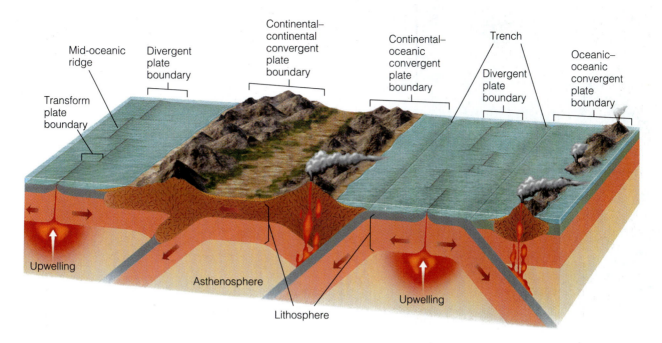

Three Principal Types of Plate Boundaries (Figure 1.13)

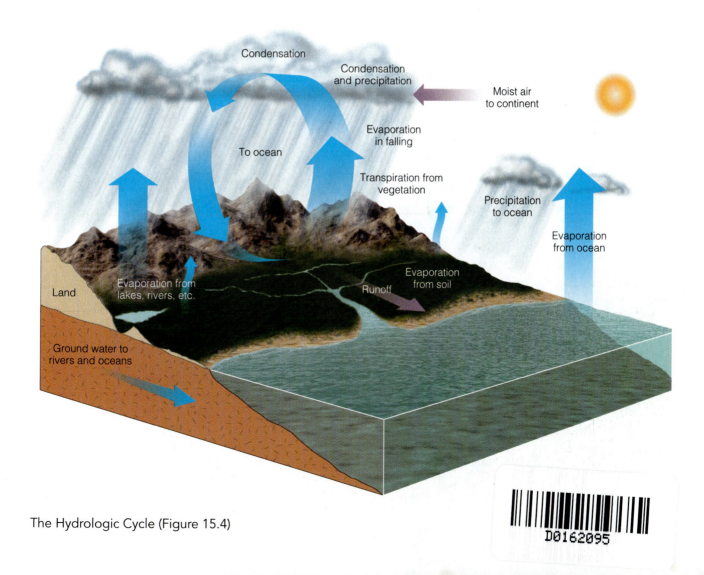

The Hydrologic Cycle (Figure 15.4)

Physical Geology

EXPLORING THE EARTH

5th EDITION

James S. Monroe

Professor Emeritus
Central Michigan University

Reed Wicander

Central Michigan University

THOMSON

BROOKS/COLE

Australia • Canada • Mexico • Singapore • Spain
United Kingdom • United States

THOMSON

★ ™

BROOKS/COLE

Earth Sciences Editor: *Keith Dodson*
Development Editor: *Alyssa White*
Assistant Editor: *Carol Ann Benedict*
Editorial Assistant: *Melissa Newt*
Technology Project Manager: *Ericka Yeoman-Saler*
Marketing Manager: *Kelley McAllister*
Marketing Assistant: *Leyla Jowza*
Advertising Project Manager: *Nathaniel Bergson-Michelson*
Project Manager, Editorial Production: *Hal Humphrey*
Senior Art Director: *Vernon T. Boes*
Print/Media Buyer: *Judy Inouye*
Permissions Editor: *Kiely Sexton*

Production Service: *Nancy Shammas, New Leaf Publishing Services*
Manuscript Editor: *Carol Reitz*
Indexer: *Kay Banning*
Text Designer: *The Davis Group*
Photo Researcher: *Kathleen Olson*
Illustrator: *Precision Graphics*
Cover Designer: *Denise Davidson*
Cover Image: *Dead Horse Point State Park, Utah;* © 2004 *David Carriere/Index Stock Imagery*
Compositor: *Carlisle Communications, Ltd.*
Printer: *Quebecor World/Dubuque*

For more information about our products, contact us at:
Thomson Learning Academic Resource Center
1-800-423-0563
For permission to use material from this text or product, submit a request online at http://www.thomsonrights.com. Any additional questions about permissions can be submitted by e-mail to thomsonrights@thomson.com.

Library of Congress Control Number: 2004100756

Student Edition: ISBN 0-534-39987-8

Instructor's Edition: ISBN 0-534-39988-6

Brooks/Cole—Thomson Learning
10 Davis Drive
Belmont, CA 94002
USA

Asia
Thomson Learning
5 Shenton Way #01-01
UIC Building
Singapore 068808

Australia/New Zealand
Thomson Learning
102 Dodds Street
Southbank, Victoria 3006
Australia

Canada
Nelson
1120 Birchmount Road
Toronto, Ontario M1K 5G4
Canada

Europe/Middle East/Africa
Thomson Learning
High Holborn House
50/51 Bedford Row
London WC1R 4LR
United Kingdom

Latin America
Thomson Learning
Seneca, 53
Colonia Polanco
11560 Mexico D.F.
Mexico

Spain/Portugal
Paraninfo
Calle Magallanes, 25
28015 Madrid, Spain

About the Authors

Photo courtesy of Sue Monroe

Photo courtesy of Melanie Wicander

James S. Monroe is professor emeritus of geology at Central Michigan University where he taught physical geology, historical geology, prehistoric life, and stratigraphy and sedimentology from 1975 until he retired in 1997. He has co-authored several textbooks with Reed Wicander and has interests in Cenozoic geology and geologic education.

Reed Wicander is a geology professor at Central Michigan University where he teaches physical geology, historical geology, prehistoric life, and invertebrate paleontology. He has co-authored several geology textbooks with James S. Monroe. His main research interests involve various aspects of Paleozoic palynology, specifically the study of acritarchs, on which he has published many papers. He is a past president of the American Association of Stratigraphic Palynologists and currently a councillor of the International Federation of Palynological Societies.

About the Cover

Your textbook cover features a spectacular image of the vast canyon eroded by the Colorado River and its tributaries as seen from Dead Horse Point in Dead Horse Point State Park near Moab, Utah. The sedimentary rocks exposed in the 600-m-deep canyon are composed mostly of mud (shale), sand (sandstone), and gravel (conglomerate) that were deposited during the Late Paleozoic and Mesozoic eras.

Dead Horse Point itself is a narrow promontory surrounded by cliffs. In the past cowboys used it as a natural horse coral, needing only a 9-m-long fence across the promontory's neck. On one occasion, a fence gate was left open but the horses remained on the promontory where they died of thirst, hence the name of the point and the state park.

When we stand at Dead Horse Point and look down into the canyon, we are actually looking far back into Earth's history. The oldest rocks, those exposed adjacent to the river, formed during the Permian Period (245–286 million years ago), whereas the youngest ones formed between 140 and 180 million years ago during the Jurassic Period. Detailed studies of these rocks indicate that they originated in a variety of environments such as stream channels and their adjacent flood plains and deserts.

In addition to Dead Horse Point State Park's scenic appeal, the rocks here also illustrate some of the basic principles geologists use to decipher Earth's history, such as the principle of superposition, which holds that in an undisturbed sequence of rock layers the oldest is at the bottom. Furthermore, these canyons show how running water and gravity-driven processes modify Earth's surface. The canyon started forming millions of years ago but even now the landscape of bold cliffs and gentle slopes of many-hued rocks continues to evolve.

Brief Contents

Contents

6 Sediment and Sedimentary Rocks   148

5 Weathering, Erosion, and Soil   120

7 Metamorphism and Metamorphic Rocks   182

8 Geologic Time: Concepts and Principles   206

9 Earthquakes 240

10 Earth's Interior 276

11 The Seafloor 302

12 Plate Tectonics: A Unifying Theory 330

14

Mass Wasting   396

13

Deformation, Mountain Building, and the Evolution of Continents   362

15 Running Water   430

16 Groundwater   468

17 Glaciers and Glaciation   500

18 The Work of Wind and Deserts   534

19 Shorelines and Shoreline Processes   560

20 Physical Geology in Perspective   592

Appendices

Answers

Preface

Earth is a dynamic planet that has changed continuously during its 4.6 billion years of existence. The size, shape, and geographic distribution of the continents and ocean basins have changed through time, as have the atmosphere and biota. As scientists and concerned citizens, we have become increasingly aware of how fragile our planet is and, more importantly, how interdependent all of its various systems and subsystems are. We have also learned that we cannot continually pollute our environment and that our natural resources are limited and, in most cases, nonrenewable. Furthermore, we are coming to realize how central geology is to our everyday lives. For these and other reasons, geology is one of the most important college or university courses a student can take.

Physical Geology: Exploring the Earth is designed for a one-semester introductory course in geology that serves both majors and nonmajors in geology and the Earth sciences. One of the problems with any introductory science course is that students are overwhelmed by the amount of material that must be learned. Furthermore, most of the material does not seem to be linked by any unifying theme and does not always appear to be relevant to their lives. This book, however, is written with students in mind in that it shows, in its easy-to-read style, that geology is an exciting and ever-changing science, and one in which new discoveries and insights are continually being made.

The goals of this book are to provide students with a basic understanding of geology and its processes and, more importantly, with an understanding of how geology relates to the human experience: that is, how geology affects not only individuals, but society in general. With these goals in mind, we introduce the major themes of the book in the first chapter to provide students with an overview of the subject and to enable them to see how the various systems and subsystems of Earth are interrelated. We also discuss the economic and environmental aspects of geology throughout the book rather than treating these topics in separate chapters. In this way, students can see, through relevant and interesting examples, how geology impacts our lives.

NEW FEATURES IN THE FIFTH EDITION

The fifth edition has undergone considerable rewriting and updating to produce a book that is not only easy to read, but has a high level of current information, many new photographs, a completely revamped art program, and various new features to help students maximize their learning and understanding of Earth and its systems. Drawing on the comments and suggestions of reviewers, we have incorporated many new features into this edition. Perhaps the most noticeable change is that the chapter on the History of the Universe, the Solar System, and the Planets has now been incorporated into the first chapter to give students a complete view of Earth's earliest development and relation to the rest of the planets in our Solar System. In addition, the final chapter now provides students with an overview of the concepts presented throughout the book and ties together the various themes covered.

The former *Prologue* and *Introduction* in each chapter are now combined into a new *Introduction*. These new *Introductions* begin each chapter with a story related to the chapter material as well as addressing the question of why each chapter's material is relevant and important to the student's overall understanding of the topic.

Concept Art Spreads are found in all but two chapters. These two-page art pieces are designed to enhance students' interest in the chapter material by visual learning. Some of the topics covered include the Burren Region of

Ireland, Rock Art, and the Many Uses of Marble, to name a few.

Another new feature, *Geology in Unexpected Places*, discusses interesting geology or geological phenomena in unusual places. We think this feature will be particularly appealing to students by relating geology to the human experience. Geology of the Great Wall of China, a bit of Egypt in Central Park, New York, and geology in cemeteries are just a few of the topics covered.

We also have a powerful new interactive media program called Physical GeologyNow, which has been seamlessly integrated with the text, enhancing students' understanding of important geological processes. It brings geology alive with animated figures, media-enhanced activities, tutorials, and personalized learning plans. And like other features in our new edition, it encourages students to be curious, to think about geology in new ways, and to connect their new found knowledge of the world around them to their own lives.

Many of the popular *What Would You Do?* boxes have been redone with new topics and questions posed. These boxes are designed to encourage students to think critically about what they're learning. They incorporate material from each chapter and ask open-ended questions to elicit discussion and have the student formulate reasoned responses to particular situations.

The previous edition's *Perspectives* have been replaced by *Geo-Focus* sections on a variety of either new topics or updated previous ones. The *Guest Essays* have been replaced by *Geo-Profiles*. There are eight *Geo-Profiles* in this edition, six of which are new. These new *Geo-Profiles* emphasize careers that students may not associate with geology, such as sculpturing, the politics of geology, and paleontology combined with movie making.

Many photographs in the fourth edition have been replaced, including most of the chapter opening photographs. In addition, a number of photographs within the chapters have been enlarged to enhance their visual impact.

The art program has been completely revamped to provide the most accurate and visually stimulating figures possible.

We feel the rewriting and updating done in the text as well as the addition of new photographs and newly rendered art greatly improves the fifth edition by making it easier to read and comprehend, as well as a more effective teaching tool. Additionally, improvements have been made in the ancillary package that accompanies the book.

TEXT ORGANIZATION

Plate tectonic theory is the unifying theme of geology and this book. This theory has revolutionized geology because it provides a global perspective of Earth and allows geologists to treat many seemingly unrelated geologic phenomena as part of a total planetary system. Because plate tectonic theory is so important, it is introduced in Chapter 1 and is discussed in most subsequent chapters in terms of the subject matter of that chapter.

Another theme of this book is that Earth is a complex, dynamic planet that has changed continually since its origin some 4.6 billion years ago. We can better understand this complexity by using a systems approach in the study of Earth and emphasizing this approach through the book.

We have organized *Physical Geology: Exploring the Earth* into several informal sections. Chapter 1 is an introduction to geology and Earth systems, its relevance to the human experience, the origin of the solar system and Earth's place in it, plate tectonic theory, the rock cycle, and geologic time and uniformitarianism. Chapters 2–7 examine Earth's materials (minerals and igneous, sedimentary, and metamorphic rocks) and the geologic processes associated with them, including the role of plate tectonics in their origin and distribution. Chapter 8 discusses geologic time, introduces several dating methods, and explains how geologists correlate rocks. Chapters 9–13 deal with the related topics of Earth's interior, the seafloor, earthquakes, deformation and mountain building, and plate tectonics. Chapters 14–19 cover Earth's surface processes. Chapter 20 summarizes and synthesizes the concepts, themes, and major topics covered in this book.

We have found that presenting the material in this order works well for most students. We know, however, that many instructors prefer an entirely different order of topics, depending on the emphasis in their course. We have therefore written this book so instructors can present the chapters in any order that suits the needs of their course.

CHAPTER ORGANIZATION

All chapters have the same organizational format. Each chapter has a photograph relating to the chapter material, an *Outline* that engages students by having many of the headings as questions, a *Chapter Objectives* outline that serves to alert students to the learning outcome objectives of the chapter, followed by a new *Introduction* that is intended to stimulate interest in the chapter by discussing some aspect of the material and show students how the chapter material fits into the larger geologic perspective.

The text is written in a clear, informal style, making it easy for students to comprehend. Numerous newly rendered color diagrams and photographs complement the text, providing a visual representation of the concepts and information presented.

Each chapter contains at least one *Geo-Focus* that presents a brief discussion of an interesting aspect of geology or geological research. *What Would You Do?* boxes, usually two per chapter, are designed to encourage thinking by students as they attempt to solve a hypothetical problem or issue relating to the chapter material. Mineral and energy resources are discussed in the final sections of a number of chapters to provide interesting, relevant information in the context of the chapter topics.

Geology in Unexpected Places are found in each chapter. This new feature is designed to focus on interesting geology in unusual places, or geology in a setting you might not have thought about.

The end-of-chapter materials begin with a concise review of important concepts and ideas in the *Review Workbook.* The *Important Terms,* which are printed in boldface type in the chapter text, are listed at the end of each chapter for easy review as well as the page number where they are first defined. A full *Glossary* of important terms appears at the end of the text. The *Review Questions* are another important feature of this book; they include multiple-choice questions with answers as well as short answer, essay, and thought-provoking and quantitative questions. Many new questions have been added in each chapter of this edition. Each chapter concludes with *World Wide Web Activities* that provides students with the URL for this book. At the Brooks/Cole website, students can assess their understanding of each chapter's topics, take quizzes, and participate in comprehensive interactivities as well as access up-to-date Web links and find additional readings

ANCILLARY MATERIALS

We are pleased to offer a full suite of text and multimedia products to accompany the fifth edition of *Physical Geology: Exploring the Earth.*

For Instructors

Instructor's Edition of Physical Geology The Instructor's Edition provides a visual preface and additional information for the instructor, including a complete resource integration guide to all print and media resources accompanying the text.

Instructor's Manual with Test Bank This manual contains chapter-by-chapter outlines, learning objectives, summaries, lecture suggestions, enrichment topics, and a complete test bank with answers.

ExamView® The ExamView assessment and tutorial system allows you to create, deliver, and customize tests and study guides, both print and online.

Transparency Acetates This full color set of 100 transparencies contains images taken directly from the text.

Slide Set This full color set of slides contains photos and images from the text.

Multimedia Manager with Living Lecture™ Tools. Create fluid multimedia lectures using your own lecture notes and clips, traditional art from all of our Earth sciences texts, and three-dimensional, animated models of *Active Figures* from the text itself.

WebTutor ToolBox to accompany Physical Geology Available free with this text (if requested), WebTutor ToolBox is available via PIN code. WebTutor ToolBox pairs all the content of the text's companion Web site with the course management of a WebCT or Blackboard product. Students have access only to student resources, while instructors can access password-protected instructor resources

Physical GeologyNow Physical GeologyNow is the first assessment-centered student learning tool for physical geology. It is tied to your lectures and the text through Living Lecture Tools that bring geologic processes to life. Physical GeologyNow is Web-based and free with every new copy of the text.

The Brooks/Cole Earth Sciences Resource Center
http://earthscience.brookscole.com

Book Companion Web Site
http://earth science.brookscole.com/physgeo5e
The Brooks/ Cole Earth Sciences Resource Center and the Book Companion Web Site feature a rich array of learning resources for your students. The text-specific companion Web site includes quizzing and other Web-based activities that will help students explore the concepts presented in the text.

For Students

Study Guide to Accompany Physical Geology This student study guide contains Chapter Overviews, Important Terms, Study Questions and Answers, Activities, and more. ISBN: 0534-399894

Earth Lab: Exploring the Earth Sciences By Owen, Pirie, and Draper. The experiments featured in this lab manual teach and reinforce core skills that characterize the Earth sciences and illustrate how the scientific method works. ISBN: 0534-379532

The Brooks/Cole Earth Sciences Resource Center
http://earthscience.brookscole.com

Book Companion Web Site

http://earthscience.brookscole.com/physgeo5e
The Brooks/Cole Earth Sciences Resource Center and the Book Companion Web Site feature a rich array of learning resources. The text-specific companion Web site includes quizzing and other Web-based activities that will help students explore the concepts that are presented in the text.

Physical GeologyNow

Physical GeologyNow is available through the book's Companion Web Site at *http://earthscience.brookscole.com/physgeo5e*. In addition to Physical GeologyNow, students using the Web site have access to maps, Web links, Internet and Info-Trac® College Edition exercises, learning objectives, discussion questions, chapter outlines, and much more.

WebTutor ToolBox to accompany Physical Geology

Available free with this text (if requested), WebTutor ToolBox is preloaded with content and available via PIN code. Students have access to an array of student resources including the text's Book Companion Web Site.

GIS Investigations for the Earth Sciences

Exploring Tropical Cyclones: GIS Investigations for the Earth Sciences. ISBN: 0534-391478. *Exploring Water Resources: GIS Investigations for the Earth Sciences*. ISBN: 0534-391567. *Exploring the Dynamic Earth: GIS Investigations for the Earth Sciences*. ISBN: 0534-391389.

These three groundbreaking guides by Hall-Wallace et al. let even novice users tap the power of GIS to explore, manipulate, and analyze large data sets. The guides come with all the software and data sets needed to complete the exercises.

InfoTrac® College Edition

Every new copy of the text is accompanied by four months of free access to InfoTrac College Edition, the online library. The new and improved InfoTrac College Edition puts cutting-edge research and the latest headlines at your students' fingertips, offering more than 10 million articles from nearly 5,000 diverse sources, such as academic journals, newsletters, and up-to-the-minute periodicals including *Environment, Journal of Geology, Science News*, and more. New—Students now also gain instant access to critical thinking and paper writing tools through InfoWrite. ISBN: 0534-409059

ACKNOWLEDGMENTS

As authors, we are, of course, responsible for the organization, style, and accuracy of the text, and any mistakes, omissions, or errors are our responsibility. The finished product is the culmination of many years of work during which we received numerous comments and advice from many geologists who reviewed parts of the text. We wish to express our sincere appreciation to the reviewers who reviewed the fourth edition and made many helpful and useful comments that led to the improvements in this fifth edition:

William J. Frazier, *Columbus State University*
Thomas J. Leonard, *William Paterson University*
Donald Lovejoy, *Palm Beach Atlantic University*
Peter Bower, *Barnard College/Columbia University*
Tom Shoberg, *Pittsburg State University*
Bethany D. Rinard, *Tarleton State University*
Thom Wilch, *Albion College*
Glen Merrill, *University of Houston*
Jon C. Crawley, *Roanoke College*
Neil Johnson, *Appalachian State University*
Richard. L. Mauger, *East Carolina University*
Eddie B. Robertson, *Reinhardt College*
John Tacinelli, *Rochester Community and Technical College*
Ntungwa Maasha, *Coastal Georgia Community College*
Cathy Baker, *Arkansas Tech University*
David P. Lawrence, *East Carolina University*
Nicholas A. Gioppo, *Mohawk Valley Community College*
Stan P. Dunagan, *University of Tennessee at Martin*
Dave Thomas, *Washtenaw Community College*
Ronald A. Johnston, *Fayetteville State University*
Steve Mattox, *Grand Valley State University*
John Leland, *Glendale Community College*
Jim Van Alstine, *University of Minnesota*
Ed van Hees, *Wayne State University*
Ed Wehling, *Anoka-Ramsey Community College*

Our thanks also to the third edition reviewers, whose comments improved the fourth edition.

Steven R. Dent, *Northern Kentucky University*
Michael R. Forrest, *Rio Hondo Community College*
René De Hon, *Northeast Louisiana University*
Robert B. Jorstad, *Eastern Illinois University*
David T. King, *Auburn University*
Peter L. Kresan, *University of Arizona*
Gary D. Rosenberg, *Indiana University–Purdue University at Indianapolis*
Darrel W. Schmitz, *Mississippi State University*
David G. Towell, *Indiana University*

We would also like to thank the reviewers of the second edition whose comments and suggestions greatly improved the third edition.

Gary Allen, *University of New Orleans*
Richard Beck, *Miami University*
Roger Bilham, *University of Colorado, Boulder*
John P. Buchanan, *Eastern Washington University*
Jeffrey B. Connelly, *University of Arkansas at Little Rock*
Peter Copeland, *University of Houston*
Rachael Craig, *Kent State University*

P. Johnathan Patchett, *University of Arizona*

Gary Rosenberg, *Indiana University–Purdue University at Indianapolis*

Chris Sanders, *Southeast Missouri State University*

Paul B. Tomascak, *University of Maryland at College Park*

Harve S. Waff, *University of Oregon*

In addition, we thank those individuals who reviewed the first edition and made many useful and insightful comments that were incorporated into the second edition of this book.

Theodore G. Benitt, *Nassau Community College*

Bruce A. Blackerby, *California State University–Fresno*

Gerald F. Brem, *California State University–Fullerton*

Ronald E. Davenport, *Louisiana Tech University*

Isabella M. Drew, *Ramapo College of New Jersey*

Jeremy Dunning, *Indiana University*

Norman K. Grant, *Miami University*

Norris W. Jones, *University of Wisconsin–Oshkosh*

Patricia M. Kenyon, *University of Alabama*

Peter L. Kresan, *University of Arizona*

Dave B. Loope, *University of Nebraska–Lincoln*

David N. Lumsden, *Memphis State University*

Steven K. Reid, *Morehead State University*

M.J. Richardson, *Texas A&M University*

Charles J. Ritter, *University of Dayton*

Gary D. Rosenberg, *Indiana University–Purdue University at Indianapolis*

J. Alexander Speer, *North Carolina State University*

James C. Walters, *University of Northern Iowa*

Lastly, we would like to provide thanks to the individuals who reviewed the first edition of this book in manuscript form. Their comments helped make the book a success for students and instructors alike.

Gary C. Allen, *University of New Orleans*

R. Scott Babcock, *Western Washington University*

Kennard Bork, *Denison University*

Thomas W. Broadhead, *University of Tennessee at Knoxville*

Anna Buising, *California State University at Hayward*

F. Howard Campbell III, *James Madison University*

Larry E. Davis, *Washington State University*

Noel Eberz, *California State University at San Jose*

Allan A. Ekdale, *University of Utah*

Stewart S. Farrar, *Eastern Kentucky University*

Richard H. Fluegeman, Jr., *Ball State University*

William J. Fritz, *Georgia State University*

Kazuya Fujita, *Michigan State University*

Norman Gray, *University of Connecticut*

Jack Green, *California State University at Long Beach*

David R. Hickey, *Lansing Community College*

R. W. Hodder, *University of Western Ontario*

Cornelis Klein, *University of New Mexico*

Lawrence W. Knight, *William Rainey Harper College*

Martin B. Lagoe, *University of Texas at Austin*

Richard H. Lefevre, *Grand Valley State University*

I. P. Martini, *University of Guelph, Ontario*

Michael McKinney, *University of Tennessee at Knoxville*

Robert Merrill, *California State University at Fresno*

Carleton Moore, *Arizona State University*

Alan P. Morris, *University of Texas at San Antonio*

Harold Pelton, *Seattle Central Community College*

James F. Petersen, *Southwest Texas State University*

Katherine H. Price, *DePauw University*

William D. Romey, *St. Lawrence University*

Gary Rosenberg, *Indiana University–Purdue University at Indianapolis*

David B. Slavsky, *Loyola University of Chicago*

Edward F. Stoddard, *North Carolina State University*

Charles P. Thornton, *Pennsylvania State University*

Samuel B. Upchurch, *University of South Florida*

John R. Wagner, *Clemson University*

We also wish to thank Kathy Benison, R. V. Dietrich (Professor Emeritus), David J. Matty, Jane M. Matty, Wayne E. Moore (Professor Emeritus), and Sven Morgan of the Geology Department, and Bruce M. C. Pape of the Geography Department of Central Michigan University, as well as Eric Johnson (Hartwick College, New York) and Stephen D. Stahl (St. Bonaventure, New York) for providing us with photographs and answering our questions concerning various topics. We also thank Arlene Sharp-Wilson of the Geology Department, whose general efficiency and good humor was invaluable during the preparation of this book. Stephanie Buck and Julie Hotchkiss, also of the Geology Department, helped with some of the clerical tasks associated with this book. We are also grateful for the generosity of the various agencies and individuals from many countries who provided photographs.

Special thanks must go to Keith Dodson, Earth Sciences Editor at Brooks/Cole, who initiated this fifth edition, and to Alyssa White, Development Editor, who saw this edition through to completion. We are equally indebted to our production manager, Nancy Shammas of New Leaf Publishing Services, for all her help. Her attention to detail and consistency as well as general good nature is greatly appreciated. We would also like to thank Carol Reitz for her copyediting skills. We appreciate her help in improving our manuscript. We thank Kathleen Olsen for her invaluable help in locating appropriate photos and checking on permissions and Kathy Joneson for updating our art style while keeping it consistent. We would also like to thank Lisa Torri for her design and contribution to the two-page art spreads. Because geology is largely a visual science, we extend special thanks to the artists at Precision Graphics, who were responsible for rendering and updating much of the art program. They all did an excellent job, and we enjoyed working with them.

As always, our families were very patient and encouraging when much of our spare time and energy were devoted to this book. We again thank them for their continued support and understanding.

James S. Monroe
Reed Wicander

Developing Critical Thinking and Study Skills

INTRODUCTION

College is a demanding and important time, a time when your values will be challenged, and you will try out new ideas and philosophies. You will make personal and career decisions that will affect your entire life. One of the most important lessons you can learn in college is how to balance your time among work, study, and recreation. If you develop good time management and study skills early in your college career, you will find that your college years will be successful and rewarding.

This section offers some suggestions to help you maximize your study time and develop critical thinking and study skills that will benefit you, not only in college, but throughout your life. While mastering the content of a course is obviously important, learning how to study and to think critically is, in many ways, far more important. Like most things in life, learning to think critically and study efficiently will initially require additional time and effort, but once mastered, these skills will save you time in the long run.

You may already be familiar with many of the suggestions and may find that others do not directly apply to you. Nevertheless, if you take the time to read this section and apply the appropriate suggestions to your own situation, we are confident that you will become a better and more efficient student, find your classes more rewarding, have more time for yourself, and get better grades. We have found that the better students are usually also the busiest. Because these students are busy with work or extracurricular activities, they have had to learn to study efficiently and manage their time effectively.

One of the keys to success in college is avoiding procrastination. While procrastination provides temporary satisfaction because you have avoided doing something you did not want to do, in the long run it leads to stress. While a small amount of stress can be beneficial, waiting until the last minute usually leads to mistakes and a subpar performance. By setting clear, specific goals and working toward them on a regular basis, you can greatly reduce the temptation to procrastinate. It is better to work efficiently for short periods of time than to put in long, unproductive hours on a task, which is usually what happens when you procrastinate.

Another key to success in college is staying physically fit. It is easy to fall into the habit of eating junk food and never exercising. To be mentally alert, you must be physically fit. Try to develop a program of regular exercise. You will find that you have more energy, feel better, and study more efficiently.

GENERAL STUDY SKILLS

Most courses, and geology in particular, build upon previous material, so it is extremely important to keep up with the coursework and set aside regular time for study in each of your courses. Try to follow these hints, and you will find you do better in school and have more time for yourself:

- Develop the habit of studying on a daily basis.
- Set aside a specific time each day to study. Some people are day people, and others are night people. Determine when you are most alert and use that time for study.
- Have an area dedicated for study. It should include a well-lighted space with a desk and the study materials you need, such as a dictionary, thesaurus, paper, pens, and pencils, and a computer if you have one.
- Study for short periods and take frequent breaks, usually after an hour of study. Get up and move

around and do something completely different. This will help you stay alert, and you'll return to your studies with renewed vigor.

- Try to review each subject every day or at least the day of the class. Develop the habit of reviewing lecture material from a class the same day.
- Become familiar with the vocabulary of the course. Look up any unfamiliar words in the glossary of your textbook or in a dictionary. Learning the language of the discipline will help you learn the material.

GETTING THE MOST FROM YOUR NOTES

If you are to get the most out of a course and do well on exams, you must learn to take good notes. Taking good notes does not mean you should try to write down every word your professor says. Part of being a good note taker is knowing what is important and what you can safely leave out.

Early in the semester, try to determine whether the lecture will follow the textbook or be predominantly new material. If much of the material is covered in the textbook, your notes do not have to be as extensive or detailed as when the material is new. In any case, the following suggestions should make you a better note taker and enable you to derive the maximum amount of information from a lecture:

- Regardless of whether the lecture discusses the same material as the textbook or supplements the reading assignment, read or scan the chapter the lecture will cover *before* class. This way you will be somewhat familiar with the concepts and can listen critically to what is being said rather than trying to write down everything. Later a few key words or phrases will jog your memory about what was said.
- Before each lecture, briefly review your notes from the previous lecture. Doing this will refresh your memory and provide a context for the new material.
- Develop your own style of note taking. Do not try to write down every word. These are notes you're taking, not a transcript. Learn to abbreviate and develop your own set of abbreviations and symbols for common words and phrases: for example, w/o (without), w (with), = (equals), ^ (above or increases), v (below or decreases), < (less than), > (greater than), & (and), u (you).
- Geology lends itself to many abbreviations that can increase your note-taking capability: for example, pt (plate tectonics), ig (igneous), meta (metamorphic), sed (sedimentary), rx (rock or rocks), ss (sandstone), my (million years), and gts (geologic time scale).

- Rewrite your notes soon after the lecture. Rewriting your notes helps reinforce what you heard and gives you an opportunity to determine whether you understand the material.
- By learning the vocabulary of the discipline before the lecture, you can cut down on the amount you have to write—you won't have to write down a definition if you already know the word.
- Learn the mannerisms of the professor. If he or she says something is important or repeats a point, be sure to write it down and highlight it in some way. Students have told me (RW) that when I stated something twice during a lecture, they knew it was important and probably would appear on a test. (They were usually right!)
- Check any unclear points in your notes with a classmate or look them up in your textbook. Pay particular attention to the professor's examples, which usually elucidate and clarify an important point and are easier to remember than an abstract concept.
- Go to class regularly and sit near the front of the class if possible. It is easier to hear and see what is written on the board or projected onto the screen, and there are fewer distractions.
- If the professor allows it, tape record the lecture, but don't use the recording as a substitute for notes. Listen carefully to the lecture and write down the important points; then fill in any gaps when you replay the tape.
- If your school allows it, and if they are available, buy class lecture notes. These are usually taken by a graduate student who is familiar with the material; typically they are quite comprehensive. Again use these notes to supplement your own.
- Ask questions. If you don't understand something, ask the professor. Many students are reluctant to do this, especially in a large lecture hall, but if you don't understand a point, other people are probably confused as well. If you can't ask questions during a lecture, talk to the professor after the lecture or during office hours.

GETTING THE MOST OUT OF WHAT YOU READ

The old adage that "you get out of something what you put into it" is true when it comes to reading textbooks. By carefully reading your text and following these suggestions, you can greatly increase your understanding of the subject:

- Look over the chapter outline to see what the material is about and how it flows from topic to topic. If you have time, skim through the chapter before you start to read in depth.

- Pay particular attention to the tables, charts, and figures. They contain a wealth of information in abbreviated form and illustrate important concepts and ideas. Geology, in particular, is a visual science, and the figures and photographs will help you visualize what is being discussed in the text and provide actual examples of features such as faults or unconformities.
- As you read your textbook, highlight or underline key concepts or sentences, but make sure you don't highlight everything. Make notes in the margins. If you don't understand a term or concept, look it up in the glossary.
- Read the chapter summary carefully. Be sure you understand all the key terms, especially those in boldface or italic type. Because geology builds on previous material, it is imperative that you understand the terminology.
- Go over the end-of-chapter questions. Write your answers as if you were taking a test. Only when you see your answer in writing will you know if you really understood the material.
- Access the latest geologic information on the Internet. The end-of-chapter World Wide Web Activities will enhance your understanding of the chapter concepts and the way geologic information is disseminated today. Knowing how to search the Internet is an essential skill.

DEVELOPING CRITICAL THINKING SKILLS

Few things in life are black and white, and it is important to be able to examine an issue from all sides and come to a logical conclusion. One of the most important things you will learn in college is to think critically and not accept everything you read and hear at face value. Thinking critically is particularly important in learning new material and relating it to what you already know. Although you can't know everything, you can learn to question effectively and arrive at conclusions consistent with the facts. Thus, these suggestions for critical thinking can help you in all your courses:

- Whenever you encounter new facts, ideas, or concepts, be sure you understand and can define all of the terms used in the discussion.
- Determine how the facts or information was derived. If the facts were derived from experiments, were the experiments well executed and free of bias? Can they be repeated? The controversy over cold fusion is an excellent example. Two scientists claimed to have produced cold fusion reactions using simple experimental laboratory apparatus, yet other scientists have never been able to achieve the same reaction by repeating the experiments.
- Do not accept any statement at face value. What is the source of the information? How reliable is the source?
- Consider whether the conclusions follow from the facts. If the facts do not appear to support the conclusions, ask questions and try to determine why they don't. Is the argument logical or is it somehow flawed?
- Be open to new ideas. After all, the underlying principles of plate tectonic theory were known early in this century yet were not accepted until the 1970s despite overwhelming evidence.
- Look at the big picture to determine how various elements are related. For example, how will constructing a dam across a river that flows to the sea affect the stream's profile? What will be the consequences to the beaches that will be deprived of sediment from the river? One of the most important lessons you can learn from your geology course is how interrelated the various systems of Earth are. When you alter one feature, you affect numerous other features as well.

IMPROVING YOUR MEMORY

Why do you remember some things and not others? The reason is that the brain stores information in different ways and forms, making it easy to remember some things and difficult to remember others. Because college requires that you learn a vast amount of information, any suggestions that can help you retain more material will help you in your studies:

- Pay attention to what you read or hear. Focus on the task at hand and avoid daydreaming. Repetition of any sort will help you remember material. Review the previous lecture before going to class, or look over the last chapter before beginning the next. Ask yourself questions as you read.
- Use mnemonic devices to help you learn unfamiliar material. For example, the order of the Paleozoic periods (Cambrian, Ordovician, Silurian, Devonian, Mississippian, Pennsylvanian, and Permian) of the geologic time scale can be remembered by the phrase, **Campbell's Onion Soup Does Make Peter Pale**, or the order of the Cenozoic Epochs (Paleocene, Eocene, Oligocene, Miocene, Pliocene, and Pleistocene) can be remembered by the phrase, **Put Eggs On My Plate Please**. Using rhymes can also be helpful.

- Look up the roots of important terms. If you understand where a word comes from, its meaning will be easier to remember. For example, *pyroclastic* comes from *pyro,* meaning "fire," and *clastic,* meaning "broken pieces." Hence a pyroclastic rock is one formed by volcanism and composed of pieces of other rocks. We have provided the roots of many important terms throughout this text to help you remember their definitions.

- Outline the material you are studying. This practice will help you see how the various components are interrelated. Learning a body of related material is much easier than learning unconnected and discrete facts. Looking for relationships is particularly helpful in geology because so many things are interrelated. For example, plate tectonics explains how mountain building, volcanism, and earthquakes are all related. The rock cycle relates the three major groups of rocks to each other and to subsurface and surface processes (Chapter 1).

- Use deductive reasoning to tie concepts together. Remember that geology builds on what you learned previously. Use that material as your foundation and see how the new material relates to it.

- Draw a picture. If you can draw a picture and label its parts, you probably understand the material. Geology lends itself very well to this type of memory device because so much is visual. For example, instead of memorizing a long list of glacial terms, draw a picture of a glacier and label its parts and the type of topography it forms.

- Focus on what is important. You can't remember everything, so focus on the important points of the lecture or the chapter. Try to visualize the big picture and use the facts to fill in the details.

PREPARING FOR EXAMS

For most students, tests are the critical part of a course. To do well on an exam, you must be prepared. These suggestions will help you focus on preparing for examinations:

- The most important advice is to study regularly rather than try to cram everything into one massive study session. Get plenty of rest the night before an exam, and stay physically fit to avoid becoming susceptible to minor illnesses that sap your strength and lessen your ability to concentrate on the subject at hand.

- Set up a schedule so that you cover small parts of the material on a regular basis. Learning some concrete examples will help you understand and remember the material.

- Review the chapter summaries. Construct an outline to make sure you understand how everything fits together. Drawing diagrams will help you remember key points. Make flash cards to help you remember terms and concepts.

- Form a study group, but make sure your group focuses on the task at hand, not on socializing. Quiz each other and compare notes to be sure you have covered all the material. We have found that students dramatically improved their grades after forming or joining a study group.

- Write the answers to all the Review Questions. Before doing so, however, become thoroughly familiar with the subject matter by reviewing your lecture notes and reading the chapter. Otherwise, you will spend an inordinate amount of time looking up answers.

- If you have any questions, visit the professor or teaching assistant. If review sessions are offered, be sure to attend. If you are having problems with the material, ask for help as soon as you have difficulty. Don't wait until the end of the semester.

- If old exams are available, look at them to see what is emphasized and what types of questions are asked. Find out whether the exam will be all objective or all essay or a combination. If you have trouble with a particular type of question (such as multiple choice or essay), practice answering questions of that type—your study group or a classmate may be able to help.

TAKING EXAMS

The most important thing to remember when taking an exam is not to panic. This, of course, is easier said than done. Almost everyone suffers from test anxiety to some degree. Usually, it passes as soon as the exam begins, but in some cases, it is so debilitating that an individual does not perform as well as he or she could. If you are one of those people, get help as soon as possible. Most colleges and universities have a program to help students overcome test anxiety or at least keep it in check. Don't be afraid to seek help if you suffer test anxiety. Your success in college depends to a large extent on how well you perform on exams, so by not seeking help, you are only hurting yourself. In addition, the following suggestions may be helpful:

- First of all, relax. Then look over the exam briefly to see its format and determine which questions are worth the most points. If it helps, quickly jot down any information you are afraid you might forget or particularly want to remember for a question.

- Answer the questions that you know the best first. Make sure, however, that you don't spend too much time on any one question or on one that is worth only a few points.

- If the exam is a combination of multiple choice and essay, answer the multiple-choice questions first. If you are not sure of an answer, go on to the next one. Sometimes the answer to one question can be found in another question. Furthermore, the multiple-choice questions may contain many of the facts needed to answer some of the essay questions.

- Read the question carefully and answer only what it asks. Save time by not repeating the question as your opening sentence to the answer. Get right to the point. Jot down a quick outline for longer essay questions to make sure you cover everything.

- If you don't understand a question, ask the examiner. Don't assume anything. After all, it is your grade that will suffer if you misinterpret the question.

- If you have time, review your exam to make sure you covered all the important points and answered all the questions.

- If you have followed our suggestions, by the time you finish the exam, you should feel confident that you did well and will have cause for celebration.

CONCLUDING COMMENTS

We hope that the suggestions we have offered will be of benefit to you, not only in this course but throughout your college career. Though it is difficult to break old habits and change a familiar routine, we are confident that following these suggestions will make you a better student. Furthermore, many of the suggestions will help you work more efficiently, not only in college, but also throughout your career. Learning is a lifelong process that does not end when you graduate. The critical thinking skills that you learn now will be invaluable throughout your life, both in your career and as an informed citizen.

Physical Geology

Understanding Earth: A Dynamic and Evolving Planet

CHAPTER 1
OUTLINE

PHYSICAL Geology ⇌ Now *This icon, appearing throughout the book, indicates an opportunity to explore interactive tutorials, animations, or practice problems available on the Physical GeologyNow Web site at http://earthscience.brookscole.com/physgeo5e.*

OBJECTIVES
At the end of this chapter, you will have learned that

- Geology is the study of Earth.

- Earth is a complex, integrated system of interconnected components that interact and affect one another in various ways.

- Geology plays an important role in the human experience and affects us as both individuals and members of society and nation-states.

- The universe is thought to have originated about 15 billion years ago with a Big Bang. The solar system and planets evolved from a turbulent, rotating cloud of material surrounding the embryonic Sun.

- Theories are based on the scientific method.

- Plate tectonic theory revolutionized geology.

- The rock cycle illustrates the interrelationships between Earth's internal and external processes and shows how and why the three major rock groups are related.

- An appreciation of geologic time and the principle of uniformitarianism is central to understanding the evolution of Earth and its biota.

- Geology is an integral part of our lives.

Satellite-based image of Earth. North America can clearly be seen in the center of this view as well as Central America and the northern part of South America. The present locations of continents and ocean basins are the result of plate movements. The interaction of plates through time has affected the physical and biological history of Earth. Source: NASA

Introduction

A major benefit of the space age has been the ability to look back from space and view our planet in its entirety. Every astronaut has remarked in one way or another on how Earth stands out as an inviting oasis in the otherwise black void of space (see the chapter opening photo). We are able to see not only the beauty of our planet but also its fragility and the role humans play in shaping and modifying its environment. We can also decipher Earth's long and frequently turbulent history by reading the clues preserved in the geologic record.

A major theme of this book is that Earth is a complex, dynamic planet that has changed continuously since its origin some 4.6 billion years ago. These changes and the present-day features we observe result from the interactions among Earth's various internal and external systems, subsystems, and cycles. Earth is unique among the planets of our solar system in that it supports life and has oceans of water, a hospitable atmosphere, and a variety of climates. It is ideally suited for life as we know it because of a combination of factors, including its distance from the Sun and the evolution of its interior, crust, oceans, and atmosphere. Life processes have, over time, influenced the evolution of Earth's atmosphere, oceans, and to some extent its crust. In turn, these physical changes have affected the evolution of life.

By viewing Earth as a whole—that is, thinking of it as a system—we not only see how its various components are interconnected, but also better appreciate its complex and dynamic nature. The system concept makes it easier to study a complex subject such as Earth because it divides the whole into smaller components we can easily understand, without losing sight of how the components all fit together as a whole.

A **system** is a combination of related parts that interact in an organized fashion (■ Figure 1.1). Information, materials, and energy entering the system from the outside are *inputs*, whereas information, materials, and energy that leave the system are *outputs*. An automobile is a good example of a system. Its various subsystems include engine, transmission, steering, and brakes. These subsystems are interconnected in such a way that a change in any one of them affects the others. The main input into the automobile system is gasoline, and its outputs are movement, heat, and pollutants.

We can examine Earth in the same way we view an automobile—that is, as a system of interconnected components that interact and affect each other in many ways. The principal subsystems of Earth are the *atmosphere, biosphere, hydrosphere, lithosphere, mantle,* and *core* (■ Figure 1.2). The complex interactions among these subsystems result in a dynamically changing body that exchanges matter and energy and recycles them into different forms (Table 1.1). The rock cycle is an excellent example of how the interaction between Earth's internal and external processes recycles Earth materials to form the three major rock groups (see Figure 1.14). Likewise, the movement of plates has profoundly affected the formation of landscapes, the distribution of mineral resources, and atmospheric and oceanic circulation patterns, which in turn have affected global climate changes.

We must also not forget that humans are part of the Earth system, and our presence alone affects this system to some extent. Accordingly, we must understand that actions we take can produce changes with wide-ranging consequences of which we might not initially be aware. For this reason, an understanding of geology, and science in general, is of great importance. If the human species is to survive, we must understand how the various Earth systems work and interact, and how our actions affect the delicate balance between these systems.

When people discuss and debate such environmental issues as acid rain, the greenhouse effect and global warming, and the depleted ozone layer, it is important to

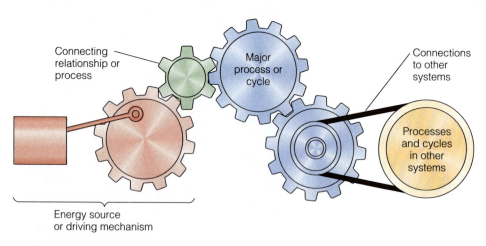

Connecting relationship or process

Major process or cycle

Connections to other systems

Processes and cycles in other systems

Energy source or driving mechanism

■ **Figure 1.1**

A series of gears can be used to illustrate how some of Earth's systems and processes interact. Pistons and driving rods represent energy sources or driving mechanisms, large gears represent important processes or cycles, and small gears represent connecting processes or relationships. Pulleys are used to show relationships to other systems. Although gears are a useful way to represent systems diagrammatically, remember that real Earth systems are far more complex.

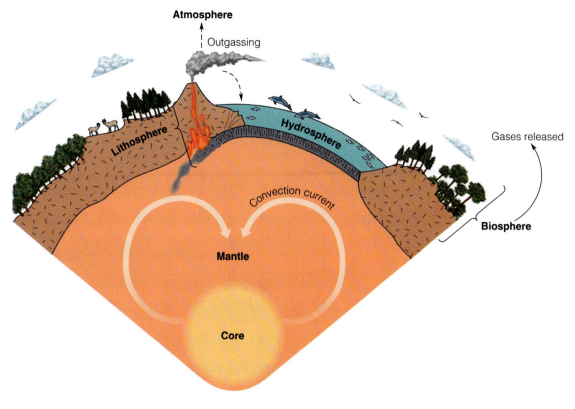

■ Figure 1.2

The atmosphere, biosphere, hydrosphere, lithosphere, mantle, and core can all be thought of as subsystems of Earth. The interactions among these subsystems are what make Earth a dynamic planet, which has evolved and changed since its origin 4.6 billion years ago.

Table 1.1

Interactions Among Earth's Principal Subsystems

	Atmosphere	Hydrosphere	Biosphere	Lithosphere
Atmosphere	Interaction among various air masses	Surface currents driven by wind Evaporation	Gases for respiration Dispersal of spores, pollen, and seed by wind	Weathering by wind erosion Transport of water vapor for precipitation of rain and snow
Hydrosphere	Input of water vapor and stored solar heat	Hydrologic cycle	Water for life	Precipitation Weathering and erosion
Biosphere	Gases from respiration	Removal of dissolved materials by organisms	Global ecosystems Food cycles	Modification of weathering and erosion processes Formation of soil
Lithosphere	Input of stored solar heat Landscapes affect air movements	Source of solid and dissolved materials	Source of mineral nutrients Modification of ecosystems by plate movements	Plate tectonics

remember that these effects are not isolated, but part of the larger Earth system. Furthermore, remember that Earth goes through time cycles that are much longer than humans are used to. Although they may have disastrous short-term effects on the human species, global warming and cooling are also part of a longer-term cycle that has resulted in many glacial advances and retreats during the past 1.6 million years. Because of their geologic perspective, geologists can make vital contributions to the debate on global warming.

They can study long-term trends by analyzing deep-sea sediments, ice cores, changes in sea level during the geologic past, and the distribution of plants and animals through time.

As you read this book, keep in mind that the different topics you study are parts of a system of interconnected components, and not isolated pieces of information. Examined in this manner, the continuous evolution of Earth and its life is not a series of isolated and unrelated events, but a dynamic interaction among it various subsystems.

WHAT IS GEOLOGY?

What is geology and what do geologists do? **Geology,** from the Greek *geo* and *logos,* is defined as the study of Earth. It is generally divided into two broad areas: physical geology and historical geology. *Physical geology* is the study of Earth materials, such as minerals and rocks, as well as the processes operating within Earth and on its surface. *Historical geology* examines the origin and evolution of Earth, its continents, oceans, atmosphere, and life.

The discipline of geology is so broad that it is subdivided into many different fields or specialties. Table 1.2 shows many of the diverse fields of geology and their relationship to the sciences of astronomy, biology, chemistry, and physics.

Nearly every aspect of geology has some economic or environmental relevance. Many geologists are involved in exploration for mineral and energy resources, using their specialized knowledge to locate the natural resources on which our industrialized society is based. As the demand for these nonrenewable resources increases, geologists apply the basic principles of geology in increasingly sophisticated ways to help focus their attention on areas with a high potential for economic success.

Although some geologists work on locating mineral and energy resources, an extremely important role, other geologists use their expertise to help solve environmental problems. Some geologists find groundwater for the ever-burgeoning needs of communities and industries or monitor surface and underground water pollution and suggest ways to clean it up. Geologic engineers help find safe locations for dams, waste-disposal sites, and power plants and design earthquake-resistant buildings.

Table 1.2

Specialties of Geology and Their Broad Relationship to the Other Sciences

Specialty	Area of Study	Related Science
Geochronology	Time and history of Earth	Astronomy
Planetary geology	Geology of the planets	
Paleontology	Fossils	Biology
Economic geology	Mineral and energy resources	
Environmental geology	Environment	
Geochemistry	Chemistry of Earth	Chemistry
Hydrogeology	Water resources	
Mineralogy	Minerals	
Petrology	Rocks	
Geophysics	Earth's interior	Physics
Structural geology	Rock deformation	
Seismology	Earthquakes	
Geomorphology	Landforms	
Oceanography	Oceans	
Paleogeography	Ancient geographic features and locations	
Stratigraphy/sedimentology	Layered rocks and sediments	

Geologists also make short- and long-range predictions about earthquakes and volcanic eruptions and the potential destruction that may result. In addition, they work with civil defense planners to help draw up contingency plans should such natural disasters occur.

As this brief survey illustrates, geologists pursue a wide variety of careers and roles. As the world's population increases and makes greater demands on Earth's limited resources, we will depend even more on geologists and their expertise.

HOW DOES GEOLOGY RELATE TO THE HUMAN EXPERIENCE?

Many people are surprised at the extent to which we depend on geology in our everyday lives and also at the numerous references to geology in the arts, music, and literature. Many sketches and paintings represent rocks and landscapes realistically. Examples by famous artists include Leonardo da Vinci's *Virgin of the Rocks* and *Virgin and Child with Saint Anne,* Giovanni Bellini's *Saint Francis in Ecstasy* and *Saint Jerome,* and Asher Brown Durand's *Kindred Spirits* (■ Figure 1.3).

In the field of music, Ferde Grofé's *Grand Canyon Suite* was no doubt inspired by the grandeur and timelessness of Arizona's Grand Canyon and its vast rock exposures (■ Figure 1.4). The rocks on the island of Staffa in the Inner Hebrides provided the inspiration for Felix Mendelssohn's famous *Hebrides* Overture.

References to geology abound in *The German Legends of the Brothers Grimm.* Jules Verne's *Journey to the Center of the Earth* describes an expedition into Earth's interior. On one level, the poem "Ozymandias" by Percy B. Shelley deals with the fact that nothing lasts forever and even solid rock eventually disintegrates under the ravages of time and weathering. Even comics contain references to geology. Two of the best known are *B.C.* by Johnny Hart and *The Far Side* by Gary Larson (■ Figure 1.5).

Geology has also played an important role in history. Wars have been fought for the control of such natural resources as oil, gas, gold, silver, diamonds, and other valuable minerals. Empires throughout history have risen and fallen on the distribution and exploitation of natural resources. The configuration of Earth's surface, or its *topography,* which is

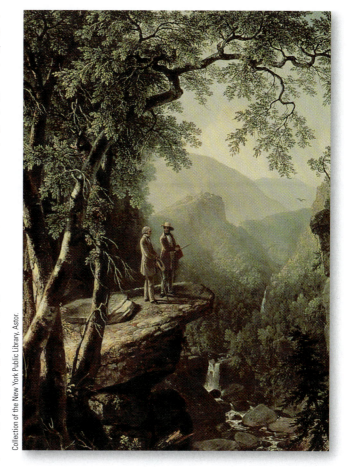

Collection of the New York Public Library, Astor.

■ **Figure 1.3**

Kindred Spirits by Asher Brown Durand (1849) realistically depicts the layered rocks along gorges in the Catskill Mountains of New York State. Durand was one of numerous artists of the 19th-century Hudson River School, which was known for realistic landscapes. This painting was done to show Durand conversing with the recently deceased Thomas Cole, the original founding force of the Hudson River School.

■ **Figure 1.4**

Ferde Grofé's *Grand Canyon Suite* pays homage to the grandeur of Arizona's Grand Canyon.

Reed Wicander

"You know, I used to like this hobby. ... But shoot!
Seems like *everybody*'s got a rock collection."

■ Figure 1.5

References to geology are frequently found in comics, as this *Far Side* cartoon by Gary Larson illustrates.

shaped by geologic agents, plays a critical role in military tactics. Natural barriers, such as mountain ranges and rivers have frequently served as political boundaries.

HOW DOES GEOLOGY AFFECT OUR EVERYDAY LIVES?

Most readers of this book will not become professional geologists. Everyone, however, should have a basic understanding of the geologic processes that ultimately affect all of us. We can trace many connections between geology and various aspects of our lives. Natural events or disasters, by their sheer magnitude, provide perhaps the most obvious connection. Less apparent, but equally significant, are the connections between geology and economic, social, and political issues.

Natural Events

Events such as destructive volcanic eruptions, devastating earthquakes, disastrous landslides, gigantic sea waves, floods, and droughts make headlines and affect many people in obvious ways. Although we cannot prevent most of these natural disasters from happening, the more knowledge we have about what causes them, the better we will be able to predict, and possibly control, the severity of their impact.

Economics and Politics

Equally important, but not always as well understood or appreciated, is the connection between geology and economic and political power. Mineral and energy resources are not equally distributed and no country is self-sufficient in all of them. Throughout history, people have fought wars to secure these resources. We need look no farther than 1990–1991 to see that the United States was involved in the Gulf War largely because it needed to protect its oil interests in that region. Mineral and energy availability and needs in many cases shape foreign policy. The sanctions imposed by the United States on South Africa in 1986, for example, did not include most of the important minerals we had been importing and needed for our industrialized society such as platinum-group minerals. Many foreign policies and treaties develop from the need to acquire and maintain adequate supplies of mineral and energy resources.

Our Role As Decision Makers

You may become involved in geologic decisions in various ways—for instance, as a member of a planning board or as a property owner with mineral rights. In such cases, you must have a basic knowledge of geology to make informed decisions. Furthermore, many professionals must deal with geologic issues as part of their jobs. For example, lawyers are becoming more involved in issues ranging from ownership of natural resources to how development activities affect the environment. As government plays a greater role in environmental issues and regulations, members of Congress have increased the number of staff devoted to studying issues related to the environment and geology.

Consumers and Citizens

Most people are unaware of the extent to which geology affects their lives. If issues like nonrenewable energy resources, waste disposal, and pollution seem simply too far removed or too complex to be fully appreciated, consider for a moment just how dependent we are on geology in our daily routines.

Much of the electricity for our appliances comes from the burning of coal, oil, or natural gas or from uranium consumed in nuclear-generating plants. It is geologists who locate the coal, petroleum, and uranium. The copper or other metal wires through which electricity travels are manufactured from materials found as the result of mineral exploration. The buildings we live and work in owe their very existence to geologic resources. Consider the concrete foundation (concrete is a mixture of clay, sand, or gravel, and limestone), the drywall (made largely from the mineral gypsum), the windows (the mineral quartz is the principal ingredient in the manufacture of glass), and the metal or plastic plumbing fixtures inside the building (the metals are from ore deposits, and the plastics are most likely manufactured from petroleum distillates of crude oil).

When we go to work, the car or public transportation we use is powered and lubricated by some type of petroleum by-product and is constructed of metal alloys and plastics. And the roads or rails we ride over come from geologic materials, such as gravel, asphalt, concrete, or steel. All these items are the result of processing geologic resources.

As individuals and societies, we enjoy a standard of living that is obviously directly dependent on the consumption of geologic materials. Therefore we need to be aware of geology and of how our use and misuse of geologic resources may affect the delicate balance of nature and irrevocably alter our culture as well as our environment.

Sustainable Development

The concept of *sustainable development* has received increasing attention, particularly since the United Nations Conference on Environment and Development met in Rio de Janeiro, Brazil, during the summer of 1992. This important concept puts satisfying basic human needs side by side with safeguarding our environment to ensure continued economic development. By redefining "wealth" to include such natural capital as clean air and water, as well as productive land, we can take appropriate measures to ensure that future generations have sufficient natural resources to maintain and improve their standard of living.

If we are to have a world in which poverty is not widespread, then we must develop policies that encourage management of our natural resources along with continuing economic development. A growing global population will mean increased demand for food, water, and natural resources, particularly nonrenewable mineral and energy resources. Geologists will play an important role in meeting these demands by locating the needed resources and ensur-

What Would You Do ?

The concept of sustainable development links satisfying basic human needs with safeguarding our environment to ensure continued economic development. The standard of living we enjoy depends directly on the consumption of geologic materials. As the president of a large multinational mining company, discuss how you might balance the need to extract valuable ore deposits and turn a profit for your company with the need to safeguard the environment, particularly if these deposits are located in an underdeveloped country with no environmental protection laws.

ing protection of the environment for the benefit of future generations.

GLOBAL GEOLOGIC AND ENVIRONMENTAL ISSUES FACING HUMANKIND

Most scientists would argue that the greatest environmental problem facing the world today is overpopulation (■ Figure 1.6). With the world's population surpassing 6 billion in 1999, projections indicate that this number will grow by at least another billion during the next two

■ Figure 1.6

A bathing fair at Pushkar, Rajasthan, India. Overpopulation is the greatest environmental problem facing the world today. Until the world's increasing population is brought under control, people will continue to strain Earth's limited resources.

GEOPROFILE

DAVID R. WUNSCH

Geology and Public Policy

David R. Wunsch is the State Geologist and Director of the New Hampshire Geological Survey. He is also an adjunct professor at the University of New Hampshire and a visiting scholar at Dartmouth College. Previously he spent 15 years in the Water Section of the Kentucky Geological Survey, where he served as the coordinator of the Coal-Field Hydrology Program. Dr. Wunsch earned a B.A. in geology from SUNY Oneonta, an M.S. in geology from the University of Akron, and a Ph.D. in hydrogeology from the University of Kentucky.

Unlike many of my friends, I consider myself very lucky because I seemed to know from a very early age what I wanted to do with my life—become a scientist. My first scientific interest was in meteorology; I aspired to become a severe storm chaser in the 1970s, *way before* the Weather Channel made it cool. However, the old adage that teachers change lives rang true for me because a geology professor I had as an undergraduate student influenced me to instead study geology. I was captivated by his command of this intriguing topic, along with his wit, artistic talents, and comfortable lecture style.

I became a geologist also because of my innate curiosity about Earth and a yearning to work at least part of the time outdoors, and because it seemed to be a profession that could satisfy my quest for excitement. Throughout my college years I held a series of summer jobs that gave me a bit of geologically related experience

while also providing excitement at times. For example, one of my more pleasant summer employment experiences was as a park naturalist, where I led nature hikes and geology tours within a beautiful state park. However, these tranquil memories are balanced by

some harrowing ones, such as almost being asphyxiated by methane while working as an oil-field roughneck. My colleagues and I were trying to cap a highly pressurized gas well that was blowing crude oil and gas after being rammed by an errant bulldozer.

As I look back at the academic and professional choices I made early in my career, it is clear that my family, friends, and even the geography of my hometown had a profound influence on my decisions and career track. For example, my mother spent most of her career as a dental assistant with two prominent local dentists. One advocated strongly for adding fluoride to drinking water because of its supposed beneficial dental effects, and the other was avidly against it. The fluoride debate interested me such that I jumped at the opportunity to study the natural occurrence of fluoride in groundwater as the topic of my master's thesis. I continued my research on fluoride in groundwater, as well as other aspects of groundwater chemistry, for my Ph.D. dissertation. To further demonstrate the influence of my background on later decisions in life, I can point to the fact that my hometown in upstate New York has the rare distinction of being located on an Indian reservation. The affairs of the local, state, and federal government, especially as they pertained to our

decades, bringing Earth's human population to more than 7 billion. Though this may not seem to be a geologic problem, remember that these people must be fed, housed, and clothed, and all with a minimal impact on the environment. Some of this population growth will be in areas that are already at risk from such geologic hazards as earthquakes, volcanic eruptions, and mass wasting. Safe and adequate water supplies must be found and kept from being polluted. More oil, gas, coal,

and alternative energy resources must be discovered and used to provide the energy to fuel the economies of nations with ever-increasing populations. New mineral resources must be located. In addition, ways to reduce usage and reuse materials must be devised so as to decrease dependence on new sources of these materials.

The problems of overpopulation and how it affects the global ecosystem vary from country to country. For many poor and nonindustrialized countries, the problem is too

Dr. David Wunsch, State Geologist of New Hampshire, collects a rock sample from a boulder that was part of the *Old Man of the Mountain,* the New Hampshire landmark that collapsed from its perch on Profile Mountain on May 3, 2003. Dr. Wunsch is a member of the governor's special task force charged with developing a plan to revitalize the memory of the Old Man.

unique hometown, were regular topics at the dinner table. Probably as influential to my budding interest in politics was my father's involvement in the local political scene. His actions piqued my interest in public policy and served as an example to get involved.

I spent most of my professional and scientific career at the Kentucky Geological Survey, where I specialized in water, environmental, and geotechnical issues in the Kentucky coalfields. In 1998 I was awarded a Congressional Science Fellowship spon-

sored by the American Geological Institute. The fellowship program, coordinated through the American Association for the Advancement of Science, selects scientists to work as science advisors for the U.S. Congress and other federal agencies in Washington, DC. I ended up working on the congressional staff of the House of Representatives' Subcommittee on Energy and Mineral Resources. My geological background allowed me to immediately contribute to the committee, which oversees agencies in the Department of the Interior and has jurisdiction over issues related to mining, energy, and the use of federal lands. My fellowship year was one of the best years of my life. Although the work was challenging and stressful at times, there was a great deal of personal satisfaction because I was able to present scientific considerations to the policymakers, which led to better government overall. My fellowship also presented me with a wonderful mid-career change and gave me the opportunity to learn and gain new expertise (such as an in-depth understanding of the legislative process). This experience rekindled my basic thirst for science because my multiple exposures to leading scientists, institutes, and agencies in the nation's capitol has expanded my knowledge base.

My practice at integrating science and policy has been an absolute benefit to me in my current position as State Geologist. Based on my experience, the time is ripe for young people to pursue scientific degrees and then look into nontraditional careers in public policy, government affairs, and outreach. In this way, they can grasp an exciting opportunity to utilize their scientific and technical background while giving something back to this wonderful country in which we live.

many people and not enough food. For the more developed and industrialized countries, it is too many people rapidly depleting both the nonrenewable and renewable natural resource base. And in the most industrially developed countries, it is people producing more pollutants than the environment can safely recycle on a human time scale. The common thread tying all of these varied situations together is the environmental imbalance created by a human population exceeding Earth's carrying capacity.

One result of environmental imbalance and an excellent example of the interrelationships among Earth's systems and subsystems is global warming caused by the greenhouse effect. Carbon dioxide is produced as a by-product of respiration and the burning of organic material. As such, it is a component of the global ecosystem and is constantly being recycled as part of the carbon cycle. The concern in recent years over the increase in atmospheric carbon dioxide relates to its role in the greenhouse effect.

GEOFOCUS

1.1

The Kyoto Protocol

At the 1997 Climate Treaty Meeting in Kyoto, Japan, a group of nations took the first steps in reducing greenhouse-causing emissions by negotiating a treaty calling for the United States and 37 other industrialized nations to reduce their greenhouse emissions by an average of 5.2% below their 1990 levels between 2008 and 2012. Whereas this might not seem like a significant decrease, it should be kept in mind that the industrialized nations, which comprise only 20% of the world's population, have to date emitted 90% of all human-produced carbon emissions.

The Kyoto Protocol requires the largest polluters to make the greatest sacrifices. The United States must cut its greenhouse gas emissions to 7% below 1990 levels, and the European Union and Japan must reduce their 1990 emissions by 8% and 6%, respectively. The greenhouse-causing gases that must be reduced are carbon dioxide, methane, nitrous oxide, hydrofluorocarbons, perfluorocarbons, and sulphur hexafluoride. It was also agreed that further reductions will be phased in beginning after 2012.

It should be noted that developing countries are not required to reduce their greenhouse-causing gases unless they want to, and that countries with extensive forests and jungles, which act as carbon dioxide "sinks," can receive credit toward their emission-reduction goals. Research indicates that deforestation of large areas, particularly in the tropics, is also a cause of increased levels of carbon dioxide. This is because trees and plants use carbon dioxide in photosynthesis, thus acting as carbon dioxide "sinks" by removing it from the atmosphere. With a decrease in the global vegetation cover, less carbon dioxide is removed from the atmosphere by photosynthesis, and carbon dioxide levels can increase.

Whereas a reduction in greenhouse-causing emissions is certainly desirable, the economic impact on industrialized countries must also be considered. It is estimated that meeting the Kyoto reductions could cause as much as a 2% drop in the U.S. gross domestic production, which is equivalent to about $227 billion. Furthermore, the economic impact would not be spread evenly throughout the population, and it could have devastating effects on many industries, particularly those that produce and use large quantities of energy. Thus the possible benefits to be gained by reducing greenhouse-causing emissions must be weighed against the possible economic hardships that meeting such reductions would entail.

The recycling of carbon dioxide between the crust and the atmosphere is an important climatic regulator because carbon dioxide, as well as other gases such as methane, nitrous oxide, chlorofluorocarbons, and water vapor, allows sunlight to pass through them but trap the heat reflected back from Earth's surface. Heat is thus retained, causing the temperature of Earth's surface and, more important, the atmosphere to increase, producing the greenhouse effect.

With industrialization and its accompanying burning of tremendous amounts of fossil fuels, carbon dioxide levels in the atmosphere have been steadily increasing since about 1880, causing many scientists to conclude that a global warming trend has already begun and will result in severe global climatic shifts (see Geo-Focus 1.1). Most computer models based on the current rate of increase in greenhouse gases show Earth warming as a whole by as much as 5°C during the next 100 years. Such a temperature change will be uneven, however, with the greatest warming occurring in the higher latitudes. As a consequence of this warming, rainfall patterns will shift dramatically, which will have a major effect on the largest grain-producing areas of the world, such as the American Midwest. Drier and hotter conditions will intensify the severity and frequency of droughts, leading to more crop failure and higher food prices. With such shifts in climate, Earth's deserts may grow, which will remove valuable crop and grazing lands.

With continued global warming, mean sea level will also rise as icecaps and glaciers melt and contribute their water to the world's oceans. It is predicted that by the 2050s, sea level will rise 21 cm, increasing the number of people at risk from flooding in coastal areas by approximately 20 million.

We would be remiss, however, if we did not point out that many other scientists are not convinced that the global warming trend is the direct result of increased human activity related to industrialization. They point out that although the amount of greenhouse gases has increased, we are still uncertain about their rate of generation and rate of removal, and whether the rise in global temperature during the past century resulted from normal climatic variations through time or from human activity. Furthermore, they point out that even if there is a general global warming during the next 100 years, it is not certain that the dire predictions made by proponents of global warming will come true.

Earth, as we know, is a remarkably complex system, with many feedback mechanisms and interconnections throughout its various subsystems and cycles. It is very difficult to predict all of the consequences that global warming would have for atmospheric and oceanic circulation patterns.

PHYSICAL
Geology⇌Now   Click Geology Interactive to work through an activity on absorbers.

ORIGIN OF THE UNIVERSE AND SOLAR SYSTEM, AND EARTH'S PLACE IN THEM

How did the universe begin? What has been its history? Is it infinite? What is its eventual fate? These are just some of the basic questions people have asked and wondered about since they first looked into the nighttime sky and saw the vastness of the universe beyond Earth.

Origin of the Universe—Did It Begin with a Big Bang?

Most scientists think that the universe originated about 15 billion years ago in what is popularly called the **Big Bang.** In a region infinitely smaller than an atom, both time and space were set at zero. Therefore there is no "before the Big Bang," only what occurred after it. The reason is that space and time are unalterably linked to form a space–time continuum demonstrated by Einstein's theory of relativity. Without space, there can be no time.

How do we know the Big Bang took place approximately 15 billion years ago? Why couldn't the universe have always existed as we know it today? Two fundamental phenomena indicate that the Big Bang occurred. First, the universe is expanding. When astronomers look beyond our own solar system, they observe that everywhere in the universe galaxies are apparently moving away from each other at tremendous speeds. By measuring this expansion rate, astronomers can calculate how long ago the galaxies were all together at a single point. Second, everywhere in the universe there is a pervasive background radiation of 2.7°K above absolute zero (absolute zero equals $-273°C$). This background radiation is thought to be the faint afterglow of the Big Bang.

According to the currently accepted theory, matter as we know it did not exist at the moment of the Big Bang, and the universe consisted of pure energy. During the first second following the Big Bang, the four basic forces— *gravity* (the attraction of one body toward another), *electromagnetic force* (combines electricity and magnetism into one force and binds atoms into molecules), *strong nuclear force* (binds protons and neutrons together), and *weak nuclear force* (responsible for the breakdown of an atom's nucleus, producing radioactive decay)—separated and the universe experienced enormous expansion. About 300,000 years later, the universe was cool enough for complete atoms of hydrogen and helium atoms to form, and photons (the energetic particles of light) separated from matter and light burst forth for the first time.

During the next 200 million years, as the universe continued expanding and cooling, stars and galaxies began to form and the chemical makeup of the universe changed. Initially the universe was 100% hydrogen and helium, whereas today it is 98% hydrogen and helium and 2% all other elements by weight. How did such a change in the universe's composition occur? Throughout their life cycle, stars undergo many nuclear reactions by which lighter elements are converted into heavier elements by nuclear fusion. When a star dies, often explosively, the heavier elements that were formed in its core are returned to interstellar space and are available for inclusion in new stars. In this way, the composition of the universe is gradually enhanced in heavier elements.

Our Solar System—Its Origin and Evolution

Our solar system, which is part of the Milky Way Galaxy, consists of a Sun, 9 planets, 101 known moons or satellites (although this number keeps changing with the discovery of new moons and satellites surrounding

What Would You Do?

An important environmental issue facing the world today is global warming. How can this problem be approached from a global systems perspective? What are the possible consequences of global warming, and can we really do anything about it? Are there ways to tell whether global warming occurred in the geologic past?

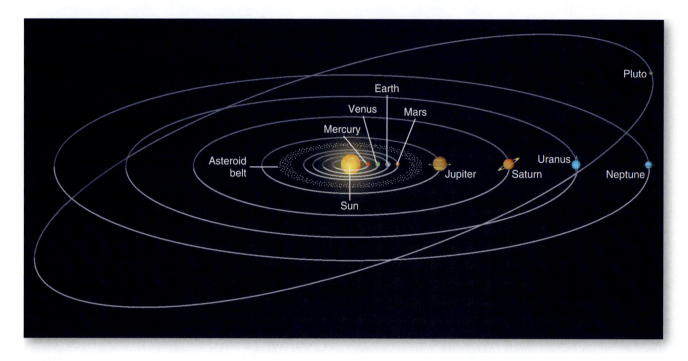

■ Figure 1.7

Diagrammatic representation of the solar system, showing the planets and their orbits around the Sun.

the Jovian planets), a tremendous number of asteroids—most of which orbit the Sun in a zone between Mars and Jupiter—and millions of comets and meteorites as well as interplanetary dust and gases (■ Figure 1.7). Any theory formulated to explain the origin and evolution of our solar system must therefore take into account its various features and characteristics.

Various scientific theories of the origin of the solar system have been suggested, modified, and discarded since the French scientist and philosopher René Descartes first proposed, in 1644, that the solar system formed from a gigantic whirlpool within a universal fluid. Today the **solar nebula theory** for the origin of our solar system involves the condensation and collapse of interstellar material in a spiral arm of the Milky Way Galaxy.

The collapse of this cloud of gases and small grains into a counterclockwise rotating disk concentrated about 90% of the material in the central part of the disk and formed an embryonic Sun, around which swirled a rotating cloud of material called a *solar nebula*. Within this solar nebula were localized eddies in which gases and solid particles condensed. During the condensation process, gaseous, liquid, and solid particles began accreting into ever-larger masses called *planetesimals* (■ Figure 1.8) that collided and grew in size and mass until they eventually became planets.

The composition and evolutionary history of the planets are a consequence, in part, of their distance from the Sun (see "The Terrestrial and Jovian Planets" on pages 16 and 17). The **terrestrial planets**—Mercury, Venus, Earth, and Mars—so named because they are

similar to *terra*, Latin for "earth," are all small and composed of rock and metallic elements that condensed at the high temperatures of the inner nebula. The **Jovian planets**—Jupiter, Saturn, Uranus, and Neptune—so named because they resemble Jupiter (the roman god was also named Jove), all have small central rocky cores compared to their overall size, and are composed mostly of hydrogen, helium, ammonia, and methane, which condense at low temperatures.

While the planets were accreting, material that had been pulled into the center of the nebula also condensed, collapsed, and was heated to several million degrees by gravitational compression. The result was the birth of a star, our Sun.

During the early accretionary phase of the solar system's history, collisions between various bodies were common, as indicated by the craters on many planets and moons. Asteroids probably formed as planetesimals in a localized eddy between what eventually became Mars and Jupiter in much the same way that other planetesimals formed the terrestrial planets. The tremendous gravitational field of Jupiter, however, prevented this material from ever accreting into a planet. Comets, which are interplanetary bodies composed of loosely bound rocky and icy material, are thought to have condensed near the orbits of Uranus and Neptune.

The solar nebula theory of the formation of the solar system thus accounts for most of the characteristics of the planets and their moons, the differences in composition between the terrestrial and Jovian planets, and the presence of the asteroid belt. Based on the available

■ **Figure 1.8**

At the stage of development shown here, planetesimals have formed in the inner solar system, and large eddies of gas and dust remain at great distances from the protosun.

data, the solar nebula theory best explains the features of the solar system and provides a logical explanation for its evolutionary history.

PHYSICAL
Geology ⇌ Now   Click Geology Interactive to work through an activity on the origin of a star in the spiral arm of a galaxy.

Earth—Its Place in Our Solar System

Some 4.6 billion years ago, various planetesimals in our solar system gathered enough material together to form Earth and eight other planets. Scientists think that this

early Earth was probably cool, of generally uniform composition and density throughout, and composed mostly of silicate compounds, iron and magnesium oxides, and smaller amounts of all the other chemical elements (■ Figure 1.9a). Subsequently, when the combination of meteorite impacts, gravitational compression, and heat from radioactive decay increased the temperature of Earth enough to melt iron and nickel, this homogeneous composition disappeared (Figure 1.9b) and was replaced by a series of concentric layers of differing composition and density, resulting in a differentiated planet (Figure 1.9c).

This differentiation into a layered planet is probably the most significant event in Earth history. Not only did it lead to the formation of a crust and eventually to continents, but it also was probably responsible for the emission of gases from the interior, which eventually led to the formation of the oceans and atmosphere.

WHY IS EARTH A DYNAMIC AND EVOLVING PLANET?

Earth is a dynamic planet that has continuously changed during its 4.6-billion-year existence. The size, shape, and geographic distribution of continents and ocean basins have changed through time, the composition of the atmosphere has evolved, and life-forms existing today differ from those that lived during the past. Mountains and hills have been worn away by erosion, and landscapes have been changed by

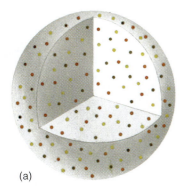

(a)

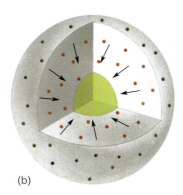

(b)

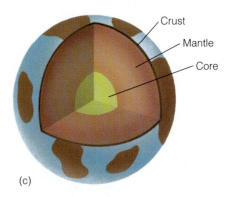

Crust
Mantle
Core

(c)

■ **Figure 1.9**

Homogeneous accretion theory for the formation of a differentiated Earth. (a) Early Earth was probably of uniform composition and density throughout. (b) Heating of early Earth reached the melting point of iron and nickel, which, being denser than silicate minerals, settled to Earth's center. At the same time, the lighter silicates flowed upward to form the mantle and the crust. (c) In this way, a differentiated Earth formed, consisting of a dense iron–nickel core, an iron-rich silicate mantle, and a silicate crust with continents and ocean basins.

The Terrestrial and Jovian Planets

The planets of our solar system can be divided into two major groups that are quite different indicating that the two underwent very different evolutionary histories. The four inner planets—Mercury, Venus, Earth, and Mars—are the terrestrial planets; they are small and dense (composed of a metallic core and silicate mantle-crust), ranging from no atmosphere (Mercury) to an oppressively thick one (Venus). The outer four planets (excluding Pluto, which some astronomers don't regard as a planet at all)—Jupiter, Saturn, Uranus, and Neptune—are the Jovian planets; they are large, ringed, low-density planets with liquid interiors surrounded by thick atmospheres.

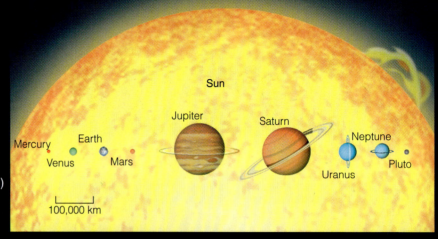

The relative sizes of the planets and the Sun. (Distances between planets are not to scale.)

100,000 km

Venus is surrounded by an oppressively thick atmosphere that completely obscures its surface. However, radar images from orbiting spacecraft reveal a wide variety of terrains, including volcanic features, folded mountain ranges, and a complex network of faults.

The **Moon** is one-fourth the diameter of Earth, has a low density relative to the terrestrial planets, and is extremely dry. Its surface is divided into low-lying dark colored plains, and light-colored highlands that are heavily cratered, attesting to a period of massive meteorite bombardment in our solar system more than 4 billion years ago. The hypothesis that best accounts for the origin of the Moon has a giant planetesimal, the size of Mars or larger, crashing into Earth 4.6 to 4.4 billion years ago, causing ejection of a large quantity of hot material that cooled and formed the Moon.

JPL/NASA

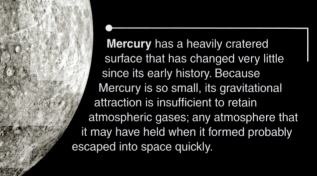

Mercury has a heavily cratered surface that has changed very little since its early history. Because Mercury is so small, its gravitational attraction is insufficient to retain atmospheric gases; any atmosphere that it may have held when it formed probably escaped into space quickly.

Earth is unique among our solar system's planets in that it has a hospitable atmosphere, oceans of water, variety of climates, and it supports life.

Mars has a thin atmos▢▢little water, and distinct seasons. Its southern hemisphere is heavily cratered like the surfaces of Mercury and the Moon. The northern hemisphere has large smooth plains, fewer craters and evidence of extensive volcanism. The largest volcano in the solar system is found in the northern hemisphere as are huge canyons, the largest of which, if present on Earth would stretch from San Francisco to New York!

Jupiter is the largest of the Jovian planets. With its moons, rings, strong magnetic field, and intense radiation belts, Jupiter is the most complex and varied planet in our solar system. Jupiter's cloudy and violent atmosphere is divided into a series of different colored bands and a variety of spots (the Great Red Spot) that interact in incredibly complex motions.

Saturn's most conspicuous feature is its ring system, consisting of thousands of rippling, spiraling bands of countless particles. The width of Saturn's rings would just reach from Earth to the Moon.

Uranus is the only planet that lies on its side, that is, its axis of rotation nearly parallels the plane of the ecliptic. Some scientists think that a collision with an Earth-sized body early in its history may have knocked Uranus on its side. Like the other Jovian planets, Uranus has a ring system, albeit a faint one.

Neptune is a dynamic stormy planet with an atmosphere similar to those of the other Jovian planets. Winds up to 2000 km/h blow over the planet, creating tremendous storms, the largest of which, the Great Dark Spot, seen in the center, is nearly as big as Earth and is similar to the Great Red Spot on Jupiter.

the forces of wind, water, and ice. Volcanic eruptions and earthquakes reveal an active interior, and folded and fractured rocks indicate the tremendous power of Earth's internal forces.

Earth consists of three concentric layers: the core, the mantle, and the crust (■ Figure 1.10). This orderly division results from density differences between the layers as a function of variations in composition, temperature, and pressure.

The **core** has a calculated density of 10 to 13 grams per cubic centimeter (g/cm^3) and occupies about 16% of Earth's total volume. Seismic (earthquake) data indicate that the core consists of a small, solid inner region and a larger, apparently liquid, outer portion. Both are thought to consist largely of iron and a small amount of nickel.

The **mantle** surrounds the core and comprises about 83% of Earth's volume. It is less dense than the core (3.3–5.7 g/cm^3) and is thought to be composed largely of *peridotite*, a dark, dense igneous rock containing abundant iron and magnesium. The mantle can be divided into three distinct zones based on physical characteristics. The lower mantle is solid and forms most of the volume of Earth's interior. The **asthenosphere** surrounds the lower mantle. It has the same composition as the lower mantle but behaves plastically and slowly flows. Partial melting within the asthenosphere generates *magma* (molten material), some of which rises to the surface because it is less dense than the rock from which it was derived. The upper mantle surrounds the asthenosphere. The solid upper mantle and the overlying crust constitute the **lithosphere,** which is broken into numerous individual pieces called **plates** that move over the asthenosphere as a result of underlying *convection cells* (■ Figure 1.11). Interactions of these plates are responsible for such phenomena as earthquakes, volcanic eruptions, and the formation of mountain ranges and ocean basins.

The **crust,** Earth's outermost layer, consists of two types. *Continental crust* is thick (20–90 km), has an average density of 2.7 g/cm^3, and contains considerable silicon and aluminum. *Oceanic crust* is thin (5–10 km), denser than continental crust (3.0 g/cm^3), and is composed of dark-colored igneous rock.

PHYSICAL Geology⇌Now Click Geology Interactive to work though an activity on core studies through Earth's layers.

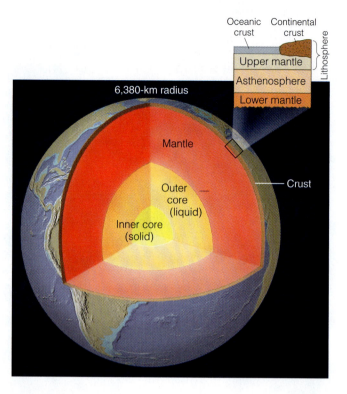

■ Figure 1.10

A cross section of Earth, illustrating the core, mantle, and crust. The enlarged portion shows the relationship between the lithosphere (composed of the continental crust, oceanic crust, and solid upper mantle) and the underlying asthenosphere and lower mantle.

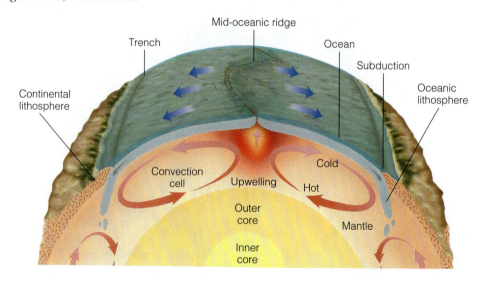

■ Figure 1.11

Earth's plates are thought to move as a result of underlying mantle convection cells in which warm material from deep within Earth rises toward the surface, cools, and then, upon losing heat, descends back into the interior. The movement of these convection cells is thought to be the mechanism responsible for the movement of Earth's plates, as shown in this diagrammatic cross section.

GEOLOGY AND THE FORMULATION OF THEORIES

The term **theory** has various meanings. In colloquial usage, it means a speculative or conjectural view of something—hence, the widespread belief that scientific theories are little more than unsubstantiated wild guesses. In scientific usage, however, a theory is a coherent explanation for one or several related natural phenomena supported by a large body of objective evidence. From a theory are derived predictive statements that can be tested by observations and/or experiments so that their validity can be assessed. The law of universal gravitation is an example of a theory describing the attraction between masses (an apple and Earth in the popularized account of Newton and his discovery).

Theories are formulated through the process known as the **scientific method.** This method is an orderly, logical approach that involves gathering and analyzing the facts or data about the problem under consideration. Tentative explanations, or **hypotheses,** are then formulated to explain the observed phenomena. Next the hypotheses are tested to see if what they predicted actually occurs in a given situation. Finally, if one of the hypotheses is found, after repeated tests, to explain the phenomena, then the hypothesis is proposed as a theory. Remember, however, that in science even a theory is still subject to further testing and refinement as new data become available.

The fact that a scientific theory can be tested and is subject to such testing separates science from other forms of human inquiry. Because scientific theories can be tested, they have the potential of being supported or even proved wrong. Accordingly, science must proceed without any appeal to beliefs or supernatural explanations, not because such beliefs or explanations are necessarily untrue but because we have no way to investigate them. For this reason, science makes no claim about the existence or nonexistence of a supernatural or spiritual realm.

Each scientific discipline has certain theories that are of particular importance for that discipline. In geology, the formulation of plate tectonic theory has changed the way geologists view Earth. Geologists now view Earth from a global perspective in which all its subsystems and cycles are interconnected, and Earth history is seen to be a continuum of interrelated events that are part of a global pattern of change.

PLATE TECTONIC THEORY

The recognition that the lithosphere is divided into rigid plates that move over the asthenosphere forms the foundation of **plate tectonic theory** (■ Figure 1.12). Zones of volcanic activity,

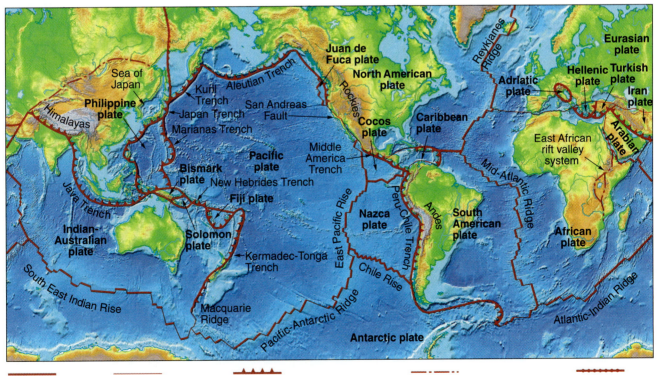

Ridge axis
Divergent boundary

Transform

Subduction zone
Convergent boundary

Zones of extension within continents

Uncertain plate boundary

■ **Figure 1.12**

Earth's lithosphere is divided into rigid plates of various sizes that move over the asthenosphere.

earthquakes, or both mark most plate boundaries. Along these boundaries plates diverge, converge, or slide sideways past each other (■ Figure 1.13).

The acceptance of plate tectonic theory is recognized as a major milestone in the geologic sciences, comparable to the revolution Darwin's theory of evolution caused in biology. Plate tectonics has provided a framework for interpreting the composition, structure, and internal processes of Earth on a global scale. It has led to the realization that the continents and ocean basins are part of a lithosphere–asthenosphere–hydrosphere system that evolved together with Earth's interior (Table 1.3).

A revolutionary concept when it was proposed in the 1960s, plate tectonic theory has had significant and far-reaching consequences in all fields of geology because it provides the basis for relating many seemingly unrelated phenomena. Besides being responsible for the major features of Earth's crust, plate movements also affect the formation and occurrence of Earth's natural resources, as well as influencing the distribution and evolution of the world's biota.

The impact of plate tectonic theory has been particularly notable in the interpretation of Earth history. For example, the Appalachian Mountains in eastern North America and the mountain ranges of Greenland, Scotland, Norway, and Sweden are not the result of unrelated mountain-building episodes but, rather, are part of a larger mountain-building event that involved the closing of an ancient "Atlantic Ocean" and the formation of the supercontinent Pangaea about 245 million years ago.

THE ROCK CYCLE

A rock is an aggregate of **minerals,** which are naturally occurring, inorganic, crystalline solids that have definite physical and chemical properties. Minerals are composed of elements such as oxygen, silicon, and aluminum, and elements are made up of atoms, the smallest particles of matter that retain the characteristics of an element. More than 3500 minerals have been identified and described, but only about a dozen make up the bulk of the rocks in Earth's crust (see Table 2.4).

Geologists recognize three major groups of rocks—*igneous, sedimentary,* and *metamorphic*—each of which is characterized by its mode of formation. Each group contains a variety of individual rock types that differ from one another on the basis of composition or texture (the size, shape, and arrangement of mineral grains).

The **rock cycle** provides a way of viewing the interrelationships between Earth's internal and external processes (■ Figure 1.14). It relates the three rock groups to each other; to superficial processes such as weathering, transportation, and deposition; and to internal processes such as magma generation and metamorphism. Plate movement is the mechanism

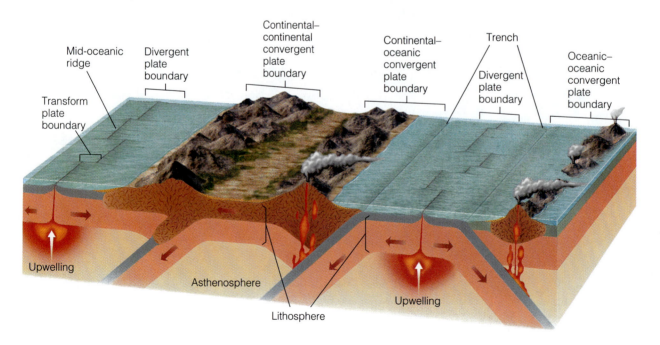

■ **Figure 1.13**

An idealized cross section illustrating the relationship between the lithosphere and the underlying asthenosphere and the three principal types of plate boundaries: divergent, convergent, and transform.

Table 1.3

Plate Tectonics and Earth Systems

Solid Earth

Plate tectonics is driven by convection in the mantle and in turn drives mountain-building and associated igneous and metamorphic activity.

Atmosphere

Arrangement of continents affects solar heating and cooling, and thus winds and weather systems. Rapid plate spreading and hot-spot activity may release volcanic carbon dioxide and affect global climate.

Hydrosphere

Continental arrangement affects ocean currents. Rate of spreading affects volume of mid-oceanic ridges and hence sea level. Placement of continents may contribute to onset of ice ages.

Biosphere

Movement of continents creates corridors or barriers to migration, the creation of ecological niches, and transport of habitats into more or less favorable climates.

Extraterrestrial

Arrangement of continents affects free circulation of ocean tides and influences tidal slowing of Earth's rotation.

Source: Adapted by permission from Stephen Dutch, James S. Monroe, and Joseph Moran, *Earth Science* (Minneapolis/St. Paul: West Publishing Co., 1997).

responsible for recycling rock materials and therefore drives the rock cycle.

Igneous rocks result when magma crystallizes or volcanic ejecta such as ash accumulate and consolidate. As magma cools, minerals crystallize, and the resulting rock is characterized by interlocking mineral grains. Magma that cools slowly beneath the surface produces *intrusive igneous rocks* (■ Figure 1.15a); magma that cools at the surface produces *extrusive igneous rocks* (Figure 1.15b).

Rocks exposed at Earth's surface are broken into particles and dissolved by various weathering processes. The particles and dissolved materials may be transported by wind, water, or ice and eventually deposited as *sediment*. This sediment may then be compacted or cemented into sedimentary rock.

Sedimentary rocks form in one of three ways: consolidation of rock fragments, precipitation of mineral matter from solution, or compaction of plant or animal remains (Figure 1.15c, d). Because sedimentary rocks form at or near Earth's surface, geologists can make inferences about the environment in which they were deposited, the transporting agent, and perhaps even something about the source from which the sediments were derived (see Chapter 6). Accordingly, sedimentary rocks are especially useful for interpreting Earth history.

Metamorphic rocks result from the alteration of other rocks, usually beneath the surface, by heat, pressure, and the chemical activity of fluids. For example,

marble, a rock preferred by many sculptors and builders, is a metamorphic rock produced when the agents of metamorphism are applied to the sedimentary rock limestone or dolostone. Metamorphic rocks are either *foliated* (Figure 1.15e) or *nonfoliated* (Figure 1.15f). Foliation, the parallel alignment of minerals due to pressure, gives the rock a layered or banded appearance.

How Are the Rock Cycle and Plate Tectonics Related?

Interactions between plates determine, to some extent, which of the three rock groups will form (■ Figure 1.16). For example, when plates converge, heat and pressure generated along the plate boundary may lead to igneous activity and metamorphism within the descending oceanic plate, thus producing various igneous and metamorphic rocks.

Some of the sediments and sedimentary rocks on the descending plate are melted, whereas other sediments and sedimentary rocks along the boundary of the nondescending plate are metamorphosed by the heat and pressure generated along the converging plate boundary. Later, the mountain range or chain of volcanic islands formed along the convergent plate boundary will be weathered and eroded, and the new sediments will be transported to the ocean to begin yet another cycle.

The interrelationship between the rock cycle and plate tectonics is just one example of how Earth's various

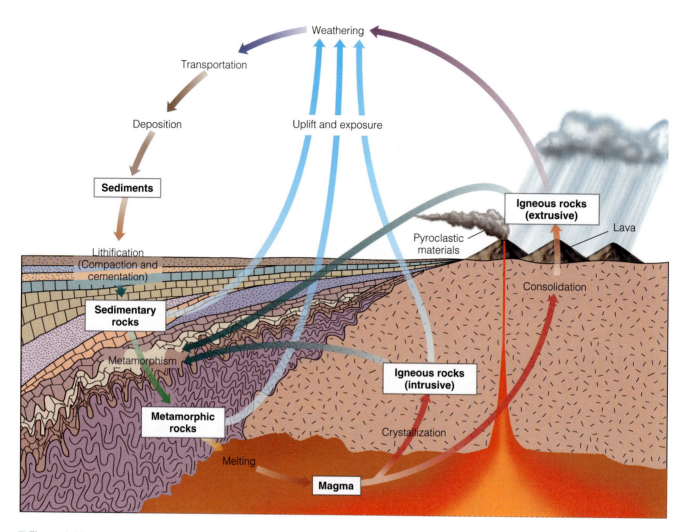

■ **Figure 1.14**

The rock cycle showing the interrelationships between Earth's internal and external processes and how the three major rock groups are related.

subsystems and cycles are all interrelated. Heating within Earth's interior results in convection cells that power the movement of plates, and also in magma, which forms intrusive and extrusive igneous rocks. Movement along plate boundaries may result in volcanic activity, earthquakes, and in some cases mountain building. The interaction between the atmosphere, hydrosphere, and biosphere contributes to the weathering of rocks exposed on Earth's surface. Plates descending back into Earth's interior are subjected to increasing heat and pressure, which may lead to metamorphism as well as the generation of magma, and yet another recycling of materials.

PHYSICAL
Geology⇌Now Click Geology Interactive to work through an activity on rock cycles.

GEOLOGIC TIME AND UNIFORMITARIANISM

An appreciation of the immensity of geologic time is central to understanding the evolution of Earth and its biota. Indeed, time is one of the main aspects that sets geology apart from the other sciences, except astronomy. Most people have difficulty comprehending geologic time because they tend to think in terms of the human perspective—seconds, hours, days, and years. Ancient history is what occurred hundreds or even thousands of years ago. When geologists talk of ancient geologic history, however, they are referring to events that happened hundreds of millions or even billions of years ago. To a geologist, recent geologic events are those that occurred within the last million years or so.

(a) Granite

(b) Basalt

(c) Conglomerate

(d) Limestone

(e) Gneiss

(f) Quartzite

■ **Figure 1.15**

Hand specimens of common igneous (a, b), sedimentary (c, d), and metamorphic (e, f) rocks. (a) Granite, an intrusive igneous rock. (b) Basalt, an extrusive igneous rock. (c) Conglomerate, a sedimentary rock formed by the consolidation of rock fragments. (d) Limestone, a sedimentary rock formed by the extraction of mineral matter from seawater by organisms or by the inorganic precipitation of the mineral calcite from seawater. (e) Gneiss, a foliated metamorphic rock. (f) Quartzite, a nonfoliated metamorphic rock.

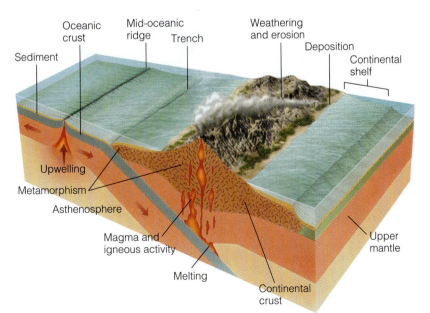

■ **Figure 1.16**

Plate tectonics and the rock cycle. The cross section shows how the three major rock groups—igneous, metamorphic, and sedimentary—are recycled through both the continental and oceanic regions.

GEOLOGY
IN UNEXPECTED PLACES

A Bit of Egypt in Central Park, New York City, and London, England

Most tourists visiting Central Park in New York City probably don't expect to see a relic of Egypt's past rising above the trees. But that is exactly what you will find if you visit Cleopatra's Needle, located behind the Metropolitan Museum of Art.

The ruler of Egypt, the Khedive Ismail Pasha, promised the United States in 1869 an obelisk in appreciation for its aid in the construction of the Suez Canal. The obelisk and its companion, which is now in London, were originally erected in front of the great temple of Heliopolis during the reign of Pharaoh Tuthmosis III (1504–1450 B.C.). They were later moved to Alexandria, where they remained until one was transported to London in 1878 and erected along the Thames on the Victoria Embankment and the other was transported to New York in 1880 and erected at its present site in 1881. Both are dubbed "Cleopatra's Needle," even though they have nothing to do with

Cleopatra and their inscriptions celebrate the pharaohs of ancient Egypt.

Henry Gorridge, a lieutenant commander of the U.S. Navy, was assigned the task of transporting the second of the twin obelisks and its pedestal to New York in 1880. The obelisk arrived at the Quarantine Station in New York in July 1880, and then it took 112 days to transport it to its final destination in Central Park, where it was raised in early 1881 in front of more than 10,000 cheering New Yorkers.

What of the geology of the obelisks? Both are made from the red granite of Syene and have suffered the visages of weathering. The New York obelisk is 21.4 m tall and weighs 193 tons, whereas the London obelisk is 21.1 m high and weighs 187 tons. Because of air pollution, the New York obelisk is not as well preserved as the one in London, and the inscriptions on the New York "Needle" are readable on only two sides.

■ Figure 1

A jogger runs by "Cleopatra's Needle," which rises above the trees of New York's Central Park. This obelisk, a gift from the Khedive Ismail, stands sentry behind the Metropolitan Museum of Art.

■ Figure 2

View of two sides of the New York obelisk, clearly showing three columns of hieroglyphic inscriptions on one side and a nearly smooth adjacent side. Chemical weathering, resulting from air pollution, has visibly damaged the obelisk. Also note more weathering and damage to the lower portion of the obelisk on both sides.

■ Figure 3

The London obelisk, located on the Victoria Embankment along the Thames River.

It is also important to remember that Earth goes through cycles of much longer duration than the human perspective of time. Although they may have disastrous effects on the human species, global warming and cooling are part of a larger cycle that has resulted in numerous glacial advances and retreats during the past 1.6 million years. Because of their geologic perspective on time and how the various Earth subsystems and cycles are interrelated, geologists can make valuable contributions to many of the current environmental debates, such as those involving global warming and sea-level changes.

The **geologic time scale** resulted from the work of many 19th-century geologists who pieced together information from numerous rock exposures and constructed a sequential chronology based on changes in Earth's biota through time. Subsequently, with the discovery of radioactivity in 1895 and the development of various radiometric dating techniques, geologists have been able to assign absolute age dates in years to the subdivisions of the geologic time scale (■ Figure 1.17).

One of the cornerstones of geology is the **principle of uniformitarianism,** which is based on the premise that present-day processes have operated throughout geologic time. Therefore, to understand and interpret geologic events from evidence preserved in rocks, we must first understand present-day processes and their results. In fact, uniformitarianism fits in completely with the system approach we are following for the study of Earth.

Uniformitarianism is a powerful principle that allows us to use present-day processes as the basis for interpreting the past and for predicting potential future events. We should keep in mind, however, that uniformitarianism does not exclude sudden or catastrophic events such as volcanic eruptions, earthquakes, landslides, or flooding. These are processes that shape our modern world, and some geologists view Earth history as a series of such short-term or punctuated events. This view is certainly in keeping with the modern principle of uniformitarianism.

Furthermore, uniformitarianism does not require that the rates and intensities of geologic processes be constant through time. We know that volcanic activity was more intense in North America 5 to 10 million years ago than it is today, and that glaciation has been more prevalent during the last several million years than in the previous 300 million years.

What uniformitarianism means is that even though the rates and intensities of geologic processes have varied during the past, the physical and chemical laws of nature have remained the same. Although Earth is in a dynamic state of change and has been ever since it formed, the processes that have shaped it during the past are the same ones operating today.

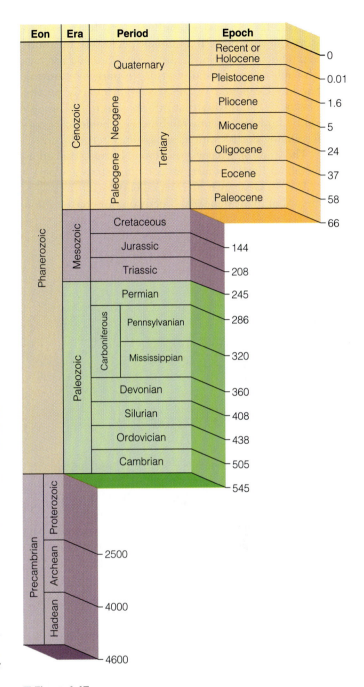

■ **Figure 1.17**

The geologic time scale. Numbers to the right of the columns are ages in millions of years before the present.

HOW DOES THE STUDY OF GEOLOGY BENEFIT US?

The most meaningful lesson to learn from the study of geology is that Earth is an extremely complex planet in which interactions among its various subsystems are taking place and have been occurring for the past 4.6 billion years. If we want to ensure the survival of the human species, we must understand how the various subsystems work and interact with each other. We can do this, in part, by studying what has happened in the past, particularly on the global scale, and using that information to try and determine how our actions might affect the delicate balance among Earth's various subsystems in the future.

The study of geology goes beyond learning numerous facts about Earth. In fact, we don't just study geology—we "live" it. Geology is an integral part of our lives. Our standard of living depends directly on our consumption of natural resources, resources that formed millions and billions of years ago. However, the way we consume natural resources and interact with the environment, as individuals and as a society, also determines our ability to pass on this standard of living to the next generation.

As you study the various topics covered in this book, keep in mind the themes discussed in this chapter and how, like the parts of a system, they are interrelated. By relating each chapter's topic to its place in the entire Earth system, you will gain a greater appreciation of why geology is so integral to our lives.

REVIEW
WORKBOOK

Chapter Summary

- Earth can be viewed as a system of interconnected components that interact and affect each other. The principal subsystems of Earth are the atmosphere, hydrosphere, biosphere, lithosphere, mantle, and core. Earth is considered a dynamic planet that changes continuously because of the interactions among its various subsystems and cycles.

- Geology, the study of Earth, is divided into two broad areas. Physical geology is the study of Earth materials as well as the processes that operate within and on Earth's surface; historical geology examines the origin and evolution of Earth, its continents, oceans, atmosphere, and life.

- Geology is part of the human experience. We can find references to it in the arts, music, and literature. A basic understanding of geology is also important for dealing with the many environmental problems and issues facing society.

- Geologists engage in a variety of occupations, the main one being exploration for mineral and energy resources. They are also becoming increasingly involved in environmental issues and making short- and long-range predictions of the potential dangers from such natural disasters as volcanic eruptions and earthquakes.

- The universe began with a Big Bang approximately 15 billion years ago. Astronomers have deduced this age by observing that celestial objects are moving away from each other in what appears to be an ever-expanding universe. Furthermore, the universe has a background radiation of 2.7° above absolute zero, radiation that is thought to be the faint afterglow of the Big Bang.

- About 4.6 billion years ago, the solar system formed from a rotating cloud of interstellar matter. As this cloud condensed, it eventually collapsed under the influence of gravity and flattened into a counterclockwise rotating disk. Within this rotating disk, the Sun, planets, and moons formed from the turbulent eddies of nebular gases and solids.

- Earth formed from a swirling eddy of nebular material 4.6 billion years ago. It probably accreted as a solid body and then soon underwent differentiation during a period of internal heating.

- Earth is differentiated into layers. The outermost layer is the crust, which is divided into continental and oceanic portions. The crust and underlying solid part of the upper mantle, also known as the lithosphere, overlie the asthenosphere, a zone that slowly flows. The asthenosphere is underlain by the solid lower mantle. Earth's core consists of an outer liquid portion and an inner solid portion.

- The lithosphere is divided into a series of plates that diverge, converge, and slide sideways past one another.

■ The scientific method is an orderly, logical approach that involves gathering and analyzing facts about a particular phenomenon, formulating hypotheses to explain the phenomenon, testing the hypotheses, and finally proposing a theory. A theory is a testable explanation for some natural phenomenon that has a large body of supporting evidence.

■ Plate tectonic theory provides a unifying explanation for many geologic features and events. The interaction between plates is responsible for volcanic eruptions, earthquakes, the formation of mountain ranges and ocean basins, and the recycling of rock materials.

■ The three major rock groups are igneous, sedimentary, and metamorphic. Igneous rocks result from the crystallization of magma or the consolidation of volcanic ejecta. Sedimentary rocks are formed mostly by the consolidation of rock fragments, precipitation of mineral matter from solution, or compaction of plant or animal remains. Metamorphic rocks are produced from other rocks, generally beneath Earth's surface, by heat, pressure, and chemically active fluids.

■ The rock cycle illustrates the interactions between internal and external Earth processes and shows how the three rock groups are interrelated.

■ Time sets geology apart from the other sciences except astronomy, and an appreciation of the immensity of geologic time is central to understanding Earth's evolution. The geologic time scale is the calendar geologists use to date past events.

■ The principle of uniformitarianism is basic to the interpretation of Earth history. This principle holds that the laws of nature have been constant through time and that the same processes operating today have operated in the past, though at different rates.

■ Geology is an integral part of our lives. Our standard of living depends directly on our consumption of natural resources, resources that formed millions and billions of years ago.

Important Terms

asthenosphere (p. 18)
Big Bang (p. 13)
core (p. 18)
crust (p. 18)
geologic time scale (p. 25)
geology (p. 6)
hypothesis (p. 19)
igneous rock (p. 21)
Jovian planets (p. 14)

lithosphere (p. 18)
mantle (p. 18)
metamorphic rock (p. 21)
mineral (p. 20)
plate (p. 18)
plate tectonic theory (p. 19)
principle of uniformitarianism
 (p. 25)
rock (p. 20)

rock cycle (p. 20)
scientific method (p. 19)
sedimentary rock (p. 21)
solar nebula theory (p. 14)
system (p. 4)
terrestrial planets (p. 14)
theory (p. 19)

Review Questions

1. Rocks that result from the alteration of other rocks, usually beneath the surface, by heat, pressure, and the chemical activity of fluids are:
   a. _____ igneous; b. _____ sedimentary;
   c. _____ metamorphic; d. _____ volcanic;
   e. _____ answers a and d.

2. A combination of related parts interacting in an organized fashion is called:
   a. _____ a cycle; b. _____ a theory; c. _____ uniformitarianism; d. _____ a hypothesis;
   e. _____ a system.

3. The composition of the universe has been changing since the Big Bang, yet 98% of it by weight still consists of the elements:
   a. _____ hydrogen and nitrogen; b. _____ hydrogen and helium; c. _____ hydrogen and carbon; d. _____ helium and carbon; e. _____ carbon and nitrogen.

4. According to the currently accepted theory for the origin of the solar system:
   a. _____ a huge nebula collapsed under its own gravitational attraction; b. _____ the

nebula formed a disc with the Sun in the center; c. _____ planetesimals accreted from gaseous, liquid, and solid particles; d. _____ all of the preceding answers; e. _____ none of the preceding answers.

5. The study of Earth materials is called:

a. _____ economic geology; b. _____ physical geology; c. _____ historical geology; d. _____ structural geology; e. _____ environmental geology.

6. It is thought that plate movement results from:

a. _____ gravitational forces; b. _____ density differences between the mantle and the core; c. _____ rotation of the mantle around the core; d. _____ convection cells; e. _____ the Coriolis effect.

7. Interaction between the atmosphere, hydrosphere, and biosphere is a major contributor to:

a. _____ mountain building; b. _____ the generation of magma; c. _____ weathering of Earth materials; d. _____ metamorphism; e. _____ plate movement.

8. What two observations led scientists to conclude that the Big Bang occurred approximately 15 billion year ago?

a. _____ a steady-state universe and 2.7° above absolute zero background radiation; b. _____ a steady-state universe and opaque background radiation; c. _____ an expanding universe and opaque background radiation; d. _____ an expanding universe and 2.7° above absolute zero background radiation; e. _____ a shrinking universe and opaque background radiation.

9. Which of the following statements about a scientific theory is *not* true?

a. _____ it is an explanation for some natural phenomenon; b. _____ predictive statements can be derived from it; c. _____ it is a conjecture or guess; d. _____ it has a large body of supporting evidence; e. _____ it is testable.

10. The lithosphere consists of:

a. _____ the crust and solid portion of the upper mantle; b. _____ the asthenosphere and the solid portion of the upper mantle; c. _____ the crust and asthenosphere; d. _____ continental and oceanic crust only; e. _____ the core and mantle.

11. The premise that present-day processes have operated throughout geologic time is the principle of:

a. _____ fossil succession; b. _____ uniformitarianism; c. _____ continental drift; d. _____ plate tectonics; e. _____ scientific deduction.

12. Which layer has the same composition as the mantle but behaves plastically?

a. _____ continental crust; b. _____ oceanic crust; c. _____ outer core; d. _____ inner core; e. _____ asthenosphere.

13. Why is an accurate geologic time scale particularly important for geologists in examining changes in global temperatures during the past?

14. Why is it important that everyone have a basic understanding of geology, even if they aren't going to become geologists?

15. Describe how you would use the scientific method to formulate a hypothesis explaining the similarity of mountain ranges on the east coast of North America and those in England, Scotland, and the Scandinavian countries. How would you test your hypothesis?

16. Discuss how the three major layers of Earth differ from each other and why the differentiation into a layered planet is probably the most significant event in Earth history.

17. Explain how the principle of uniformitarianism allows for catastrophic events.

18. Discuss why plate tectonic theory is a unifying theory of geology.

19. Explain the advantage of using a system approach to the study of Earth.

20. Explain how a knowledge of geology would be useful in planning a military campaign against another country.

World Wide Web Activities

PHYSICAL Geology⇌Now Assess your understanding of this chapter's topics with additional quizzing and comprehensive interactivities at

http://earthscience.brookscole.com/physgeo5e

as well as current and up-to-date weblinks, additional readings, and InfoTrac College Edition exercises.

Minerals—The Building Blocks of Rocks

CHAPTER 2
OUTLINE

PHYSICAL Geology⇌Now *This icon, appearing throughout the book, indicates an opportunity to explore interactive tutorials, animations, or practice problems available on the Physical GeologyNow Web site at http://earthscience.brookscole.com/physgeo5e.*

OBJECTIVES

At the end of this chapter, you will have learned that

- All matter, including minerals, is made up of atoms that bond to form elements and compounds.

- Geologists have a very specific definition for the term *mineral.*

- You can distinguish minerals from other naturally occurring and manufactured substances.

- Minerals are incredibly varied, yet only a few are particularly common.

- Geologists use physical properties such as color, hardness, and density to identify minerals.

- Minerals originate in various ways and under varied conditions.

- Some minerals, designated rock-forming minerals, are particularly common in rocks, whereas others are found in minor quantities.

- Various minerals and rocks are important natural resources that are essential to industrialized societies.

These black pearls valued at about $13,000 are on display at Maui Pearls in the town of Avarua on the Island of Rarotanga, which is part of the Cook Islands in the South Pacific. Pearls are composed mostly of the mineral aragonite. Source: Sue Monroe

Introduction

The term *mineral* is commonly used for dietary substances we need for good nutrition, such as calcium, iron, and magnesium, but these are actually chemical elements, not minerals, at least in the geologic sense. The term also designates substances that are neither animal nor vegetable, implying that minerals are inorganic, which is correct, but not all inorganic substances are minerals. For example, water and water vapor are inorganic, but neither is a mineral. Yet ice is a mineral. Thus minerals must be solids as opposed to liquids or gases. So, **minerals** are naturally occurring inorganic solids that are further characterized as *crystalline,* meaning that their atoms are arranged in a specific way, as opposed to glass, which has no such ordered internal structure.Minerals also have a narrowly defined chemical composition and characteristic physical properties such as color, hardness, and density. We will examine all parts of this rather lengthy definition later in this chapter.

The importance of minerals in many human endeavors cannot be overstated. The ore deposits we rely on to sustain our standard of living and our industrialized societies are simply natural concentrations of minerals and rocks of economic importance. Iron ore, industrial minerals for abrasives, glass and cement, as well as minerals and rocks needed for animal feed supplements and fertilizers are essential to our economic well-being. The United States and Canada owe much of their economic success to the availability of abundant natural resources, although they must import some important commodities, thus accounting for political and economic ties with various other nations.

An important reason to study minerals is that they are the building blocks of rocks, so rocks, with few exceptions, are combinations of one or more minerals. Granite, for instance, is made up of specified percentages of minerals known as quartz and feldspars along with other minerals in minor quantities. In several of the following chapters we will have much more to say about the mineral composition of various rocks as well as the importance of minerals in rock identification and classification.

Some minerals are attractive and eagerly sought by private collectors and for museum displays (■ Figure 2.1). Other minerals are known as *gemstones*—that is, any precious or

Los Angeles County Museum specimen, © Harold and Erica Van Pelt

(a) A spectacular example of tourmaline and quartz (colorless) from the Himalaya Mine, San Diego County, California.

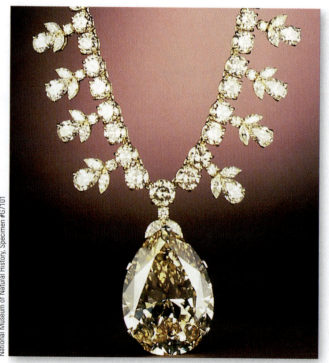

National Museum of Natural History, Specimen #67101

(b) The pendent in this necklace is the Victoria Transvaal diamond from South Africa. It is in the Smithsonian Institution.

■ **Figure 2.1**

Museum-quality mineral specimens. Some minerals are attractive and valued as precious and semiprecious gemstones, whereas many others are important natural resources.

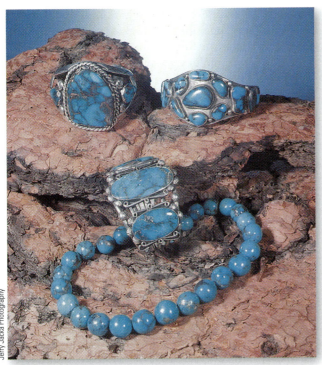

(c) Turquoise—a sky-blue, blue-green, or light green hydrated copper aluminum phosphate—is a semiprecious stone used for jewelry and as a decorative stone.

(d) Two copper minerals, azurite (blue) and malachite (green), in the Flagg Collection, Arizona Mining and Mineral Museum.

■ **Figure 2.1 (Continued)**

semiprecious mineral or rock used for decorative purposes, especially jewelry. As their name implies, the precious gemstones such as diamond, ruby, sapphire, and emerald are most desirable and most expensive. Many people have small precious gemstones and perhaps some semiprecious ones, such as garnet, peridot, and turquoise (Figure 2.1). The lore associated with gemstones, such as relating them to one's birth month, makes them even more appealing to many people.

Amber and pearl are included among the semiprecious gemstones, but are they really minerals? Amber is hardened resin (sap) from coniferous trees and thus an organic substance and not a mineral, but it is nevertheless prized as a decorative "stone" (■ Figure 2.2). It is best known from the Baltic Sea region of Europe, where sun-worshiping cultures, noting its golden translucence resembling the Sun's rays, thought it possessed mystical powers.

Pearls form when mollusks, such as clams or oysters, deposit successive layers of tiny mineral crystals around some irritant, perhaps a sand grain. Most pearls are lustrous white, but some are silver gray, green, or black (see the chapter opening photo). Unlike other gemstones, pearls need no shaping or polishing before they are used in jewelry; that is, they are essentially ready to use when found.

From our discussion so far we have a formal definition of the term *mineral* and we know that minerals are the basic constituents of rocks. Now let's delve deeper into what minerals are made of by considering matter, atoms, elements, and bonding.

PHYSICAL Geology⇌Now   Click Geology Interactive to work through an activity on mineral labs.

■ **Figure 2.2**

Insect preserved in amber. Although amber is an organic substance and thus not a mineral, it is nevertheless valued as a semiprecious gemstone. Recall that amber played a pivotal role in the book and movie *Jurassic Park*.

MATTER—WHAT IS IT?

Anything that has mass and occupies space is *matter.* Accordingly, water, plants, animals, the atmosphere, and minerals and rocks are composed of matter. Physicists recognize three states or phases of matter:* *liquids, gases,* and *solids.* Liquids, such as surface water and groundwater, as well as atmospheric gases are important in our considerations of several surface processes, such as running water and wind, but here our main concern is with solids because by definition minerals are solids. So, the next question is, What is matter made of?

Atoms and Elements

Matter is made up of chemical **elements,** which in turn are composed of tiny particles known as **atoms** (■ Figure 2.3). Atoms are the smallest units of matter that retain the characteristics of a particular element. That is, they cannot be split into substances of different composition, except in radioactive decay (discussed in Chapter 8). Thus an element is made up of atoms all of which have the same properties. Scientists have discovered 92 naturally occurring elements, some of which are listed

---

*Actually, scientists recognize a fourth state of matter known as *plasma,* an ionized gas as in fluorescent and neon lights and matter in the Sun and stars.

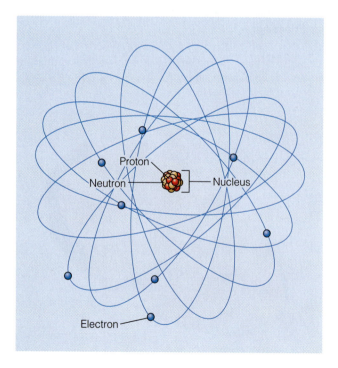

■ **Figure 2.3**

The structure of an atom. The dense nucleus consisting of protons and neutrons is surrounded by a cloud of orbiting electrons.

in Table 2.1, and several others have been made in laboratories (see Appendix A). All naturally occurring elements and most artificial ones have a name and a symbol—for example, oxygen (O), aluminum (Al), and potassium (K).

At the center of an atom is a tiny **nucleus** made up of one or more particles known as **protons,** which have a positive electrical charge, and **neutrons,** which are electrically neutral (Figure 2.3). The nucleus is only about 1/100,000 of the diameter of an atom, yet it contains virtually all of the atom's mass. **Electrons,** particles with a negative electrical charge, orbit rapidly around the nucleus at specific distances in one or more **electron shells.** The electrons determine how an atom interacts with other atoms, but the nucleus determines how many electrons an atom has, because the positively charged protons attract and hold negatively charged electrons in their orbits.

The number of protons in its nucleus determines an atom's identity and its **atomic number.** Hydrogen (H), for instance, has 1 proton in its nucleus and thus has an atomic number of 1. The nuclei of helium (He) atoms possess 2 protons, whereas those of carbon (C) have 6, and uranium (U) have 92, so their atomic numbers are 2, 6, and 92, respectively. Atoms are also characterized by their **atomic mass number,** which is the sum of protons and neutrons in the nucleus (electrons contribute negligible mass to atoms). However, atoms of the same chemical element might have different atomic mass numbers because the number of neutrons can vary. All carbon (C) atoms have 6 protons—otherwise they would not be carbon—but the number of neutrons can be 12, 13, or 14. Thus we recognize three types of carbon, each with a different atomic mass number, or what are known as *isotopes* (■ Figure 2.4).

These isotopes of carbon, or those of any other element, behave the same chemically; carbon 12 and carbon 14 are both present in carbon dioxide (CO_2), for example. However, some isotopes are radioactive, meaning that they spontaneously decay or change to other stable elements. Carbon 14 is radioactive, whereas both carbon 12 and carbon 13 are stable. Radioactive isotopes are important for determining the absolute ages of rocks (see Chapter 8).

Bonding and Compounds

Interactions among electrons around atoms can result in two or more atoms joining together, a process known as **bonding.** If atoms of two or more elements bond, the resulting substance is a **compound.** Gaseous oxygen consists of only oxygen atoms and is thus an element, whereas the mineral quartz, consisting of silicon and oxygen atoms, is a compound. Most minerals are compounds, although gold, platinum, and several others are important exceptions.

To understand bonding, it is necessary to delve deeper into the structure of atoms. Recall that negatively charged electrons orbit the nuclei of atoms in electron shells. With the exception of hydrogen, which has only one proton and one electron, the innermost electron shell of an atom contains only two electrons. The other shells contain various numbers of electrons, but the outermost shell never has more than eight (Table 2.1). The electrons in the outermost shell are those that are usually involved in chemical bonding.

Two types of chemical bonds, *ionic* and *covalent*, are particularly important in minerals, and many minerals contain both types of bonds. Two other types of chemi-

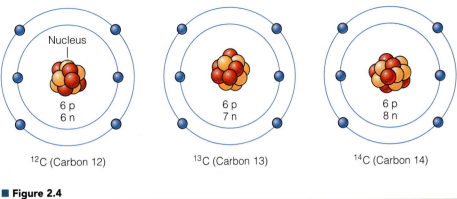

^{12}C (Carbon 12) ^{13}C (Carbon 13) ^{14}C (Carbon 14)

■ **Figure 2.4**

Schematic representation of the isotopes of carbon. Carbon has an atomic number of 6 and an atomic mass number of 12, 13, or 14, depending on the number of neutrons (n) in its nucleus.

cal bonds, *metallic* and *van der Waals*, are much less common but extremely important in determining the properties of some useful minerals.

Ionic Bonding   Notice in Table 2.1 that most atoms have fewer than eight electrons in their outermost electron

Table 2.1

Symbols, Atomic Numbers, and Electron Configurations for Some of the Naturally Occurring Elements

Element	Symbol	Atomic Number	Number of Electrons in Each Shell			
			1	2	3	4
Hydrogen	H	1	1			
Helium	He	2	2			
Lithium	Li	3	2	1		
Beryllium	Be	4	2	2		
Boron	B	5	2	3		
Carbon	C	6	2	4		
Nitrogen	N	7	2	5		
Oxygen	O	8	2	6		
Fluorine	F	9	2	7		
Neon	Ne	10	2	8		
Sodium	Na	11	2	8	1	
Magnesium	Mg	12	2	8	2	
Aluminum	Al	13	2	8	3	
Silicon	Si	14	2	8	4	
Phosphorus	P	15	2	8	5	
Sulfur	S	16	2	8	6	
Chlorine	Cl	17	2	8	7	
Argon	Ar	18	2	8	8	
Potassium	K	19	2	8	8	1
Calcium	Ca	20	2	8	8	2

shell. However, some elements, including neon and argon, have complete outer shells containing eight electrons; because of this electron configuration, these elements, known as the *noble gases,* do not react readily with other elements to form compounds. Interactions among atoms tend to produce electron configurations similar to those of the noble gases. That is, atoms interact so that their outermost electron shell is filled with eight electrons, unless the first shell (with two electrons) is also the outermost electron shell, as in helium.

One way that the noble gas configuration can be attained is by the transfer of one or more electrons from one atom to another. Common salt is composed of the elements sodium (Na) and chlorine (Cl), each of which is poisonous, but when combined chemically they form the compound sodium chloride (NaCl), better known as the mineral halite. Notice in Figure 2.5a that sodium has 11 protons and 11 electrons; thus the positive electrical charges of the protons are exactly balanced by the negative charges of the electrons, and the atom is electrically neutral. Likewise, chlorine with 17 protons and 17 electrons is electrically neutral (■ Figure 2.5a). But

neither sodium nor chlorine has 8 electrons in its outermost electron shell; sodium has only 1, whereas chlorine has 7. To attain a stable configuration, sodium loses the electron in its outermost electron shell, leaving its next shell with 8 electrons as the outermost one (Figure 2.5a). Sodium now has one fewer electron (negative charge) than it has protons (positive charge), so it is an electrically charged **ion** and is symbolized Na^{+1}.

The electron lost by sodium is transferred to the outermost electron shell of chlorine, which had 7 electrons to begin with. The addition of one more electron gives chlorine an outermost electron shell of 8 electrons, the configuration of a noble gas. But its total number of electrons is now 18, which exceeds by 1 the number of protons. Accordingly, chlorine also becomes an ion, but it is negatively charged (Cl^{-1}). An **ionic bond** forms between sodium and chlorine because of the attractive force between the positively charged sodium ion and the negatively charged chlorine ion (Figure 2.5a).

In ionic compounds, such as sodium chloride (the mineral halite), the ions are arranged in a three-dimensional framework that results in overall electrical neutrality. In halite, sodium ions are bonded to chlorine ions on all sides, and chlorine ions are surrounded by sodium ions (Figure 2.5b).

Covalent Bonding Co-**valent bonds** form between atoms when their electron shells overlap and they share electrons. For example, atoms of the same element, such as oxygen in oxygen gas, cannot bond by transferring electrons from one atom to another. And carbon (C), which forms the minerals graphite and diamond, has four electrons in its outermost electron shell (■ Figure 2.6a). If these four electrons were transferred to another carbon atom, the atom receiving the electrons would have the noble gas configuration of eight electrons in its outermost electron shell, but the atom contributing the electrons would not.

In such situations, adjacent atoms share electrons by overlapping their electron shells. A carbon atom in diamond, for instance, shares all

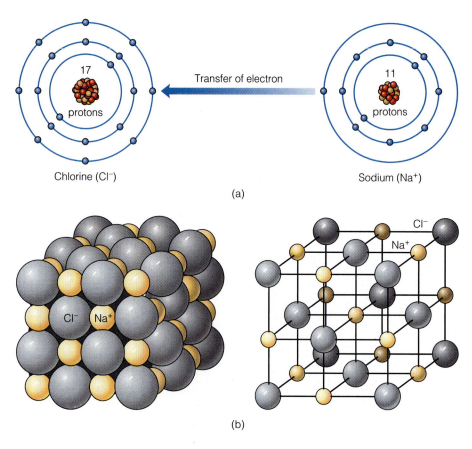

Transfer of electron

Chlorine (Cl⁻)   Sodium (Na⁺)

(a)

(b)

■ **Figure 2.5**

(a) Ionic bonding. The electron in the outermost shell of sodium is transferred to the outermost electron shell of chlorine. Once the transfer has occurred, sodium and chlorine are positively and negatively charged ions, respectively. (b) The crystal structure of sodium chloride, the mineral halite. The diagram on the left shows the relative sizes of sodium and chlorine ions, and the diagram on the right shows the locations of the ions in the crystal structure.

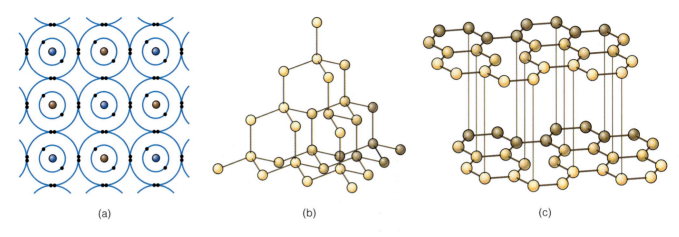

(a)   (b)   (c)

■ **Figure 2.6**

(a) Covalent bonds formed by adjacent atoms sharing electrons in diamond. (b) Carbon atoms in diamond are covalently bonded to form a three-dimensional framework. (c) Covalent bonding is also found in graphite, but here the carbon atoms are bonded together to form sheets that are held to one another by van der Waals bonds. The sheets themselves are strong, but the bonds between sheets are weak.

four of its outermost electrons with a neighbor to produce a stable noble gas configuration (Figure 2.6a).

Covalent bonds are not restricted to substances composed of atoms of a single kind. Among the most common minerals, the silicates (discussed later in this chapter), the element silicon forms partly covalent and partly ionic bonds with oxygen.

Metallic and van der Waals Bonds   *Metallic bonding* results from an extreme type of electron sharing. The electrons of the outermost electron shell of metals such as gold, silver, and copper readily move about from one atom to another. This electron mobility accounts for the fact that metals have a metallic luster (their appearance in reflected light), provide good electrical and thermal conductivity, and can be easily reshaped. Only a few minerals possess metallic bonds, but those that do are very useful; copper, for example, is used for electrical wiring because of its high electrical conductivity.

Some electrically neutral atoms and molecules* have no electrons available for ionic, covalent, or metallic bonding. They nevertheless have a weak attractive force between them when in proximity. This weak attractive force is a *van der Waals* or *residual bond*. The carbon atoms in the mineral graphite are covalently bonded to form sheets, but the sheets are weakly held together by van der Waals bonds (Figure 2.6c). This type of bonding makes graphite useful for pencil leads; when a pencil is moved across a piece of paper, small pieces of graphite flake off along the planes held together by van der Waals bonds and adhere to the paper.

PHYSICAL
Geology ⇌ Now   Click Geology Interactive to work through an activity on atoms and crystals.

WHAT ARE MINERALS?

n the Introduction we defined a mineral as an inorganic, naturally occurring, crystalline solid with a narrowly defined chemical composition and characteristic physical properties. Furthermore, we know from the previous section on Bonding and Compounds that most minerals are compounds of two or more chemically bonded elements as in quartz (SiO_2). In the following sections we will examine each part of the formal definition of the term *mineral*.

Naturally Occurring Inorganic Substances

The criterion *naturally occurring* excludes from minerals all substances manufactured by humans. Accordingly, most geologists do not regard synthetic diamonds and rubies and other artificially synthesized substances as minerals. This criterion is particularly important to those who buy and sell gemstones, most of which are minerals, because some human-made substances are very difficult to distinguish from natural gem minerals.

Some geologists think the term *inorganic* in the mineral definition is superfluous. It does remind us that animal matter and vegetable matter are not minerals. Nevertheless, some organisms, including corals, clams, and a number of other animals and plants, construct their shells of the compound calcium carbonate

---

*A molecule is the smallest unit of a substance that has the properties of that substance. A water molecule (H_2O), for example, possesses two hydrogen atoms and one oxygen atom.

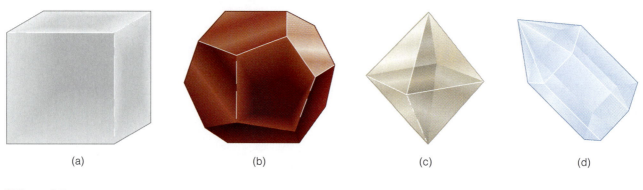

(a)   (b)   (c)   (d)

■ **Figure 2.7**

Mineral crystals develop in a variety of shapes. (a) Cubic crystals are typical of the minerals halite, galena, and pyrite. (b) Dodecahedron crystals such as those of garnet have 12 sides. (c) Diamond has octahedral, or 8-sided, crystals. (d) A prism terminated by a pyramid is found in quartz.

($CaCO_3$), which is either the mineral aragonite or calcite, or their shells are made of silicon dioxide (SiO_2), as in the mineral quartz.

PHYSICAL
Geology⇌Now   Click Geology Interactive to work through an activity on atomic behavior.

Mineral Crystals

By definition minerals are **crystalline solids,** in which the constituent atoms are arranged in a regular, three-dimensional framework, as in the mineral halite (Figure 2.5b). Under ideal conditions, such as in a cavity, mineral crystals grow and form perfect crystals that possess planar surfaces (crystal faces), sharp corners, and straight edges (■ Figure 2.7). In other words, the regular geometric shape of a well-formed mineral crystal is the exterior manifestation of an ordered internal atomic arrangement. Not all rigid substances are crystalline solids; natural and manufactured glass lack the ordered arrangement of atoms and are said to be *amorphous,* meaning "without form."

In the preceding paragraph we used the terms *crystalline* and *crystal*. Keep in mind that *crystalline* refers to a solid with a regular three-dimensional internal framework of atoms, whereas a **crystal** is a geometric shape with planar faces (crystal faces), sharp corners, and straight edges. Thus a crystal is the external manifestation of a crystalline structure. Minerals are by definition crystalline solids, but crystalline solids do not necessarily always yield well-formed crystals. The reason is that when crystals form, they may grow in proximity to form an interlocking mosaic in which individual crystals are not apparent or easily discerned (■ Figure 2.8).

So how do we know that the mass of minerals in Figure 2.8b is actually crystalline? X-ray beams and light transmitted through mineral crystals or crystalline masses behave in a predictable manner, providing compelling evidence for an internal orderly structure. Another way we

can determine that minerals with no obvious crystals are actually crystalline is by their *cleavage,* the property of breaking or splitting repeatedly along smooth, closely spaced planes. Not all minerals have cleavage planes, but many do, and such regularity certainly indicates that splitting is controlled by internal structure.

As early as 1669, the Danish scientist Nicolas Steno determined that the angles of intersection of equivalent crystal faces on different specimens of quartz are identical. Since then, this *constancy of interfacial angles* has been demonstrated for many other minerals, regardless of their size, shape, age, or geographic occurrence (Figure 2.8c). Steno postulated that mineral crystals are made up of very small, identical building blocks, and that the arrangement of these building blocks determines the external form of mineral crystals. In short, he proposed that external form results from internal structure, a proposal that has since been verified.

Chemical Composition of Minerals

Mineral composition is shown by a chemical formula, which is a shorthand way of indicating the numbers of atoms of different elements that make up a mineral. The mineral quartz consists of one silicon (Si) atom for every two oxygen (O) atoms and thus has the formula SiO_2; the subscript number indicates the number of atoms. Orthoclase is composed of one potassium, one aluminum, three silicon, and eight oxygen atoms, so its formula is $KAlSi_3O_8$. A few minerals known as **native elements** consist of a single element and include such minerals as graphite and diamond, both of which are composed of carbon (C), silver (Ag), platinum (Pt), and gold (Au).

The definition of a mineral contains the phrase *a narrowly defined chemical composition* because some minerals actually have a range of compositions. For many minerals, the chemical composition does not vary. Quartz is always composed of silicon and oxygen (SiO_2), and halite contains only sodium and chlorine (NaCl).

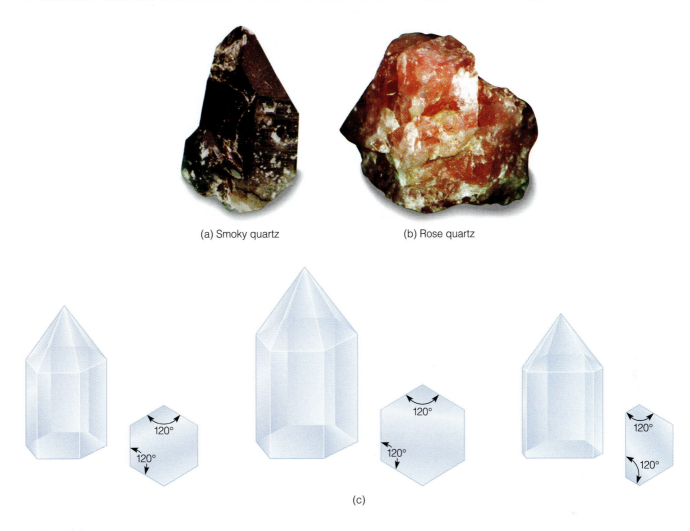

(a) Smoky quartz

(b) Rose quartz

(c)

■ **Figure 2.8**

(a) Well-shaped crystal of smoky quartz. (b) Specimen of rose quartz in which no obvious crystals can be discerned. (c) Side views and cross sections of quartz crystals showing the constancy of interfacial angles. A well-shaped crystal (left), a larger well-shaped crystal (middle), and a poorly shaped crystal (right). The angles formed between equivalent crystal faces on different specimens of the same mineral are the same regardless of the size, shape, age, or geographic occurrence of the specimens.

Other minerals have a range of compositions because one element can substitute for another if the atoms of two or more elements are nearly the same size and the same charge. Notice in ■ Figure 2.9 that iron and magnesium atoms are about the same size; therefore they can substitute for each other. The chemical formula for the mineral olivine is $(Mg,Fe)_2SiO_4$, meaning that, in addition to silicon and oxygen, it may contain only magnesium, only iron, or a combination of both. As a matter of fact, the term *olivine* is usually applied to minerals that contain both iron and magnesium, whereas forsterite is olivine with only magnesium (Mg_2SiO_4) and olivine with only iron is fayalite (Fe_2SiO_4). A number of other minerals also have ranges of compositions, so these are actually mineral groups with several members.

Physical Properties of Minerals

The last criterion in our definition of a mineral, *characteristic physical properties*, refers to such properties as hardness, color, and crystal form. These properties are controlled by composition and structure. We will have more to say about physical properties of minerals later in this chapter.

HOW MANY MINERALS ARE THERE?

Geologists have identified and described more than 3500 minerals, but only a few—perhaps two dozen—are particularly common. Considering that 92 naturally occurring elements are known, one might think that an extremely large number of minerals could be formed, but several factors limit the number possible. For one thing, many combinations of elements simply do not occur; no

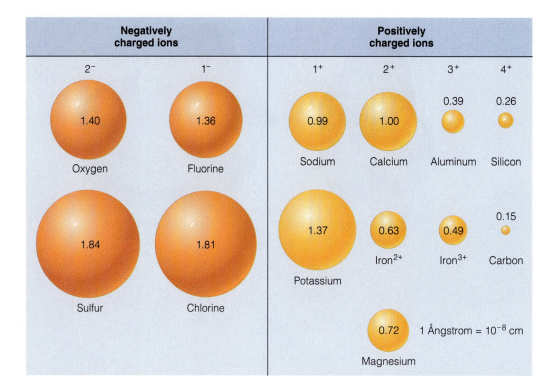

■ **Figure 2.9**

Electrical charges and relative sizes of ions common in minerals. The numbers within the ions are the radii shown in Ångstrom units.

compounds are composed of only potassium and sodium or of silicon and iron, for example. Another important factor restricting the number of common minerals is that the bulk of Earth's crust is made up of only eight chemical elements (■ Figure 2.10). Oxygen and silicon constitute more than 74% (by weight) of the crust and nearly 84% of the atoms available to form compounds. By far the most common minerals in the crust consist of silicon and oxygen combined with one or more of the other elements listed in Figure 2.10.

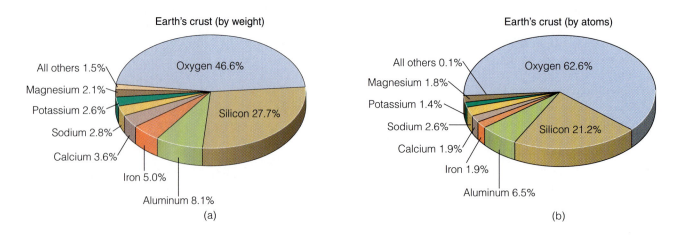

■ **Figure 2.10**

Common elements in Earth's crust. (a) Percentage of crust by weight, and (b) percentage of crust by atoms. Source: (a) From Miller, G. T. 1996. *Living in the Environment: Principles, Concepts and Solutions.* Wadsworth Publishing. Figure 8.3.

GEOLOGY
IN UNEXPECTED PLACES

The Queen's Jewels

Because of their beauty and scarcity, gemstones have fascinated people for thousands of years. Indeed, ancient people used various minerals, rocks, and fossils for their presumed mystical powers or simply because they were attractive. One of the most impressive collections of gemstones is the Crown Jewels housed in the Tower of London in England. The Tower of London is a formidable stone structure on the Thames River that has served as a fortification, the residence for kings and queens, and a prison for such notable people as Sir Walter Raleigh, who was incarcerated there for 13 years. Construction on the Tower of London began during the reign of William the Conqueror (1066–1087), but it was enlarged and modified until about 1300, and since then it has remained much the same.

Within the Tower, the Waterloo Barracks, originally built for 1000 soldiers, have housed the British Crown Jewels since the beginning of the 14th century. Only during World War II (1939–1945) were the Crown Jewels removed to a secret location for safekeeping and then they were returned to the Waterloo Barracks. Among the Crown Jewels is the crown made for the coronation of George VI in 1937 and later modified for Queen Elizabeth II's coronation in 1953. It is set with 2868 diamonds, 17 sapphires, 11 emeralds, 5 rubies, and 273 pearls (■ Figure 1). In addition to various other crowns, the Crown Jewels comprise

■ **Figure 1**

The Imperial State Crown was made for the coronation of George VI in 1937 and altered for Her Majesty Queen Elizabeth II in 1953.

gold plates, christening fonts, and scepters, including the Scepter with Cross with the 530-carat First Star of Africa diamond mounted in its head, the largest cut diamond in the world. Actually, the First Star of Africa diamond is the largest of nine stones cut from the much larger Cullinan Diamond from Africa.

MINERAL GROUPS RECOGNIZED BY GEOLOGISTS

Geologists recognize mineral classes or groups, each with members that share the same negatively charged ion or ion group (Table 2.2). We mentioned in a previous section that ions are atoms that have either a positive or negative electrical charge resulting from the loss or gain of electrons in their outermost shell. In addition to ions, some minerals contain tightly bonded, complex groups of different atoms known as *radicals* that act as single units within minerals. A good example is the carbonate radical, consisting of a carbon atom bonded to three oxygen atoms and thus having the formula CO_3 and a -2 electrical charge. Other common radicals and their charges are sulfate (SO_4, -2), hydroxyl (OH_2, -1), and silicate (SiO_2, -4) (■ Figure 2.11).

Table 2.2

Mineral Groups Recognized by Geologists

Mineral Group	Negatively Charged Ion or Radical	Examples	Composition
Carbonate	$(CO_3)^{-2}$	Calcite	$CaCO_3$
		Dolomite	$CaMg(CO_3)_2$
Halide	Cl^{-1}, F^{-1}	Halite	$NaCl$
		Fluorite	CaF_2
Hydroxide	$(OH)^{-1}$	Brucite	$Mg(OH)_2$
Native element	—	Gold	Au
		Silver	Ag^*
		Diamond	C
Phosphate	$(PO_4)^{-3}$	Apatite	$Ca_5(PO_4)_3(F,Cl)$
Oxide	O^{-2}	Hematite	Fe_2O_3
		Magnetite	Fe_3O_4
Silicate	$(SiO_4)^{-4}$	Quartz	SiO_2
		Potassium feldspar	$KAlSi_3O_8$
		Olivine	$(Mg, Fe)_2SiO_4$
Sulfate	$(SO_4)^{-2}$	Anhydrite	$CaSO_4$
		Gypsum	$CaSO_4 \cdot 2H_2O$
Sulfide	S^{-2}	Galena	PbS
		Pyrite	FeS_2
		Argentite	Ag_2S^*

*Note that silver is found as a native element and as a sulfide mineral.

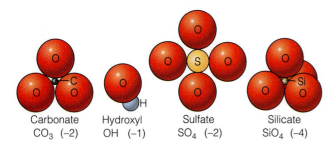

■ Figure 2.11

Many minerals contain radicals, which are complex groups of atoms tightly bonded together. The silicate and carbonate radicals are particularly common in many minerals, such as quartz (SiO_2) and calcite ($CaCO_3$).

The Silicate Minerals

Because silicon and oxygen are the two most abundant elements in Earth's crust, it is not surprising that many minerals contain these elements. A combination of silicon and oxygen is known as **silica,** and minerals that contain silica are **silicates.** Quartz (SiO_2) is pure silica because it is composed entirely of silicon and oxygen. But most silicates have one or more additional elements, as in orthoclase ($KAlSi_3O_8$) and olivine [$(Mg,Fe)_2SiO_4$]. Silicate minerals include about one third of all known minerals, but their abundance is even more impressive when one considers that they make up perhaps 95% of Earth's crust.

The basic building block of all silicate minerals is the **silica tetrahedron,** consisting of one silicon atom and four oxygen atoms (■ Figure 2.12a); the silica radical is shown in Figure 2.11. These atoms are arranged so that the four oxygen atoms surround a silicon atom, which occupies the space between the oxygen atoms, thus forming a four-faced pyramidal structure. The silicon atom has a positive charge of 4, and each of the four oxygen atoms has a negative charge of 2, resulting in a radical with a total negative charge of 4 $(SiO_4)^{-4}$.

Because the silica tetrahedron has a negative charge, it does not exist in nature as an isolated ion group; rather, it combines with positively charged ions or shares its oxygen atoms with other silica tetrahedra. In the simplest silicate minerals, the silica tetrahedra exist as single units bonded to positively charged ions.

			Formula of negatively charged ion group	Example
(b)	Isolated tetrahedra	△	$(SiO_4)^{-4}$	Olivine
(c)	Continuous chains of tetrahedra	← ⟁⟁ →	$(SiO_3)^{-2}$	Pyroxene group (augite)
		← ⟁⟁⟁ →	$(Si_4O_{11})^{-6}$	Amphibole group (hornblende)
(d)	Continuous sheets	↑ ← ⟁⟁⟁ → ↓	$(Si_4O_{10})^{-4}$	Micas (muscovite)
(e)	Three-dimensional networks	Too complex to be shown by a simple two-dimensional drawing	$(SiO_2)^0$ $(Si_3AlO_8)^{-1}$ $(Si_2Al_2O_8)^{-2}$	Quartz Orthoclase feldspars Plagioclase feldspars

$(SiO_4)^{-4}$

$$O^{-2} - Si^{4+}$$ with O^{-2} bonds

(a)

PHYSICAL Geology⇌Now ■ **Active Figure 2.12**

(a) Model of the silica tetrahedron, showing the unsatisfied negative charges at each oxygen. (b)–(e) Structures of the common silicate minerals shown by various arrangements of the silica tetrahedra. (b) Isolated tetrahedra. (c) Continuous chains. (d) Continuous sheets. (e) Networks. The arrows adjacent to single-chain, double-chain, and sheet silicates indicate that these structures continue indefinitely in the directions shown.

In minerals containing isolated tetrahedra, the silicon to oxygen ratio is 1:4, and the negative charge of the silica ion is balanced by positive ions (Figure 2.12b). Olivine [$(Mg,Fe)_2SiO_4$], for example, has either two magnesium (Mg^{+2}) ions, two iron (Fe^{+2}) ions, or one of each to offset the -4 charge of the silica ion.

Silica tetrahedra may also join together to form chains of indefinite length (Figure 2.12c). Single chains, as in the pyroxene minerals, form when each tetrahedron shares two of its oxygens with an adjacent tetrahedron, resulting in a silicon to oxygen ratio of 1:3. Enstatite, a pyroxene-group mineral, reflects this ratio in its chemical formula $MgSiO_3$. Individual chains, however, possess a net -2 electrical charge, so they are balanced by positive ions, such as Mg^{+2}, that link parallel chains together (Figure 2.12c).

The amphibole group of minerals is characterized by a double-chain structure in which alternate tetrahedra in two parallel rows are cross-linked (Figure 2.12c). The formation of double chains results in a silicon to oxygen ratio of 4:11, so each double chain possesses a -6 electrical charge. Mg^{+2}, Fe^{+2}, and Al^{+2} are usually involved in linking the double chains together.

In sheet structure silicates, three oxygens of each tetrahedron are shared by adjacent tetrahedra (Figure 2.12d). Such structures result in continuous sheets of silica tetrahedra with silicon to oxygen ratios of 2:5. Continuous sheets also possess a negative electrical charge satisfied by positive ions located between the sheets. This particular structure accounts for the characteristic sheet structure of the *micas*, such as biotite and muscovite, and the *clay minerals*.

Three-dimensional networks of silica tetrahedra form when all four oxygens of the silica tetrahedra are shared by adjacent tetrahedra (Figure 2.12e). Such sharing of oxygen atoms results in a silicon to oxygen ratio of 1:2, which is electrically neutral. Quartz is a common framework silicate.

Ferromagnesian Silicates   Some silicate minerals contain iron (Fe), magnesium (Mg), or both, and are known as **ferromagnesian silicates.** These minerals are commonly dark and more dense than nonferromagnesian silicates. Some of the common ferromagnesian silicate minerals are olivine, the pyroxenes, the amphiboles,

Sue Monroe

Olivine

Augite

Hornblende

Biotite mica

■ **Figure 2.13**

Examples of common ferromagnesian silicate minerals.

and biotite (■ Figure 2.13). Olivine, an olive green mineral, is common in some igneous rocks but uncommon in most other rock types. The pyroxenes and amphiboles are actually mineral groups, but the varieties augite and hornblende are the most common. Biotite mica is a common, dark ferromagnesian silicate with a distinctive sheet structure (Figure 2.12d).

Nonferromagnesian Silicates The **nonferromagnesian silicates,** as their name implies, lack iron and magnesium, are generally light colored, and are less dense than ferromagnesian silicates (■ Figure 2.14).

Sue Monroe

Quartz

Orthoclase

Plagioclase

Muscovite mica

■ **Figure 2.14**

Examples of common nonferromagnesian silicate minerals.

The most common minerals in Earth's crust are nonferromagnesian silicates known as *feldspars.* Feldspar is a general name, however, and two distinct groups are recognized, each of which includes several species. The *potassium feldspars,* represented by microcline and orthoclase ($KAlSi_3O_8$), are common in igneous, metamorphic, and some sedimentary rocks. Like all feldspars, microcline and orthoclase have two internal planes of weakness along which they break or cleave.

The second group of feldspars, the *plagioclase feldspars,* range from calcium-rich ($CaAl_2Si_2O_8$) to sodium-rich ($NaAlSi_3O_8$) varieties. They possess the characteristic feldspar cleavage and typically are white or cream to medium gray. Plagioclase cleavage surfaces commonly show numerous distinctive, closely spaced, parallel lines called *striations.*

Quartz (SiO_2), a very abundant nonferromagnesian silicate, is common in the three major rock groups, especially in such rocks as granite, gneiss, and sandstone. A framework silicate, it can usually be recognized by its glassy appearance and hardness (Figure 2.14).

Another fairly common nonferromagnesian silicate is muscovite, which is a mica. Like biotite it is a sheet silicate, but muscovite is typically nearly colorless (Figure 2.14), whereas biotite is black. Various clay minerals also possess the sheet structure typical of the micas, but their crystals are so small that they can be seen only with extremely high magnification. These clay minerals are important constituents of several types of rocks and are essential components of soils (see Chapter 5).

Carbonate Minerals

Carbonate minerals, those containing the negatively charged carbonate radical $(CO_3)^{-2}$ include calcium carbonate ($CaCO_3$) as the minerals *aragonite* or *calcite* (■ Figure 2.15a). Aragonite is unstable and commonly changes to calcite, the main constituent of the sedimentary rock *limestone.* A number of other carbonate minerals are known, but only one of these need concern us: *Dolomite* [$CaMg(CO_3)_2$] forms by the chemical alteration of calcite by the addition of magnesium. Sedimentary rock composed of the mineral dolomite is *dolostone* (see Chapter 6).

Other Mineral Groups

In addition to silicates and carbonates, geologists recognize several other mineral groups (Table 2.2). And even though minerals from these groups are less common than silicates and carbonates, many are found in rocks in small quantities, and others are very important resources. In the oxides, an element combines with oxygen, as in hematite

(a) Calcite

(b) Galena

(c) Gypsum

Sue Monroe

(d) Halite

■ Figure 2.15

Representative minerals from four groups. (a) Calcite ($CaCO_3$) is the most common carbonate mineral. (b) The sulfide mineral galena (PbS) is the ore of lead. (c) Gypsum ($CaSO_4 \cdot 2H_2O$) is a common sulfate mineral. (d) Halite ($NaCl$) is a good example of a halide mineral.

(Fe_2O_3) and magnetite (Fe_3O_4). Rocks with high concentrations of these minerals in the Lake Superior region of Canada and the United States are sources of iron ores for the manufacture of steel. The related hydroxides form mostly by the chemical alteration of other minerals.

We have already noted that the *native elements* are minerals composed of a single element. Examples are diamond and graphite (C) and the precious metals gold (Au), silver (Ag), and platinum (Pt), two of which are featured in "The Precious Metals" on pages 46 and 47. Some elements such as silver and copper are found both as native elements and as compounds, and are thus also included in other mineral groups—the silver sulfide argentite (Ag_2S), for example.

Several minerals and rocks containing the phosphate radical $(PO_4)^{-3}$ are important sources of phosphorus for fertilizers. The sulfides such as the mineral galena (PbS), the ore of lead, have a positively charged ion combined with sulfur (S^{-2}) (Figure 2.15b), whereas the sulfates have an element combined with the complex radical $(SO_4)^{-2}$ as in gypsum ($CaSO_4 \cdot 2H_2O$) (Figure 2.15c). The halides contain the halogen elements, fluorine (F^{-1}) and chlorine (Cl^{-1}); examples are the minerals halite ($NaCl$) (Figure 2.15d) and fluorite (CaF_2).

HOW ARE MINERALS IDENTIFIED?

nternal structure and chemical composition determine the characteristic physical properties of all minerals. Many physical properties are remarkably constant for a given mineral species, but some, especially color, may vary. Although professional geologists use sophisticated techniques to study and identify minerals, most common minerals can be identified by using the physical properties described next (see Appendix C).

Luster and Color

Luster (not to be confused with *color*) is the quality and intensity of light reflected from a mineral's surface. Geologists define two basic types of luster: metallic, having the appearance of a metal, and nonmetallic. Notice that of the four minerals shown in Figure 2.15 only galena has a metallic luster. Among the several types of nonmetallic luster are glassy or vitreous (as in quartz), dull or earthy, waxy, greasy, and brilliant (as in diamond).

The Precious Metals

The discovery of gold by James Marshall at Sutter's Mill near Coloma in 1848 sparked the California gold rush (1849–1853) during which $200 million in gold was recovered.

National Museum of Natural History, Specimen #R12197

Specimen of gold from Grass Valley, California. Gold is too heavy and too soft for tools and weapons, so it has been prized for jewelry and as a symbol of wealth, but it is also used in glass making, electrical circuitry, gold plating, the chemical industry, and dentistry.

NEVADA

Carson City
Lake Tahoe

Placerville
Sacramento

San Francisco

Stockton

Sierra Nevada

Yosemite National Park

CALIFORNIA

PACIFIC OCEAN

Colorado River

☐ Gold rush belt

```
0        100        200
0    100    200    300 km
```

A miner pans for gold (foreground) by swirling water, sand, and gravel in a broad, shallow pan. The heavier gold sinks to the bottom. At the far left a miner washes sediment in a cradle. As in panning, the cradle separates heavier gold from other materials.

Bettmann/Corbis

Gold miners on the American River near Sacramento, California. Most of the gold came from placer deposits in which running water separated and concentrated minerals and rock fragments by their density.

Bettmann/Corbis

Hydraulic mining in California in which strong jets of water washed gold-bearing sand and gravel into sluices. In this image taken in 1905 at Junction City, California, water is directed through a monitor onto a hillside. Hydraulic mining was efficient from the mining point of view but caused considerable environmental damage.

Reports in 1876 of gold in the Black Hills of South Dakota resulted in a flood of miners that led to hostilities with the Sioux Indians, and the annihilation of Lt. Col. George Armstrong Custer and 260 of his men at the Battle of the Little Big Horn in Montana. This view shows the headworks (upper right) of the Homestake Mine at Lead, South Dakota in 1900. The headworks is the cluster of buildings near the opening to a mine.

J.C.H. Grabill/Corbis

Like gold, silver is found as a native element as in this specimen, but it also occurs as a compound in the sulfide mineral argentite (Ag_2S). Silver is used in North America for silver halide film, jewelry, flatware, surgical instruments, and backing for mirrors.

Ken Lucas/Visuals Unlimited

This image shows the headworks of the Yellowjacket Mine at Gold Hill, Nevada, and the inset shows silver-bearing quartz (white) in volcanic rock. This largest silver discovery in North America, called the Comstock Lode, was responsible for bringing Nevada into the Union in 1864 during the Civil War, even though it had too few people to qualify for statehood. The Comstock Lode was mined for silver and gold from 1859 until 1898.

James S. Monroe

Sue Monroe

Beginning students are distressed by the fact that the color of some minerals varies considerably, making the most obvious physical property of little use for their identification. Geologists know that color or lack of color in minerals is caused by how the various wavelengths of visible light are absorbed or transmitted, but they know little about what causes such differences. In any case, we can make some generalizations about color that are helpful in mineral identification. Ferromagnesian silicates are typically black, brown, or dark green, although olivine is olive green (Figure 2.13). Nonferromagnesian silicates, on the other hand, vary considerably in color but are rarely very dark. White, cream, colorless, and shades of pink and pale green are more typical (Figure 2.14).

Another helpful generalization is that the color of minerals that have a metallic luster is more consistent than it is for nonmetallic minerals. For example, galena is always lead-gray (Figure 2.15b), whereas pyrite is in-variably brassy yellow. In contrast, quartz, a nonmetallic mineral, may be colorless, smoky brown to almost black, rose, yellow-brown, milky white, blue, or violet to purple (Figure 2.8a, b).

Crystal Form

As previously noted, mineral crystals are not common, so many mineral specimens you encounter will not show the perfect crystal form typical of that mineral species (■ Figures 2.7 and 2.16). Keep in mind, however, that even though crystals may not be apparent, minerals nevertheless possess the atomic structure that would have yielded well-formed crystals if they had developed under ideal conditions.

Some minerals do typically occur as crystals. For example, 12-sided crystals of garnet are common, as are 6- and 12-sided crystals of pyrite. Minerals that grow in cavities or are precipitated from circulating hot water (hydrothermal solutions) in cracks and crevices in rocks also commonly occur as crystals (see Geo-Focus 2.1).

Crystal form can be a useful characteristic for mineral identification, but a number of minerals have the same crystal form. Pyrite (FeS_2), galena (PbS), and halite (NaCl) all occur as cubic crystals, but they can be easily identified by other properties such as color, luster, hardness, and density.

Cleavage and Fracture

Not all minerals possess **cleavage,** but those that do tend to break, or split, along a smooth plane or planes of weakness determined by the strength of the bonds within a mineral crystal. Cleavage is characterized in

(a) Fluorite

(b) Calcite

(c) Barite

Sue Monroe

■ **Figure 2.16**

Mineral crystals. (a) Cubic crystals of fluorite. (b) Calcite crystal. (c) Blade-shaped crystals of barite.

GEOFOCUS

Mineral Crystals

Certainly the most alluring aspect of minerals are crystals, most of which are rather small, measuring a few millimeters to centimeters long, but some reach gigantic proportions. Spodumene crystals up to 14 m long were mined in South Dakota for their lithium content, quartz crystals weighing several metric tons have been found in Russia, and sheets of muscovite measuring more than 2.4 m across come from mines in Ontario. Invariably, such large crystals grow in cavities where their growth is unrestricted, or they are found in *pegmatites,* a type of igneous rock with especially large minerals (see Chapter 3).

The most remarkable recent find of giant crystals took place in April 2000 in a silver and lead mine in Chihuahua, Mexico. A cavern there is lined with hundreds of gypsum crystals more than 1 m long and what one author called "crystal moonbeams" made of gypsum crystals 1.2 m in diameter and up to 15.2 m long (■ Figure 1). Perhaps these are the largest mineral crystals anywhere in the world. Fearing vandalism, the company that owned the mine kept the crystals a secret for some time, but the 65°C temperature and 100% humidity in the crystal-filled cavern would keep out all but the most determined vandals.

For many centuries, crystals and minerals were desired for their alleged healing powers and mystical properties. In fact, many minerals, especially mineral crystals, as well as some rocks and fossils have served as religious symbols and talismans,

or have been carried, worn, applied externally, or ingested for their presumed mystical or curative powers. Diamond, according to one legend, wards off evil spirits, sickness, and floods, whereas topaz was thought to avert mental disorders, and ruby was believed to preserve its owner's health and warn of imminent bad luck. Indeed, even today ads touting the healing qualities of various crystals and claims that they enhance emotional stability and clear thinking are seen in magazines and tabloids. Unfortunately for those purchasing crystals for these purposes, they provide no more benefit than artificial ones. In short, wishful thinking and the placebo effect are responsible for any perceived beneficial results.

One reason some people think crystals have favorable attributes is the curious property called the piezoelectric effect. When some crystals are compressed or an electrical current is applied, these minerals produce an electrical charge that enables them to be accurate timekeepers. For example, the electrical current from a watch's battery causes a quartz crystal to expand and contract very rapidly and regularly (about 100,000 times per second). Quartz crystal clocks were first developed in 1928, and now quartz clocks and watches are commonplace. Even inexpensive quartz timepieces are very accurate, and precision-manufactured quartz clocks used in astronomy do not gain or lose more than 1 second in 10 years.

An interesting historical note is that during World War II (1939–1945) the United States had

Richard D. Fisher

■ Figure 1

Some of these gypsum crystals in a cavern in Chihuahua, Mexico, measure up to 15.2 m long and may be the world's largest crystals. They were discovered in April 2000.

difficulty obtaining Brazilian quartz crystals needed for making radios. This shortage prompted the development of artificially synthesized quartz, and now most quartz used in watches and clocks is synthetic. So even though the piezoelectric effect imparts no healing or protective powers to crystals, it is essential in a number of applications in which precise measurements of time, pressure, or acceleration are needed. And of course many people are intrigued by crystals simply because they are so attractive.

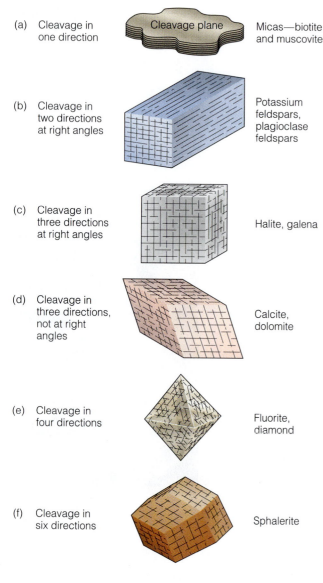

(a) Cleavage in one direction — Cleavage plane — Micas—biotite and muscovite

(b) Cleavage in two directions at right angles — Potassium feldspars, plagioclase feldspars

(c) Cleavage in three directions at right angles — Halite, galena

(d) Cleavage in three directions, not at right angles — Calcite, dolomite

(e) Cleavage in four directions — Fluorite, diamond

(f) Cleavage in six directions — Sphalerite

■ **Figure 2.17**

Several types of mineral cleavage. (a) One direction. (b) Two directions at right angles. (c) Three directions at right angles. (d) Three directions, not at right angles. (e) Four directions. (f) Six directions.

terms of quality (perfect, good, poor), direction, and angles of intersection of cleavage planes. Biotite, a common ferromagnesian silicate, has perfect cleavage in one direction (■ Figure 2.17a). The fact that biotite preferentially cleaves along a number of closely spaced, parallel planes is related to its structure; it is a sheet silicate with the sheets of silica tetrahedra weakly bonded to one another by iron and magnesium ions (Figure 2.12d).

Feldspars possess two directions of cleavage that intersect at right angles (Figure 2.17b), and the mineral halite has three directions of cleavage, all of which in-

tersect at right angles (Figure 2.17c). Calcite also possesses three directions of cleavage, but none of the intersection angles is a right angle, so cleavage fragments of calcite are rhombohedrons (Figure 2.17d). Minerals with four directions of cleavage include fluorite and diamond (Figure 2.17e). Ironically, diamond, the hardest mineral, can be easily cleaved. A few minerals such as sphalerite, an ore of zinc, have six directions of cleavage (Figure 2.17f).

Cleavage is an important diagnostic property of minerals, and its recognition is essential in distinguishing between some minerals. The pyroxene mineral augite and the amphibole mineral hornblende, for example, look much alike; both are dark green to black, have the same hardness, and possess two directions of cleavage. But the cleavage planes of augite intersect at about 90 degrees, whereas the cleavage planes of hornblende intersect at angles of 56 degrees and 124 degrees (■ Figure 2.18).

In contrast to cleavage, *fracture* is mineral breakage along irregular surfaces. Any mineral can be fractured if enough force is applied, but the fracture surfaces are commonly uneven or conchoidal (curved) rather than smooth.

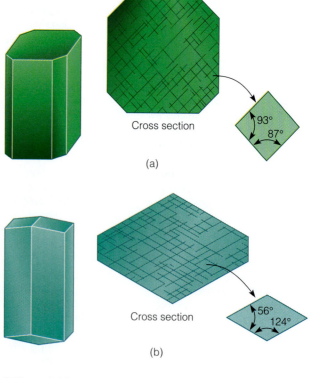

Cross section

93°
87°

(a)

Cross section

56°
124°

(b)

■ **Figure 2.18**

Cleavage in augite and hornblende. (a) Augite crystal and cross section of crystal showing cleavage. (b) Hornblende crystal and cross section of crystal showing cleavage.

Table 2.3

Mohs Hardness Scale

Hardness	Mineral	Hardness of Some Common Objects
10	Diamond	
9	Corundum	
8	Topaz	
7	Quartz	
		Steel file (6½)
6	Orthoclase	
		Glass (5½–6)
5	Apatite	
4	Fluorite	
3	Calcite	Copper penny (3)
		Fingernail (2½)
2	Gypsum	
1	Talc	

Hardness

An Austrian geologist, Friedrich Mohs, devised a relative hardness scale for 10 minerals. He arbitrarily assigned a hardness value of 10 to diamond, the hardest mineral known, and lesser values to the other minerals. Relative hardness is easily determined by the use of Mohs hardness scale (Table 2.3). Quartz will scratch fluorite but cannot be scratched by fluorite, gypsum can be scratched by a fingernail, and so on. So **hardness** is defined as a mineral's resistance to abrasion and is controlled mostly by internal structure. For example, both graphite and diamond are composed of carbon, but the former has a hardness of 1 to 2, whereas the latter has a hardness of 10.

Specific Gravity (Density)

Specific gravity and density are two separate concepts, but here we will use them more or less as synonyms. A mineral's **specific gravity** is the ratio of its weight to the weight of an equal volume of pure water. Thus a mineral with a specific gravity of 3.0 is three times as heavy as water. Like all ratios, specific gravity is not expressed in units such as grams per cubic centimeters; it is a dimensionless number. **Density,** in contrast, is a mineral's mass (weight) per unit of volume expressed in grams per cubic centimeters. So the specific gravity of galena (Figure 2.15b) is 7.58 and its density is 7.58 g/cm^3. In most instances we will refer to a mineral's density, and in some of the following chapters we will mention the density of various rocks.

Structure and composition control a mineral's specific gravity and density. Because ferromagnesian silicates contain iron, magnesium, or both, they tend to be denser than nonferromagnesian silicates. In general, the metallic minerals, such as galena and hematite, are denser than nonmetals. Pure gold with a density of 19.3 g/cm^3 is about two and one half times as dense as lead. Diamond and graphite, both of which are composed of carbon (C), illustrate how structure controls specific gravity or density. The specific gravity of diamond is 3.5, whereas that of graphite varies from 2.09 to 2.33.

Other Useful Mineral Properties

Other physical properties characterize some minerals. Talc has a distinctive soapy feel, graphite writes on paper, halite tastes salty, and magnetite is magnetic (■ Figure 2.19). Calcite possesses the property of *double refraction,* meaning that an object when viewed through

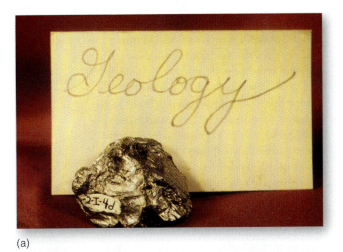

(a)

James S. Monroe

(b)

■ **Figure 2.19**

Various mineral properties. Graphite (a), the mineral used to make pencil "lead," writes on paper, whereas magnetite (b), an important iron ore, is magnetic.

a transparent piece of calcite will have a double image. Some sheet silicates are plastic and, when bent into a new shape, will retain that shape; others are flexible and, if bent, will return to their original position when the forces that bent them are removed.

A simple chemical test to identify the minerals calcite and dolomite involves applying a drop of dilute hydrochloric acid to the mineral specimen. If the mineral is calcite, it will react vigorously with the acid and release carbon dioxide, which causes the acid to bubble or effervesce. Dolomite, in contrast, will not react with hydrochloric acid unless it is powdered.

WHERE AND HOW DO MINERALS FORM?

Thus far we have discussed the composition, structure, and physical properties of minerals but have not fully addressed how they originate. One phenomenon that accounts for the origin of minerals is the cooling of molten rock material known as *magma* (magma that reaches the surface is called *lava*). As magma or lava cools, minerals crystallize and grow, thereby determining the mineral composition of various igneous rocks such as basalt (dominated by ferromagnesian silicates) and granite (dominated by nonferromagnesian silicates) (see Chapter 3). Hot water solutions derived from magma commonly invade cracks and crevasses in adjacent rocks, and from these solutions a variety of minerals crystallize, some of economic importance. Minerals also originate when water in hot springs cools (see Chapter 16), and when hot, mineral-rich water discharges onto the seafloor at hot springs known as black smokers (see Chapter 11).

Dissolved materials in seawater, more rarely lake water, combine to form minerals such as halite (NaCl), gypsum ($CaSO_4 \cdot 2H_2O$), and several others when the water evaporates. Aragonite and/or calcite, both varieties of calcium carbonate ($CaCO_3$), might also form from evaporating water, but most originates when organisms such as clams, oysters, corals, and floating microorganisms use this compound to construct their shells. And a few plants and animals use silicon dioxide (SiO_2) for their skeletons, which accumulate as mineral matter on the seafloor when the organisms die (see Chapter 6).

Some clay minerals form when chemical processes compositionally and structurally alter other minerals, such as

feldspars (see Chapter 5), and others originate when rocks are changed during metamorphism (see Chapter 7). In fact, the agents that cause metamorphism—heat, pressure, and chemically active fluids—are responsible for the origin of many minerals. A few minerals even originate when gasses such as hydrogen sulfide (H_2S) and sulfur dioxide (SO_2) react at volcanic vents to produce sulfur.

WHAT ARE ROCK-FORMING MINERALS?

Geologists use the term **rock** for a solid aggregate of one or more minerals, but the term also refers to masses of mineral-like matter as in the natural glass obsidian (see Chapter 3) and masses of solid organic matter as in coal (see Chapter 6). Granite, consisting of certain percentages of potassium feldspars and quartz, is a rock, as is limestone although it is composed of only calcite. And even though some rocks might contain many minerals, only a few, designated **rock-forming minerals**, are sufficiently common to be used in

Table 2.4

Important Rock-Forming Minerals

Mineral	Primary Occurrence
Ferromagnesian silicates	
Olivine	Igneous, metamorphic rocks
Pyroxene group	
Augite most common	Igneous, metamorphic rocks
Amphibole group	
Hornblende most common	Igneous, metamorphic rocks
Biotite	All rock types
Nonferromagnesian silicates	
Quartz	All rock types
Potassium feldspar group	
Orthoclase, microcline	All rock types
Plagioclase feldspar group	All rock types
Muscovite	All rock types
Clay mineral group	Soils, sedimentary rocks, some metamorphic rocks
Carbonates	
Calcite	Sedimentary rocks
Dolomite	Sedimentary rocks
Sulfates	
Anhydrite	Sedimentary rocks
Gypsum	Sedimentary rocks
Halides	
Halite	Sedimentary rocks

Sue Monroe

Granite   Potassium   Quartz   Biotite
                       feldspar

■ **Figure 2.20**

The igneous rock granite is made up of mostly potassium feldspar and quartz, both of which are common rock-forming minerals. Granite also usually contains a small amount of the sheet silicate biotite.

rock identification and classification (Table 2.4). Others, known as *accessory minerals,* are present in such small quantities that they can be disregarded. In addition to potassium feldspars and quartz, granite commonly contains conspicuous but minor quantities of biotite mica and other minerals in even lesser amounts, all of which are nonessential for identifying granite (■ Figure 2.20).

We have emphasized that silicate minerals are by far the most common minerals in Earth's crust, so it follows that most rocks are composed of these minerals. Indeed, feldspar minerals (plagioclase feldspars and potassium feldspars) and quartz make up more than 60% of Earth's crust. But even among the hundreds of silicates only a few are particularly common in rocks, although many others are present as accessory minerals.

The most common nonsilicate rock-forming minerals are the carbonates calcite ($CaCO_3$) and dolomite [$CaMg(CO_3)_2$], the main constituents of the sedimentary rocks limestone and dolostone, respectively (see Chapter 6). Among the sulfates and halides, gypsum ($CaSO_4 \cdot 2H_2O$) in rock gypsum and halite (NaCl) in rock salt (see Chapter 6) are common enough to qualify as rock-forming minerals. But even though these minerals, and their corresponding rocks, might be common in some areas, their overall abundance is limited compared to the silicate and carbonate rock-forming minerals.

rally occurring solid, liquid, or gaseous material in or on Earth's crust in such form and amount that economic extraction of a commodity from the concentration is currently or potentially feasible."

Natural resources are mostly valuable concentrations of minerals, rocks, or both, but liquid petroleum and natural gas are also included. In fact, we refer to a variety of resources, some of which are *metallic resources* (copper, tin, iron ore, etc.), *nonmetallic resources* (sand and gravel, crushed stone, salt, sulfur, etc.), and *energy resources* (petroleum, natural gas, coal, and uranium). All of the foregoing are indeed resources, but we must make a distinction between a resource, the total amount of a commodity whether discovered or undiscovered, and a **reserve,** which is only that part of the resource base that is known and can be economically recovered. Aluminum can be extracted from aluminum-rich igneous and sedimentary rocks, but at present that cannot be done economically.

The distinction between a resource and a reserve is simple enough in principle, but in practice it depends on several factors, not all of which remain constant. Geographic location may be the determining factor in evaluating a resource versus a reserve. For instance, a resource in a remote region might not be mined because transportation costs are too high, and what might be deemed a resource rather than a reserve in the United

NATURAL RESOURCES AND RESERVES

The United States and Canada, both highly industrialized nations, have enjoyed considerable economic success because they have abundant natural resources. But what are resources, how and where do they form, and how are they found and exploited? Geologists at the U.S. Geological Survey use this definition: A **resource** is "a concentration of natu-

What Would You Do

The distinction between minerals and rocks is not easy for beginning students to understand. As a teacher, you know that minerals are made up of chemical elements and that rocks consist of one or more minerals, but despite your best efforts to clearly define them, your students commonly mistake one for the other. Can you think of analogies that might help students understand the difference between minerals and rocks?

TERRY S. MOLLO

Sculpture Out of Stone

Terry S. Mollo is a full-time sculptor living in New York. She regularly exhibits her work in New York's Hudson Valley, New York City, and New Jersey. She has a B.A. in Journalism and Communication Arts from Pace University and currently studies sculpture at the Sculpture/Fine Arts Studio in Pomona, New York, under Martin Glick and at the Art Students League of New York with Gary Sussman.

It wasn't until 1997 that I started working with various stones, giving them new form, but I recall being drawn to the beauty of these natural formations at an early age. In 1954, when my grade school class studied the California gold rush, my friends and I formed a geology exploration club, which involved "borrowing" hammers, small chisels, and various other tools from our family garages to break apart pretty "rocks" we collected around the neighborhood. Black and silver mica was one of our favorite specimens; it was easy to find and lots of fun to crumble. We pulverized it with our hammers and chisels, poured it through makeshift strainers into jars, then weighed it, marked it "gem-dust," and stashed it away— a secret, sparkling treasure.

Now, many years later, as a full-time sculptor, I have returned to these remarkable stones that I sought out in childhood, but the tools I use today are specifically designed to file, chip, and carve stone into a new work of art. In my work, I use a method called the "subtractive" method, which is to say that I remove something to create something else. By removing stone and creating areas of depth and shadow, I strive to create an aesthetic work of art. Hundreds of tools are now available for stone carving (manual, pneumatic, electric, and industrial).

Despite recent innovations and variations in tools, many sculptors still depend on the same basic handful of tools that Michelangelo used—hammers and chisels. In fact, popular hand tools like files, rasps, and rifflers, which are hand-forged so no two are identical, come from Milan, Italy, where great masters like Michelangelo worked.

When an artist is blocking out a piece in its early stages, large amounts of excess stone can be removed quickly with pneumatic chisels as well as a flexible shaft used for cutting, grinding, and contouring. For fine details, a sculptor often depends on that small traditional group of manual files, rasps, and rifflers to finish a piece. At the end, sanding and polishing are done with silicon carbide, diamond wet/dry sandpapers, or electric machines.

Great strides have been made technologically during the last 10 to 20 years, and now it is actually possible to start with a clay or plaster maquette and use a digitally based carving process called a CNC (Computer Numerically Controlled) carving machine. CNC uses a 3-D scanned image of the original work, digital technology, and an enormous five-axis milling machine. Even with these modern inventions, sculpting stone is a slow and arduous art, and unfortunately a dying one. Those of us who are still active sculptors are often told that we must teach all we know and pass it down, so it doesn't go away.

I have worked with several types of stone, including marble, alabaster, agate, and limestone. There is a large variety of material to choose from, and within each category the colors, shades, shapes, and veining are very exciting to an artist, each lending itself to a different kind of piece, be it abstract, realistic, allegorical—figurative or nonfigurative. Many sculptors today prefer the alabasters, since they offer the

This flower sculture by Terry Mollo is titled *Hu Yi Fang* (Breathe Comfortable Fragrance) and is made from white/gold alabaster.

widest variety of colors and interesting veining. Alabaster (as well as steatite or soapstone) is a softer stone and is more easily carved as opposed to marble, which is metamorphic, or granite, which is igneous and extremely hard. Marble and granite are both dense and difficult to carve; nonetheless, most sculptors at some point in their career seek out a beautiful piece of Italian marble to carve.

Stones are available from quarries or through suppliers both in the United States and abroad. Domestic alabasters, limestone, and good marbles come from many quarries in the United States in Vermont, Georgia, and Colorado. Italy has amazing

marble, agate, alabaster, and travertine. There are also Portuguese pink marble, Belgian black marble, Persian travertine, and African wonderstone, which is a beautiful deep gray and black, although it is very difficult to carve and is one of the most expensive stones. In the New York area current stone prices range from about $0.75 per pound to $4 or $5 per pound. Indiana limestone is quite inexpensive. A piece of Italian white alabaster a little bigger than a football may cost $60.

The most common question people ask me about my work as a sculptor is whether or not I see a piece of stone I admire and bring it to the studio where I decide what to do with it, or whether I have a particular idea and then go out after the perfect piece of stone to create that vision. Well, the truth is that I have experienced the process both ways. Many of the ideas I have with regard to stone involve organic shapes such as leaves, flowers, and trees. For example, one morning I had an idea to create a big flower in an upright position and went out in search of the perfect stone to carve into the flower I imagined. I walked around a gigantic supply basement filled with thousands of pieces that varied in size and color from all over the world. Finally, I chose a white/gold alabaster that weighed about 350 pounds. It had a very nice shape and size for the vision floating in my head. I could see my flower deep within the stone. Then, as I waited for the piece to be loaded into my car, I saw the most beautiful piece of honey-brown-colored Italian agate with beige and pale yellow veining, so I bought it. I had no idea what I would do with it, however, so it sat in my basement for well over a year waiting for my inspiration. Then one evening I came across a framed black and white Ansel Adams photograph of a wave . . . and that was it. I took my Italian brown agate into the studio and carved away. The beige at the tips became sea froth, the opaque, dark chocolate section the undercurrent, and the veins running diagonally my shoreline.

States and Canada may be mined in a developing country where labor costs are low. The commodity in question is also important. Gold or diamonds in sufficient quantity can be mined profitably just about anywhere, whereas all sand and gravel deposits must be close to their market areas.

Obviously the market price is important in evaluating any resource. From 1935 until 1968, the U.S. government maintained the price of gold at $35 per troy ounce (1 troy ounce = 31.1 g). When this restriction was removed, demand determined the market price and gold prices rose, reaching an all-time high of $843 per troy ounce in 1980. As a result, many marginal deposits became reserves and a number of abandoned mines were reopened.

The status of a resource is also affected by changes in technology. By the time of World War II (1939–1945), the richest iron ores of the Great Lakes region in the United States and Canada had been mostly depleted. But the development of a method for separating the iron from unusable rock and shaping it into pellets ideal for use in blast furnaces made it profitable to mine rocks with less iron. As a matter of fact, a large part of the mineral revenue of Newfoundland and Quebec, Canada, and Minnesota and Michigan comes from mining iron ore.

Most people know that industrialized societies depend on a variety of natural resources but have little knowledge about their occurrence, methods of recovery, and economics. Geologists are, of course, essential in finding and evaluating deposits, but extraction involves engineers and chemists, not to mention many people in support industries that supply mining equipment. Ultimately, though, the decision about whether a deposit should be mined or not is made by people trained in business and economics. In short, extraction must yield a profit. The extraction of natural resources, other than oil, natural gas, and coal, amounted to more than $40 billion during 2002 in the United States, and in Canada the extraction of nonfuel resources during the same year was nearly $18 billion (Canadian dollars). We will have much more to say about energy resources in later chapters, especially Chapter 6.

Everyone is aware of the importance of resources such as petroleum, gold, and ores of iron, copper, and lead. However, some quite common minerals are also essential. For example, pure quartz sand is used to manufacture glass and optical instruments as well as sandpaper and steel alloys. Clay minerals are needed to make ceramics and paper, and feldspars are used for porcelain, ceramics, enamel, and glass. Micas are used in a variety of products including lipstick, glitter, and eye

■ **Figure 2.21**

This magnesite mine at Gabbs, Nevada, employs 96 of the town's 667 residents. Magnesite is used to manufacture high-temperature–resistant construction materials, refractory bricks, oxychloride cement, medicines, and cosmetics.

shadow as well as the lustrous paints on appliances and automobiles. Phosphate-bearing rock used in fertilizers mined in Florida accounts for a large part of that state's mineral production, and a magnesite ($MgCO_3$) mine is the only employer of any consequence in the small town of Gabbs, Nevada (■ Figure 2.21).

Access to many resources is essential for industrialization and the high standard of living enjoyed in many countries. The United States and Canada are fortunate to be resource-rich nations, but resources are used much faster than they form so they are *nonrenewable*, meaning that once a resource has been depleted, new deposits or suitable substitutes, if available, must either be found or be imported from elsewhere. For some essential resources, the United States is totally dependent on imports—no cobalt was mined in this country during 2002. Yet the United States, the world's largest consumer of cobalt, uses this essential metal in gas-turbine aircraft engines and magnets and for corrosion- and wear-resistant alloys. Obviously all cobalt is imported, as is all manganese, an element essential for making steel.

In addition to cobalt and manganese, the United States imports all the aluminum ore it uses as well as all or some of many other resources (■ Figure 2.22). Canada, in contrast, is more self-reliant, meeting most of its domestic mineral and energy needs. Nevertheless, it must import phosphate, chromium, manganese, and aluminum ore. Canada also produces more crude oil and natural gas than it uses, and it is among the world leaders in producing and exporting uranium.

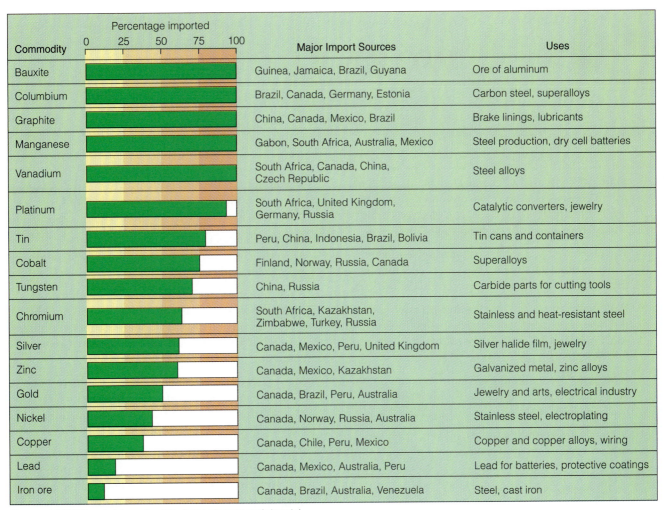

Commodity	Percentage imported 0 25 50 75 100	Major Import Sources	Uses
Bauxite		Guinea, Jamaica, Brazil, Guyana	Ore of aluminum
Columbium		Brazil, Canada, Germany, Estonia	Carbon steel, superalloys
Graphite		China, Canada, Mexico, Brazil	Brake linings, lubricants
Manganese		Gabon, South Africa, Australia, Mexico	Steel production, dry cell batteries
Vanadium		South Africa, Canada, China, Czech Republic	Steel alloys
Platinum		South Africa, United Kingdom, Germany, Russia	Catalytic converters, jewelry
Tin		Peru, China, Indonesia, Brazil, Bolivia	Tin cans and containers
Cobalt		Finland, Norway, Russia, Canada	Superalloys
Tungsten		China, Russia	Carbide parts for cutting tools
Chromium		South Africa, Kazakhstan, Zimbabwe, Turkey, Russia	Stainless and heat-resistant steel
Silver		Canada, Mexico, Peru, United Kingdom	Silver halide film, jewelry
Zinc		Canada, Mexico, Kazakhstan	Galvanized metal, zinc alloys
Gold		Canada, Brazil, Peru, Australia	Jewelry and arts, electrical industry
Nickel		Canada, Norway, Russia, Australia	Stainless steel, electroplating
Copper		Canada, Chile, Peru, Mexico	Copper and copper alloys, wiring
Lead		Canada, Mexico, Australia, Peru	Lead for batteries, protective coatings
Iron ore		Canada, Brazil, Australia, Venezuela	Steel, cast iron

Sources: USGS Minerals Information: http://minerals.usgs.gov/minerals/
USGS Mineral Commodity Summaries 2003: http://usgs.gov/minerals/pubs/mcs/2003.pdf

■ **Figure 2.22**

The dependence of the United States on imports of various mineral commodities is apparent from this chart. The lengths of the green bars correspond to the amounts of resources imported.

To ensure continued supplies of essential minerals and energy resources, geologists as well as other scientists, government agencies, and leaders in business and industry continually assess the status of resources in view of changing economic and political conditions and changes in science and technology. The U.S. Geological Survey, for instance, keeps detailed statistical records of mine production, imports, and exports, and regularly publishes reports on the status of numerous commodities. Similar reports appear regularly in the *Canadian Minerals Yearbook*. In several of the following chapters we will discuss the geologic occurrence of several natural resources.

What Would You Do

Let's say that some reputable businesspeople tell you of opportunities to invest in natural resources. Two ventures look promising: a gold mine and a sand and gravel pit. Given that gold sells for about $290 per ounce, whereas sand and gravel are worth about $4 or $5 per ton, would it be more prudent to invest in the gold mine? Explain not only how market price would influence your decision but also what other factors you might need to consider.

REVIEW WORKBOOK

Chapter Summary

- Matter is composed of chemical elements, each of which consists of atoms. Protons and neutrons are present in an atom's nucleus, and electrons orbit around the nucleus in electron shells.

- The number of protons in an atom's nucleus determines its atomic number. The atomic mass number is the number of protons plus neutrons in the nucleus.

- Bonding results when atoms join with other atoms; if different elements bond, a compound forms. Most minerals are compounds, although there are a few exceptions.

- Ionic and covalent bonds are most common in minerals, but metallic and van der Waals bonds are found in some.

- Minerals are crystalline solids, meaning they possess an ordered internal arrangement of atoms.

- Mineral composition is indicated by a chemical formula, such as SiO_2 for quartz.

- Some minerals have a range of compositions because different elements substitute for one another if the atoms are of about the same size and the same electrical charge.

- More than 3500 minerals are known, most of which are silicates. The two types of silicates are ferromagnesian and nonferromagnesian.

- In addition to silicates, geologists recognize carbonates, native elements, hydroxides, oxides, phosphates, halides, sulfates, and sulfides.

- Structure and composition control the physical properties of minerals, such as crystal form, hardness, color, and cleavage.

- A few minerals, designated rock-forming minerals, are common enough in rocks to be essential in their identification and classification. Most rock-forming minerals are silicates, but some carbonates are also common.

- Many resources are concentrations of minerals or rocks of economic importance. They are further characterized as metallic resources, nonmetallic resources, and energy resources.

- Reserves are that part of the resource base that can be extracted profitably. The status of a resource versus a reserve depends on market price, labor costs, geographic location, and developments in science and technology.

- The United States must import many resources to maintain its industrial capacity. Canada is more self-reliant, but it too must import some commodities.

Important Terms

atom (p. 34)
atomic mass number (p. 34)
atomic number (p. 34)
bonding (p. 34)
carbonate mineral (p. 44)
cleavage (p. 48)
compound (p. 34)
covalent bond (p. 36)
crystal (p. 38)
crystalline solid (p. 38)
density (p. 51)
electron (p. 34)

electron shell (p. 34)
element (p. 34)
ferromagnesian silicate (p. 43)
hardness (p. 51)
ion (p. 36)
ionic bond (p. 36)
luster (p. 45)
mineral (p. 32)
native element (p. 38)
neutron (p. 34)
nonferromagnesian silicate
  (p. 44)

nucleus (p. 34)
proton (p. 34)
reserve (p. 53)
resource (p. 53)
rock (p. 52)
rock-forming mineral (p. 52)
silica (p. 42)
silica tetrahedron (p. 42)
silicate (p. 42)
specific gravity (p. 51)

Review Questions

1. A common rock-forming silicate mineral is _____, whereas the most common carbonate mineral is _____.
   a. _____ olivine/gypsum; b. _____ quartz/calcite; c. _____ hematite/galena; d. _____ halite/biotite; e. _____ muscovite/hornblende.

2. In what type of chemical bonding are electrons shared by adjacent atoms?
   a. _____ van der Waals; b. _____ silicate; c. _____ octahedral; d. _____ spherical; e. _____ covalent.

3. The two most abundant elements in Earth's crust are:
   a. _____ oxygen and silicon; b. _____ iron and potassium; c. _____ aluminum and calcium; d. _____ granite and basalt; e. _____ magnesium and iridium.

4. An atom with 6 protons and 8 neutrons in its nucleus has an atomic mass number of:
   a. _____ 6; b. _____ 8; c. _____ 14; d. _____ 48; e. _____ 2.

5. Any mineral composed of an element combined with sulfur (S^{-2}) as in galena (PbS) is a(n):
   a. _____ oxide; b. _____ sulfide; c. _____ carbonate; d. _____ silicate; e. _____ hydroxide.

6. The atoms of the noble gases do not react to form compounds because they have:
   a. _____ eight electrons in their outermost electron shell; b. _____ more positive charges than negative charges; c. _____ three directions of cleavage intersecting at right angles; d. _____ atomic mass numbers exceeding 92; e. _____ too much silica and not enough calcium.

7. A rock-forming mineral is any mineral:
   a. _____ found in rocks; b. _____ containing the $(CO_3)^{-2}$ radical; c. _____ in which oxygen combines with iron; d. _____ essential for classification of rocks; e. _____ from the silicate group.

8. A mineral known as a native element is one in which:
   a. _____ one element can substitute for another; b. _____ composition is determined by reactions between oxygen and iron; c. _____ atoms bond to form continuous sheets; d. _____ at least silicon and oxygen are found; e. _____ only one chemical element is present.

9. Minerals that possess the property known as cleavage:
   a. _____ are denser than minerals lacking this property; b. _____ exhibit double refraction; c. _____ break along smooth internal planes of weakness; d. _____ include obsidian and coal; e. _____ are composed mostly of the noble gases.

10. The ferromagnesian silicate olivine has the chemical formula $(Mg,Fe)_2SiO_4$, which means that:
    a. _____ silicon and oxygen may or may not be present; b. _____ magnesium and iron can substitute for each other; c. _____ magnesium and iron are less abundant in Earth's crust than silicon and oxygen; d. _____ olivine contains either magnesium or iron but not both; e. _____ ferromagnesian silicates are darker than nonferromagnesian silicates.

11. Explain the distinction between rock-forming minerals and accessory minerals. Also, name some of the most common silicate rock-forming minerals and one carbonate rock-forming mineral.

12. Why must the United States, a resource-rich nation, import most or all of some of the resources it needs? What are some of the problems such a dependence on imports creates?

13. What is cleavage in minerals? How can it be used in "cutting" gemstones?

14. How do minerals characterized as silicates differ from carbonates and oxides?

15. Briefly discuss three ways in which minerals originate.

16. Compare ionic and covalent bonding.

17. Under what conditions do well-formed mineral crystals originate? Why are well-formed crystals not very common?

18. What accounts for the fact that some minerals, such as plagioclase feldspars, have a range of chemical compositions? Give an example from the ferromagnesian silicates.

19. How would the color and density of a rock composed mostly of ferromagnesian silicates differ from one made up primarily of nonferromagnesian silicates?

20. What is the basic distinction between minerals and rocks?

World Wide Web Activities

PHYSICAL Geology⟳Now Assess your understanding of this chapter's topics with additional quizzing and comprehensive interactivities at

http://earthscience.brookscole.com/physgeo5e

as well as current and up-to-date weblinks, additional readings, and InfoTrac College Edition exercises.

Igneous Rocks and Intrusive Igneous Activity

CHAPTER 3

OUTLINE

PHYSICAL Geology ⇌ Now *This icon, appearing throughout the book, indicates an opportunity to explore interactive tutorials, animations, or practice problems available on the Physical GeologyNow Web site at http://earthscience.brookscole.com/physgeo5e.*

OBJECTIVES

At the end of this chapter, you will have learned that

- With few exceptions magma is composed of silicon and oxygen with lesser amounts of several other chemical elements.

- Temperature and especially composition are the most important controls on the mobility of magma and lava.

- Most magma originates within the upper mantle or lower crust at or near divergent and convergent plate boundaries.

- Several processes bring about chemical changes in magma, so magma may evolve from one kind into another.

- All igneous rocks form when magma or lava cools and crystallizes or by the consolidation of pyroclastic materials ejected during explosive eruptions.

- Geologists use texture and composition to classify igneous rocks.

- Intrusive igneous bodies called plutons form when magma cools below Earth's surface. The origin of the largest plutons is not fully understood.

These granitic rocks in Yosemite National Park in California are part of the Sierra Nevada batholith that measures 640 km long and up to 110 km wide. This near-vertical cliff is El Capitan, meaning "The Chief." It rises more than 900 m above the valley floor, making it the highest unbroken cliff in the world. Source: Courtesy of Richard L. Chambers

Introduction

We mentioned that the term *rock* applies to a solid aggregate of one or more minerals as well as mineral-like matter as in natural glass and solid masses of organic matter as in coal. Furthermore, in Chapter 1 we briefly discussed the three main families of rocks: igneous, sedimentary, and metamorphic. Recall that *igneous rocks* form when molten rock material known as *magma* or *lava* cools and crystallizes to form a variety of minerals, or when particulate matter called *pyroclastic materials* become consolidated. We are most familiar with igneous rocks formed from lava flows and pyroclastic materials because they are easily observed at the surface, but you should be aware that most magma never reaches Earth's surface. Indeed, much of it cools and crystallizes far underground and thus forms *plutons,* igneous bodies of various shapes and sizes.

Granite and several similar appearing rocks, collectively called *granitic rocks,* are the most common ones encountered in the larger plutons, such as those in the Sierra Nevada of California (see the chapter opening photo), in Acadia National Park, Maine, or in Denali (Mount McKinley) in Alaska. The images of Presidents Lincoln, Roosevelt, Jefferson, and Washington at Mount Rushmore National Memorial in South Dakota as well as the nearby Crazy Horse Memorial (under construction) are in the 1.7-billion-year-old Harney Peak Granite consisting of a number of plutons (■ Figure 3.1). These huge bodies of granitic rock formed far below the surface, but subsequent uplift and deep erosion exposed them in their present form. Some granitic rocks are quite attractive, especially when sawed and polished. They are used for tombstones, mantlepieces, kitchen counters, facing stones on

buildings, pedestals for statues, and statuary itself. More important, though, is the fact that fluids emanating from plutons account for many important ore minerals of important metals such as copper in adjacent rocks.

The origin of plutons, or intrusive igneous activity, and volcanism involving eruption of lava flows and pyroclastic materials are closely related topics even though we discuss them in separate chapters. The same kinds of magmas are involved in both processes, but magma varies in its mobility, which accounts for the fact that only some reaches the surface. Furthermore, plutons typically lie beneath areas of volcanism and, in fact, are the source of the overlying lava flows and pyroclastic materials. Plutons and most volcanoes are found at or near divergent and convergent plate boundaries, so igneous rocks serve as some of the criteria for recognizing ancient plate boundaries and also help us unravel the complexities of mountain-building episodes (see Chapter 13).

One important reason to study igneous rocks and intrusive igneous activity is that igneous rocks are one of the three main families of rocks. In addition, igneous rocks make up large parts of all continents and nearly all of the oceanic crust, which is formed continuously by igneous activity at divergent plate boundaries. And, as already mentioned, important mineral deposits are found adjacent to many plutons.

In this chapter our main concerns are (1) the origin, composition, textures, and classification of igneous rocks, and (2) the origin, significance, and types of plutons. In the following chapter we will consider volcanism, volcanoes, and associated phenomena that result from magma reaching Earth's surface. Remember, though, that the origin of plutons and volcanism are related topics.

(a)   (b)

■ **Figure 3.1**

(a) The presidents' images at Mount Rushmore, South Dakota, were carved in the Harney Peak Granite. The 18-m-high images were carved between 1927 and 1941 and are now the primary attraction at Mount Rushmore National Memorial. (b) The nearby Crazy Horse Memorial, also in the Harney Peak Granite, is still under construction.

THE PROPERTIES AND BEHAVIOR OF MAGMA AND LAVA

n Chapter 2 we noted that one process that accounts for the origin of minerals, and thus of rocks, is the cooling and crystallization of molten rock material known as *magma* and *lava*. **Magma** is simply molten rock below the surface; the same material at the surface is called **lava.** No other distinction is necessary, so the two terms simply tell us the location of the molten rock. Any magma is less dense than the rock from which it was derived, so it tends to move toward the surface. However, much of it cools and solidifies deep underground, thus accounting for the origin of various plutons. Magma reaching the surface either erupts as **lava flows** or is forcefully ejected into the atmosphere as particles known as **pyroclastic materials** (from the Greek *pyro,* "fire," and *klastos,* "broken"). Lava flows and eruptions of pyroclastic materials are the most awe-inspiring manifestations of all processes related to magma, but result from only a small percentage of all magma that forms.

All **igneous rocks** derive from magma, but two separate processes account for their origin. They form when (1) magma or lava cools and crystallizes to form minerals, or (2) pyroclastic materials such as volcanic ash are consolidated, forming solid masses from the previously loose particles. Igneous rocks that result from cooling lava flows and consolidation of pyroclastic materials are further characterized as **volcanic rocks** or **extrusive igneous rocks,** whereas those that form when magma cools and crystallizes below the surface are known as **plutonic rocks** or **intrusive igneous rocks.**

This brief introduction to magma and lava is enough to clarify what these substances are and what kinds of rocks are derived from them. However, let's explore magma, the source of all igneous rocks, a bit further and consider its composition, temperature, and resistance to flow, or what is called *viscosity*.

Composition of Magma

In Chapter 2 we noted that by far the most abundant minerals in Earth's crust are silicates such as quartz, feldspars, and several ferromagnesian silicates, all composed of silicon, oxygen, and other elements shown in Figure 2.10. As a result, melting of crustal rocks yields mostly silica-rich magmas that also contain considerable aluminum, calcium, sodium, iron, magnesium, potassium, and several other elements in lesser quantities. Another source of magma is melting of rocks in Earth's upper mantle, which are composed largely of ferromagnesian silicates. Thus magma derived from this source contains comparatively less silica and more iron and magnesium.

Silica is the primary constituent of most magmas, but it varies enough to distinguish magmas characterized as felsic, intermediate, and mafic (Table 3.1). **Felsic magma,** with more than 65% silica, is silica rich and contains considerable sodium, potassium, and aluminum, but little calcium, iron, and magnesium. **Mafic magma,** in contrast, contains less than 52% silica, but proportionately more calcium, iron, and magnesium. As its name implies, **intermediate magma** has a composition intermediate between felsic and mafic magma (Table 3.1).

How Hot Are Magma and Lava?

Whether you have witnessed a lava flow or not, you know that lava is very hot. But how hot is hot? Erupting lavas generally have temperatures in the range of 1000° to 1200°C, although a temperature of 1350°C was recorded above Hawaiian lava lakes where volcanic gases reacted with the atmosphere. Magma must be even hotter, but no direct measurements of magma temperatures have been made.

Most temperature measurements are taken where volcanoes show little or no explosive activity, so our best information comes from mafic lava flows such as those issuing from the Hawaiian volcanoes (■ Figure 3.2). In contrast, eruptions of felsic lava flows are not as common, and, in fact, volcanoes erupting silica-rich lava tend to be explosive and thus cannot be approached safely. Nevertheless, the temperatures of some bulbous masses of felsic lava in lava domes have been measured at a distance with an instrument called an optical

Table 3.1

The Most Common Types of Magmas and Their Characteristics

Type of Magma	Silica Content (%)	Sodium, Potassium, and Aluminum	Calcium, Iron, and Magnesium
Ultramafic	<45		Increase
Mafic	45–52		
Intermediate	53–65		
Felsic	>65	Increase	

P. Mouginis-Mark

■ **Figure 3.2**

A geologist uses a thermocouple to determine the temperature of a lava flow in Hawaii.

Silica content strongly controls magma and lava viscosity. With increasing silica content, numerous networks of silica tetrahedra form and retard flow because for flow to take place, the strong bonds of the networks must be ruptured. Mafic magma and lava with 45–52% silica have fewer silica tetrahedra networks and as a result are more mobile than felsic magma and lava flows. One mafic flow in 1783 in Iceland flowed about 80 km, and geologists traced some ancient flows in Washington State for more than 500 km. Felsic magma, in contrast, because of its higher viscosity, does not reach the surface as commonly as mafic magma. And when felsic lava flows do occur, they tend to be slow moving and thick and to move only short distances. A thick, pasty lava flow that erupted in 1915 from Lassen Peak in California flowed only about 300 m before it ceased moving.

pyrometer. The surfaces of these domes are as hot as 900°C, but their interiors must surely be even hotter.

When Mount St. Helens erupted in 1980, it ejected felsic magma as particulate matter in pyroclastic flows. Two weeks later, these flows still had temperatures between 300°C and 420°C, and a steam explosion took place more than a year later when water encountered some of the still-hot deposits. The reason magma and lava retain heat so well is that rock conducts heat so poorly. Accordingly, the interiors of thick lava flows may remain hot for months or years, whereas plutons, depending on their size and depth, may not completely cool for thousands to millions of years.

Viscosity—Resistance to Flow

All liquids have the property of **viscosity,** or simply resistance to flow. For many liquids, such as water, viscosity is very low, so they are highly fluid and flow readily. For other liquids, though, viscosity is so high that they flow much more slowly. Good examples are cold motor oil and syrup, both of which are quite viscous and thus flow only with difficulty. But when these same liquids are heated, their viscosity is much lower and they flow more easily. That is, they become more fluid with increasing temperature. Accordingly, you might suspect that temperature controls the viscosity of magma and lava, and this inference is partly correct. We can generalize and say that hot magma or lava moves more readily than cooler magma or lava, but we must qualify this statement by noting that temperature is not the only control of viscosity.

HOW DOES MAGMA ORIGINATE AND CHANGE?

Most of us have not witnessed a volcanic eruption, but we have nevertheless seen news reports or documentaries showing magma issuing forth as lava flows or pyroclastic materials. If you want to actually see a lava flow in progress, you need only visit Hawaii Volcanoes National Park where Kilauea is now in its 20th consecutive year of eruption. You can even see pyroclastic materials erupted at lava fountains in Hawaii, but the truly explosive eruptions that yield huge quantities of particulate matter are episodic and quite dangerous. In any case, we are familiar with some aspects of igneous activity, but most people are unaware of how and where magma originates, how it rises from its place of origin, and how it might change. Indeed, many believe the misconception that lava comes from a continuous layer of molten rock beneath the crust or that it comes from Earth's molten core.

First, let us address how and where magma originates. We know that the atoms in a solid are in constant motion, and that to remain solid the forces binding atoms together must exceed the energy of motion. If a solid is heated, though, the energy of motion exceeds the binding forces and the solid melts. We are all familiar with this phenomenon, and we are also aware that not all solids melt at the same temperature. Once magma forms, it tends to rise because it is less dense than the rock that melted, and some actually makes it to the surface. Its upward mobility depends largely on its viscosity; fluid mafic magma reaches the surface more commonly than viscous felsic magma.

Magma may come from depths of 100 to 300 km, but most forms at much shallower depths in the upper mantle or lower crust and accumulates in reservoirs known as **magma chambers.** Beneath spreading ridges, where the crust is thin, magma chambers exist at a depth of only a few kilometers, but along convergent plate boundaries, magma chambers are commonly a few tens of kilometers deep. The volume of a magma chamber ranges from a few to many hundreds of cubic kilometers of molten rock within the otherwise solid lithosphere. Some simply cools and crystallizes within Earth's crust, thus accounting for the origin of various plutons, whereas some rises to the surface and is erupted as lava flows or pyroclastic materials.

perature range is reached in which a given mineral begins to crystallize. A previously formed mineral reacts with the remaining liquid magma (the melt) so that it forms the next mineral in the sequence. For instance, olivine $[(Mg,Fe)_2SiO_4]$ is the first ferromagnesian silicate to crystallize. As the magma continues to cool, it reaches the temperature range at which pyroxene is stable; a reaction occurs between the olivine and the remaining melt, and pyroxene forms.

With continued cooling, a similar reaction takes place between pyroxene and the melt, and the pyroxene structure is rearranged to form amphibole. Further cooling causes a reaction between the amphibole and the melt, and its structure is rearranged so that the sheet structure of biotite mica forms. Although the reactions just described tend to convert one mineral to the next in the series, the reactions are not always complete. Olivine, for example, might have a rim of pyroxene, indicating an incomplete reaction. If magma cools rapidly

Bowen's Reaction Series

During the early part of the last century, N. L. Bowen hypothesized that mafic, intermediate, and felsic magmas could all derive from a parent mafic magma. He knew that minerals do not all crystallize simultaneously from a cooling magma, but rather crystallize in a predictable sequence. Based on his observations and laboratory experiments, Bowen proposed a mechanism, now called **Bowen's reaction series,** to account for the derivation of intermediate and felsic magmas from mafic magma. Bowen's reaction series consists of two branches: a *discontinuous branch* and a *continuous branch* (■ Figure 3.3). As the temperature of magma decreases, minerals crystallize along both branches simultaneously, but for convenience we will discuss them separately.

In the discontinuous branch, which contains only ferromagnesian silicates, one mineral changes to another over specific temperature ranges (Figure 3.3). As the temperature decreases, a tem-

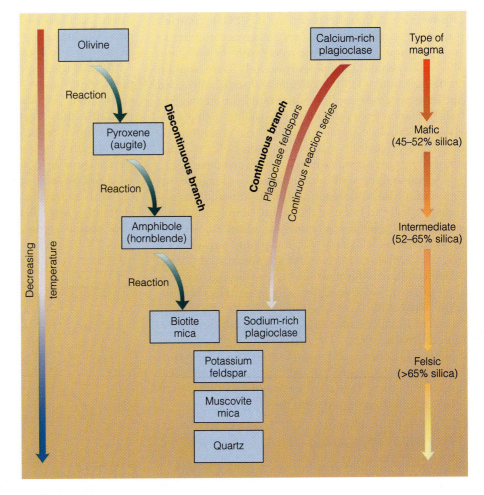

■ **Figure 3.3**

Bowen's reaction series consists of a discontinuous branch along which a succession of ferromagnesian silicates crystallize as the magma's temperature decreases, and a continuous branch along which plagioclase feldspars with increasing amounts of sodium crystallize. Notice also that the composition of the initial mafic magma changes as crystallization takes place along the two branches.

What Would You Do

Let's say you are a high school science teacher interested in developing experiments to show your students that (1) composition and temperature affect the viscosity of a lava flow, and (2) when magma/lava cools, some minerals crystallize before others. Describe the experiments you might devise to illustrate these points.

enough, the early-formed minerals do not have time to react with the melt, and thus all the ferromagnesian silicates in the discontinuous branch can be in one rock. In any case, by the time biotite has crystallized, essentially all magnesium and iron present in the original magma have been used up.

Plagioclase feldspars, which are nonferromagnesian silicates, are the only minerals in the continuous branch of Bowen's reaction series (Figure 3.3). Calcium-rich plagioclase crystallizes first. As the magma continues to cool, calcium-rich plagioclase reacts with the melt, and plagioclase containing proportionately more sodium crystallizes until all of the calcium and sodium are used up. In many cases cooling is too rapid for a complete transformation from calcium-rich to sodium-rich plagioclase. Plagioclase forming under these conditions is *zoned*, meaning that it has a calcium-rich core surrounded by zones progressively richer in sodium.

As minerals crystallize simultaneously along the two branches of Bowen's reaction series, iron and magnesium are depleted because they are used in ferromagnesian silicates, whereas calcium and sodium are used up in plagioclase feldspars. At this point, any leftover magma is enriched in potassium, aluminum, and silicon, which combine to form orthoclase ($KAlSi_3O_8$), a potassium feldspar, and if water pressure is high the sheet silicate muscovite forms. Any remaining magma is enriched in silicon and oxygen (silica) and forms the mineral quartz (SiO_2). The crystallization of orthoclase and quartz is not a true reaction series as is the crystallization of ferromagnesian silicates and plagioclase feldspars, because they form independently rather than by a reaction of orthoclase with the melt.

The Origin of Magma at Spreading Ridges

One fundamental observation we can make regarding the origin of magma is that Earth's temperature increases with depth. Known as the *geothermal gradient*, this temperature increase averages about 25°C/km. Accordingly, rocks at depth are hot but remain solid because their melting temperature rises with increasing pressure (■ Figure 3.4a). However, beneath spreading ridges the temperature locally exceeds the melting tem-

perature, at least in part, because pressure decreases. That is, plate separation at ridges probably causes a decrease in pressure on the already hot rocks at depth, thus initiating melting (Figure 3.4a). In addition, the presence of water decreases the melting temperature beneath spreading ridges because water aids thermal energy in breaking the chemical bonds in minerals (Figure 3.4b).

Localized, cylindrical plumes of hot mantle material, called *mantle plumes*, rise beneath spreading ridges and spread out in all directions (■ Figure 3.5). Perhaps concentrations of radioactive minerals within the crust and upper mantle decay and provide the heat necessary to melt rocks and thus generate magma.

Magma formed beneath spreading ridges is invariably mafic (45–52% silica). But the upper mantle rocks from which this magma is derived are characterized as ultramafic (<45% silica), consisting largely of ferromagnesian silicates and lesser amounts of nonferromagnesian silicates. To explain how mafic magma originates from ultramafic rock, geologists propose that the magma forms from source rock that only partially melts. This phenomenon of partial melting takes place because not all of the minerals in rocks melt at the same temperature.

Recall the sequence of minerals in Bowen's reaction series (Figure 3.3). The order in which these minerals melt is the opposite of their order of crystallization. Accordingly, rocks made up of quartz, potassium feldspar, and sodium-rich plagioclase begin melting at lower temperatures than those composed of ferromagnesian silicates and the calcic varieties of plagioclase. So when ultramafic rock starts to melt, the minerals richest in silica melt first, followed by those containing less silica. Therefore, if melting is not complete, mafic magma containing proportionately more sil-

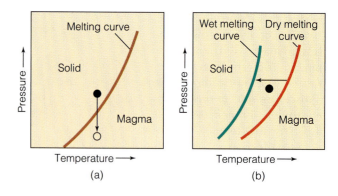

■ Figure 3.4

The effects of pressure and water on melting. (a) As pressure decreases, even when temperature remains constant, melting takes place. The black circle represents rock at high temperature. The same rock (open circle) melts at lower pressure. (b) If water is present, the melting curve shifts to the left because water provides an additional agent to break chemical bonds. Accordingly, rocks melt at a lower temperature (green melting curve) if water is present.

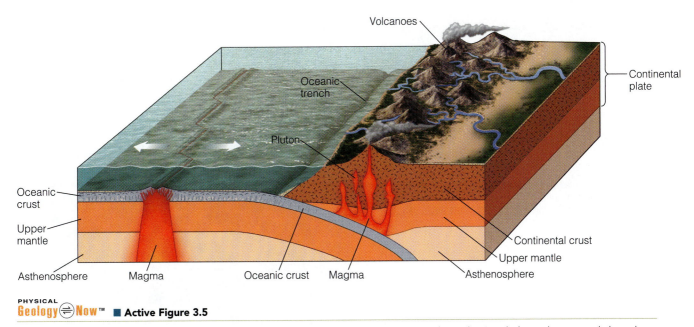

PHYSICAL
Geology ⇌ Now™ ■ **Active Figure 3.5**

Both intrusive and extrusive igneous activity take place at divergent plate boundaries (spreading ridges) and where plates are subducted at convergent plate boundaries. Oceanic crust is composed largely of dark igneous rocks in plutons that cooled from submarine lava flows. Magma forms where an oceanic plate is subducted beneath another oceanic plate or beneath a continental plate as shown here. Much of the magma forms plutons, but some is erupted to form volcanoes (see Chapter 4).

ica than the source rock results. Once this mafic magma forms, some of it rises to the surface, where it forms lava flows, and some simply cools beneath the surface forming various plutons.

Subduction Zones and the Origin of Magma

Another fundamental observation regarding magma is that, where an oceanic plate is subducted beneath either a continental plate or another oceanic plate, a belt of volcanoes and plutons is found near the leading edge of the overriding plate (Figure 3.5). It would seem, then, that subduction and the origin of magma must be related in some way, and indeed they are. Furthermore, magma at these convergent plate boundaries is mostly intermediate (53–65% silica) or felsic (>65% silica).

Once again, geologists invoke the phenomenon of partial melting to explain the origin and composition of magma at subduction zones. As a subducted plate descends toward the asthenosphere, it eventually reaches the depth where the temperature is high enough to initiate partial melting. In addition, the wet oceanic crust descends to a depth at which dewatering takes place, and as the water rises into the overlying mantle, it enhances melting and magma forms (Figure 3.4b).

Recall that partial melting of ultramafic rock at spreading ridges yields mafic magma. Similarly, partial melting of mafic rocks of the oceanic crust yields intermediate (53–65% silica) and felsic (>65% silica) magmas, both of which are richer in silica than the source rock. Moreover, some of the silica-rich sediments and

sedimentary rocks of continental margins are probably carried downward with the subducted plate and contribute their silica to the magma. Also, mafic magma rising through the lower continental crust must be contaminated with silica-rich materials, which changes its composition (see the next section).

PHYSICAL
Geology ⇌ Now   Click Geology Interactive to work through activities on these animations:
- rock labs
- circulation
- fault types

Processes That Bring About Compositional Changes in Magma

Once magma forms, its composition may change by **crystal settling,** involving the physical separation of minerals by crystallization and gravitational settling (■ Figure 3.6). Olivine, the first ferromagnesian silicate to form in the discontinuous branch of Bowen's reaction series, has a density greater than that of the remaining magma and tends to sink in the melt. Accordingly, the remaining melt becomes comparatively rich in silica, sodium, and potassium because much of the iron and magnesium were removed when minerals containing these elements crystallized.

Although crystal settling does take place in magmas, it does not do so on a scale that would yield very much felsic magma. In some thick, sheetlike, intrusive igneous bodies called *sills,* the first-formed minerals in the reaction series are indeed concentrated. The lower parts of these bodies contain more olivine and pyroxene than the

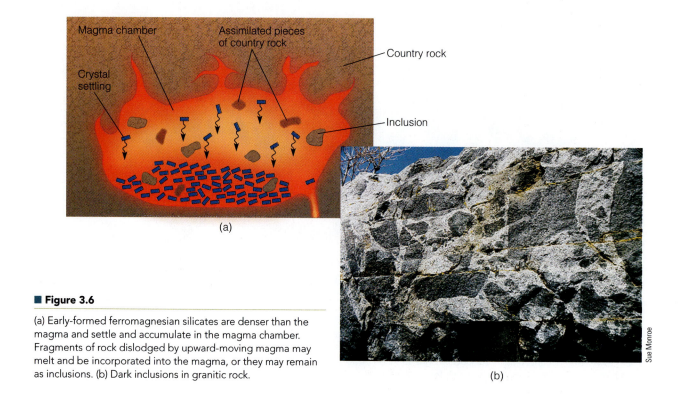

■ **Figure 3.6**

(a) Early-formed ferromagnesian silicates are denser than the magma and settle and accumulate in the magma chamber. Fragments of rock dislodged by upward-moving magma may melt and be incorporated into the magma, or they may remain as inclusions. (b) Dark inclusions in granitic rock.

upper parts, which are less mafic. But even in these bodies, crystal settling has yielded little felsic magma from an original mafic magma.

If felsic magma could be derived on a large scale from mafic magma, there should be far more mafic magma than felsic magma. To yield a particular volume of granite (a felsic igneous rock), about 10 times as much mafic magma would have to be present initially for crystal settling to yield the volume of granite in question. If this were so, then mafic intrusive igneous rocks should be much more common than felsic ones. However, just the opposite is the case, so it appears that mechanisms other than crystal settling must account for the large volume of felsic magma. Partial melting of mafic oceanic crust and silica-rich sediments of continental margins during subduction yields magma richer in silica than the source rock. Furthermore, magma rising through the continental crust absorbs some felsic materials and becomes more enriched in silica.

The composition of magma also changes by **assimilation,** a process by which magma reacts with preexisting rock, called **country rock,** with which it comes in contact (Figure 3.6). The walls of a volcanic conduit or magma chamber are, of course, heated by the adjacent magma, which may reach temperatures of 1300°C. Some of these rocks partly or completely melt, provided their melting temperature is lower than that of the magma. Because the assimilated rocks seldom have the same composition as the magma, the composition of the magma changes.

The fact that assimilation occurs is indicated by *inclusions,* incompletely melted pieces of rock that are fairly common within igneous rocks. Many inclusions were simply wedged loose from the country rock as magma forced its way into preexisting fractures (Figure 3.6). No one doubts that assimilation takes place, but its effect on the bulk composition of magma must be slight. The reason is that the heat for melting comes from the magma itself, and this has the effect of cooling the magma. Only a limited amount of rock can be assimilated by magma, and that amount is insufficient to bring about a major compositional change.

Neither crystal settling nor assimilation can produce a significant amount of felsic magma from a mafic one. But both processes, if operating concurrently, can bring about greater changes than either process acting alone. Some geologists think that this is one way that intermediate magma forms where oceanic lithosphere is subducted beneath continental lithosphere.

The fact that a single volcano can erupt lavas of different composition indicates that magmas of differing composition are present. It seems likely that some of these magmas would come into contact and mix with one another. If this is the case, we would expect that the composition of the magma resulting from **magma mixing** would be a modified version of the parent magmas. Suppose rising mafic magma mixes with felsic magma of about the same volume (■ Figure 3.7). The resulting "new" magma would have a more intermediate composition.

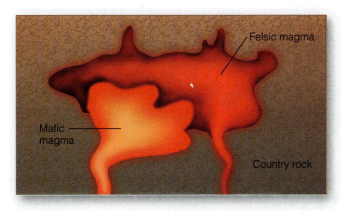

■ Figure 3.7

Magma mixing. Two magmas mix and produce magma with a composition different from either of the parent magmas. In this case, the resulting magma has an intermediate composition.

IGNEOUS ROCKS—THEIR CHARACTERISTICS AND CLASSIFICATION

We have defined *plutonic* or *intrusive igneous rocks* and *volcanic* or *extrusive igneous rocks*. Here we will have considerably more to say about the texture, composition, and classification of these rocks, which constitute one of the three major rock families depicted in the rock cycle (see Figure 1.14).

Igneous Rock Textures

The term *texture* refers to the size, shape, and arrangement of mineral grains composing igneous rocks. Size is the most important because mineral crystal size is related to the cooling history of magma or lava and generally indicates whether an igneous rock is intrusive or extrusive. The atoms in magma and lava are in constant motion, but when cooling begins, some atoms bond to form small nuclei. As other atoms in the liquid chemically bond to these nuclei, they do so in an orderly geometric arrangement and the nuclei grow into crystalline *mineral grains,* the individual particles that make up igneous rocks.

During rapid cooling, as takes place in lava flows, the rate at which mineral nuclei form exceeds the rate of growth and an aggregate of many small mineral grains is formed. The result is a fine-grained or **aphanitic texture,** in which individual minerals are too small to be seen without magnification (■ Figure 3.8a, b). With slow cooling, the rate of growth exceeds the rate of nuclei formation, and relatively large min-

eral grains form, thus yielding a coarse-grained or **phaneritic texture,** in which minerals are clearly visible (Figure 3.8c, d). Aphanitic textures generally indicate an extrusive origin, whereas rocks with phaneritic textures are usually intrusive. However, shallow plutons might have an aphanitic texture, and the rocks that form in the interiors of thick lava flows might be phaneritic.

Another common texture in igneous rocks is one termed **porphyritic,** in which minerals of markedly different size are present in the same rock. The larger minerals are *phenocrysts* and the smaller ones collectively make up the *groundmass,* which is simply the grains between phenocrysts (Figure 3.8e, f). The groundmass can be either aphanitic or phaneritic; the only requirement for a porphyritic texture is that the phenocrysts be considerably larger than the minerals in the groundmass. Igneous rocks with porphyritic textures are designated *porphyry,* as in basalt porphyry. These rocks have more complex cooling histories than those with aphanitic or phaneritic textures that might involve, for example, magma partly cooling beneath the surface followed by its eruption and rapid cooling at the surface.

Lava may cool so rapidly that its constituent atoms do not have time to become arranged in the ordered, three-dimensional frameworks of minerals. As a consequence, *natural glass* such as *obsidian* forms (Figure 3.8g). Even though obsidian with its glassy texture is not composed of minerals, geologists classify it as an igneous rock.

Some magmas contain large amounts of water vapor and other gases. These gases may be trapped in cooling lava where they form numerous small holes or cavities known as **vesicles;** rocks with many vesicles are termed *vesicular,* as in vesicular basalt (Figure 3.8h).

A **pyroclastic** or **fragmental texture** characterizes igneous rocks formed by explosive volcanic activity (Figure 3.8i). For example, ash discharged high into the atmosphere eventually settles to the surface where it accumulates; if consolidated, it forms pyroclastic igneous rocks.

Composition of Igneous Rocks

Most igneous rocks, just as the magma from which they originate, are characterized as mafic (45–52% silica), intermediate (53–65% silica), or felsic (>65% silica). A few are referred to as ultramafic (<45% silica), but these are probably derived from mafic magma by a process discussed later. The parent magma plays an important role in determining the mineral composition of igneous rocks, yet it is possible for the same magma to yield a variety of igneous rocks because its composition can change as a result of crystal settling, assimilation, magma mixing, and the sequence in which minerals crystallize (see Figures 3.3, 3.6, and 3.7).

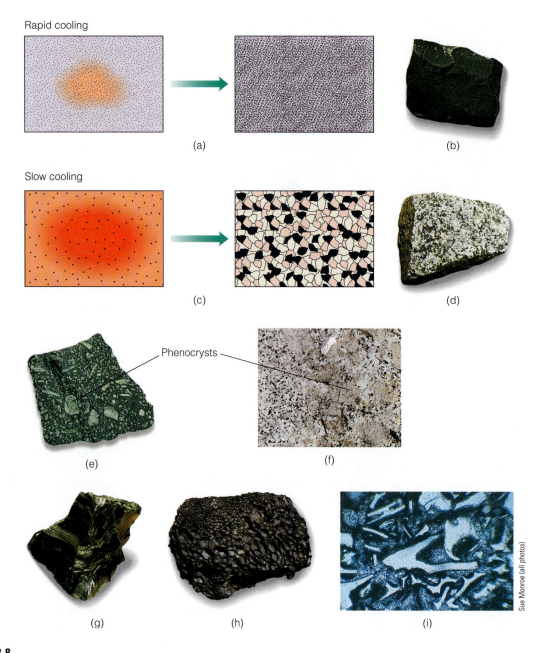

Rapid cooling

(a)

(b)

Slow cooling

(c)

(d)

Phenocrysts

(e)

(f)

(g)

(h)

(i)

Sue Monroe (all photos)

■ **Figure 3.8**

The various textures of igneous rocks. Texture is one criterion used to classify igneous rocks. (a,b) Rapid cooling as in lava flows results in many small minerals and an aphanitic (fine-grained) texture. (c,d) Slower cooling in plutons yields a phaneritic texture. (e,f) These porphyritic textures indicate a complex cooling history. (g) Obsidian has a glassy texture because magma cooled too quickly for mineral crystals to form. (h) Gases expand in lava and yield a vesicular texture. (i) Microscopic view of an igneous rock with a fragmental texture. The colorless, angular objects are pieces of volcanic glass measuring up to 2 mm.

Classifying Igneous Rocks

With few exceptions, geologists use texture and composition to classify igneous rocks. Notice in ■ Figure 3.9 that all rocks except peridotite constitute pairs; the members of a pair have the same composition but different textures. Basalt and gabbro, andesite and diorite, and rhyolite and granite are compositional (mineralogical) equivalents, but basalt, andesite, and rhyolite are aphanitic and most commonly extrusive, whereas gabbro, diorite, and granite have phaneritic textures that generally indicate an intrusive origin. The extrusive and intrusive members of each pair can usually be differentiated by texture, but many shallow intrusive rocks have textures that cannot readily be distinguished from those of extrusive igneous rocks. In other words, they all exist in a textural continuum.

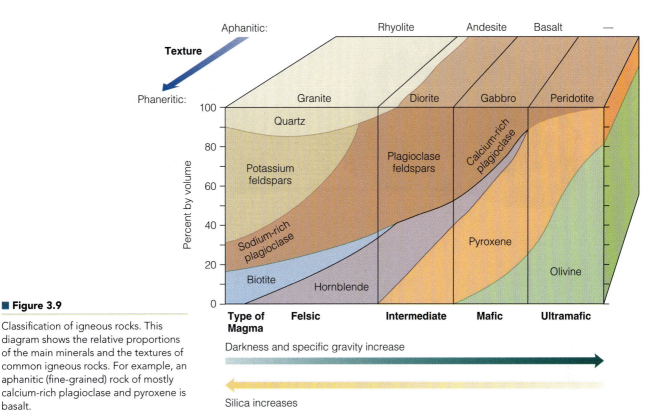

■ Figure 3.9

Classification of igneous rocks. This diagram shows the relative proportions of the main minerals and the textures of common igneous rocks. For example, an aphanitic (fine-grained) rock of mostly calcium-rich plagioclase and pyroxene is basalt.

The igneous rocks shown in Figure 3.9 are also differentiated by composition. Reading across the chart from rhyolite to andesite to basalt, for example, we see that the relative proportions of nonferromagnesian and ferromagnesian silicates differ. The differences in composition are gradual, however, so that a compositional continuum exists. In other words, rocks exist with compositions intermediate between rhyolite and andesite, and so on.

Ultramafic Rocks   Ultramafic rocks (<45% silica) are composed largely of ferromagnesian silicates. The ultramafic rock *peridotite* contains mostly olivine, lesser amounts of pyroxene, and usually a little plagioclase feldspar (■ Figures 3.9 and 3.10). Another ultramafic rock (pyroxenite) is composed predominately of pyroxene. Because these minerals are dark, the rocks are generally black or dark green. Peridotite is likely the rock type that makes up the upper mantle (see Chapter 10). Ultramafic rocks probably originate by concentration of the early-formed ferromagnesian minerals that separated from mafic magmas.

Ultramafic lava flows are known in rocks older than 2.5 billion years, but younger ones are rare or absent. The reason is that to erupt, ultramafic lava must have a near-surface temperature of about 1600°C; the surface temperatures of present-day mafic

lava flows are between 1000° and 1200°C. During early Earth history, though, more radioactive decay heated the mantle to as much as 300°C hotter than now and ultramafic lavas could erupt onto the surface. Because the amount of heat has decreased over time, Earth has cooled, and eruptions of ultramafic lava flows ceased.

Basalt-Gabbro   *Basalt* and *gabbro* are the fine-grained and coarse-grained rocks, respectively, that crystallize from mafic magma (45–52% silica) (■ Figure 3.11). Thus both have the same composition—mostly calcium-rich plagioclase and pyroxene, with smaller amounts of olivine and amphibole (Figure 3.9). Because they contain a large proportion of ferromagnesian silicates, basalt and gabbro are dark; those that are porphyritic typically contain calcium plagioclase or olivine phenocrysts.

Basalt is the most common extrusive igneous rock. Extensive basalt lava flows cover vast areas in Washington, Oregon, Idaho, and northern California (see Chapter 4). Oceanic islands such as Iceland,

■ Figure 3.10

This specimen of the ultramafic rock peridotite is made up mostly of olivine. Notice in Figure 3.9 that peridotite is the only phaneritic rock that does not have an aphanetic counterpart. Peridotite is rare at Earth's surface but is very likely the rock making up the mantle. Source: Sue Monroe

(a) Basalt

(b) Gabbro

■ **Figure 3.11**

Mafic igneous rocks. (a) Basalt is aphanetic. (b) Gabbro is phaneritic. Notice the light reflected from crystal faces.

the Galápagos, the Azores, and the Hawaiian Islands are composed mostly of basalt, and basalt makes up the upper part of the oceanic crust.

Gabbro is much less common than basalt, at least in the continental crust or where it can be easily observed. Small intrusive bodies of gabbro are present in the continental crust, but intermediate to felsic intrusive rocks such as diorite and granite are much more common. The lower part of the oceanic crust is composed of gabbro, however.

Andesite-Diorite   Intermediate composition magma (53–65% silica) crystallizes to form *andesite* and *diorite,* which are compositionally equivalent fine- and coarse-grained igneous rocks (■ Figure 3.12). Andesite and diorite are composed predominately of plagioclase feldspar, with the typical ferromagnesian component being amphibole or biotite (Figure 3.9). Andesite is generally medium to dark gray, but diorite has a salt-and-pepper appearance because of its white to light gray plagioclase and dark ferromagnesian silicates (Figure 3.12b).

Andesite is a common extrusive igneous rock formed from lava erupted in volcanic chains at convergent plate margins. The volcanoes of the Andes Mountains of South America and the Cascade Range in western North America are composed in part of andesite. Intrusive bodies of diorite are fairly common in the continental crust but are not nearly as abundant as granitic rocks.

Rhyolite-Granite   *Rhyolite* and *granite* (>65% silica) crystallize from felsic magma and are therefore silica-rich rocks (■ Figure 3.13). They consist largely of potassium feldspar, sodium-rich plagioclase, and quartz, with perhaps some biotite and rarely amphibole (Figure 3.9). Because nonferromagnesian silicates predominate, rhyolite and granite are typically light colored. Rhyolite is fine grained, although most often it contains phenocrysts of potassium feldspar or quartz, and granite is coarse grained. Granite porphyry is also fairly common.

Rhyolite lava flows are much less common than andesite and basalt flows. Recall that the greatest control of magma viscosity is silica content. Thus, if felsic magma rises to the surface, it begins to cool, the pressure on it decreases, and gases are released explosively, usually yielding rhyolitic pyroclastic materials. The rhyolitic lava flows that do occur are thick and highly viscous and move only short distances.

Granite is a coarsely crystalline igneous rock with a composition corresponding to that of the field shown in Figure 3.9. Strictly speaking, not all rocks in this field are granites. For example, a rock with a composition close to the line separating granite and diorite is called *granodiorite.* To avoid the confusion that might result from introducing more rock names, we will follow the practice of referring to rocks to the left of the granite-diorite line in Figure 3.9 as *granitic.*

(a) Andesite

(b) Diorite

■ **Figure 3.12**

Intermediate igneous rocks. (a) Andesite. This specimen has hornblende phenocrysts and is thus andesite porphyry. (b) Diorite has a salt-and-pepper appearance because it contains light-colored nonferromagnesian silicates and dark-colored ferromagnesian silicates.

Sue Monroe

(a) Rhyolite

Sue Monroe

(b) Granite

■ **Figure 3.13**

Felsic igneous rocks. (a) Rhyolite and (b) granite are typically light-colored because they contain mostly nonferromagnesian silicate minerals. The dark spots in the granite specimen are biotite mica. The white and pinkish minerals are feldspars, whereas the glassy-appearing minerals are quartz.

Granitic rocks are by far the most common intrusive igneous rocks, although they are restricted to the continents. Most granitic rocks were intruded at or near convergent plate margins during mountain-building episodes. When these mountainous regions are uplifted and eroded, the vast bodies of granitic rocks forming their cores are exposed. The granitic rocks of the Sierra Nevada of California form a composite body measuring about 640 km long and 110 km wide, and the granitic rocks of the Coast Ranges of British Columbia, Canada, are even more voluminous.

Pegmatite   The term *pegmatite* refers to a particular texture rather than a specific composition, but most pegmatites are composed largely of quartz, potassium feldspar, and sodium-rich plagioclase, thus corresponding closely to granite. A few pegmatites are mafic or intermediate in composition and are appropriately called *gabbro* and *diorite pegmatites*. The most remarkable feature of pegmatites is the size of their minerals, which measure at least 1 cm across, and in some pegmatites they measure tens of centimeters or meters (■ Figure 3.14). Many pegmatites are associated with large granite intrusive bodies and are composed of minerals that formed from the water-rich magma that remained after most of the granite crystallized.

When magma cools and forms granite, the remaining water-rich magma has properties that differ from the magma from which it separated. It has a lower density and viscosity and commonly invades the adjacent rocks where minerals crystallize. This water-rich magma also contains a number of elements that rarely enter into the common minerals that form granite. Pegmatites that are essentially very coarsely crystalline granite are simple pegmatites, whereas those with minerals containing elements such as lithium, beryllium, cesium, boron, and several others are complex pegmatites. Some complex pegmatites contain 300 different mineral species, a few of which are important economically. In addition, several gem minerals such as emerald and aquamarine, both of which are varieties of the silicate mineral beryl, and tourmaline (Figure 3.14c) are found in some pegmatites. Many rare minerals of lesser value and well-formed crystals of common minerals, such as quartz, are also mined and sold to collectors and museums.

The formation and growth of mineral-crystal nuclei in pegmatites are similar to those processes in other magmas but with one critical difference: The water-rich magma from which pegmatites crystallize inhibits the formation of nuclei. However, some nuclei do form, and because the appropriate ions in the liquid can move easily and attach themselves to a growing crystal, individual minerals have the opportunity to grow very large.

Other Igneous Rocks   A few igneous rocks, including tuff, volcanic breccia, obsidian, pumice, and scoria, are identified primarily by their textures (■ Figure 3.15). Much of the fragmental material erupted by volcanoes is *ash,* a designation for pyroclastic materials that measure less than 2.0 mm, most of which consists of broken pieces or shards of volcanic glass (Figure 3.8i). The consolidation of ash forms the pyroclastic rock *tuff* (■ Figure 3.16a). Most tuff is silica rich and light colored and is appropriately called *rhyolite tuff.* Some ash flows are so hot that as they come to rest, the ash particles fuse together and form a *welded tuff.* Consolidated deposits of larger pyroclastic materials, such as cinders, blocks, and bombs, are *volcanic breccia* (Figure 3.15).

Courtesy of Steve Stahl

(a)

Sue Monroe

(b)

Wendell E. Wilson

(c)

■ **Figure 3.14**

(a) This pegmatite, the light-colored rock, is exposed in the Black Hills of South Dakota. (b) Closeup view of a specimen from a pegmatite with minerals measuring 2 to 3 cm across. (c) Tourmaline from the Dunton Pegmatite in Maine.

Both *obsidian* and *pumice* are varieties of volcanic glass (Figure 3.16b, c). Obsidian may be black, dark gray, red, or brown, depending on the presence of iron. Obsidian breaks with the conchoidal (smoothly curved) fracture typical of glass. Analyses of many samples indicate that most obsidian has a high silica content and is compositionally similar to rhyolite.

Pumice is a variety of volcanic glass containing numerous vesicles that develop when gas escapes through lava and forms a froth (Figure 3.16c). Some pumice forms as particles erupted from explosive volcanoes. If pumice falls into water, it can be carried great distances because it is so porous and light that it floats. Another vesicular rock is *scoria*, which has more vesicles than solid rock (Figure 3.16d).

Composition	Felsic ←————→ Mafic		
Texture	Vesicular	Pumice	Scoria
	Glassy	Obsidian	
	Pyroclastic or Fragmental	←—— Volcanic Breccia ——→ Tuff/welded tuff	

■ **Figure 3.15**

Igneous rocks for which texture is the main consideration in classification. Composition is shown, but it is not essential for naming these rocks.

What Would You Do

As the only member of your community with any geology background, you are considered the local expert on minerals and rocks. Suppose one of your friends brings you a rock specimen with the following features—composition: mostly potassium feldspar and plagioclase feldspar with about 10% quartz and minor amounts of biotite; texture: minerals average 3 mm across, but several potassium feldspars are up to 3 cm. Give the specimen a rock name and tell your friend as much as you can about the rock's history. Why are the minerals so large? If the magma from which the rock crystallized reached the surface, what kind of rock would have formed?

(a) Tuff

(b) Obsidian

(c) Pumice

(d) Scoria

■ **Figure 3.16**

Examples of igneous rocks classified primarily by their texture. (a) Tuff is composed of pyroclastic materials such as those in Figure 3.8i. (b) The natural glass obsidian. (c) Pumice is glassy and extremely vesicular. (d) Scoria is also vesicular, but it is darker, heavier, and more crystalline than pumice.

GEOLOGY
IN UNEXPECTED PLACES

Little Rock, Big Story

More than a million people visit Plymouth Rock at Pilgrim Memorial State Park in Plymouth, Massachusetts, each year, but most are surprised to find out that it is a rather small boulder. Legend holds that it is the landing place where the Pilgrims first set foot in the New World in 1620. In fact, the Pilgrims first landed near Provincetown on Cape Cod and then went on to the Plymouth area, but even then probably landed north of Plymouth, not at Plymouth Rock.

Actually the boulder that is Plymouth Rock was much larger when the Pilgrims landed, but an attempt to move the symbolic stone to the town square in 1774 split it in two. The lower half of the rock remained near the seashore. In 1880 the two pieces of the rock were reunited, and in 1921 the boulder was moved once again, this time to a stone canopy erected over the original site.

Plymouth Rock has great symbolic value, but otherwise is a rather ordinary rock, although it is geologically interesting. The stone itself is from the 600+-million-year-old Dedham Granodiorite, an intrusive igneous rock similar to granite, which was carried to its present site and deposited by a glacier during the Ice Age (1.6 million to 10,000 years

ago). The Dedham Granodiorite is part of an association of rocks in New England that geologists think represent a chain of volcanic islands similar to the Aleutian Islands. These volcanic islands were incorporated into North America when plates collided and caused a period of mountain building called the Taconic orogeny about 450 million years ago.

Marcus Kazmierczak, www.mkaz.com

■ **Figure 1**

Plymouth Rock

PLUTONS—THEIR CHARACTERISTICS AND ORIGINS

Unlike volcanism and the origin of volcanic rocks, both of which can be observed, we can study intrusive igneous activity only indirectly. Intrusive igneous bodies known as **plutons** form when magma cools and crystallizes within the crust (see "Plutons" on pages 78–79). So plutons can be observed only after erosion has exposed them at the surface. Fur-

thermore, geologists cannot duplicate the conditions under which plutons form except in small laboratory experiments. Accordingly, geologists face a much greater challenge in interpreting the mechanisms whereby plutons form. Magma that cools to form plutons is emplaced mostly at divergent and convergent plate boundaries, which are also areas of active volcanism.

Geologists recognize several types of plutons based on their geometry (three-dimensional shape) and relationships to the country rock. In terms of their geometry, plutons are massive (irregular), tabular, cylindrical, or mushroom shaped. Plutons are also either **concor-**

dant, meaning they have boundaries that parallel the layering in the country rock, or discordant, with boundaries that cut across the country rock's layering (see "Plutons" on pages 78–79).

Dikes and Sills

Dikes and sills are tabular or sheetlike plutons, differing only in that dikes are discordant whereas sills are concordant (see "Plutons" on pages 78–79). Dikes are quite common. Most are small bodies measuring 1 or 2 m across, but they range from a few centimeters to more than 100 m thick. Invariable they are emplaced within preexisting fractures or where fluid pressure is great enough for them to form their own fractures.

Erosion of the Hawaiian volcanoes exposes dikes in rift zones, the large fractures that cut across these volcanoes. The Columbia River basalts in Washington (discussed in Chapter 4) issued from long fissures, and magma that cooled in the fissures formed dikes. Some of the large historic fissure eruptions are underlain by dikes; for example, dikes underlie both the Laki fissure eruption of 1783 in Iceland and the Eldgja fissure, also in Iceland, where eruptions occurred in A.D. 950 from a fissure nearly 30 km long.

Concordant sheetlike plutons are sills, many of which are a meter or less thick, although some are much thicker. A well-known sill in the United States is the Palisades sill that forms the Palisades along the west side of the Hudson River in New York and New Jersey (see Figure 12.17). It is exposed for 60 km along the river and is up to 300 m thick. Most sills were intruded into sedimentary rocks, but eroded volcanoes also reveal that sills are commonly injected into piles of volcanic rocks. In fact, some inflation of volcanoes preceding eruptions may be caused by the injection of sills (see Chapter 4).

In contrast to dikes, which follow zones of weakness, sills are emplaced when the fluid pressure is so great that the intruding magma actually lifts the overlying rocks. Because emplacement requires fluid pressure exceeding the force exerted by the weight of the overlying rocks, many sills are shallow intrusive bodies, but some were emplaced deep in the crust.

Laccoliths

Laccoliths are similar to sills in that they are concordant, but instead of being tabular, they have a mushroomlike geometry (see "Plutons" on pages 78–79). They tend to have a flat floor and are domed up in their central part. Like sills, laccoliths are rather shallow intrusive bodies that actually lift up the overlying rocks when magma is intruded. In this case, however, the rock layers are arched up over the pluton. Most lacco-

liths are rather small bodies. Well-known laccoliths in the United States are in the Henry Mountains of southeastern Utah, and several buttes in Montana are eroded laccoliths.

Volcanic Pipes and Necks

A volcano has a cylindrical conduit known as a volcanic pipe that connects its crater with an underlying magma chamber. Through this structure magma rises to the surface. When a volcano ceases to erupt, its slopes are attacked by water, gases, and acids and it erodes, but the magma that solidified in the pipe is commonly more resistant to alteration and erosion. Consequently, much of the volcano is eroded but the pipe remains as a remnant called a volcanic neck (see "Plutons" on pages 78–79). Several volcanic necks are found in the southwestern United States, especially in Arizona and New Mexico, and others are recognized elsewhere (see Geo-Focus 3.1).

Batholiths and Stocks

By definition a batholith, the largest of all plutons, must have at least 100 km² of surface area, and most are far larger. A stock, in contrast, is similar but smaller. Some stocks are simply parts of large plutons that once exposed by erosion are batholiths (see "Plutons" on pages 78–79). Both batholiths and stocks are generally discordant, although locally they may be concordant, and batholiths, especially, consist of multiple intrusions. In other words, a batholith is a large composite body produced by repeated, voluminous intrusions of magma in the same region. The coastal batholith of Peru, for instance, was emplaced during a period of 60 to 70 million years and is made up of as many as 800 individual plutons.

The igneous rocks composing batholiths are mostly granitic, although diorite may also be present. Batholiths and stocks are emplaced mostly near convergent plate boundaries during episodes of mountain building. One example is the Sierra Nevada batholith of California (see the chapter opening photo), which formed over millions of years during a mountain-building episode known as the Nevadan orogeny. Later uplift and erosion exposed this huge composite pluton at the surface. Other large batholiths in North America include the Idaho batholith and the Coast Range batholith in British Columbia, Canada.

A number of mineral resources are found in rocks of batholiths and stocks and in the adjacent country rocks. Granitic rocks are the primary source of gold, which forms from mineral-rich solutions moving through cracks and fractures of the igneous body. The copper deposits at Butte, Montana, are in rocks near

Plutons

Intrusive bodies called plutons are common, but we see them at the surface only after deep erosion. Notice that they vary in geometry and their relationships to the country rock.

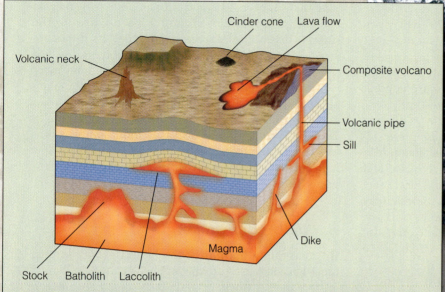

Block diagram showing various plutons. Some plutons cut across the layering in country rock and are discordant, whereas others parallel the layering and are concordant.

Part of the Sierra Nevada batholith in Yosemite National Park, California. The batholith, consisting of multiple intrusions of granitic rock, is more than 600 km long and up to 110 km wide. To appreciate the scale in this image, the waterfall has a descent of 435 m.

A volcanic neck in Monument Valley Tribal Park, Arizona. This landform is 457 m high. Most of the original volcano was eroded, leaving only this remnant.

Granitic rocks of a small stock at Castle Crags State Park, California.

The dark materials in this image are igneous rocks, whereas the light layers are sedimentary. Notice that the sill parallels the layering, so it is concordant. The dike, though, clearly cuts across the layering and is discordant. Sills and dikes have sheetlike geometry, but in this view we can see them in only two dimensions.

Sill

Dike

Martin G. Miller/ Visuals Unlimited

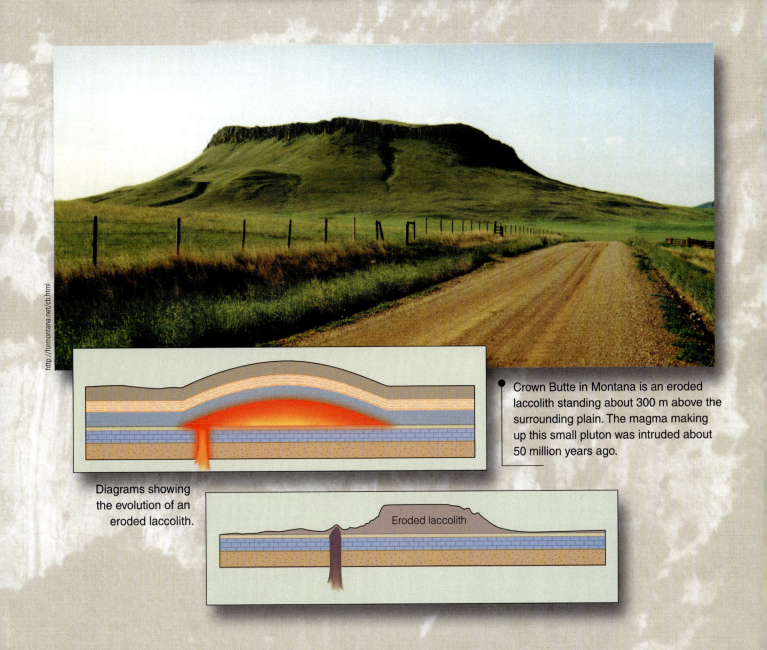

http://formontana.net/cb.html

Diagrams showing the evolution of an eroded laccolith.

Eroded laccolith

Crown Butte in Montana is an eroded laccolith standing about 300 m above the surrounding plain. The magma making up this small pluton was intruded about 50 million years ago.

GEO**FOCUS**

Some Remarkable Volcanic Necks

We mentioned in the text that as an extinct volcano weathers and erodes, a remnant of the original mountain may persist as a volcanic neck. The origin of volcanic necks is well known, but these isolated monoliths rising above otherwise rather flat land are scenic, awe inspiring, and the subject of legends. They are found in many areas of recently active volcanism. A rather small volcanic neck rising only 79 m above the surface in the town of Le Puy, France, is the scenic site of the 11th-century chapel of Saint Michel d'Aiguilhe (■ Figure 1). It is so steep that materials and tools used in its construction had to be hauled up in baskets.

Perhaps the most famous volcanic neck in the United States is Shiprock, New Mexico, which rises nearly 550 m above the surrounding plain and is visible from 160 km away. Radiating outward from this conical structure are three vertical dikes that stand like walls above the adjacent countryside (■ Figure 2). According to one legend, Shiprock, or *Tsae-bidahi*, meaning "winged rock," represents a giant bird that brought the Navajo people from the north. The same legend holds that the dikes are snakes that turned to stone.

An absolute age determined for one of the dikes indicates that Shiprock is about 27 million years old. When the original volcano formed, apparently during explosive eruptions, rising magma penetrated various rocks including the Mancos Shale, the rock unit now exposed at the surface adjacent to Shiprock. The rock that makes up Shiprock itself is tuff-breccia, consisting of fragmented volcanic debris as well as pieces of metamorphic, sedimentary, and igneous rocks.

Geologists agree that Devil's Tower in northwestern Wyoming cooled from a small body of magma and that erosion has exposed it in its present form (■ Figure 3). However, opinion is divided on whether it is a volcanic neck or an eroded laccolith. In either case, the rock that makes up Devil's Tower is 45 to 50 million years old, and President Theodore Roosevelt designated this impressive landform as our first national monument in 1906. At 260 m high, Devil's Tower is visible from 48 km away and served as a landmark for early travelers in this area. It achieved further distinction in 1977 when it was featured in the film *Close Encounters of the Third Kind*.

The Cheyenne and Sioux Indians call Devil's Tower *Mateo Tepee*, meaning "Grizzly Bear Lodge." It was also called the "Bad God's Tower," and reportedly "Devil's Tower" is a translation from this phrase. The tower's most conspicuous features are the near-vertical lines that according to Cheyenne legends are scratch

---

the margins of the granitic rocks of the Boulder batholith (■ Figure 3.17). Near Salt Lake City, Utah, copper is mined from the mineralized rocks in the Bingham stock, a composite pluton composed of granite and granite porphyry.

As noted earlier, batholiths are emplaced at convergent plate boundaries during mountain building. However, large exposures of granitic rocks are also present within the interiors of continents where mountains are absent. A large area in Canada is underlain by extensive granitic rocks as well as by other rock types. These granites were emplaced during mountain-building episodes during early Earth history. The mountains have long since been eroded; the remaining rocks represent the eroded "roots" of these ancient mountains.

HOW ARE BATHOLITHS INTRUDED INTO EARTH'S CRUST?

Geologists realized long ago that the emplacement of batholiths posed a space problem. What happened to the rock that was once

marks made by a gigantic grizzly bear. One legend holds that the bear made the scratches while pursuing a group of children. Another tells of six brothers and a woman also pursued by a grizzly bear. One brother carried a rock, and when he sang a song it grew into Devil's Tower, safely carrying the brothers and woman out of the bear's reach.

Although not nearly as interesting as the Cheyenne legends, the origin for the "scratch marks" is well understood. These lines actually formed at the intersections of *columnar joints*, fractures that form in response to cooling and contraction that occur in some plutons and lava flows (see Chapter 4). The columns outlined by these fractures are up to 2.5 m across, and the pile of rubble at the tower's base is simply an accumulation of collapsed columns.

PHYSICAL
Geology⇌Now   Click Geology Interactive to work through an activity on rock labs.

Richard List/Corbis

■ Figure 1

This volcanic neck in Le Puy, France, rises 79 m above the surface of the town. Workers on the Chapel of Saint Michel d'Aiguilhe had to haul building materials and tools up in baskets.

Courtesy of Frank Hanna

■ Figure 2

Shiprock, a volcanic neck in northwestern New Mexico, rises nearly 550 m above the surrounding plain. One of the dikes radiating from Shiprock is in the foreground.

James S. Monroe

■ Figure 3

Devil's Tower in northeastern Wyoming rises about 260 m above its base. It may be a volcanic neck or an eroded laccolith. The vertical lines result from intersections of fractures called columnar joints. According to Cheyenne legend, though, a gigantic grizzly bear made the deep scratches.

in the space now occupied by a batholith? One proposed answer was that no displacement had occurred, but rather that batholiths formed in place by alteration of the country rock through a process called *granitization.* According to this view, granite did not originate as magma but rather from hot, ion-rich solutions that simply altered the country rock and transformed it into granite. Granitization is a solid-state phenomenon, so it is essentially an extreme type of metamorphism (see Chapter 7).

Granitization is no doubt a real phenomenon, but most granitic rocks show clear evidence of an igneous origin. For one thing, if granitization had taken place,

we would expect the change from country rock to granite to take place gradually over some distance. However, in almost all cases no such gradual change can be detected. In fact, most granitic rocks have what geologists refer to as sharp contacts with adjacent rocks. Another feature indicating an igneous origin for granitic rocks is the alignment of elongate minerals parallel with their contacts, which must have occurred when magma was injected. A few granitic rocks lack sharp contacts and gradually change in character until they resemble the adjacent country rock. These probably did originate by granitization. In the opinion of most geologists, only small quantities of granitic rock could form by this

James S. Monroe

■ **Figure 3.17**

More than one and a half billion tons of copper ore were mined from the Berkeley Pit at Butte, Montana, during its operation from 1955 until 1982. The copper-bearing minerals are in rocks adjacent to the Boulder batholith. This photo was taken during the early 1970s. The excavation is more than 540 m deep.

process, so it cannot account for the huge volume of granitic rocks of batholiths. Accordingly, geologists conclude that an igneous origin for almost all granite is clear, but they still must deal with the space problem.

One solution is that these large igneous bodies melted their way into the crust. In other words, they simply assimilated the country rock as they moved upward (Figure 3.6). The presence of inclusions, especially near the tops of some plutons, indicates that assimilation does occur. Nevertheless, as we noted previously, assimilation is a limited process because magma cools as country rock is assimilated; calculations indicate that far too little heat is available in magma to assimilate the huge quantities of country rock necessary to make room for a batholith.

Geologists now generally agree that batholiths were emplaced by *forceful injection* as magma moved upward. Recall that granite is derived from viscous felsic magma and therefore rises slowly. It appears that the magma deforms and shoulders aside the country rock, and as it rises farther, some of the country rock fills the space beneath the magma (■ Figure 3.18). A somewhat analogous situation was discovered in which large masses of sedimentary rock known as *rock salt* rise through the overlying rocks to form *salt domes* (see Figure 6.24c).

Salt domes are recognized in several areas of the world, including the Gulf Coast of the United States. Layers of rock salt exist at some depth, but salt is less dense than most other types of rock materials. When

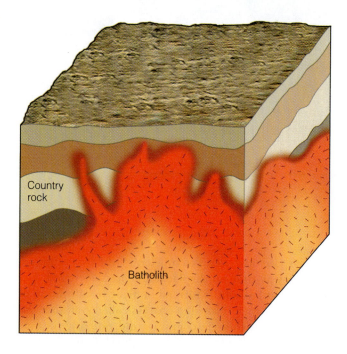

■ **Figure 3.18**

Emplacement of a hypothetical batholith. As the magma rises, it shoulders aside and deforms the country rock.

under pressure, it rises toward the surface even though it remains solid, and as it moves up, it pushes aside and deforms the country rock. Natural examples of rock salt

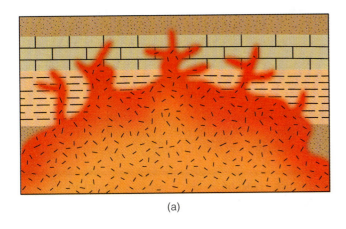

(a)

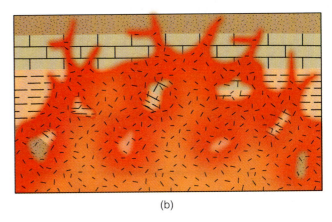

(b)

■ **Figure 3.19**

Emplacement of a batholith by stoping. (a) Magma is injected into fractures and planes between layers in the country rock. (b) Blocks of country rock are detached and engulfed in the magma, thus making room for the magma to rise farther. Some of the engulfed blocks might be assimilated, and some might remain as inclusions (see Figure 3.6).

flowage are known, and it can easily be demonstrated experimentally. In the arid Middle East, for example, salt moving up in the manner described actually flows out at the surface.

Some batholiths do indeed show evidence of having been emplaced forcefully by shouldering aside and deforming the country rock. This mechanism probably occurs in the deeper parts of the crust where temperature and pressure are high and the country rocks are easily deformed in the manner described. At shallower depths,

the crust is more rigid and tends to deform by fracturing. In this environment, batholiths may be emplaced by **stoping,** a process in which rising magma detaches and engulfs pieces of country rock (■ Figure 3.19). According to this concept, magma moves up along fractures and the planes separating layers of country rock. Eventually, pieces of country rock detach and settle into the magma. No new room is created during stoping; the magma simply fills the space formerly occupied by country rock (Figure 3.19).

3 REVIEW WORKBOOK

Chapter Summary

- *Magma* is the term for molten rock below the surface, whereas the same material at the surface is *lava.*

- Silica content distinguishes among mafic (45–52% silica), intermediate (53–65% silica), and felsic (>65% silica) magmas.

- Magma and lava viscosity depends on temperature and especially on composition; the more silica, the greater the viscosity.

- Minerals crystallize from magma and lava when small crystal nuclei form and grow.

- Rapid cooling accounts for the aphanitic textures of volcanic rocks, whereas comparatively slow cooling yields the phaneritic textures of plutonic rocks. Igneous rocks with markedly different sized minerals are porphyritic.

- Igneous rock composition is determined largely by the composition of the parent magma, but magma composition can change so that the same magma may yield more than one kind of igneous rock.

- According to Bowen's reaction series, cooling mafic magma yields a sequence of minerals each of which is stable within specific temperature ranges.
  a. Only ferromagnesian silicates are found in the discontinuous branch of Bowen's reaction series.
  b. The continuous branch of the reaction series yields only plagioclase feldspars that become increasingly enriched in sodium as cooling occurs.

- A chemical change in magma may take place as early-formed ferromagnesian silicates form and, because of their density, settle in the magma.

- Compositional changes also take place in magma when it assimilates country rock or when one magma mixes with another.

- Geologists recognize two broad categories of igneous rocks: volcanic or extrusive and plutonic or intrusive.

- Texture and composition are the criteria used for igneous rock classification, although a few are defined only by texture.

- Crystallization from water-rich magma results in very large minerals in rocks known as pegmatite. Most pegmatite has an overall composition similar to granite.

- Intrusive igneous bodies known as plutons vary in their geometry and their relationships to country rock; some are concordant, whereas others are discordant.

- The largest plutons, known as batholiths, consist of multiple intrusions of magma over long periods of time.

- Most plutons, including batholiths, are found at or near divergent and convergent plate boundaries.

Important Terms

aphanitic texture (p. 69)
assimilation (p. 68)
batholith (p. 77)
Bowen's reaction series (p. 65)
concordant pluton (p. 76)
country rock (p. 68)
crystal settling (p. 67)
dike (p. 77)
discordant pluton (p. 77)
felsic magma (p. 63)
igneous rock (p. 63)
intermediate magma (p. 63)

laccolith (p. 77)
lava (p. 63)
lava flow (p. 63)
mafic magma (p. 63)
magma (p. 63)
magma chamber (p. 65)
magma mixing (p. 68)
phaneritic texture (p. 69)
pluton (p. 76)
plutonic (intrusive igneous) rock (p. 63)
porphyritic texture (p. 69)

pyroclastic materials (p. 63)
pyroclastic (fragmental) texture (p. 69)
sill (p. 77)
stock (p. 77)
stoping (p. 83)
vesicle (p. 69)
viscosity (p. 64)
volcanic neck (p. 77)
volcanic pipe (p. 77)
volcanic (extrusive igneous) rock (p. 63)

Review Questions

1. A dike is a discordant pluton, whereas a _____ is concordant.
   a. _____ batholith; b. _____ volcanic neck;
   c. _____ laccolith; d. _____ stock;
   e. _____ ash fall.

2. An aphanitic igneous rock composed mostly of pyroxenes and calcium-rich plagioclase is:
   a. _____ granite; b. _____ obsidian;
   c. _____ rhyolite; d. _____ diorite;
   e. _____ basalt.

3. The size of the mineral grains that make up an igneous rock is a useful criterion for determining whether the rock is _____ or _____.

   a. _____ volcanic/plutonic; b. _____ discordant/concordant; c. _____ vesicular/fragmental;
   d. _____ porphyritic/felsic; e. _____ ultramafic/igneous.

4. Magma characterized as intermediate:
   a. _____ flows more readily than mafic magma; b. _____ has between 53% and 65% silica; c. _____ crystallizes to form granite and rhyolite; d. _____ cools to form rocks making up most of the oceanic crust; e. _____ is one from which ultramafic rocks are derived.

5. The phenomenon by which pieces of country rock are detached and engulfed by rising magma is known as:

a. _____ stoping; b. _____ assimilation;
c. _____ magma mixing; d. _____ Bowen's
reaction series; e. _____ crystal settling.

6. An igneous rock that has minerals large
enough to see without magnification is said to
have a(n) _____ texture and is probably _____.
a. _____ laccolithic/pegmatite; b. _____ frag-
mental/felsic; c. _____ isometric/magmatic;
d. _____ phaneritic/plutonic; e. _____ frag-
mental/obsidian.

7. Which one of the following statements regarding batholiths is *false*?
a. _____ They consist of multiple voluminous
intrusions; b. _____ They form mostly at convergent plate boundaries during mountain
building; c. _____ They consist of a variety of
volcanic rocks, but especially basalt; d. _____
They must have at least 100 km² of surface
area; e. _____ Although locally concordant,
they are mostly discordant.

8. An igneous rock characterized as a porphyry
is one:
a. _____ that formed by crystal settling and
assimilation; b. _____ possessing minerals of
markedly different sizes; c. _____ made up
largely of potassium feldspar and quartz;
d. _____ that forms when pyroclastic materials
are consolidated; e. _____ resulting from very
rapid cooling.

9. Which pair of the following igneous rocks has
the same texture?
a. _____ basalt-andesite; b. _____ granite-
rhyolite; c. _____ pumice-obsidian;
d. _____ tuff-diorite; e. _____ scoria-lapilli.

10. One process by which magma changes composition is:
a. _____ crystal settling; b. _____ rapid cooling; c. _____ explosive volcanism; d. _____
fracturing; e. _____ plate convergence.

11. Two aphanitic igneous rocks have the following compositions. Specimen 1: 15% biotite;
15% sodium-rich plagioclase, 60% potassium
feldspar, and 10% quartz. Specimen 2:
10% olivine, 55% pyroxene, 5% hornblende,
and 30% calcium-rich plagioclase. Use
Figure 3.9 to classify these rocks. Which
would be the darkest and most dense?

12. How does a sill differ from a dike? A diagram
would be helpful.

13. How do crystal settling and assimilation bring
about compositional changes in magma?
Also, give evidence that these processes actually take place.

14. Describe or diagram the sequence of events
leading to the origin of a volcanic neck.

15. Describe a porphyritic texture and explain
how it might originate.

16. Why are felsic lava flows so much more viscous than mafic ones?

17. How does a pegmatite form, and why are
their mineral crystals so large?

18. Compare the continuous and discontinuous
branches of Bowen's reaction series. Why are
potassium feldspar and quartz not part of
either branch?

19. You analyze the mineral composition of a
thick sill and find that it has a mafic lower
part but its upper part is more intermediate.
Given that it all came from a single magma
injected at one time, how can you account for
its compositional differences?

20. What kind(s) of evidence would you look for
to determine whether the granite in a
batholith crystallized from magma or originated by granitization?

World Wide Web Activities

PHYSICAL
Geology⇌Now   Assess your understanding of this
chapter's topics with additional quizzing and comprehensive interactivities at

http://earthscience.brookscole.com/physgeo5e

as well as current and up-to-date weblinks, additional
readings, and InfoTrac College Edition exercises.

Volcanism and Volcanoes

CHAPTER 4
OUTLINE

PHYSICAL
Geology⇌Now *This icon, appearing throughout the book, indicates an opportunity to explore interactive tutorials, animations, or practice problems available on the Physical GeologyNow Web site at http://earthscience.brookscole.com/physgeo5e.*

OBJECTIVES

At the end of this chapter, you will have learned that

- In addition to lava flows, erupting volcanoes eject pyroclastic materials, especially ash, and various gases.

- Geologists identify the basic types of volcanoes by their eruptive style, composition, and shape.

- Although all volcanoes are unique, most are identified as shield volcanoes, cinder cones, or composite volcanoes.

- Volcanoes characterized as lava domes tend to erupt explosively and thus are dangerous.

- There are active volcanoes in the United States in Hawaii, Alaska, and the Cascade Range of the Pacific Northwest.

- Eruptions in Hawaii and Alaska are commonplace, but only two eruptions have occurred in the continental United States during the 1900s. Canada has had no eruptions during historic time.

- Some eruptions yield vast sheets of lava or pyroclastic materials rather than volcanoes.

- Geologists have devised the volcanic explosivity index as a measure of an eruption's size.

- Some volcanoes are carefully monitored to help geologists anticipate eruptions.

- Most volcanoes are located in belts at or near divergent and convergent plate boundaries.

Mount Vesuvius has erupted 80 times since A.D. 79, most recently in 1944. Naples, Italy, and other communities are either on the flanks of the volcano or nearby. The Bay of Naples is on the right. Source: Stone/Getty Images

Introduction

No other geologic phenomenon has captured the public imagination more than volcanism. Eruptions featured in television documentaries and movies portray the destruction wrought by lava flows and explosive volcanism. Depictions of lava flows in movies notwithstanding, however, humans have little to fear from these incandescent streams of molten rock, although in both 1977 and 2002 lava flows killed dozens of people in the Democratic Republic of the Congo. Lava flows may destroy buildings and roadways and cover otherwise productive agricultural land. And some volcanic eruptions, especially those at convergent plate boundaries, are explosive and pose considerable danger to nearby populated areas.

One of the best-known catastrophic eruptions ever recorded was the A.D. 79 eruption of Mount Vesuvius that destroyed the thriving Roman communities of Pompeii, Herculaneum, and Stabiae in what is now Italy (see the chapter opening photo and ■ Figure 4.1). Fortunately for us, Pliny

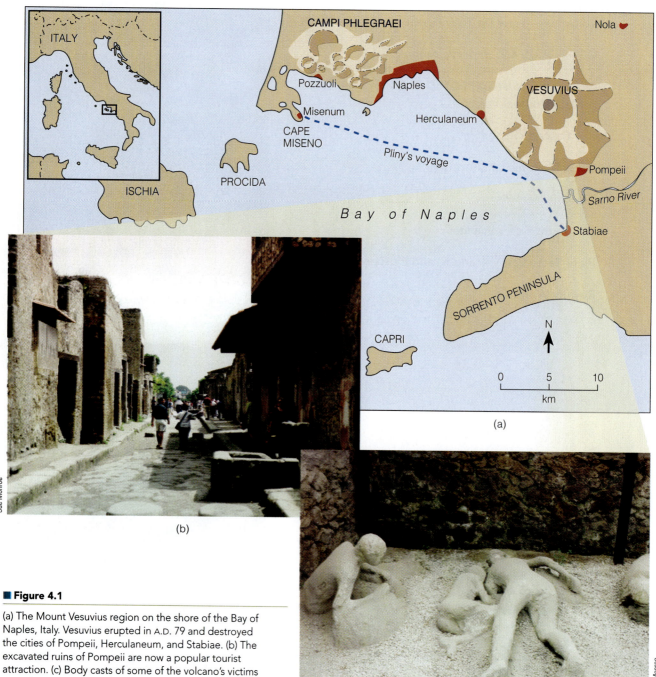

■ Figure 4.1

(a) The Mount Vesuvius region on the shore of the Bay of Naples, Italy. Vesuvius erupted in A.D. 79 and destroyed the cities of Pompeii, Herculaneum, and Stabiae. (b) The excavated ruins of Pompeii are now a popular tourist attraction. (c) Body casts of some of the volcano's victims in Pompeii.

Table 4.1

Some Notable Volcanic Eruptions

Date	Volcano	Deaths
Apr. 10, 1815	Tambora, Indonesia	117,000; includes deaths from eruption and famine and disease
Oct. 8, 1822	Galunggung, Java	Pyroclastic flows and mudflows killed 4011
Mar. 2, 1856	Awu, Indonesia	2806 died in pyroclastic flows
Aug. 27, 1883	Krakatau, Indonesia	More than 36,000 died; most killed by tsunami
June 7, 1892	Awu, Indonesia	1532 died in pyroclastic flows
May 8, 1902	Mount Pelée, Martinique	Nuée ardente engulfed St. Pierre and killed 28,000
Oct. 24, 1902	Santa Maria, Guatemala	5000 died during eruption
May 19, 1919	Kelut, Java	Mudflows devastated 104 villages and killed 5110
Jan. 21, 1951	Lamington, New Guinea	Pyroclastic flows killed 2942
Mar. 17, 1963	Agung, Indonesia	1148 perished during eruption
May 18, 1980	Mount St. Helens, Washington	63 killed; 600 km^2 of forest devastated
Mar. 28, 1982	El Chichón, Mexico	Pyroclastic flows killed 1877
Nov. 13, 1985	Nevado del Ruiz, Colombia	Minor eruption triggered mudflows that killed 23,000
Aug. 21, 1986	Oku volcanic field, Cameroon	Cloud of CO_2 released from Lake Nyos killed 1746
June 15, 1991	Mount Pinatubo, Philippines	~281 killed during eruption; 83 died in later mudflows; 358 died of illness
July 1999	Soufrière Hills, Montserrat	19 killed; 12,000 evacuated
Jan. 17, 2002	Nyiragongo, Zaire	Lava flow killed 80–100 in Goma

the Younger recorded the event in detail; his uncle, Pliny the Elder, died while trying to investigate the eruption. In fact, Pliny the Younger's account is so vivid that this and similar eruptions during which huge quantities of pumice are blasted into the air are still called *plinian*.

Pompeii, a city of about 20,000 people and only 9 km downwind from the volcano, was buried in nearly 3 m of pyroclastic materials that covered all but the tallest buildings (Figure 4.1). About 2000 victims have been discovered in the city, but certainly far more than this were killed. Pompeii was covered by volcanic debris rather gradually, but surges of incandescent volcanic materials in glowing avalanches swept through Herculaneum, quickly burying the town to a depth of about 20 m. Since A.D. 79, Mount Vesuvius has erupted 80 times, most violently in 1631 and 1906; it last erupted in 1944. Continuing volcanic and seismic activity in this area poses a continuing threat to the many cities and towns along the shores of the Bay of Naples (Figure 4.1).

One very good reason to study volcanism is that the complex interactions among Earth's systems are well illustrated by volcanic eruptions. Volcanism, especially the emission of gases and pyroclastic materials, has an immediate and profound impact on the atmosphere, hydrosphere, and biosphere, at least in the vicinity of an eruption. And in some cases the effects are worldwide, as they were following the eruptions of Tambora in 1815, Krakatau in 1883, and

Pinatubo in 1991. Furthermore, the fact that lava flows and explosive eruptions cause property damage, injuries, and fatalities (Table 4.1) and at least short-term atmospheric changes indicates that these are catastrophic events, at least from the human perspective.

Ironically, though, when considered in the context of Earth history, volcanism is actually a constructive process. The atmosphere and surface waters most likely resulted from emission of gases during Earth's early history, and oceanic crust is continuously produced by volcanism at spreading ridges. Oceanic islands such as the Hawaiian Islands, Iceland, and the Azores owe their existence to volcanism, and weathering of lava and pyroclastic materials in tropical areas such as Indonesia converts lava flows, pyroclastic materials, and volcanic mudflows to productive soils.

Most of us will never experience an eruption, but such events are commonplace in some parts of the world. Residents of the islands of Hawaii, the Philippines, Japan, and Iceland are well aware of eruptions, but eruptions in the continental United States have occurred only twice since 1914, both in the Cascade Range, which stretches from northern California through Oregon and Washington and into southern British Columbia, Canada. Canada has had no eruptions during historic time. Ancient and ongoing volcanism in the western United States has yielded interesting features, several of which are featured in this chapter.

VOLCANISM

The term **volcanism** immediately brings to mind lava flows, and it does in fact encompass such activity, but the term refers specifically to those processes whereby lava and its contained gases as well as pyroclastic materials are extruded onto the surface or into the atmosphere. And of course volcanism yields several distinctive landforms, particularly volcanoes, as well as volcanic (extrusive) igneous rocks. About 550 volcanoes are presently *active*; that is, they have erupted during historic time. Only 12 or so are erupting at any one time. Most of this activity is minor, although large eruptions are certainly not uncommon.

All of the terrestrial planets and Earth's moon were volcanically active during their early histories, but now volcanoes are known only on Earth and on one or two other bodies in the solar system. Triton, one of Neptune's moons, probably has active volcanism, and Jupiter's moon Io is by far the most volcanically active body in the solar system (■ Figure 4.2). Many of its 100 or so volcanoes are erupting at any given time.

In addition to active volcanoes, Earth has numerous *dormant* volcanoes that have not erupted during historic time but may do so in the future. Prior to its eruption in

A.D. 79, Mount Vesuvius had not been active in human memory. The largest volcanic outburst in the last 50 years took place when Mount Pinatubo in the Philippines erupted in 1991 after lying dormant for 600 years. Some volcanoes have not erupted in historic time and show no signs of doing so again; thousands of these *extinct* or *inactive* volcanoes are known.

Volcanic Gases

Samples taken from present-day volcanoes indicate that 50–80% of all volcanic gases are water vapor. Lesser amounts of carbon dioxide; nitrogen; sulfur gases, especially sulfur dioxide and hydrogen sulfide; and very small amounts of carbon monoxide, hydrogen, and chlorine are also emitted. In areas of recent volcanism, such as Lassen Volcanic National Park in California, one cannot help but notice the rotten-egg odor of hydrogen sulfide gas (■ Figure 4.3a).

When magma rises toward the surface, pressure is reduced and the contained gases begin to expand. In highly viscous felsic magma, expansion is inhibited and gas pressure increases. Eventually, the pressure may become great enough to cause an explosion and produce pyroclastic materials such as ash. In contrast, low-viscosity mafic magma allows gases to expand and escape easily. Accordingly, mafic magma generally erupts rather quietly.

Most gases released during eruptions quickly dissipate in the atmosphere and pose little danger to humans, but on several occasions they have caused numerous fatalities. In 1783 toxic gases, probably sulfur dioxide, erupted from Laki fissure in Iceland had devastating effects. About 75% of the nation's livestock died, and the haze resulting from the gas caused lower temperatures and crop failures; about 24% of Iceland's population died as a result of the ensuing Blue Haze Famine. The country suffered its coldest winter in 225 years in 1783–1784, with temperatures 4.8°C below the long-term average. The eruption also produced what Benjamin Franklin called a "dry fog" that was responsible for dimming the intensity of sunlight in Europe. The severe winter of 1783–1784 in Europe and eastern North America is attributed to the presence of this "dry fog" in the upper atmosphere.

The particularly cold spring and summer of 1816 are attributed to the 1815 eruption of Tambora in Indonesia, the largest and most deadly eruption during historic time. The eruption of Mayon volcano in the Philippines during the previous year may have contributed to the cool spring and summer of 1816 as well. Another large historic eruption that had widespread climatic effects was the eruption of Krakatau in 1883.

In 1986, in the African nation of Cameroon, 1746 people died when a cloud of carbon dioxide engulfed them. The gas accumulated in the waters of Lake Nyos, which occupies a volcanic caldera. Scientists disagree

■ **Figure 4.2**

You can see one of Io's erupting volcanoes on the left side of this image. Unlike volcanoes on Earth, Io's volcanoes erupt mostly sulfur compounds from an interior heated by "kneading" caused by the gravitational attraction of the nearby massive planet Jupiter.

JSC/NASA

Sue Monroe

Sue Monroe

(b)

Sue Monroe

(c)

(a)

■ **Figure 4.3**

Volcanic gases. (a) Gases emitted at the Sulfur Works in Lassen Volcanic National Park, California. (b) Trees killed by carbon dioxide rising from beneath Mammoth Mountain volcano in California. (c) The CO_2 gas at Mammoth Mountain has not affected humans, but this sign warns of the potential danger.

about what caused the gas to suddenly burst forth from the lake, but once it did, it flowed downhill along the surface because it was denser than air. In fact, the density and velocity of the gas cloud were great enough to flatten vegetation, including trees, a few kilometers from the lake. Unfortunately, thousands of animals and many people, some as far as 23 km from the lake, were asphyxiated.

In 1990 forest rangers noticed that trees began dying in an area of about 170 acres on the south side of Mammoth Mountain volcano in eastern California (Figure 4.3b, c). Previously the trees had been thriving, but earthquakes in 1991 apparently triggered the release of carbon dioxide (CO_2) gas from beneath the volcano. Investigations showed that CO_2 gas accounted for 20–90% of the gas content of the soil. During photosynthesis, plant leaves produce oxygen (O_2) from CO_2, but their roots must absorb O_2. Diminished quantities of O_2 in the soil were responsible for the tree deaths.

Residents of the island of Hawaii have coined the term *vog* for volcanic smog. Kilauea volcano has been erupting continuously since 1983, releasing small amounts of lava, copious quantities of carbon dioxide, and about 1800 tons of sulfur dioxide per day. Carbon dioxide has been no problem, but sulfur dioxide produces a haze and the unpleasant odor of sulfur. Vog

probably poses little or no risk for tourists, but a long-term threat exists for residents of the west side of the island where vog is most common.

Lava Flows

Lava flows are portrayed in movies and on television as fiery streams of incandescent rock that pose a great danger to humans. Actually, lava flows are the least dangerous manifestation of volcanism, although on occasion they have caused fatalities (see Geo-Focus 4.1). Most lava flows do not move particularly fast, and because they are fluid, they follow existing low areas. Thus, once a flow erupts from a volcano, determining the path it will take is fairly easy, and anyone in areas likely to be affected can be evacuated.

Even low-viscosity lava flows generally do not move very rapidly. Flows can move much faster, though, when their margins cool to form a channel, and especially when insulated on all sides as in a *lava tube,* where a speed of more than 50 km/hr has been recorded. A conduit known as a **lava tube** within a lava flow forms when the margins and upper surface of the flow solidify. Thus confined and insulated, the flow moves rapidly and over great distances. As an eruption ceases, the tube drains,

GEOFOCUS

4.1

Lava Flows Pose Little Danger to Humans—Usually

In the text we made the point that incandescent streams of molten rock are impressive and commonly portrayed in movies as a danger to humans. We also mentioned that lava flows are actually the least dangerous manifestation of volcanic activity, although they may destroy buildings, highways, and cropland. We should note, though, that on some occasions they have been directly or indirectly responsible for fatalities.

The most recent fatalities caused by lava flows took place during January 2002 in the city of Goma in the Democratic Republic of Congo (formerly Zaire). Nyiragongo, the volcano from which the flows issued, has erupted 19 times since 1884, and during two of these eruptions lava flows caused fatalities (■ Figure 1). One of several volcanoes in Africa's Virunga Volcanic Chain, Nyiragongo is a composite volcano along the East African Rift. A lava lake in its summit caldera

was active for decades, and in 1977 the lake suddenly drained through fissures and covered several square kilometers with fluid lava flows. Unfortunately, about 70 people (300 in another estimate) and a herd of elephants perished in the flows that moved at speeds of 60 km/hr.

More recently, a massive eruption at Nyiragongo took place on January 17, 2002, when a huge plume of ash rose above the mountain and three fast-moving lava flows descended its western and eastern flanks. Fourteen villages near the volcano were destroyed, and within one day one of the flows sliced though the city of Goma, 19 km south of Nyiragongo, destroying everything in a 60-m-wide path. The lava ignited many fires, and huge explosions took place where it came into contact with gasoline storage tanks.

The number of fatalities in Goma is uncertain, but most estimates are

between 80 and 100. Explosions triggered by the lava flow rather than the flow itself were responsible for most of these deaths. About 400,000 people were evacuated from the city for three days. Although of little consolation to the survivors of those who died, the death toll was actually quite low considering that a lava flow moved quickly though a large, densely populated city.

Our statement that "lava flows are the least dangerous manifestation of volcanic activity" is correct, even though there are some exceptions. Far greater dangers are posed by explosive eruptions during which huge quantities of pyroclastic materials and gases are ejected into the atmosphere and form volcanic mudflows (lahars). Fatalities from these eruptions and associated activity might be in the thousands or tens of thousands (Table 4.1).

Images of the World

(a)

AFP/Corbis

(b)

■ **Figure 1**

(a) Nyiragongo is a composite volcano in central Africa that has erupted 19 times since 1884. It stands 3470 m high. (b) Part of one of the January 17, 2002, lava flows that killed dozens of people in Goma, Democratic Republic of Congo. The lava flow is moving across the runway at Goma airport.

(a)

Hawaii Volcanoes National Park, USGS

(b)

J. B. Judd/USGS

(c)

Sue Monroe

■ **Figure 4.4**

(a) The hollow beneath a lava flow is a lava tube. (b) Part of this lava tube's roof has collapsed, forming a skylight through which the active flow can be seen. (c) Entrance to one of the nearly 200 lava tubes in Lava Beds National Monument, California.

leaving an empty tunnel-like structure (■ Figure 4.4a). Part of the roof of a lava tube may collapse, forming a *skylight* through which an active flow can be observed (Figure 4.4b), or access can be gained to an inactive lava tube. In Hawaii, lava moves through lava tubes many kilometers long and in some cases discharges into the sea.

Geologists define two types of lava flows, both named for Hawaiian flows. A **pahoehoe** (pronounced *pah-hoy-hoy*) flow has a ropy surface much like taffy (■ Figure 4.5a, b). The surface of an **aa** (pronounced *ah-ah*) flow is characterized by rough, jagged, angular blocks and fragments (Figure 4.5c). Pahoehoe flows are less viscous than aa flows; indeed, the latter are viscous enough to break up into blocks and move forward as a wall of rubble (Figure 4.5d).

Pressure on the partly solidified crust of a still-moving lava flow causes the surface to buckle into *pressure ridges* (■ Figure 4.6a). Gases escaping from a flow hurl globs of lava into the air, which fall back to the surface and adhere to one another, thus forming small,

steep-sided *spatter cones* or *spatter ramparts* if they are elongate (Figure 4.6b). Spatter cones a few meters high are common on lava flows in Hawaii, and you can see ancient ones in Craters of the Moon National Monument in Idaho.

Columnar joints are common in many lava flows, especially mafic flows, but they also occur in other kinds of flows and in some intrusive igneous rocks. Once a lava flow ceases moving, it contracts as it cools, thus producing forces that cause fractures called *joints* to open. On the surface of a flow, these joints commonly form polygonal (often six-sided) cracks (■ Figure 4.7). These cracks extend downward into the flow, thus forming parallel columns with their long axes perpendicular to the principal cooling surface. Excellent examples of columnar joints are found in many areas.

Lava flows are often perched on top of ridges or form the cap of flat-topped hills (■ Figure 4.8a). But we know these flows were fluid and flowed into low areas such as valleys, so why do they now stand high above the

(a)

(b)

(c)

(d)

J. D. Griggs/USGS

J. D. Griggs/USGS

Robert Tilling/USGS

James S. Monroe

■ **Figure 4.5**

(a) A pahoehoe lava flow in Hawaii. Notice the smooth lobes of lava, but in the right center you can see the smooth, folded texture of the flow's surface. (b) Excellent closeup view of the taffylike appearance of pahoehoe in Hawaii. (c) An aa flow advances over an older pahoehoe flow in Hawaii. (d) This pile of rubble is actually the margin of an aa flow that formed in about 1650 in California. It is about 10 m high.

T. J. Takahashi/USGS

J. D. Griggs/USGS

(a)   (b)

■ **Figure 4.6**

(a) Pressure ridge on a 1982 lava flow in Hawaii. (b) Spatter rampart that formed in 1984 in Hawaii. Spatter ramparts are elongate, whereas spatter cones are cone shaped.

James S. Monroe

James S. Monroe

(a)   (b)

■ **Figure 4.7**

(a) Columnar joints in a 60-million-year-old lava flow at the Giant's Causeway in Northern Ireland. As the lava cooled, fractures formed and intersected to form mostly five- and six-sided columns. (b) Surface view of the columns at the "other end" of the Giant's Causeway on the island of Staffa in Scotland about 85 km from the view seen in (a).

(a)

(b)

(c)

■ **Figure 4.8**

(a) Lava flows into a small valley, where it cools and crystallizes. (b) The adjacent higher areas erode more rapidly than the lava flow, producing an inversion of topography. That is, what was a valley is now a ridge top. (c) The black rock at the top of this small hill is basalt that once filled a valley.

rapidly and the former valley becomes a ridge or hilltop (Figure 4.8b, c).

Much of the igneous rock in the upper part of the oceanic crust is a distinctive type consisting of bulbous masses of basalt resembling pillows, hence the name **pillow lava.** It was long recognized that pillow lava forms when lava is rapidly chilled beneath water, but its formation was not observed until 1971. Divers near Hawaii saw pillows form when a blob of lava broke through the crust of an underwater lava flow and cooled almost instantly, forming a pillow-shaped structure with a glassy exterior. The remaining fluid inside then broke through the crust of the pillow, repeating the process and resulting in an accumulation of interconnected pillows (■ Figure 4.9).

general surface? When a lava flow cools, it forms rock that is commonly harder and more resistant to erosion than the adjacent rocks. Accordingly, as erosion proceeds, the rocks along the lava flow's margin erode more

PHYSICAL
Geology ⇌ Now   Click Geology Interactive to work through an activity of Volcanic Landforms.

(a)

(b)

Pillow lava

■ **Figure 4.9**

(a) These bulbous masses of pillow lava form when magma is erupted underwater. (b) Ancient pillow lava on land in Marin County, California. Two complete pillows and many broken ones are visible.

Sue Monroe

(a)

■ Figure 4.10

(a) Pyroclastic materials. The large object on the left is a volcanic bomb about 20 cm long. The granular materials in the upper right are pyroclastic materials known as lapilli. The pile of gray-white material on the lower right is ash. (b) A huge cloud of ash and steam erupted from Mount Pinatubo in the Philippines on June 15, 1991.

Pyroclastic Materials

In addition to lava flows, erupting volcanoes eject pyroclastic materials, especially **ash,** a designation for pyroclastic particles measuring less than 2.0 mm (■ Figure 4.10). In some cases ash is ejected into the atmosphere and settles to the surface as an *ash fall.* In 1947 ash erupted from Mount Hekla in Iceland fell 3800 km away on Helsinki, Finland. In contrast to an ash fall, an *ash flow* is a cloud of ash and gas that flows along or close to the land surface. Ash flows can move faster than 100 km/hr, and some cover vast areas.

In populated areas adjacent to volcanoes, ash falls and ash flows pose serious problems, and volcanic ash in the atmosphere is a hazard to aviation. Since 1980, about 80 aircraft have been damaged when they encountered clouds of volcanic ash. The most serious incident took place in 1989 when ash from Redoubt Volcano in Alaska caused all four jet engines to fail on KLM Flight 867. The plane carrying 231 passengers nearly crashed when it fell more than 3 km before the crew could restart the engines. The plane landed safely in Anchorage, Alaska, but it required $80 million in repairs.

In addition to ash, volcanoes erupt *lapilli,* consisting of pyroclastic materials measuring from 2 to 64 mm, and *blocks* and *bombs,* both of which are larger than 64 mm (Figure 4.10a). Bombs have a twisted, streamlined shape, indicating they were erupted as globs of magma that cooled and solidified during their flight through the air. Blocks, in contrast, are angular pieces of rock ripped from a volcanic conduit or pieces of a solidified crust of a lava flow. Because of their size, lapilli, bombs, and blocks are confined to the immediate area of an eruption.

Reuters/Corbis

(b)

PHYSICAL
Geology ⇌ Now Click Geology Interactive to work through an activity of Magma Chemistry and Explosivity.

VOLCANOES—HOW DO THEY FORM?

S imply put, a **volcano** is a hill or mountain that forms around a vent where lava, pyroclastic materials, and gases erupt. Some volcanoes are conical, but others are simply bulbous, steep-sided masses of magma, and some resemble an inverted shield lying on the ground. In all cases, though, volcanoes have a conduit or conduits leading to a magma chamber beneath the surface. The Roman deity of fire, Vulcan, was the inspiration for calling these mountains volcanoes, and because of their danger and obvious connection to Earth's interior they have been held in awe by many cultures.

Probably no other geologic phenomenon, except perhaps earthquakes, has so much lore associated with it. In Hawaiian legends the volcano goddess Pele resides in the crater of Kilauea on Hawaii. During one of her frequent rages, Pele causes earthquakes and lava flows,

(a)

(b)

(c)

(d)

Wizard Island   Crater Lake

(e)

James S. Monroe

■ **Figure 4.11**

Events leading to the origin of Crater Lake, Oregon. (a,b) Ash clouds and ash flows partly drain the magma chamber beneath Mount Mazama. (c) The collapse of the summit and formation of the caldera. (d) Postcaldera eruptions partly cover the caldera floor, and the small cinder cone known as Wizard Island forms. (e) View from the rim of Crater Lake showing Wizard Island. Source: Figure 4.11(a–d): From Howell Williams, *Crater Lake: The Story of Its Origin* (Berkeley, Calif. University of California Press): Illustrations from p. 84. © 1941 Regents of the University of California, © renewed 1969, Howell Williams.

and she may hurl flaming boulders at those who offend her. Native Americans in the Pacific Northwest tell of a titanic battle between the volcano gods Skel and Llao to account for huge eruptions that took place about 6600 years ago in Oregon and California. Pliny the Elder (A.D. 23–79), mentioned in the Introduction, believed that before eruptions "the air is extremely calm and the sea quiet, because the winds have already plunged into the earth and are preparing to reemerge."*

Geologists recognize several major types of volcanoes, but one must realize that each volcano is unique in its history of eruptions and development. For instance, the frequency of eruptions varies considerably;

the Hawaiian volcanoes and Mount Etna on Sicily have erupted repeatedly, whereas Pinatubo in the Philippines erupted in 1991 for the first time in 600 years (Figure 4.10b). And some volcanoes are complex mountains that have characteristics of more than one type of volcano.

Most volcanoes have a circular depression known as a **crater** at their summit, or on their flanks, that forms by explosions or collapse. Craters are generally less than 1 km across, whereas much larger rimmed depressions on volcanoes are **calderas.** In fact, some volcanoes have a summit crater within a caldera. Calderas are huge structures that form following voluminous eruptions during which part of a magma chamber drains and the mountain's summit collapses into the vacated space below. An excellent example is misnamed Crater Lake in Oregon (■ Figure 4.11). Crater Lake is actually a steep-rimmed caldera that formed about

*Quoted from M. Krafft, *Volcanoes: Fire from the Earth* (New York: Harry N. Abrams, 1993), p. 40.

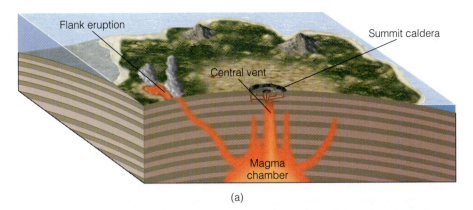

(a)

Robert Tilling/USGS

(b)

James S. Monroe

(c)

■ **Figure 4.12**

Shield volcano. Each layer shown consists of numerous thin basalt lava flows. (b) View of Mauna Loa, an active shield volcano on Hawaii, with its upper 1.5 km covered by snow. (c) Crater Mountain is an extinct shield volcano in Lassen County, California. It is about 10 km across and stands 460 m high. the depression at the summit is a 2-km-wide crater. Notice in the illustration and the images that shield volcanoes are not very steep.

6600 years ago in the manner just described; it is more than 1200 m deep and measures 9.7 × 6.5 km. As impressive as Crater Lake is, it is not nearly as large as some other

calderas, such as the Toba caldera in Sumatra, which measures 100 km long and 30 km wide, and the Yellowstone caldera in Wyoming (see Geo-Focus 4.2).

Shield Volcanoes

Shield volcanoes resemble the outer surface of a shield lying on the ground, with the convex side up (■ Figure 4.12). They have low, rounded profiles with gentle slopes ranging from about 2 to 10 degrees; they are composed mostly of low-viscosity mafic lava flows, so the flows spread out and form thin layers. Eruptions from shield volcanoes, sometimes called *Hawaiian-type eruptions,* are quiet compared with those of volcanoes such as Mount St. Helens. Lava most commonly rises to the surface with little explosive activity, so it usually poses little danger to humans. Lava fountains, some up to 400 m high, contribute some pyroclastic materials to shield volcanoes, but otherwise they are composed of basalt lava flows; flows comprise more than 99% of the Hawaiian volcanoes above sea level.

Although eruptions of shield volcanoes tend to be rather quiet, some of the Hawaiian volcanoes have, on occasion, produced sizable explosions when magma comes in contact with groundwater, causing it to vaporize instantly. One such explosion occurred in 1790 while Chief Keoua was leading about 250 warriors across the summit of Kilauea volcano to engage a rival chief in battle. About 80 of Keoua's warriors were killed by a cloud of hot volcanic gases.

The current activity at Kilauea is impressive for another reason; it has been erupting continuously since January 3, 1983, making it the longest recorded eruption. During these 21 years, more than 2.3 km³ of molten rock has flowed out at the surface, much of it reaching the sea and forming 2.2 km² of new property on the island of Hawaii. Unfortunately, lava flows from

GEOFOCUS

4.2

Supervolcanoes

Geologists have no formal definition for the term *supervolcano,* but we can take it to mean an eruption that results in the explosive ejection of hundreds of cubic kilometers of pyroclastic materials and the origin of a huge caldera. Fortunately, such voluminous eruptions are rare, and none have occurred during historic time. One supervolcano of particular interest in North America is the one that yielded the Yellowstone caldera in Wyoming. Continuing earthquakes and hydrothermal activity (hot springs and geysers) remind us that renewed volcanism in this area is possible.

The Yellowstone caldera lies within the confines of Yellowstone National Park, the first area in the United States set aside as a national park. Yellowstone National Park is noted for its scenery, wildlife, boiling mud pots, hot springs, and geysers, especially Old Faithful (see Chapter 16), but few tourists are aware of the region's volcanic history. The fact that a large body of magma is present beneath the surface is accepted among geologists,

many of whom are convinced that the Yellowstone caldera remains active.

On three separate occasions supervolcano eruptions followed the accumulation of rhyolitic magma beneath the surface of the Yellowstone region. Each eruption yielded a widespread blanket of volcanic ash and pumice as well as collapse of the surface and origin of a large caldera. We can summarize Yellowstone's volcanic history by noting that supervolcano eruptions took place 2 million years ago, 1.3 million years ago, and 600,000 years ago. It was during this last huge eruption that the present-day Yellowstone caldera originated, which is actually part of a larger composite caldera resulting from the three cataclysmic events noted above (■ Figure 1a).

The magnitude of these supervolcano eruptions is difficult to imagine. In fact, the pyroclastic flows and airborne ash cover not only the areas within and adjacent to the Yellowstone region but also a large part of the western United States and northern Mexico. Nothing in the immediate areas of the

eruptions could have survived the intense heat and choking clouds of volcanic ash.

Since the last supervolcano eruption, rhyolitic magma has continued to accumulate beneath the caldera, thereby elevating parts of its floor in what are known as *resurgent domes.* Precise leveling shows that the caldera floor continues to rise; more than 80 cm of uplift has taken place since 1923. And between 150,000 and 75,000 years ago, an additional 1000 km^3 of pyroclastic materials was erupted within the caldera. An excellent example is the Yellowstone Tuff into which the picturesque Grand Canyon of the Yellowstone River is incised (Figure 1b).

Many geologists are convinced that a *mantle plume,* a cylindrical mass of magma rising from the mantle, underlies the area. As this rising magma nears the surface, it triggers volcanic eruptions, and because the magma is rhyolitic the eruptions are particularly explosive. The Yellowstone hot spot, as it is called, is one of only a few dozen hot spots recognized on Earth.

Kilauea have also destroyed about 200 homes and caused some $61 million in damages.

Shield volcanoes are most common in the ocean basins, such as the Hawaiian Islands and Iceland, but some are also present on the continents—for example, in east Africa. The island of Hawaii consists of five huge shield volcanoes, two of which, Kilauea and Mauna Loa, are active much of the time. Mauna Loa at nearly 100 km across its base and more than 9.5 km above the surrounding seafloor is the largest

volcano in the world (Figure 4.12b). Its volume is estimated at about 50,000 km^3. By contrast, a very large volcano in the continental United States, Mount Shasta in northern California, has a volume of only about 350 km^3.

Cinder Cones

Cinder cones are small, steep-sided volcanoes made up of pyroclastic materials resembling cinders (■ Figure 4.13).

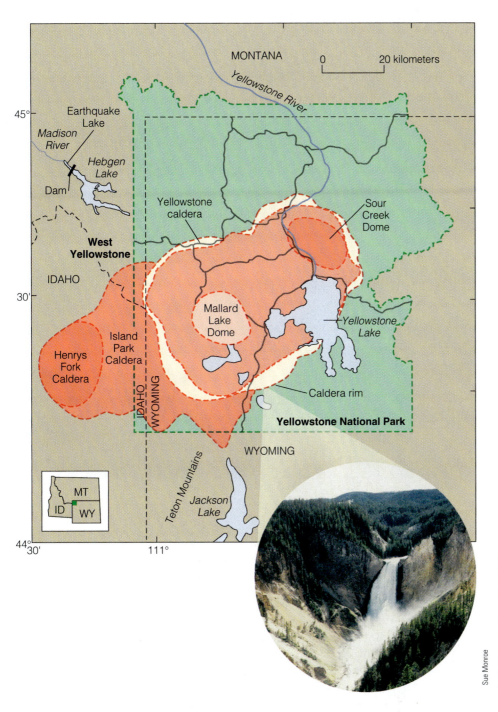

■ **Figure 1**

(a) The Yellowstone caldera formed following huge eruptions about 600,000 years ago. It is part of a larger composite caldera. (b) The walls of the Grand Canyon of the Yellowstone River are made up of the hydrothermally altered Yellowstone Tuff that partly fills the Yellowstone caldera.

Sue Monroe

They form when pyroclastic materials accumulate around a vent from which they were erupted. Cinder cones are quite small, rarely exceeding 400 m high, with slope angles up to 33 degrees, depending on the angle that can be maintained by the angular pyroclastic materials. Many of these small volcanoes have a large, bowl-shaped crater, and if they issue any lava flows, they usually break through the base of lower flanks of the mountains. Although all cinder cones are rather conical, their symmetry varies from those almost perfectly symmetrical to those that formed when prevailing winds caused pyroclastic materials to build up higher on the downwind side of the vent.

Many cinder cones form on the flanks or within the calderas of larger volcanoes and represent the final stages of activity, particularly in areas of basalt lava flows. Wizard Island in Crater Lake, Oregon, is a small cinder cone that formed after the summit of Mount Mazama collapsed to form a caldera (Figure 4.11).

■ **Figure 4.13**

(a) A 230-m-high cinder cone in Lassen Volcanic National Park, California. (b) The large, bowl-shaped crater at the summit of the cinder cone shown in (a). (c) View from the top of the cinder cone in (a) showing an aa lava flow that broke through the base of the volcano during the 1650s. The light-colored material is volcanic ash that fell on the still-hot lava flow. (d) Eldfel, a cinder cone in Iceland, began erupting in 1973 and in two days grew to 100 m high. Another cinder cone known as Helgafel is also visible.

Cinder cones are common in the southern Rocky Mountain states, particularly New Mexico and Arizona, and many others are found in California, Oregon, and Washington.

Eruptive activity at cinder cones is rather short-lived. For instance, on February 20, 1943, a farmer in Mexico noticed fumes emanating from a crack in his cornfield, and a few minutes later ash and cinders were erupted. Within a month, a 300-m-high cinder cone had formed. Lava flows broke through the flanks of the new volcano, which was later named Paricutín, and covered two nearby towns. Activity ceased in 1952.

In 1973, on the Icelandic island of Heimaey, the town of Vestmannaeyjar was threatened by a new cinder cone. The initial eruption began on January 23, and within two days a cinder cone, later named Eldfell, rose to about 100 m above the surrounding area (Figure 4.13d). Pyroclastic materials from the volcano buried parts of the town, and by February a massive aa lava flow

was advancing toward the town. The flow's leading edge ranged from 10 to 20 m thick, and its central part was as much as 100 m thick. The residents of Vestmannaeyjar sprayed the leading edge of the flow with seawater in an effort to divert it from the town. The flow was in fact diverted, but how effective the efforts of the townspeople were is not clear; they may simply have been lucky.

Composite Volcanoes (Stratovolcanoes)

Pyroclastic layers as well as lava flows, both of intermediate composition, are found in **composite volcanoes,** which are also called *stratovolcanoes* (■ Figure 4.14). As the lava flows cool, they typically form andesite. Recall that intermediate lava flows are more viscous than mafic ones that yield basalt. Geologists use the term **lahar** for volcanic mudflows, which are common on composite volcanoes. A lahar may form when rain falls on unconsolidated pyroclastic materials and creates a muddy

Pyroclastic layers
Lava flows

(a)

(b)

R. Solkowski/Consulting Geologists, Vancouver, WA

Sue Monroe

(c)

Courtesy of Wayne E. Moore

(d)

■ **Figure 4.14**

(a) Composite volcanoes, or stratrovolcanoes, are composed mostly of lava flows and pyroclastic materials of intermediate composition. (b) Mayon volcano in the Philippines is one of the most nearly symmetrical composite volcanoes in the world. It erupted during 1999 for the 13th time this century. (c, d) Two views of Mount Shasta, a large composite volcano in northern California. Notice in (d), a view from the north, that a cone known as Shastina is present on the flank of the larger mountain.

slurry that moves downslope (■ Figure 4.15). On November 13, 1985, a minor eruption of Nevado del Ruiz in Colombia melted snow and ice on the volcano, causing lahars that killed about 23,000 people (Table 4.1).

Composite volcanoes differ from shield volcanoes and cinder cones in composition as noted above, but their overall shape differs, too. Remember that shield volcanoes are very low slopes, whereas cinder cones are small, steep-sided, conical mountains. In marked contrast, composite volcanoes are steep-sided near their summits, perhaps as much as 30 degrees, but the slope decreases toward the base, where it may be no more than 5 degrees. Mayon volcano in the Philippines is one of the most nearly symmetrical composite volcanoes anywhere (Figure 4.14b). It erupted in 1999 for the 13th time during the 1900s.

When most people think of volcanoes, they picture the graceful profiles of composite volcanoes, which are the typical large volcanoes found on the continents and island arcs. And some of these volcanoes are indeed large: Mount Shasta in northern California is made up of about 350 km^3 of material and measures about 20 km across its base (Figure 4.14c, d). Indeed, it is so large that it dominates the skyline when one approaches from any direction. Other familiar composite volcanoes are several in the Cascade Range of the Pacific Northwest as well as Fujiyama in Japan and Mount Vesuvius in Italy (see the chapter opening photo).

Mount Pinatubo in the Philippines (Figure 4.10b) erupted violently on June 15, 1991. Huge quantities of gas and an estimated 3–5 km^3 of ash were discharged into the atmosphere, making this the world's largest

GEOLOGY
IN UNEXPECTED PLACES

A Most Unusual Volcano

Perhaps the East African Rift is not an unexpected place for volcanoes, but one known as Oldoinyo Lengai or Ol Doinyo Lengai in Tanzania is certainly among Earth's most peculiar volcanoes. Oldoinyo Lengai, which means "Mountain of God" in the Masai language, is an active composite volcano (it last erupted in August 2002), standing about 2890 m high. It is one in an east–west belt of about 20 volcanoes near the southern part of the East African Rift. Its most notable feature, however, is its eruptions of magma that cools to form carbonitite, an igneous rock with at least 50% carbonate minerals, mostly calcite ($CaCO_3$) and dolomite [$CaMg(CO_3)_2$]. In fact, carbonitite looks much like the metamorphic rock marble (see Chapter 7).

Remember that based on silica content most magma varies from mafic to felsic, but only rarely does magma have a significant amount of carbonate minerals. At Oldoinyo Lengai the carbonitite magma typically has very low viscosity and is fluid at temperatures of only 540° to 595°C, reflecting the low melting temperatures of carbonate minerals. As a result it flows rather quickly and it is not incandescent but rather looks like black mud. However, its minerals are chemically unstable, so they react with water in the atmosphere and their color changes to pale gray very quickly (■ Figure 1).

Courtesy of Frederick A. Belton

■ **Figure 1**

Lava spattering from the top of a small cone on the crater floor on August 8, 1990. The black lava on the right side of the cone is fresh (erupted within the last hour); the white material in the foreground was black when first erupted and became white over several months.

---

eruption since 1912. Fortunately, warnings of an impending eruption were heeded, and 200,000 people were evacuated from around the volcano; yet, the eruption was still responsible for 722 deaths (Table 4.1).

Lava Domes

Although some volcanoes show features of more than one of the types discussed so far, most can be classified as shield volcanoes, cinder cones, or composite volcanoes. One more type of volcano deserves our attention, however. **Lava domes,** also known as *volcanic domes* and *plug domes,* are steep-sided, bulbous mountains that form when viscous felsic magma, occasionally intermediate magma, is forced toward the surface (■ Figure 4.16).

Because felsic magma is so viscous, it moves upward very slowly and only when the pressure from below is great.

Beginning in 1980, a number of lava domes were emplaced in the crater of Mount St. Helens in Washington; most of these were destroyed during subsequent eruptions. Since 1983, Mount St. Helens has been characterized by sporadic dome growth. In June 1991 a lava dome in Japan's Unzen volcano collapsed under its own weight, causing the flow of debris and hot ash that killed 43 people in a nearby town.

Lava dome eruptions are some of the most violent and destructive. In 1902 viscous magma accumulated beneath the summit of Mount Pelée on the island of Martinique. Eventually the pressure increased until the side of the mountain blew out in a tremendous explo-

(a)   (b)

■ Figure 4.15

Volcanic mudflows, or lahars, in the Philippines and Colombia. (a) Homes partly buried by a lahar on June 15, 1991, in the Philippines following the eruption of Mount Pinatubo. (b) Air view of Armero, Colombia, where at least 23,000 people died in lahars that inundated the area.

sion, ejecting a mobile, dense cloud of pyroclastic materials and a glowing cloud of gases and dust called a **nuée ardente** (French for "glowing cloud"). The pyroclastic flow followed a valley to the sea, but the nuée ardente jumped a ridge and engulfed the city of St. Pierre (■ Figure 4.17).

A tremendous blast hit St. Pierre and leveled buildings; hurled boulders, trees, and pieces of masonry down the streets; and moved a 3-ton statue 16 m. Accompanying the blast was a swirling cloud of incandescent ash and gases with an internal temperature of 700°C that incinerated everything in its path. The nuée ardente passed through St. Pierre in 2 or 3 minutes, only to be followed by a firestorm as combustible materials burned and casks

of rum exploded. But by then most of the 28,000 residents of the city were already dead. In fact, in the area covered by the nuée ardente, only 2 survived!* One survivor was on the outer edge of the nuée ardente, but even there he was terribly burned and his family and neighbors were all killed. The other survivor, a stevedore incarcerated the night before for disorderly conduct, was in a windowless cell partly below ground level. He remained in his cell badly burned for four days after the eruption

*Although it is commonly reported that only 2 people survived the eruption, at least 69 and possibly as many as 111 people survived beyond the extreme margins of the nuée ardente and on ships in the harbor. Many, however, were badly injured.

(a)   (b)

■ Figure 4.16

(a) Chaos Crags in the distance are made up of at least four lava domes that formed less than 1200 years ago in Lassen Volcanic National Park, California. The debris in the foreground is Chaos Jumbles, which resulted from partial collapse of some of the domes. (b) This steep-sided lava dome lies atop Novarupta in Katmai National Park and Preserve in Alaska.

(a)

■ Figure 4.17

(a) St. Pierre, Martinique, after it was destroyed by a nuée ardente erupted from Mount Pelée in 1902. Only 2 of the city's 28,000 inhabitants survived. (b) A large nuée ardente from Mount Pelée a few months after the one that destroyed St. Pierre.

(b)

until rescue workers heard his cries for help. He later became an attraction in the Barnum & Bailey Circus, where he was advertised as "the only living object that survived in the 'Silent City of Death' where 40,000 beings were suffocated, burned or buried by one belching blast of Mont Pelée's terrible volcanic eruption."*

PHYSICAL Geology⇌Now Click Geology Interactive to work through an activity on Volcanic Watch, USA through Volcanic Landforms.

THE CASCADE RANGE

Large volcanic mountains dominate the **Cascade Range** in the Pacific Northwest (see "Cascade Range Volcanoes" on pages 108–109). During the summer of 1914, Lassen Peak in California began to erupt and culminated with the "Great Hot Blast," a huge steam explosion on May 22, 1915. The blast devastated a large area, but because the area was so sparsely settled, there were no human fatalities. Following Lassen Peak's eruption, the Cascade vol-

canoes remained quiet for 63 years. Then, on March 16, 1980, Mount St. Helens in Washington showed signs of renewed activity, and on May 18 it erupted violently, causing the most fatalities for any eruption in U.S. history (Table 4.1).

What Kinds of Volcanoes Are Present?

Most of the large volcanoes in the Cascade Range are composite volcanoes, such as Mount St. Helens and Mount Rainier in Washington, Mount Hood in Oregon, and Mount Shasta in California (Figure 4.14c, d). Lassen Peak in northern California, however, is the world's largest lava dome; actually it is a comparatively small volcanic mountain that developed 27,000 years ago on the flank of a much larger, deeply eroded composite volcano. It has been quiet since 1917, but a source of heat beneath the surface is obvious from ongoing hydrothermal activity (Figure 4.3a).

What was once a nearly symmetrical composite volcano changed markedly when Mount St. Helens erupted explosively in 1980, killing about 63 people, wiping out thousands of animals, and leveling some 600 km² of forest. Geologists, citing Mount St. Helens's history, warned that it was the most likely volcano in the range to erupt violently. The huge explosion caused much of the damage and fatalities, but snow and ice on the

*Quoted from A. Scarth, *Vulcan's Fury: Man Against the Volcano* (New Haven, CT: Yale University Press, 1999), p. 177.

mountain melted and pyroclastic materials displaced water in lakes and streams, causing lahars and extensive flooding.

Other than Lassen Peak, all of the large volcanoes in the Cascade Range are composite volcanoes. However, Medicine Lake Volcano in northern California and Newberry Volcano in Oregon, both lying just east of the main trend of the Cascades, are two large shield volcanoes. A distinctive feature at Newberry Volcano is a 1600-year-old obsidian flow and casts of trees formed when lava flowed around them and solidified.

Cinder cones are also common in the range. We have already mentioned Wizard Island in Crater Lake, Oregon (Figure 4.11e). Cinder Cone in Lassen Volcanic National Park, at about 230 m high, erupted aa lava flows as recently as the 1650s (Figure 4.13a–c).

Will the Cascade Volcanoes Erupt Again?

Although Mount Garibaldi in British Columbia may be extinct, many of the other volcanoes in the Cascade Range are dormant and will no doubt erupt again. Future eruptions of Mount Shasta in California would threaten several nearby communities, and Mount Hood in Oregon lies less than 65 km from the densely populated Portland area. But the most dangerous volcano is probably Mount Rainier in Washington.

Rather than lava flows, ash falls, or even a colossal explosion, the greatest threat from Mount Rainier is volcanic mudflows or lahars similar to the one in Figure 4.15. Of the 60 large flows that have occurred during the last 10,000 years, the largest, consisting of 4 km^3 of debris, covered an area now occupied by more than 120,000 people. No one knows when the next flow will take place, but at least one community has taken the threat seriously enough to formulate an emergency evacuation plan. Unfortunately, the residents would have only 1 or 2 hours to carry out an evacuation.

OTHER VOLCANIC LANDFORMS

The term *volcano* immediately brings to mind the magnificent composite volcanoes such as those in Figure 4.14. However, as noted earlier, even though some volcanoes are the typical mountains envisioned, numerous volcanoes with other shapes are found in many areas (Figures 4.12 and 4.13). In fact, in some areas of volcanism, volcanoes fail to develop at all. For instance, during *fissure eruptions* fluid lava pours out and simply builds up rather flat-lying areas, whereas huge explosive eruptions might yield *pyroclastic sheet deposits,* which, as their name implies, have a sheetlike geometry.

What Would You Do?

No one doubts that some of the Cascade Range volcanoes will erupt again, but no one knows when or how large these eruptions will be. A job transfer takes you to a community in Oregon with several large volcanoes nearby. You have some concerns about possible future eruptions, so what kinds of information would you seek out before buying a home in this area? In addition, as a concerned citizen, could you make any suggestions about what should be done in the case of a large eruption?

Fissure Eruptions

Some 164,000 km^2 of eastern Washington and parts of Oregon and Idaho were covered by overlapping basalt lava flows between about 17 and 5 million years ago. Now known as the Columbia River basalts, they are well exposed in canyons eroded by the Snake and Columbia Rivers (■ Figure 4.18a–c). Rather than being erupted from a central vent, these flows issued from long cracks or fissures and are thus known as **fissure eruptions.** Lava erupted from these fissures was so fluid (had such low viscosity) that it simply spread out, covering vast areas and building up a **basalt plateau,** which is simply a broad, flat, elevated area underlain by lava flows.

The Columbia River basalt flows have an aggregate thickness of about 1000 m, and some individual flows cover huge areas—for example, the Roza flow, which is 30 m thick, advanced along a front about 100 km wide and covered 40,000 km^2.

Fissure eruptions and basalt plateaus are not common, although several large areas with these features are known. Currently, this kind of activity occurs only in Iceland. A number of volcanic mountains are present in Iceland, but the bulk of the island is composed of basalt flows erupted from fissures. Two large fissure eruptions, one in A.D. 930 and the other in 1783, account for about half of the magma erupted in Iceland during historic time. The 1783 eruption issued from the Laki fissure, which is more than 30 km long; about 12.5 km^3 of lava flowed several tens of kilometers from the fissure; the lava covered an area of more than 560 km^2 and in one place filled a valley to a depth of about 200 m.

Pyroclastic Sheet Deposits

More than 100 years ago, geologists were aware of vast areas covered by felsic volcanic rocks a few meters to hundreds of meters thick. It seemed improbable that these could have formed as vast lava flows, but it seemed equally unlikely that they were ash fall deposits. Based on observations of historic pyroclastic

Cascade Range Volcanoes

Several large volcanoes and hundreds of smaller volcanic vents are in the Cascade Range, which stretches from northern California through Oregon and Washington and into southern British Columbia, Canada. Medicine Lake Volcano and Newberry Volcano are shield volcanoes that lie just east of the main trend of the Cascade Range.

Mount St. Helens, a composite volcano, as it appeared from the east in 1978.

D. R. Crandel / USGS

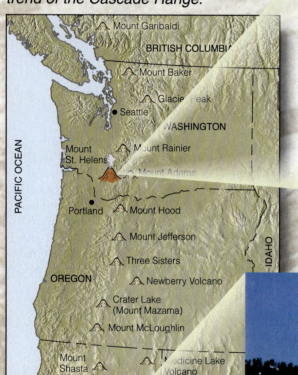

Lassen Peak today. This most southerly peak in the Cascade Range is made up of 2 km³ of material including the bulbous masses of rock visible in this image.

Sue Monroe

- Of the Cascade Range volcanoes, only Lassen Peak and Mount St. Helens have erupted during the 1900s.

- Lassen Peak is a lava dome that formed about 27,000 years ago on the northeast flank of a deeply eroded composite volcano known as Mount Tehama.

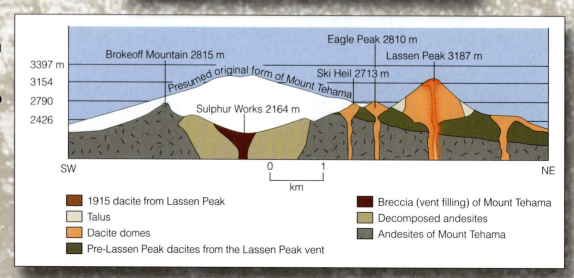

Brokeoff Mountain 2815 m
Eagle Peak 2810 m
Ski Heil 2713 m
Lassen Peak 3187 m
Sulphur Works 2164 m
Presumed original form of Mount Tehama

3397 m
3154
2790
2426

SW   0   1   km   NE

- 1915 dacite from Lassen Peak
- Talus
- Dacite domes
- Pre-Lassen Peak dacites from the Lassen Peak vent
- Breccia (vent filling) of Mount Tehama
- Decomposed andesites
- Andesites of Mount Tehama

Mount St. Helens. The lateral blast on May 18, 1980, took place when a bulge on the volcano's north face collapsed, reducing the pressure on gas-charged magma.

Courtesy of Keith Ronnholm

PhotoDisc/ Getty Images

Mount St. Helens' lateral blast killed about 63 people and leveled 600km² of forest. These trees in Smith Creek Valley are in part of the blast zone. Two geologists in the lower right part of the image provide scale.

Lynn Topinka/ USGS

Shortly after the lateral blast, this 19-km-high ash and steam cloud erupted from Mount St. Helens.

Lassen Peak erupted numerous times from 1914 to 1917. This eruption took place in 1915.

A huge steam explosion called the Great Hot Blast leveled the area in the foreground in 1915. In the 88 years since the eruption, trees are becoming reestablished in this area known as the Devastated Area.

Sue Monroe

Frank Kujawa, University of Central Florida/GeoPhoto

(a)

Earlier flows

(b)

Fissures

W. E. Scott/USGS

(c)

■ **Figure 4.18**

(a) About 20 lava flows of the Columbia River basalts, exposed in the canyon of the Grand Ronde River in Washington. (b) Block diagram showing fissure eruptions and the origin of a basalt plateau. (c) Pyroclastic flow deposits that issued from Mount Pinatubo on June 15, 1991, in the Philippines. Some of the flows moved 16 km from the volcano and filled this valley to depths of 50 to 200 m.

flows, such as the nuée ardente erupted by Mount Pelée in 1902, it seems that these ancient rocks originated as pyroclastic flows—hence, the name **pyroclastic sheet deposits** (Figure 4.18c). They cover far greater areas than any observed during historic time and apparently erupted from long fissures rather than

from a central vent. The pyroclastic materials of many of these flows were so hot that they fused together to form *welded tuff*.

Geologists now think that major pyroclastic flows issued from fissures formed during the origin of calderas (see Geo-Focus 4.2). For instance, pyroclastic flows were

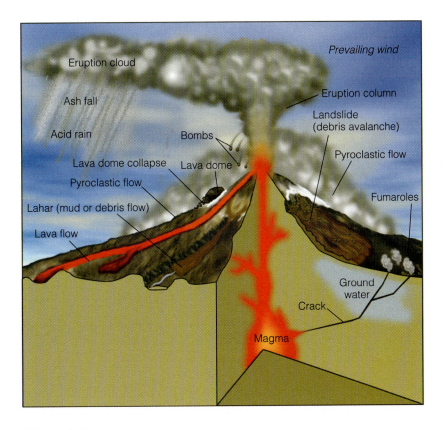

■ Figure 4.19

Volcanoes produce a variety of hazards that kill or injure people and destroy property. Some hazards, such as landslides and lahars, may occur even when a volcano is not erupting. This illustration shows a typical volcano in Alaska and the western United States, but volcanoes in Hawaii also pose some hazards.

erupted during the formation of a large caldera in the area of present-day Crater Lake, Oregon (Figure 4.11). Similarly, the Bishop Tuff of eastern California was erupted shortly before the formation of the Long Valley caldera. Interestingly, earthquake activity in the Long Valley caldera and nearby areas beginning in 1978 may indicate that magma is moving up beneath part of the caldera. Thus the possibility of future eruptions in that area cannot be discounted.

VOLCANIC HAZARDS

We have discussed several volcanic hazards such as lava flows, nuée ardentes, gases, and lahars in this chapter. Lava flows and nuée ardentes obviously are threats only during an eruption, but lahars and landslides may take place even when no eruption has occurred for a long time (■ Figure 4.19). Certainly the most vulnerable areas in the United States are Alaska, Hawaii, California, Oregon, and Washington, but some other places in the western part of the country might experience renewed volcanism, too.

How Large Is an Eruption and How Long Do Eruptions Last?

Geologists have devised several ways of expressing the size of a volcanic eruption. One, called the destructiveness index, is based on the area covered by lava or pyroclastic materials during an eruption. Geologists also rank eruptions in terms of their intensity and magnitude, but these have both been incorporated into the more widely used **volcanic explosivity index (VEI)** (■ Figure 4.20). Unlike the Richter Magnitude Scale for earthquakes (see Chapter 9), the VEI is only semiquantitative, being based partly on subjective criteria.

The volcanic explosivity index (VEI) ranges from 0 (nonexplosive) to 8 and is based on several aspects of an eruption, such as the volume of material explosively ejected and the height of the eruption plume. However, the volume of lava, fatalities, and property damage are not considered. For instance, the 1985 eruption of Nevado del Ruiz in Colombia killed 23,000 people, yet has a VEI of only 3. In contrast, the huge eruption (VEI = 6) of Novarupta in Alaska in 1912 caused no fatalities or injuries. Since A.D. 1500, only the 1815 eruption of Tambora had a value of 7; it was both large and deadly (Table 4.1). Nearly 5700 eruptions during the last 10,000 years have been assigned VEI numbers, but none has exceeded 7, and most (62%) were assigned a value of 2.

The duration of eruptions varies considerably. Fully 42% of about 3300 historic eruptions lasted less than one month. About 33% erupted for one to six months, but some 16 volcanoes have been active more or less continuously for more than 20 years. Stromboli and Mount Etna in Italy and Erta Ale in Ethiopia are good examples. In some explosive volcanoes, the time from the onset of their eruptions to the climactic event is weeks or months. A case in point is the colossal explosive eruption of Mount St. Helens on May 18, 1980, that occurred two months after eruptive activity began. Unfortunately, many volcanoes give little or no warning of such large-scale events; of 252 explosive eruptions, 42% erupted most violently during their first day of activity. As one might imagine, predicting eruptions is complicated by those volcanoes that give so little warning of impending activity (see the next section).

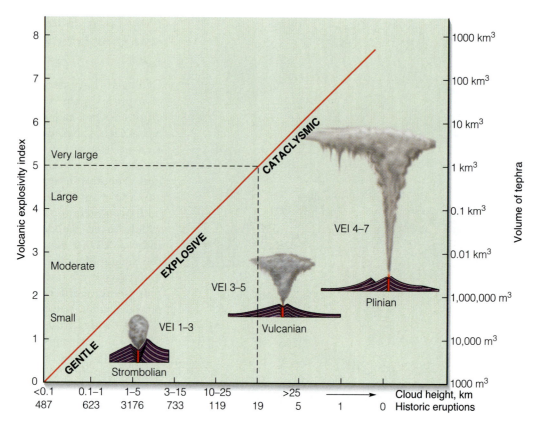

■ Figure 4.20

The volcanic explosivity index (VEI). In this example, an eruption with a VEI of 5 has an eruption cloud up to 25 km high and ejects at least 1 km³ of tephra, a collective term for all pyroclastic materials. Geologists characterize eruptions as Hawaiian (nonexplosive), Strombolian, Vulcanian, and Plinian.

Is It Possible to Forecast Eruptions?

Most of the dangerous volcanoes are at or near the margins of tectonic plates, especially at convergent plate boundaries. At any one time about a dozen volcanoes are erupting, but most eruptions cause little or no property damage, injuries, or fatalities. Unfortunately, some do. The 1815 eruption of Tambora in Indonesia and a rather minor eruption of Nevado del Ruiz in Colombia in 1985 are good examples (Table 4.1).

Only a few of Earth's potentially dangerous volcanoes are monitored, including some in Japan, Italy, Russia, New Zealand, and the United States. Four facilities in the United States are devoted to volcano monitoring; Hawaiian Volcano Observatory on Kilauea Volcano; the David A. Johnston Cascades Volcano Observatory in Vancouver, Washington; Alaska Volcano Observatory in Fairbanks, Alaska; and the Long Valley Observatory in Menlo Park, California. Many of the methods now used to monitor volcanoes were developed at the Hawaiian Volcano Observatory.

Volcano monitoring involves recording and analyzing various physical and chemical changes at volcanoes. Tiltmeters detect changes in the slopes of a volcano as it inflates when magma rises beneath it, whereas a geodimeter uses a laser beam to measure horizontal distances, which

also change as a volcano inflates (■ Figure 4.21). Geologists also monitor gas emissions, changes in groundwater level and temperature, hot springs activity, and changes in local magnetic and electrical fields. Even the accumulating snow and ice, if any, are evaluated to anticipate hazards from floods should an eruption take place.

Of critical importance in volcano monitoring and warning of an imminent eruption is the detection of **volcanic tremor,** the continuous ground motion lasting for minutes to hours as opposed to the sudden, sharp jolts produced by most earthquakes. Volcanic tremor, also known as *harmonic tremor,* indicates that magma is moving beneath the surface.

To more fully anticipate the future activity of a volcano, its eruptive history must also be known. Accordingly, geologists study the record of past eruptions preserved in rocks. Detailed studies before 1980 indicated that Mount St. Helens, Washington, had erupted explosively 14 or 15 times during the last 4500 years, so geologists concluded that it was one of the most likely Cascade Range volcanoes to erupt again. In fact, maps they prepared showing areas in which damage from an eruption could be expected were helpful in determining which areas should have restricted access and evacuations once an eruption did take place.

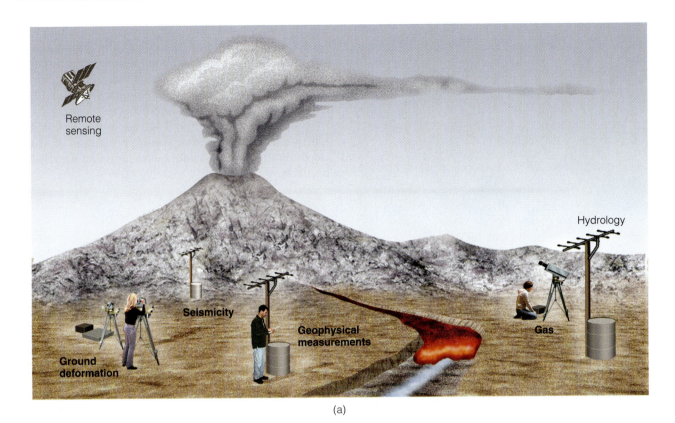

Remote sensing

Hydrology

Seismicity

Geophysical measurements

Gas

Ground deformation

(a)

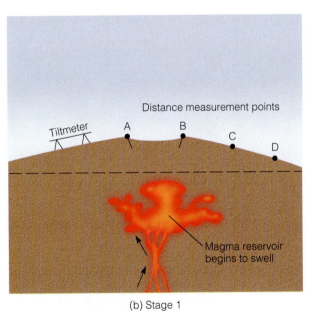

Distance measurement points

Tiltmeter

A   B   C   D

Magma reservoir begins to swell

(b) Stage 1

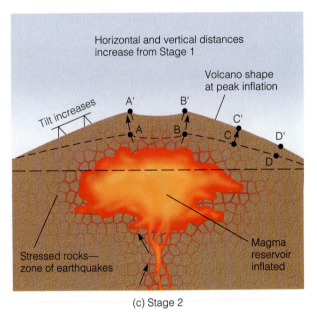

Horizontal and vertical distances increase from Stage 1

Volcano shape at peak inflation

Tilt increases

A'   B'   C'
A   B   C   D'
D

Stressed rocks— zone of earthquakes

Magma reservoir inflated

(c) Stage 2

■ **Figure 4.21**

(a) Techniques used to monitor volcanoes. (b, c) Detection of ground deformation by tiltmeters and measurements of horizontal and vertical distances. As a volcano inflates when magma moves beneath it, volcanic tremor is also detected.

Geologists successfully gave timely warnings of impending eruptions of Mount St. Helens, Washington, and Mount Pinatubo in the Philippines, but in both cases the climactic eruptions were preceded by eruptive activity of lesser intensity. In some cases, however, the warning signs are much more subtle and difficult to interpret. Numerous small earthquakes and other warning signs indicated to geologists of the U.S. Geological Survey (USGS) that magma was moving beneath the surface of the Long Valley caldera in eastern California, so in 1987 they issued a low-level warning, and then nothing happened.

What Would You Do ?

You are an enthusiast of natural history and would like to share your interests with your family. Accordingly, you plan a vacation to see some of the volcanic features in our national parks and monuments. Let's assume your planned route will take you through Wyoming, Idaho, Washington, Oregon, and California. What specific areas might you visit, and what kinds of volcanic features would you see in these areas? Are there any other parts of the United States you might visit in the future to see additional evidence for volcanism?

area in the country) and property values plummeted. Monitoring continues in the Long Valley caldera, and the signs of renewed volcanism, including earthquake swarms, trees being killed by carbon dioxide gas apparently emanating from magma (Figure 4.3b), and hot spring activity, cannot be ignored.

For the better monitored volcanoes it is now possible to make accurate short-term predictions of eruptions. But for many volcanoes, little or no information is available.

Volcanic activity in the Long Valley caldera occurred as recently as 250 years ago, and there is every reason to think it will occur again, but when renewed activity will take place remains an unanswered question. Unfortunately, the local populace was largely unaware of the geologic history of the region, the USGS did a poor job in communicating its concerns, and premature news releases caused more concern than was justified. In any case, local residents where outraged because the warnings caused a decrease in tourism (Mammoth Mountain on the margins of the caldera is the second largest ski

DISTRIBUTION OF VOLCANOES

Most of the world's active volcanoes are in well-defined zones or belts rather than being randomly distributed. The **circum-Pacific belt,** with more than 60% of all active volcanoes, includes those in the Andes of South America; the volcanoes of Central America, Mexico, and the Cascade Range of North America; as well as the Alaskan volcanoes and those in Japan, the Philippines, Indonesia, and New Zealand (■ Figure 4.22). Also included in

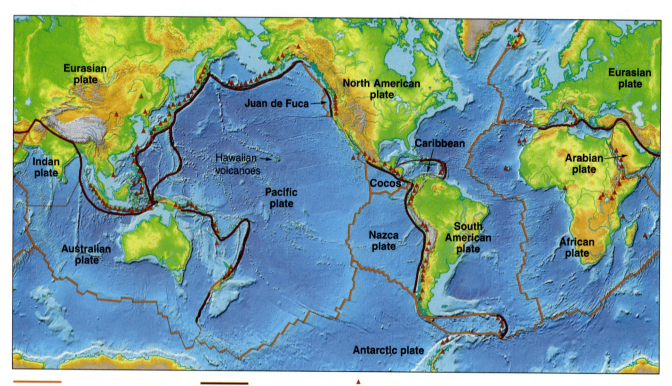

Divergent plate boundary
(some transform plate boundaries)   Convergent boundary   Volcano

■ **Figure 4.22**

Most volcanoes are at or near convergent and divergent plate boundaries. The two major volcano belts are the circum-Pacific belt, with about 60% of all active volcanoes, and the Mediterranean belt, with 20% of active volcanoes. Most of the rest are near mid-oceanic ridges.

the circum-Pacific belt are the southernmost active volcanoes at Mount Erebus in Antarctica, and a large caldera at Deception Island that erupted most recently during 1970. In fact, this belt nearly encircling the Pacific Ocean basin is popularly called the Ring of Fire, alluding to the fact that it contains so many active volcanoes.

The second common area of active volcanism is the **Mediterranean belt** (Figure 4.22). About 20% of all active volcanism takes place in this belt, where the famous Italian volcanoes such as Mount Etna and Vesuvius and the Greek volcano Santorini are found. Mount Etna has issued lava flows 190 times since 1500 B.C., when activity was first recorded. A particularly violent eruption of Santorini in 1390 B.C. might be the basis for the myth regarding the lost continent of Atlantis (see Chapter 11), and in A.D. 79 an eruption of Vesuvius destroyed Pompeii and other nearby cities (see the Introduction).

Nearly all the remaining active volcanoes are at or near mid-oceanic ridges or the extensions of these ridges onto land (Figure 4.22). These include the East Pacific Rise and the longest of all mid-oceanic ridges, the Mid-Atlantic Ridge. The latter is located near the center of the Atlantic Ocean basin, accounting for the volcanism in Iceland and elsewhere. It continues around the southern tip of Africa where it connects with the Indian Ridge. Branches of the Indian Ridge extend into the Red Sea and East Africa, where such volcanoes as Kilamanjaro in Tanzania, Nyiragongo in Zaire (see Geo-Focus 4.1), and Erta Ale in Ethiopia with its continuously active lava lake are found.

Anyone familiar with volcanoes will have noticed that no mention has been made so far of the Hawaiian volcanoes. This is not an oversight; they are the notable exceptions to the distribution of volcanoes in well-defined belts. Their location and significance are discussed in the following section.

PLATE TECTONICS, VOLCANOES, AND PLUTONS

n Chapter 3 we discussed the origin and evolution of magma and concluded that (1) mafic magma is generated beneath spreading ridges and (2) intermediate and felsic magma forms where an oceanic plate is subducted beneath another oceanic plate or a continental plate. Accordingly, most of Earth's volcanism and emplacement of plutons take place at divergent and convergent plate boundaries.

Divergent Plate Boundaries and Igneous Activity

Much of the mafic magma that forms beneath spreading ridges is emplaced at depth as vertical dikes and gab-bro plutons. But some rises to the surface, where it forms submarine lava flows and pillow lava (Figure 4.9). Indeed, the oceanic crust is composed largely of gabbro and basalt. Much of this submarine volcanism goes undetected, but researchers in submersible craft have observed the results of these eruptions.

Pyroclastic materials are not common in this environment because mafic lava is very fluid, allowing gases to escape easily, and at great depth, water pressure prevents gases from expanding. Accordingly, the explosive eruptions that yield pyroclastic materials are not common. Should an eruptive center along a ridge build above sea level, however, pyroclastic materials may be erupted at lava fountains, but most of the magma issues forth as fluid lava flows that form shield volcanoes.

Excellent examples of divergent plate boundary volcanism are found along the Mid-Atlantic Ridge, particularly where it takes place above sea level as in Iceland (Figure 4.22). In November 1963 a new volcanic island, later name Surtsey, rose from the sea just south of Iceland. The East Pacific Rise and the Indian Ridge are areas of similar volcanism. Not all divergent plate boundaries are beneath sea level as in the previous examples. For instance, divergence and igneous activity are taking place in Africa at the East African Rift system (Figure 4.22).

PHYSICAL Geology Now Click Geology Interactive to work through an activity on Distribution of Volcanism.

Igneous Activity at Convergent Plate Boundaries

Nearly all of the large active volcanoes in both the circum-Pacific and Mediterranean belts are composite volcanoes near the leading edges of overriding plates at convergent plate boundaries (Figure 4.22). The overriding plate, with its chain of volcanoes, may be oceanic as in the case of the Aleutian Islands, or it may be continental as is, for instance, the South American plate with its chain of volcanoes along its western edge.

As previously noted, these volcanoes at convergent plate boundaries consist largely of lava flows and pyroclastic materials of intermediate to felsic composition. Remember that when mafic oceanic crust partially melts, some of the magma generated is emplaced near plate boundaries as plutons and some is erupted to build up composite volcanoes. More viscous magmas, usually of felsic composition, are emplaced as lava domes, thus accounting for the explosive eruptions that typically occur at convergent plate boundaries.

In previous sections, we alluded to several eruptions at convergent plate boundaries. Good examples are the explosive eruptions of Mount Pinatubo and Mayon volcano in the Philippines, both of which are situated near a plate boundary beneath which an oceanic plate is subducted. Mount St. Helens, Washington, is similarly situated, but it is on a continental rather than an oceanic plate. Mount Vesuvius in Italy, one of several active volcanoes in that region, lies on a plate that the northern margin of the African plate is subducted beneath.

Intraplate Volcanism

Mauna Loa and Kilauea on the island of Hawaii and Loihi just 32 km to the south are within the interior of a rigid plate far from any divergent or convergent plate boundary (Figure 4.22). The magma is derived from the upper mantle, as it is at spreading ridges, and accordingly is mafic so it builds up shield volcanoes. Loihi is particularly interesting because it represents an early stage in the origin of a new Hawaiian island. It is a submarine volcano that rises more than 3000 m above the adjacent seafloor, but its summit is still about 940 m below sea level.

Even though the Hawaiian volcanoes are not at a spreading ridge near a subduction zone, their evolution is related to plate movements. Notice in Figure 12.24 that the ages of the rocks that make up the various Hawaiian Islands increase toward the northwest; Kauai formed 3.8 to 5.6 million years ago, whereas Hawaii began forming less than 1 million years ago, and Loihi began to form even more recently. Continuous movement of the Pacific plate over the hot spot, now beneath Hawaii and Loihi, has formed the islands in succession.

4 REVIEW
WORKBOOK

Chapter Summary

- Volcanism encompasses those processes by which magma rises to the surface as lava flows and pyroclastic materials, and associated gases are released into the atmosphere.

- Only a few percent by weight of magma consists of gases, most of which is water vapor, but sulfur gases may have far-reaching climatic effects.

- Aa lava flows have surfaces of jagged, angular blocks, whereas the surfaces of pahoehoe flows are smoothly wrinkled.

- Several other features of lava flows are spatter cones, pressure ridges, lava tubes, and columnar joints. Lava erupted underwater typically forms bulbous masses known as pillow lava.

- Volcanoes are found in various shapes and sizes, but all form where lava and pyroclastic materials are erupted from a vent.

- Shield volcanoes have low, rounded profiles and are composed mostly of mafic flows that cool and form basalt. Small, steep-sided cinder cones form around a vent where pyroclastic materials are erupted and accumulate. Composite volcanoes are made up of lava flows and pyroclastic materials of intermediate composition and volcanic mudflows.

- Viscous bulbous masses of lava, generally of felsic composition, form lava domes, which are dangerous because they erupt explosively.

- The summits of volcanoes have either a crater or a much larger caldera. Most calderas form following voluminous eruptions and the volcanic peak collapses into a partially drained magma chamber.

- Fluid mafic lava from fissure eruptions spreads over large areas to form a basalt plateau.

- Pyroclastic sheet deposits result when huge eruptions of ash and other pyroclastic materials take place, particularly when calderas form.

- Geologists have devised a volcanic explosivity index (VEI) to give a semiquantitative measure of the size of an eruption. Volume of material erupted and height of an eruption plume are criteria used to determine VEI; fatalities and property damage are not considered.

- To effectively monitor volcanoes, geologists evaluate several physical and chemical aspects of volcanic regions. Of particular importance in monitoring volcanoes and forecasting eruptions is detecting volcanic tremor and determining the eruptive history of a volcano.

- About 80% of all volcanic eruptions take place in the circum-Pacific belt and the Mediterranean belt, mostly at convergent plate boundaries. Most of the rest of the eruptions occur along mid-oceanic ridges or their extensions onto land.

- The two active volcanoes on the island of Hawaii and one just to the south apparently lie above a hot spot over which the Pacific plate moves.

Important Terms

aa (p. 93)
ash (p. 97)
basalt plateau (p. 107)
caldera (p. 98)
Cascade Range (p. 106)
cinder cone (p. 100)
circum-Pacific belt (p. 114)
columnar joint (p. 93)
composite volcano
  (stratovolcano) (p. 102)

crater (p. 98)
fissure eruption (p. 107)
lahar (p. 102)
lava dome (p. 104)
lava tube (p. 91)
Mediterranean belt (p. 115)
nuée ardente (p. 105)
pahoehoe (p. 93)
pillow lava (p. 96)

pyroclastic sheet deposit (p. 110)
shield volcano (p. 99)
volcanic explosivity index (VEI)
  (p. 111)
volcanic tremor (p. 112)
volcanism (p. 90)
volcano (p. 97)

Review Questions

1. One of the warning signs of an impending volcanic eruption is volcanic tremor, which is:
   a. _____ inflation of a volcano as magma rises; b. _____ changes in groundwater temperature; c. _____ ground shaking lasting for minutes or hours; d. _____ cooling and shrinkage in lava to form columnar joints; e. _____ eruptions of fluid lava from long fissures.

2. Pillow lava forms when:
   a. _____ pyroclastic materials accumulate in thick layers; b. _____ lava erupts underwater; c. _____ globs of lava stick together on a lava flow's surface; d. _____ pressure within a flow causes buckling; e. _____ a volcano's summit collapses.

3. An incandescent cloud of gas and particles erupted from a volcano is a:
   a. _____ nuée ardente; b. _____ pahoehoe; c. _____ spatter cone; d. _____ lapilli; e. _____ caldera.

4. The fact that _____ have slopes less than 10 degrees is because they are composed of low-viscosity lava flows.
   a. _____ lava tubes; b. _____ pyroclastic sheet deposits; c. _____ basalt plateaus; d. _____ volcanic bombs; e. _____ shield volcanoes.

5. Basalt plateaus form as a result of:
   a. _____ repeated eruptions of felsic lava; b. _____ erosion of composite volcanoes; c. _____ inflation of a volcano as magma rises; d. _____ eruptions of fluid lava from fissures; e. _____ volcanic mudflows on cinder cones.

6. Most active volcanoes are found in the:
   a. _____ mid-oceanic ridge volcanic zone; b. _____ Sierra Nevada–Cascade volcanic province; c. _____ circum-Pacific belt; d. _____ eastern Atlantic subduction zone; e. _____ Mediterranean divergent boundary.

7. The summits of some volcanoes have very wide, steep-sided depressions known as _____, most of which are formed by _____.
   a. _____ explosion pits/fissure eruptions; b. _____ calderas/summit collapse; c. _____ lava domes/forceful injection; d. _____ basalt plateaus/eruptions of cinders; e. _____ parsitic cones/lava flows.

8. Pahoehoe is a kind of lava flow with a:
   a. _____ large component of pyroclastic materials; b. _____ lava tube; c. _____ smooth ropy surface; d. _____ mass of interconnected pillows; e. _____ fracture pattern outlining polygons.

9. Volcanoes emit several gases, but the most common one is:

   a. _____ water vapor; b. _____ carbon dioxide; c. _____ hydrogen sulfide; d. _____ methane; e. _____ chlorine.

10. An area of active volcanism in the Pacific Northwest of the United States is the:

    a. _____ Appalachian Mountains; b. _____ Cascade Range; c. _____ southern Rocky Mountains; d. _____ Marathon Mountains; e. _____ Teton Range.

11. Suppose you find rocks on land consisting of layers of pillow lava overlain by deep-sea sedimentary rocks. Where did the pillow lava originate, and what type of rock would you expect to lie beneath the pillow lava?

12. Why are most composite volcanoes at convergent plate boundaries, whereas most shield volcanoes are at or near divergent plate boundaries?

13. Explain how a caldera forms. Where would you go to see an example of a caldera?

14. What kinds of information do geologists evaluate when they monitor volcanoes and warn of imminent eruptions?

15. How do columnar joints and spatter cones form? Where are good places to see each?

16. What geologic events would have to occur for a chain of volcanoes to form along the east coast of the United States and Canada?

17. What is a lava dome, and why are lava dome eruptions dangerous?

18. How do aa and pahoehoe lava flows differ? What accounts for their differences?

19. How does a crater differ from a caldera? Also, how does each form?

20. Explain why eruptions of mafic lava are rather quiet, whereas eruptions of felsic lava are commonly explosive.

World Wide Web Activities

PHYSICAL **Geology⇌Now**   Assess your understanding of this chapter's topics with additional quizzing and comprehensive interactivities at

http://earthscience.brookscole.com/physgeo5e

as well as current and up-to-date weblinks, additional readings, and InfoTrac College Edition exercises.

Weathering, Erosion, and Soil

CHAPTER 5
OUTLINE

PHYSICAL Geology⇌Now *This icon, appearing throughout the book, indicates an opportunity to explore interactive tutorials, animations, or practice problems available on the Physical GeologyNow Web site at http://earthscience, brookscole.com/physgeo5e.*

OBJECTIVES
At the end of this chapter, you will have learned that

- Weathering causes physical and chemical changes in Earth materials, whereas erosion removes materials from the weathering site.

- Mechanical weathering processes, such as water freezing and thawing in cracks and the activities of organisms, break rocks into smaller pieces but do not change their composition.

- Chemical weathering processes include solution and oxidation and bring about a change in the composition of rocks and minerals.

- The rate at which chemical weathering proceeds depends on climate, particle size, and the activities of organisms.

- Weathering, both mechanical and chemical, is responsible for the origin of soils, and it yields the materials that make up sedimentary rocks.

- The type of soil that develops and its fertility depend on several factors, especially climate.

- Soil erosion and physical and chemical deterioration of soils cause many problems, including decreases in productivity.

- Mechanical and chemical weathering processes are important in the origin or concentrations of some natural resources.

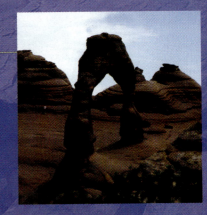

Weathering and erosion along parallel fractures in sedimentary rocks have yielded the arches and other features in Arches National Park, Utah. Delicate Arch shown here measures 9.7 m across and 14 m high.
Source: James S. Monroe

Introduction

Weathering, defined as the physical breakdown (disintegration) and chemical alteration (decomposition) of rocks and minerals at or near the surface, is such a pervasive phenomenon that many people completely overlook it. Nevertheless, it takes place continuously at variable rates on all Earth materials, including those used in construction, and on all rocklike substances, such as materials used for roadways as well as concrete in sidewalks, foundations, and bridges (■ Figure 5.1).

Actually, weathering is a group of physical and chemical processes that alter Earth materials so that they are more nearly in equilibrium with a new set of environmental conditions. For instance, many rocks form within the crust, where pressures and temperatures are high and where little or no water or oxygen is present. These same rocks at the surface are exposed to low pressures and temperatures, atmospheric gases, water, acids, and the activities of organisms. In short, interactions of Earth materials with the atmosphere, hydrosphere, and biosphere bring about changes.

There are good reasons to study these phenomena that alter Earth materials. First, the **parent material**—that is, the material being weathered—is disaggregated to form smaller particles, and some of the parent material's constituent minerals are dissolved and removed from the weathering site. This removal of weathered materials, known as **erosion,** is accomplished by gravity, running water, glaciers, wind, and waves on shorelines, and the same process may then **transport** the eroded materials some distance and deposit it as sediment, the raw materials for *sedimentary rocks* (see Chapter 6). In short, weathering, erosion, transport, and deposition are essential parts of the rock cycle (see Figure 1.14).

A second reason to study the phenomenon of weathering is that some of the weathered material is further altered to form *soil*. So weathering provides the raw materials for sedimentary rocks and for soil. A third reason to study weathering is that it is responsible for the origin of some natural resources

(a)

(b)

■ **Figure 5.1**

(a) Heavily weathered granitic rock. Most of the exposed rocks have been so thoroughly weathered that only a few rounded masses of rock appear unaltered. The small cones in the right and left foreground consist of individual mineral grains and granitic rock fragments. (b) Closeup view of some of the weathered material. These particles retain their original composition, so mechanical weathering processes have predominated.

James S. Monroe

(a)

(b)

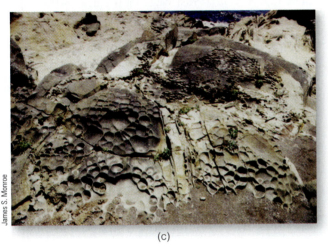

(c)

■ **Figure 5.2**

The effects of differential weathering and erosion. (a) These spires and pillars in Bryce Canyon National Park in Utah are known as hoodoos. (b) Differential weathering and erosion of limestone in Spain. (c) Honeycomb weathering of rocks at Pebble Beach, California. In coastal areas this kind of weathering results from dissagregation of granular rocks. The partitions between cavities are protected by coatings of microscopic algae.

such as aluminum ore, and it enriches others by removing soluble materials. In the several chapters on surface processes, such as running water, glaciers, and wind, we will have much more to say about erosion, transport, and deposition.

Several geologic processes such as volcanism, deformation, and glaciation have yielded many areas of exceptional scenery, but so have weathering and erosion. We marvel at the intricately sculpted landscape at Bryce Canyon National Park, Utah (■ Figure 5.2a); the rugged shoreline at Acadia National Park, Maine; and the badlands at Dinosaur Provincial Park in Alberta, Canada. Weathering and erosion of fractured rocks in Arches National Park, Utah, have yielded a landscape of isolated spires and balanced rocks as well as the famous arches for which the park was named (see the chapter opening photo).

We have mentioned that weathering is responsible for our productive soils, but in some areas erosion takes place faster than soil-forming processes operate, thus accounting for soil losses and decreased agricultural productivity. Certainly some soil losses are natural as ongoing evolution of the land takes place, but human activities have certainly compounded the problem in many areas.

HOW ARE EARTH MATERIALS ALTERED?

Of course, weathering is a surface or near-surface process, but the rocks it acts on are not structurally and compositionally homogeneous throughout, accounting for **differential weathering.** That is, weathering takes place at different rates even in the same area, which commonly results in uneven surfaces. Differential weathering and *differential erosion*—that is, variable rates of erosion—combine to yield some unusual and even bizarre features, such as hoodoos, arches, and pockmarked surfaces (Figure 5.2).

The two recognized types of weathering, *mechanical* and *chemical*, both proceed simultaneously on parent material as well as on materials in transport and those deposited as sediment. In short, all surface or near-surface materials weather, although one type of weathering may predominate over the other depending on such variables as climate and rock type. We will discuss mechanical and chemical weathering separately in the following sections only for convenience.

MECHANICAL WEATHERING— DISAGGREGATION OF EARTH MATERIALS

Mechanical weathering takes place when physical forces break Earth materials into smaller pieces that retain the chemical composition of the parent material. Granite, for instance, may weather mechanically to produce smaller pieces of granite, or disintegration liberates individual minerals (Figure 5.1). Thus mechanically weathered granite yields granite rock fragments and mineral fragments of quartz, potassium feldspars, plagioclase feldspars, and several other minerals in lesser quantities, all of which have the same chemical composition as they did in the parent material. The physical processes responsible for mechanical weathering include frost action, pressure release, thermal expansion and contraction, salt crystal growth, and the activities of organisms.

Frost Action

Frost action involving repeated freezing and thawing of water in cracks and pores in rocks is particularly effective where temperatures commonly fluctuate above and below freezing. In the high mountains of the western United States and Canada, frost action is effective even during summer months. As one would expect, it is of little or no importance in the tropics or where water is permanently frozen.

When water seeps into a crack and freezes, it expands by about 9% and exerts great force on the walls of the crack, thereby widening and extending it by **frost wedging.** As a result of repeated freezing and thawing, pieces of rock eventually detach from the parent material (■ Figure 5.3a). Frost wedging is particularly effective if the crack is convoluted.

The debris produced by frost wedging and other weathering processes in mountains commonly accumulates as large cones of **talus** lying at the bases of slopes (Figure 5.3b). The debris forming talus is simply angular pieces of rock from a larger body that has been mechanically and to a lesser degree chemically weathered. Most rocks have a system of fractures called *joints* along which frost action is particularly effective. Water seeps along the joint surfaces and eventually wedges pieces of

(a)

(b)

James S. Monroe

■ **Figure 5.3**

(a) Frost wedging takes place when water seeps into cracks and expands as it freezes. Angular pieces of rock are pried loose by repeated freezing and thawing. (b) Accumulation of talus (foreground) at the base of a slope. The parent material is highly fractured and quite susceptible to frost wedging, although other weathering processes also help break the rock into smaller pieces.

rock loose, which then tumble downslope to accumulate with other loosened rocks.

In the phenomenon known as **frost heaving,** a mass of sediment or soil undergoes freezing, expansion, and actual lifting, followed by thawing, contraction, and lowering of the mass. Frost heaving is particularly evident where water freezes beneath roadways and sidewalks.

Pressure Release

The mechanical weathering process called **pressure release** is especially evident in rocks that formed as deeply buried intrusive bodies such as batholiths, but it occurs in other types of rocks as well. When a batholith forms, magma crystallizes under tremendous pressure (the weight of the overlying rock) and the resulting rock is stable under these pressure conditions. If the batholith is uplifted and the overlying rock

eroded, the pressure is reduced. The now-exposed rock contains energy that is released by outward expansion and the formation of **sheet joints,** large fractures that more or less parallel the rock surface (■ Figure 5.4a). Slabs of rock bounded by sheet joints may slip, slide, or spall (break) off of the host rock—a process known as **exfoliation**—and accumulate as talus. Large rounded domes of rock resulting from this process, called **exfoliation domes,** are found in Yosemite National Park, California, and at Stone Mountain, Georgia (Figure 5.4b, c).

That solid rock expands and produces fractures is counterintuitive but is nevertheless a well-known phenomenon. In deep mines, masses of rock suddenly detach from the sides of the excavation, often with explosive violence. Spectacular examples of these *rock bursts* take place in deep mines, where they and the related but less violent phenomenon called *popping* pose a danger to mine workers. In South Africa about 20 miners are killed by rock bursts every year.

(a)

(b)

(c)

Sue Monroe

James S. Monroe

Mark Gibson/Visuals Unlimited

■ **Figure 5.4**

(a) Slabs of granitic rock bounded by sheet joints in the Sierra Nevada of California. Notice that these slabs are inclined down toward the roadway visible in the lower part of the image. (b) A small exfoliation dome in the Sierra Nevada near Donner Pass in California. (c) Stone Mountain is a large exfoliation dome in Georgia.

■ **Figure 5.5**

Sheet joint formed by expansion in the Mount Airy Granite in North Carolina. The hammer is about 30 cm long.

Courtesy of W. D. Lowry

ficient to overcome the internal strength of a rock? Experiments in which rocks were heated and cooled repeatedly to simulate years of such activity indicate that thermal expansion and contraction is not an important agent of mechanical weathering.[†] Despite these experimental results, some rocks in deserts do indeed appear to show the effects of this process.

Daily temperature variation is the most common cause of alternating expansion and contraction, but these changes take place over periods of hours. In contrast, fire causes very rapid expansion. During a forest fire, rocks may heat very rapidly, especially near the surface, because they conduct heat so poorly. The heated surface layer expands more rapidly than the interior, and thin sheets paralleling the rock surface become detached.

In some quarrying operations,[*] the removal of surface materials to a depth of only 7 or 8m has led to the formation of sheet joints in the underlying rock (■ Figure 5.5). At quarries in Vermont and Tennessee, the excavation of marble exposed rocks that were formerly buried and under great pressure. When the overlying rock was removed, the marble expanded and sheet joints formed. Some slabs of rock bounded by sheet joints burst so violently that quarrying machines weighing more than a ton were thrown from their tracks, and some quarries had to be abandoned because fracturing rendered the stone useless.

Thermal Expansion and Contraction

During **thermal expansion and contraction,** the volume of rocks changes in response to heating and cooling. In a desert, where the temperature may vary as much as 30°C in one day, rocks expand when heated and contract as they cool. Rock is a poor conductor of heat, so its outside heats up more than its inside; the surface expands more than the interior, producing stresses that may cause fracturing. Furthermore, dark minerals absorb heat faster than light-colored ones, so differential expansion occurs even between the minerals in some rocks.

Repeated thermal expansion and contraction is a common phenomenon, but are the forces generated suf-

Growth of Salt Crystals

Under some circumstances, salt crystals that form from solution cause disaggregation of rocks. Growing crystals exert enough force to widen cracks and crevices or dislodge particles in porous, granular rocks such as sandstone. Even in crystalline rocks such as granite, **salt crystal growth** may pry loose individual minerals. To the extent that salt crystal growth produces forces that expand openings in rocks, it is similar to frost wedging. Most salt crystal growth occurs in hot, arid areas, although it probably affects rocks in some coastal regions as well.

Organisms

Animals, plants, and bacteria all participate in the mechanical and chemical alteration of rocks. Burrowing animals, such as worms, termites, reptiles, rodents, and many others, constantly mix soil and sediment particles and bring material from depth to the surface where further weathering occurs. Even materials ingested by worms are further reduced in size, and animal burrows give gases and water easier access to greater depths. The roots of plants, especially large bushes and trees, wedge themselves into cracks in rocks and further widen them (■ Figure 5.6a). Tree roots growing under or through sidewalks and foundations do considerable damage.

---

*A *quarry* is a surface excavation, generally for the extraction of building stone.

[†]Thermal expansion and contraction may be more significant on the Moon, where extreme temperature changes occur quickly.

Sue Monroe

(a)

Sue Monroe

(b)

■ **Figure 5.6**

Organisms and weathering. (a) This tree near Anchorage, Alaska, is growing in a crack in the rocks and thus contributes to mechanical weathering. (b) The irregular orange masses on these rocks on a small island in the Irish Sea are lichens (composite organisms of fungi and algae). Lichens derive their nutrients from the rock and contribute to chemical weathering.

CHEMICAL WEATHERING—DECOMPOSITION OF EARTH MATERIALS

Chemical weathering includes those processes by which rocks and minerals are decomposed by alteration of parent material. In contrast to mechanical weathering, chemical weathering results in a change in the composition of weathered materials. For example, several clay minerals (sheet silicates) form by the chemical and structural alteration of other minerals such as potassium feldspars and plagioclase feldspars, both of which are framework silicates. Other minerals are completely decomposed during chemical weathering as their ions are taken into solution, but some chemically stable minerals are simply liberated from the parent material.

Important agents of chemical weathering include atmospheric gases, especially oxygen, water, and acids.

Organisms also play an important role. Rocks with lichens (composite organisms consisting of fungi and algae) on their surfaces undergo more extensive chemical alteration than lichen-free rocks (Figure 5.6b). In addition, plants remove ions from soil water and reduce the chemical stability of soil minerals, and plant roots release organic acids. Other chemical weathering processes include solution, oxidation, and hydrolysis.

Solution

During **solution** the ions of a substance separate in a liquid and the solid substance dissolves. Water is a remarkable solvent because its molecules have an asymmetric shape, consisting of one oxygen atom with two hydrogen atoms arranged so that the angle between the two hydrogens is about 104 degrees (■ Figure 5.7a). Because of this asymmetry, the oxygen end of the molecule retains a slight negative electrical charge, whereas the hydrogen end retains a slight positive charge. When a soluble substance such as the mineral halite ($NaCl$) comes in contact with a water molecule, the positively

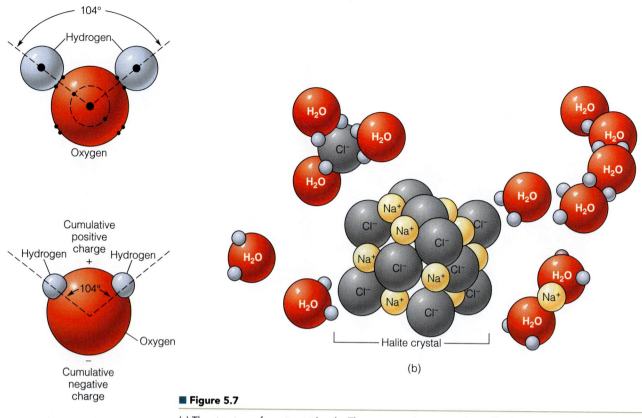

■ Figure 5.7

(a) The structure of a water molecule. The asymmetric arrangement of hydrogen atoms causes the molecule to have a slight positive electrical charge at its hydrogen end and a slight negative charge at its oxygen end. (b) Solution of sodium chloride (NaCl), the mineral halite, in water. Note that the sodium atoms are attracted to the oxygen end of a water molecule, whereas chloride ions are attracted to the hydrogen end of the molecule.

charged sodium ions are attracted to the negative end of the water molecule, and the negatively charged chloride ions are attracted to the positively charged end of the water molecule (Figure 5.7b). Thus ions are liberated from the crystal structure, and the solid dissolves.

Most minerals are not very soluble in pure water because the attractive forces of water molecules are not sufficient to overcome the forces between particles in minerals. The mineral calcite ($CaCO_3$), the major constituent of the sedimentary rock limestone and the metamorphic rock marble, is practically insoluble in pure water, but it rapidly dissolves if a small amount of acid is present. An easy way to make water acidic is by dissociating the ions of carbonic acid. That is, the ions become separated as carbonic acid breaks down into other substances.

$$H_2O \;+\; CO_2 \;\rightleftharpoons\; H_2CO_3 \;\rightleftharpoons\; H^+ \;+\; HCO_3{}^-$$

WATER   CARBON   CARBONIC   HYDROGEN   BICARBONATE
         DIOXIDE   ACID   ION   ION

According to this chemical equation, water and carbon dioxide combine to form *carbonic acid,* a small amount of which dissociates to yield hydrogen and bicarbonate ions. The concentration of hydrogen ions determines the acidity of a solution; the more hydrogen ions present, the stronger the acid.

Carbon dioxide from several sources combines with water and reacts to form acid solutions. The atmosphere is mostly nitrogen and oxygen, but about 0.03% is carbon dioxide, causing rain to be slightly acidic. Human activities have added gases to the atmosphere that contribute to the problem of acid rain (see Geo-Focus 5.1). Decaying organic matter and the respiration of organisms produce carbon dioxide in soils, so groundwater in humid areas is slightly acidic. Arid regions have sparse vegetation and few soil organisms, so groundwater has a limited supply of carbon dioxide and tends to be alkaline rather than acidic—that is, it has a low concentration of hydrogen ions.

Whatever the source of carbon dioxide, once an acidic solution is present, calcite rapidly dissolves according to the following reaction:

$$CaCO_3 + H_2O + CO_2 \;\rightleftharpoons\; Ca^{+2} + 2HCO_3{}^-$$

CALCITE   WATER   CARBON   CALCIUM   BICARBONATE
                      DIOXIDE   ION   ION

The dissolution of the calcite in limestone and marble has had dramatic effects in many places, ranging from small cavities to large caverns such as Mammoth Cave in Kentucky and Carlsbad Caverns in New Mexico (see Chapter 16). The varying solubility of limestone and other rocks with the carbonate radical ($CO_3{}^{-2}$) such as dolostone is

■ **Figure 5.8**

Exposure of the Bighorn Dolomite in Wyoming. Limestone and dolostone are resistant to chemical weathering in the semiarid to arid West where groundwater is alkaline. The same rocks in more humid climates tend to have more subdued exposures.

well demonstrated by the bold cliffs they form in semiarid and arid regions where groundwater is alkaline (■ Figure 5.8). Subdued exposures of these rocks are more common in humid regions where groundwater is acidic.

Oxidation

The term **oxidation** has a variety of meanings for chemists, but in chemical weathering it refers to reactions with oxygen to form an oxide (one or more metallic elements combined with oxygen) or, if water is present, a hydroxide (a metallic element or radical combined with OH). For example, iron rusts when it combines with oxygen and forms the iron oxide hematite:

$$4Fe + 3O_2 \rightarrow 2Fe_2O_3$$

IRON   OXYGEN   IRON OXIDE
(HEMATITE)

Of course, atmospheric oxygen is abundantly available for oxidation reactions, but oxidation is generally a slow process unless water is present. Most oxidation is carried out by oxygen dissolved in water.

Oxidation is important in the alteration of ferromagnesian silicates such as olivine, pyroxenes, amphiboles, and biotite. Iron in these minerals combines with oxygen to form the reddish iron oxide hematite (Fe_2O_3) or the yellowish or brown hydroxide limonite [$FeO(OH) \cdot nH_2O$]. The yellow, brown, and red colors of many soils and sedimentary rocks result from the presence of small amounts of hematite or limonite (see Chapter 6).

The oxidation of iron sulfides such as the mineral pyrite (FeS_2) is of particular concern in some areas. Pyrite is commonly associated with coal, so in mine tailings* pyrite oxidizes to form sulfuric acid (H_2SO_4) and

---

Tailings are the rock debris of mining; they are considered too poor for further processing and are left as heaps on the surface.

iron oxide. Acid soils and waters in coal-mining areas are produced in this manner and present a serious environmental hazard (■ Figure 5.9).

■ **Figure 5.9**

The oxidation of pyrite in mine tailings forms acid water, as in this small stream. More than 11,000 km of U.S. streams, mostly in the Appalachian region, are contaminated by abandoned coal mines that leak sulfuric acid.

GEOFOCUS 5.1

Industrialization and Acid Rain

One result of industrialization is atmospheric pollution, which causes smog, possible disruption of the ozone layer, global warming, and acid rain. Acidity, a measure of hydrogen ion concentration, is measured on the pH scale (■ Figure 1a). A pH value of 7 is neutral, whereas acidic conditions correspond to values less than 7, and values greater than 7 denote alkaline, or basic, conditions. Normal rain has a pH value of about 5.6, making it slightly acidic, but acid rain has a pH of less than 5.0. In addition to acid rain, some areas experience acid snow, and even acid fog with a pH as low as 1.7 occurs in some industrialized areas.

Several natural processes, including soil bacteria metabolism and volcanism, release gases into the atmosphere that contribute to acid rain. Human activities also produce added atmospheric stress, especially burning fossil fuels that release carbon dioxide and nitrogen oxide from internal combustion engines. Both of these gases add to acid rain, but the greatest culprit is sulfur dioxide released mostly by burning coal that contains sulfur that oxidizes to form sulfur dioxide (SO_2):

$$S \text{ (in coal)} + O_2 \text{ (gas)} \rightarrow SO_2 \text{ (gas)}$$

As sulfur dioxide rises into the atmosphere from coal-burning power plants, it reacts with oxygen and forms sulfur trioxide:

$$2SO_2 \text{ (gas)} + O_2 \text{ (gas)} \rightarrow 2SO_3 \text{ (gas)}$$

And finally, sulfur trioxide reacts with water droplets in the atmosphere and forms sulfuric acid (H_2SO_4), the main component of acid rain:

$$SO_3 \text{ (gas)} + H_2O \text{ (liquid)} \rightarrow H_2SO_4$$

Robert Angus Smith first recognized acid rain in England in 1872, but not until 1961 did it become an environmental concern when it was realized that acid rain is corrosive and irritating, kills vegetation, and has a detrimental effect on surface waters. Since then, the effects of acid rain are apparent in Europe (especially in eastern Europe where so much coal is burned) and the eastern part of North America, where the problem has been getting worse for the last three decades (Figure 1b). In recent years, several of the developed countries have reduced their emissions at coal-burning facilities, but many developing nations have increased theirs.

The areas affected by acid rain invariably lie downwind from plants that emit sulfur gases, but the effects of acid rain in these areas may be modified by the local geology. For instance, if the area is underlain by limestone or alkaline soils, acid rain tends to be neutralized, but granite has little or no modifying effect. Small lakes lose their ability to neutralize acid rain and become more and more acidic until various types of organisms disappear, and in some cases all life-forms eventually die.

Acid rain also causes increased chemical weathering of limestone and marble and, to a lesser degree, sandstone. The effects are especially evident on buildings, monuments, and tombstones as in Gettysburg National Military Park in Pennsylvania, which lies in an area that receives some of the most acidic rain in the country.

The devastation caused by sulfur gases on vegetation near coal-burning plants is apparent, but some have questioned whether acid rain has much effect on forests and crops distant from these sources. Nevertheless, many forests in the eastern United States show signs of stress than cannot be attributed to other causes. In Germany's Black Forest, the needles of firs, spruce, and pines are turning yellow and falling off.

Millions of tons of sulfur dioxide are released yearly into the atmosphere in the United States. Power

Hydrolysis

Hydrolysis is the chemical reaction between the hydrogen (H^+) ions and hydroxyl (OH^-) ions of water and a mineral's ions. In hydrolysis, hydrogen ions actually replace positive ions in minerals, thus changing the composition of minerals and liberating soluble compounds and iron that then may be oxidized.

Consider the chemical alteration of feldspars by hydrolysis. Potassium feldspars such as orthoclase ($KAlSi_3O_8$) are common in many rocks, as are the plagioclase feldspars (which vary in composition from $CaAl_2Si_2O_8$ to $NaAlSi_3O_8$). All feldspars are framework silicates, but when altered, they yield materials in solution and clay minerals, such as kaolinite, which are sheet silicates.

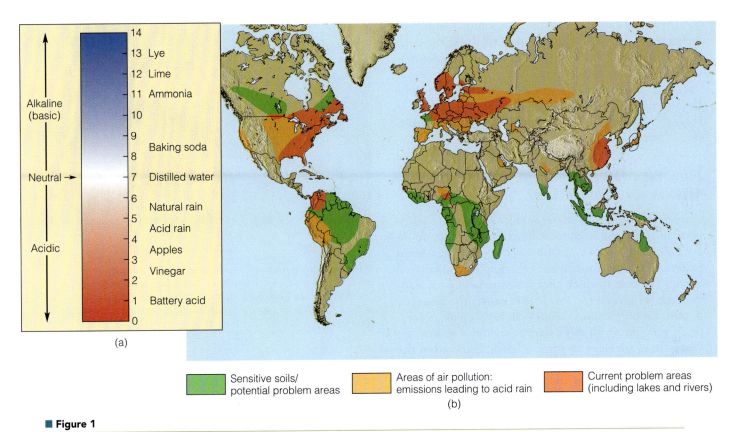

Figure 1

(a) Values less than 7 on the pH scale indicate acidic conditions, whereas those greater than 7 are alkaline. This is a logarithmic scale, so a decrease of one unit is a 10-fold increase in acidity. (b) Areas where acid rain is now a problem, and areas where the problem may develop.

plants built before 1975 have no emission controls, but the problems they pose must be addressed if emissions are to be reduced to an acceptable level. The most effective way to reduce emissions from these older plants is with flue-gas desulfurization, a process that removes up to 90% of the sulfur dioxide from exhaust gases.

Flue-gas desulfurization has some drawbacks. One is that some plants are simply too old to be profitably upgraded. Other problems include disposal of sulfur wastes, the lack of control on nitrogen gas emissions, and reduced efficiency of the power plant, which must burn more coal to make up the difference.

Other ways to control emissions are burning low-sulfur coal, fluidized bed combustion, and conservation of electricity. Natural gas contains practically no sulfur, but converting to this alternative energy source would require the installation of expensive new furnaces in existing plants.

Acid rain, like global warming, is a worldwide problem that knows no national boundaries. Wind may blow pollutants from the source in one country to another where the effects are felt. For instance, much of the acid rain in eastern Canada actually comes from sources in the United States.

Developed nations have the economic resources to reduce emissions, but many underdeveloped nations cannot afford to do so. Furthermore, many nations have access to only high-sulfur coal and cannot afford to install flue-gas desulfurization devices. Nevertheless, acid rain can be controlled only by the cooperation of all nations contributing to the problem.

The chemical weathering of potassium feldspar by hydrolysis occurs as follows:

$$2KAlSi_3O_8 + 2H^+ + 2HCO_3^- + H_2O \rightarrow$$

ORTHOCLASE — HYDROGEN ION — BICARBONATE ION — WATER

$$Al_2Si_2O_5(OH)_4 + 2K^+ + 2HCO_3^- + 4SiO_2$$

CLAY (KAOLINITE) — POTASSIUM ION — BICARBONATE ION — SILICA

In this reaction, hydrogen ions attack the ions in the orthoclase structure, and some liberated ions are incorporated in a developing clay mineral, while others simply go into solution. On the right side of the equation is excess silica that would not fit into the crystal structure of the clay mineral. This dissolved silica is an important source of cement that binds together the particles in some sedimentary rocks (see Chapter 6).

Plagioclase feldspars are also altered by hydrolysis, but they yield soluble calcium and sodium compounds rather than potassium compounds. Otherwise the reaction is the same. In fact, these dissolved compounds, particularly calcium compounds, are what make water hard. Calcium in water is a problem because it inhibits the reaction of detergents with dirt and precipitates as scaly mineral matter in water pipes and water heaters (see Chapter 16).

How Fast Does Chemical Weathering Take Place?

Chemical weathering operates on the surface of particles, so rocks and minerals alter from the outside inward. In fact, rocks commonly have a rind of weathered material near the surface but are completely unaltered inside. The rate at which chemical weathering proceeds depends on several factors. One is simply the presence or absence of fractures, because fluids seep along fractures, accounting for more intense chemical weathering along these surfaces (■ Figure 5.10). Thus, given the same rock type under similar conditions, the more fractures, the more rapid the chemical weathering. Of course, other factors also control the rate of chemical weathering, including particle size, climate, and parent material.

How Does Particle Size Affect the Rate of Chemical Weathering?

Because chemical weathering affects particle surfaces, the greater the surface area, the more effective the weathering. It is important to realize that small particles have larger surface areas compared to their volume than do large particles. Notice in ■ Figure 5.11 that a block measuring 1 m on a side has a total surface area of 6 m², but when the block is broken into particles measuring 0.5 m on a side, the total surface area increases to 12 m². And if these particles are all reduced to 0.25 m on a side, the total surface area increases to 24 m² but the total volume remains the same at 1 m³.

We can make two important statements regarding the block in Figure 5.11. First, as it is divided into a number of smaller blocks, its total surface area increases. Second, the smaller any single block is, the more surface area it has compared to its volume. We can conclude that mechanical weathering, which reduces the size of particles, contributes to chemical weathering by exposing more surface area.

Your own experiences with particle size verify our contention regarding surface area and volume. Because of its very small particle size, powdered sugar gives an intense burst of sweetness as the tiny pieces dissolve rapidly, but otherwise it is the same as the granular sugar we use on our cereal or in our coffee. As an experiment, see how rapidly crushed ice and an equal volume of block ice melt, or determine the time it takes to boil an entire potato as opposed to one cut into small pieces.

Climate and Chemical Weathering

Chemical processes proceed more rapidly at high temperatures and in the presence of liquids, so chemical weathering is more effective in the tropics than in arid and arctic regions because temperatures and rainfall are high and evaporation rates are low (■ Figure 5.12). In addition, vegetation and animal life are much more abundant in

Sue Monroe

■ **Figure 5.10**

Fluids seep along fractures in rocks where chemical weathering is more intense than it is in unfractured parts of the same rock. Notice too that a narrow white band stands out in relief near the left side of the image. This is a quartz vein that is more resistant to chemical weathering than its granitic host rock.

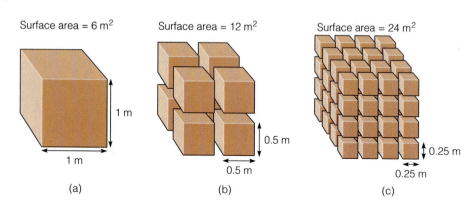

Surface area = 6 m²   Surface area = 12 m²   Surface area = 24 m²

1 m
1 m
(a)

0.5 m
0.5 m
(b)

0.25 m
0.25 m
(c)

■ **Figure 5.11**

Particle size and chemical weathering. As a rock is divided into smaller and smaller particles, its surface area increases but its volume remains the same. In (a) the surface area is 6 m², in (b) it is 12 m², and in (c) 24 m², but the volume remains the same at 1 m³. Small particles have more surface area compared to their volume than do large particles.

GEOLOGY
IN UNEXPECTED PLACES

Gravestones

Cemeteries might not seem like good places to study geology, but they provide an excellent opportunity to see the effects of weathering and other geologic phenomena. Any readily available stone can be used for headstones, but one of the most popular today is granite or similar rocks. ■ Figure 1 shows a small cemetery in Deerfield, Massachusetts, where some of the headstones date back to the 1600s.

The two headstones shown here are made of sandstone, but the lettering on one is clearly legible whereas nothing can be discerned on the other. What accounts for the differences is their age; the one on the right is older. Although you cannot see any detail on the gravestones in the background, some made of slate date from the 1660s and yet their inscriptions are easily read. Why? This slate simply does not weather as rapidly as the sandstone. Another observation is that some of the gravestones in the background are at odd angles—that is, tilted from their original vertical position. Assuming no vandalism, how would you account for this observation?

When you visit a cemetery, notice that some gravestones have growths or protrusions that in many cases obscure the inscriptions. These are lichens (composite organisms of fungi and algae) similar to those shown in Figure 5.6b. Such organisms derive their nutrients from the rock and thus contribute to chemical weathering.

James S. Monroe

■ **Figure 1**

Headstones in a cemetery in Deerfield, Massachusetts.

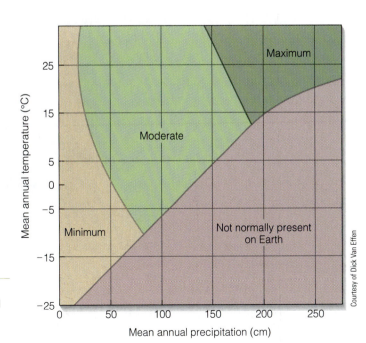

Courtesy of Dick Van Effen

■ **Figure 5.12**

Relationships of chemical weathering rates and climate. Chemical weathering is at a maximum where temperature and rainfall are high, and at a minimum in arid environments whether hot or cold.

Table 5.1

Stability of Silicate Minerals

	Ferromagnesian Silicates	Nonferromagnesian Silicates
Increasing Stability ↓	Olivine	Calcium plagioclase
	Pyroxene	
	Amphibole	Sodium plagioclase
	Biotite	Potassium feldspar
		Muscovite
		Quartz

the tropics. Consequently, the effects of weathering extend to depths of several tens of meters, but commonly extend only centimeters to a few meters deep in arid and arctic regions. You should realize, though, that chemical weathering goes on everywhere, except perhaps where earth materials are permanently frozen.

The Importance of Parent Material Some rocks are more resistant to chemical alteration than others and are not altered as rapidly. The metamorphic rock quartzite, composed of quartz (SiO_2), is an extremely stable substance that alters very slowly compared with most other rock types. In contrast, basalt, with its large amount of calcium-rich plagioclase and pyroxene minerals, decomposes rapidly because these minerals are chemically unstable. In fact, the stability of common minerals is just the opposite of their order of crystallization in Bowen's reaction series (Table 5.1). The minerals that form last in this series are chemically stable, whereas those that form early are more easily altered by chemical processes.

One manifestation of chemical weathering is **spheroidal weathering** (■ Figure 5.13). In spheroidal weathering, a stone, even a rectangular one, weathers into a more spherical shape because that is the most stable shape it can assume. The reason? On a rectangular stone, the corners are attacked by weathering processes from three sides, and the edges are attacked from two sides, but the flat surfaces weather more or less uniformly (Figure 5.13). Consequently, the corners and edges alter more rapidly, the material sloughs off them, and a more spherical shape develops. Once a spherical shape is present, all surfaces weather at the same rate.

The effects of spheroidal weathering are obvious in many rock bodies, particularly those that have a rectangular pattern of fractures (joints) similar to those in Figure 5.10. Fluids seep along the joint surfaces, resulting

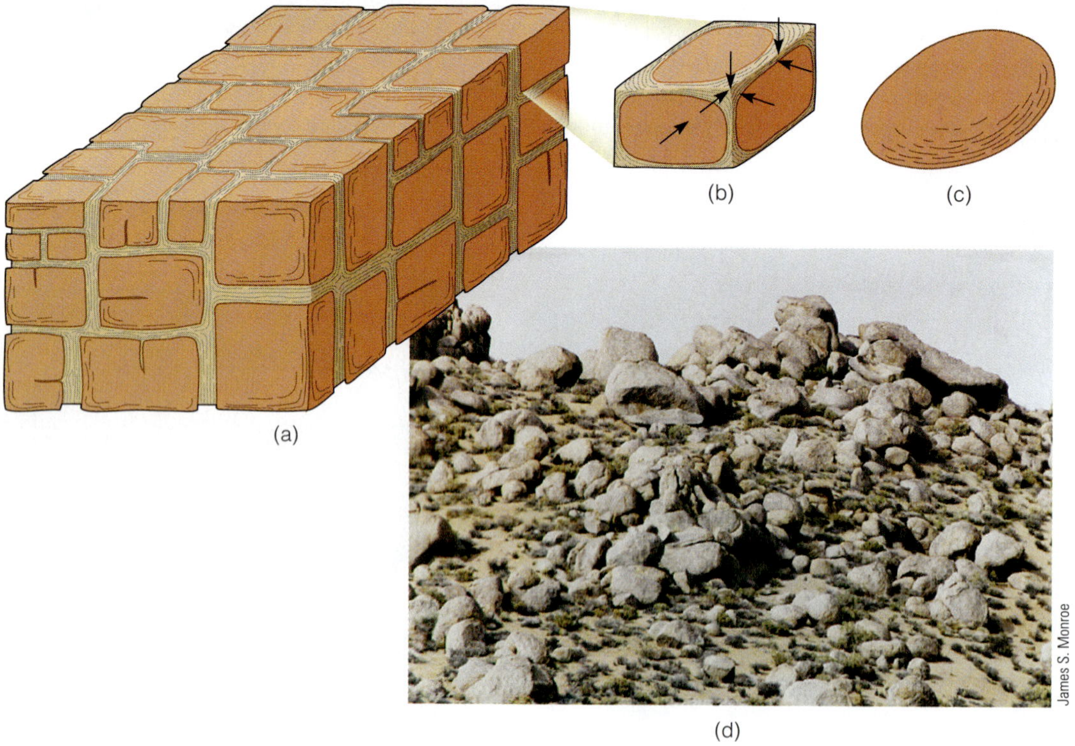
(a)
(b)
(c)
(d)
James S. Monroe

■ Figure 5.13

Spheroidal weathering. (a) The rectangular blocks outlined by fractures are attacked by chemical weathering processes, much like those in Figure 5.10, but (b) the corners and edges are weathered most rapidly. (c) When a block has weathered so that its shape is more nearly spherical, its surface is weathered evenly, and no further change in shape takes place. (d) Exposure of granitic rocks reduced to spherical boulders.

in more intense weathering at the edges and corners of the rectangular blocks, thus yielding more nearly spherical objects. Fractured granitic rocks are especially susceptible to spheroidal weathering (Figure 5.13d), but good examples can be found in all rock types.

SOIL—ONE PRODUCT OF WEATHERING

Most of Earth's land surface is covered by **regolith,** a collective term for sediment regardless of how it was deposited, as well as layers of pyroclastic materials and the residue formed in place by weathering. Regolith that we call **soil** consists of weathered materials, air, water, and organic matter and supports vegetation. Soil is an essential link between the parent material below and life above. Almost all land-dwelling organisms depend on soil for their existence. Plants grow in soil from which they derive their nutrients and most of their water, whereas land-dwelling animals depend directly or indirectly on plants for nutrients.

Only about 45% of a good soil for farming or gardening is composed of weathered material, mostly sand, silt, and clay. Much of the remaining 55% is simply void spaces filled with either air or water and a small but important amount of humus. **Humus** consists of carbon derived by bacterial decay of organic matter and is highly resistant to further decay. Even a fertile soil might contain as little as 5% humus, but humus is nevertheless an important source of plant nutrients and it enhances moisture retention. Furthermore, it gives the upper layers of many soils their dark color; soils with little humus are lighter colored and not as productive.

Some weathered materials in soils are simply sand- and silt-sized mineral grains, especially quartz, but other weathered materials may be present as well. These solid particles hold soil particles apart, allowing oxygen and water to circulate more freely. Clay minerals are also important constituents of soils and aid in the retention of water as well as supply nutrients to plants. Soils with excess clay minerals, however, drain poorly and are sticky when wet and hard when dry.

If a body of rock weathers and the weathering residue accumulates over it, the soil so formed is *residual,* meaning that it formed in place (■ Figure 5.14a). In contrast, *transported soil* develops on weathered material that was eroded and transported from the weathering site and deposited elsewhere, such as on a stream's floodplain (Figure 5.14b). Many fertile transported soils of the Mississippi River valley and the Pacific Northwest developed on deposits of wind-blown dust called *loess* (see Chapter 16).

The Soil Profile

Observed in vertical cross section, a soil has distinct layers, or **soil horizons,** that differ from one another in texture, structure, composition, and color (■ Figure 5.15). Starting from the top, the horizons are designated

(a)

(b)

■ **Figure 5.14**

(a) Residual soil on bedrock in California and (b) transported soil in a small gully in Nevada. Notice that in (a) horizons A, B, and C are easily seen, but in (b) horizon C is covered by debris.

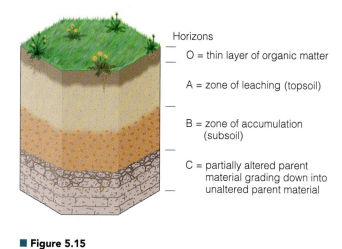

Horizons

— O = thin layer of organic matter

— A = zone of leaching (topsoil)

— B = zone of accumulation (subsoil)

— C = partially altered parent material grading down into unaltered parent material

■ Figure 5.15

The soil horizons in a fully developed soil.

O, A, B, and C, but the boundaries between horizons are transitional rather than sharp. Because soil-forming processes begin at the surface and work downward, the upper soil layer is more altered from the parent material than are the layers below.

Horizon O, which is only a few centimeters thick, consists of organic matter. The remains of plants are clearly recognizable in the upper part of horizon O, but its lower part consists of humus.

Horizon A, called *topsoil*, contains more organic matter than horizons B and C, and it is also characterized by intense biological activity because plant roots, bacteria, fungi, and animals such as worms are abundant. Threadlike soil bacteria give freshly plowed soil its earthy aroma. In soils developed over a long period of time, horizon A consists mostly of clays and chemically stable minerals such as quartz. Water percolating down through horizon A dissolves soluble minerals and carries them away or down to lower levels in the soil by a process called *leaching*, so horizon A is also known as the **zone of leaching** (Figure 5.15).

Horizon B, or *subsoil*, contains fewer organisms and less organic matter than horizon A. Horizon B is known as the **zone of accumulation** because soluble minerals leached from horizon A accumulate as irregular masses. If horizon A is stripped away by erosion, leaving horizon B exposed, plants do not grow as well; and if horizon B is clayey, it is harder when dry and stickier when wet than other soil horizons.

Horizon C, the lowest soil layer, consists of partially altered parent material grading down into unaltered parent material (Figure 5.15). In horizons A and B, the composition and texture of the parent material have been so thoroughly altered that it is no longer recognizable. In contrast, rock fragments and mineral grains of the parent material retain their identity in horizon C. Horizon C contains little organic matter.

What Factors Are Important in Soil Formation?

All soils form by mechanical and chemical weathering, but they differ in texture, color, thickness, and fertility. Accordingly, we are interested in the factors that control the attributes and locations of various soils as well as how rapidly soil-forming processes operate. Climate, parent material, organic activity, relief and slope, and time are the critical factors in soil formation. Complex interactions among these determine soil type, thickness, and fertility (■ Figure 5.16).

Climate and Soil   Soil scientists know that climate is the single most important factor influencing soil type and depth. Intense chemical weathering in the tropics yields deep soils from which most of the soluble minerals have been removed by leaching. In arctic and desert climates, soils tend to be thin, contain significant quantities of soluble minerals, and are composed mostly of materials derived by mechanical weathering (Figure 5.16).

A very general classification recognizes three major soil types characteristic of different climatic settings. Soils that develop in humid regions, such as the eastern United States and much of Canada, are **pedalfers,** a name derived from the Greek word *pedon*, meaning "soil," and from the chemical symbols for aluminum (Al) and iron (Fe). Because pedalfers form where abundant moisture is present, most of the soluble minerals have been leached from horizon A. Although it may be gray, horizon A is commonly dark because of abundant organic matter, and aluminum-rich clays and iron oxides tend to accumulate in horizon B.

Soils found in much of the arid and semiarid western United States, especially the Southwest, are **pedocals.** Pedocal derives its name in part from the first three letters of *calcite*. These soils contain less organic matter than pedalfers, so horizon A is lighter colored and contains more unstable minerals because of less intense chemical weathering. As soil water evaporates, calcium carbonate leached from above precipitates in horizon B, where it forms irregular masses of *caliche* (■ Figure 5.17a). Precipitation of sodium salts in some desert areas where soil water evaporation is intense yields *alkali soils* that are so alkaline that they support few plants (Figure 5.17b).

Laterite is a soil formed in the tropics where chemical weathering is intense and leaching of soluble minerals is complete. These soils are red, commonly extend to depths of several tens of meters, and are composed largely of aluminum hydroxides, iron oxides, and clay minerals; even quartz, a chemically stable mineral, is leached out (■ Figure 5.18a).

Although laterites support lush vegetation, as in tropical rain forests, they are not very fertile. The native vegetation is sustained by nutrients derived mostly from the surface layer of organic matter. When these soils are

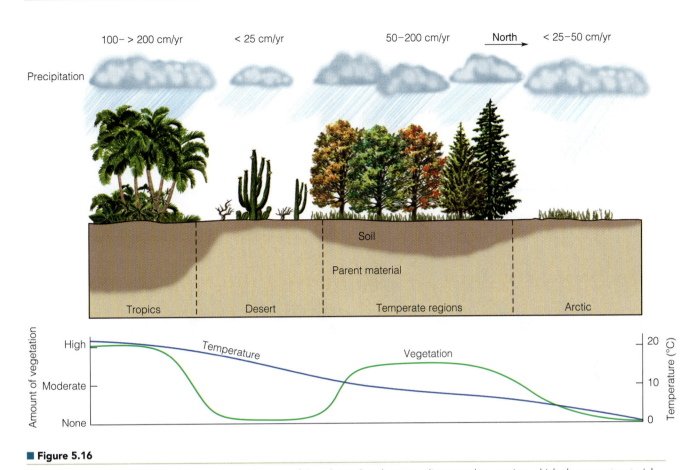

■ **Figure 5.16**

Generalized diagram showing soil formation as a function of the relationships between climate and vegetation, which alter parent material over time. Soil-forming processes operate most vigorously where precipitation and temperatures are high, as in the tropics.

cleared of their native vegetation, the surface accumulation of organic matter rapidly oxidizes, and there is little to replace it. Consequently, when societies practicing slash-and-burn agriculture clear these soils, they can raise crops for only a few years at best (Figure 5.18b, c). Then the soil is depleted of plant nutrients, the clay-rich laterite bakes brick hard in the tropical sun, and the farmers move on to another area, where the process is repeated.

■ **Figure 5.17**

(a) This boulder has been turned over to show the scaly white material known as *caliche* that formed on its underside. Irregular masses of caliche are common in horizon B of many pedocals. (b) An alkali soil in California. The white material is sodium carbonate or potassium carbonate. Notice that only a few hardy plants grow in this area.

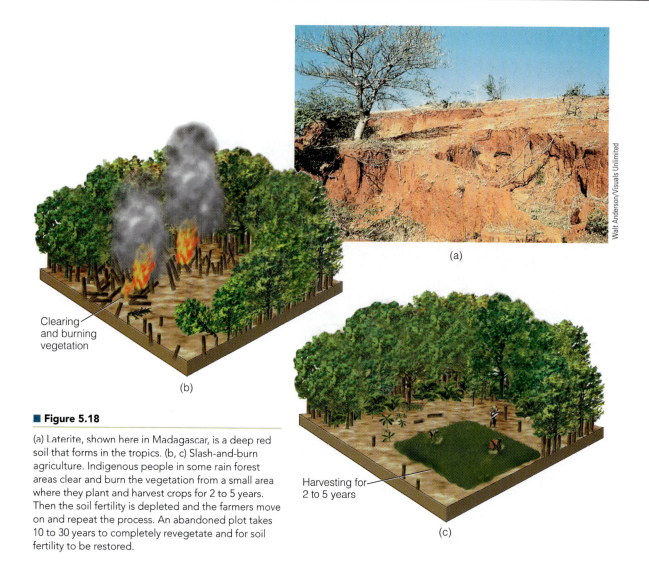

Walt Anderson/Visuals Unlimited

(a)

Clearing and burning vegetation

(b)

Harvesting for 2 to 5 years

(c)

■ **Figure 5.18**

(a) Laterite, shown here in Madagascar, is a deep red soil that forms in the tropics. (b, c) Slash-and-burn agriculture. Indigenous people in some rain forest areas clear and burn the vegetation from a small area where they plant and harvest crops for 2 to 5 years. Then the soil fertility is depleted and the farmers move on and repeat the process. An abandoned plot takes 10 to 30 years to completely revegetate and for soil fertility to be restored.

Parent Material   The same rock type can yield different soils in different climatic regimes, and in the same climatic regime the same soils can develop on different rock types. Thus it seems that climate is more important than parent material in determining the type of soil. Nevertheless, rock type does exert some control. For example, the metamorphic rock quartzite will have a thin soil over it because it is chemically stable, whereas an adjacent body of granite will have a much deeper soil (■ Figure 5.19a).

Soil that develops on basalt will be rich in iron oxides because basalt contains abundant ferromagnesian silicates, but rocks lacking these minerals will not yield an iron oxide-rich soil no matter how thoroughly they are weathered. Also, weathering of a pure quartz sandstone yields no clay, whereas weathering of clay yields no sand.

Activities of Organisms   Soil not only depends on organisms for its fertility but also provides a suitable habitat for organisms ranging from microscopic, single-celled bacteria to burrowing animals such as ground squirrels and gophers. Earthworms—as many as a million per acre—ants, sowbugs, termites, centipedes, millipedes, and nematodes, along with various types of fungi, algae, and single-celled animals, make their homes in soil. All contribute to the formation of soils and provide humus when they die and are decomposed by bacterial action.

Much of the humus in soils is provided by grasses or leaf litter that microorganisms decompose to obtain food. In so doing, they break down organic compounds within plants and release nutrients back into the soil. In addition, organic acids produced by decaying soil organisms are important in further weathering of parent materials and soil particles.

Burrowing animals constantly churn and mix soils, and their burrows provide avenues for gases and water. Soil organisms, especially some types of bacteria, are extremely important in changing atmospheric nitrogen into a form of soil nitrogen suitable for use by plants.

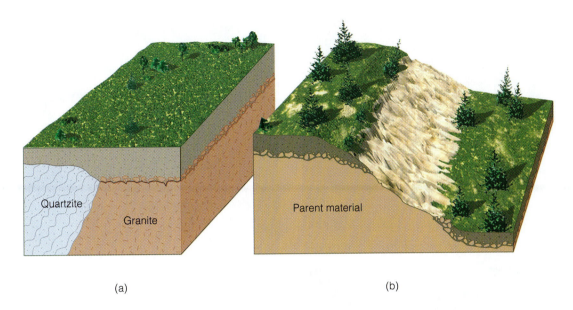

■ **Figure 5.19**

(a) The influence of parent material on soil development. Quartzite is resistant to chemical weathering, whereas granite alters more quickly.
(b) The effect of slope on soil formation. Where slopes are steep, erosion occurs faster than soil can form.

The Lay of the Land—Relief and Slope *Relief* is the difference in elevation between high and low points in a region. In some mountainous areas relief is measured in hundreds of meters, whereas it rarely exceeds a few meters in, for instance, the Great Plains of the United States and Canada. Because climate is such an important factor in soil formation and climate changes with elevation, areas with considerable relief have different soils in mountains and adjacent lowlands.

Slope influences soil formation in two ways. One is simply *slope angle:* Steep slopes have little or no soil because weathered materials erode faster than soil-forming processes operate (Figure 5.19b). The other slope factor is *slope direction*—that is, the direction a slope faces. In the Northern Hemisphere, north-facing slopes receive less sunlight than south-facing slopes. In fact, steep north-facing slopes may not receive any sunlight at all. Accordingly, north-facing slopes have cooler internal temperatures, support different vegetation, and, if in a cold climate, remain snow covered or frozen longer.

Time Soil-forming processes begin at the surface and work downward, so horizon A has been altered longer than the other horizons, and thus parent material is no longer recognizable. Even in horizon B, parent material is usually not discernable, but it is in horizon C. In fact, soil properties are determined by climate and organisms altering parent material through time (Figure 5.16), so the longer the processes have operated, the more fully developed a soil will be. If weathering takes place for an extended period, especially in humid climates, soil fertility decreases as

plant nutrients are leached out, unless new materials are delivered. For instance, agricultural lands adjacent to rivers such as the Nile in Egypt have their soils replenished yearly during floods. In areas of active tectonism, erosion of uplifted areas provides fresh materials that are transported to nearby lowlands, where they contribute to soils.

How much time is needed to develop a centimeter of soil or a fully developed soil a meter or so deep? We can give no definitive answer because weathering proceeds at vastly different rates depending on climate and parent material, but an overall average might be about 2.5cm per century. However, a lava flow a few centuries old in Hawaii may have a well-developed soil on it, whereas a flow the same age in Iceland has considerably less soil. Given the same climatic conditions, soil develops faster on unconsolidated sediment than it does on bedrock.*

Under optimum conditions, soil-forming processes operate rapidly in the context of geologic time. From the human perspective, though, soil formation is a slow process; consequently, soil is a nonrenewable resource.

Expansive Soils

News reports commonly cover geologic events that cause fatalities, injuries, and property damage such as floods, earthquakes, volcanic eruptions, and landslides.

Bedrock is a general term for the rock underlying soil or unconsolidated sediment.

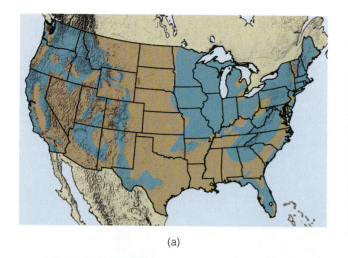

(a)

(b)

Ed Nuhfer, Director, Teaching Effectiveness

■ **Figure 5.20**

(a) The brown pattern shows the distribution of expansive soils in the United States. Many areas of expansive soils are also present in the blue areas but are too small to show at this scale. (b) The undulations in this sidewalk near Dallas, Texas, were caused by expansive soil.

These more sensational events overshadow the fact that processes that rarely make the news cause more property damage. One of these, soil creep, is considered in Chapter 14, but here we are concerned with **expansive soils,** soils containing clay minerals that increase in volume when wet and shrink when they dry out. Soils with clays that expand 6% are considered highly expansive, and some are even more expansive. About $6 billion in damage to foundations, roadways, sidewalks, and other structures takes place each year in the United States, mostly in the Rocky Mountain states, the Southwest,

and some of the states along the Gulf of Mexico (■ Figure 5.20a).

When soil expands and contracts, overlying structures are first uplifted and then subside, thus experiencing forces that usually are not equally applied. Part of a house or sidewalk might be uplifted more than an adjacent part of the same structure (Figure 5.20b). Avoiding areas of expansive soils is the best way to prevent damage, but what can be done to minimize the damage to existing structures? In some cases the soil is removed, or mixed with chemicals that change the way it reacts with water, or covered by a layer of nonexpansive fill, all expensive but perhaps necessary remedies. Also, soil should be kept as dry as possible to inhibit expansion, and specialized building methods can be employed, such as placing structures on piers or reinforced foundations designed to minimize the effects of expansion.

SOIL DEGRADATION

Given that a fully mature soil takes centuries to thousands of years to form, any soil losses that exceed the rate of formation are viewed with alarm. Likewise, any decrease in soil productivity is cause for concern, especially in those parts of the world where soils provide only a marginal existence. Any loss of soil to erosion or decrease in soil productivity resulting from physical and chemical phenomena is collectively called **soil degradation.**

According to studies by the World Resources Institute, 17% of the world's soils were degraded to some extent by human activities between 1945 and 1990. The estimate for North America is only 5.3%, whereas the figures for other parts of the world are much higher. The three general types of soil degradation are *erosion, chemical deterioration,* and *physical deterioration,* each with several separate but related processes.

Soil Erosion

Wind and running water are responsible for most soil erosion. Of course, erosion is a natural, ongoing process, but it is usually slow enough for soil-forming processes to keep pace. Unfortunately, several human activities create problems by introducing elements into the system that would not otherwise be present. Examples of these human activities are the removal of natural vegetation by agricultural practices such as plowing and overgrazing, and overexploitation for firewood and deforestation by logging operations. Exposed soil is simply more easily eroded than vegetation-covered soil.

Just how easily soil pulverized by plowing can be eroded by wind was clearly demonstrated during the 1930s when a large area of the western United States lost millions of tons of topsoil. Several states in the Great Plains were particularly hard hit by drought, and wind eroded the exposed soils. Indeed, in the Dust Bowl, as it is called (■ Figure 5.21a), agricultural productivity plummeted, and tens of thousands of people were left destitute. The migrant farm workers immortalized in John Steinbeck's novel *The Grapes of Wrath* came from the Dust Bowl area. Furthermore, the area had huge dust storms (Figure 5.21b); more than 140 were recorded during 1936 and 1937. One storm in 1934 covered more than 3.5 million km² and blew millions of tons of soil far to the east, where it settled along the East Coast and on ships as far as 480 km out in the Atlantic Ocean.

Wind is certainly effective in some areas, especially on exposed soils, but running water is capable of much greater erosion. Some soil is removed by **sheet erosion,** which erodes thin layers of soil over an extensive, gently sloping area by water not confined to channels. In contrast, **rill erosion** takes place when running water scours small, troughlike channels. If these channels are shal-

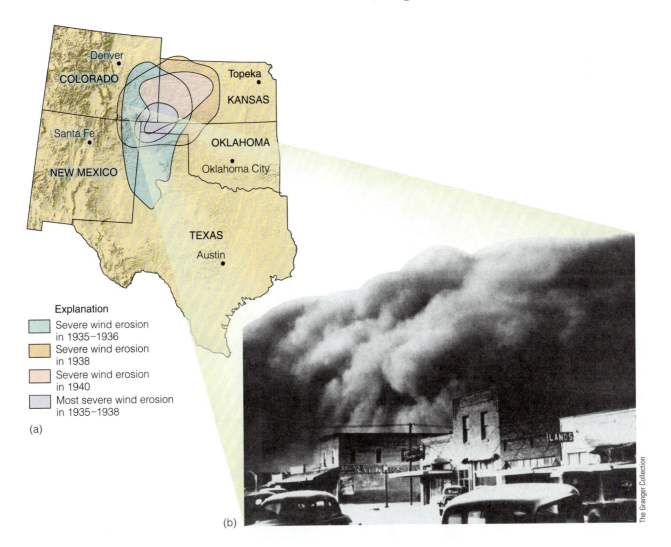

Explanation

- Severe wind erosion in 1935–1936
- Severe wind erosion in 1938
- Severe wind erosion in 1940
- Most severe wind erosion in 1935–1938

(a)

(b)

The Granger Collection

■ **Figure 5.21**

The Dust Bowl of the 1930s. (a) Drought conditions extended far beyond the boundaries shown here, but this area was particularly hard hit by drought, dust storms, and wind erosion. (b) This huge dust storm was photographed at Lamar, Colorado, in 1934. Source: From *Dust Bowl: The Southern Plains in the 1930s,* by Donald Worster. Copyright © 1979 by Oxford University Press, Inc. Reprinted by permission.

low enough to be eliminated by plowing, they are called *rills,* but if deeper than about 30 cm, they cannot be plowed over and are called *gullies* (■ Figure 5.22a, b). Where *gullying* is extensive, croplands can no longer be tilled and must be abandoned.

Increased soil erosion invariable follows clearing of tropical rain forests for agriculture or logging operations (Figure 5.22c). In either case, more rainwater runs off at the surface, gullying becomes more common, and flooding is more frequent because trees with their huge water-holding capacity are no longer present. Clearing woodlands in less humid regions such as much of the eastern United States also results in accelerated erosion, at least initially. Studies of lake deposits in several areas indicate that years of rapid erosion follow clearing and then fall off markedly when the land is covered by crops. Nevertheless, erosion rates may be 10 times greater than they were before clearing took place.

If soil losses to erosion are minimal, soil-forming processes keep pace and the soil remains productive, unless, of course, the soil is chemically or physically degraded, too. If the loss rate exceeds the rate of renewal, though, the most productive soil layer, horizon A, is removed, exposing horizon B, which is much less productive. High rates of soil erosion are obviously problems, but there are additional consequences. For one thing, eroded soil is transported elsewhere, perhaps onto roadways or into streams and rivers. Also, sediment accumulates in canals and irrigation ditches, and fertilizers and pesticides are carried into waterways and lakes.

Soil erosion problems experienced during the past, especially during the 1930s, motivated federal and state agencies to develop practices to minimize soil erosion on agricultural lands (Table 5.2). Crop rotation, contour plowing and strip-cropping (■ Figure 5.23), and terracing have all proved effective and are now used routinely. No-till planting, in which the residue from a harvested crop is left on the ground to inhibit erosion, has also been effective.

(a)

(b)

(c)

James S. Monroe

H. H. Walsron/USGS

Frank Lembrecht/Visuals Unlimited

■ **Figure 5.22**

(a) Rill erosion in a field in Michigan during a rainstorm. The rill was later plowed over. (b) A large gully in the upper basin of the Rio Reventado in Costa Rica. (c) Accelerated soil erosion on a bare surface in Madagascar that was once covered by lush forest.

Table 5.2

Soil Conservation Practices

Terracing	Creating flat areas on sloping ground. One of the oldest and most effective ways of preserving soil and water.
Strip-cropping	Growing of different crops on alternate, parallel strips of ground to minimize wind and water erosion. Alternating strips of corn and alfalfa, for instance.
Crop Rotation	Yearly alternation of crops on the same land. Significant reduction in soil erosion when soil-depleting crops are alternated with soil-enriching crops.
Contour Plowing	Plowing along a slope's contours so that furrows and ridges are perpendicular to the slope.
No-till Planting	Planting seeds through the residue of a previously harvested crop.
Windbreaks	Planting trees or large shrubs along the margins of a field. Especially effective in reducing wind erosion.

Chemical and Physical Soil Degradation

Soil undergoes chemical deterioration when its nutrients are depleted and its productivity decreases. Loss of soil nutrients is most notable in many of the populous developing nations where soils are overused to maintain high levels of agricultural productivity. Chemical deterioration is also caused by insufficient use of fertilizers and by clearing soils of their natural vegetation. Examples of chemical deterioration are found everywhere but it is most prevalent in South America, where it accounts for 30% of all soil degradation.

Other types of chemical deterioration are pollution and *salinization,* which occurs when the concentration of salts increases in soil, making it unfit for agriculture. Pollution can be caused by improper disposal of domestic, industrial, and mining wastes, oil and chemical spills, and the concentration of insecticides and pesticides in soils. Soil pollution is a particularly severe problem in eastern Europe.

Physical deterioration of soils results when soil particles are compacted under the weight of heavy machinery and livestock, especially cattle. Compacted soils are more costly to plow, and plants have a more difficult time emerging from them. Furthermore, water does not readily infiltrate, so more runoff occurs, which in turn accelerates the rate of water erosion.

Soil degradation is a serious problem in many parts of the world. In North America, with local exceptions, it is moderate but nevertheless of some concern. The rich prairie soils of the midwestern United States and the Great Plains of the United States and Canada remain

Visuals Unlimited

■ **Figure 5.23**

Contour plowing and strip-cropping are two soil conservation practices used on this farm. Contour plowing involves plowing parallel to the contours of the land to inhibit runoff and soil erosion. In strip-cropping row crops such as corn alternate with other crops such as alfalfa.

productive, although their overall productivity has decreased somewhat over the last several decades. In any case, lessons learned during the past have convinced farmers, government agencies, and the public in general that they can no longer take soil for granted or regard it as a resource that needs no nurturing.

WEATHERING AND NATURAL RESOURCES

We have already discussed various aspects of soils, which are certainly one of our most precious natural resources. Indeed, if it were not for soils, food production on Earth would be vastly different and capable of supporting far fewer people. In addition, other aspects of soils are important economically. We discussed the origin of laterites in response to intense chemical weathering in the tropics, and noted further that they are not very productive. However, if the parent material is rich in aluminum, the ore of aluminum called *bauxite* accumulates in horizon B. Some bauxite is found in Arkansas, Alabama, and Georgia, but at present it is cheaper to import it rather than mine these deposits, so both the United States and Canada depend on foreign sources of aluminum ore.

Bauxite and other accumulations of valuable minerals formed by the selective removal of soluble substance during chemical weathering are known as *residual concentrations*. Certainly bauxite is the best known example of a residual concentration, but other deposits that formed in a similar fashion include ores of iron, manganese, clays, nickel, phosphate, tin, diamonds, and gold. Some of the sedimentary iron deposits in the Lake Superior region of the United States and Canada were enriched by chemical weathering when soluble parts of the deposits were carried away.

Most of the economic deposits of various clay minerals formed by sedimentary processes or by hydrothermal alteration of granitic rocks, but some are residual concentrations. A number of kaolinite deposits in the southern United States formed when chemical weathering altered feldspars in pegmatites or as residual concentrations of clay-rich limestones and dolostones. Kaolinite is a type of clay mineral used in the manufacture of paper and ceramics.

Chemical weathering is also responsible for gossans and ore deposits that lie beneath them. A *gossan* is a yellow to red deposit made up mostly of hydrated iron oxides that formed by oxidation and leaching of sulfide minerals such as pyrite (FeS_2). The dissolution of pyrite and other sulfides forms sulfuric acid, which causes other metallic minerals to dissolve, and these tend to be carried down toward the groundwater table, where the descending solutions form minerals containing copper, lead, and zinc. And below the water table the concentration of sulfide minerals forms the bulk of the ore minerals. Gossans have been mined for iron, but they are far more important as indicators of underlying ore deposits.

5 REVIEW WORKBOOK

Chapter Summary

- Mechanical and chemical weathering disintegrate and decompose parent material so that it is more nearly in equilibrium with new physical and chemical conditions.

- The products of weathering include rock fragments and minerals liberated from parent material as well as soluble compounds and ions in solution.

- Weathering yields materials that may become soil or sedimentary rock.

- Mechanical weathering processes include frost action, pressure release, thermal expansion and contraction, salt crystal growth, and the activities of organisms. The particles yielded retain the composition of the parent material.

- Chemical weathering by solution, hydrolysis, and oxidation results in a chemical change in parent material and proceeds most rapidly in hot, wet environments.

- Mechanical weathering contributes to chemical weathering by breaking parent material into smaller pieces, thereby exposing more surface area.

- Soils possess horizons designated, in descending order, as O, A, B, and C, which differ from one another in texture, composition, structure, and color.

- The important factors controlling soil formation are climate, parent material, organic activity, relief and slope, and time.

- Soils in humid regions are broadly referred to as pedalfers, whereas those of semiarid regions are known as pedocals. Laterite is soil that forms in the topics where chemical weathering is intense.

- Soil degradation results from erosion as well as from physical and chemical deterioration. Human activities such as construction, agriculture, deforestation, waste disposal, and chemical spills contribute to soil degradation.

- Chemical weathering is responsible for the origin of some mineral deposits such as residual concentrations of iron, lead, manganese, and clay.

- Gossans and their underlying ores result from chemical weathering.

Important Terms

chemical weathering (p. 127)
differential weathering (p. 123)
erosion (p. 122)
exfoliation (p. 125)
exfoliation dome (p. 125)
expansive soil (p. 140)
frost action (p. 124)
frost heaving (p. 125)
frost wedging (p. 124)
humus (p. 135)
hydrolysis (p. 130)
laterite (p. 136)

mechanical weathering (p. 124)
oxidation (p. 129)
parent material (p. 122)
pedalfer (p. 136)
pedocal (p. 136)
pressure release (p. 125)
regolith (p. 135)
rill erosion (p. 141)
salt crystal growth (p. 126)
sheet erosion (p. 141)
sheet joint (p. 125)
soil (p. 135)

soil degradation (p. 140)
soil horizon (p. 135)
solution (p. 127)
spheroidal weathering (p. 134)
talus (p. 124)
thermal expansion and
   contraction (p. 126)
transport (p. 122)
weathering (p. 122)
zone of accumulation (p. 136)
zone of leaching (p. 136)

Review Questions

1. An essential component of soils is decomposed organic matter known as:

   a. _____ humus; b. _____ regolith; c. _____ talus; d. _____ gossan; e. _____ carbonic acid.

2. If a small amount of carbonic acid is present in groundwater, _____ dissolves rapidly.

   a. _____ pedalfer; b. _____ exfoliation domes; c. _____ limestone; d. _____ horizon A; e. _____ manganese.

3. The process by which hydrogen and hydroxyl ions of water replace ions in minerals is:

   a. _____ solution; b. _____ differential weathering; c. _____ residual replacement; d. _____ supergene enrichment; e. _____ hydrolysis.

4. Which one of the following is *not* a mechanical weathering process?

   a. _____ salt crystal growth; b. _____ frost wedging; c. _____ oxidation; d. _____ pressure release; e. _____ thermal expansion and contraction.

5. The soil typical of semiarid regions is _____, whereas _____ is much more common in more humid areas.

   a. _____ pedocal/pedalfer; b. _____ regolith/laterite; c. _____ caliche/talus; d. _____ residual/compound; e. _____ exfoliated/carbonized.

6. The accumulation of angular pieces of rock at the base of a slope is known as:

   a. _____ parent material; b. _____ talus; c. _____ soil; d. _____ laterite; e. _____ caliche.

7. Horizon C differs from other soil horizons in that it:

   a. _____ is the most fertile; b. _____ has weathered the longest; c. _____ is made up of sodium sulfate; d. _____ grades down into parent material; e. _____ contains the most humus.

8. Pressure release is the primary process responsible for:

   a. _____ spheroidal weathering; b. _____ exfoliation domes; c. _____ residual ores; d. _____ frost heaving; e. _____ soil degradation.

9. Spheroidal weathering takes place because:

   a. _____ oxidation changes limestone to clay; b. _____ naturally occurring rocks are spherical to begin with; c. _____ thermal expansion and contraction are so effective; d. _____ aluminum oxides are nearly insoluble; e. _____ corners and edges of stones weather faster than flat surfaces.

10. Any loss of soil to erosion or decrease in productivity is known as:

    a. _____ expansive soils; b. _____ salt crystal growth; c. _____ frost heaving; d. _____ soil degradation; e. _____ strip-cropping.

11. Explain how residual concentrations form and why they are important.

12. Notice in Figure 6.23b that both vertical cliffs and gentle slopes are present. How can you account for these differences?

13. How does mechanical weathering differ from and contribute to chemical weathering?

14. Draw soil profiles for semiarid and humid regions and list the characteristics of each.

15. Why is groundwater in humid regions acidic, whereas in arid areas it is alkaline?

16. Explain how exfoliation domes form. In what kinds of rocks do they develop, and where would you go to see examples?

17. How do the factors of climate, parent material, and time determine the depth and fertility of soil?

18. Describe the types of soil degradation. What practices are used to prevent or at least decrease soil erosion?

19. What are differential weathering and differential erosion and why do they occur?

20. Explain why particle size, climate, and parent material control the rate of chemical weathering.

World Wide Web Activities

PHYSICAL
Geology⇌Now Assess your understanding of this chapter's topics with additional quizzing and comprehensive interactivities at

http://earthscience.brookscole.com/physgeo5e

as well as current and up-to-date weblinks, additional readings, and InfoTrac College Edition exercises.

Sediment and Sedimentary Rocks

CHAPTER 6
OUTLINE

PHYSICAL Geology⇌Now *This icon, appearing throughout the book, indicates an opportunity to explore interactive tutorials, animations, or practice problems available on the Physical GeologyNow Web site at http://earthscience.brookscole.com/physgeo5e.*

OBJECTIVES

After reading this chapter, you will have learned that

- Sediment comes from weathering of rock and is transported and deposited by a variety of processes.

- Compaction and cementation convert unconsolidated sediment into sedimentary rocks.

- Texture and composition are the criteria used to classify sedimentary rocks.

- Individual layers of sediment differ in texture, composition, or both.

- Many widespread layers of sedimentary rocks were deposited during marine transgressions and regressions, both of which can be recognized by the evidence they leave in the geologic record.

- Sedimentary rocks possess a variety of structures that formed when they were deposited, and some also contain fossils.

- Several geologic processes lead to the preservation of fossils.

- Structures and fossils in sedimentary rocks are used to determine depositional environments—that is, to read the story told by sedimentary rocks.

- Some sedimentary rocks contain resources or may be resources themselves.

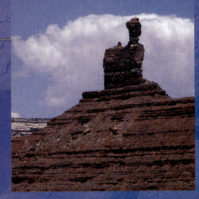

Sedimentary rocks in the Valley of the Gods in southern Utah. Notice that the layering in the rocks is horizontal and that erosion has exposed them in their present form. The red color results from iron oxide cement in the rocks, which are mostly sandstone. Source: Sue Monroe

Introduction

In Chapter 5 we emphasized that weathering, erosion, transport, and deposition are essential parts of the rock cycle, and the reason is that these processes are responsible for the origin and deposition of *sediment* that may become *sedimentary rock*. The term **sediment** refers to (1) all solid particles of preexisting rocks yielded by mechanical and chemical weathering, (2) minerals derived from solutions containing materials dissolved during chemical weathering, and (3) minerals extracted from seawater by organisms to build their shells. **Sedimentary rock** is simply any rock composed of sediment.

Perhaps some examples will clarify what we mean by sediment as opposed to sedimentary rock. The gravel in a stream channel, sand on beaches (■ Figure 6.1a), and mud on the seafloor are examples of sediment, loose aggregates of solid particles derived by weathering. If these sediments become solid aggregates—that is, the loose particles are held together—they are sedimentary rocks (Figure 6.1b). We will discuss the processes by which sediment is converted to sedimentary rock later in this chapter.

Of course, interactions among various Earth systems are responsible for the sediment that eventually becomes sedimentary rock. Recall, though, that one aspect of the rock cycle (see Figure 1.14) is that all three major rock families are interrelated in complex ways, and this being the case we

might expect that some rocks would show characteristics of more than one rock-forming process. Some of the rocks found in John Day Fossil Beds National Monument in central Oregon provide a good example.

Geologists agree that many of the rocks in the monument are indeed sedimentary or volcanic, but some are characterized as *volcanoclastic*, meaning they contain large amounts of volcanic rock fragments and pyroclastic materials. Many of these rocks have few of the internal structures so typical of sedimentary rocks, and yet geologists who study igneous rocks and processes mostly ignored them, thinking they were sedimentary. Now, however, the rocks have been thoroughly studied and reveal information about both igneous and sedimentary processes that took place in central Oregon between 6 and 54 million years ago (■ Figure 6.2). The rocks also contain numerous fossils, especially of land-dwelling mammals and plants, both of which tell us something about the biosphere and the ancient climate in this part of North America.

Earth's crust is composed mostly of *crystalline rock*, a term that refers loosely to metamorphic rocks and most igneous rocks (those composed of pyroclastic materials excepted). Nevertheless, sediment and sedimentary rocks, comprising perhaps 5% of the crust, are by far the most common in surface exposures and in the shallow subsurface.

(a)   (b)

■ **Figure 6.1**

(a) The gravel on this beach is an example of sediment, a loose aggregate of solids derived from preexisting rocks. (b) The gravel in this exposure was lithified—that is, converted to sedimentary rock—when chemical cement formed in pore spaces and bound the loose particles together. Accordingly, it is the sedimentary rock known as conglomerate.

Sue Monroe

(a)

Rocks in John Day Fossil Beds National Monument in central Oregon provide a record of events that took place between 6 and 54 million years ago. (a) At Sheep Rock the rocks are mostly sedimentary, but a remnant of lava flow is present on the hilltop. (b) Another exposure of sedimentary rocks in the monument. (c) Mammals that lived in the area between 37 and 55 million years ago include (1) titanotheres, (2) a carnivore, (3) ancient horses, (4) tapirs, and (5) rhinoceroses. The climate was subtropical then, but now it is semiarid.

Sue Monroe

(b)

The Field Museum

(c)

They cover about two thirds of the continents and most of the seafloor, except spreading ridges. All rocks are important for deciphering Earth history, but sedimentary rocks play a special role because they preserve evidence of the surface processes responsible for them (Figure 6.2). Accordingly, geologists study sedimentary rocks to determine the past distribution of streams, lakes, deserts, glaciers, and shorelines, and they also make inferences about ancient climates and the biosphere. In addition, some sediments and sedimentary rocks are themselves natural resources, or they are the host for resources such as petroleum and natural gas.

SEDIMENT TRANSPORT AND DEPOSITION

Recall from Chapter 5 that erosion involves removing weathered materials from their source area by running water, wind, glaciers, gravity, and waves, and that the same processes transport sediment some distance and eventually deposit it. Our task here is to understand transport mechanisms, depositional processes, and environments of deposition.

Table 6.1

Size Classification of Sedimentary Particles

Size	Sediment Name	
>2 mm	Gravel	
1/16–2 mm	Sand	
1/256–1/16 mm	Silt	Mud*
<1/256 mm	Clay	

*Mud is a mixture of silt and clay-sized particles.

Transport of Sediment

Weathering and erosion are fundamental processes in the origin of sediment, but so is *sediment transport*— that is, the movement of sediment by natural processes. One important criterion for classifying sedimentary particles is their size, particularly for *detrital sediment*, which consists of all solid particles derived from any kind of weathering (Table 6.1). Granite, for instance, may yield small pieces of granite (rock fragments) and individual mineral grains of quartz, feldspars, and perhaps biotite. Any detrital particle that measures more than 2 mm, regardless of its composition, is *gravel*, whereas sand measures $\frac{1}{16}$–2 mm. Mixtures of silt, measuring between $\frac{1}{256}$ and $\frac{1}{16}$ mm, and clay ($>\frac{1}{256}$ mm) are collectively referred to as *mud*.

Chemical sediment, on the other hand, is made up of solid particles derived by inorganic chemical processes, such as evaporation of seawater, as well as the activities of organisms. Clams, oysters, corals and a number of other animals, and some plants construct their skeletons of minerals, especially aragonite or calcite ($CaCO_3$) or silicon dioxide (SiO_2).

Most sediment, be it detrital or chemical, accumulates as loose solids that are later converted to sedimentary rock, but a few skip the unconsolidated sediment stage and form directly as solids. Wave-resistant structures known as *reefs* are made up of skeletons of corals and mollusks as well as encrusting algae. Reefs are solids when they form, and they constitute a type of *limestone*. However, if a reef is ripped up during a storm and pieces of reef material are deposited on the seafloor, the pieces are sedimentary particles, and these too are characterized as gravel-, sand-, silt-, and clay-sized particles.

All detrital sediment is transported some distance from its source, whereas chemical sediment forms in the area where it is deposited. Thus the latter may be transported, but only locally where waves or tidal currents move sediment short distances. During the transport of detrital sediment, *abrasion* reduces the size of particles and the sharp corners and edges are worn smooth, a process called **rounding**, as pieces of gravel and sand collide with one another (■ Figure 6.3). Various processes also bring about **sorting**, which refers to the size distribution of the particles in sediment or sedimentary rock (Figure 6.3). Sediment is well sorted if all the particles are about the same size, and poorly sorted if the range of sizes is wide. Rounding and sorting may seem rather unimportant, but both have significant implications for other aspects of sediment and sedimentary rocks, such as how groundwater, liquid petroleum, and gas move through them. They are also useful for determining how deposition occurred, a topic covered more fully in a later section.

Deposition and Depositional Environments

The transport distance for sediment is variable, but regardless of transport mechanism or distance, all sediment is eventually deposited; that is, it accumulates in a specific area. Silt and clay may be transported to a lake

(a)   (b)   (c)

■ **Figure 6.3**

Rounding and sorting in sediments. (a) Beginning students often mistake rounding to mean ball-shaped. All three of these stones are well rounded; that is, their sharp corners and edges have been worn smooth. (b) Deposit of well-rounded, well-sorted gravel. The particles on average measure about 5 cm across. (c) Angular, poorly sorted gravel. Note the quarter to provide scale.

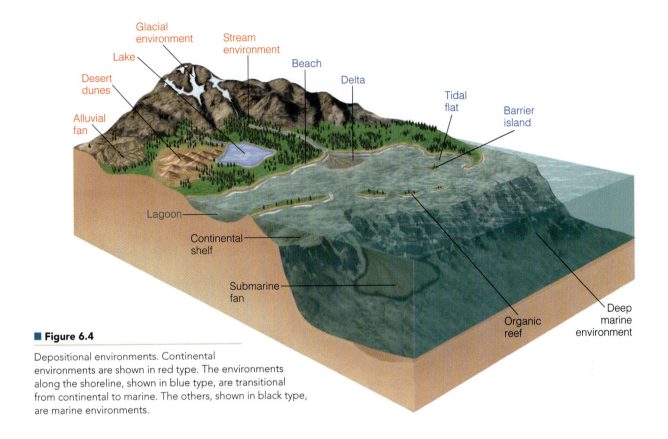

■ Figure 6.4

Depositional environments. Continental environments are shown in red type. The environments along the shoreline, shown in blue type, are transitional from continental to marine. The others, shown in black type, are marine environments.

or a lagoon or a river's floodplain, where the lack of currents and fluid turbulence allows them to settle, thus forming a layer of mud. Larger particles of sand and gravel commonly accumulate in stream channels or on beaches, whereas chemical sediment may be deposited in shallow seawater where organisms use dissolved substances to make their skeletons.

Any geographic area in which sediment accumulates is a **depositional environment**, where physical, chemical, and biological processes impart distinctive characteristics to sedimentary deposits. A completely satisfactory classification of sedimentary environments does not exist, but geologists recognize three broad depositional settings: continental, transitional, and marine, each with several specific depositional environments (■ Figure 6.4). Deserts, lakes, streams and rivers and their adjacent floodplains, and areas covered by and adjacent to glaciers are the main continental environments. That is, in these cases deposition takes place on the land. Deltas, beaches, tidal flats, and barrier islands are transitional environments because processes operating on land as well as those in the marine realm affect the deposits. A delta, for example, is deposited by running water (a stream or river), but waves and tides modify the deposit. Seaward of the transitional environments are those where only marine processes are important as on the continental shelf and beyond (Figure 6.4).

PHYSICAL
Geology ⇌ Now Click Geology Interactive to work through an activity on Rock Lab.

HOW DOES SEDIMENT BECOME SEDIMENTARY ROCK?

When detrital sediment is deposited, it consists of a loose aggregate of particles (Figure 6.1a). Mud accumulating in lakes and sand and gravel in stream channels or on beaches are good examples. To convert these aggregates of particles into sedimentary rocks requires **lithification** by compaction, cementation, or both (■ Figure 6.5).

Compaction

Compaction and cementation are certainly responsible for lithification, but the importance of one or the other varies depending on the type of sediment. To illustrate this point, consider two detrital sedimentary deposits: one made up of mud and the other of sand. In both cases, the sediment consists of solid particles and *pore*

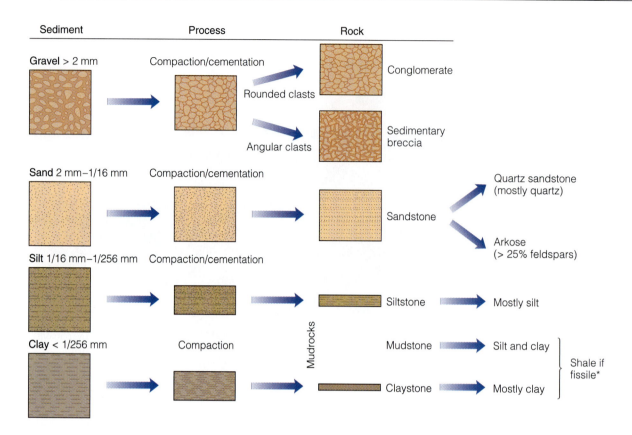

*Fissile refers to rocks capable of splitting along closely spaced planes.

■ Figure 6.5

Lithification and classification of detrital sedimentary rocks. Notice that little compaction takes place in sand and gravel.

spaces, the voids between particles. These deposits are subjected to **compaction** from their own weight and the weight of any additional sediment deposited on top of them, thereby reducing the amount of pore space and the volume of the deposit. Our hypothetical mud deposit may have 80% water-filled pore space, but after compaction its volume is reduced by as much as 40% (Figure 6.5). The sand deposit with as much as 50% pore space is also compacted, but far less than the mud deposit, so that the grains fit more tightly together (Figure 6.5).

Cementation

Compaction alone is sufficient for lithification of mud, but for sand and gravel **cementation** involving the precipitation of minerals in pore spaces is also necessary. The cement in the form of calcium carbonate ($CaCO_3$), silicon dioxide (SiO_2), and iron oxides and hydroxides, such as hematite (Fe_2O_3) and limonite ($FeO(OH) \cdot nH_2O$), respectively, are the most common chemical cements. Recall from Chapter 5 that calcium carbonate readily dissolves in water containing a small amount of carbonic acid, and that chemical weathering of feldspars and other minerals yields silica in solution.

Circulating water that contains these compounds precipitates minerals in the pore spaces of sediment, thereby binding the loose particles together.

Iron oxide and hydroxide cements are much less common than the others mentioned above, but their presence accounts for the red, yellow, and brown sedimentary rocks found in many areas (■ Figure 6.6). Much of the iron in these rocks was derived by oxidation of ferromagnesian silicates in the original deposit, but circulating groundwater carried some of it in.

Our discussion so far has been adequate to explain lithification of detrital sediments but has not addressed this process in chemical sediments. By far the most common sediments in this category are calcium carbonate mud and sand- and gravel-sized accumulations of calcium carbonate grains, such as shells and shell fragments. Compaction and cementation also take place in these sediments, converting them into various types of limestone, but compaction is generally less effective because cementation takes place soon after deposition. In any case, the cement is calcium carbonate derived by partial solution of some of the particles in the deposit.

PHYSICAL Geology ⬛Now Click Geology Interactive to work through an activity on Rock Cycle.

■ Figure 6.6

These layers of sandstone at Jemez Pueblo, New Mexico, are red because they contain iron oxide cement.

TYPES OF SEDIMENTARY ROCKS

In the Introduction we noted that sedimentary rocks make up only a small part of Earth's crust, but along with sediments they cover most of the seafloor and about two thirds of the continents. Accordingly, they are the most commonly encountered Earth materials. We have discussed the origin of sediment, its transport, deposition, and lithification, so now we will turn to a consideration of the types of sedimentary rocks and their classification. Geologists recognize two broad categories of sedimentary rocks, *detrital* and *chemical*, which, with the exception of coal, are derived from preexisting rocks (■ Figure 6.7).

■ Figure 6.7

The derivation of sediment from preexisting rocks. Whether derived by chemical or mechanical weathering, solid particles and materials in solution are transported and deposited as sediment, which if lithified becomes detrital or chemical sedimentary rock. This illustration simply shows part of the rock cycle in greater detail (see Figure 1.14).

Detrital Sedimentary Rocks

Detrital sedimentary rocks are made up of *detritus*, the solid particles such as gravel, sand, and mud derived from preexisting rocks by mechanical and chemical weathering. They have a *clastic texture*, meaning they are composed of particles or fragments known as *clasts* (■ Figure 6.8a). The several varieties of detrital sedimentary rocks are classified primarily by the size of their constituent particles, although composition is used to modify some of the rock names (Figure 6.5).

Conglomerate and Sedimentary Breccia

Conglomerate and *sedimentary breccia* consist of gravel-sized particles—that is, detrital particles measuring more than 2 mm (■ Figures 6.5 and 6.9). Composition is not a consideration in naming these rocks; both are made up mostly of rock fragments of various sizes, but some large, individual mineral grains may also be present. The only difference between the two rocks is the shape of their gravel-sized particles; conglomerate has rounded gravel, whereas sedimentary breccia has angular gravel called *rubble*.

Conglomerate is fairly common, but sedimentary breccia is rare because gravel-sized particles become rounded quickly during sediment transport or by processes operating in environments where gravel is

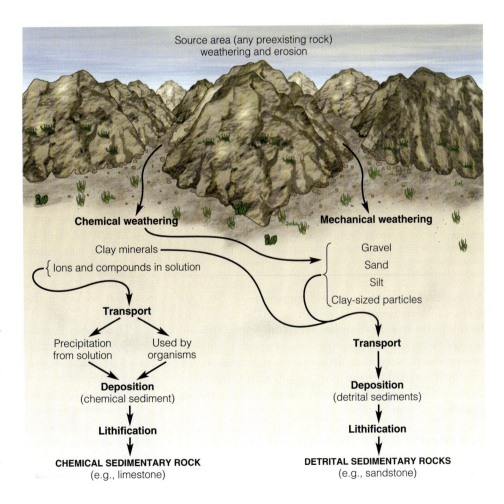

Source area (any preexisting rock) weathering and erosion

Chemical weathering　　　　　　　　　**Mechanical weathering**

Clay minerals

{ Ions and compounds in solution

Gravel
Sand
Silt
Clay-sized particles

Transport

Precipitation　　　　Used by
from solution　　　　organisms

Transport

Deposition
(chemical sediment)

Deposition
(detrital sediments)

Lithification

Lithification

CHEMICAL SEDIMENTARY ROCK
(e.g., limestone)

DETRITAL SEDIMENTARY ROCKS
(e.g., sandstone)

James S. Monroe

(a)

James S. Monroe

(b)

■ **Figure 6.8**

(a) Microscopic view of sandstone showing its clastic texture consisting of fragments of minerals, mostly quartz, averaging about 0.5 mm across. How would you characterize the sorting and rounding? (b) Microscopic view of the crystalline texture of limestone showing a mosaic of calcite crystals that measure about 1 mm across.

(a) Conglomerate

(b) Sedimentary breccia

■ **Figure 6.9**

Detrital sedimentary rocks composed of gravel. (a) Rounded gravel is present in conglomerate. The particles in this specimen average 4 or 5 mm across. (b) Sedimentary breccia consists of angular gravel.
Source: Sue Monroe

deposited. For instance, gravel transported a few kilometers in a stream or moved to and fro by waves on a beach is invariably rounded. Thus, if you encounter sedimentary breccia, you can conclude that its angular gravel experienced very little transport. Considerable energy is needed to transport gravel, so it tends to be deposited in high-energy environments such as stream channels and on beaches.

Sandstone   *Sand* is simply a designation for detrital particles measuring between $\frac{1}{16}$ and 2 mm regardless of composition, so any mineral or rock fragment can be in

sandstone. Although composition is not a consideration for sandstone classification, geologists recognize varieties based on composition (■ Figures 6.5 and 6.10). *Quartz sandstone* is the most common and, as the name implies, is composed mostly of quartz grains. Another variety of sandstone called *arkose* contains at least 25% feldspar minerals.

Given that Earth's crust is made up of an estimated 51% feldspars (39% plagioclase and 12% potassium feldspars), 24% ferromagnesian silicates (olivine, pyroxene, amphiboles, and biotite), 12% quartz, and 13% other minerals, why is quartz by far the most common constituent of sandstones? The chance that any mineral will end up in a sedimentary rock depends on its availability, mechanical durability, and chemical stability. Feldspars and quartz are both abundant, but quartz is hard, lacks cleavage, and is chemically stable, whereas feldspars are not quite as hard, possess two directions of cleavage along which they break easily, and are more susceptible to chemical weathering. Other mechanically durable and chemically stable minerals such as zircon and tourmaline are rare in sedimentary rocks simply because they are uncommon in source rocks.

The only other particles of any consequence in sandstones are grains of chert derived from rock composed of microscopic quartz crystals and the micas (muscovite and biotite), although these rarely comprise more than a few percent of any sandstone. But what about the other ferromagnesian silicates, such as olivine, pyroxenes, and amphiboles, which make up about 24% of Earth's crust? Although common, with the exception of biotite they are scarce in sandstones because of their chemical instability (see Table 5.1).

(a) Quartz sandstone

(b) Arkose

■ **Figure 6.10**

(a) Quartz sandstone is the most common variety of sandstone, but (b), arkose a feldspar-rich sandstone, is also fairly abundant. Iron oxides are responsible for the red parts of the rock. Source: Sue Monroe

Sandstone forms in a variety of depositional environments, such as stream channels, beaches and barrier islands, deltas, and the continental shelf. It is a particularly common sedimentary rock and easily recognized by its gritty appearance and feel.

Mudrock   *Mudrock* is a general term that encompasses all detrital rocks composed of silt- and clay-sized particles (■ Figures 6.5 and 6.11). Varieties of mudrocks include *siltstone*, composed mostly of silt-sized particles, *mudstone*, a mixture of silt- and clay-sized particles, and *claystone*, composed primarily of clay-sized particles. The term *clay* actually has two meanings; it refers to sheet silicates known as *clay minerals*, and it is a size designation for particles smaller than $\frac{1}{256}$ mm. And even though mudstones and claystones contain significant amounts of clay minerals, their identity is not important in classifying these rocks. Some mudstones and claystones are designated as *shale* if they are fissile, meaning that they break along closely spaced parallel planes (Figure 6.11a).

About 40% of all detrital sedimentary rocks are mudrocks, making them more common than sandstone or conglomerate. Because silt and clay particles are so small, they can be transported by weak currents and kept suspended in water by minor turbulence. Consequently deposition takes place where turbulence is at a minimum, as in the quiet offshore waters of lakes, in lagoons, and on river floodplains.

Sue Monroe

(a)

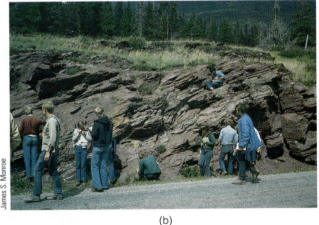

James S. Monroe

(b)

■ **Figure 6.11**

(a) Exposure of shale in Tennessee. Notice that the rock breaks along closely spaced planes, so it is fissile. (b) Mudstone in Glacier National Park, Montana. What do you think makes this rock red?

Table 6.2

Classification of Chemical and Biochemical Sedimentary Rocks

CHEMICAL SEDIMENTARY ROCKS		
Texture	**Composition**	**Rock Name**
Varies	Calcite ($CaCO_3$)	Limestone ⎫
Varies	Dolomite [$CaMg(CO_3)_2$]	Dolostone ⎬ Carbonate rocks
Crystalline	Gypsum ($CaSO_4 \cdot 2H_2O$)	Rock gypsum ⎫
Crystalline	Halite (NaCl)	Rock salt ⎬ Evaporites
BIOCHEMICAL SEDIMENTARY ROCKS		
Clastic	Calcite ($CaCO_3$) shells	Limestone (various types such as chalk and coquina)
Usually crystalline	Altered microscopic shells of SiO_2	Chert (various color varieties)
—	Carbon from altered land plants	Coal (lignite, bituminous, anthracite)

Chemical and Biochemical Sedimentary Rocks

Various compounds and ions taken into solution during chemical weathering are the raw materials for **chemical sedimentary rocks** (Table 6.2). They are so named because chemical processes are responsible for their origin, as when minerals form as the result of either inorganic chemical processes or the chemical activities of organisms. Some of these rocks have a *crystalline texture*, meaning they are composed of an interlocking mosaic of mineral crystals (Figure 6.8b). Others, however, have a clastic texture; some limestones, for instance, are composed of fragments of seashells. Organisms play an important role in the origin of those chemical sedimentary rocks designated **biochemical sedimentary rocks.**

Limestone and Dolostone   Limestone and dolostone are **carbonate rocks** because each is made up of minerals containing the carbonate radical (CO_3) (see Figure 2.11); limestone consists of calcite ($CaCO_3$) and dolostone is made up of dolomite [$CaMg(CO_3)_2$]. In Chapter 5 we noted that calcite rapidly dissolves in acidic water, but the chemical reaction leading to dissolution is reversible, so calcite can precipitate from solution under some circumstances. Some limestone, although probably not very much, forms by inorganic chemical precipitation. Travertine, a finely crystalline type of limestone precipitated around hot springs, is one example. And some limestones consist of small spherical grains known as *ooids* (■ Figure 6.12a), which form as layers of calcite chemically precipitate around some kind of nucleus such as a sand grain or shell frag-

ment. Lithified deposits of ooids are known as *oolitic limestones.*

Limestone is a common sedimentary rock; most has a large component of calcite that was extracted from seawater by organisms (Table 6.2). Corals, clams, oysters, snails, algae, and a number of other marine organisms construct their shells of aragonite, an unstable form of calcium carbonate that alters to calcite. When the organisms die, their skeletons or fragments of skeletons from clay- to gravel-size particles accumulate as sediment, which when lithified is limestone. The limestone known as *coquina* consists entirely of shell fragments cemented by calcium carbonate, and *chalk* is a variety of limestone made up of microscopic shells of organisms (Figure 6.12b and c). Because organisms play such an important role in their origin, most limestones are *biochemical sedimentary rocks* (Figure 6.12d).

Dolostone is similar to limestone, but most or all of it probably formed secondarily by the alteration of limestone. The consensus among geologists is that dolostone forms when magnesium replaces some of the calcium in calcite, thereby converting $CaCO_3$ to dolomite $CaMg(CO_3)_2$. One way this might take place is in a lagoon where evaporation rates are high, and much of the calcium in seawater is used in calcite ($CaCO_3$) and gypsum ($CaSO_4 \cdot 2H_2O$). Magnesium (Mg) becomes concentrated in the water, which then becomes denser, sinks, and permeates preexisting limestone, converting it to dolostone by replacing some of the calcium with magnesium.

Evaporites   In Chapter 5 we discussed the fact that during chemical weathering some minerals are taken into solution. Some of these dissolved substances are used by organisms and account for the origin of several varieties of limestone, but others precipitate from evaporating water by chemical processes, forming rocks known as **evaporites** (Table 6.2). *Rock salt,* composed of halite (NaCl), and *rock gypsum,* composed of gypsum ($CaSO_4 \cdot 2H_2O$), are the most common evaporites (■ Figure 6.13). Both have a crystalline texture.

All groundwater, water in lakes and streams, and seawater contains dissolved substances. If evaporation occurs, as it does in some lakes and restricted, marginal parts of seas, the volume of water decreases, thereby increasing the amount of dissolved mineral matter in relation to the volume of the fluid. If evaporation continues, the fluid eventually reaches the saturation point, the point at which it can no longer contain all the dissolved material and precipitation of minerals occurs. Rock salt is simply sodium chloride, and rock gypsum is calcium sulfate, both formed by the process just outlined.

Compared to mudrocks, sandstone, and limestone, evaporites are not very common but nevertheless are significant deposits in some areas. Vast deposits of rock salt and rock gypsum underlie parts of Michigan, Ohio, and New York; the Louann Salt of the Gulf Coast

region underlies a large area; and similar deposits are present in Canada and on other continents. In addition to rock salt and rock gypsum, other evaporite minerals and rocks are known, but most are rare. Some are important, though, as resources. Sylvite, for instance, a potassium chloride (KCl), is used in manufacturing fertilizers, dyes, and soaps.

Chert   *Chert* is a hard rock composed of microscopic crystals of quartz (SiO_2) (Table 6.2; ■ Figure 6.14). Color varieties of chert include *flint,* which is black because of inclusions of organic matter, and *jasper,* which is red or brown due to iron oxides. Because chert is hard and lacks cleavage, it can be shaped to form sharp cutting edges. Many cultures used chert to manufacture tools, spear points, and arrowheads.

Some chert is found as irregular masses or *nodules* within other rocks, especially limestones, and as distinct layers of *bedded chert* (Figure 6.14b). Chert nodules in limestone are clearly secondary; that is, they have replaced part of the host rock, apparently being precipitated from solution. Some bedded chert may precipitate from seawater, but because so little silica is dissolved in seawater, a biochemical origin is more likely. It seems that bedded chert

(b) Coquina

(a) Ooids

(d) Limestone with fossils

■ **Figure 6.12**

(a) Present-day ooids up to 2 mm across from the Bahamas. A rock made up of lithified ooids is oolitic limestone. Three types of limestone (b, c, and d).
(b) Coquina is composed of broken shells. (c) Chalk cliffs in Denmark. Chalk is made up of microscopic shells.
(d) Limestone with numerous fossil shells.

(c) Chalk

forms from deposits of shells of silica-secreting, single-celled organisms such as radiolarians and diatoms. Unfortunately, these shells are easily altered, so the evidence for a biochemical origin of bedded cherts is obscured.

Coal Coal, a biochemical sedimentary rock, consists of compressed, altered remains of land plants (Table 6.2). It forms in swamps and bogs where the water is oxygen deficient or where organic matter accumulates faster than it decomposes. In oxygen-deficient swamps and bogs, the bacteria that decompose vegetation can live without oxygen, but their wastes must be oxidized, and because little or no oxygen is present, wastes accumulate and kill the bacteria. Thus bacterial decay ceases and the vegetation does not completely decompose. These partly altered plant remains form organic muck, which, when buried and compressed, becomes *peat*, which looks rather like coarse pipe tobacco (■ Figure 6.15a). Where peat is

(a) Rock salt

(b) Rock gypsum

■ **Figure 6.13**

Evaporites form when minerals precipitate from evaporating water. (a) Core of rock salt from an oil well in Michigan. (b) Rock gypsum. Both rock salt and rock gypsum have many uses.

(a) Chert

(b) Bedded chert

■ **Figure 6.14**

(a) Chert is a dense, hard rock composed of microscopic crystals of quartz. (b) Thin layer of bedded chert.

abundant, as in Ireland and Scotland, it is used for fuel.

If peat is more deeply buried and compressed, and especially if it is heated too, it is converted to dull black or brown coal called *lignite,* in which plant remains are clearly visible (Figure 6.15b). During the change from organic muck to lignite, the easily vaporized or volatile elements of the vegetation such as oxygen, nitrogen, and hydrogen are driven off, enriching the residue in carbon; lignite contains about 70% carbon, whereas only about 50% is present in peat.

Bituminous coal, which contains about 80% carbon, is dense, black, and so thoroughly altered that plant remains can only rarely be seen (Figure 6.15c). It is a higher-grade coal than lignite because it burns more efficiently, but the highest-grade coal is *anthracite,* a metamorphic type of coal (see Chapter 7). It contains up to 98% carbon and, when burned, yields more heat per unit volume than other types of coal.

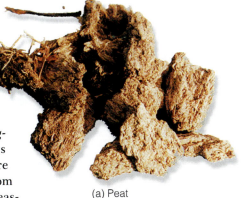

(a) Peat

(b) Lignite

(c) Bituminous coal

■ **Figure 6.15**

(a) Peat is partly decomposed plant material. It represents the first stage in the origin of coal. (b) Lignite is a dull variety of coal in which plant remains are still visible. (c) Bituminous coal is shinier and darker than lignite and only rarely are plant remains visible. Source: Sue Monroe

SEDIMENTARY FACIES

f you trace a layer of sediment or sedimentary rock laterally, it generally shows changes in composition, texture, or both, resulting from the simultaneous operation of different processes in adjacent depositional environments. For example, sand may be deposited in a high-energy nearshore marine environment, whereas mud and carbonate sediments accumulate simultaneously in the laterally adjacent low-energy offshore environments (■ Figure 6.16a). Deposition in each of these environments produces **sedimentary facies,** bodies of sediment that possess distinctive physical, chemical, and biological attributes.

Any aspect of sedimentary rocks that makes them recognizably different from adjacent rocks of the same age, or approximately the same age, can be used to establish a sedimentary facies. Figure 6.16a illustrates three sedimentary facies: a sandstone facies, a shale facies, and a limestone facies.

Marine Transgressions and Regressions

Many sedimentary rocks in the interiors of continents show clear evidence of having been deposited in transitional and marine environments. The rocks in Figure 6.16d, for instance, consist of a sandstone facies that was deposited in a nearshore marine environment overlain by shale and limestone facies deposited in offshore environments. We explain this vertical sequence of facies by deposition during a time when sea level rose with respect to the continent and the shoreline moved inland, thereby giving rise to a **marine transgression** (Figure 6.16). As the shoreline advanced inland, the depositional environments parallel to the shoreline did likewise. As a result of a marine transgression, the facies that formed in the offshore environments were superposed over the facies deposited in the nearshore environment, thus accounting for the vertical succession of sedimentary facies in Figure 6.16d.

Another important aspect of marine transgressions is that individual facies are deposited over a huge geographic area. Even though the nearshore environment is long and narrow at any particular time, deposition takes place continuously as the environment migrates

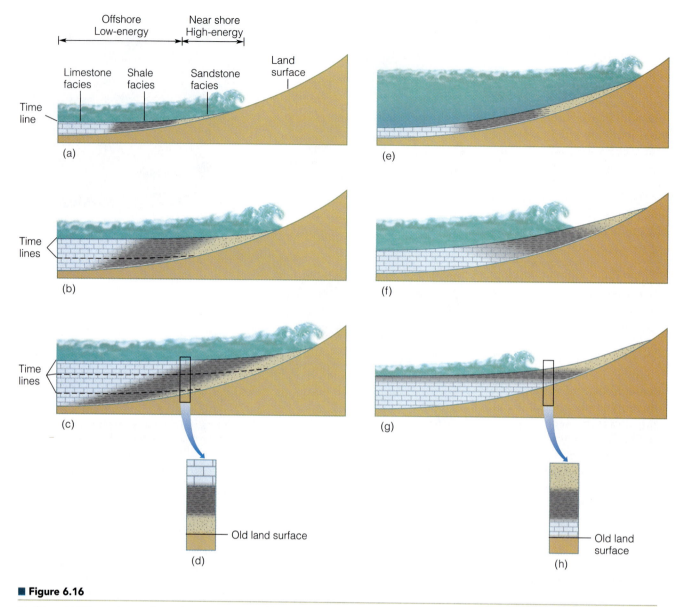

■ Figure 6.16

(a), (b), and (c) Three stages of a marine transgression. (d) Diagrammatic representation of the vertical sequence of facies resulting from a transgression. (e), (f), and (g) Three stages of a marine regression. (h) Diagrammatic representation of the vertical sequence of facies resulting from a regression.

landward. The sand deposit may be tens to hundreds of meters thick but has horizontal dimensions of length and width measured in hundreds of kilometers.

The opposite of a marine transgression is a **marine regression.** If sea level falls with respect to a continent, the shoreline and environments that parallel the shoreline move seaward. The vertical sequence produced by a marine regression has facies of the nearshore environment superposed over facies of offshore environments (Figure 6.16h). Marine regressions also account for the deposition of a facies over a large geographic area.

Causes of Marine Transgressions and Regressions

Uplift or subsidence of the continents, rates of seafloor spreading, and the amount of seawater contained in glaciers are sufficient to cause marine transgressions and regressions. If a continent is uplifted, it rises with respect to sea level, the shoreline moves seaward, and a regression ensues. The opposite type of movement, subsidence, results in a transgression.

Seafloor spreading causes these events by changing the volumes of the ocean basins. During comparatively rapid episodes of seafloor spreading, greater heat beneath the mid-oceanic ridges causes them to expand, displacing seawater onto the continents. When seafloor spreading is slower, the ridges subside and the volume of the ocean basins increases, and the seas retreat from the continents.

Changes in sea level related to the addition or removal of water from the oceans during glacial episodes also cause transgressions and regressions. During a widespread glacial episode, large quantities of seawater are on land as glacial ice, and consequently sea level falls with respect to continents. When glaciers melt, the water returns to the seas, and sea level rises.

What Would You Do ?

You live in the continental interior where flat-lying sedimentary rock layers are well exposed. Some local residents tell you of a location nearby where sandstone and mudstone with dinosaur fossils are overlain first by a seashell-bearing sandstone, followed upward by shale and finally limestone containing the remains of clams, oysters, and corals. How would you explain the presence of fossils, especially marine fossils so far from the sea, and how this vertical sequence of rocks came to be deposited?

READING THE STORY IN SEDIMENTARY ROCKS

The only record of prehistoric events is found in the rocks of Earth's crust, especially in sedimentary rocks, because they preserve evidence of the surface conditions that prevailed at the time of deposition. In short, sedimentary rocks acquired many of their properties as a result of the physical, chemical, and biological processes that operated in the depositional environment. Accordingly, geologists evaluate several features of these rocks such as textures (sorting and rounding) and composition. Wind-blown dune deposits, for instance, tend to be well sorted and well rounded, whereas glacial deposits are typically poorly sorted. In most cases, composition tells little about depositional processes, although we can infer from evaporites that they formed in arid environments. Other features of these rocks are *sedimentary structures* and *fossils*, both of which are very important in environmental analyses.

Sedimentary Structures

When sediment is deposited, it has a variety of features known as **sedimentary structures** that formed as a result of physical and biological processes operating in the depositional environment. One of the most common of these sedimentary structures is distinct layers known as **strata**, or **beds** (■ Figure 6.17). Beds vary in thickness from less than a millimeter up to many meters. Many sedimentary rock layers are separated from one another by surfaces known as *bedding planes*, above and below which the rocks differ markedly in composition, grain size, color, or a combination of features. Such abrupt changes indicate rapid changes in the type of sediment that accumulated or perhaps a period of nondeposition or erosion followed by renewed deposition. In contrast, a layer may grade upward from one rock type into another, thus indicating a gradual change in deposition. Almost all sedimentary rocks show some kind of stratification or bedding; a few, such as limestones that formed as coral reefs, lack this feature.

Many sedimentary rocks are characterized by **cross-bedding**, in which layers are arranged at an angle to the surface on which they are deposited (Figure 6.17). Cross-bedding is common in desert dunes and in sediments in stream channels and shallow marine environments. Invariably, cross-beds result from transport by wind or water and deposition on the downcurrent side of dune-like structures. Cross-beds are inclined downward, or dip, in the direction of flow and so are good indicators of ancient current directions, or **paleocurrents** (Figure 6.17).

In **graded bedding**, grain size decreases upward within a single bed (■ Figure 6.18). Most graded bedding formed from turbidity currents, underwater flows of sediment–water mixtures that are denser than sediment-free water. These flows move downslope along the bottom of the sea or a lake until they reach the relatively level seafloor or lake floor. There, they rapidly slow down and begin depositing transported sediment, the largest particles first followed by progressively smaller particles (Figure 6.18a, b). Although turbidity currents account for most graded bedding, some forms in stream channels during the waning stages of floods (Figure 6.18c).

In sand deposits, one commonly sees small-scale, ridgelike **ripple marks** on the surfaces of beds. One type of ripple mark is asymmetric in cross section, with a gentle upstream slope and a steep downstream slope. Known as *current ripple marks* (■ Figure 6.19b, c), they form as a result of currents that move in one direction as in a stream channel. Like cross-bedding, current ripple marks are good paleocurrent indicators. In contrast, the to-and-fro motion of waves produces ripples that tend to be symmetric in cross section.

(a)

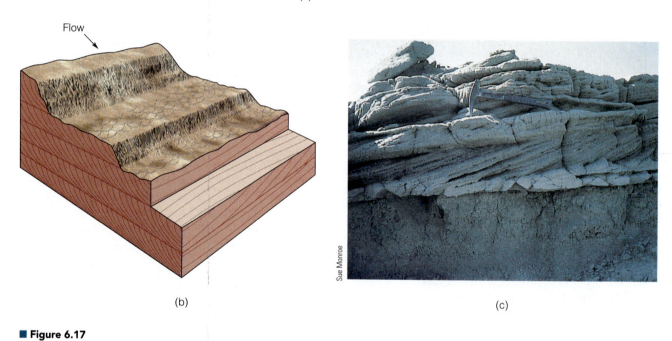

Flow

(b)

(c)

■ **Figure 6.17**

(a) Bedding or stratification is obvious in these alternating layers of mudrock (shale in this case) and sandstone. (b) Origin of cross-bedding by deposition on a sloping surface in a stream channel in this example. (c) Cross-beds in ancient sandstone in Montana. The hammer is about 30 cm long. Can you tell which direction the current flowed that produced this cross-bedding?

Known as *wave-formed ripple marks,* they form mostly in the shallow, nearshore waters of oceans and lakes (Figure 6.19d, e).

When clay-rich sediment dries, it shrinks and forms intersecting fractures known as **mud cracks** (■ Figure 6.20). These features in ancient sedimentary rocks indicate that the sediment was deposited where periodic drying was possible, as on a river floodplain, near a lakeshore, or where muddy deposits are exposed on marine shorelines at low tide.

Geologists routinely use the sedimentary structures we have discussed when they interpret depositional environments of sedimentary rocks. There are, however,

some sedimentary structures that form long after deposition and tell nothing of the original conditions of sedimentation. Nevertheless, some are interesting and attractive (■ Figure 6.21).

Fossils

Fossils, the remains or traces of ancient organisms, are mostly the hard skeletal parts such as shells, bones, and teeth, but under exceptional conditions even the soft-part anatomy may be preserved. Several frozen woolly mammoths have been discovered in Alaska and Siberia with hair, flesh, and internal organs preserved.

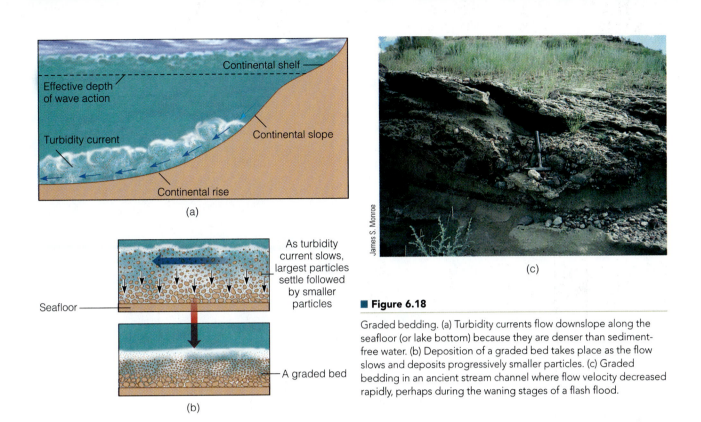

■ Figure 6.18

Graded bedding. (a) Turbidity currents flow downslope along the seafloor (or lake bottom) because they are denser than sediment-free water. (b) Deposition of a graded bed takes place as the flow slows and deposits progressively smaller particles. (c) Graded bedding in an ancient stream channel where flow velocity decreased rapidly, perhaps during the waning stages of a flash flood.

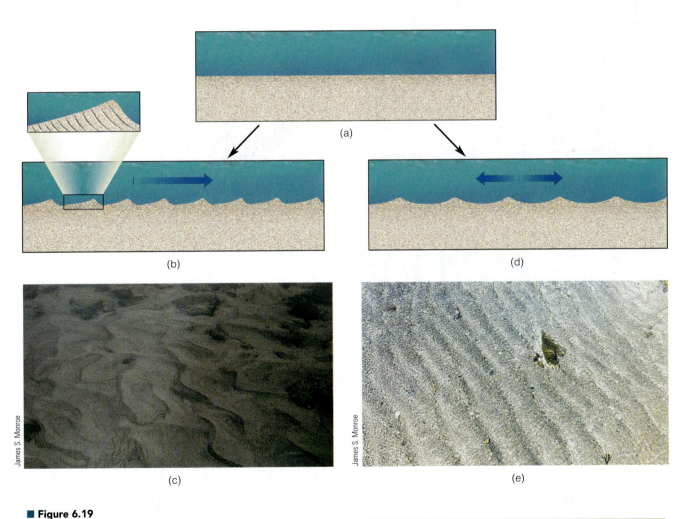

■ Figure 6.19

Ripple marks. (a) Undisturbed layer of sand. (b) Current ripple marks form in response to flow in one direction, as in a stream channel. The enlargement of one ripple shows its internal structure. Note that individual layers within the ripple are inclined, showing an example of cross-bedding. (c) Current ripples that formed in a small stream channel; flow was from right to left. (d) The to-and-fro currents of waves in shallow water deform the surface of the sand layer into wave-formed ripple marks. (e) Wave-formed ripple marks in sand in shallow seawater.

Richard V. Dietrich

Jeff Mayo/GeoPhoto Publishing

(a)   (b)

■ **Figure 6.20**

(a) Mud cracks form in clay-rich sediments when they dry and shrink. (b) Mud cracks in ancient rocks in Glacier National Park, Montana. Notice that the cracks have been filled by sediment.

Sue Monroe

■ **Figure 6.21**

Some sedimentary structures form long after deposition and tell nothing about the original depositional environment. This 50-cm-long geode formed by partial filling of a cavity by color-banded agate, which conforms to the walls of the cavity, and by quartz crystals pointing inward toward the center of the cavity.

The actual remains of organisms are known as *body fossils* to distinguish them from *trace fossils* such as tracks, trails, nests, and burrows, which are indications of ancient organic activity (see "Fossilization" on pages 168–169).

For any potential fossil to be preserved, it must escape the ravages of destructive processes such as running water, waves, scavengers, exposure to the atmosphere, and bacterial decay. Obviously, the soft parts of organisms are devoured or decomposed most rapidly, but even the hard skeletal parts are destroyed unless buried and protected in mud, sand, or volcanic ash. Even if buried, bones and shells may be dissolved by groundwater or destroyed by alteration of the host rock during metamorphism. Nevertheless, fossils are common. The remains of microscopic plants and animals are the most common, but these require specialized methods of recovery, preparation, and study and are not sought out by casual fossil collectors. Shells of marine animals are also common and easily collected in many areas, and even the bones and teeth of dinosaurs are much more common than most people realize (see Geo-Focus 6.1 on pages 170–171).

Some fossils retain their original composition and structure and are preserved as unaltered remains, but many have been altered in some way. Dissolved minerals may precipitate in the pores of bones, teeth, and shells or can fill the spaces within cells of wood. Wood may be preserved by silica replacing the woody tissues; it then is referred to as *petrified*, a term that means "to become stone." Silicon dioxide (SiO_2) or iron sulfide (FeS_2) can completely replace the calcium carbonate ($CaCO_3$) shells of marine animals. Insects and the leaves, stems, and roots of plants are commonly preserved as thin carbon films that show the details of the original organism. Shells and bones in sediment may be dissolved, leaving

GEOLOGY
IN UNEXPECTED PLACES

Sandstone Lion

The 9-m-long Lion Monument in Lucerne, Switzerland, was chiseled into sandstone in 1821 as a memorial to the approximately 850 Swiss soldiers who died during the French Revolution of 1792 in Paris (■ Figure 1a). Lukas Ahorn chiseled the monument into the sandstone wall of a quarry; the inscription above the lion pays honor to the "loyalty and courage of the Swiss." An officer on leave from the army at the time of the battle in Paris took the first steps to set up the monument.

Notice that the sandstone layers are inclined downward, or dip, to the left at about 50 degrees. We could postulate that (1) the original layers were horizontal and simply tilted 50 degrees into this position, or (2) perhaps they were rotated 140 degrees from their original position so that the layers are now upside down, or overturned in geologic parlance. To resolve this problem, you must determine which of the layers shown was at the top of the original sequence of beds and thus the youngest. In Figure 1b, notice that the cross-beds have a sharp angular contact with the younger rocks above them, whereas they are nearly parallel with the older rocks below, as in Figure 6.17. Accordingly, we conclude that the youngest rock layer is the one toward the upper left and the rock layers have not been overturned.

Having determined which layer is oldest and which is youngest, we now know that any rocks exposed to the right of the image are older than the ones shown and, of course, any to the left are younger. However, it is important to note that we have determined *relative ages* only—that is, which layers are older versus younger. Nothing in this image tells us the absolute age in number of years before the present. We consider relative and absolute ages more fully in Chapter 8.

Sue Monroe

(a)

Cross-bedding

Sue Monroe

(b)

■ **Figure 1**

(a) Lion Monument in Lucerne, Switzerland. (b) Cross-bedding shows sharp, angular contact with younger rocks above and nearly parallel contact with older rocks below.

Fossilization

Although the chance that any one organism will be preserved in the fossil record is slight, fossils are nevertheless common because so many billions of organisms lived during the past several hundred million years. Hard skeletal parts of organisms living where burial was likely are most common.

The bones of this dinosaur and the shells of these marine animals called ammonites have had minerals added to their pores, making them more durable.

Sue Monroe

Sue Monroe

Trace fossils do not include actual remains—only tracks, burrows, nests, and droppings.

A tiny amphibian track (left) and a coprolite (fossilized feces, above) from a carnivorous mammal. The coprolite is about 5.5 cm long.

Sue Monroe

This object looks like a clam, but it is simply sediment that filled a space formed when a clam shell dissolved. The material filling the space, called a mold, is a cast.

Sue Monroe

Fossil insect replaced by silicon dioxide (SiO_2).

Reed Wicander

Fossilized tree stump at Florissant Fossil Beds National Monument, Colorado. The woody tissue has been replaced by silicon dioxide. Mudflows 3 to 6 m deep buried the lower parts of many trees at this site.

A palm frond and an insect preserved by carbonization.

Insects preserved in amber, hardened resin secreted by coniferous trees.

This 6- or 7-month old frozen baby mammoth was found in Siberia in 1977. It measures 1.15 m long and 1 m high. Most of its hair has fallen out except around the feet. The carcass is about 40,000 years old.

This insect from the La Brea Tar Pits in Los Angeles is just one of hundreds of species of animals found in the asphaltlike substance in the pits.

Sue Monroe

Sovfoto/V.Khristoforov

J. Koivula/Science Source/Photo Researchers, Inc.

Sue Monroe

Sue Monroe

GEO**FOCUS**

6.1

Fossils and Fossilization

Shells of marine invertebrate animals such as clams and corals are the most easily collected fossils, but even fossils of some vertebrate animals—fish, amphibians, reptiles, birds, and mammals—are far more common than most people realize. For example, thousands of fossil fish are found on single surfaces within the Green River Formation of Wyoming. Cretaceous rocks in Montana contain an estimated 10,000 duck-billed dinosaurs in an area measuring only about 2 by 0.4 km; they were overcome by volcanic gases and subsequently buried in ash. A deposit in Jurassic rocks measuring only 18 by 15 m in Wyoming has yielded 4000 bones from about 20 large dinosaurs that became mired in mud and died (■ Figure 1). Hundreds of rhinoceroses, three-toed horses, saber-toothed deer, and other animals have been recovered from Miocene-aged ash in Nebraska (■ Figure 2).

Perhaps such remarkable concentrations of fossils make one think that some kinds of unknown processes operated in the past. In nearly all cases, though, we can find present-day examples to help us understand the conditions for fossilization. During the 1830s in Uruguay, flooding rivers buried thousands of horses and cattle that had died during a previous drought. Similarly, many of the fossil mammals in the White River

Reproduced from Bones for Barnum Brown: Adventures

■ Figure 1

Howe Quarry in Wyoming, where more than 4000 dinosaurs bones were recovered.

University of Nebraska State Museum

■ Figure 2

Paleontologists excavating rhinoceros (foreground) and horse (background) skeletons from volcanic ash near Orchard, Nebraska.

a cavity called a *mold* shaped like the shell or bone. If a mold is filled in, it becomes a *cast*. (See "Fossilization" on pages 168–169.)

If it were not for fossils, we would have no knowledge of trilobites, dinosaurs, and other extinct organisms. Thus fossils constitute our only record of ancient life. But they are not simply curiosities. In many geologic studies, it is necessary to correlate or determine age equivalence of sedimentary rocks in different areas. Such correlations are most commonly demonstrated

with fossils; we will discuss correlation more fully in Chapter 8. Fossils are also useful in determining environments of deposition.

Figuring Out the Environment of Deposition

Geologists rely on textures, sedimentary structures, and fossils to make interpretations about how a particular

■ Figure 3

Mural by Charles R. Knight showing Pleistocene mammals at the La Brea Tar Pits in Los Angeles, California, including a giant ground sloth, saber-toothed cats, a huge vulture, and a herd of mammoths in the background.

George C. Page Museum

Badlands of South Dakota were buried in floodplain deposits. Although a rare occurrence, African elephants become mired in mud and perish as did Jurassic dinosaurs in Wyoming. In 1984 thousands of caribou died in the Caniapiskau River of Quebec, Canada, and in Africa wildebeests die by the hundreds during river crossings. A bone bed in Canada with the remains of hundreds of horned dinosaurs appears to have formed in a similar fashion.

Scientists in a submersible saw the remains of a whale in the Gulf of Mexico that was slowly being covered by sediment. In fact, the sediment is similar to deposits in southern California from which fossil whales have been recovered. When Mount St. Helens erupted in 1980, mudflows transported and buried logs, stumps, and thousands of trees still in their original growth position. Following this event, stream erosion exposed trees buried in growth position in 1885. Fossil forests in what is now Yellowstone National Park, Wyoming, and Florissant Fossil Beds National Monument, Colorado, were buried in a similar manner.

The famous La Brea Tar Pits in downtown Los Angeles, California, contained numerous fossil saber-toothed cats, horses, camels, vultures, and armored mammals called glyptodonts (■ Figure 3). The tar (actually an asphaltlike substance) in the tar pits is simply the sticky residue that forms when the easily vaporized constituents of oil are lost. Of course, the area no longer supports saber-toothed cats and so on, but small animals are still trapped exactly as were their ancient counterparts.

The burial conditions for fossilization may be very rapid, as when mudflows engulf trees or animals, but they may just as well be quite slow, as in the case of the whale in the Gulf of California. In all cases, though, we have every reason to think that organisms in the fossil record were buried just as organic remains are buried today.

sedimentary rock body was deposited. Furthermore, they compare the features seen in ancient rocks with those in deposits forming today. But are we justified in using present-day processes and environments to make inferences about what happened when no human observers were present? Perhaps some examples will clarify this point.

Notice in ■ Figure 6.22a that the present-day deposits in and adjacent to this stream are similar to the ancient deposits nearby (Figure 6.22b) and to those in Figure 6.22c, which are more than 1 billion years old. Thus we conclude that the ancient deposits formed just as the recent ones have. If you visit a lakeshore or seashore, you might see small wave-formed ripple marks in sand near the shoreline. This observation indicates that the to-and-fro motion of water in waves deforms sand into nearly symmetrical ridges with intervening troughs (Figure 6.19d, e). In this example, as well as the previous one, we simply apply the principle of uniformitarianism (see Chapter 1) and conclude that the

■ **Figure 6.22**

(a) Gravel deposited by a swiftly flowing stream. Most transport and deposition take place when the stream is much higher than it is in this image. (b) This nearby deposit also formed when gravel was deposited in a similar stream channel. (c) These rocks in Michigan are about 1 billion years old, yet they show features similar to those in (a) and (b). Accordingly, we are justified in concluding that this conglomerate also resulted from deposition by running water.

same processes were responsible for both present-day and ancient deposits and ripple marks.

And, of course, fossils might also help in our endeavor to interpret depositional environments. It is true that fossils of land animals and plants are occasionally washed into transitional and marine environments, but most of them are preserved in deposits that accumulated on land. Fossils of marine-dwelling creatures obviously indicate transitional and marine environments.

Particles such as ooids in limestone (Figure 6.12a) form in shallow marine environments where currents are vigorous, and large-scale cross-bedding is typical of but not restricted to desert dunes. Glacial deposits are typically poorly sorted, lack discernable laying, and are

found with other features indicating glacial transport and deposition.

The Navajo Sandstone of the southwestern United States is an ancient desert dune deposit that formed when the prevailing winds blew from the northeast. But what evidence justifies this conclusion? This 300-m-thick sandstone is made up of well-sorted, well-rounded sand grains measuring 0.2 to 0.5 mm in diameter. Furthermore, it has cross-beds up to 30 m high (■ Figure 6.23a) and current ripple marks, both typical of desert dunes. Some of the sand layers have preserved dinosaur tracks and tracks of other land-dwelling animals, ruling out the possibility of a marine origin. In short, the Navajo Sandstone possesses several features that indicate a desert dune depositional

James S. Monroe

(a)

Alan Mayo/GeoPhoto

Muav Limestone

Bright Angel Shale

Tapeats Sandstone

(b)

■ Figure 6.23

Ancient sedimentary rocks and their interpretation. (a) The Jurassic-aged Navajo Sandstone in Zion National Park, Utah, is a wind-blown dune deposit. Vertical fractures intersect cross-beds, giving this cliff its checkerboard appearance—hence, the name Checkerboard Mesa. (b) View of the Tapeats Sandstone, Bright Angel Shale, and Muav Limestone in the Grand Canyon in Arizona. These rocks were deposited during a marine transgression. Compare with the vertical sequence of rocks in Figure 6.16d.

What Would You Do ?

No one was present millions of years ago to record data about the climate, geography, and geologic processes. So how is it possible to decipher unobserved past events? In other words, what features in rocks, and especially sedimentary rocks, would you look for to determine what happened in the far distant past? Can you think of any economic reasons to decipher Earth history from the record preserved in rocks?

In the Grand Canyon of Arizona several formations are well exposed; a *formation* is simply a widespread unit of rock, especially sedimentary rock, recognizably different from the rocks above and below. A vertical sequence consisting of the Tapeats Sandstone, Bright Angel Shale, and Muav Limestone is present in the lower part of the canyon (Figure 6.23b), all of which contain features, including fossils, clearly indicating that they were deposited in transitional and marine environments. As a matter of fact, all three were forming simultaneously in different adjacent environments, and during a marine transgression they were deposited in the vertical sequence now seen. That is, they conform closely to the sequence shown in Figure 6.16d.

Although determining depositional environments satisfies our curiosity about Earth history, such endeavors are not of only academic interest. Quite the contrary. Knowing how sedimentary rocks were deposited might be helpful in locating hydrocarbon reservoirs or finding and recovering adequate supplies of groundwater. Indeed, the professional journals of such organizations as the *American Association of Petroleum Geologists* often contain articles about depositional environments of various resource-bearing rocks.

IMPORTANT RESOURCES IN SEDIMENTS AND SEDIMENTARY ROCKS

Sediments and sedimentary rocks or the materials they contain have a variety of uses. Sand and gravel are essential to the construction industry, pure clay deposits are used for ceramics, and limestone is used in the manufacture of cement and in blast furnaces where iron ore is refined to make steel. Evaporites are the source of table salt as well as a number of chemical compounds, and rock gypsum is used to manufacture wallboard. Sand composed mostly of

environment. And, finally, the cross-beds are inclined downward toward the southwest, indicating that the prevailing winds were from the northeast.

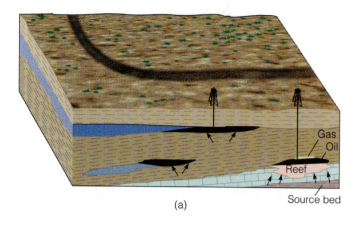

(a)

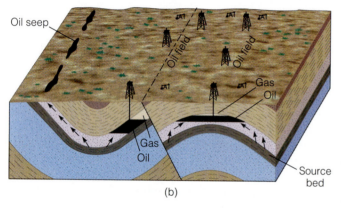

(b)

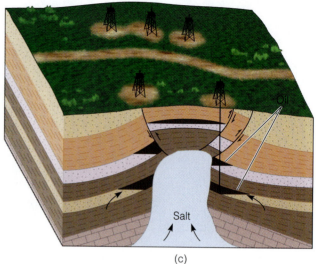

(c)

■ **Figure 6.24**

Oil and natural gas traps. The arrows in the diagrams indicate the migration of hydrocarbons. (a) Two examples of stratigraphic traps. (b) Two examples of structural traps, one formed by folding, the other by faulting. (c) An example of structures adjacent to a salt dome where oil and natural gas might be trapped.

quartz, or what is called *silica sand,* has a variety of uses, including the manufacture of glass, refractory bricks for blast furnaces, and molds for casting iron, aluminum, and copper alloys.

Placer deposits are surface accumulations resulting from the separation and concentration of materials of greater density from those of less density in streams and on beaches. Much of the gold recovered during the California gold rush (1848–1853) was mined from placer deposits, and placers of a variety of other minerals such as diamonds and tin are important.

The tiny island nation of Nauru, with one of the highest per capita incomes in the world, has an economy based almost entirely on mining and exporting phosphate-bearing sedimentary rock used in fertilizers. More than half of Florida's mineral value comes from mining phosphate rock (see Chapter 11). Dolostones in Missouri are the host rocks for ores of lead and zinc. Diatomite, a lightweight, porous sedimentary rock composed of the microscopic silica (SiO_2) shells of single-celled plants, is used in gas purification and to filter a number of fluids such as molasses, fruit juices, water, and sewage.

Petroleum and Natural Gas

Petroleum and natural gas are *hydrocarbons,* meaning they are composed solely of hydrogen and carbon. Hydrocarbons form when the remains of microorganisms settle to the seafloor or lake floor where little oxygen is available to decompose them. As they are buried under layers of sediment, they are heated and transformed into petroleum and natural gas.

The rock in which hydrocarbons formed is the *source rock,* but for them to accumulate in economic quantities, they must migrate into and become trapped in some kind of *reservoir rock* beneath cap rock. Otherwise, they would migrate upward and eventually seep out at the surface (■ Figure 6.24). Effective reservoir rocks contain considerable pore space where appreciable quantities of hydrocarbons accumulate. Furthermore, reservoir rocks must possess high *permeability,* or the capacity to transmit fluids; otherwise, hydrocarbons cannot be extracted in reasonable quantities.

Many oil and gas traps are called *stratigraphic traps* (Figure 6.24a), because they owe their existence to

SUSAN M. LANDON

Exploring for Oil and Natural Gas

Susan M. Landon began her career in 1974 with Amoco Production Company and, in 1989, opened her own consulting office in Denver, Colorado. She is currently a partner in the exploration group, Thomasson Partner Associates. In 1990 she was elected president of the American Institute of Professional Geologists and, in 1992, treasurer of the American Association of Petroleum Geologists.

I am an independent petroleum geologist. I specialize in applying geologic principles to oil and gas exploration in frontier areas—places where little or no exploration has occurred and few or no hydrocarbons have been discovered. It is very much like solving a mystery. The Earth provides a variety of clues—rock type, organic content, stratigraphic relationships, structure, and the like—that geologists must piece together to determine the potential for the presence of hydrocarbons.

For example, as part of my work for Amoco, I explored the Precambrian Midcontinent Rift frontier in the north central portion of the United States. Some rifts, like the Gulf of Suez and the North Sea, are characterized by significant hydrocarbon reserves, and the presence of an unexplored rift basin in the center of North America is intriguing. A copper mine in the Upper Peninsula of Michigan, the White Pine Mine, has historically been plagued by oil bleeding out of fractures in the shale. For many years, this had been documented as academically interesting because the rocks are much older than those that typically have been associated with hydrocarbon production.

Field and laboratory work documented that the copper-bearing shale at the White Pine Mine contained adequate organic material to be the source of the oil. The thermal history of the basin was modeled to determine the timing of hydrocarbon generation. If hydrocarbons had been generated prior to deposition of an effective seal and formation of a trap, the hydrocarbons would have leaked out naturally into the atmosphere.

Further work identified sandstones with enough porosity to serve as reservoirs for hydrocarbons. Seismic data were acquired and interpreted to identify specific traps. We then had to convince Amoco management that this prospect had high enough potential to contain hydrocarbon reserves to offset the significant risks and costs. In this case, management agreed that the risk was offset by the potential for a very large accumulation of hydrocarbons, and a well was authorized. Amoco drilled a 5441-m well in Iowa to test the prospect, at a cost of nearly $5 million. The well was dry (economically unsuccessful), but the geologic information obtained as a result of drilling the well will be used to continue to define prospective drilling sites in the Midcontinent Rift.

My career in the petroleum industry began with Amoco Production Company, and after 15 years, I made the decision to leave the company to work independently. For several years, I consulted for a variety of companies, assisting them in exploration projects. I am now involved in a partnership with several other exploration geologists and geophysicists, and we are actively exploring for oil and natural gas in the United States. We develop ideas, defining areas that we believe may be prospective, and attract other companies as partners to assist us in further exploration and drilling. Currently, I am working on projects located in Wyoming, southern Illinois, southern Michigan, Iowa, and Minnesota. I was Manager of Exploration Training when I left Amoco, and as a result, I currently teach a few courses (such as Petroleum Geology for Engineers), which allows me to travel to places like Cairo, Egypt; Quito, Ecuador; and Houston, Texas.

variations in the rock layers or strata. Ancient coral reefs are good stratigraphic traps, and some of the oil in the Persian Gulf region is trapped in ancient reefs. *Structural traps* result when rocks are deformed by folding, fracturing, or both (Figure 6.24b). In sedimentary rocks deformed into a series of folds, hydrocarbons migrate to and accumulate in the high parts of these structures. Displacement of rocks along faults (fractures along which movement has occurred) also yields situations conducive to trapping hydrocarbons (Figure 6.24b).

In the Gulf Coast region, hydrocarbons are commonly found in structures adjacent to salt domes. A vast layer of ancient rock salt is present, which is a low-density sedimentary rock. When deeply buried beneath more dense sediments such as sand and mud, it rises toward the surface in pillars known as *salt domes*. As the rock salt rises, it penetrates and deforms the overlying rock layers, forming structures along its margins that may trap petroleum and gas (Figure 6.24c).

Although large concentrations of petroleum are present in many areas of the world, more than 50% of all proven reserves are in the Persian Gulf region! Furthermore, some of the oil fields are gigantic; at least 20 are expected to yield more than 5 billion barrels of oil each, and several have already surpassed this figure.

Many nations are heavily dependent on imports of Persian Gulf oil. The United States is the third largest oil producer and yet must import about 9.2 million barrels daily, just over half of what it uses, much of it from the Middle East, although it also receives large quantities from Venezuela and Mexico. Most geologists think that all the truly gigantic oil fields have already been found but concede that some significant discoveries are yet to be made. One must view these potential discoveries in the proper perspective, however. For example, the discovery of an oil field comparable to that of the North Slope of Alaska (about 10 billion barrels) (■ Figure 6.25) constitutes about a 2-year supply for the United States at the current consumption rate.

Oil Shale and Tar Sands

Another source of petroleum is *oil shale,* a fine-grained, thinly layered sedimentary rock containing an organic substance called *kerogen* from which liquid oil and combustible gases can be extracted. During the Middle Ages (from about A.D. 600 to 1350), Europeans used oil shale as a solid fuel for heating. Small oil shale industries existed in the eastern United States during the 1850s but were discontinued when drilling and pumping liquid oil began in 1858.

About two thirds of all known oil shale is found in the United States, most of it in the Green River Formation of Wyoming, Colorado, and Utah, but large deposits are also present in South America. No liquid oil or combustible gases are currently produced from oil shale be-

cause the process of extracting them is more expensive than conventional drilling and pumping. Nevertheless, oil shale represents a huge untapped resource; according to one estimate, 80 billion barrels of oil could be recovered from the Green River Formation with present technology.

If the Green River Formation were mined for its oil shale, large-scale excavations would be necessary as would huge quantities of water, in a region where water is already in short supply. Overall, a venture of this magnitude would have a profound impact on the environment. And given the current and expected future rates of consumption, even if the full potential of oil shale could be realized, it would still not solve all our energy needs.

In addition to oil shale, liquid petroleum can be derived from *tar sand,* a type of sandstone with its pore spaces filled with viscous, asphaltlike hydrocarbons, the sticky residue of once-liquid petroleum from which the volatile elements have been lost. The Athabaska tar sands in Alberta, Canada, is one of the largest deposits of this type, but the United States has few tar sands and cannot look to this source as a significant future energy resource. Large-scale extraction of oil is currently taking place from the Athabaska tar sands, which contain an estimated 300 billion barrels of recoverable petroleum.

Coal

Historically, most coal mined in the United States was bituminous coal from Kentucky, Ohio, West Virginia, and Pennsylvania that formed in coastal swamps during the Pennsylvanian Period between 286 and 320 million years ago. Now, though, huge lignite and subbituminous coal deposits in the western United States are becoming increasingly important. During 2001, more than a billion metric tons of coal were mined in the United States, more than half from mines in Wyoming, West Virginia, and Kentucky.

Anthracite coal is an especially desirable resource because it burns hot with a smokeless flame. Unfortunately, it is the least common type of coal, so most coal used for heating buildings and for generating electrical energy is bituminous (see Figure 6.15c). *Coke* is a hard, gray substance consisting of the fused ash of bituminous coal. It is prepared by heating coal and driving off the volatile matter and is used to fire blast furnaces for steel production. Synthetic oil and gas and a number of other products are also made from bituminous coal and lignite.

Uranium

Most uranium used in nuclear reactors in North America comes from the complex potassium-, uranium-, vanadium-bearing mineral *carnotite* found in some sed-

(a)

(b)

■ **Figure 6.25**

(a) The 1290-km-long Alaska Pipeline carries oil from Prudhoe Bay to Valdez, where it is then shipped by oil tankers. The pipeline, seen here near Fairbanks, Alaska, is above ground where it crosses areas of permafrost and underground elsewhere. (b) The red line encircles the producing oil fields in the Prudhoe Bay region on Alaska's North Slope, but discoveries have also been made in adjacent areas. This is the largest oil field in North America. It was discovered in 1968.

imentary rocks. Some uranium is also derived from *uraninite* (UO_2), a uranium oxide found in granitic rocks and hydrothermal veins. Uraninite is easily oxidized and dissolved in groundwater, transported elsewhere, and chemically reduced and precipitated in the presence of organic matter.

The richest uranium ores in the United States are in the Colorado Plateau area of Colorado and adjoining parts of Wyoming, Utah, Arizona, and New Mexico. These ores, consisting of fairly pure masses and encrustations of carnotite, are associated with plant remains including petrified trees in sandstones that formed in ancient stream channels.

Large reserves of low-grade uranium ore also are found in the Chattanooga Shale. The uranium is finely disseminated in this black, organic-rich mudrock that underlies large parts of several states, including Illinois,

Indiana, Ohio, Kentucky, and Tennessee. Canada is the world's largest producer and exporter of uranium.

Banded Iron Formation

Banded iron formation (BIF), consisting of alternating thin layers of chert and iron minerals, mostly the iron oxides hematite and magnetite, is a chemical sedimentary rock of great economic importance (■ Figure 6.26). Banded iron formations are present on all the continents and account for most of the iron ore mined in the world today. Vast BIF are present in the Lake Superior region of the United States and Canada and in the Labrador trough of eastern Canada.

The origin of BIF is not fully understood, and none are currently forming. Fully 92% of all BIF were deposited in shallow seas during the Proterozoic Eon between 2.5 and 2.0 billion years ago. A highly reactive element, iron in the presence of oxygen combines to form rustlike oxides that are not readily soluble in water. During early Earth history, little oxygen was present in the atmosphere, so little was dissolved in seawater. However, soluble reduced iron (Fe^{+2}) and silica were present in seawater.

Geologic evidence indicates that abundant photosynthesizing organisms were present about 2.5 billion years ago. These organisms, such as bacteria, release oxygen as a by-product of respiration; thus they released oxygen into seawater and caused large-scale precipitation of iron oxides and silica as banded iron formations.

(a)

(b)

■ **Figure 6.26**

(a) The Empire Mine at Palmer, Michigan, where iron ore from the Negaunee Iron Formation is mined. (b) Banded iron formation at Ishpeming, Michigan. At this location the rocks are alternating layers of red chert and silver-colored iron minerals. (c) The iron ore in this region is processed and shaped into pellets about 1 cm across, and then either shipped directly to steel mills elsewhere or stored for later shipment.

(c)

6 REVIEW
WORKBOOK

Chapter Summary

- Detrital sediment consists of weathered solid particles, whereas chemical sediment consists of minerals extracted from solution by inorganic chemical processes and the activities of organisms.

- Sedimentary particles are designated in order of decreasing size as gravel, sand, silt, and clay.

- During transport, sedimentary particles are rounded and sorted, although the degree of rounding and sorting depends on particle size, transport distance, and depositional process.

- Any area where sediment is deposited is a depositional environment. Major depositional settings are continental, transitional, and marine, each of which includes several specific depositional environments.

- Lithification takes place when sediments are compacted and cemented and thus converted into sedimentary rock. Silica and calcium carbonate are the most common chemical cements, but iron oxide and iron hydroxide cements are important in some rocks.

- Sedimentary rocks are classified as detrital or chemical:

  a. Detrital sedimentary rocks consist of solid particles derived from preexisting rocks.

  b. Chemical sedimentary rocks are derived from substances in solution by inorganic chemical processes or the activities of organisms. A subcategory called biochemical sedimentary rocks is recognized.

- Carbonate rocks contain minerals with the carbonate radical $(CO_3)^{-2}$ as in limestone and dolostone. Dolostone forms when magnesium partly replaces the calcium in limestone.

- Evaporites include rock salt and rock gypsum, both of which form by inorganic precipitation of minerals from evaporating water.

- Coal is a type of biochemical sedimentary rock composed of the altered remains of land plants.

- Sedimentary facies are bodies of sediment or sedimentary rock that are recognizably different from adjacent sediments or rocks.

- Vertical sequences of rocks with offshore facies overlying nearshore facies form when sea level rises with respect to the land, causing a marine transgression. A rise in the land relative to sea level causes a marine regression, which results in nearshore facies overlying offshore facies.

- Sedimentary structures such as bedding, cross-bedding, and ripple marks form in sediments during or shortly after deposition. These features help geologists determine ancient current directions and depositional environments.

- Fossils provide the only record of prehistoric life and are useful for correlation and environmental interpretations.

- Depositional environments of ancient sedimentary rocks are determined by studying sedimentary textures and structures, examining fossils, and making comparisons with present-day sediments deposited by known processes.

- Many sediments and sedimentary rocks, including sand, gravel, evaporites, coal, and banded iron formations, are important natural resources. Most oil and natural gas are found in sedimentary rocks.

Important Terms

bed (bedding) (p. 163)
biochemical sedimentary rock
  (p. 158)
carbonate rock (p. 158)
cementation (p. 154)
chemical sedimentary rock
  (p. 158)

compaction (p. 154)
cross-bedding (p. 163)
depositional environment
  (p. 153)
detrital sedimentary rock (p. 155)
evaporite (p. 159)
fossil (p. 166)

graded bedding (p. 163)
lithification (p. 153)
marine regression (p. 163)
marine transgression (p. 162)
mud crack (p. 164)
paleocurrent (p. 163)
ripple mark (p. 164)

rounding (p. 152)   sedimentary rock (p. 150)   sorting (p. 152)
sediment (p. 150)   sedimentary structure (p. 163)   strata (stratification) (p. 163)
sedimentary facies (p. 161)

Review Questions

1. A vertical sequence of sedimentary rocks in which nearshore facies overlie offshore facies resulted from:

   a. _____ deposition by turbidity currents; b. _____ a marine regression; c. _____ meandering stream deposition; d. _____ compaction and cementation of evaporites; e. _____ granitization.

2. Although there are local exceptions, most detrital sand is made up largely of the mineral:

   a. _____ biotite; b. _____ plagioclase; c. _____ augite; d. _____ calcite; e. _____ quartz.

3. Cross-bedding preserved in sedimentary rocks is a good indication of:

   a. _____ the intensity of organic activity; b. _____ ancient current directions; c. _____ the amount of silica cement; d. _____ how old the containing rocks are; e. _____ whether or not the rocks contain important resources.

4. Dolostone forms from limestone when:

   a. _____ limestone loses some of its water; b. _____ evaporite deposition takes place in a lagoon; c. _____ organic matter accumulates in a swamp; d. _____ sand is deposed over a layer of mud; e. _____ some of the calcium in limestone is replaced by magnesium.

5. A deposit of detrital sediment characterized as poorly sorted has:

   a. _____ a large amount of calcium carbonate cement; b. _____ cross-bedding and current ripple marks; c. _____ particles of markedly different sizes; d. _____ more ferromagnesian silicates than nonferromagnesian silicates; e. _____ iron oxide cement.

6. Arkose is a type of sedimentary rock:

   a. _____ with at least 25% feldspar minerals; b. _____ in which the common mica is biotite; c. _____ with a high percentage of fossil shells; d. _____ in which calcium has replaced magnesium; e. _____ that is fissile.

7. Lithification involves cementation and:

   a. _____ replacement; b. _____ compaction; c. _____ inversion; d. _____ granitization; e. _____ stoping.

8. Which one of the following is a common evaporite rock?

   a. _____ rock gypsum; b. _____ siltstone; c. _____ chert; d. _____ claystone; e. _____ bituminous coal.

9. Which one of the following is a body fossil?

   a. _____ clam shell; b. _____ elephant footprint; c. _____ worm burrow; d. _____ dinosaur nest; e. _____ reptile footprint.

10. Traps for petroleum and natural gas formed by folding and fracturing of rocks are known as _____ traps.

    a. _____ lithologic; b. _____ compaction; c. _____ stratigraphic; d. _____ compositional; e. _____ structural.

11. Organisms play a significant role in the origin of _____ sedimentary rocks.

    a. _____ evaporite; b. _____ biochemical; c. _____ detrital; d. _____ lithified; e. _____ crystalline.

12. The most common sedimentary rocks are:

    a. _____ sandstone, mudstone, limestone; b. _____ rock salt, siltstone, coal; c. _____ arkose, sedimentary breccia, claystone; d. _____ shale, dolostone, chert; e. _____ lignite, conglomerate, shale.

13. In one of our national parks you observe a vertical sequence of rocks with sandstone at the base followed upward by shale and limestone, each of which contains fossil clams and corals. Give an account of the history of these rocks. That is, how were they deposited and how did they come to be superposed in the sequence observed?

14. Describe two ways that limestone forms.

15. In what fundamental way(s) do detrital sedimentary rocks differ from chemical sedimentary rocks?

16. Describe the processes leading to the lithification of deposits of sand and mud.

17. The United States used about 860 million metric tons of coal per year from its reserve of 243 billion metric tons. Assuming that all of this coal can be mined, how long will it last at the current rate of consumption? Why

is it improbable that all of this reserve can be mined?

18. Illustrate and describe two sedimentary structures that can be used to determine ancient current directions.

19. Fossils are quite common, yet very few of the organisms that ever lived were actually fossilized. How can you explain this apparent contradiction?

20. How does coal form, and what varieties of coal do geologists recognize? Which of these varieties makes the best fuel?

21. Explain what structural and stratigraphic traps are and how they differ from one another.

22. What distinction(s) do geologists make to differentiate siltstone, mudstone, and claystone? Also, explain what shale is.

23. How does dolostone differ from limestone, and what is one possible way that dolostone forms?

24. You encounter a sandstone with well-sorted, well-rounded sand grains, large-scale cross-bedding, and reptile tracks. What can you infer about the depositional environment?

World Wide Web Activities

PHYSICAL Geology⇌Now Assess your understanding of this chapter's topics with additional quizzing and comprehensive interactivities at

http://earthscience.brookscole.com/physgeo5e

as well as current and up-to-date weblinks, additional readings, and InfoTrac College Edition exercises.

Metamorphism and Metamorphic Rocks

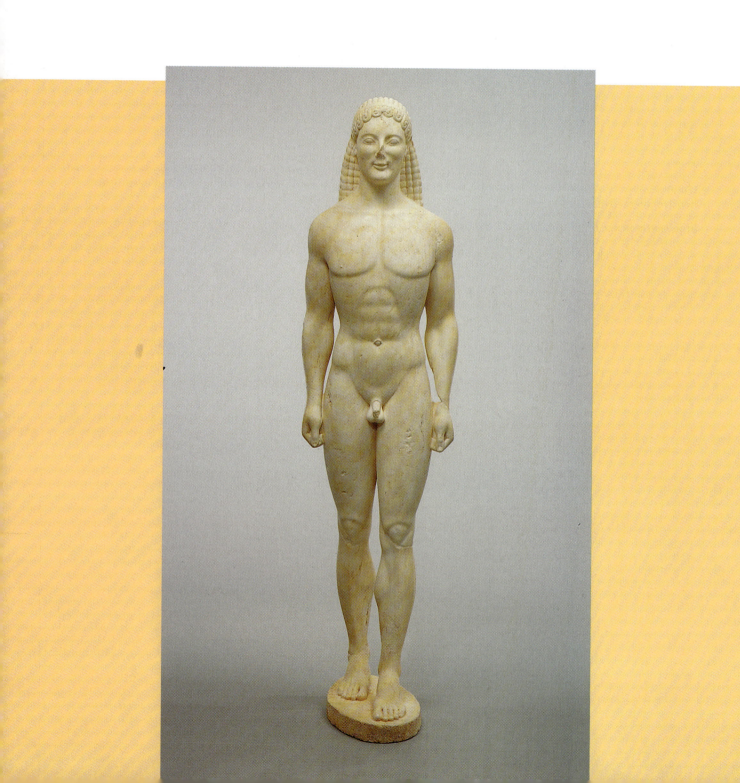

CHAPTER 7
OUTLINE

PHYSICAL Geology⇌Now *This icon, appearing throughout the book, indicates an opportunity to explore interactive tutorials, animations, or practice problems available on the Physical GeologyNow Web site at http://earthscience, brookscole.com/physgeo5e.*

OBJECTIVES

At the end of this chapter, you will have learned that

- Metamorphic rocks result from the transformation of other rocks by various processes occurring beneath Earth's surface.

- Heat, pressure, and fluid activity are the three agents of metamorphism.

- Contact, dynamic, and regional metamorphism are the three types of metamorphism.

- Metamorphic rocks are typically divided into two groups, primarily on the basis of texture.

- Metamorphic rocks with a foliated texture include slate, phyllite, schist, gneiss, and amphibolite.

- Metamorphic rocks with a nonfoliated texture include marble, quartzite, greenstone, and hornfels.

- Metamorphic rocks can be grouped into metamorphic zones based on the presence of index minerals that form under specific temperature and pressure conditions.

- The appearance of particular index minerals results from increasing metamorphic intensity.

- Metamorphism is associated with all three types of plate boundaries, but it is most widespread along convergent plate boundaries.

- Many metamorphic minerals and rocks are valuable metallic ores, building materials, and gemstones.

This Greek kouros, which stands 206 cm tall, has been the object of an intensive authentication study by the Getty Museum. Using a variety of geologic tests, scientists have determined that the kouros was carved from dolomitic marble that probably came from the Cape Vathy quarries on the island of Thasos. Source: Garry Hobart/GeoImagery

Introduction

Its homogeneity, softness, and various textures have made marble, a metamorphic rock formed from limestone or dolostone, a favorite rock of sculptors throughout history. As the value of authentic marble sculptures has increased over the years, the number of forgeries has also increased. With the price of some marble sculptures in the millions of dollars, private collectors and museums need some means of ensuring the authenticity of the work they are buying. Aside from the monetary considerations, it is important that forgeries not become part of the historical and artistic legacy of human endeavor.

Experts have traditionally relied on artistic style and weathering characteristics to determine whether a marble sculpture is authentic or a forgery. Because marble is not very resistant to weathering, however, forgers have been able to produce the weathered appearance of an authentic work. Using newly developed techniques, geologists can now distinguish a naturally weathered marble surface from one that has been artificially altered. Yet, there are examples in which expert opinion is still divided on whether a sculpture is authentic or not.

One of the best examples is the Greek kouros (a sculpted figure of a Greek youth) the J. Paul Getty Museum in Malibu, California, purchased for a reputed price of $7 million in 1984 (see the chapter opening photo). Because some of its stylistic features caused some experts to question its authenticity, the museum had a variety of geochemical and mineralogical tests performed in an effort to determine the authenticity of the kouros.

Although numerous scientific tests have not unequivocally proven authenticity, they have shown that the weathered surface layer of the kouros bears more similarities to naturally occurring weathered surfaces of dolomitic marble than to known artificially produced surfaces. Furthermore, no evidence indicates that the surface alteration of the kouros is of modern origin.

Unfortunately, despite intensive study by scientists, archaeologists, and art historians, opinion is still divided on the authenticity of the Getty kouros. Most scientists accept that the kouros was carved sometime around 530 B.C. Pointing to inconsistencies in its style of sculpture for that period, other art historians think it is a modern forgery.

Regardless of whether the Getty kouros will be proven to be authentic or a forgery, geologic testing to authenticate marble sculptures is now an important part of many museums' curatorial functions. To help geologists in the

What Would You Do ?

As the director of a major museum, you have the opportunity to purchase, for a considerable sum of money, a newly discovered marble bust by a famous ancient sculptor. You want to be sure it is not a forgery. What would you do to ensure that the bust is authentic and not a clever forgery? After all, you are spending a large sum of the museum's money. As a nonscientist, how would you go about making sure the proper tests are being performed to authenticate the bust?

authentication of marble sculptures, a large body of data about the characteristics and origin of marble is being amassed as more sculptures and marble quarries are analyzed.

Metamorphic rocks (from the Greek *meta*, "change," and *morpho*, "shape") constitute the third major group of rocks. They result from the transformation of other rocks by metamorphic processes that usually occur beneath Earth's surface (see Figure 1.14). During metamorphism, rocks are subjected to sufficient heat, pressure, and fluid activity to change their mineral composition, texture, or both, thus forming new rocks. These transformations take place below the melting temperature of the rock; otherwise, an igneous rock would result.

A useful analogy for metamorphism is baking a cake. Just like a metamorphic rock, the resulting cake depends on the ingredients, their proportions, how they are mixed together, how much water or milk is added, and the temperature and length of time used for baking the cake.

Except for marble and slate, most people are not familiar with metamorphic rocks. Students frequently ask us, Why is it important to study metamorphic rocks and processes? Our answer to that is: Look around you.

A large portion of Earth's continental crust is composed of metamorphic and igneous rocks. Together they form the crystalline basement rocks underlying the sedimentary rocks of a continent's surface. These basement rocks are widely exposed in regions of the continents known as *shields*, which have been very stable during the past 600 million years

(■ Figure 7.1). Metamorphic rocks also constitute a sizable portion of the crystalline core of large mountain ranges. Some of the oldest known rocks, dated at 3.96 million years from the Canadian Shield, are metamorphic, indicating they formed from even older rocks!

Metamorphic rocks such as marble and slate are used as building materials, and certain metamorphic minerals are economically important. Garnets, for example, are used as gemstones or abrasives; talc is used in cosmetics, in manufacturing paint, and as a lubricant; and kyanite is used in producing heat-resistant materials in sparkplugs. Therefore knowledge of metamorphic rocks and processes is of economic value.

Asbestos, a metamorphic mineral, is used for insulation and fireproofing and is widespread in buildings and building materials. There are different forms of asbestos, however, and they do not all pose the same health hazards. Recognizing this fact would have been useful during the debates over the dangers posed to the public's health by asbestos (see Geo-Focus 7.1).

What Would You Do

The problem of removing asbestos from public buildings is an important national health and political issue. The current policy of the Environmental Protection Agency (EPA) mandates that all forms of asbestos are treated as identical hazards. Yet studies indicate that only one form of asbestos is a known health hazard. Because the cost of asbestos removal has been estimated to be as high as $100 billion, many people are questioning whether it is cost effective to remove asbestos from all public buildings where it has been installed.

As a leading researcher on the health hazards of asbestos, you have been asked to testify before a congressional committee on whether it is worthwhile to spend so much money for asbestos removal. How would you address this issue in terms of formulating a policy that balances the risks and benefits of removing asbestos from public buildings? What role would geologists play in formulating this policy?

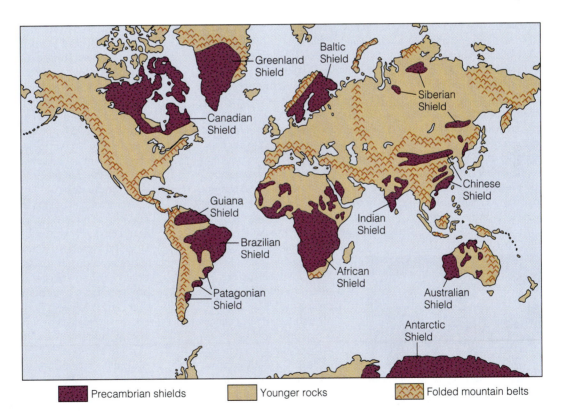

■ **Figure 7.1**

Metamorphic rock occurrences. Shields are the exposed portions of the crystalline basement rocks underlying each continent; these areas have been very stable during the past 600 million years. Metamorphic rocks also constitute the crystalline core of major mountain belts.

GEOFOCUS

7.1

Asbestos: Good or Bad?

Asbestos (from the Latin, meaning "unquenchable") is a general term applied to any silicate mineral that easily separates into flexible fibers. The combination of such features as noncombustibility and flexibility makes asbestos an important industrial material of considerable value. In fact, asbestos has more than 3000 known uses, including brake linings, fireproof fabrics, and heat insulators.

Asbestos is divided into two broad groups, serpentine and amphibole. *Chrysotile* is the fibrous form of serpentine asbestos (■ Figure 1); it is the most valuable type and constitutes the bulk of all commercial asbestos. Its strong, silky fibers are easily spun and can withstand temperatures of up to 2750°C.

The vast majority of chrysotile asbestos is in serpentine, a type of rock formed by the alteration of ultramafic igneous rocks such as peridotite under low- and medium-grade metamorphic conditions. Other chrysotile results when the metamorphism of magnesium limestone or dolostone produces discontinuous serpentine bands within the carbonate beds.

Among the varieties of amphibole asbestos, *crocidolite* is the most common. Also known as blue asbestos, crocidolite is a long, coarse, spinning fiber that is stronger but more brittle than chrysotile and also less resistant to heat. Crocidolite is found in such metamorphic rocks as slates and schists and is thought to form by the solid-state alteration of other minerals as a result of deep burial.

Despite the widespread use of asbestos, the U.S. Environmental Protection Agency (EPA) instituted a gradual ban on all new asbestos products. The ban was imposed because some forms of asbestos can cause lung cancer and scarring of the lungs if fibers are inhaled. Because the EPA apparently paid little attention to the issue of risks versus benefits when it enacted this rule, the U.S. Fifth Circuit Court of Appeals overturned the EPA ban on asbestos in 1991.

The threat of lung cancer has also resulted in legislation mandating the removal of asbestos already in place in all public buildings, including all public and private schools. However, important questions have been raised concerning the threat posed by asbestos and the additional potential hazards that may arise from its improper removal.

Current EPA policy mandates that all forms of asbestos are to be treated as identical hazards. Yet studies indicate that only the amphibole forms constitute a known health hazard. Chrysotile, whose fibers tend to be curly, does not become lodged in the lungs. Furthermore, its fibers are generally soluble and disappear in tissue. In contrast, crocidolite has long, straight, thin fibers that penetrate the lungs and stay there. These fibers irritate the lung tissue and over a long period of time can lead to lung cancer. Thus crocidolite, and not chrysotile, is overwhelmingly responsible for asbestos-related lung cancer. Because about 95% of the asbestos in place in the United States is chrysotile, many people question whether the dangers from asbestos are exaggerated.

Removing asbestos from buildings where it has been installed could cost as much as $100 billion. Unless the material containing the asbestos is disturbed, asbestos does not shed fibers and thus does not contribute to airborne asbestos that can be inhaled. Furthermore, improper removal of asbestos can lead to contamination. In most cases of improper removal, the concentration of airborne asbestos fibers is far higher than if the asbestos had been left in place.

The problem of asbestos contamination is a good example of how geology affects our lives and why a basic knowledge of science is important.

Collection of the J. Paul Getty Museum

■ **Figure 1**

Specimen of chrysotile from Thetford, Quebec, Canada. Chrysotile is the fibrous form of serpentine asbestos.

WHAT ARE THE AGENTS OF METAMORPHISM?

The three agents of metamorphism are heat, pressure, and fluid activity. During metamorphism, the original rock undergoes change to achieve equilibrium with its new environment. The changes may result in the formation of new minerals and/or a change in the texture of the rock, brought about by the reorientation of the original minerals. In some instances, the change is minor, and features of the original rock can still be recognized. In other cases, the rock changes so much that the identity of the original rock can be determined only with great difficulty, if at all.

Besides heat, pressure, and fluid activity, time is also important to the metamorphic process. Chemical reactions proceed at different rates and thus require different amounts of time to complete. Reactions involving silicate compounds are particularly slow, and because most metamorphic rocks are composed of silicate minerals, it is thought that metamorphism is a slow geologic process.

Heat

Heat is an important agent of metamorphism because it increases the rate of chemical reactions that may produce minerals different from those in the original rock. The heat may come from intrusive magmas or result from deep burial in the crust such as occurs during subduction along a convergent plate boundary.

When rocks are intruded by bodies of magma, they are subjected to intense heat that affects the surrounding rock; the most intense heating usually occurs adjacent to the magma body and gradually decreases with distance from the intrusion. The zone of metamorphosed rocks that forms in the country rock adjacent to an intrusive igneous body is usually rather distinct and easy to recognize.

Recall that temperature increases with depth and that Earth's geothermal gradient averages about 25°C/km. Rocks forming at the surface may be transported to great depths by subduction along a convergent plate boundary and thus subjected to increasing temperature and pressure. During subduction, some minerals may be transformed into other minerals that are more stable under the higher temperature and pressure conditions.

Pressure

When rocks are buried, they are subjected to increasingly greater **lithostatic pressure**; this pressure results from the weight of the overlying rocks and is applied equally in all directions (■ Figure 7.2a). A similar situation occurs when an object is immersed in water. For ex-

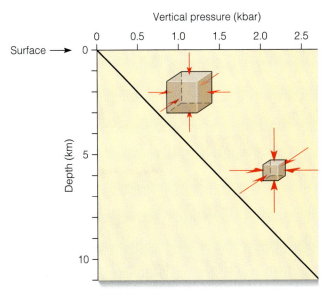

1 kilobar (kbar) = 1000 bars
Atmospheric pressure at sea level = 1 bar

(a)

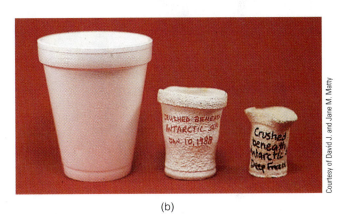

Courtesy of David J. and Jane M. Matty

(b)

■ **Figure 7.2**

(a) Lithostatic pressure is applied equally in all directions in Earth's crust due to the weight of overlying rocks. Thus pressure increases with depth, as indicated by the sloping black line. (b) A similar situation occurs when 200-ml Styrofoam cups are lowered to ocean depths of approximately 750 m and 1500 m. Increased water pressure is exerted equally in all directions on the cups, and they consequently decrease in volume while still maintaining their general shape. Source: (a): From C. Gillen, *Metamorphic Geology*, Figure 4.4, p. 73. Copyright © 1982. Reprinted with the kind permission of Kluwer Academic Publishers and C. Gillen.

ample, the deeper a Styrofoam cup is submerged in the ocean, the smaller it gets because water pressure increases with depth and is exerted on the cup equally in all directions, thereby compressing the Styrofoam (Figure 7.2b).

Just as in the Styrofoam cup example, rocks are subjected to increasing lithostatic pressure with depth such that the mineral grains within a rock become more closely packed. Under these conditions, the minerals may *recrystallize*; that is, they become smaller, denser minerals.

Eric Johnson

■ Figure 7.3

Differential pressure is pressure that is unequally applied to an object. Rotated garnets are a good example of the effects of differential pressure applied to a rock during metamorphism. This rotated garnet (center) came from a schist in northeast Sardinia.

seawater moving through hot basaltic rock of the oceanic crust transforms olivine into the metamorphic mineral serpentine:

$$2Mg_2SiO_4 + 2H_2O \rightarrow Mg_3Si_2O_5(OH)_4 + MgO$$

OLIVINE   WATER   SERPENTINE   CARRIED AWAY IN SOLUTION

The chemically active fluids important in the metamorphic process come primarily from three sources. The first is water trapped in the pore spaces of sedimentary rocks as they form. A second is the volatile fluid within magma. The third source is the dehydration of water-bearing minerals such as gypsum ($CaSO_4 \cdot 2H_2O$) and some clays.

WHAT ARE THE THREE TYPES OF METAMORPHISM?

Geologists recognize three major types of metamorphism: contact metamorphism in which magmatic heat and fluids act to produce change; dynamic metamorphism, which is principally the result of high differential pressures associated with intense deformation; and regional metamorphism, which occurs within a large area and is caused primarily by mountain-building forces. Even though we will discuss each type of metamorphism separately, the boundary between them is not always distinct and depends largely on which of the three metamorphic agents was dominant.

Contact Metamorphism

Contact metamorphism takes place when a body of magma alters the surrounding country rock. At shallow depths, intruding magma raises the temperature of the surrounding rock, causing thermal alteration. Furthermore, the release of hot fluids into the country rock by the cooling intrusion can aid in the formation of new minerals.

Important factors in contact metamorphism are the initial temperature and size of the intrusion as well as the fluid content of the magma and/or country rock. The initial temperature of an intrusion is controlled, in part, by its composition: mafic magmas are hotter than felsic magmas (see Chapter 3) and hence have a greater thermal effect on the rocks surrounding them. The size of the intrusion is also important. In the case of small intrusions, such as dikes and sills, usually only those rocks in immediate contact with the intrusion are affected. Because large intrusions, such as batholiths, take a long time to cool, the increased temperature in the surrounding rock may last long enough for a larger area to be affected.

Along with lithostatic pressure resulting from burial, rocks may also experience **differential pressures** (■ Figure 7.3). In this case, the pressures are not equal on all sides, and the rock is consequently distorted. Differential pressures typically occur during deformation associated with mountain building and can produce distinctive metamorphic textures and features.

Fluid Activity

In almost every region of metamorphism, water and carbon dioxide (CO_2) are present in varying amounts along mineral grain boundaries or in the pore spaces of rocks. These fluids, which may contain ions in solution, enhance metamorphism by increasing the rate of chemical reactions. Under dry conditions, most minerals react very slowly, but when even small amounts of fluid are introduced, reaction rates increase, mainly because ions move readily through the fluid and thus enhance chemical reactions and the formation of new minerals.

The following reaction provides a good example of how new minerals can be formed by **fluid activity**. Here,

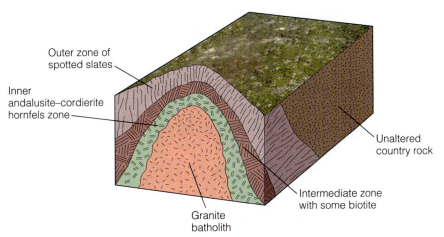

Outer zone of spotted slates

Inner andalusite–cordierite hornfels zone

Granite batholith

Intermediate zone with some biotite

Unaltered country rock

■ Figure 7.4

A metamorphic aureole typically surrounds many igneous intrusions. The metamorphic aureole associated with this idealized granite batholith contains three zones of mineral assemblages reflecting the decreases in temperature with distance from the intrusion. An andalusite–cordierite hornfels forms adjacent to the batholith. This is followed by an intermediate zone of extensive recrystallization in which some biotite develops, and farthest from the intrusion is the outer zone, which is characterized by spotted slates.

Fluids also play an important role in contact metamorphism. Many magmas are wet and contain hot, chemically active fluids that may emanate into the surrounding rock. These fluids can react with the rock and aid in the formation of new minerals. In addition, the country rock may contain pore fluids that, when heated by the magma, also increase reaction rates.

Temperatures can reach nearly 900°C adjacent to an intrusion, but they gradually decrease with distance. The effects of such heat and the resulting chemical reactions usually occur in concentric zones known as **aureoles** (■ Figure 7.4). The boundary between an intrusion and its aureole may be either sharp or transitional (■ Figure 7.5).

Metamorphic aureoles vary in width depending on the size, temperature, and composition of the intrusion as well as the mineralogy of the surrounding country rock. Typically, large intrusive bodies have several metamorphic zones, each characterized by distinctive mineral assemblages indicating the decrease in temperature with distance from the intrusion (Figure 7.4). The zone closest to the intrusion, and hence subject to the highest temperatures, may contain high-temperature metamorphic minerals (that is, minerals in equilibrium with the higher-temperature environment) such as sillimanite. The outer zones may be characterized by lower-temperature metamorphic minerals such as chlorite, talc, and epidote.

The formation of new minerals by contact metamorphism depends not only on proximity to the intrusion but also on the composition of the country rock. Shales, mudstones, impure limestones, and impure dolostones are particularly susceptible to the formation of new minerals by contact metamorphism, whereas pure sandstones or pure limestones typically are not.

Because heat and fluids are the primary agents of contact metamorphism, two types of contact metamorphic rocks are generally recognized: those resulting from baking of country rock and those altered by hot solutions. Many of the rocks resulting from contact metamorphism have the texture of porcelain; that is, they are hard and fine grained. This is particularly true for rocks with a high clay content, such as shale. Such texture results because the clay minerals in the rock are baked, just as a clay pot is baked when fired in a kiln.

During the final stages of cooling, when intruding magma begins to crystallize, large amounts of hot, watery solutions are often released. These solutions may react with the country rock and produce new metamorphic minerals. This process, which usually occurs near Earth's surface, is called *hydrothermal alteration* (from the Greek *hydro*, "water," and *therme*, "heat") and may result in valuable mineral deposits. Geologists think that many of the world's ore deposits result from the migration of metallic ions in hydrothermal solutions. Examples are copper, gold, iron ores, tin, and zinc in various localities including Australia, Canada, China, Cyprus, Finland, Russia, and the western United States.

Dynamic Metamorphism

Most **dynamic metamorphism** is associated with fault (fractures along which movement has occurred) zones where rocks are subjected to high differential pressures. The metamorphic rocks that result from pure dynamic metamorphism are called *mylonites* and are typically restricted to narrow zones adjacent to faults. Mylonites are hard, dense, fine-grained rocks, many of which are characterized by thin laminations (■ Figure 7.6). Tectonic settings where mylonites occur include the Moine Thrust Zone in northwest Scotland, the Adirondack Highlands in New York, and portions of the San Andreas fault in California (see Chapter 13).

Igneous rock   Metamorphic rock

David J. Matty

■ Figure 7.5

A sharp and clearly defined boundary (red line) occurs between the intruding light-colored igneous rock on the left and the dark metamorphosed country rock on the right. The intrusion is part of the Peninsular Ranges Batholith, east of San Diego, California.

Eric Johnson

■ Figure 7.6

Mylonite from the Adirondack Highlands, New York. Note the thin laminations.

Regional Metamorphism

Most metamorphic rocks result from **regional metamorphism,** which occurs over a large area and is usually caused by tremendous temperatures, pressures, and deformation within the deeper portions of the crust. Regional metamorphism is most obvious along convergent plate boundaries where rocks are intensely deformed and recrystallized during convergence and subduction. Within these metamorphic rocks there is usually a gradation of metamorphic intensity from areas that were subjected to the most intense pressures and/or highest temperatures to areas of lower pressures and temperatures. Such a gradation in metamorphism can be recognized by the metamorphic minerals that are present.

Regional metamorphism is not confined to only convergent margins. It also occurs in areas where plates diverge, although usually at much shallower depths because of the high geothermal gradient associated with these areas.

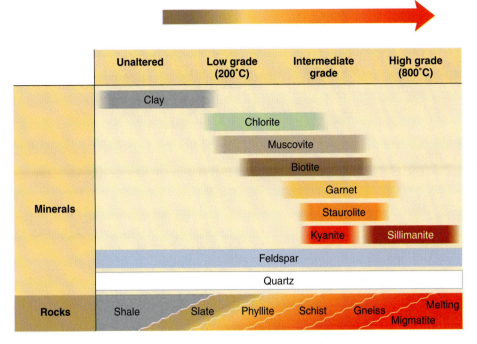

Change in mineral assemblage and rock type with increasing metamorphism of shale. When a clay-rich rock such as shale is subjected to increasing metamorphism, new minerals form, as shown by the various colored bars. The progressive appearance of particular minerals allows geologists to recognize low-, intermediate-, and high-grade metamorphic zones.

From field studies and laboratory experiments, certain minerals are known to form only within specific temperature and pressure ranges. Such minerals are known as **index minerals** because their presence allows geologists to recognize low-, intermediate-, and high-grade metamorphic zones (■ Figure 7.7).

When a clay-rich rock such as shale is metamorphosed, new minerals form as a result of metamorphic processes. The mineral chlorite, for example, is produced under relatively low temperatures of about 200°C, so its presence indicates low-grade metamorphism. As temperatures and pressures continue to increase, new minerals form that are stable under those conditions. Thus there is a progression in the appearance of new minerals from chlorite, whose presence indicates low-grade metamorphism, to sillimanite, whose presence indicates high-grade metamorphism and temperatures exceeding 500°C.

Different rock compositions develop different index minerals. When sandy dolomites are metamorphosed, for example, they produce an entirely different set of index minerals. Thus a specific set of index minerals commonly forms in specific rock types as metamorphism progresses.

Although such common minerals as mica, quartz, and feldspar can occur in both igneous and metamorphic rocks, other minerals such as andalusite, sillimanite, and kyanite generally occur only in metamorphic rocks derived from clay-rich sediments. Although these three minerals all have the same chemical formula (Al_2SiO_5), they differ in crystal structure and other physical properties because each forms under a different range of pressures and temperatures. Thus they are sometimes used as index minerals for metamorphic rocks formed from clay-rich sediments.

PHYSICAL **Geology ⇌ Now** ™ Click Geology Interactive to work through an activity on Rock Cycle.

HOW ARE METAMORPHIC ROCKS CLASSIFIED?

For purposes of classification, metamorphic rocks are commonly divided into two groups: those exhibiting a foliated (from the Latin *folium,* "leaf") texture and those with a nonfoliated texture (Table 7.1).

Foliated Metamorphic Rocks

Rocks subjected to heat and differential pressure during metamorphism typically have minerals arranged in a parallel fashion, giving them a **foliated texture** (■ Figure 7.8). The size and shape of the mineral grains determine whether the foliation is fine or coarse. If the foliation is such that the individual grains cannot be recognized without magnification, the rock is said to be slate (■ Figure 7.9a). A coarse foliation results when granular minerals such as quartz and feldspar are segregated into roughly parallel and streaky zones that differ in composition and color as in gneiss. Foliated metamorphic rocks can be arranged in order of increasingly coarse grain size and perfection of foliation.

Slate is a very fine-grained metamorphic rock that commonly exhibits *slaty cleavage* (Figure 7.9b). Slate is

Table 7.1

Classification of Common Metamorphic Rocks

Texture	Metamorphic Rock	Typical Minerals	Metamorphic Grade	Characteristics of Rocks	Parent Rock
Foliated	Slate	Clays, micas, chlorite	Low	Fine-grained, splits easily into flat pieces	Mudrocks, volcanic ash
	Phyllite	Fine-grained quartz, micas, chlorite	Low to medium	Fine-grained, glossy or lustrous sheen	Mudrocks
	Schist	Micas, chlorite, quartz, talc, hornblende, garnet, staurolite, graphite	Low to high	Distinct foliation, minerals visible	Mudrocks, carbonates, mafic igneous rocks
	Gneiss	Quartz, feldspars, hornblende, micas	High	Segregated light and dark bands visible	Mudrocks, sandstones, felsic igneous rocks
	Amphibolite	Hornblende, plagioclase	Medium to high	Dark, weakly foliated	Mafic igneous rocks
	Migmatite	Quartz, feldspars, hornblende, micas	High	Streaks or lenses of granite intermixed with gneiss	Felsic igneous rocks mixed with sedimentary rocks
Nonfoliated	Marble	Calcite, dolomite	Low to high	Interlocking grains of calcite or dolomite, reacts with HCl	Limestone or dolostone
	Quartzite	Quartz	Medium to high	Interlocking quartz grains, hard, dense	Quartz sandstone
	Greenstone	Chlorite, epidote, hornblende	Low to high	Fine-grained, green	Mafic igneous rocks
	Hornfels	Micas, garnets, andalusite, cordierite, quartz	Low to medium	Fine-grained, equidimensional grains, hard, dense	Mudrocks
	Anthracite	Carbon	High	Black, lustrous, subconcoidal fracture	Coal

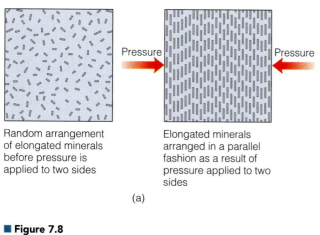

Random arrangement of elongated minerals before pressure is applied to two sides

Elongated minerals arranged in a parallel fashion as a result of pressure applied to two sides

(a)

Reed Wicander

(b)

■ **Figure 7.8**

(a) When rocks are subjected to differential pressure, the mineral grains are typically arranged in a parallel fashion, producing a foliated texture. (b) Photomicrograph of a metamorphic rock with a foliated texture showing the parallel arrangement of mineral grains.

GEOLOGY
IN UNEXPECTED PLACES

Starting Off with a Clean Slate

Slate is a common metamorphic rock that has many uses. Two familiar uses are in the playing surface of billiard tables and roofing shingles.

Although slate is abundant throughout the world, most of it is unsuitable for billiard tables. For billiard tables, the slate must have a very fine grain so it can be honed to a smooth surface, somewhat elastic so it will expand and contract with the table's wood frame, and essentially nonabsorbent. Presently Brazil, China, India, and Italy are the major exporters of billiard table–quality slate, with the best coming from the Liguarian region of northern Italy. Most quality tables use at least 1-inch-thick slate that is split into three pieces. Although using three slabs requires extra work to ensure a tight fit and smooth surface, a table with three pieces is preferred over a single piece because it is less likely to fracture. Furthermore, the slate is usually slightly larger than the playing surface so that it extends below the rails of the table, thus giving additional strength to the rails and stability to the table. In addition, a quality table will have a wood backing glued to the underside of the slate so the felt cloth that is stretched tightly over the slate's surface can be stapled to the wood to provide a smooth playing surface.

Slate has been used as a roofing material for centuries. When properly installed and maintained, slate normally lasts for 60 to 125 years; many slate roofs have been around for more than 200 years. In the United States, slate roofing tiles typically come in shades of gray, green, purple, black, and red. There are 36 common sizes of tiles, ranging from 12 to 24 inches long with the width about half the length. The typical slate tile is usually $\frac{1}{4}$ inch thick. Thicker tiles may be used, but they are harder to work with and greatly increase the weight of the roof.

The years between 1897 and 1914 witnessed the height of the U.S. roofing slate industry in both quantity and value of output. By the end of the 19th century, more than 200 slate quarries were operating in 13 states. With the introduction of asphalt shingles, which can be mass produced, easily transported, and installed at a much lower cost than slate shingles, the slate shingle industry in the United States began to decline around 1915. The renewed popularity of historic preservation and the recognition of slate's durability, however, have brought a resurgence in the slate roofing industry. It's not that unusual these days for geology to be overhead as well as underfoot.

■ **Figure 1**

Different colored slates make up the roof of this elementary school in Mount Pleasant, Michigan.

the result of low-grade regional metamorphism of shale or, more rarely, volcanic ash. Because it can easily be split along cleavage planes into flat pieces, slate is an excellent rock for roofing and floor tiles, billiard and pool table tops, and blackboards (Figure 7.9c). The different colors of most slates are caused by minute amounts of graphite (black), iron oxide (red and purple), and/or chlorite (green).

Phyllite is similar in composition to slate but coarser grained. The minerals, however, are still too small to be

(a) Slate

Phyllite

■ **Figure 7.10**

Specimen of phyllite. Note the lustrous sheen as well as the bedding (upper left to lower right) at an angle to the cleavage of the specimen.

(b)

(c)

■ **Figure 7.9**

(a) Hand specimen of slate. (b) This panel of Arvonia Slate from Albemarne Slate Quarry, Virginia, shows bedding (upper right to lower left) at an angle to the slaty cleavage. (c) Slate roof of Chalet Enzian, Switzerland.

identified without magnification. Phyllite can be distinguished from slate by its glossy or lustrous sheen (■ Figure 7.10). It represents an intermediate grain size between slate and schist.

Schist is most commonly produced by regional metamorphism. The type of schist formed depends on the intensity of metamorphism and the character of the original rock (■ Figure 7.11). Metamorphism of many rock types can yield schist, but most schist appears to have formed from clay-rich sedimentary rocks (Table 7.1).

All schists contain more than 50% platy and elongated minerals, all of which are large enough to be clearly visible. Their mineral composition imparts a *schistosity* or *schistose foliation* to the rock that usually produces a wavy type of parting when split. Schistosity is common in low- to high-grade metamorphic environments, and each type of schist is known by its most conspicuous mineral or minerals, such as mica schist, chlorite schist, and talc schist.

Gneiss is a metamorphic rock that is streaked or has segregated bands of light and dark minerals. Gneisses are composed mostly of granular minerals such as quartz and/or feldspar, with lesser percentages of platy or elongated minerals such as micas or amphiboles (■ Figure 7.12). Quartz and feldspar are the principal light-colored minerals, whereas biotite and hornblende are the typical dark minerals. Gneiss typically breaks in an irregular manner, much like coarsely crystalline nonfoliated rocks.

Most gneiss probably results from recrystallization of clay-rich sedimentary rocks during regional metamorphism (Table 7.1). Gneiss also can form from igneous rocks such as granite or older metamorphic rocks.

Another fairly common foliated metamorphic rock is *amphibolite*. A dark rock, it is composed mainly of hornblende and plagioclase. The alignment of the hornblende crystals produces a slightly foliated texture. Many amphibolites result from medium- to high-grade metamorphism of such ferromagnesian silicate-rich igneous rocks as basalt.

In some areas of regional metamorphism, exposures of "mixed rocks" having both igneous and high-grade metamorphic characteristics are present. In these rocks,

(a) Garnet–mica schist

Sue Monroe

(b) Hornblende–mica–garnet

Sue Monroe

■ **Figure 7.11**

Schist. (a) Garnet–mica schist. (b) Hornblende–mica–garnet schist.

called *migmatites,* streaks or lenses of granite are usually intermixed with high-grade ferromagnesian-rich metamorphic rocks, thereby imparting a wavy appearance to the rocks (■ Figure 7.13).

Most migmatites are thought to be the product of extremely high-grade metamorphism, and several models for their origin have been proposed. Part of the problem in determining the origin of migmatites is explaining how the granitic component formed. According to one model,

Reed Wicander

■ **Figure 7.12**

Gneiss is characterized by segregated bands of light and dark minerals. This folded gneiss is exposed at Wawa, Ontario, Canada.

the granitic magma formed in place by the partial melting of rock during intense metamorphism. Such an origin is possible provided that the host rocks contained quartz and feldspars and that water was present. Another possibility is that the granitic components formed by the redistribution of minerals by recrystallization in the solid state—that is, by pure metamorphism.

Nonfoliated Metamorphic Rocks

In some metamorphic rocks, the mineral grains do not show a discernable preferred orientation. Instead, these rocks consist of a mosaic of roughly equidimensional minerals and are characterized as having a **nonfoliated texture** (■ Figure 7.14). Most nonfoliated metamorphic rocks result from contact or regional metamorphism of rocks with no platy or elongate minerals. Frequently, the only indication that a granular rock has been metamorphosed is the large grain size resulting from recrystallization. Nonfoliated metamorphic rocks are generally of two types: those composed mainly of only one mineral— for example, marble or quartzite; and those in which the different mineral grains are too small to be seen without magnification, such as greenstone and hornfels.

Marble is a well-known metamorphic rock composed predominantly of calcite or dolomite; its grain size ranges from fine to coarsely granular (see the chapter opening photo and ■ Figure 7.15). Marble results from either contact or regional metamorphism of limestones or dolostones (Table 7.1). Pure marble is snowy white or bluish, but many color varieties exist because of the presence of mineral impurities in the original sedimentary rock. The softness of marble, its uniform texture, and its various colors have made it the favorite rock of builders and sculptors throughout history (see the Introduction and "The Many Uses of Marble" on pages 196 and 197).

Quartzite is a hard, compact rock formed from quartz sandstone under medium- to high-grade metamorphic conditions during contact or regional metamorphism (■ Figure 7.16). Because recrystallization is so complete, metamorphic quartzite is of uniform strength and therefore usually breaks across the component quartz grains rather than around them when it is struck. Pure quartzite is white, but iron and other impurities commonly impart a reddish or other color to it. Quartzite is commonly used as foundation material for road and railway beds.

The name *greenstone* is applied to any compact, dark green, altered, mafic igneous rock that formed under low- to high-grade metamorphic conditions. The green color results from the presence of chlorite, epidote, and hornblende.

Hornfels, a fine-grained, nonfoliated metamorphic rock resulting from contact metamorphism, is composed of various equidimensional mineral grains. The composition of hornfels directly depends on the composition of the original rock, and many compositional varieties

The Many Uses of Marble

Marble is a remarkable stone that has a variety of uses. Formed from limestone or dolostone by the metamorphic processes of heat and pressure, marble comes in a variety of colors and textures.

Marble has been used by sculptors and architects for many centuries in statuary, monuments, as a facing and main stone in buildings and structures, as well as for floor tiling and other ornamental and structural uses. Marble can also be found in toothpaste and as a source of lime in agricultural fertilizers.

Aphrodite of Melos, also known as *Venus de Milo*, is one of the most recognizable works of art in the world. Dated at around 150 B.C., *Venus de Milo* was created by an unknown artist during the Hellenistic period and carved from the world-famous Parian marble from Paros in the Cyclades. Today *Venus de Milo* attracts thousands of visitors a year to the Louvre Museum in Paris, where she can be viewed and appreciated.

Reed Wicander

Marble has been used extensively as a building stone through the ages and throughout the world. For example, the Greek Parthenon was constructed of white Pentelic marble from Mt. Pentelicus in Attica.

Photodisc/ Getty Images

Photodisc/ Getty Images

The Taj Mahal in India is largely constructed of Makrana marble quarried from hills just southwest of Jaipur in Rajasthan. In addition to its main use as a building material, marble was used throughout the structure in art works and intricately carved marble flowers (right). All in all, it took more than 20,000 workers 17 years to build the Taj Mahal from A.D 1631 to 1648.

Galen Rowell/ Corbis

In the United States, marble is used as a building stone in many structures and is quarried from many locations. Marble is used in a variety of buildings and monuments in Washington, D.C. The Washington Monument is built from three different kinds of marble. The first 152 feet of the monument, built between 1848 and 1854, is faced with marble from the Texas, Maryland, quarry. Following 25 years of virtual inactivity, construction resumed with four rows of white marble from Lee, Massachusetts, added above the Texas marble. This marble proved to be too costly, so the upper part of the monument was finished with Cockeysville marble from quarries at Cockeysville, Maryland. The three different marbles can be distinguished by the slight differences in color.

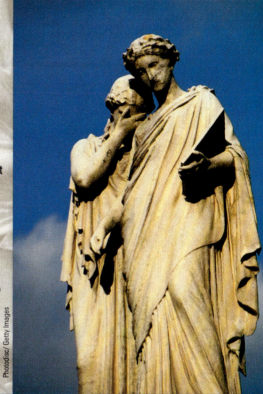

The Peace Monument at Pennsylvania Avenue on the west side of the Capitol is constructed from white marble from Carrara, Italy, a locality famous for its marble.

A marble quarry in northcentral Vermont. Vermont is known for producing some of the finest marble in the United States.

Another example of a marble building in Washington, D.C., is the Lincoln Monument, built from the Colorado Yule Marble, which is quarried at Marble, Colorado. This very pure white marble has been used not only for the Lincoln Monument but also for many other prominent buildings throughout the United States.

Ed Bartram

■ **Figure 7.13**

Migmatites consist of high-grade metamorphic rock intermixed with streaks or lenses of granite. This migmatite is exposed at Thirty Thousand Islands of Georgian Bay, Lake Huron, Ontario, Canada.

Reed Wicander

■ **Figure 7.14**

Nonfoliated textures are characterized by a mosaic of roughly equidimensional minerals, as in this photomicrograph of marble.

are known. The majority of hornfels, however, are apparently derived from contact metamorphism of clay-rich sedimentary rocks or impure dolostones.

Anthracite is a black, lustrous, hard coal that contains a high percentage of fixed carbon and a low percentage of volatile matter. It usually forms from the metamorphism of lower-grade coals by heat and pressure and is thus considered by many geologists to be a metamorphic rock.

PHYSICAL
Geology ⊜ Now™ Click Geology Interactive to work through an activity on Rock Lab.

WHAT ARE METAMORPHIC ZONES AND FACIES?

The first systematic study of metamorphic zones was conducted during the late 1800s by George Barrow and other British geologists working in the Dalradian schists of the southwestern Scottish Highlands. Here clay-rich sedimentary rocks

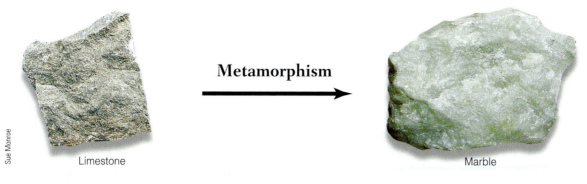

Sue Monroe

Limestone → **Metamorphism** → Marble

■ **Figure 7.15**

Marble results from the metamorphism of the sedimentary rock limestone or dolostone.

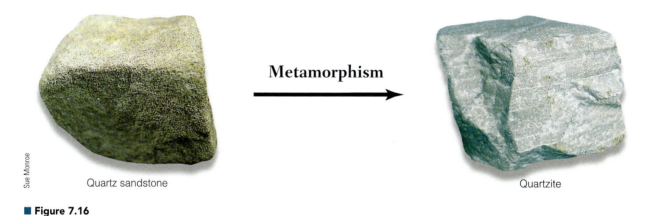

Sue Monroe

Quartz sandstone → **Metamorphism** → Quartzite

■ **Figure 7.16**

Quartzite results from the metamorphism of quartz sandstone.

have been subjected to regional metamorphism, and the resulting metamorphic rocks can be divided into different zones based on the presence of distinctive silicate mineral assemblages. These mineral assemblages, each recognized by the presence of one or more index minerals, indicate different degrees of metamorphism. The index minerals Barrow and his associates chose to represent increasing metamorphic intensity were chlorite, biotite, garnet, staurolite, kyanite, and sillimanite (Figure 7.7). Note that these are the metamorphic minerals produced from clay-rich sedimentary rocks. Other mineral assemblages and index minerals are produced from rocks with different original compositions.

The successive appearance of metamorphic index minerals indicates gradually increasing or decreasing intensity of metamorphism. Going from lower- to higher-grade zones, the first appearance of a particular index mineral indicates the location of the minimum temperature and pressure conditions needed for the formation of that mineral. When the locations of the first appearances of that index mineral are connected on a map, the result is a line of equal metamorphic intensity, or an *isograd*. The region between isograds is known as a **metamorphic zone.** By noting the occurrence of metamorphic index minerals, geologists can construct a map showing the metamorphic zones of an entire area (■ Figure 7.17).

Numerous studies of different metamorphic rocks have demonstrated that although the texture and composition of any rock may be altered by metamorphism, the overall chemical composition may be little changed. Thus the different mineral assemblages found in increasingly higher-grade metamorphic rocks derived from the same original rock result from changes in temperature and pressure.

A **metamorphic facies** is a group of metamorphic rocks characterized by particular mineral assemblages formed under the same broad temperature–pressure conditions (■ Figure 7.18). Each facies is named after its most characteristic rock or mineral. For example, the green metamorphic mineral chlorite, which forms under relatively low temperatures and pressures, yields rocks belonging to the *greenschist facies.* Under increasingly higher temperatures and pressures, other metamorphic facies, such as the *amphibolite* and *granulite facies,* develop.

Although usually applied to areas where the original rocks were clay rich, the concept of metamorphic facies can be used with modification in other situations. It cannot, however, be used in areas where the original rocks were pure quartz sandstones or pure limestones or dolostones. Such rocks would yield only quartzites and marbles, respectively.

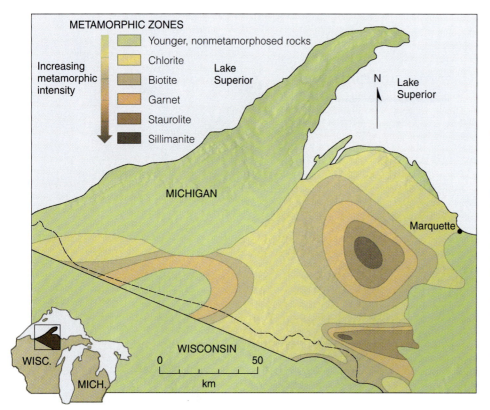

■ **Figure 7.17**

Metamorphic zones in the Upper Peninsula of Michigan. The zones in this region are based on the presence of distinctive silicate mineral assemblages resulting from the metamorphism of sedimentary rocks during an interval of mountain building and minor granitic intrusion during the Proterozoic Eon, about 1.5 billion years ago. The lines separating the different metamorphic zones are isograds. Source: From H. L. James, *G. S. A. Bulletin*, vol. 66, plate 1, page 1454, with permission of the publisher, the Geological Society of America, Boulder, Colorado. USA. Copyright © 1955 Geological Society of America.

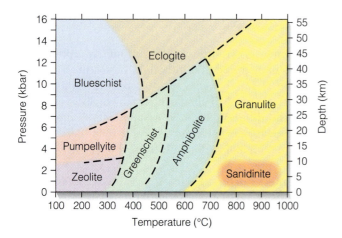

■ **Figure 7.18**

A pressure–temperature diagram showing where various metamorphic facies occur. A facies is characterized by a particular mineral assemblage that formed under the same broad temperature–pressure conditions. Each facies is named after its most characteristic rock or mineral. Source: From AGI Data Sheet 35.4, *AGI Data Sheets,* 3rd edition (1989) with the kind permission of the American Geological Institute.

HOW DOES PLATE TECTONICS AFFECT METAMORPHISM?

Although metamorphism is associated with all three types of plate boundaries (see Figure 1.13), it is most common along convergent plate boundaries. Metamorphic rocks form at convergent plate boundaries because temperature and pressure increase as a result of plate collisions.

■ Figure 7.19 illustrates the various temperature–pressure regimes produced along an oceanic–continental convergent plate boundary and the type of metamorphic facies and rocks that can result. When an oceanic plate collides with a continental plate, tremendous pressure is generated as the oceanic plate is subducted. Because rock is a poor heat conductor, the cold descending oceanic plate heats slowly, and metamorphism is caused mostly by increasing pressure with depth. Metamorphism in such an environment produces rocks typical of the *blueschist facies* (low temperature, high pressure), which is characterized by the blue-colored amphibole mineral glaucophane (Figure 7.18). Geologists use the occurrence of blueschist facies rocks as evidence of ancient subduction zones. An excellent example of blueschist metamorphism can be found in the California Coast

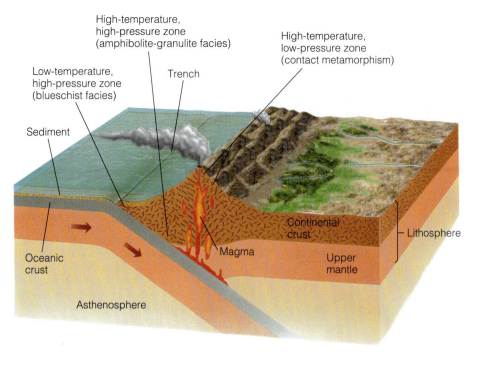

High-temperature, high-pressure zone (amphibolite-granulite facies)

Trench

High-temperature, low-pressure zone (contact metamorphism)

Low-temperature, high-pressure zone (blueschist facies)

Sediment

Continental crust

Lithosphere

Oceanic crust

Magma

Upper mantle

Asthenosphere

■ Figure 7.19

Metamorphic facies resulting from various temperature–pressure conditions produced along an oceanic–continental convergent plate boundary.

As subduction along the oceanic–continental plate boundary continues, both temperature and pressure increase with depth and yield high-grade metamorphic rocks. Eventually, the descending plate begins to melt and generates magma that moves upward. This rising magma may alter the surrounding rock by contact metamorphism, producing migmatites in the deeper portions of the crust and hornfels at shallower depths. Such an environment is characterized by high temperatures and low to medium pressures.

Although metamorphism is most common along convergent plate margins, many divergent plate boundaries are characterized by contact metamorphism. Rising magma at mid-oceanic ridges heats the adjacent rocks, producing contact metamorphic minerals and

Ranges. Here rocks of the Franciscan Complex were metamorphosed under low-temperature, high-pressure conditions that clearly indicate the presence of a former subduction zone (■ Figure 7.20).

textures. Besides contact metamorphism, fluids emanating from the rising magma—and its reaction with seawater—very commonly produce metal-bearing hydrothermal solutions that may precipitate minerals of economic value.

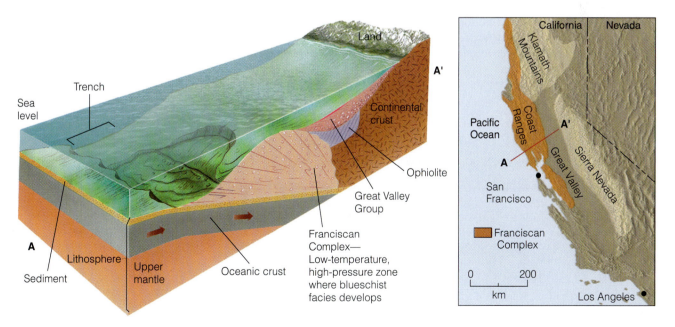

Land

Trench

A'

Sea level

Continental crust

Ophiolite

Great Valley Group

Franciscan Complex—Low-temperature, high-pressure zone where blueschist facies develops

A

Lithosphere

Sediment

Upper mantle

Oceanic crust

California   Nevada

Klamath Mountains

Pacific Ocean

Coast Ranges

A'

A

San Francisco

Sierra Nevada

Great Valley

Franciscan Complex

0   200
km

Los Angeles

■ Figure 7.20

Index map of California showing the location of the Franciscan Complex and a diagrammatic reconstruction of the environment in which it was regionally metamorphosed under low-temperature, high-pressure subduction conditions approximately 150 million years ago. The red line on the index map shows the orientation of the reconstruction to the current geography. Source: From "Effects of Late Jurassic-Early Tertiary Subduction in California," *San Joaquin Geological Society Short Course*, 1977, 66, Figure 5-9.

Peter Hulme/Ecoscene/Corbis

■ **Figure 7.21**

Slate quarry in Wales. Slate, which has a variety of uses, is the result of low-grade metamorphism of shale. These high-quality slates were formed by a mountain-building episode that occurred approximately 400 to 440 million years ago in the present-day countries of Iceland, Scotland, Wales, and Norway.

These deposits may eventually be brought to Earth's surface by later tectonic activity. The copper ores of Cyprus are a good example of such hydrothermal activity (see Chapter 12).

METAMORPHISM AND NATURAL RESOURCES

Many metamorphic rocks and minerals are valuable natural resources. Although these resources include various types of ore deposits, the two most familiar and widely used metamorphic rocks, as such, are marble and slate, which, as previously discussed, have been used for centuries in a variety of ways (■ Figure 7.21).

Many ore deposits result from contact metamorphism during which hot, iron-rich fluids migrate from igneous intrusions into the surrounding rock, thereby producing rich ore deposits. The most common sulfide ore minerals associated with contact metamorphism are bornite, chalcopyrite, galena, pyrite, and sphalerite; two common oxide ore minerals are hematite and magnetite. Tin and tungsten are also important ores associated with contact metamorphism (Table 7.2).

Table 7.2

The Main Ore Deposits Resulting from Contact Metamorphism

Ore Deposit	Major Mineral	Formula	Use
Copper	Bornite Chalcopyrite	Cu_5FeS_4 $CuFeS_2$	Important sources of copper, which is used in various aspects of manufacturing, transportation, communications, and construction
Iron	Hematite Magnetite	Fe_2O_3 Fe_3O_4	Major sources of iron for manufacture of steel, which is used in nearly every form of construction, manufacturing, transportation, and communications
Lead	Galena	PbS	Chief source of lead, which is used in batteries, pipes, solder, and elsewhere where resistance to corrosion is required
Tin	Cassiterite	SnO_2	Principal source of tin, which is used for tin plating, solder, alloys, and chemicals
Tungsten	Scheelite Wolframite	$CaWO_4$ $(Fe,Mn)WO_4$	Chief sources of tungsten, which is used in hardening metals and manufacturing carbides
Zinc	Sphalerite	$(Zn, Fe)S$	Major source of zinc, which is used in batteries and in galvanizing iron and making brass

Other economically important metamorphic minerals include talc for talcum powder; graphite for pencils and dry lubricants; garnets and corundum, which are used as abrasives or gemstones, depending on their quality; and andalusite, kyanite, and sillimanite, all of which are used in manufacturing high-temperature porcelains and temperature-resistant minerals for products such as spark plugs and the linings of furnaces.

7 REVIEW WORKBOOK

Chapter Summary

- Metamorphic rocks result from the transformation of other rocks, usually beneath Earth's surface, as a consequence of one or a combination of three agents: heat, pressure, and fluid activity.

- Heat for metamorphism comes from intrusive magmas or deep burial. Pressure is either lithostatic or differential. Fluids trapped in sedimentary rocks or emanating from intruding magmas can enhance chemical changes and the formation of new minerals.

- The three major types of metamorphism are contact, dynamic, and regional.

- Index minerals—minerals that form only within specific temperature and pressure ranges—allow geologists to recognize low-, intermediate-, and high-grade metamorphic zones.

- Metamorphic rocks are primarily classified according to their texture. In a foliated texture, platy minerals have a preferred orientation. A nonfoliated texture does not exhibit any discernable preferred orientation of the mineral grains.

- Foliated metamorphic rocks can be arranged in order of grain size and/or perfection of their foliation. Slate is fine grained, followed by (in

coarser-grained order) phyllite and schist; gneiss displays segregated bands of minerals. Amphibolite is another fairly common foliated metamorphic rock.

- Marble, quartzite, greenstone, and hornfels are common nonfoliated metamorphic rocks.

- Metamorphic zones are based on index minerals and are areas of equal metamorphic intensity. Metamorphic facies are characterized by particular assemblages of minerals that formed under specific metamorphic conditions. These facies are named for a characteristic constituent mineral or rock type.

- Metamorphism occurs along all three kinds of plate boundaries but most commonly at convergent plate margins.

- Metamorphic rocks formed near Earth's surface along an oceanic–continental plate boundary result from low-temperature, high-pressure conditions. As a subducted oceanic plate descends, it is subjected to increasingly higher temperatures and pressures that result in higher-grade metamorphism.

- Many metamorphic rocks and minerals, such as marble, slate, graphite, talc, and asbestos, are valuable natural resources.

Important Terms

aureole (p. 189)
contact metamorphism (p. 188)
differential pressure (p. 188)
dynamic metamorphism (p. 189)
fluid activity (p. 188)

foliated texture (p. 191)
heat (p. 187)
index minerals (p. 191)
lithostatic pressure (p. 187)
metamorphic facies (p. 199)

metamorphic rock (p. 184)
metamorphic zone (p. 199)
nonfoliated texture (p. 195)
regional metamorphism (p. 190)

Review Questions

1. Which of the following is *not* an agent or process of metamorphism?

    a. _____ pressure; b. _____ heat; c. _____ fluid activity; d. _____ time; e. _____ gravity.

2. The nonfoliated metamorphic rock formed from limestone or dolostone is called:

    a. _____ quartzite; b. _____ marble; c. _____ hornfels; d. _____ greenstone; e. _____ schist.

3. Metamorphic rocks can form from what type of original rock?

    a. _____ igneous; b. _____ sedimentary; c. _____ metamorphic; d. _____ volcanic; e. _____ all of the preceding answers.

4. Pressure resulting from deep burial and applied equally in all directions on a rock is:

    a. _____ directional; b. _____ differential; c. _____ lithostatic; d. _____ shear; e. _____ unilateral.

5. Magmatic heat and fluid activity are the primary agents involved in what type of metamorphism?

    a. _____ dynamic; b. _____ lithostatic; c. _____ contact; d. _____ regional; e. _____ thermodynamic.

6. Along what type of plate boundary is metamorphism most common?

    a. _____ divergent; b. _____ transform; c. _____ aseismic; d. _____ convergent; e. _____ lithospheric.

7. Concentric zones surrounding an igneous intrusion and characterized by distinctive mineral assemblages are:

    a. _____ thermodynamic rings; b. _____ hydrothermal regions; c. _____ metamorphic layers; d. _____ regional facies; e. _____ aureoles.

8. The majority of metamorphic rocks result from which type of metamorphism?

    a. _____ lithostatic; b. _____ contact; c. _____ regional; d. _____ local; e. _____ dynamic.

9. Which is the order of increasingly coarser grain size and perfection of foliation?

    a. _____ gneiss → schist → phyllite → slate; b. _____ phyllite → slate → schist → gneiss; c. _____ schist → slate → gneiss → phyllite; d. _____ slate → phyllite → schist → gneiss; e. _____ slate → schist → phyllite → gneiss.

10. Metamorphic zones:
    a. _____ reflect a metamorphic grade;
    b. _____ are characterized by distinctive mineral assemblages; c. _____ are separated from each other by isograds; d. _____ all of the preceding answers; e. _____ none of the preceding answers.

11. Discuss the role each of the three agents of metamorphism plays in transforming any rock into a metamorphic rock.

12. Why is metamorphism more common along convergent plate boundaries than along any other type of plate boundary?

13. How can aureoles be used to determine the effects of metamorphism?

14. Describe the two types of metamorphic texture and discuss how they are produced.

15. Why should the average citizen know anything about metamorphic rocks and how they form?

16. What is regional metamorphism, and under what conditions does it occur?

17. Name several economically valuable metamorphic minerals or rocks and discuss why they are valuable.

18. What specific features about foliated metamorphic rocks make them unsuitable as foundations for dams? Are there any metamorphic rocks that would make good foundations? Why?

19. Using Figure 7.18, go to a point that is represented by 450°C and 6 kilobars of pressure. What metamorphic facies is represented by those conditions? If pressure is raised to 10 kilobars, what facies is represented by the new conditions? What change in depth of burial is required to effect the pressure change of 6 to 10 kilobars?

20. If plate tectonic movement did not exist, could there be metamorphism? Do you think metamorphic rocks exist on other planets in our solar system? Why?

World Wide Web Activities

PHYSICAL Geology⇌Now Assess your understanding of this chapter's topics with additional quizzing and comprehensive interactivities at

http://earthscience.brookscole.com/physgeo5e

as well as current and up-to-date weblinks, additional readings, and InfoTrac College Edition exercises.

Geologic Time: Concepts and Principles

CHAPTER 8
OUTLINE

PHYSICAL GeologyNow *This icon, appearing throughout the book, indicates an opportunity to explore interactive tutorials, animations, or practice problems available on the Physical GeologyNow Web site at http://earthscience.brookscole.com/physgeo5e.*

OBJECTIVES
At the end of this chapter, you will have learned that

- The concept of geologic time and its measurements have changed through human history.

- The principle of uniformitarianism is fundamental to geology.

- The fundamental principles of relative dating provide a means to interpret geologic history.

- The three types of unconformities—disconformities, angular unconformities, and nonconformities—are erosional surfaces separating younger from older rocks and represent significant intervals of geologic time for which we have no record at a particular location.

- Time equivalency of rock units can be demonstrated by various correlation techniques.

- Different absolute dating methods are used to date geologic events in terms of years before present.

- The most accurate radiometric dates are obtained from igneous rocks.

- The geologic time scale evolved primarily during the 19th century through the efforts of many people.

The Grand Canyon, Arizona. Major John Wesley Powell led two expeditions down the Colorado River and through the canyon in 1869 and 1871. He was struck by the seemingly limitless time represented by the rocks exposed in the canyon walls and by the recognition that these rock layers, like the pages in a book, contain the geologic history of this region.
Source: Royalty-Free/Corbis

Introduction

n 1869 Major John Wesley Powell, a Civil War veteran who lost his right arm in the battle of Shiloh, led a group of hardy explorers down the uncharted Colorado River through the Grand Canyon. With no maps or other information, Powell and his group ran the many rapids of the Colorado River in fragile wooden boats, hastily recording what they saw. Powell wrote in his diary that "all about me are interesting geologic records. The book is open and I read as I run."

From this initial reconnaissance, Powell led a second expedition down the Colorado River in 1871. This second trip included a photographer, a surveyor, and three topographers. The expedition made detailed topographic and geologic maps of the Grand Canyon area as well as the first photographic record of the region.

Probably no one has contributed as much to the understanding of the Grand Canyon as Major Powell. In recognition of his contributions, the Powell Memorial was erected on the South Rim of the Grand Canyon in 1969 to commemorate the 100th anniversary of this history-making first expedition.

Most tourists today, like Powell and his fellow explorers in 1869, are astonished by the seemingly limitless time represented by the rocks exposed in the walls of the Grand Canyon. For most people, staring down 1.5 km at the rocks in the canyon is their only exposure to the concept of geologic time. When standing on the rim and looking down into the Grand Canyon, we are really looking far back in time, all the way back to the early history of our planet. In fact, more than 1 billion years of history are preserved in the rocks of the Grand Canyon.

Just as we read the pages of a book, we can read the rock layers of the Grand Canyon and learn that this region underwent episodes of mountain building as well as periods of advancing and retreating shallow seas. How do we know this? The answer lies in applying the principles of relative dating to the rocks we see exposed in the canyon, as well as recognizing that the geologic processes we see operating today, have operated throughout Earth history.

We begin this chapter by asking the question, What is time? We seem obsessed with time, and we organize our lives around it with the help of clocks, calendars, and appointment books. Yet most of us feel we don't have enough of it; we are always running "behind" or "out of time." Whereas physicists deal with extremely short intervals of time and geologists deal with incredibly long periods of time, most of us tend to view time from the perspective of our own existence; that is, we partition our lives into seconds, hours, days, weeks, months, and years. Ancient history is what occurred hundreds or even thousands of years ago. Yet when geologists talk of ancient geologic history, they are referring to events that happened millions or even billions of years ago!

Time sets geology apart from most of the other sciences, and an appreciation of the immensity of geologic time is fundamental to understanding the physical and biological history of our planet. In fact, understanding and accepting the magnitude of geologic time are major contributions geology has made to the sciences.

In some respects, time is defined by the methods used to measure it. Geologists use two different frames of reference when discussing geologic time. **Relative dating** involves placing geologic events in a sequential order as determined from their position in the geologic record. Relative dating will not tell us how long ago a particular event took place, only that one event preceded another. A useful analogy for relative dating is a television guide that does not list the times programs are shown. You cannot tell what time a particular program will be shown, but by watching a few shows and checking the guide, you can determine whether you have missed the show or how many shows are scheduled before the one you want to see.

The various principles used to determine relative dating were discovered hundreds of years ago, and since then they have been used to construct the *relative geologic time scale* (■ Figure 8.1). Furthermore, these principles are still widely used by geologists today.

Absolute dating results in specific dates for rock units or events expressed in years before the present. In our analogy of the television guide, the time when the programs were actually shown would be the absolute dates. In this way, you not only could determine whether you had missed a show (relative dating), but also would know how long it would be until a show you wanted to see would be shown (absolute dating).

Radiometric dating is the most common method of obtaining absolute ages. Such dates are calculated from the natural decay rates of various radioactive elements present in trace amounts in some rocks. It was not until the discovery of radioactivity near the end of the 19th century that absolute ages could be accurately applied to the relative geologic time scale. Today the geologic time scale is really a dual scale: a relative scale based on rock sequences with radiometric dates expressed as years before the present (Figure 8.1).

Besides providing an appreciation for the immensity of geologic time, why is the study of geologic time important? One of the most important lessons to be learned in this chapter is how to reason and apply the fundamental geologic principles to solve geologic problems. The logic used

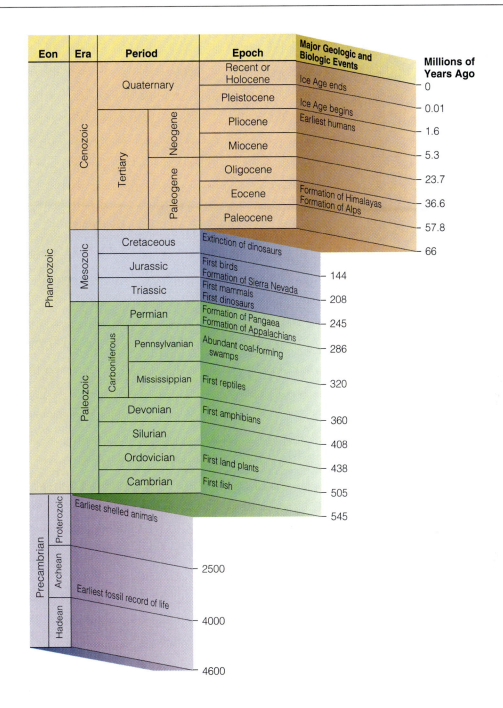

Eon	Era	Period		Epoch	Major Geologic and Biologic Events	Millions of Years Ago
Phanerozoic	Cenozoic	Quaternary		Recent or Holocene	Ice Age ends	0
				Pleistocene	Ice Age begins	0.01
		Tertiary	Neogene	Pliocene	Earliest humans	1.6
				Miocene		5.3
			Paleogene	Oligocene		23.7
				Eocene	Formation of Himalayas Formation of Alps	36.6
				Paleocene		57.8
	Mesozoic	Cretaceous			Extinction of dinosaurs	66
		Jurassic			First birds Formation of Sierra Nevada	144
		Triassic			First mammals First dinosaurs	208
	Paleozoic	Permian			Formation of Pangaea Formation of Appalachians	245
		Carboniferous	Pennsylvanian		Abundant coal-forming swamps	286
			Mississippian		First reptiles	320
		Devonian			First amphibians	360
		Silurian				408
		Ordovician			First land plants	438
		Cambrian			First fish	505
Precambrian	Proterozoic				Earliest shelled animals	545
	Archean					2500
					Earliest fossil record of life	4000
	Hadean					4600

■ **Figure 8.1**

The geologic time scale. Some of the major geologic and biologic events are indicated along the right-hand margin. Source: Modified from A. R. Palmer, "The Decade of North American Geology, 1983 Geologic Time Scale." Geology (Geological Society of America, 1983), p. 504.

in applying the principles of relative dating to interpret the geologic history of an area involves basic reasoning skills that can be transferred to and used in almost any profession.

Advances and refinements in the various absolute dating techniques during the 20th century have changed the way we view Earth in terms of when events occurred in the past and the rates of geologic change through time. The ability to accurately determine past climatic changes and

their causes has important implications for the current debate on global warming and its effects on humans (see Geo-Focus 8.1).

PHYSICAL Geology⇌Now Click Geology Interactive to work through an activity on Relative Dating and Absolute Dating through Geologic Time.

GEOFOCUS

8.1

Geologic Time and Climate Change

With all the debate concerning global warming and its possible implications, it is extremely important to be able to reconstruct past climatic regimes as accurately as possible. To model how Earth's climate system has responded to changes in the past and use that information for simulations of future climatic scenarios, we must have as precise a geologic calendar as possible.

One way to study climatic changes is to examine lake sediment or ice cores that contain organic matter. By taking closely spaced samples and dating the organic matter in the cores using the carbon 14 dating technique, geologists can construct a detailed chronology for each core examined. Changes in isotope ratios, pollen, and plant and invertebrate fossil assemblages can then be accurately dated, and the time and duration of climate changes correlated over increasingly larger areas. Without a means of precise dating, we would have no way to accurately model past climatic changes with the precision needed to predict possible future climate changes on a human time scale.

An interesting method that is becoming more common in reconstructing past climates is to analyze stalagmites from caves. Stalagmites are icicle-shaped structures rising from a cave floor and formed of calcium carbonate precipitated from evaporating water (see Figure 16.18). A stalagmite therefore records a layered history because each newly precipitated layer of calcium carbonate is younger than the previously precipitated layer. Thus a stalagmite's layers are oldest in the center at its base and progressively younger as they move outward. Using techniques developed during the past 10 years, based on high-precision ratios of uranium 234 to thorium 230, geologists can achieve very precise radiometric dates on individual layers of a stalagmite. This technique lets geologists determine the age of materials much older than they can date by carbon 14, and it is reliable back to about 500,000 years.

An interesting recent study of stalagmites from Crevice Cave in Missouri revealed a history of climatic and vegetation change in the mid-continent region of the United States during the interval between 75,000 and 25,000 years ago. By analyzing carbon 13 and oxygen 18 isotope profiles during this interval, geologists were able to deduce that average temperature fluctuations of 4°C correlated with major changes in vegetation. During the interval between 75,000 and 55,000 years ago, the climate oscillated between warm and cold, and vegetation alternated among forest, savannah, and prairie. Fifty-five thousand years ago the climate cooled, and there was a sudden change from grasslands to forest that persisted until 25,000 years ago. This corresponds to the time when global ice sheets began building and advancing.

High-precision dating techniques in stalagmite studies, using uranium 234 and thorium 230, provide an accurate chronology that allows geologists to model climate systems of the past and perhaps to determine what causes global climatic changes and their duration. Without these sophisticated dating techniques and others like them, geologists would not be able to make precise correlations and accurately reconstruct past environments and climates. By analyzing past environmental and climate changes and their duration, geologists hope they can use these data, sometime in the near future, to predict and possibly modify regional climatic changes.

HOW HAS THE CONCEPT OF GEOLOGIC TIME AND EARTH'S AGE CHANGED THROUGHOUT HUMAN HISTORY?

The concept of geologic time and its measurement have changed through human history. Many early Christian scholars and clerics tried to establish the date of creation by analyzing historical records and the genealogies found in Scripture. Based on their analyses, they generally believed that Earth and all its features were no more than about 6000 years old. The idea of a very young Earth provided the basis for most Western chronologies of Earth history prior to the 18th century.

During the 18th and 19th centuries, several attempts were made to determine Earth's age on the basis of scientific evidence rather than revelation. The French zoologist Georges Louis de Buffon (1707–1788) assumed Earth gradually cooled to its present condition from a molten beginning. To simulate this history, he melted iron balls of various diameters and allowed them to cool to the surrounding temperature. By extrapolating their cooling rate to a ball the size of Earth, he determined that Earth was at least 75,000 years old. Although this age was much older than that derived from Scripture, it was vastly younger than we now know the planet to be.

Other scholars were equally ingenious in attempting to calculate Earth's age. For example, if deposition rates could be determined for various sediments, geologists reasoned that they could calculate how long it would take to deposit any rock layer. They could then extrapolate how old Earth was from the total thickness of sedimentary rock in its crust. Rates of deposition vary, however, even for the same type of rock. Furthermore, it is impossible to estimate how much rock has been removed by erosion, or how much a rock sequence has been reduced by compaction. As a result of these variables, estimates ranged from less than 1 million years to more than 2 billion years.

Another attempt to determine Earth's age involved calculating the age of the oceans. If the ocean basins were filled very soon after the origin of the planet, then they would be only slightly younger than Earth itself. The best-known calculations for the oceans' age were made by the Irish geologist John Joly in 1899. He reasoned that Earth's ocean waters were originally fresh and their present salinity was the result of dissolved salt being carried into the ocean basins by rivers. By measuring the present amount of salt in the world's rivers, and knowing the volume of ocean water and its salinity, Joly calculated that it would have taken at least 90 million years for the oceans to reach their present salinity level. This was still much younger than the now-accepted age of 4.6 billion years for Earth, mainly because Joly had no way to calculate how much salt had been recycled or the amount of salt stored in continental salt deposits and seafloor clay deposits.

Besides trying to determine Earth's age, the naturalists of the 18th and 19th centuries were also formulating some of the fundamental geologic principles that are used in deciphering Earth history. From the evidence preserved in the geologic record, it was clear to them that Earth is very old and that geologic processes have operated over long periods of time. A good example of geologic processes operating over long periods of time to produce a spectacular landscape is the evolution of Uluru and Kata Tjuta (see "Uluru and Kata Tjuta" on pages 212 and 213).

WHY ARE JAMES HUTTON'S CONTRIBUTIONS TO GEOLOGY IMPORTANT?

The Scottish geologist James Hutton (1726–1797) is considered by many to be the founder of modern geology. His detailed studies and observations of rock exposures and present-day geologic processes were instrumental in establishing the **principle of uniformitarianism** (see Chapter 1), the concept that the same processes today have operated over vast amounts of time. Because Hutton relied on known processes to account for Earth history, he concluded that Earth must be very old and wrote that "we find no vestige of a beginning, and no prospect of an end."

Unfortunately, Hutton was not a particularly good writer, so his ideas were not widely disseminated or accepted. In 1830 Charles Lyell published a landmark book, *Principles of Geology,* in which he championed Hutton's concept of uniformitarianism. Instead of relying on catastrophic events to explain various Earth features, Lyell recognized that imperceptible changes brought about by present-day processes could, over long periods of time, have tremendous cumulative effects. Through his writings, Lyell firmly established uniformitarianism as the guiding principle of geology. Furthermore, the recognition of virtually limitless amounts of time was also necessary for, and instrumental in, the acceptance of Darwin's 1859 theory of evolution.

After finally establishing that present-day processes have operated over vast periods of time, geologists were nevertheless nearly forced to accept a very young age for Earth when a highly respected English physicist, Lord

Uluṟu and Kata Tjuṯa

Rising majestically above the surrounding flat desert of central Australia are Uluṟu and Kata Tjuṯa. Uluṟu and Kata Tjuṯa are the aboriginal names for what most people know as Ayers Rock and The Olgas. The history of Uluṟu and Kata Tjuṯa began about 550 million years ago when a huge mountain range formed in what is now central Australia. It subsequently eroded, and vast quantities of gravel were transported by streams and deposited along its base to form large alluvial fans. Marine sediments then covered the alluvial fans and the entire region was uplifted by tectonic forces between 400 and 300 million years ago and then subjected to weathering.

The spectacular and varied rock shapes of Uluṟu and Kata Tjuṯa are the result of millions of years of weathering and erosion by water and, to a lesser extent, wind acting on the fractures formed during uplift. Differences in the composition and texture of the rocks also played a role in sculpting these colorful and magnificent structures.

Aerial view of Uluṟu with Kata Tjuṯa in the background. Contrary to popular belief, Uluṟu is not a giant boulder. Rather it is the exposed portion of the nearly vertically tilted Uluṟu Arkose. The caves, caverns, and depressions visible on the northeastern side are the result of weathering.

Reed Wicander

A closeup view of the brain- and honeycomb-like small caves seen on the northeast side of Uluṟu.

Reed Wicander

Uluru at sunset. The near-vertical tilting of the sedimentary beds of the Uluru Arkose that make up Uluru can be seen clearly. Differential weathering of the sedimentary layers has produced the distinct parallel ridges and other features characteristic of Uluru.

Reed Wicander

Aerial view of Kata Tjuta with Uluru in the background. Kata Tjuta is composed of the Mount Currie Conglomerate, a coarse-grained and poorly sorted conglomerate. The sediments that were lithified into the Mount Currie Conglomerate were deposited, like the Uluru Arkose, as an alluvial fan beginning approximately 550 million years ago.

Reed Wicander

A view of the rounded domes of Kata Tjuta and typical vegetation as seen from within one of its canyons. The red color of the rocks is the result of oxidation of iron in the sediments.

Reed Wicander

An aerial closeup view of Kata Tjuta. The distinctive dome shape of the rocks is the result of weathering and erosion of the Mount Currie Conglomerate. In addition to weathering, the release of pressure on the buried rocks as they were exposed at the surface by tectonic forces contributed to the characteristic rounded shapes of Kata Tjuta.

Reed Wicander

GEOLOGY
IN UNEXPECTED PLACES

Time Marches On—The Great Wall of China

The Great Wall of China, built over many centuries as a military fortification against invasion by enemies from the north, has largely succumbed to the ravages of nature and human activity. Originally begun as a series of short walls during the Zhou Dynasty (770–476 B.C.), the wall grew as successive dynasties connected different parts, with final improvements made during the Ming Dynasty (A.D. 1368–1644). Even though the Great Wall is not continuous, it stretches more than 5000 km across northern China from the east coast to the central part of the country. Contrary to popular belief, the Great Wall of China is not the only man-made structure visible from space or the Moon. From low Earth orbit, many artificial objects are visible, such as highways, cities, and railroads. When viewed from a distance of a few thousand kilometers, no man-made objects are visible and the Great Wall can barely be seen with the naked eye from the shuttle according to NASA. In fact, China's first astronaut Yang Liwei told state television on his return from space, "I did not see the Great Wall from space."

So with that short history and debunking of an urban legend, what is the Great Wall made of? Basically, the Great Wall was constructed with whatever material was available in the area. This included sedimentary, metamorphic, and igneous rocks, bricks, sand, gravel, and even dirt and straw. Regardless of the material, the wall was built by hand by thousands of Chinese over many centuries.

In the Badaling area of Beijing, which is the part of the Great Wall most tourists visit and has been restored, it was built using the igneous and sedimentary rocks from the mountains around Badaling. The sedimentary rocks are mudstones, sandstones, and limestones, whereas the igneous rocks are granite. The sides of the walls in this area are constructed of rectangular slabs of granite, and the top or roof of the Great Wall here is paved with large gray bricks.

The average height of the Great Wall is 8.5 m, and it is 6.5 m wide along its base. The wall along the top averages 5.7 m and is wide enough for 5 horses or 10 warriors to walk side by side.

Most visitors to the Great Wall are so impressed by its size and history that they don't even notice what it is made from. Now, however, you know and should you visit this impressive structure you can tell your fellow travelers all about it.

Reed Wicander

■ **Figure 2**

The top of the Great Wall at Badaling. Note the original rocks in the lower portion of the side of the wall and the paving bricks in the foreground comprising the top of the wall.

Reed Wicander

■ **Figure 1**

The Great Wall of China winding across the top of the hills at Badaling, just outside Beijing.

Kelvin (1824–1907), claimed, in a paper written in 1866, to have destroyed the uniformitarian foundation of geology. Starting with the generally accepted belief that Earth was originally molten, Kelvin assumed that it has gradually been losing heat and that, by measuring this heat loss, he could determine its age.

Kelvin knew from deep mines in Europe that Earth's temperature increases with depth, and he reasoned that Earth is losing heat from its interior. By knowing the size of Earth, the melting temperatures of rocks, and the rate of heat loss, Kelvin was able to calculate the age at which Earth was entirely molten. From these calculations, he concluded that Earth could be no older than 400 million years or younger than 20 million years. This wide discrepancy in age reflected uncertainties in average temperature increases with depth and the various melting points of Earth's constituent materials.

After finally establishing that Earth was very old and that present-day processes operating over long periods of time account for geologic features, geologists were in a quandary. Either they had to accept Kelvin's dates and squeeze events into a shorter time frame, or they had to abandon the concept of seemingly limitless time that was the underpinning of uniformitarian geology and one of the foundations of Darwinian evolution.

Kelvin's reasoning and calculations were sound, but his basic premises were false, thereby invalidating his conclusions. Kelvin was unaware that Earth has an internal heat source, radioactivity, that has allowed it to maintain a fairly constant temperature through time.* His 40-year campaign for a young Earth ended with the discovery of radioactivity near the end of the 19th century. His calculations were no longer valid and his proof for a geologically young Earth collapsed.

Although the discovery of radioactivity destroyed Kelvin's arguments, it provided geologists with a clock that could measure Earth's age and validate what geologists had been saying all along—namely, that Earth was indeed very old!

WHAT ARE RELATIVE DATING METHODS?

Before the development of radiometric dating techniques, geologists had no reliable means of absolute dating and therefore depended solely on relative dating methods. These methods allow events to be placed in sequential order only and do not tell us how long ago an event took place. Although the principles of relative dating may now seem self-evident, their discovery was an important scientific achievement because they provided geologists with a means to interpret geologic history and develop a relative geologic time scale.

Six fundamental geologic principles are used in relative dating: superposition, original horizontality, lateral continuity, cross-cutting relationships, inclusions, and fossil succession.

Fundamental Principles of Relative Dating

The 17th century was an important time in the development of geology as a science because of the widely circulated writings of the Danish anatomist Nicolas Steno (1638–1686). Steno observed that when streams flood, they spread out across their floodplains and deposit layers of sediment that bury organisms dwelling on the floodplain. Subsequent floods produce new layers of sediments that are deposited or superposed over previous deposits. When lithified, these layers of sediment become sedimentary rock. Thus, in an undisturbed succession of sedimentary rock layers, the oldest layer is at the bottom and the youngest layer is at the top. This **principle of superposition** is the basis for relative-age determinations of strata and their contained fossils (■ Figure 8.2).

■ **Figure 8.2**

Badlands National Monument in South Dakota illustrates three of the six fundamental principles of relative dating. The sedimentary rocks of the Badlands were originally deposited horizontally by sluggish streams in a variety of continental environments (principle of original horizontality). The oldest rocks are at the bottom of this highly dissected landscape, and the youngest rocks are at the top, forming the rims (principle of superposition). The exposed rock layers extend laterally in all directions for some distance (principle of lateral continuity).

*Actually Earth's temperature has decreased through time because the original amount of radioactive materials has been decreasing and thus is not supplying as much heat. However, the temperature is decreasing at a rate considerably slower than would be required to lend any credence to Kelvin's calculations.

(a)   (b)

■ **Figure 8.3**

The principle of cross-cutting relationships. (a) A dark dike has been intruded into older light-colored granite along the north shore of Lake Superior, Ontario, Canada. (b) A small fault (arrows show direction of movement) displacing tilted beds along Templin Highway, Castaic, California.

Steno also observed that, because sedimentary particles settle from water under the influence of gravity, sediment is deposited in essentially horizontal layers, thus illustrating the **principle of original horizontality** (Figure 8.2). Therefore a sequence of sedimentary rock layers that is steeply inclined from the horizontal must have been tilted after deposition and lithification.

Steno's third principle, the **principle of lateral continuity,** states that a layer of sediment extends laterally in all directions until it thins and pinches out or terminates against the edge of the depositional basin (Figure 8.2).

James Hutton is credited with discovering the **principle of cross-cutting relationships.** Based on his detailed studies and observations of rock exposures in Scotland, Hutton recognized that an igneous intrusion or fault must be younger than the rocks it intrudes or displaces (■ Figure 8.3).

Although this principle illustrates that an intrusive igneous structure is younger than the rocks it intrudes, the association of sedimentary and igneous rocks may cause problems in relative dating. Buried lava flows and intrusive igneous bodies such as sills look very similar in a sequence of strata (■ Figure 8.4). A buried lava flow, however, is older than the rocks above it (principle of superperposition), whereas a sill is younger than all the beds below it and younger than the bed immediately above it.

To resolve such relative-age problems as these, geologists look to see if the sedimentary rocks in contact with the igneous rocks show signs of baking or alteration by heat (see the section on contact metamorphism in Chapter 7, p. 188). A sedimentary rock showing such effects must be older than the igneous rock with which it is in contact. In Figure 8.4, for example, a sill produces a zone of baking immediately above and below it because it intruded into previously existing sedimentary rocks. A lava flow, in contrast, bakes only those rocks below it.

Another way to determine relative ages is by using the **principle of inclusions.** This principle holds that inclusions, or fragments of one rock contained within a layer of another, are older than the rock layer itself. The batholith shown in ■ Figure 8.5a contains sandstone inclusions, and the sandstone unit shows the effects of baking. Accordingly, we conclude that the sandstone is older than the batholith. In Figure 8.5b, however, the sandstone contains granite rock fragments, indicating that the batholith was the source rock for the inclusions and is therefore older than the sandstone.

Fossils have been known for centuries (see Chapter 6), yet their utility in relative dating and geologic mapping was not fully appreciated until the early 19th century. William Smith (1769–1839), an English civil engineer involved in surveying and building canals in southern England, independently recognized the principle of superposition by reasoning that the fossils at the bottom of a sequence of strata are older than those at the top of the sequence. This recognition served as the basis for the **principle of fossil succession,** or the *principle of faunal and floral succession,* as it is sometimes called (■ Figure 8.6).

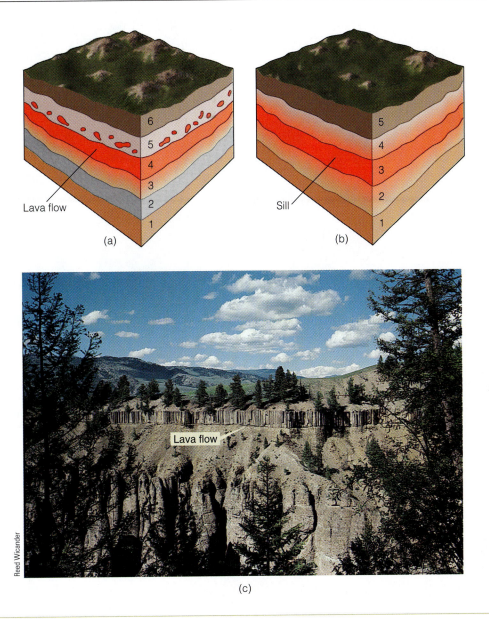

Lava flow

(a)

Sill

(b)

Lava flow

(c)

Reed Wicander

■ **Figure 8.4**

Relative ages of lava flows, sills, and associated sedimentary rocks may be difficult to determine. (a) A buried lava flow in bed 4 baked the underlying bed, and bed 5 contains inclusions of the lava flow. The lava flow is younger than bed 3 and older than beds 5 and 6. (b) The rock units above and below the sill in bed 3 have been baked, indicating that the sill is younger than beds 2 and 4, but its age relative to bed 5 cannot be determined. (c) This buried lava flow in Yellowstone National Park, Wyoming, displays columnar jointing.

According to this principle, fossil assemblages succeed one another through time in a regular and predictable order. The validity and successful use of this principle depend on three points: (1) Life has varied through time, (2) fossil assemblages are recognizably different from one another, and (3) the relative ages of the fossil assemblages can be determined. Observations of fossils in older versus younger strata clearly demonstrate that life-forms have changed. Because this is true, fossil assemblages (point 2) are recognizably different. Furthermore, superposition can be used to demonstrate the relative ages of the fossil assemblages.

Unconformities

So far we have discussed vertical relationships among conformable strata—that is, sequences of rocks in which deposition was more or less continuous. A bedding plane between strata may represent a depositional break of anywhere from minutes to tens of years, but it is inconsequential in the context of geologic time. However, in some sequences of strata, surfaces known as **unconformities** may be present, representing times of nondeposition, erosion, or both. These unconformities encompass long periods of geologic time, perhaps

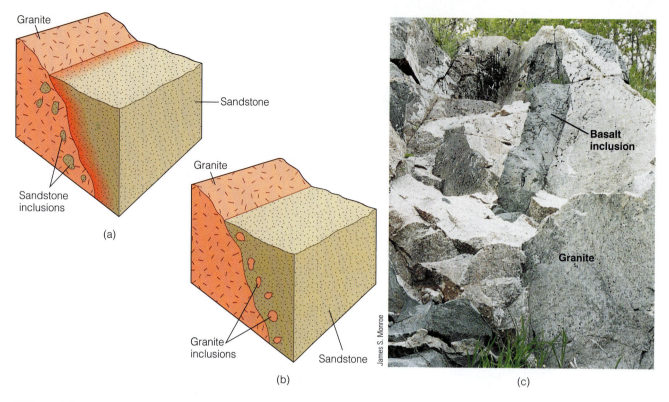

■ Figure 8.5

The principle of inclusions. (a) The batholith is younger than the sandstone because the sandstone has been baked at its contact with the granite and the granite contains sandstone inclusions. (b) Granite inclusions in the sandstone indicate that the batholith was the source of the sandstone and therefore is older. (c) Outcrop in northern Wisconsin showing basalt inclusions (black) in granite (white). Accordingly, the basalt inclusions are older than the granite.

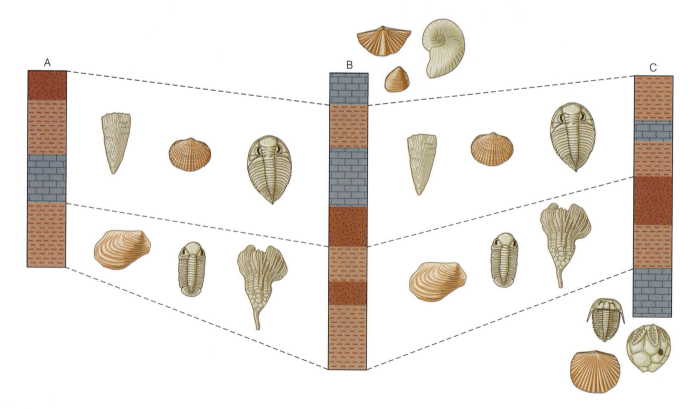

■ Figure 8.6

This generalized diagram shows how geologists use the principle of fossil succession to identify strata of the same age in different areas. The rocks in the three sections encompassed by the dashed lines contain similar fossils and are thus the same age. Note that the youngest rocks in this region are in section B, whereas the oldest rocks are in section C.

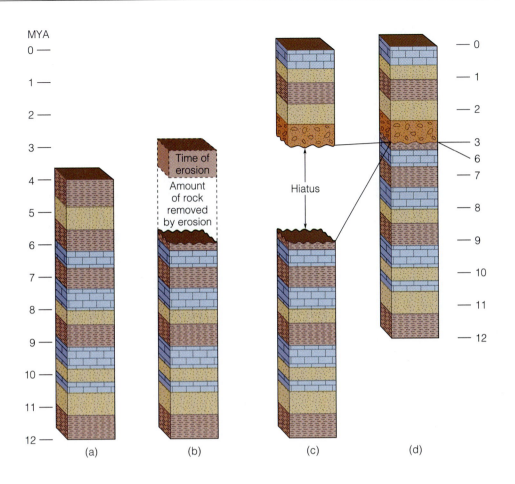

MYA

Figure 8.7

A simplified diagram showing the development of an unconformity and a hiatus. (a) Deposition began 12 million years ago (MYA) and continued more or less uninterrupted until 4 MYA. (b) A 1-million-year episode of erosion occurred, and during that time strata representing 2 million years of geologic time were eroded. (c) A hiatus of 3 million years exists between the older strata and the strata that formed during a renewed episode of deposition that began 3 MYA. (d) The actual stratigraphic record. The unconformity is the surface separating the strata and represents a major break in our record of geologic time.

millions or tens of millions of years. Accordingly, the geologic record is incomplete at that particular location, just as a book with missing pages is incomplete, and the interval of geologic time not represented by strata is called a *hiatus* (■ Figure 8.7).

The general term *unconformity* encompasses three specific types of surfaces. A **disconformity** is a surface of erosion or nondeposition separating younger from older rocks, both of which are parallel with one another (■ Figure 8.8). Unless the erosional surface separating the older from the younger parallel beds is well defined or distinct, the disconformity frequently resembles an ordinary bedding plane. Accordingly, many disconformities are difficult to recognize and must be identified on the basis of fossil assemblages.

An **angular unconformity** is an erosional surface on tilted or folded strata over which younger rocks were deposited (■ Figure 8.9). The strata below the unconformable surface generally dip more steeply than those above, producing an angular relationship.

The angular unconformity illustrated in Figure 8.9b is probably the most famous in the world. It was here at

Siccar Point, Scotland, where James Hutton realized that severe upheavals had tilted the lower rocks and formed mountains that were then worn away and covered by younger, flat-lying rocks. The erosional surface between the older tilted rocks and the younger flat-lying strata meant that a significant gap existed in the geologic record. Although Hutton did not use the term *unconformity*, he was the first to understand and explain the significance of such discontinuities in the geologic record.

A **nonconformity** is the third type of unconformity. Here an erosion surface cut into metamorphic or igneous rocks is covered by sedimentary rocks (■ Figure 8.10). This type of unconformity closely resembles an intrusive igneous contact with sedimentary rocks. The principle of inclusions is helpful in determining whether the relationship between the underlying igneous rocks and the overlying sedimentary rocks is the result of an intrusion or erosion (Figure 8.5). In the case of an intrusion, the igneous rocks are younger, but in the case of erosion, the sedimentary rocks are younger. Being able to distinguish between a nonconformity and an intrusive contact is very important because they represent different sequences of events.

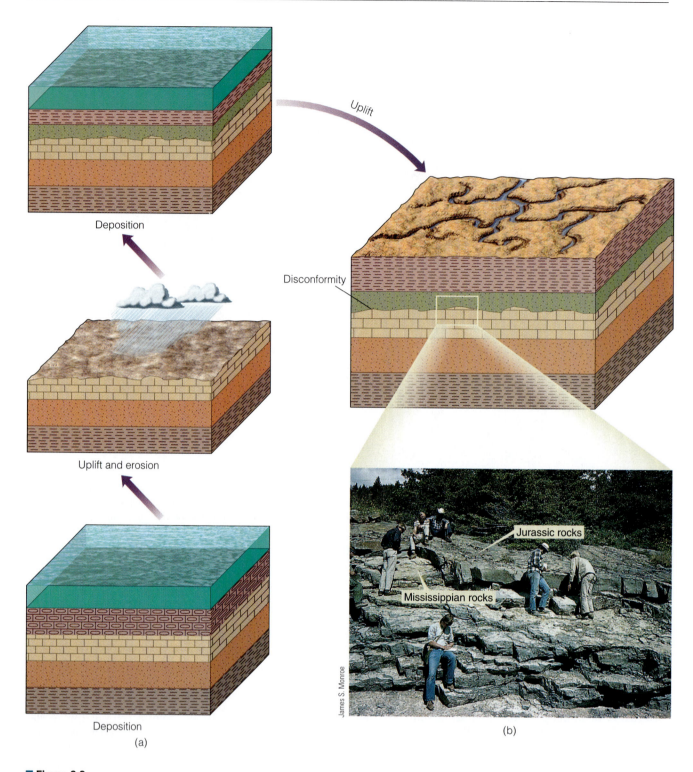

Figure 8.8

(a) Formation of a disconformity. (b) Disconformity between Mississippian and Jurassic strata in Montana. The geologist at the upper left is sitting on Jurassic strata, and his right foot is resting on Mississippian rocks.

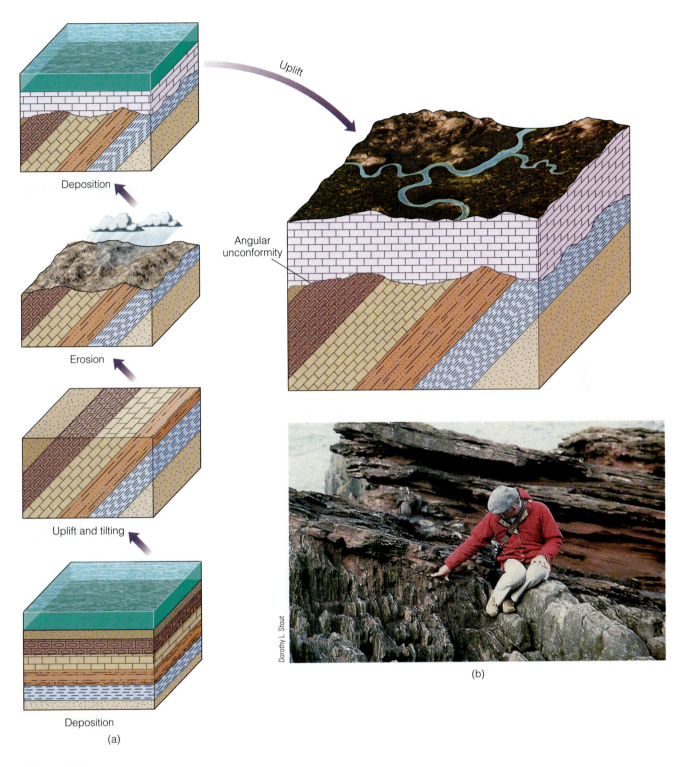

Uplift

Deposition

Erosion

Angular
unconformity

Uplift and tilting

Deposition

(a)

Dorothy L. Stout

(b)

■ **Figure 8.9**

(a) Formation of an angular unconformity. (b) Angular unconformity at Siccar Point, Scotland. James Hutton first realized the significance of unconformities at this site in 1788.

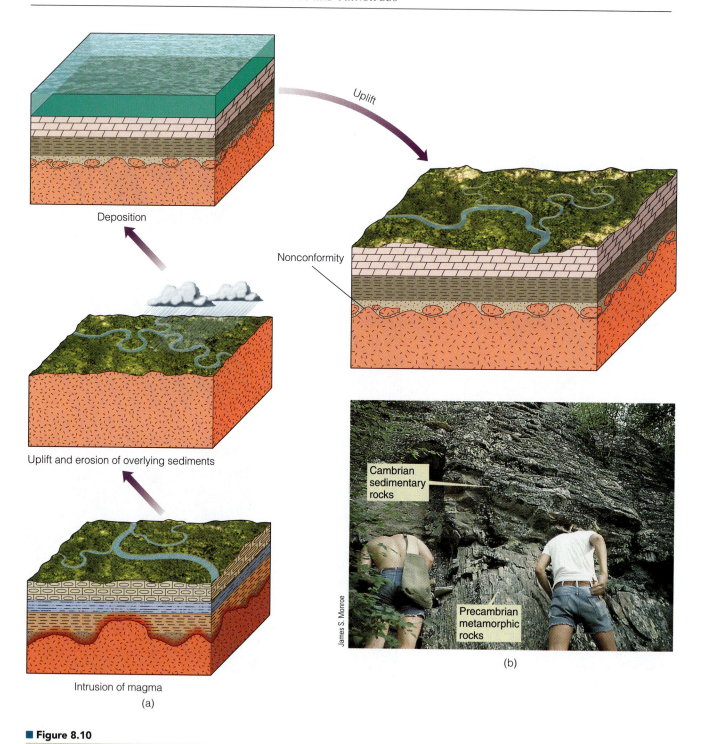

■ Figure 8.10

(a) Formation of a nonconformity. (b) Nonconformity between Precambrian metamorphic rocks and the overlying Cambrian-age Deadwood Formation, South Dakota.

Applying the Principles of Relative Dating

We can decipher the geologic history of the area represented by the block diagram in ■ Figure 8.11 by applying the various relative dating principles just discussed. The methods and logic used in this example are the same as those applied by 19th-century geologists in constructing the geologic time scale.

According to the principles of superposition and original horizontality, beds A–G were deposited horizontally; then either they were tilted, faulted (H), and eroded, or after deposition, they were faulted (H), tilted, and then eroded (■ Figure 8.12a–c). Because the fault cuts beds A–G, it must be younger than the beds according to the principle of cross-cutting relationships.

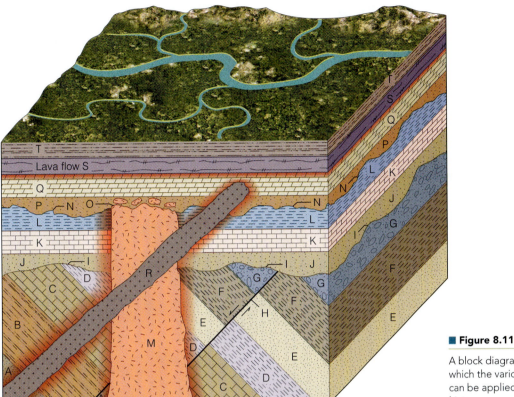

■ Figure 8.11

A block diagram of a hypothetical area in which the various relative-dating principles can be applied to determine its geologic history.

Beds J–L were then deposited horizontally over this erosional surface, producing an angular unconformity (I) (Figure 8.12d). Following deposition of these three beds, the entire sequence was intruded by a dike (M), which, according to the principle of cross-cutting relationships, must be younger than all the rocks it intrudes (Figure 8.12e).

The entire area was then uplifted and eroded; next beds P and Q were deposited, producing a disconformity (N) between beds L and P and a nonconformity (O) between the igneous intrusion M and the sedimentary bed P (Figure 8.12f, g). We know that the relationship between igneous intrusion M and the overlying sedimentary bed P is a nonconformity because of the presence of inclusions of M in P (principle of inclusions).

At this point, there are several possibilities for reconstructing the geologic history of this area. According to the principle of cross-cutting relationships, dike R must be younger than bed Q because it intrudes into it. It can have intruded anytime *after* bed Q was deposited; however, we cannot determine whether R was formed right after Q, right after S, or after T was formed. For purposes of this history, we will say that it intruded after the deposition of bed Q (Figure 8.12g, h).

Following the intrusion of dike R, lava S flowed over bed Q, followed by the deposition of bed T (Figure 8.12i, j). Although the lava flow (S) is not a sedimentary unit, the principle of superposition still applies because it flowed onto the surface, just as sediments are deposited on Earth's surface.

What Would You Do

You have been chosen to be part of the first astronaut crew to land on Mars. You were selected because you are a geologist, and therefore your primary responsibility is to map the geology of the landing site area. An important goal of the mission will be to work out the geologic history of the area. How will you go about this? Will you be able to use the principles of relative dating? How will you go about correlating the various rock units? Will you be able to determine absolute ages? How would you do this?

We have established a relative chronology for the rocks and events of this area by using the principles of relative dating. Remember, however, that we have no way of knowing how many years ago these events occurred unless we can obtain radiometric dates for the igneous rocks. With these dates, we can establish the range of absolute ages between which the different sedimentary units were deposited and also determine how much time is represented by the unconformities.

PHYSICAL
Geology⇌Now   Click Geology Interactive to work through an activity on Relative Dating through Geologic Time.

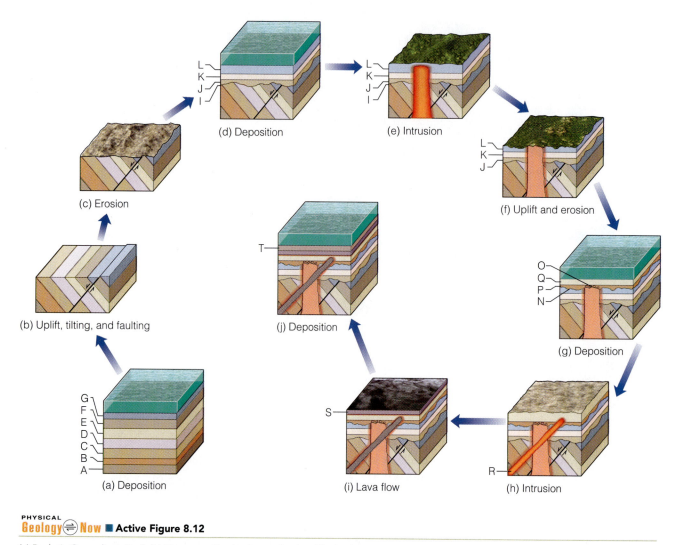

(a) Beds A–G are deposited. (b) The preceding beds are tilted and faulted. (c) Erosion. (d) Beds J–L are deposited, producing an angular unconformity I. (e) The entire sequence is intruded by a dike. (f) The entire sequence is uplifted and eroded. (g) Beds P and Q are deposited, producing a disconformity (N) and a nonconformity (O). (h) Dike R intrudes. (i) Lava (S) flows over bed Q, baking it. (j) Bed T is deposited.

HOW DO GEOLOGISTS CORRELATE ROCK UNITS?

To decipher Earth history, geologists must demonstrate the time equivalency of rock units in different areas. This process is known as **correlation.**

If surface exposures are adequate, units may simply be traced laterally (principle of lateral continuity), even if occasional gaps exist (■ Figure 8.13). Other criteria used to correlate units are similarity of rock type, position in a sequence, and key beds. *Key beds* are units, such as coal beds or volcanic ash layers, that are sufficiently distinctive to allow identification of the same unit in different areas (Figure 8.13).

Generally, no single location in a region has a geologic record of all events that occurred during its history; therefore geologists must correlate from one area to an-other in order to determine the complete geologic history of the region. An excellent example is the history of the Colorado Plateau (■ Figure 8.14). This region provides a record of events occurring over approximately 2 billion years. Because of the forces of erosion, the entire record is not preserved at any single location. Within the walls of the Grand Canyon are rocks of the Precambrian and Paleozoic eras, whereas Paleozoic and Mesozoic Era rocks are found in Zion National Park, and Mesozoic and Cenozoic Era rocks are exposed in Bryce Canyon (Figure 8.14). By correlating the uppermost rocks at one location with the lowermost equivalent rocks of another area, geologists can decipher the history of the entire region.

Although geologists can match up rocks on the basis of similar rock type and superposition, correlation of this type can be done only in a limited area where beds can be traced from one site to another. To correlate rock units over a large area or to correlate age-equivalent units of different composition, fossils and the principle of fossil succession must be used.

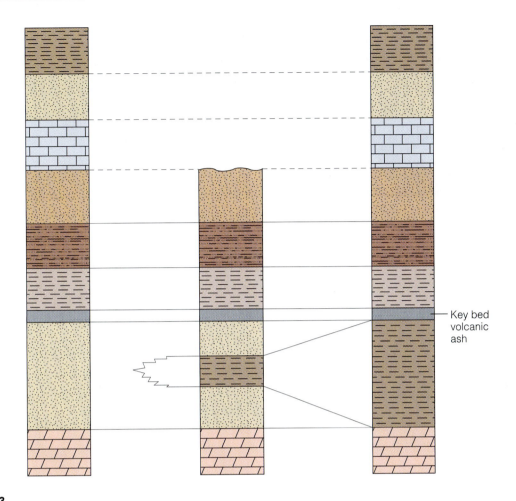

■ Figure 8.13

In areas of adequate exposures, rock units can be traced laterally, even if occasional gaps exist, and correlated on the basis of similarity in rock type and position in a sequence. Rocks can also be correlated by a key bed—in this case, volcanic ash. Source: From *History of the Earth: An Introduction to Historical Geology*, 2nd ed., by Bernhard Kummel. © 1961, 1970 by W. H. Freeman Co. Used with permission.

Fossils are useful as relative time indicators because they are the remains of organisms that lived for a certain length of time during the geologic past. Fossils that are easily identified, are geographically widespread, and existed for a rather short interval of geologic time are particularly useful. Such fossils are **guide fossils** or *index fossils* (■ Figure 8.15). The trilobite *Paradoxides* and the brachiopod *Atrypa* meet these criteria and are therefore good guide fossils. In contrast, the brachiopod *Lingula* is easily identified and widespread, but its geologic range of Ordovician to Recent makes it of little use in correlation.

Because most fossils have fairly long geologic ranges, geologists construct *concurrent range zones* to determine the age of the sedimentary rocks containing the fossils. Concurrent range zones are established by plotting the overlapping ranges of two or more fossils with different geologic ranges (■ Figure 8.16). The first and last occurrences of fossils are used to determine zone boundaries. Correlating concurrent range zones is probably the most accurate method of determining time equivalence.

Subsurface Correlation

In addition to surface geology, geologists are interested in subsurface geology because it provides additional information about geologic features beneath Earth's surface. A variety of techniques and methods are used to acquire and interpret data about the subsurface geology of an area.

When drilling is done for oil or natural gas, cores or rock chips called *well cuttings* are commonly recovered from the drill hole. These samples are studied under the microscope and reveal such important information as rock type, porosity (the amount of pore space), permeability (the ability to transmit fluids), and the presence of oil stains. In addition, the samples can be processed for a variety of microfossils that aid in determining the geologic age of the rock and the environment of deposition.

Geophysical instruments may be lowered down the drill hole to record such rock properties as electrical resistivity and radioactivity, thus providing a record or *well log* of the rocks penetrated. Cores, well cuttings, and

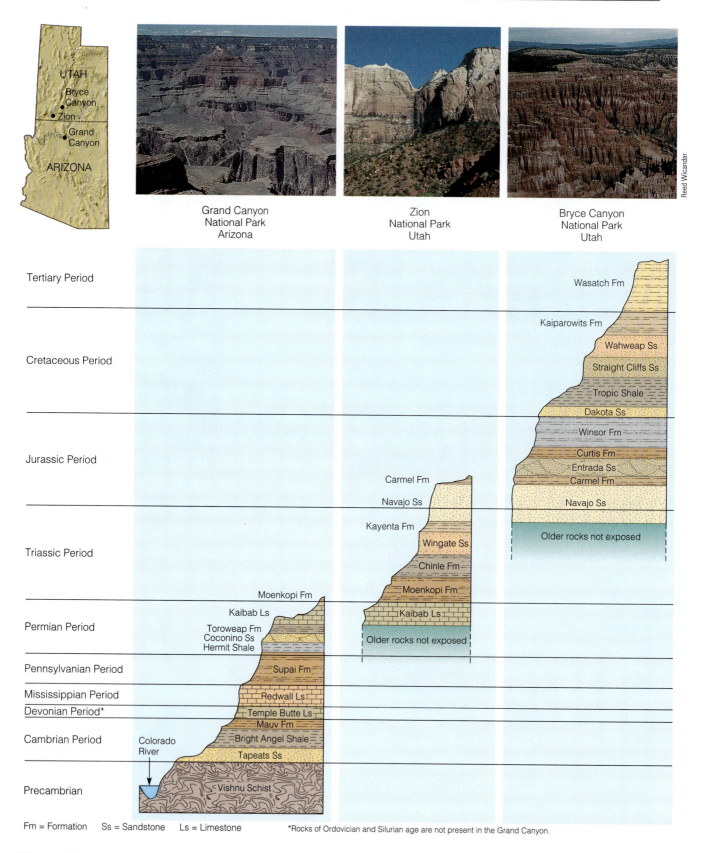

Grand Canyon
National Park
Arizona

Zion
National Park
Utah

Bryce Canyon
National Park
Utah

Reed Wicander

Fm = Formation   Ss = Sandstone   Ls = Limestone   *Rocks of Ordovician and Silurian age are not present in the Grand Canyon.

■ **Figure 8.14**

Correlation of rocks within the Colorado Plateau. From a correlation of the rocks from various locations, the history of the entire region can be deciphered.

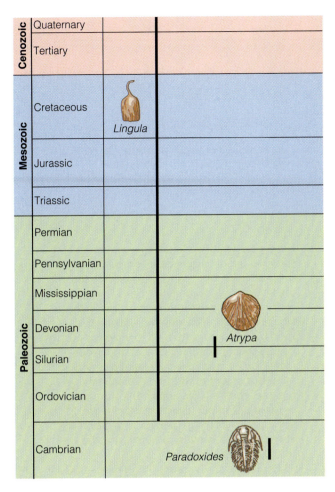

Figure 8.15

Comparison of the geologic ranges (heavy vertical lines) of three marine invertebrate animals. *Lingula* is of little use in correlation because it has such a long range. But *Atrypa* and *Paradoxides* are good guide fossils because both are widespread, easily identified, and have short geologic ranges.

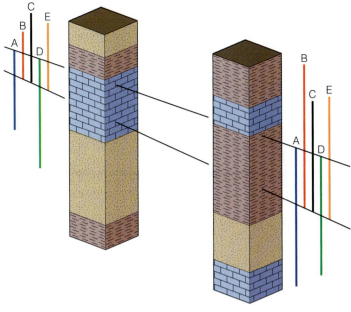

Figure 8.16

Correlation of two sections using concurrent range zones. This concurrent range zone was established by the overlapping ranges of fossils symbolized here by the letters A through E.

well logs are all extremely useful in making subsurface correlations (■ Figure 8.17).

Subsurface rock units may also be detected and traced by the study of seismic profiles. Energy pulses, such as those from explosions, travel through rocks at a velocity determined by rock density, and some of this energy is reflected from various horizons (contacts between contrasting layers) back to the surface, where it is recorded (see Figures 11.3 and 11.12). Seismic stratigraphy is particularly useful in tracing units in areas such as the continental shelves, where it is very expensive to drill holes and other techniques have limited use.

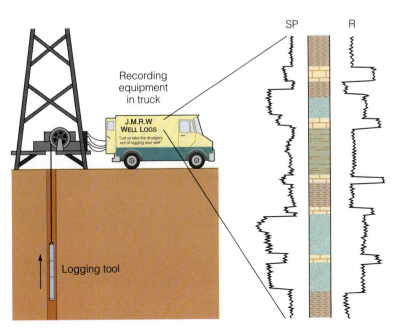

Figure 8.17

A schematic diagram showing how well logs are made. As the logging tool is withdrawn from the drill hole, data are transmitted to the surface, where they are recorded and printed as a well log. The curve labeled SP in this diagrammatic electric log is a plot of self-potential (electrical potential caused by different conductors in a solution that conducts electricity) with depth. The curve labeled R is a plot of electrical resistivity with depth. Electric logs yield information about the rock type and fluid content of subsurface formations. Electric logs are also used to correlate from well to well.

JOHN R. HORNER

Geology in Hollywood

John "Jack" Horner attended the University of Montana, where he majored in geology and zoology. In 1975 he was hired as a research assistant in the Museum of Natural History at Princeton University, where he worked until 1982. From 1982 until the present he has been Curator of Paleontology at the Museum of the Rockies in Bozeman, Montana. Jack's research teams discovered the first dinosaur egg clutches in the Western Hemisphere and six new species of dinosaurs.

Not very many scientists are likely to get a phone call from Steven Spielberg, but that's what happened to me back in the early 1990s. Steven called to ask if I would consider being a technical advisor for a movie he was working on, based on the book by Michael Crichton called *Jurassic Park*. I told Mr. Spielberg that I'd be happy to work with him on his movie, so I went to Universal Studios in Los Angeles and began working on one of the coolest movies ever made. My job was to make sure that the dinosaurs looked as accurate as possible, based on the available scientific understanding, and to help the actors like Sam Neil understand paleontology. One of my first duties was to help Stan Winston, the person whose studio made the animatronics, with the *T. rex* and velociraptors. All the animatronics in the Jurassic Park movies are life-size. They are first sculpted out of clay, then scanned in 3-D, and finally built as life-size puppet robots. When the animatronics were built and the sets finished, I was called back on several occasions to assist Steven while he shot the movie.

Celeste Horner

I was there to answer questions and make sure that the dinosaurs didn't do something they weren't supposed to, or to be sure the actors didn't say something scientifically wrong. One of my favorite scenes was in the kitchen where the children were being pursued by the raptors. It's hard to imagine, but I and a couple hundred other people were in that kitchen with piles of electronics and hundreds of feet of wire and cable. Someone did a great job of editing the film and removing all of our reflections from those shiny appliances. It gave me a great appreciation for how a camera captures what the director wants us to see.

On the day they were getting ready to shoot the sequence where the raptors first come into the kitchen, Steven said that the raptors would come in with their forked tongues waving in the air like a lizard, tasting the air for their prey. I jumped in and explained that we know dinosaurs didn't have forked tongues, and that the forked tongues would suggest that dinosaurs were cold-blooded rather than warm-blooded. Steven changed that scene so that just before they come into the kitchen, one dinosaur snorts at the window and fogs it up, revealing their warm-bloodedness. It was a

WHAT ARE ABSOLUTE DATING METHODS?

Although most of the isotopes of the 92 naturally occurring elements are stable, some are radioactive and spontaneously decay to other more stable isotopes of elements, releasing energy in the process. The discovery, in 1903 by Pierre and Marie Curie, that radioactive decay produces heat meant that geologists finally had a

mechanism for explaining Earth's internal heat that did not rely on residual cooling from a molten origin. Furthermore, geologists now had a powerful tool to date geologic events accurately and to verify the long time periods postulated by Hutton, Lyell, and Darwin.

Atoms, Elements, and Isotopes

As we discussed in Chapter 2, all matter is made up of chemical elements, each composed of extremely small

Jack Horner (second from right) on the *Jurassic Park* set with Steven Spielberg (right) and crew.

lated to velociraptor, so it's not much of an exaggeration to make them big in the movie.

One of the funny things that happened was that Steven had actually based the Alan Grant character on me, but I didn't discover this until the movie was finished. Movies are made of lots of parts, and only the director knows what it will look like in the end. It's kind of like finding a dinosaur skeleton and having to wait until it's all excavated, prepared, and mounted before everyone else knows what you found.

In the second movie, *The Lost World,* I didn't have much to do because most of the dinosaurs were the same as the ones we'd used in *Jurassic Park*. I did help with the *T. rex* nest, a lot of the footprints, and even some of the *T. rex* walking scenes.

In *Jurassic Park III* I did a great deal of work because there wasn't a book. Steven, the executive producer, Joe Johnston, the director, a group of writers, and I all got together and created the story line. I suggested bringing in the *Spinosaurus* to kill the *T. rex*, and I helped with virtually all aspects of the movie except the finishing touches. I even got to help in the editing trailer and worked very closely with the computer graphics people at Industrial Light and Magic.

For me it has been a great experience to work on these movies, but I wouldn't trade my job for any of theirs—not the director, the movie stars, or anyone else. Being a scientist and discovering new dinosaur skeletons and information that helps us unravel the geological history of our planet is the best job a person could ever have.

small change, but one that made a big difference in how we perceive the dinosaurs.

I suggested other changes to Steven as we continued to shoot the movie. Some of the suggestions he took, and others he didn't. It often depended on whether or not the suggestion was based on scientific data or was just my guess. If it was an accepted scientific idea, Steven was inclined to listen, but if it was an idea still in the process of being researched, he'd go with whatever he thought would be more exciting for the movie. Personally, I think that is what makes Steven a great director: He wasn't making a documentary, but rather a fictional movie. He wasn't tied to facts like documentaries, but instead he could exaggerate for the sake of a good movie. For example, the velociraptors in the movie are larger than any known to have lived, but there are larger dinosaurs very closely re-

particles called *atoms.* The nucleus of an atom is composed of *protons* (positively charged particles) and *neutrons* (neutral particles) with *electrons* (negatively charged particles) encircling it (see Figure 2.3). The number of protons defines an element's *atomic number* and helps determine its properties and characteristics. The combined number of protons and neutrons in an atom is its *atomic mass number.* However, not all atoms of the same element have the same number of neutrons in their nuclei. These variable forms of the same element are called *isotopes* (see Figure 2.4). Most isotopes are stable, but some are unstable and spontaneously decay to a more stable form. It is the decay rate of unstable isotopes that geologists measure to determine the absolute age of rocks.

Radioactive Decay and Half-Lives

Radioactive decay is the process by which an unstable atomic nucleus is spontaneously transformed into an atomic nucleus of a different element. Scientists recognize three types of radioactive decay, all of which result

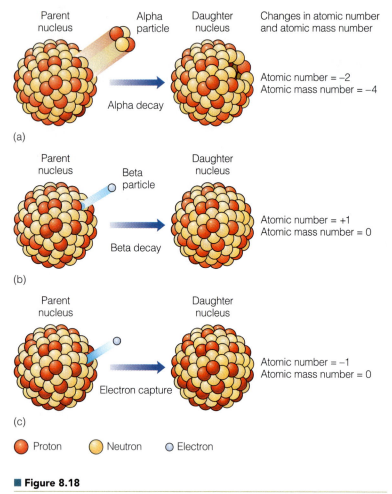

■ Figure 8.18

Three types of radioactive decay. (a) Alpha decay, in which an unstable parent nucleus emits 2 protons and 2 neutrons. (b) Beta decay, in which an electron is emitted from the nucleus. (c) Electron capture, in which a proton captures an electron and is thereby converted to a neutron.

the original unstable *parent element* to decay to atoms of a new, more stable *daughter element*. The half-life of a given radioactive element is constant and can be precisely measured. Half-lives of various radioactive elements range from less than a billionth of a second to 49 billion years.

Radioactive decay occurs at a geometric rate rather than a linear rate. Therefore a graph of the decay rate produces a curve rather than a straight line (■ Figure 8.20). For example, an element with 1,000,000 parent atoms will have 500,000 parent atoms and 500,000 daughter atoms after one half-life. After two half-lives, it will have 250,000 parent atoms (one half of the previous parent atoms, which is equivalent to one fourth of the original parent atoms) and 750,000 daughter atoms. After three half-lives, it will have 125,000 parent atoms (one half of the previous parent atoms, or one eighth of the original parent atoms) and 875,000 daughter atoms, and so on until the number of parent atoms remaining is so few that they cannot be accurately measured by present-day instruments.

By measuring the parent–daughter ratio and knowing the half-life of the parent (which has been determined in the laboratory), geologists can calculate the age of a sample containing the radioactive element. The parent–daughter ratio is usually determined by a *mass spectrometer,* an instrument that measures the proportions of atoms of different masses.

PHYSICAL
Geology⇌Now   Click Geology Interactive to work through an activity on Absolute Dating through Geologic Time.

Sources of Uncertainty

The most accurate radiometric dates are obtained from igneous rocks. As magma cools and begins to crystallize, radioactive parent atoms are separated from previously formed daughter atoms. Because they are the right size, some radioactive parent atoms are incorporated into the crystal structure of certain minerals. The stable daughter atoms, though, are a different size from the radioactive parent atoms and consequently cannot fit into the crystal structure of the same mineral as the parent atoms. Therefore a mineral crystallizing in cooling magma will contain radioactive parent atoms but no stable daughter atoms (■ Figure 8.21). Thus the time that is being measured is the time of crystallization of the mineral containing the radioactive atoms, and not the time of formation of the radioactive atoms.

in a change of atomic structure (■ Figure 8.18). In **alpha decay,** 2 protons and 2 neutrons are emitted from the nucleus, resulting in the loss of 2 atomic numbers and 4 atomic mass numbers. In **beta decay,** a fast-moving electron is emitted from a neutron in the nucleus, changing that neutron to a proton and consequently increasing the atomic number by 1, with no resultant atomic mass number change. **Electron capture** results when a proton captures an electron from an electron shell and thereby converts to a neutron, resulting in the loss of 1 atomic number but not changing the atomic mass number.

Some elements undergo only 1 decay step in the conversion from an unstable form to a stable form. For example, rubidium 87 decays to strontium 87 by a single beta emission, and potassium 40 decays to argon 40 by a single electron capture. Other radioactive elements undergo several decay steps. Uranium 235 decays to lead 207 by 7 alpha and 6 beta steps, whereas uranium 238 decays to lead 206 by 8 alpha and 6 beta steps (■ Figure 8.19).

When we discuss decay rates, it is convenient to refer to them in terms of half-lives. The **half-life** of a radioactive element is the time it takes for half of the atoms of

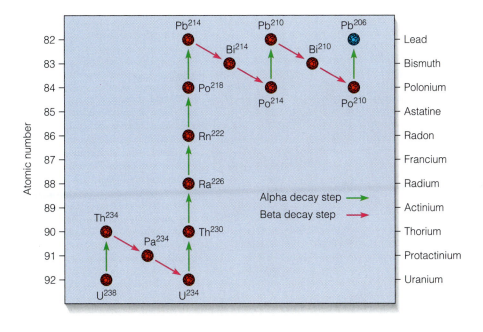

Radioactive decay series for uranium 238 to lead 206. Radioactive uranium 238 decays to its stable daughter product, lead 206, by 8 alpha and 6 beta decay steps. A number of different isotopes are produced as intermediate steps in the decay series. Source: Based on data from S. M. Richardson and H. Y. McSween, Jr., Geochemistry—*Pathways and Processes*, Prentice-Hall.

be obtained on sedimentary rocks is when the mineral glauconite is present. Glauconite is a greenish mineral containing radioactive potassium 40, which decays to argon 40 (Table 8.1). It forms in certain marine environments as a result of chemical reactions with clay minerals during the conversion of sediments to sedimentary rock. Thus glauconite forms when the sedimentary rock forms, and a radiometric date indicates the time of the sedimentary rock's origin. Being a gas, however, the daughter product argon can easily escape from a mineral. Therefore any date obtained from glauconite, or any other mineral containing the potassium 40 and argon 40 pair, must be considered a minimum age.

To obtain accurate radiometric dates, geologists must be sure that they are dealing with a *closed system*, meaning that neither parent nor daughter atoms

Except in unusual circumstances, sedimentary rocks cannot be radiometrically dated because one would be measuring the age of a particular mineral rather than the time that it was deposited as a sedimentary particle. One of the few instances in which radiometric dates can

have been added or removed from the system since crystallization and that the ratio between them results only from radioactive decay. Otherwise, an inaccurate date will result. If daughter atoms have leaked out of the mineral being analyzed, the calculated age will be too young; if

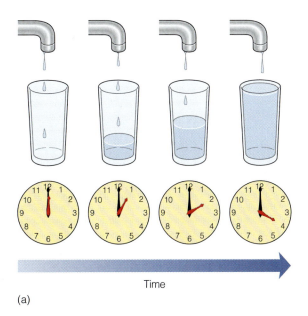

(a)

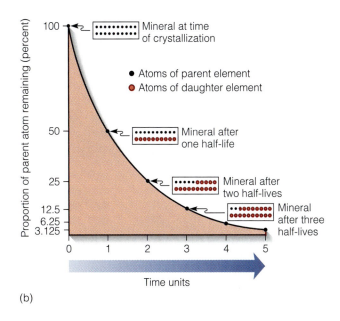

(b)

■ **Figure 8.20**

(a) Uniform, linear change is characteristic of many familiar processes. In this example, water is being added to a glass at a constant rate.
(b) Geometric radioactive decay curve, in which each time unit represents one half-life, and each half-life is the time it takes for half of the parent element to decay to the daughter element.

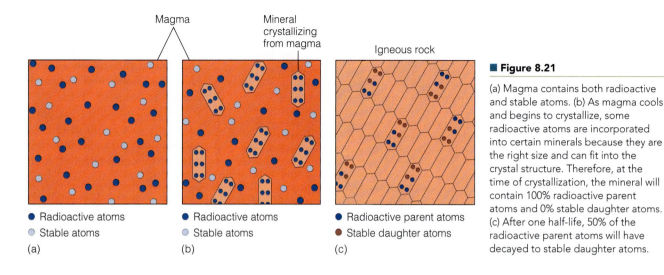

● Radioactive atoms
○ Stable atoms
(a)

● Radioactive atoms
○ Stable atoms
(b)

● Radioactive parent atoms
● Stable daughter atoms
(c)

■ **Figure 8.21**

(a) Magma contains both radioactive and stable atoms. (b) As magma cools and begins to crystallize, some radioactive atoms are incorporated into certain minerals because they are the right size and can fit into the crystal structure. Therefore, at the time of crystallization, the mineral will contain 100% radioactive parent atoms and 0% stable daughter atoms. (c) After one half-life, 50% of the radioactive parent atoms will have decayed to stable daughter atoms.

parent atoms have been removed, the calculated age will be too old.

Leakage may take place if the rock is heated or subjected to intense pressure, as can sometimes occur during metamorphism. If this happens, some of the parent or daughter atoms may be driven from the mineral being analyzed, resulting in an inaccurate age determination. If the daughter product was completely removed, then one would be measuring the time since metamorphism (a useful measurement itself), and not the time since crystallization of the mineral (■ Figure 8.22). Because heat affects the parent–daughter ratio, metamorphic rocks are difficult to date accurately. Remember that although the parent–daughter ratio may be affected by heat, the decay rate of the parent element remains constant, regardless of any physical or chemical changes.

It is sometimes possible to cross-check the radiometric date obtained by measuring the parent–daughter ratio of two different radioactive elements in the same mineral. For example, naturally occurring uranium consists of both uranium 235 and uranium 238 isotopes. Through various decay steps, uranium 235 decays to lead 207, whereas uranium 238 decays to lead 206 (Figure 8.19). If the minerals containing both uranium isotopes have remained closed systems, the ages obtained from each parent–daughter ratio should agree closely and therefore should indicate the time of crystallization of the magma. If the ages do not closely agree, other samples must be taken and ratios measured to see which, if either, date is correct.

Recent advances and the development of new techniques and instruments for measuring various isotope

Table 8.1

Five of the Principal Long-Lived Radioactive Isotope Pairs Used in Radiometric Dating

ISOTOPES		Half-Life of Parent (years)	Effective Dating Range (years)	Minerals and Rocks That Can Be Dated	
Parent	Daughter				
Uranium 238	Lead 206	4.5 billion	10 million to 4.6 billion	Zircon	
				Uraninite	
Uranium 235	Lead 207	704 million			
Thorium 232	Lead 208	14 billion			
Rubidium 87	Strontium 87	48.8 billion	10 million to 4.6 billion	Muscovite	
				Biotite	
				Potassium feldspar	
				Whole metamorphic or igneous rock	
Potassium 40	Argon 40	1.3 billion	100,000 to 4.6 billion	Glauconite	Hornblende
				Muscovite	Whole volcanic rock
				Biotite	

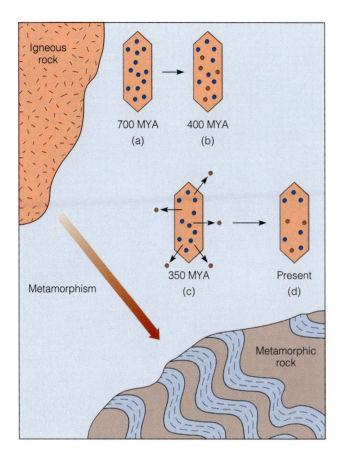

Figure 8.22

The effect of metamorphism in driving out daughter atoms from a mineral that crystallized 700 million years ago (MYA). The mineral is shown immediately after crystallization (a), then at 400 million years (b), when some of the parent atoms had decayed to daughter atoms. Metamorphism at 350 MYA (c) drives the daughter atoms out of the mineral into the surrounding rock. (d) Assuming the rock has remained a closed chemical system throughout its history, dating the mineral today yields the time of metamorphism, whereas dating the whole rock provides the time of its crystallization, 700 MYA.

ratios have enabled geologists to analyze not only increasingly smaller samples, but with a greater precision than ever before. Presently the measurement error for many radiometric dates is typically less than 0.5% of the age, and in some cases it is even better than 0.1%. Thus, for a rock 540 million years old (near the beginning of the Cambrian Period), the possible error could range from nearly 2.7 million years to less than 540,000 years.

Long-Lived Radioactive Isotope Pairs

Table 8.1 shows the five common, long-lived parent–daughter isotope pairs used in radiometric dating. Long-lived pairs have half-lives of millions or billions of years. All of these were present when Earth formed and are still present in measurable quantities. Other shorter-lived radioactive isotope pairs have decayed to the point that only small quantities near the limit of detection remain.

The most commonly used isotope pairs are the uranium–lead and thorium–lead series, which are used principally to date ancient igneous intrusives, lunar samples, and some meteorites. The rubidium–strontium pair is also used for very old samples and has been effective in dating the oldest rocks on Earth as well as meteorites. The potassium–argon method is typically used for dating fine-grained volcanic rocks from which individual crystals cannot be separated; hence, the whole rock is analyzed. Because argon is a gas, great care must be taken to ensure that the sample has not been subjected to heat, which would allow argon to escape; such a sample would yield an age that is too young. Other long-lived radioactive isotope pairs exist, but they are rather rare and are used only in special situations.

Fission Track Dating

The emission of atomic particles resulting from the spontaneous decay of uranium within a mineral damages its crystal structure. The damage appears as microscopic linear tracks that are visible only after etching the mineral with hydrofluoric acid. The age of the sample is determined on the basis of the number of fission tracks present and the amount of uranium the sample contains: the older the sample, the greater the number of tracks (■ Figure 8.23).

Fission track dating is of particular interest to archaeologists and geologists because they can use the technique to date samples ranging from only a few hundred to hundreds of millions of years in age. It is most useful for dating samples between about 40,000 and 1.5 million years ago, a period for which other dating techniques are not always particularly suitable. One of the problems in fission track dating occurs when the rocks have later been subjected to high temperatures. If this happens, the damaged crystal structures are repaired by annealing, and consequently the

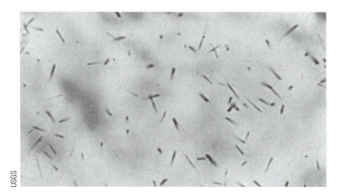

Figure 8.23

Each fission track (about 16 microns long) in this apatite crystal is the result of the radioactive decay of a uranium atom. The apatite crystal, which has been etched with hydrofluoric acid to make the fission tracks visible, comes from one of the dikes at Shiprock, New Mexico, and has a calculated age of 27 million years.

tracks disappear. In such instances, the calculated age will be younger than the actual age.

Radiocarbon and Tree-Ring Dating Methods

Carbon is an important element in nature and is one of the basic elements found in all forms of life. It has three isotopes; two of these, carbon 12 and 13, are stable, whereas carbon 14 is radioactive (see Figure 2.4). Carbon 14 has a half-life of 5730 years plus or minus 30 years. The **carbon 14 dating technique** is based on the ratio of carbon 14 to carbon 12 and is generally used to date once-living material.

The short half-life of carbon 14 makes this dating technique practical only for specimens younger than about 70,000 years. Consequently, the carbon 14 dating method is especially useful in archaeology and has greatly helped unravel the events of the latter portion of the Pleistocene Epoch. For example, carbon 14 dates of maize from the Tehuacan Valley of Mexico have forced archeologists to rethink their ideas of where the first center for maize domestication in Mesoamerica arose. Carbon 14 dating is also helping to answer the question of when humans began populating North America.

Carbon 14 is constantly formed in the upper atmosphere when cosmic rays, which are high-energy particles (mostly protons), strike the atoms of upper-atmospheric gases, splitting their nuclei into protons and neutrons. When a neutron strikes the nucleus of a nitrogen atom (atomic number 7, atomic mass number 14), it may be absorbed into the nucleus and a proton emitted. Thus the atomic number of the atom decreases by 1, while the atomic mass number stays the same. Because the atomic number has changed, a new element, carbon 14 (atomic number 6, atomic mass number 14), is formed. The newly formed carbon 14 is rapidly assimilated into the carbon cycle and, along with carbon 12 and 13, is absorbed in a nearly constant ratio by all living organisms (■ Figure 8.24). When an organism dies, however, carbon 14 is not replenished, and the ratio of carbon 14 to carbon 12 decreases as carbon 14 decays back to nitrogen by a single beta decay step (Figure 8.24).

Currently the ratio of carbon 14 to carbon 12 is remarkably constant in both the atmosphere and living organisms. There is good evidence, however, that the production of carbon 14, and thus the ratio of carbon 14 to carbon 12, has varied somewhat during the past several thousand years. This was determined by comparing ages established by carbon 14 dating of wood samples against those established by counting annual tree rings in the same samples. As a result, carbon 14 ages have been corrected to reflect such variations in the past.

Tree-ring dating is another useful method for dating geologically recent events. The age of a tree can be determined by counting the growth rings in the lower part of the stem. Each ring represents one year's growth,

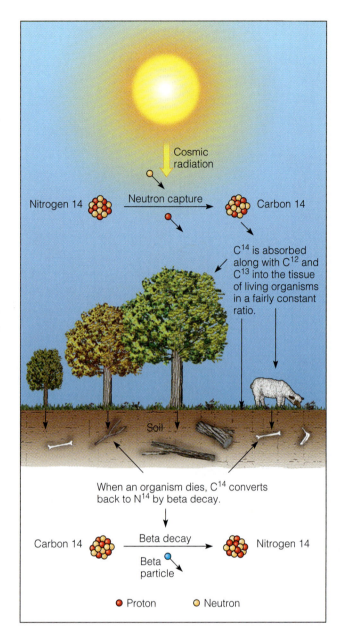

■ **Figure 8.24**

The carbon cycle showing the formation, dispersal, and decay of carbon 14.

and the pattern of wide and narrow rings can be compared among trees to establish the exact year in which the rings were formed. The procedure of matching ring patterns from numerous trees and wood fragments in a given area is referred to as *cross-dating*. By correlating distinctive tree-ring sequences from living to nearby dead trees, a time scale can be constructed that extends back to about 14,000 years ago (■ Figure 8.25). By matching ring patterns to the composite ring scale, wood samples whose ages are not known can be accurately dated.

The applicability of tree-ring dating is somewhat limited because it can be used only where continuous

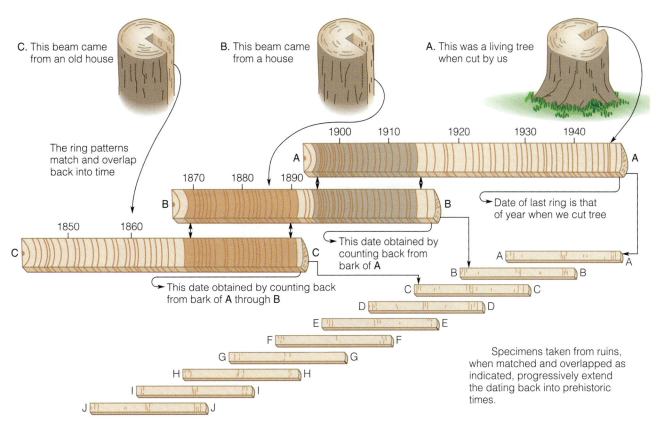

C. This beam came from an old house

B. This beam came from a house

A. This was a living tree when cut by us

The ring patterns match and overlap back into time

Date of last ring is that of year when we cut tree

This date obtained by counting back from bark of A

This date obtained by counting back from bark of A through B

Specimens taken from ruins, when matched and overlapped as indicated, progressively extend the dating back into prehistoric times.

■ **Figure 8.25**

In the cross-dating method, tree-ring patterns from different woods are matched against each other to establish a ring-width chronology backward in time. Source: From *An Introduction to Tree-Ring Dating*, by Stokes and Smiley, 1968, p. 6. Reprinted by permission of University of Chicago Press.

tree records are found. It is therefore most useful in arid regions, particularly the southwestern United States.

PHYSICAL Geology⇌Now Click Geology Interactive to work through an activity on Absolute Dating through Geologic Time.

HOW WAS THE GEOLOGIC TIME SCALE DEVELOPED?

The geologic time scale is a hierarchical scale in which the 4.6-billion-year history of Earth is divided into time units of varying duration (Figure 8.1). It was not developed by any one individual but rather evolved, primarily during the 19th century, through the efforts of many people. By applying relative-dating methods to rock outcrops, geologists in England and western Europe defined the major geologic time units without the benefit of radiometric dating techniques. Using the principles of superposition and fossil succession, they could correlate various rock exposures and piece together a composite geologic sec-

tion. This composite section is in effect a relative time scale because the rocks are arranged in their correct sequential order.

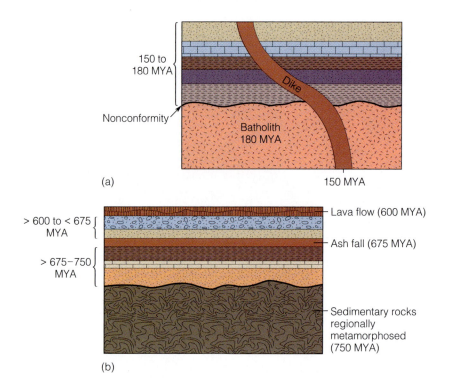

■ Figure 8.26

Absolute ages of sedimentary rocks can be determined by dating associated igneous rocks. In (a) and (b), sedimentary rocks are bracketed by rock bodies for which absolute ages have been determined.

By the beginning of the 20th century, geologists had developed a relative geologic time scale but did not yet have any absolute dates for the various time-unit boundaries. Following the discovery of radioactivity near the end of the 19th century, radiometric dates were added to the relative geologic time scale (Figure 8.1).

Because sedimentary rocks, with rare exceptions, cannot be radiometrically dated, geologists have had to rely on interbedded volcanic rocks and igneous intrusions to apply absolute dates to the boundaries of the various subdivisions of the geologic time scale (■ Figure 8.26). An ashfall or lava flow provides an excellent marker bed that is a time-equivalent surface, supplying a minimum age for the sedimentary rocks below and a maximum age for the rocks above. Ashfalls are particularly useful because they may fall over both marine and nonmarine sedimentary environments and can provide a connection between these different environments.

Thousands of absolute ages are now known for sedimentary rocks of known relative ages, and these absolute dates have been added to the relative time scale. In this way, geologists have been able to determine the absolute ages of the various geologic periods and to determine their durations (Figure 8.1).

REVIEW
WORKBOOK

Chapter Summary

- Relative dating involves placing geologic events in a sequential order as determined from their position in the geologic record. Absolute dating results in specific dates for events, expressed in years before the present.

- During the 18th and 19th centuries, attempts were made to determine Earth's age based on scientific evidence rather than revelation. Although some attempts were ingenious, they yielded a variety of ages that now are known to be much too young.

- James Hutton thought that present-day processes operating over long periods of time could explain all the geologic features of Earth. His observations were instrumental in establishing the principle of uniformitarianism.

- Uniformitarianism, as articulated by Charles Lyell, soon became the guiding principle of geology. It holds that the laws of nature have been constant through time and that the same processes operating today have operated in the past, though not necessarily at the same rates.

- Besides uniformitarianism, the principles of superposition, original horizontality, lateral continuity, cross-cutting relationships, inclusions, and fossil succession are basic for determining relative geologic ages and for interpreting Earth history.

- Surfaces of discontinuity encompassing significant amounts of geologic time are common in the geologic record. Such surfaces are unconformities and result from times of nondeposition, erosion, or both.

- Correlation is the practice of demonstrating equivalency of units in different areas. Time equivalence is most commonly demonstrated by correlating strata that contain similar fossils.

- Radioactivity was discovered during the late 19th century, and soon thereafter radiometric dating techniques allowed geologists to determine absolute ages for geologic events.

- Absolute ages for rocks are usually obtained by determining how many half-lives of a radioactive parent element have elapsed since the sample originally crystallized. A half-life is the time it takes for half of the radioactive parent element to decay to a stable daughter element.

- The most accurate radiometric dates are obtained from long-lived radioactive isotope pairs in igneous rocks. The most reliable dates are those obtained by using at least two different radioactive decay series in the same rock.

- Carbon 14 dating can be used only for organic matter such as wood, bones, and shells and is effective back to about 70,000 years ago. Unlike the long-lived isotope pairs, the carbon 14 dating technique determines age by the ratio of radioactive carbon 14 to stable carbon 12.

- Through the efforts of many geologists applying the principles of relative dating, a relative geologic time scale was established by the mid 1800s.

- Most absolute ages of sedimentary rocks and their contained fossils are obtained indirectly by dating associated metamorphic or igneous rocks.

Important Terms

absolute dating (p. 208)
alpha decay (p. 230)
angular unconformity (p. 219)
beta decay (p. 230)
carbon 14 dating technique (p. 234)

correlation (p. 224)
disconformity (p. 219)
electron capture (p. 230)
fission track dating (p. 233)
guide fossil (p. 225)
half-life (p. 230)

nonconformity (p. 219)
principle of cross-cutting relationships (p. 216)
principle of fossil succession (p. 216)
principle of inclusions (p. 216)

principle of lateral continuity (p. 216)

principle of original horizontality (p. 216)

principle of superposition (p. 215)

principle of uniformitarianism (p. 211)

radioactive decay (p. 229)

relative dating (p. 208)

tree-ring dating (p. 234)

unconformity (p. 217)

Review Questions

1. Pacing geologic events in sequential or chronological order as determined by their position in the geologic record is called:

   a. _____ absolute dating; b. _____ correlation; c. _____ historical dating; d. _____ relative dating; e. _____ uniformitarianism.

2. Who is generally considered to be the founder of modern geology?

   a. _____ Werner; b. _____ Lyell; c. _____ Steno; d. _____ Cuvier; e. _____ Hutton.

3. If a radioactive element has a half-life of 8 million years, the amount of parent material remaining after 24 million years of decay will be what fraction of the original amount?

   a. _____ 1/32; b. _____ 1/16; c. _____ 1/8; d. _____ 1/4; e. _____ 1/2.

4. If a feldspar grain within a sedimentary rock (such as a sandstone) is radiometrically dated, the date obtained will indicate when:

   a. _____ the feldspar crystal formed; b. _____ the sedimentary rock formed; c. _____ the parent radioactive isotope formed; d. _____ the daughter radioactive isotope(s) formed; e. _____ none of the preceding answers.

5. In which type of radioactive decay does a neutron change to a proton in the nucleus by the emission of an electron?

   a. _____ alpha decay; b. _____ beta decay; c. _____ electron capture; d. _____ fission track; e. _____ none of the preceding answers.

6. Demonstrating the time equivalency of rock units in different areas is called:

   a. _____ relative dating; b. _____ historical dating; c. _____ correlation; d. _____ absolute dating; e. _____ none of the preceding answers.

7. How many half-lives are required to yield a mineral with 1250 atoms of U^{235} and 18,750 atoms of Pb^{206}?

   a. _____ 1; b. _____ 2; c. _____ 4; d. _____ 8; e. _____ 16.

8. In radiocarbon dating, what isotopic ratio decreases as carbon 14 decays back to nitrogen?

   a. _____ nitrogen 14 to carbon 14; b. _____ carbon 14 to carbon 12; c. _____ carbon 13 to carbon 12; d. _____ nitrogen 14 to carbon 12; e. _____ none of the preceding answers.

9. Considering the half-life of potassium 40, which is 1.3 billion years, what fraction of the original potassium 40 can be expected within a given mineral crystal after 3.9 billion years?

   a. _____ 1/2; b. _____ 1/4; c. _____ 1/8; d. _____ 1/16; e. _____ 1/32.

10. The atomic mass number of an element is determined by the number of _____ in its nucleus.

    a. _____ protons; b. _____ neutrons; c. _____ electrons; d. _____ protons and neutrons; e. _____ protons and electrons.

11. What is being measured in radiometric dating?

    a. _____ the time when a radioactive isotope formed; b. _____ the time of crystallization of a mineral containing an isotope; c. _____ the amount of the parent isotope only; d. _____ when the dated mineral became part of a sedimentary rock; e. _____ when the stable daughter isotope was formed.

12. In some places, where disconformities are particularly difficult to discern from a physical point of view, use of the principle of fossil succession helps us delineate such unconformities. How do you suppose using fossils could help us find such hard-to-see disconformities?

13. If a rock or mineral were radiometrically dated using two or more radioactive isotope pairs (e.g., uranium 238 to lead 206 and rubidium 87 to thorium 87) and the analysis for those isotope pairs yielded distinctly different results, what possible explanation could be offered to explain how this happened? How can one rock have two correct ages?

14. If you wanted to calculate the absolute age of an intrusive body, what information would you need?

15. Where did Lord Kelvin go wrong? Explain the rationale for his age-of-Earth calculations and what discovery subsequently showed that Kelvin's calculations were, in fact, in error.

16. How does metamorphism affect the potential for accurate radiometric dating using any and all techniques discussed in this chapter? How would such radiometric dates be affected by metamorphism, and why?

17. Describe the principle of uniformitarianism according to Hutton and Lyell. What is the significance of this principle?

18. What uncertainties are associated with trying to radiometrically date any sedimentary rock?

19. Describe how the principle of inclusions would be important in recognizing a nonconformity.

20. Explain the concept of a concurrent range zone and how it relates to guide or index fossils.

World Wide Web Activities

PHYSICAL Geology⇌Now Assess your understanding of this chapter's topics with additional quizzing and comprehensive interactivities at

http://earthscience.brookscole.com/physgeo5e

as well as current and up-to-date weblinks, additional readings, and InfoTrac College Edition exercises.

Earthquakes

CHAPTER 9
OUTLINE

PHYSICAL
Geology⇌Now *This icon, appearing throughout the book, indicates an opportunity to explore interactive tutorials, animations, or practice problems available on the Physical GeologyNow Web site at http://earthscience.brookscole.com/physgeo5e.*

OBJECTIVES
At the end of this chapter, you will have learned that

- Earthquakes result from the sudden release of energy.

- Energy is stored in rocks and is released when they fracture, thus producing various types of waves that travel outward in all directions from their source.

- Most earthquakes take place in well-defined zones at transform, divergent, and convergent plate boundaries.

- An earthquake's epicenter is found by analyzing earthquake waves at no fewer than three seismic stations.

- Intensity is a qualitative assessment of the damage done by an earthquake. The Richter Magnitude Scale and Moment Magnitude Scale are used to express the amount of energy released during an earthquake.

- Great hazards are associated with earthquakes, such as ground shaking, fire, tsunami, and ground failure.

- Adequate preparation, building practices, and monitoring in earthquake-prone areas can minimize the destructive effects of earthquakes.

- Efforts by scientists to make accurate, short-term earthquake predictions have thus far met with only limited success.

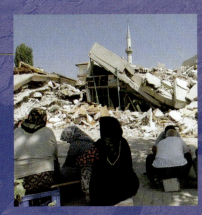

Damage at Izmit, Turkey, resulting from the earthquake of August 17, 1999, in which an estimated 17,000 people died and more than 240,000 buildings were damaged. Source: AFP/Corbis

241

Introduction

A t 3:02 A.M. on August 17, 1999, violent shaking from an earthquake awakened millions of people in Turkey. When the earthquake was over, an estimated 17,000 people were dead, at least 50,000 were injured, and tens of thousands of survivors were left homeless. The amount of destruction this earthquake caused is staggering. More than 150,000 buildings were moderately to heavily damaged, and another 90,000 suffered slight damage. Collapsed buildings were everywhere, streets were strewn with rubble, and all communications were knocked out. And if this weren't enough, the same area was struck again only three months later on November 12, 1999, by an aftershock nearly as large as the original earthquake, which killed an additional 374 people and injured about 3000 more. All in all, this was a disaster of epic proportions. Yet it was not the first, nor will it be the last major devastating earthquake in this region or other parts of the world.

As one of nature's most frightening and destructive phenomena, earthquakes have always aroused a sense of fear and as such, have been the subject of numerous myths and legends. What makes an earthquake so frightening is that when it begins, there is no way to tell how long it will last or how violent it will be. About 13 million people have died in earthquakes during the past 4000 years, with about 2.7 million of these deaths occurring during the last century alone (Table 9.1).

Geologists define an **earthquake** as the shaking or trembling caused by the sudden release of energy, usually as a result of faulting, which involves displacement of rocks along fractures (we will discuss the different types of faults in Chap-

ter 13). After an earthquake, continuing adjustments along a fault may generate a series of earthquakes known as *aftershocks*. Most of these are smaller than the main shock, but they can still cause considerable damage to already weakened structures, as we just saw in the 1999 Turkey earthquake.

Although the geologic definition of an earthquake is accurate, it is not nearly as imaginative or colorful as the explanations many people held in the past. Many cultures attributed the cause of earthquakes to movements of some kind of animal on which Earth rested. In Japan it was a giant catfish, in Mongolia a giant frog, in China an ox, in South America a whale, and to the Algonquin of North America an immense tortoise. And a legend from Mexico holds that earthquakes occur when the devil, El Diablo, rips open Earth's crust so he and his friends can reach the surface.

If earthquakes are not the result of animal movement or the devil ripping open the crust, what does cause earthquakes? Geologists know that most earthquakes result from energy released along plate boundaries, and as such, earthquakes are a manifestation of Earth's dynamic nature and the fact Earth is an internally active planet.

Why should you study earthquakes? The obvious answer is because they are destructive and cause many deaths and injuries to the people living in earthquake-prone areas. They also affect the economies of many countries in terms of cleanup costs, lost jobs, and lost business revenues. From a purely personal standpoint, you someday may be caught in an earthquake. Even if you don't plan to live in an area subject to earthquakes, you probably will someday travel where

Table 9.1

Some Significant Earthquakes

Year	Location	Magnitude (estimated before 1935)	Deaths (estimated)
1556	China (Shanxi Province)	8.0	1,000,000
1755	Portugal (Lisbon)	8.6	70,000
1906	USA (San Francisco, California)	8.3	700
1923	Japan (Tokyo)	8.3	143,000
1976	China (Tangshan)	8.0	242,000
1985	Mexico (Mexico City)	8.1	9,500
1988	Armenia	7.0	25,000
1990	Iran	7.3	40,000
1993	India	6.4	30,000
1995	Japan (Kobe)	7.2	5,000+
1998	Afghanistan	6.1	5,000+
1999	Turkey	7.4	17,000
2003	Iran	6.3	30,000

there is the threat of earthquakes, and you should know what to do if you experience one.

In this chapter you will learn what you can do to make your home more earthquake resistant, precautions to take if you live where earthquakes are common, and what to do during and after an earthquake to minimize your chances of serious injury or even death!

WHAT IS THE ELASTIC REBOUND THEORY?

Based on studies conducted after the 1906 San Francisco earthquake, H. F. Reid of Johns Hopkins University proposed the **elastic rebound theory** to explain how energy is released during earthquakes. Reid studied three sets of measurements taken across a portion of the San Andreas fault that had broken during the 1906 earthquake. The measurements revealed that points on opposite sides of the fault had moved 3.2 m during the 50-year period prior to breakage in 1906, with the west side moving northward (■ Figure 9.1).

According to Reid, rocks on opposite sides of the San Andreas fault had been storing energy and bending slightly for at least 50 years before the 1906 earthquake. Any straight line such as a fence or road that crossed the

San Andreas fault would gradually be bent, because rocks on one side of the fault moved relative to rocks on the other side (Figure 9.1). Eventually, the strength of the rocks was exceeded, the rocks on opposite sides of the fault rebounded or "snapped back" to their former undeformed shape, and the energy stored was released as earthquake waves radiating out from the break.

Additional field and laboratory studies conducted by Reid and others have confirmed that elastic rebound is the mechanism by which energy is released during earthquakes. In laboratory studies, rocks subjected to forces equivalent to those occurring in the crust initially change their shape. As more force is applied, however, they resist further deformation until their internal strength is exceeded. At that point, they break and snap back to their original undeformed shape, releasing internally stored energy.

The energy stored in rocks undergoing elastic deformation is analogous to the energy stored in a tightly wound watch spring. The tighter the spring is wound, the more energy is stored, thus making more energy available for release. If the spring is wound so tightly that it breaks, then the stored energy is released as the spring rapidly unwinds and partially regains its original shape. Perhaps an even more meaningful analogy is simply bending a long, straight stick over your knee. As the stick bends, it deforms and eventually reaches the point at which it breaks. When this happens, the two pieces

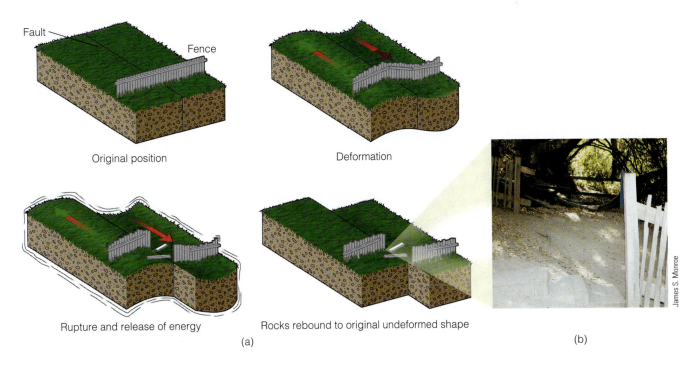

Original position

Deformation

Rupture and release of energy

Rocks rebound to original undeformed shape

(a)

(b)

James S. Monroe

PHYSICAL Geology⇌Now ■ Active Figure 9.1

(a) According to the elastic rebound theory, when rocks are deformed, they store energy and bend. When the internal strength of the rocks is exceeded, they fracture, releasing the energy as they rebound to their former undeformed shape. This sudden release of energy causes an earthquake. (b) During the 1906 San Francisco earthquake, this fence in Marin County was displaced nearly 5 m.

GEOPROFILE

JOHN C. LAHR

Listening to the Earth

John Lahr graduated with a B.S. in Physics from Rensselaer Polytechnic Institute. He attended graduate school at Columbia University and did his graduate research at Lamont–Doherty Earth Observatory. In 1971 he moved to Menlo Park, California, to begin work at the U.S. Geological Survey on their new seismic network in southern Alaska. From 1978 through 1997 John was Project Chief, Alaska Seismic Studies Project, which established a network of seismograph stations that extended along the southern coast of Alaska.

From a young age I knew that I wanted to be a scientist, but I had no idea this would lead to earthquakes! My major in college was physics primarily because that was the one subject that I could do fairly well in. I enjoyed mathematics too, but the advanced courses became too abstract for me. During my senior year at RPI, I applied to graduate schools in physics. When a friend mentioned that he might switch to geophysics in graduate school, I went to the library to see what the journals in geophysics looked like. I remember looking at a copy of the *Bulletin of the Seismological Society of America* and wondering if I would ever be able to understand such complex articles. In any case, I went ahead and applied to Columbia University in geophysics. I was accepted to both Columbia and other graduate programs and have to admit that one of the reasons I chose Columbia was that my girlfriend lived in the Bronx!

As it turned out, I was assigned to work with a seismology professor at Lamont. Life as a graduate student was busy but exciting. I worked for Robert Page, a Lamont postdoc, during the summers on his Alaskan research, which involved both recording earthquakes with portable seismometers and surveying across active faults, looking for fault creep. I was also able to go with a small team to study the 1971 San Fernando earthquake in California.

I'll always remember one incident that occurred in San Fernando. We packed loads of gear and flew to California within a day or so of the earthquake. At that time excess baggage was not a problem and security screening was not involved. When we got off the plane, it was dark. We rented a car and drove right to the area that had reported the most damage. We couldn't get very close to this area, however, because the residents had been evacuated and police were blocking all traffic. We got out of our car and tried to walk into the area but still ran into the police. We asked one policeman how we could get permission to go farther in order to study the earthquake. He gave us the address of temporary police "headquarters" where we could get a pass. This sounded good at first, until we realized this address was within the blockaded area! Thinking that all was lost, I told the policeman, "Do you realize that this (pointing to Dr. Chris Scholz) is a prominent, world-renowned seismologist?" (That was and is still a true statement.) The policeman retorted, "If he was a world-renowned seismologist, he'd have his pass!"

By the summer of 1971 I was ready for a break from school, so I applied to work for the USGS. I had not given much thought about where I might find work as a seismologist, but I was happy when it turned out to require moving to California! What a pleasant change from the weather in Buffalo, Troy, and New York.

Work at the USGS was challenging but fun. This was a very exciting time to be in seismology because from 1966 through 1975 there was a revolution in the understanding of tectonics and earthquakes. The global network of seismometers installed in the early 1960s for the purpose of nuclear monitoring turned out to provide a wealth of information on the distribution of earthquakes and their style of faulting. This evidence, along with the magnetic stripes found in the ocean crust, forced the abandonment of the view that the continents were fixed and led to the now-accepted theory of plate tectonics.

Alaska, being the most seismically active state of the United States, was an ideal

Jan Lahr

Phil Dawson

John Lahr atop Mt. Redoubt volcano.

place to study the implications of this new theory. The great 1964 Alaska earthquake, the second largest earthquake ever recorded, was still fresh in people's minds. Using our network of stations around Cook Inlet, we soon determined the location and shape of the Pacific plate that was thrusting beneath Alaska. One of my challenges was to locate these events, which ranged in depth from the surface to 200 km. We tried to use the location program developed for California, where all of the earthquakes are less than 15 km deep, but it would not work well for us. In the end a new program, called Hypoellipse, was developed by Peter Ward and me. This program had the flexibility to locate earthquakes at any depth and would also provide better error estimates.

It was very satisfying to use this new tool, along with our new regional network, to study the events in southern Alaska. One of the discoveries that we made was a weakly active dipping zone of seismicity beneath the Wrangell volcanoes. This zone could not be seen in the earthquake locations made with distant stations because the earthquakes were too small. It demonstrated beyond a doubt that these volcanoes, like most others around the Pacific's Ring of Fire, had resulted from subduction of a tectonic plate.

One of the most exciting times of my career was during the eruption of Mt. Re-

doubt volcano from 1989 to 1990. We were running a few stations near Mt. Spurr, so we were in a great position to study the relationship of the seismic signals from the volcano to its eruptions. We found that some relatively long-period signals tended to precede eruptions, and that very small, high-frequency events clearly resolved the narrow, steeply dipping conduit along which the magma was transported. Luckily nobody was injured by the eruptions, but there were some close calls. One day between eruptions we had word from someone working at the oil terminal located on the flank of the volcano that a helicopter had flown right up to the top of Mt. Redoubt and that the fresh dome of lava looked stable and hard. The next day that dome exploded, sending clouds of steam and pulverized rock (called ash) into the air and down the slopes of the volcano. This was a close call, to be sure.

Although my career in seismology was not planned well in advance, I've always been thankful for the events that led me in this direction. I've enjoyed working in a field that has contributed to the reduction of the hazards posed by earthquakes by providing information on the probable locations and magnitudes of future earthquakes. This information is incorporated into building codes that have proved effective in saving lives and reducing property damage. My one frustration is that these codes are not always adopted, as adoption is a decision made at the state or local level of government. After the 1989 Loma Prieta earthquake in the Bay Area of California, the USGS phones rang for weeks with people asking simple questions about earthquakes and their hazards. It became clear to me that the scientific community had failed in the area of public education. Without an educated public, earthquake preparedness was bound to be inadequate and there would be little support for strong building codes, especially in areas outside of California where earthquakes are rare but the risk is nonetheless significant. Since that time I have focused as much as possible on education and outreach, and in retirement I will continue to work on educating people about all that has been learned about earthquake hazards during the past 40 years.

of the original stick snap back into their original straight position. Likewise, rocks subjected to intense forces bend until they break and then return to their original position, releasing energy in the process.

WHAT IS SEISMOLOGY?

Seismology, the study of earthquakes, emerged as a true science during the 1880s with the development of **seismographs**, instruments that detect, record, and measure the vibrations produced by an earthquake (■ Figure 9.2). The record made by a seismograph is a *seismogram*. Although most seismographs today employ electronic sensors, computer printouts have largely replaced the strip-chart seismograms of earlier seismographs.

When an earthquake occurs, energy in the form of *seismic waves* radiates out from the point of release (■ Figure 9.3). These waves are somewhat analogous to the ripples that move out concentrically from the point where a stone is thrown into a pond. Unlike waves on a pond, however, seismic waves move outward in all directions from their source.

Earthquakes take place because rocks are capable of storing energy but their strength is limited, so if enough force is present, they rupture and thus release their stored energy. In other words, most earthquakes result when movement occurs along fractures (faults), most of which are related to plate movements. Once a fracture begins, it moves along the fault at several kilometers per second for as long as conditions for failure exist. The longer the fracture along which movement occurs, the more time it takes for the stored energy to be released, and therefore the longer the ground will shake. During some very large earthquakes, the ground might shake for 3 minutes, a seemingly brief period but interminable if you happen to be experiencing it.

The Focus and Epicenter of an Earthquake

The point within Earth where fracturing begins—that is, the point at which energy is first released—is an earthquake's **focus,** or *hypocenter.* What we usually hear in news reports, however, is the location of the **epicenter,** the point on Earth's surface directly above the focus (Figure 9.3). For instance, according to a report by the U.S. Geological Survey, the August 1999 earthquake in Turkey (see the Introduction) had an epicenter about 11 km southeast of the city of Izmit, and its focal depth was about 17 km.

Seismologists recognize three categories of earthquakes based on focal depth. *Shallow-focus* earthquakes

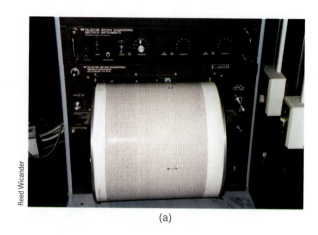

(a)

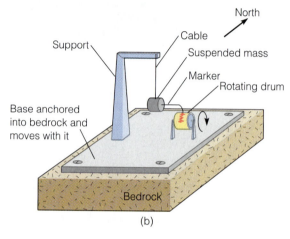

(b)

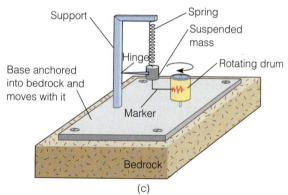

(c)

■ **Figure 9.2**

Modern seismographs record earthquake waves electronically. (a) Earthquakes recorded by a seismograph. (b) A horizontal-motion seismograph. Because of its inertia, the heavy mass that contains the marker remains stationary while the rest of the structure moves along with the ground during an earthquake. As long as the length of the arm is not parallel to the direction of ground movement, the marker will record the earthquake waves on the rotating drum. This seismograph would record waves from west or east, but to record waves from the north or south another seismograph at right angles to this one is needed. (c) A vertical-motion seismograph. This seismograph operates on the same principle as a horizontal-motion instrument and records vertical ground movement.

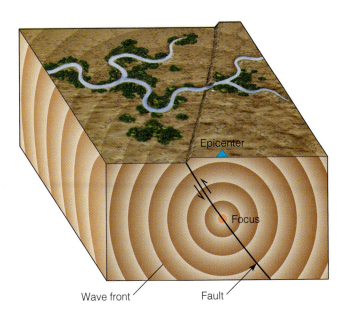

Epicenter

Focus

Wave front

Fault

PHYSICAL
Geology ⇌ Now ■ **Active Figure 9.3**

The focus of an earthquake is the location where rupture begins and energy is released. The place on the surface vertically above the focus is the epicenter. Seismic wave fronts move out in all directions from their source, the focus of an earthquake.

have focal depths of less than 70 km from the surface, whereas those with foci between 70 and 300 km are *intermediate focus*, and the foci of those characterized as *deep focus* are more than 300 km deep. Earthquakes are not evenly distributed among these three categories. Approximately 90% of all earthquake foci are at depths of less than 100 km, whereas only about 3% of all earthquakes are deep. Shallow-focus earthquakes are, with few exceptions, the most destructive.

An interesting relationship exists between earthquake foci and plate boundaries. Earthquakes generated along divergent or transform plate boundaries are invariably shallow focus, whereas many shallow- and nearly all intermediate- and deep-focus earthquakes occur along convergent margins (■ Figure 9.4). Furthermore, a pattern emerges when the focal depths of earthquakes near island arcs and their adjacent ocean trenches are plotted. Notice in ■ Figure 9.5 that the focal depth increases beneath the Tonga Trench in a narrow, well-defined zone that dips approximately 45 degrees. Dipping seismic zones, called *Benioff zones*, are common along convergent plate boundaries where one plate is subducted beneath another. Such dipping seismic zones indicate the angle of plate descent along a convergent plate boundary.

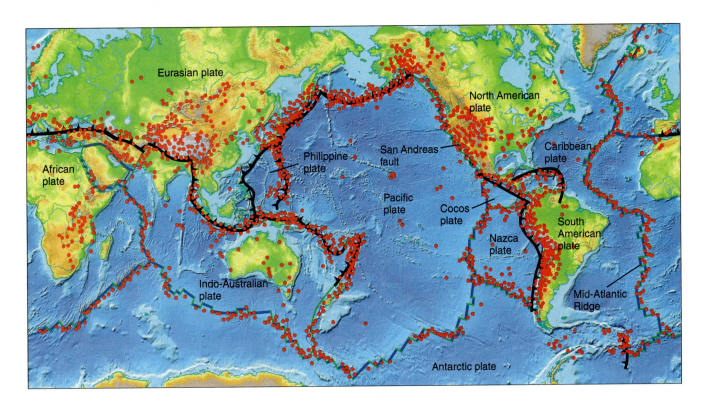

■ **Figure 9.4**

The relationship between earthquake epicenters and plate boundaries. Approximately 80% of earthquakes occur within the circum-Pacific belt, 15% within the Mediterranean–Asiatic belt, and the remaining 5% within plate interiors or along oceanic spreading ridges. Each dot represents a single earthquake epicenter. Source: Data from National Oceanic and Atmospheric Administration.

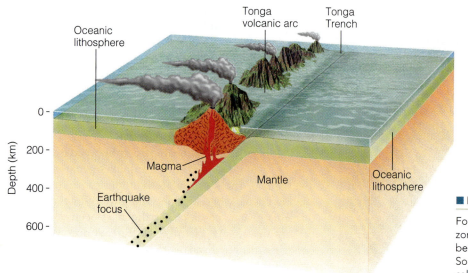

Depth (km)

0 —

200 —

400 —

600 —

Oceanic
lithosphere

Tonga
volcanic arc

Tonga
Trench

Magma

Mantle

Oceanic
lithosphere

Earthquake
focus

■ **Figure 9.5**

Focal depth increases in a well-defined
zone that dips approximately 45 degrees
beneath the Tonga volcanic arc in the
South Pacific. Dipping seismic zones are
called Benioff zones.

WHERE DO EARTHQUAKES OCCUR, AND HOW OFTEN?

No place on Earth is immune to earthquakes, but almost 95% take place in seismic belts corresponding to plate boundaries where plates converge, diverge, and slide past each other. Earthquake activity distant from plate margins is minimal but can be devastating when it occurs. The relationship between plate margins and the distribution of earthquakes is readily apparent when the locations of earthquake epicenters are superimposed on a map showing the boundaries of Earth's plates (Figure 9.4).

The majority of all earthquakes (approximately 80%) occur in the *circum-Pacific belt,* a zone of seismic activity nearly encircling the Pacific Ocean basin. Most of these earthquakes result from convergence along plate margins, as in the case of the 1995 Kobe, Japan, earthquake (■ Figure 9.6a). The earthquakes along the North American Pacific Coast, especially in California, are also in this belt, but here plates slide past one another rather than converge. The October 17, 1989, Loma Prieta earthquake in the San Francisco area (Figure 9.6b) and the January 17, 1994, Northridge earthquake (Figure 9.6c) happened along this plate boundary.

The second major seismic belt, accounting for 15% of all earthquakes, is the *Mediterranean–Asiatic belt.* This belt extends westerly from Indonesia through the Himalayas, across Iran and Turkey, and westerly through the Mediterranean region of Europe. The devastating 1990 earthquake in Iran that killed 40,000 people, the 1999 Turkey earthquake that killed about 17,000 people, the 2001 India earthquake that killed more than 20,000 people, and the 2003 Iran earthquake that killed 30,000 people are recent examples of the destructive earthquakes that strike this region (see the Introduction).

The remaining 5% of earthquakes occur mostly in the interiors of plates and along oceanic spreading-ridge systems. Most of these earthquakes are not strong, although several major intraplate earthquakes are worthy of mention. For example, the 1811 and 1812 earthquakes near New Madrid, Missouri, killed approximately 20 people and nearly destroyed the town. So strong were these earthquakes that they were felt from the Rocky Mountains to the Atlantic Ocean and from the Canadian border to the Gulf of Mexico. In addition, the earthquakes caused church bells to ring as far away as Boston, Massachusetts (1600 km). Within the immediate area, numerous buildings were destroyed and forests were flattened; the land sank several meters in some areas, causing flooding; and reportedly the Mississippi River reversed its flow during the shaking and changed its course slightly. Eyewitnesses described the scene at New Madrid as follows:

> The earth was observed to roll in waves a few feet high with visible depressions between. By and by these swells burst throwing up large volumes of water, sand, and coal.
>
> . . . Undulations of the earth resembling waves, increasing in elevations as they advanced, and when they attained a certain fearful height the earth would burst.
>
> The shocks were clearly distinguishable into two classes, those in which the motion was horizontal and those in which it was perpendicular. The latter was attended with explosions and the terrible mixture of noises, . . . but they were by no means as destructive as the former.
>
> Cpt. Sarpy tied up at this [small] island on the evening of the 15th of December, 1811. In looking around they found that a party of river pirates occupied part of the island and were expecting Sarpy

Dennis Fox

(a)

Bunyo Ishikawa/Sygma/Corbis

(b)

Roger Ressmeyer/Corbis

(c)

■ **Figure 9.6**

Earthquake damage in the circum-Pacific belt. (a) Some of the damage in Kobe, Japan, caused by the January 1995 earthquake in which more than 5000 people died. (b) Damage in Oakland, California, resulting from the October 1989 Loma Prieta earthquake. The columns supporting the upper deck of Interstate 880 failed, causing the upper deck to collapse onto the lower one. (c) View of the severe exterior damage to the Northridge Meadows apartments in which sixteen people were killed as a result of the January 1994 Northridge, California, earthquake.

with the intention of robbing him. As soon as Sarpy found that out he quietly dropped lower down the river. In the night the earthquake came and next morning when the accompanying haziness disappeared the island could no longer be seen. It had been utterly destroyed as well as its pirate inhabitants.*

Another major intraplate earthquake struck Charleston, South Carolina, on August 31, 1886, killing 60 people and causing $23 million in property damage (■ Figure 9.7). In December 1988 a large intraplate earthquake struck near Tennant Creek in Australia's Northern Territory.

The cause of intraplate earthquakes is not well understood, but geologists think they arise from localized stresses caused by the compression that most plates experience along their margins. A useful analogy might be that of moving a house. Regardless of how careful the movers are, moving something so large without its internal parts shifting slightly is impossible. Similarly, plates are not likely to move without some internal stresses that occasionally cause earthquakes. Interestingly, many intraplate earthquakes are associated with ancient and presumed inactive faults that are reactivated at various intervals.

More than 150,000 earthquakes strong enough to be felt are recorded every year by the worldwide network of seismograph stations. In addition, an estimated 900,000 earthquakes are recorded annually by seismographs but are too small to be individually cataloged. These small earthquakes result from the energy released as continuous adjustments take place between the various plates.

*C. Officer and J. Page, *Tales of the Earth* (New York: Oxford University Press, 1993), pp. 49–50.

PHYSICAL
Geology⇌Now Click Geology Interactive to work through an activity on Earthquakes in Space and Time plus Seismic Risk USA through Earthquakes and Tsunami.

South Carolina Library, University of South Carolina

■ **Figure 9.7**

Damage done to Charleston, South Carolina, by the earthquake of August 31, 1886. This earthquake is the largest reported in the eastern United States.

WHAT ARE SEISMIC WAVES?

Many people have experienced an earthquake but are probably unaware that the shaking they feel and the damage to structures are caused by the arrival of various *seismic waves,* a general term encompassing all waves generated by an earthquake. When movement on a fault takes place, energy is released in the form of two kinds of waves that radiate outward in all directions from an earthquake's focus. **Body waves,** so called because they travel through the solid body of Earth, are somewhat like sound waves, and **surface waves,** which travel along the ground surface, are analogous to undulations or waves on water surfaces.

Body Waves

An earthquake generates two types of body waves: P-waves and S-waves (■ Figure 9.8). **P-waves,** or *primary waves,* are the fastest seismic waves and can travel through solids, liquids, and gases. P-waves are compressional, or push–pull, waves and are similar to sound waves in that they move material forward and backward along a line in the same direction that the waves themselves are moving (Figure 9.8b). Thus the material through which P-waves travel is expanded and compressed as the wave moves through it and returns to its

original size and shape after the wave passes by. In fact, some P-waves emerging from within Earth are transmitted into the atmosphere as sound waves that at certain frequencies can be heard by humans and animals.

S-waves, or *secondary waves,* are somewhat slower than P-waves and can travel only through solids. S-waves are *shear waves* because they move the material perpendicular to the direction of travel, thereby producing shear stresses in the material they move through (Figure 9.8c). Because liquids (as well as gases) are not rigid, they have no shear strength, and S-waves cannot be transmitted through them.

The velocities of P- and S-waves are determined by the density and elasticity of the materials through which they travel. For example, seismic waves travel more slowly through rocks of greater density but more rapidly through rocks with greater elasticity. *Elasticity* is a property of solids, such as rocks, and means that once they have been deformed by an applied force, they return to their original shape when the force is no longer present. Because P-wave velocity is greater than S-wave velocity in all materials, P-waves always arrive at seismic stations first.

Surface Waves

Surface waves travel along the surface of the ground, or just below it, and are slower than body waves. Unlike the sharp jolting and shaking that body waves cause, surface waves generally produce a rolling or swaying motion, much like the experience of being in a boat.

Several types of surface waves are recognized. The two most important are **Rayleigh waves (R-waves)** and **Love waves (L-waves),** named after the British scientists who discovered them, Lord Rayleigh and A. E. H. Love. Rayleigh waves are generally the slower of the two and behave like water waves in that they move forward while the individual particles of material move in an elliptic path within a vertical plane oriented in the direction of wave movement (■ Figure 9.9b).

The motion of a Love wave is similar to that of an S-wave, but the individual particles of the material move only back and forth in a horizontal plane perpendicular to the direction of wave travel (Figure 9.9c). This type of lateral motion can be particularly damaging to building foundations.

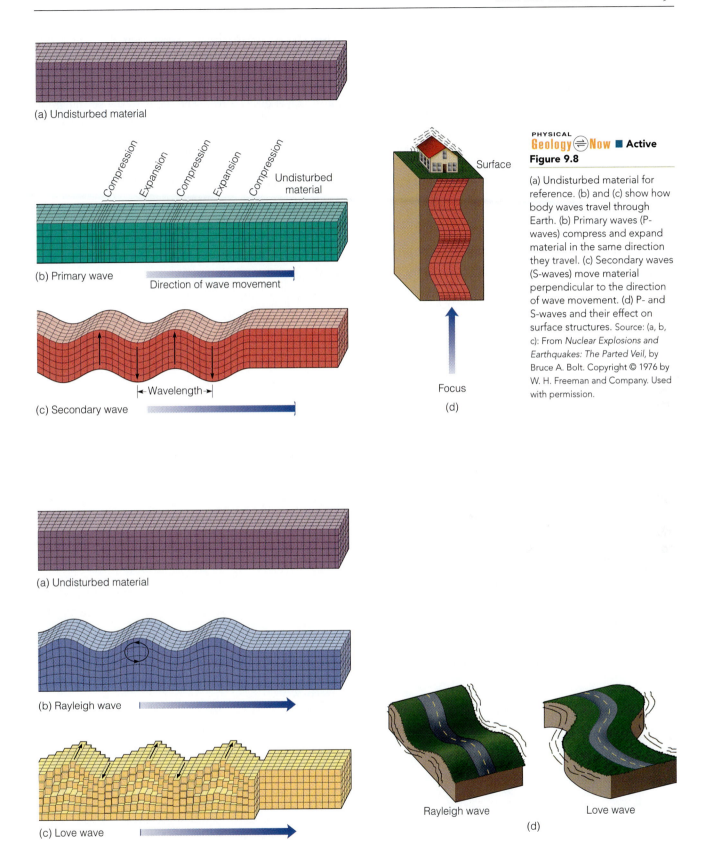

Figure 9.8

(a) Undisturbed material for reference. (b) and (c) show how body waves travel through Earth. (b) Primary waves (P-waves) compress and expand material in the same direction they travel. (c) Secondary waves (S-waves) move material perpendicular to the direction of wave movement. (d) P- and S-waves and their effect on surface structures. Source: (a, b, c): From *Nuclear Explosions and Earthquakes: The Parted Veil*, by Bruce A. Bolt. Copyright © 1976 by W. H. Freeman and Company. Used with permission.

(a) Undisturbed material

Compression Expansion Compression Expansion Compression

Undisturbed material

(b) Primary wave

Direction of wave movement

Wavelength

(c) Secondary wave

Surface

Focus

(d)

(a) Undisturbed material

(b) Rayleigh wave

(c) Love wave

Rayleigh wave

Love wave

(d)

■ **Figure 9.9**

Surface waves. (a) Undisturbed material for reference. (b) and (c) show how surface waves travel along Earth's surface or just below it. (b) Rayleigh waves (R-waves) move material in an elliptical path in a plane oriented parallel to the direction of wave movement. (c) Love waves (L-waves) move material back and forth in a horizontal plane perpendicular to the direction of wave movement. (d) The arrival of R- and L-waves causes the surface to undulate and shake from side to side. Source: (a, b, c) From *Nuclear Explosions and Earthquakes: The Parted Veil*, by Bruce A. Bolt. Copyright © 1976 by W. H. Freeman and Company. Used with permission.

HOW IS AN EARTHQUAKE'S EPICENTER LOCATED?

Previously we mentioned that news articles commonly report an earthquake's epicenter, but just how is the location of an epicenter determined? Once again, geologists rely on the study of seismic waves. We already know that P-waves travel faster than S-waves, nearly twice as fast in all substances, so P-waves arrive at a seismograph station first, followed some time later by S-waves. Both P- and S-waves travel directly from the focus to the seismograph station through Earth's interior, but L- and R-waves arrive last because they are the slowest, and they also travel the longest route along the surface (■ Figure 9.10a, b). L- and R-waves cause much of the damage during earthquakes, but only P- and S-waves need con-

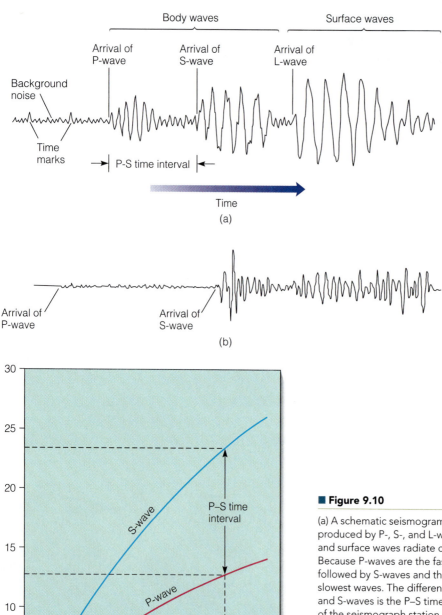

■ **Figure 9.10**

(a) A schematic seismogram showing the arrival order and pattern produced by P-, S-, and L-waves. When an earthquake occurs, body and surface waves radiate out from the focus at the same time. Because P-waves are the fastest, they arrive at a seismograph first, followed by S-waves and then by surface waves, which are the slowest waves. The difference between the arrival times of the P- and S-waves is the P–S time interval; it is a function of the distance of the seismograph station from the focus. (b) Seismogram for the 1906 San Francisco earthquake, recorded 14,668 km away in Göttingen, Germany. The total record represents about 26 minutes, so considerable time passed between the arrival of the P-waves and the slower-moving S-waves. The arrival of surface waves, not shown here, caused the instrument to go off the scale. (c) A time–distance graph showing the average travel times for P- and S-waves. The farther away a seismograph station is from the focus of an earthquake, the longer the interval between the arrivals of the P- and S-waves, and hence the greater the distance between the curves on the time–distance graph as indicated by the P–S time interval. Source: (c) Based on data from C. F. Richter, *Elementary Seismology*, 1958. W. H. Freeman and Company.

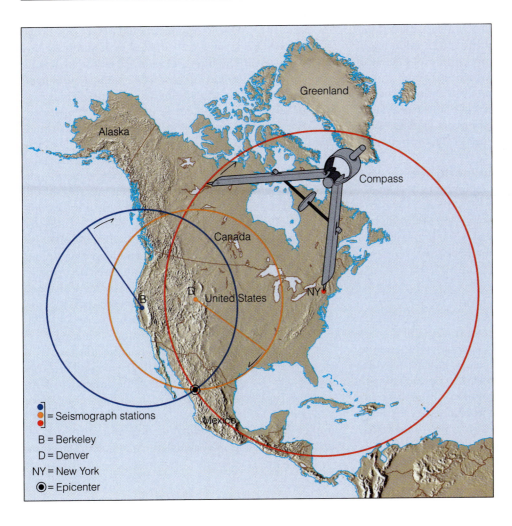

PHYSICAL
Geology⇌Now ■ **Active Figure 9.11**

Three seismograph stations are needed to locate the epicenter of an earthquake. The P–S time interval is plotted on a time–distance graph for each seismograph station to determine the distance that station is from the epicenter. A circle with that radius is drawn from each station, and the intersection of the three circles is the epicenter of the earthquake.

and a line is drawn straight down to the distance axis of the graph, thus giving the distance from the focus to each seismic station (Figure 9.10c). Next, a circle whose radius equals the distance shown on the time–distance graph from each of the seismic stations is drawn on a map (Figure 9.11). The intersection of the three circles is the location of the earthquake's epicenter. It should be obvious from Figure 9.11 that P–S time intervals from at least three seismic stations are needed. If only one were used, the epicenter could be at any location on the circle drawn around that station, and using two stations would give two possible locations for the epicenter.

Determining the focal depth of an earthquake is much more difficult and considerably less precise than finding its epicenter. The focal depth is usually found by making computations based on several assumptions, using comparisons with the results obtained at other seismic stations, and recalculating and approximating the depth as closely as possible. Even so, the results are not highly accurate, but they do tell us that most earthquakes, probably 75%, have foci no deeper than 10 to 15 km and that a few are as deep as 680 km.

cern us here because they are the ones important in determining an epicenter.

Seismologists, geologists who study seismology, have accumulated a tremendous amount of data over the years and now know the average speeds of P- and S-waves for any specific distance from their source. These P- and S-wave travel times are published in *time–distance graphs* illustrating that the difference between the arrival times of the two waves is a function of distance between a seismograph and an earthquake's focus (Figure 9.10a). That is, the farther the waves travel, the greater the *P–S time interval* or simply the time difference between the arrivals of P- and S-waves (Figure 9.10a, c).

If the P–S time intervals are known from at least three seismograph stations, then the epicenter of any earthquake can be determined (■ Figure 9.11). Here is how it works. Subtracting the arrival time of the first P-wave from the arrival time of the first S-wave gives the P–S time interval for each seismic station. Each of these time intervals is then plotted on a time–distance graph,

HOW ARE THE SIZE AND STRENGTH OF AN EARTHQUAKE MEASURED?

Following any earthquake that causes extensive damage, fatalities, and injuries, graphic reports of the quake's violence and human suffering are common. Headlines tell us that thousands died, many more were injured or homeless, and property damage is in the millions and possibly billions of dollars. Even though some of the information might be exaggerated, few other natural processes account for such tragic consequences, hurricanes coupled with coastal flooding being a notable exception. And although descriptions of fatalities and damage give some indication of the size of an earthquake, geologists are interested in more reliable methods of determining an earthquake's size.

Two measures of an earthquake's strength are commonly used. One is *intensity*, a qualitative assessment of the kinds of damage done by an earthquake. The other, *magnitude*, is a quantitative measurement of the amount of energy released by an earthquake. Each method provides important information that can be used to prepare for future earthquakes.

Intensity

Intensity is a subjective measure of the kind of damage done by an earthquake, as well as people's reaction to it. Since the mid-19th century, geologists have used intensity as a rough approximation of the size and strength of an earthquake. The most common intensity scale used in the United States is the **Modified Mercalli Intensity Scale,** which has values ranging from I to XII (Table 9.2).

Intensity maps can be constructed for regions hit by earthquakes by dividing the affected region into various intensity zones. The intensity value given for each zone is the maximum intensity that the earthquake produced for that zone. Even though intensity maps are not precise because of the subjective nature of the measurements, they do provide geologists with a rough approximation of the location of the earthquake, the kind and extent of the damage done, and the effects of local geology on different types of building construction (■ Figure 9.12). Because intensity is a measure of the kind of damage done by an earthquake, insurance companies still classify earthquakes on the basis of intensity.

Generally, a large earthquake will produce higher intensity values than a small earthquake, but many other factors besides the amount of energy released by an earthquake also affect its intensity. These include distance from the epicenter, focal depth of the earthquake, population density and local geology of the area, type of building construction employed, and duration of shaking.

A comparison of the intensity map for the 1906 San Francisco earthquake and a geologic map of the area shows a strong correlation between the amount of damage done and the underlying rock and soil conditions (■ Figure 9.13). Damage was greatest in those areas underlain by poorly consolidated material or artificial fill because the effects of shaking are amplified in these ma-

Table 9.2

Modified Mercalli Intensity Scale

I	Not felt except by a very few under especially favorable circumstances.
II	Felt by only a few people at rest, especially on upper floors of buildings.
III	Felt quite noticeably indoors, especially on upper floors of buildings, but many people do not recognize it as an earthquake. Standing automobiles may rock slightly.
IV	During the day felt indoors by many, outdoors by few. At night some awakened. Sensation like heavy truck striking building, standing automobiles rocked noticeably.
V	Felt by nearly everyone, many awakened. Some dishes, windows, etc. broken, a few instances of cracked plaster. Disturbance of trees, poles, and other tall objects sometimes noticed.
VI	Felt by all, many frightened and run outdoors. Some heavy furniture moved, a few instances of fallen plaster or damaged chimneys. Damage slight.
VII	Everybody runs outdoors. Damage negligible in buildings of good design and construction; slight to moderate in well-built ordinary structures; considerable in poorly built or badly designed structures; some chimneys broken. Noticed by people driving automobiles.
VIII	Damage slight in specially designed structures; considerable in normally constructed buildings with possible partial collapse; great in poorly built structures. Fall of chimneys, monuments, walls. Heavy furniture overturned. Sand and mud ejected in small amounts.
IX	Damage considerable in specially designed structures. Buildings shifted off foundations. Ground noticeably cracked. Underground pipes broken.
X	Some well-built wooden structures destroyed; most masonry and frame structures with foundations destroyed; ground badly cracked. Rails bent. Landslides considerable from river banks and steep slopes. Water splashed over river banks.
XI	Few, if any (masonry) structures remain standing. Bridges destroyed. Broad fissures in ground. Underground pipelines completely out of service.
XII	Damage total. Waves seen on ground surfaces. Objects thrown upward into the air.

Source: U.S. Geological Survey.

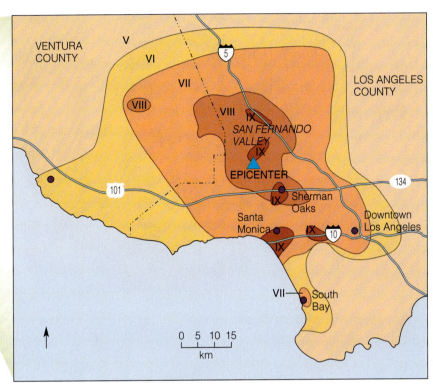

■ **Figure 9.12**

Preliminary Modified Mercalli Intensity map for the 1994 Northridge, California, earthquake, showing the region divided into intensity zones based on the kind of damage done. This earthquake had a magnitude of 6.7.

terials, whereas damage was less in areas of solid bedrock. The correlation between the geology and the amount of damage done by an earthquake was further reinforced by the 1989 Loma Prieta earthquake when many of the same areas that were extensively damaged in the 1906 earthquake were once again heavily damaged.

Magnitude

If earthquakes are to be compared quantitatively, we must use a scale that measures the amount of energy released and is independent of intensity. Such a scale was developed in 1935 by Charles F. Richter, a seismologist at the California Institute of Technology. The **Richter Magnitude Scale** measures earthquake **magnitude,** which is the total amount of energy released by an earthquake at its source. It is an open-ended scale with values beginning at 1. The largest magnitude recorded has been 8.6, and though values greater than 9 are theoretically possible, they are highly improbable because rocks are not able to store the energy necessary to generate earthquakes of that magnitude.

The magnitude of an earthquake is determined by measuring the amplitude of the largest seismic wave as recorded on a seismogram (■ Figure 9.14). To avoid large numbers, Richter used a conventional base-10 logarithmic scale to convert the amplitude of the largest recorded seismic wave to a numeric magnitude value (Figure 9.14). Therefore each whole-number increase in magnitude represents a 10-fold increase in wave amplitude. For example, the amplitude of the largest seismic wave for an earthquake of magnitude 6 is 10 times that produced by an earthquake of magnitude 5, 100 times as large as a magnitude 4 earthquake, and 1000 times that of an earthquake of magnitude 3 ($10 \times 10 \times 10 = 1000$).

A common misconception about the size of earthquakes is that an increase of one unit on the Richter Magnitude Scale—a 7 versus a 6, for instance—means a 10-fold increase in size. It is true that each whole-number increase in magnitude represents a 10-fold increase in the wave amplitude, but each magnitude increase corresponds to a roughly 30-fold increase in the amount of energy released (actually it is 31.5, but 30 is close enough for our purposes). What this means is that it

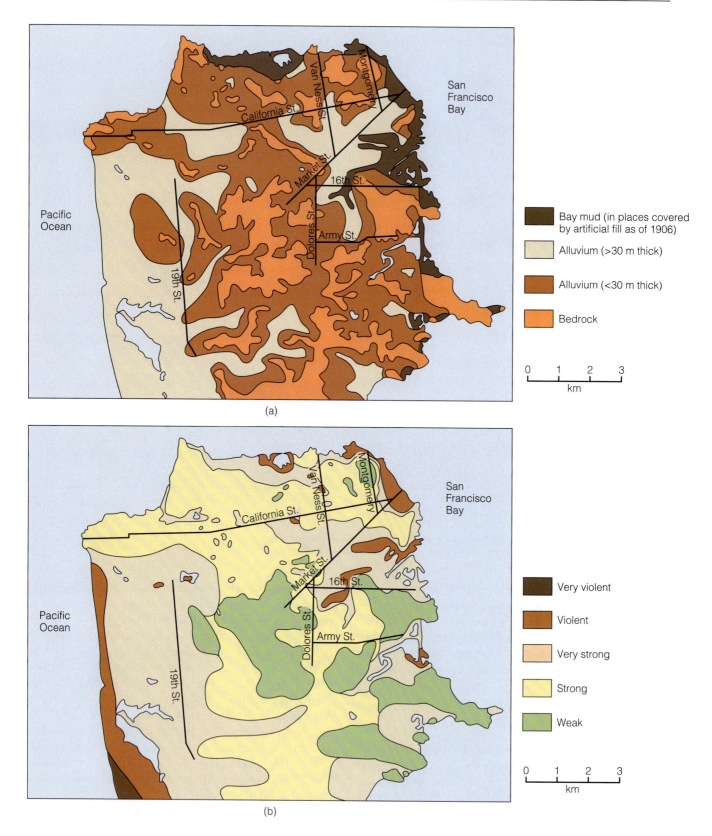

■ Figure 9.13

Comparison between (a) the general geology of the San Francisco area and (b) a Modified Mercalli Intensity map of the same area for the 1906 earthquake. A close correlation exists between the geology and intensity. Areas underlain by bedrock correspond to the lowest intensity values, followed by areas underlain by thin alluvium (sediment) and thick alluvium. Bay mud and/or artificial fill lie beneath the areas shaken most violently.

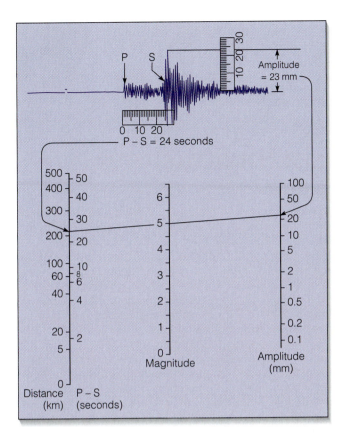

■ Figure 9.14

The Richter Magnitude Scale measures the total amount of energy released by an earthquake at its source. The magnitude is determined by measuring the maximum amplitude of the largest seismic wave and marking it on the right-hand scale. The difference between the arrival times of the P- and S-waves (recorded in seconds) is marked on the left-hand scale. When a line is drawn between the two points, the magnitude of the earthquake is the point at which the line crosses the center scale. Source: *From Earthquakes, by Bruce A. Bolt. Copyright © 1988 by W. H. Freeman and Company. Used with permission.*

Table 9.3

Average Number of Earthquakes of Various Magnitudes per Year Worldwide

Magnitude	Effects	Average Number per Year
<2.5	Typically not felt but recorded	900,000
2.5–6.0	Usually felt; minor to moderate damage to structures	31,000
6.1–6.9	Potentially destructive, especially in populated areas	100
7.0–7.9	Major earthquakes; serious damage results	20
>8.0	Great earthquakes; usually result in total destruction	1 every 5 years

Source: Modified from *Earthquake Information Bulletin*, and B. Gutenberg and C. F. Richter, *Seismicity of the Earth and Associated Phenomena* (Princeton: Princeton University Press, 1949).

would take about 30 earthquakes of magnitude 6 to equal the energy released in one with a magnitude of 7. The 1964 Alaska earthquake with a magnitude of 8.6 released almost 900 times more energy than the 1994 Northridge, California, earthquake of magnitude 6.7! And the Alaska earthquake released more than 27,000 times as much energy as one with a magnitude of 5.6 would have.

We have already mentioned that more than 900,000 earthquakes are recorded around the world each year. These figures can be placed in better perspective by reference to Table 9.3, which shows that the vast majority of earthquakes have a Richter magnitude of less than 2.5 and that great earthquakes (those with a magnitude greater than 8.0) occur, on average, only once every five years.

The Richter Magnitude Scale was devised to measure earthquake waves on a particular seismograph and at a specific distance from an earthquake. One of its limitations is that it underestimates the energy of very large

earthquakes because it measures the highest peak on a seismogram, which represents only an instant during an earthquake. For large earthquakes, though, the energy might be released over several minutes and along hundreds of kilometers of a fault. For instance, during the 1857 Fort Tejon, California, earthquake, the ground shook for more than 2 minutes and energy was released for 360 km along the fault. Despite its shortcomings, Richter magnitudes still commonly appear in news releases. More recently, seismologists developed a *Seismic-Moment Magnitude Scale* that considers the area of a fault along which rupture occurred and the amount of movement of rocks adjacent to the fault. Seismologists are confident they now have a scale with which they not only can more effectively compare different-sized earthquakes, but also can evaluate the size of earthquakes that occurred before instruments were available to record them.

WHAT ARE THE DESTRUCTIVE EFFECTS OF EARTHQUAKES?

C ertainly earthquakes are one of nature's most destructive phenomena. Little or no warning precedes earthquakes, and once they begin, little or nothing can be done to minimize their effects,

GEO**FOCUS**

9.1

Designing Earthquake-Resistant Structures

One way to reduce property damage, injuries, and loss of life is to design and build structures that are as earthquake resistant as possible. Many things can be done to improve the safety of current structures and of new buildings.

To design earthquake-resistant structures, engineers must understand the dynamics and mechanics of earthquakes, including the type and duration of the ground motion and how rapidly the ground accelerates during an earthquake. An understanding of the area's geology is also important because certain ground materials such as water-saturated sediments or landfill can lose their strength and cohesiveness during an earthquake (see Figure 9.15). Finally, engineers must be aware of how different structures behave under different earthquake conditions.

With the level of technology currently available, a well-designed, properly constructed building should be able to withstand small,

short-duration earthquakes of less than 5.5 magnitude with little or no damage. In moderate earthquakes (5.5–7.0 magnitude), the damage suffered should not be serious and should be repairable. In a major earthquake of greater than 7.0 magnitude, the building should not collapse, although it may later have to be demolished.

Many factors enter into the design of an earthquake-resistant structure, but the most important is that the building be tied together; that is, the foundation, walls, floors, and roof should all be joined together to create a structure that can withstand both horizontal and vertical shaking (■ Figure 1). Almost all the structural failures resulting from earthquake ground movement occur at weak connections, where the various parts of a structure are not securely tied together. Buildings with open or unsupported first stories are particularly susceptible to damage. Some reinforcement must be done, or collapse is a distinct possibility (Figure 1).

Tall buildings, such as skyscrapers, must be designed so that a certain amount of swaying or flexing can occur, but not so much that they touch neighboring buildings during swaying. If a building is brittle and does not give, it will crack and fail. Besides designed flexibility, engineers must ensure that a building does not vibrate at the same frequency as the ground does during an earthquake. When that happens, the force applied by the seismic waves at ground level is multiplied several times by the time they reach the top of the building.

Damage to high-rise structures can also be minimized or prevented by using diagonal steel beams to help prevent swaying. In addition, tall buildings in earthquake-prone areas are now commonly placed on layered steel and rubber structures and devices similar to shock absorbers that help decrease the amount of sway.

What about structures built many years ago? Just as in new buildings, the most important thing that can be done to increase

although planning before an earthquake can help. However, earthquake prediction may become a reality in the future (discussed in a later section). The destructive effects of earthquakes include ground shaking, fire, seismic sea waves, and landslides, as well as panic, disruption of vital services, and psychological shock. In some cases, rescue attempts are hampered by inadequate resources or planning, existing conditions of civil unrest, or simply the magnitude of the disaster.

The number of deaths and injuries as well as the amount of property damage depend on several factors.

Generally speaking, earthquakes during working hours and school hours in densely populated urban areas are the most destructive and cause the most fatalities and injuries. However, magnitude, duration of shaking, distance from the epicenter, geology of the affected region, and type of structures are also important considerations. Given these variables, it should not be surprising that a comparatively small earthquake can have disastrous effects, whereas a much larger one might go largely unnoticed, except perhaps by seismologists.

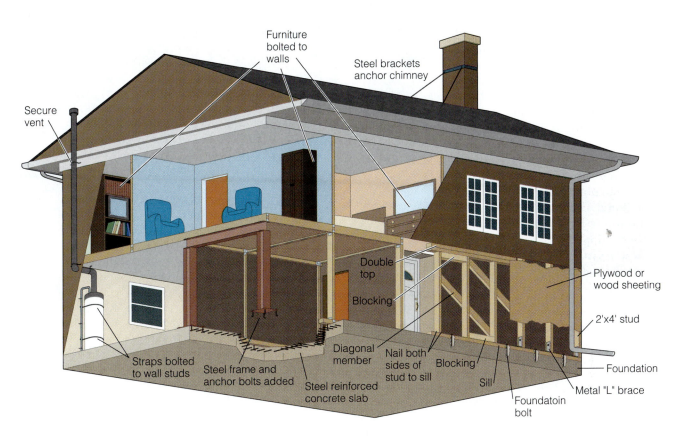

Furniture bolted to walls

Steel brackets anchor chimney

Secure vent

Double top

Blocking

Plywood or wood sheeting

2'x4' stud

Straps bolted to wall studs

Steel frame and anchor bolts added

Steel reinforced concrete slab

Diagonal member

Nail both sides of stud to sill

Blocking

Sill

Foundation

Metal "L" brace

Foundatoin bolt

■ Figure 1

This illustration shows some of the things a homeowner can do to reduce damage to a building because of ground shaking during an earthquake. Notice that the structure must be solidly attached to its foundation, and bracing the walls helps prevent damage from horizontal motion.

the stability and safety of older structures is to tie together the different components of each building. This can be done by adding a steel frame to unreinforced parts of a building such as a garage, bolting the walls to the foundation, adding reinforced beams to the exterior, and using beam and joist connectors whenever possible. Although such modifications are expensive, they are usually cheaper than having to replace a building that was destroyed by an earthquake.

Ground Shaking

Ground shaking, the most obvious and immediate effect of an earthquake, varies depending on magnitude, distance from the epicenter, and the type of underlying materials in the area—unconsolidated sediment or fill versus bedrock, for instance. Certainly ground shaking is terrifying, and it might be violent enough for fissures to open in the ground. Nevertheless, contrary to popular myth, fissures do not swallow up people and buildings and then close on them. And although California will no doubt have big earth-

quakes in the future, rocks cannot store enough energy to displace a landmass as large as California into the Pacific Ocean, as the tabloids sometimes suggest will happen.

The effects of ground shaking, such as collapsing buildings, falling building facades and window glass, and toppling monuments and statues, cause more damage and result in more loss of life and injuries than any other earthquake hazard. Structures built on solid bedrock generally suffer less damage than those built on poorly consolidated material such as water-saturated sediments or artificial fill (see Geo-Focus 9.1).

Structures built on poorly consolidated or water-saturated material are subjected to ground shaking of longer duration and greater S-wave amplitude than those on bedrock (■ Figure 9.15). In addition, fill and water-saturated sediments tend to liquefy, or behave like a fluid, a process known as *liquefaction*. When shaken, the individual grains lose cohesion and the ground flows. Dramatic examples of damage resulting from liquefaction include Niigata, Japan, where large apartment buildings were tipped on their sides after the water-saturated soil of the hillside collapsed (■ Figure 9.16), and Turnagain Heights, Alaska, where many homes were destroyed when the Bootlegger Cove Clay lost all of its strength when it was shaken by the 1964 earthquake (see Figure 14.21).

Besides the magnitude of an earthquake and the underlying geology, the material used and the type of construction also affect the amount of damage done (see Geo-Focus 9.1). Adobe and mud-walled structures are the weakest of all and almost always collapse during an earthquake. Unreinforced brick structures and poorly built concrete structures are also particularly susceptible to collapse. For example, the 1976 earthquake in Tangshan, China, completely leveled the city because hardly any structures were built to resist seis-

National Geophysical Data Center

■ Figure 9.16

The effects of ground shaking on water-saturated soil are dramatically illustrated by the collapse of these buildings in Niigata, Japan, during a 1964 earthquake. The buildings were designed to be earthquake resistant and fell over on their sides intact.

mic forces. In fact, most had unreinforced brick walls, which have no flexibility, and consequently they collapsed during the shaking (■ Figure 9.17a). The 6.4 magnitude earthquake that struck India in 1993 killed about 30,000 whereas the 6.7 magnitude Northridge earthquake resulted in only 61 deaths. Both earthquakes occurred in densely populated regions, but in India the brick and stone buildings could not withstand ground shaking; most collapsed, entombing their occupants (Figure 9.17b).

Fire

In many earthquakes, particularly in urban areas, fire is a major hazard. Almost 90% of the damage done in the 1906 San Francisco earthquake was caused by fire. The shaking severed many of the electrical and gas lines, which touched off flames and started fires all over the city. Because water mains were ruptured by the earthquake, there was no effective way to fight the fires that raged out of control for three days, destroying much of the city.

Eighty-three years later, during the 1989 Loma Prieta earthquake, a fire broke out in the Marina district of San Francisco (■ Figure 9.18). This time, however, it was contained within a small area because San Francisco had a system of valves throughout its water and gas pipeline system so that lines could be isolated from breaks (see "The San Andreas Fault" on pages 262 and 263).

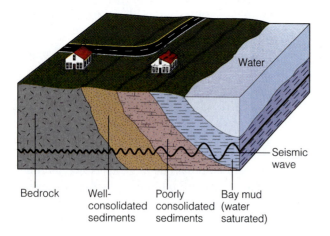

Bedrock   Well-consolidated sediments   Poorly consolidated sediments   Bay mud (water saturated)   Water   Seismic wave

■ Figure 9.15

The amplitude and duration of seismic waves generally increase as they pass from bedrock to poorly consolidated or water-saturated material. Thus structures built on weaker material typically suffer greater damage than similar structures built on bedrock.

(a)　　　　　　　　　　　　　　　　　　　　　　　　　　　　(b)

■ **Figure 9.17**

(a) Many of the approximately 242,000 people who died in the 1976 earthquake in Tangshan, China, were killed by collapsing structures. Many buildings were constructed from unreinforced brick, which has no flexibility and quickly fell during the quake. (b) In 1993 India experienced its worst earthquake in more than 50 years. Thousands of brick and stone houses collapsed, killing at least 30,000 people.

■ **Figure 9.18**

San Francisco Marina district fire caused by broken gas lines during the 1989 Loma Prieta earthquake.

The San Andreas Fault

The circum-Pacific belt is well known for its volcanic activity and earthquakes. Indeed, about 60% of all volcanic eruptions and 80% of all earthquakes take place in this belt that nearly encircles the Pacific Ocean basin (see Figure 9.4).

One well-known and well-studied segment of the circum-Pacific belt is the 1300-km-long San Andreas fault extending from the Gulf of California north through coastal California until it terminates at the Mendocino fracture zone off California's north coast. In plate tectonic terminology, it marks a transform plate boundary between the North American and Pacific plates (see Chapter 12).

Earthquakes along the San Andreas and related faults will continue to occur. But other segments of the circum-Pacific belt as well as the Mediterranean-Asiatic belt are also quite active, and will continue to experience earthquakes.

Aerial view of the San Andreas fault.

View across the San Andreas fault at Tomales Bay, north of San Francisco. The low area occupied by the bay is underlain by shattered rocks of the San Andreas fault zone. Rocks underlying the hills in the distance are on the North American plate, whereas those at the point where this photograph was taken are on the Pacific plate.

This shop in Olema, California, is rather whimsically called The Epicenter, alluding to the fact that it is in the San Andreas fault zone.

THE EPICENTER

Rocks on opposite sides of the San Andreas fault periodically lurch past one another, generating large earthquakes. The most famous one destroyed San Francisco on April 18, 1906. It resulted when 465 km of the fault ruptured, causing about 6 m of horizontal displacment in some areas (see Figure 9.1b). It is estimated that 3000 people died. The shaking lasted nearly 1 minute and caused property damage estimated at $400 million in 1906 dollars! About 28,000 buildings were destroyed, many of them by the three-day fire that raged out of control and devastated about 12 km^2 of the city.

San Francisco following the 1906 earthquake. This view along Sacramento Street shows damaged buildings and the approaching fire.

Since 1906 the San Andreas fault and its subsidiary faults have spawned many more earthquakes; one of the most tragic was centered at Northridge, California, a small community north of Los Angeles. During the early morning hours of January 17, 1994, Northridge and surrounding areas were shaken for 40 seconds. When it was over, 61 people were dead and thousands injured; an oil main and at least 250 gas lines had ruptured, igniting numerous fires; nine freeways were destroyed; and thousands of homes and other buildings were damaged or destroyed.

Severe damage caused by ground shaking during the 1994 Northridge earthquake. Sixteen died in this building.

A portion of Interstate 5 Golden State Freeway collapsed during the 1994 Northridge earthquake. Fortunately, no one was killed at this location.

A spectacular fire on Balboa Boulevard, Northridge, was caused by a gas-main explosion during the earthquake.

The Bishop Museum

■ Figure 9.19

As a tsunami crashes into the street behind them, residents of Hilo, Hawaii, run for their lives. This tsunami was generated by an earthquake in the Aleutian Islands on April 1, 1946, and resulted in massive property damage to Hilo and the deaths of 159 people.

During the September 1, 1923, earthquake in Japan, fires destroyed 71% of the houses in Tokyo and practically all the houses in Yokohama. In all, 576,262 houses were destroyed by fire, and 143,000 people died, many as a result of the fire. A horrible example occurred in Tokyo where thousands of people gathered along the banks of the Sumida River to escape the raging fires. Suddenly, a firestorm swept over the area, killing more than 38,000 people. The fires from this earthquake were particularly devastating because most of the buildings were constructed of wood; many fires were started by chemicals and fanned by 20 km/hr winds.

Tsunami: Killer Waves

On April 1, 1946, the residents of Hilo, Hawaii, were completely unaware of an earthquake more than 3500 km away in the Aleutian Islands, but this earthquake ended up killing 159 people in that city and causing $25 million in property damage (■ Figure 9.19). Thus earthquakes can kill at a considerable distance, in some cases even much farther away than in this example. This earthquake generated what is popularly called a "tidal wave" but more correctly termed a *seismic sea wave* or **tsunami,** a Japanese term meaning "harbor wave." The term *tidal wave* nevertheless persists in popular literature and some news accounts, but these waves are not caused by or related to tides. Indeed, tsunami are de-

structive sea waves generated when large amounts of energy are rapidly released into a body of water. Many result from submarine earthquakes, but volcanoes at sea or submarine landslides can also cause them. For example, the 1883 eruption of Krakatau between Java and Sumatra generated a large sea wave that killed 36,000 people on nearby islands.

Once a tsunami is generated, it can travel across an entire ocean and cause devastation far from its source. In the open sea, tsunami travel at several hundred kilometers per hour, and commonly go unnoticed as they pass beneath ships because they are usually less than 1 m high and the distance between wave crests is typically hundreds of kilometers. When they enter shallow water, however, the wave slows down and water piles up to heights anywhere from a meter or two to many meters high. The 1946 tsunami that struck Hilo, Hawaii, was 16.5 m high! In any case, the tremendous energy possessed by a tsunami is concentrated on a shoreline when it hits either as a large breaking wave or, in some cases, as what appears to be a very rapidly rising tide.

The Hawaiian Islands are especially vulnerable to tsunami, but other areas around the Pacific Rim have also been devastated by these waves. Japan and parts of the coast of South America have experienced a number of tsunami, and following the 1964 Alaska earthquake the waterfronts at Anchorage, Alaska (■ Figure 9.20), and Crescent City, California, were heavily damaged and a number of people were swept away.

A common popular belief is that a tsunami is a single large wave that crashes onto a shoreline. Any tsunami consists of a series of waves that pour onshore for as long as 30 minutes followed by an equal time during which water rushes back to sea. Furthermore, after the first wave hits, more waves follow at 20- to 60-minute intervals. About 80 minutes after the 1755 Lisbon, Portugal, earthquake, the first of three tsunami, the largest more than 12 m high, destroyed the waterfront area and killed numerous people. Following the arrival of a 2-m-high tsunami in Crescent City, California, in 1964, curious people went to the waterfront to inspect the damage.

and instruments that detect earthquake-generated waves. Whenever a strong earthquake takes place anywhere within the Pacific basin, its location is determined, and instruments are checked to see if a tsunami has been generated. If it has, a warning is sent out to evacuate people from low-lying areas that may be affected. Nevertheless, tsunami remain a threat to people in coastal areas, especially around the Pacific Ocean.

PHYSICAL Geology ⇌ Now Click Geology Interactive to work through an activity on Tsunami through Earthquakes and Tsunami.

Ground Failure

Earthquake-triggered landslides are particularly dangerous in mountainous regions and have been responsible for tremendous amounts of damage and many deaths. The 1959 earthquake in Madison Canyon, Montana, for example, caused a huge rock slide (■ Figure 9.22), while the 1970 Peru earthquake caused an avalanche that destroyed the town of Yungay and killed an estimated 66,000 people. Most of the 100,000 deaths from the 1920 earthquake in Gansu, China, resulted when cliffs composed of loess (wind-deposited silt) collapsed. More than 20,000 people were killed when two thirds of the town of Port Royal, Jamaica, slid into the sea following an earthquake on June 7, 1692.

PHYSICAL Geology ⇌ Now Click Geology Interactive to work through an activity on Seismic Case Study, Alaska, 1964, through Earthquakes and Tsunami.

© Pierre Mion

■ **Figure 9.20**

This painting is an artist's concept of the tsunami that hit the Anchorage, Alaska, waterfront following the 1964 earthquake. It is based on eyewitness accounts.

Unfortunately, 10 were killed by a following 4-m-high wave!

One of nature's warning signs of an approaching tsunami is that some are preceded by a sudden withdrawal of the sea from a coastal region. In fact, the sea might withdraw so far that it cannot be seen and the seafloor is laid bare over a huge area. On more than one occasion, people have rushed out to inspect exposed reefs or collect fish and shells, only to be swept away when the tsunami arrived.

Following the tragic 1946 tsunami that hit Hilo, Hawaii, the U.S. Coast and Geodetic Survey established a Tsunami Early Warning System in Honolulu, Hawaii (■ Figure 9.21). This system combines seismographs

CAN EARTHQUAKES BE PREDICTED?

A successful prediction must include a time frame for the occurrence of an earthquake, its location, and its strength. Despite the tremendous amount of information geologists have gathered about the cause of earthquakes, successful predictions are still rare. Nevertheless, if reliable predictions can be made, they can greatly reduce the number of deaths and injuries.

GEOLOGY
IN UNEXPECTED PLACES

It's Not My Fault

Are you the type of person who finds fault with others, or are you a geologist who just finds faults? Faults can be seen in some rather unexpected places. And what does finding a fault have to do with earthquakes? Remember that most earthquakes are caused by movement along a fault. This movement releases energy in the form of seismic waves, which are what we feel during an earthquake. So when we find faults or evidence of fault movement, the energy released along that fault may have generated an earthquake.

The most direct evidence of a fault is a fault scarp. Imagine my surprise when I was visiting a tranquil lake in Nanjing, China, many years ago and, while walking along a trail, noticed the afternoon sun reflecting off an ancient fault scarp (■ Figure 1). That discovery certainly ranks as one of my all-time great "geology in unexpected places" experiences.

Another time, while hiking in the Gallatin National Forest in Montana, I came across the approximately 6-m-high Red Canyon Fault scarp. Vertical movement along this fault on August 17, 1959, produced a magnitude 7.3 earthquake that triggered a landslide that blocked the Madison River in Montana and created Earthquake Lake (see Figure 9.22). Because of the magnitude of the earthquake that occurred, I was not surprised to find a fault scarp in the area, but the height of the scarp and its preservation 18 years later were unexpected.

Sometimes we can fault others for not recognizing a potential problem when it presents itself. Such is the case in Dana Point, California, where houses have been built next to and over an ancient fault (■ Figure 2). Although it is unlikely that the fault is active and movement will occur along it, erosion is proceeding along the fault at a faster rate than it is elsewhere in the area, causing the homeowners to have to take remedial action to slow the rate of erosion.

■ Figure 2

This inactive fault can be recognized by the fact there are pink colored rocks (left) juxtaposed against white colored rocks (right). Because movement along this fault has pulverized the rocks, erosion has proceeded more rapidly along the fault, forming a narrow valley. Note the retaining wall in the upper right hand part of the photo and house behind it. Both the retaining wall and house are built directly over this fault.

■ Figure 1

Light reflecting off a fault scarp along a lake trail in Nanjing, China. Note the highly polished scarp surface, indicating movement along this surface as rocks moved past each other.

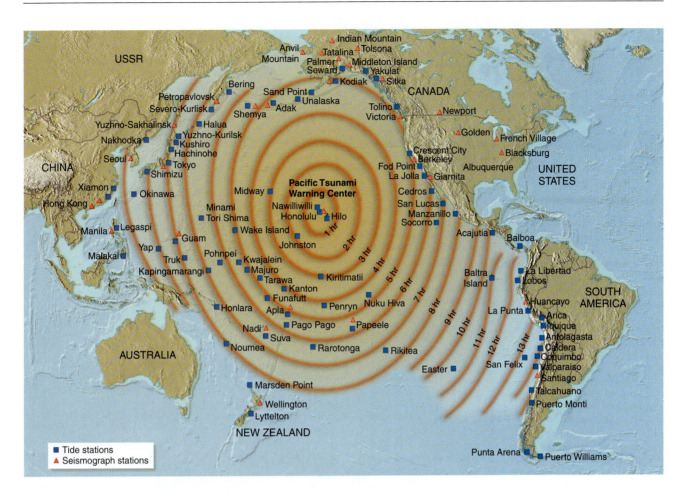

PHYSICAL
Geology⇌Now ■ **Active Figure 9.21**

The Pacific Tsunami Early Warning System. Reporting stations and tsunami travel time to Honolulu, Hawaii. Source: Data from National Oceanic and Atmospheric Administration.

From an analysis of historic records and the distribution of known faults, **seismic risk maps** can be constructed that indicate the likelihood and potential severity of future earthquakes based on the intensity of past earthquakes. An international effort by scientists from several countries resulted in the publication of the first Global Seismic Hazard Assessment Map in December 1999 (■ Figure 9.23). Although such maps cannot be used to predict when an earthquake will take place in any particular area, they are useful in anticipating future earthquakes and helping people prepare for them.

Earthquake Precursors

Studies conducted during the past several decades indicate that most earthquakes are preceded by both short-term and long-term changes within Earth. Such changes are called *precursors*.

One long-range prediction technique used in seismically active areas involves plotting the location of major earthquakes and their aftershocks to detect areas that have had major earthquakes in the past but are currently inactive. Such regions are locked and not releas-

> ## What Would You Do
>
> Your city has experienced moderate to large earthquakes in the past, and as a result, the local planning committee, of which you are a member, has been charged with making recommendations as to how your city can best reduce damage as well as potential injuries and fatalities resulting from future earthquakes. You are told to consider zoning regulations; building codes for private dwellings, hospitals, public buildings, and high-rise structures; and emergency contingency plans. What kinds of recommendations would you make, and what and who would you ask for professional guidance?

ing energy. Nevertheless, pressure is continuing to accumulate in these regions due to plate motions, making these *seismic gaps* prime locations for future earthquakes. Several seismic gaps along the San Andreas fault have the potential for future major earthquakes

(a)

Reed Wicander

Source of landslide

Landslide deposit

(b)

Reed Wicander

■ **Figure 9.22**

On August 17, 1959, an earthquake with a Richter magnitude of 7.3 shook southwestern Montana and a large area in adjacent states. (a) The fault scarp in this image was produced when the block on the right moved up several meters compared to the one on the left. (b) The earthquake triggered a landslide (visible in the distance) that blocked the Madison River in Montana and created Earthquake Lake (foreground). The slide entombed about 26 people in a campground at the valley bottom.

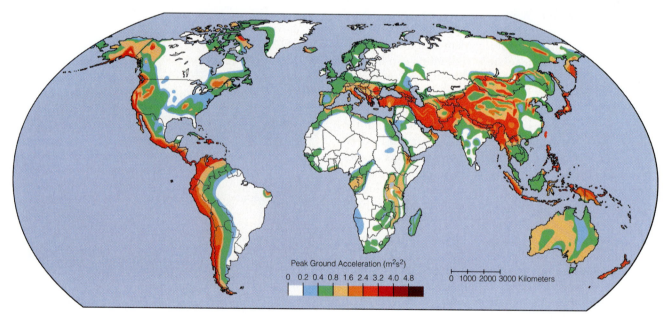

Peak Ground Acceleration (m²s²)

0 0.2 0.4 0.8 1.6 2.4 3.2 4.0 4.8

0 1000 2000 3000 Kilometers

■ **Figure 9.23**

The Global Seismic Hazard Assessment Program published this seismic hazard map showing peak ground accelerations. The values are based on a 90% probability that the indicated horizontal ground acceleration during an earthquake is not likely to be exceeded in 50 years. The higher the number, the greater the hazard. As expected, the greatest seismic risks are in the circum-Pacific belt and the Mediterranean–Asiatic belt.

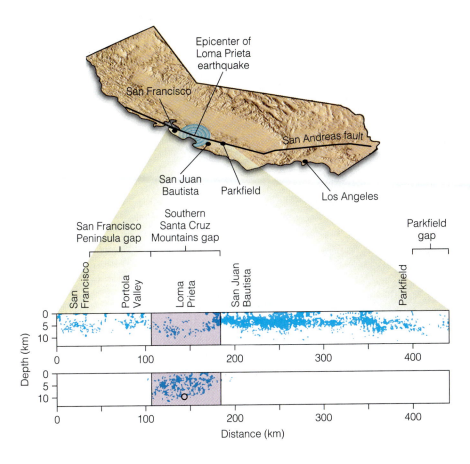

■ Figure 9.24

Three seismic gaps are evident in this cross section along the San Andreas fault from north of San Francisco to south of Parkfield. The first is between San Francisco and Portola Valley, the second near Loma Prieta Mountain, and the third is southeast of Parkfield. The top section shows the epicenters of earthquakes between January 1969 and July 1989. The bottom section shows the southern Santa Cruz Mountains gap after it was filled by the October 17, 1989, Loma Prieta earthquake (open circle) and its aftershocks. Source: Data from *The Loma Prieta Earthquake of October 17, 1989.* U.S. Geological Survey.

(■ Figure 9.24). A major earthquake that damaged Mexico City in 1985 occurred along a seismic gap in the convergence zone along the west coast of Mexico.

Changes in elevation and tilting of the land surface have frequently preceded earthquakes and may be warnings of impending quakes. Extremely slight changes in the angle of the ground surface can be measured by *tiltmeters*. Tiltmeters have been placed on both sides of the San Andreas fault to measure tilting of the ground surface that is thought to result from increasing pressure in the rocks. Data from measurements in central California indicate significant tilting immediately preceding small earthquakes. Furthermore, extensive tiltmeter work performed in Japan prior to the 1964 Niigata earthquake clearly showed a relationship between increased tilting and the main shock. Although more research is needed, such changes appear to be useful in making short-term earthquake predictions.

Other earthquake precursors include fluctuations in the water level of wells and local changes in Earth's magnetic field and the electrical resistance of the ground. These fluctuations are thought to result from changes in the amount of pore space in rocks due to increasing pressure.

The Chinese used the precursors just mentioned, except seismic gaps, to successfully predict a large earthquake in Haicheng on February 4, 1975. The earthquake had a magnitude of 7.5 and destroyed hundreds of buildings but claimed very few lives because most people had been evacuated from the buildings and were outdoors when it occurred. Unfortunately, visits by U.S. scientists revealed that the prediction resulted from unique circumstances that really could not be applied to earthquake prediction elsewhere. And for that matter, the Chinese failed to predict a 1976 earthquake that killed 242,000 people (Figure 19.17a).

Many of the precursors just discussed can be related to the *dilatancy model*, which is based on changes occurring in rocks subjected to very high pressures. Laboratory experiments have shown that rocks undergo an increase in volume, known as *dilatancy*, just before rupturing. As pressure builds in rocks along faults, numerous small cracks are produced that alter the physical properties of the rocks. Water enters the cracks and increases the fluid pressure; this further increases the volume of the rocks and decreases their inherent strength until failure eventually occurs, producing an earthquake.

The dilatancy model is consistent with many earthquake precursors (■ Figure 9.25). Although additional research is needed, this model has the potential for predicting earthquakes under certain circumstances.

Earthquake Prediction Programs

Currently, only four nations—the United States, Japan, Russia, and China—have government-sponsored earthquake prediction programs. These programs include laboratory and field studies of rock behavior before, during, and after large earthquakes, as well as monitoring activity along major active faults.

The Chinese have perhaps one of the most ambitious earthquake prediction programs in the world,

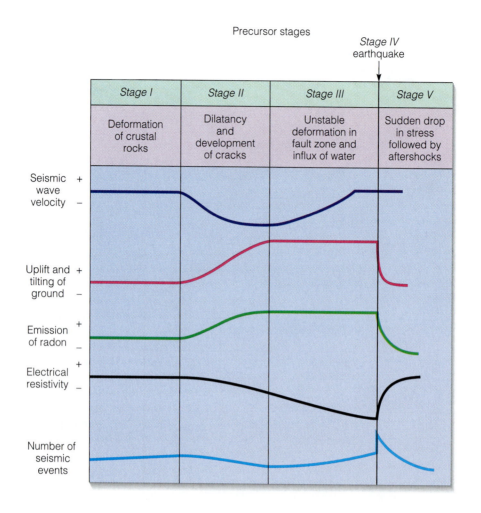

■ Figure 9.25

The relationship between dilatancy and various other earthquake precursors. The onset of dilatancy matches a change in each of the precursors illustrated. For example, a drop in seismic wave velocity corresponds to the onset of dilatancy and the development of cracks. Source: From *Predicting Earthquakes: A Scientific and Technical Evaluation—With Implications for Society*, p. 41. Copyright © 1976 National Academy Press. Reprinted by permission.

which is understandable considering their long history of destructive earthquakes. Their earthquake prediction program was initiated soon after two large earthquakes occurred at Xingtai (300 km southwest of Beijing) in 1966. This program includes extensive study and monitoring of all possible earthquake precursors. In addition, the Chinese emphasize changes in phenomena that can be observed and heard without the use of sophisticated instruments. They successfully predicted the 1975 Haicheng earthquake, as we already noted, but failed to predict the devastating 1976 Tangshan earthquake that killed at least 242,000 people (Figure 9.17a).

Progress is being made toward dependable, accurate earthquake predictions, and studies are under way to assess public reactions to long-, medium-, and short-term earthquake warnings. However, unless short-term warnings are actually followed by an earthquake, most people will probably ignore the warnings, as they frequently do now for hurricanes, tornadoes, and tsunami. Perhaps the best we can hope for is that people in seismically active areas will take measures to minimize their risk from the next major earthquake (Table 9.4).

CAN EARTHQUAKES BE CONTROLLED?

Reliable earthquake prediction is still in the future, but can anything be done to control or at least partly control earthquakes? Because of the tremendous energy involved, it seems unlikely that humans will ever be able to prevent earthquakes. However, it may be possible to gradually release the energy stored in rocks, thus decreasing the probability of a large earthquake and extensive damage.

During the early to mid-1960s, Denver, Colorado, experienced numerous small earthquakes. This was sur-

Table 9.4

What You Can Do to Prepare for an Earthquake

Anyone who lives in an area that is subject to earthquakes or who will be visiting or moving to such an area can take certain precautions to reduce the risks and losses resulting from an earthquake.

Before an earthquake:

1. Become familiar with the geologic hazards of the area where you live and work.
2. Make sure your house is securely attached to the foundation by anchor bolts and that the walls, floors, and roof are all firmly connected together.
3. Heavy furniture such as bookcases should be bolted to the walls; semiflexible natural gas lines should be used so that they can give without breaking; water heaters and furnaces should be strapped and the straps bolted to wall studs to prevent gas-line rupture and fire. Brick chimneys should have a bracket or brace that can be anchored to the roof.
4. Maintain a several-day supply of freshwater and canned foods, and keep a fresh supply of flashlight and radio batteries as well as a fire extinguisher.
5. Maintain a basic first-aid kit and have a working knowledge of first-aid procedures.
6. Learn how to turn off the various utilities at your house.
7. Above all, have a planned course of action for when an earthquake strikes.

During an earthquake:

1. Remain calm and avoid panic.
2. If you are indoors, get under a desk or table if possible, or stand in an interior doorway or room corner as these are the structurally strongest parts of a room; avoid windows and falling debris.
3. In a tall building, do not rush for the stairwells or elevators.
4. In an unreinforced or other hazardous building, it may be better to get out of the building rather than to stay in it. Be on the alert for fallen power lines and the possibility of falling debris.
5. If you are outside, get to an open area away from buildings if possible.
6. If you are in an automobile, stay in the car, and avoid tall buildings, overpasses, and bridges if possible.

After an earthquake:

1. If you are uninjured, remain calm and assess the situation.
2. Help anyone who is injured.
3. Make sure there are no fires or fire hazards.
4. Check for damage to utilities and turn off gas valves if you smell gas.
5. Use your telephone only for emergencies.
6. Do not go sightseeing or move around the streets unnecessarily.
7. Avoid landslide and beach areas.
8. Be prepared for aftershocks.

---

prising because Denver had not been prone to earthquakes in the past. In 1962 geologist David M. Evans suggested that Denver's earthquakes were directly related to the injection of contaminated wastewater into a disposal well 3674 m deep at the Rocky Mountain Arsenal, northeast of Denver (■ Figure 9.26a). The U.S. Army initially denied that a connection existed, but a USGS study concluded that the pumping of waste fluids into fractured rocks beneath the disposal well decreased the friction on opposite sides of fractures and, in effect, lubricated them so that movement occurred, causing the earthquakes that Denver experienced.

Figure 9.26b shows the relationship between the average number of earthquakes in Denver per month and the average amount of contaminated fluids injected into the disposal well per month. Obviously, a high degree of correlation between the two exists, and the correlation is particularly convincing considering that during the time when no waste fluids were injected, earthquake activity decreased dramatically.

Experiments conducted in 1969 at an abandoned oil field near Rangely, Colorado, confirmed the arsenal hy-

What Would You Do ?

Some geologists think that by pumping liquids into locked segments of active faults, they can generate small- to moderate-sized earthquakes. These earthquakes would relieve the buildup of pressure along a fault and thus prevent very large earthquakes from taking place. What do you think of this proposal? What kind of social, political, and economic consequences would there be, and do you think such an effort will ever actually reduce the threat of earthquakes?

pothesis. Water was pumped into and out of abandoned oil wells, the pore-water pressure in these wells was measured, and seismographs were installed in the area to measure any seismic activity. Monitoring showed that small earthquakes were occurring in the area when fluid was injected and that earthquake activity declined when

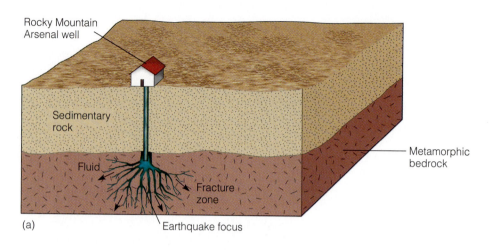

(a)

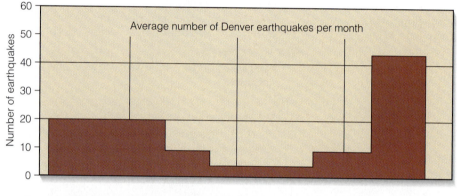

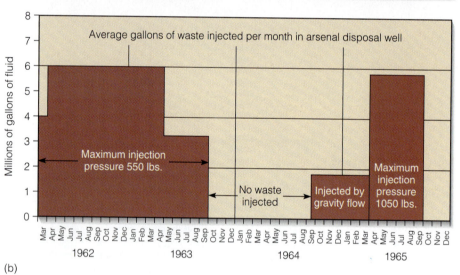

(b)

■ Figure 9.26

(a) A block diagram of the Rocky Mountain Arsenal well and the underlying geology. (b) A graph showing the relationship between the amount of wastewater injected into the well per month and the average number of Denver earthquakes per month. There have not been any significant earthquakes in Denver since injection of wastewater into the disposal well ceased in 1965. Source: From Figure 6, page 17, *Geotimes Vol. 10, No. 9* (1966) with the kind permission of the American Geological Institute.

the fluids were pumped out. What the geologists were doing was starting and stopping earthquakes at will, and the relationship between pore-water pressures and earthquakes was established.

Based on these results, some geologists have proposed that fluids be pumped into the locked segments or seismic gaps of active faults to cause small- to moderate-sized earthquakes. They think this would relieve the pressure on the fault and prevent a major earthquake from occurring. Although this plan is intriguing, it also

has many potential problems. For instance, there is no guarantee that only a small earthquake might result. Instead, a major earthquake might occur, causing tremendous property damage and loss of life. Who would be responsible? Certainly, a great deal more research is needed before such an experiment is performed, even in an area of low population density.

It appears that until such time as earthquakes can be accurately predicted or controlled, the best means of defense is careful planning and preparation (Table 9.4).

9 REVIEW
WORKBOOK

Chapter Summary

- Earthquakes are vibrations of Earth caused by the sudden release of energy, usually along a fault.

- The elastic rebound theory holds that pressure builds in rocks on opposite sides of a fault until the strength of the rocks is exceeded and rupture occurs. When the rocks rupture, stored energy is released as they snap back to their original position.

- Seismology is the study of earthquakes. Earthquakes are recorded on seismographs, and the record of an earthquake is a seismogram.

- The point where energy is released is an earthquake's focus, whereas its epicenter is vertically above the focus on the surface.

- Approximately 80% of all earthquakes occur in the circum-Pacific belt, 15% within the Mediterranean–Asiatic belt, and the remaining 5% mostly in the interior of the plates or along oceanic spreading-ridge systems.

- The two types of body waves are P-waves and S-waves. P-waves travel through all materials, whereas S-waves do not travel through liquids. P-waves are the fastest waves and are compressional, whereas S-waves are shear.

- The two types of surface waves are Rayleigh and Love waves, and they travel along or just below the surface.

- The epicenter of an earthquake is located by the use of a time–distance graph of the P- and S-waves from any given distance. Three seismographs are needed to locate the epicenter.

- Intensity is a measure of the kind of damage done by an earthquake and is expressed by values from I to XII in the Modified Mercalli Intensity Scale.

- Magnitude measures the amount of energy released by an earthquake and is expressed in the Richter Magnitude Scale. Each increase in the magnitude number represents about a 30-fold increase in energy released.

- The Seismic-Moment Magnitude Scale more accurately estimates the energy released during very large earthquakes.

- Ground shaking is the most destructive of all earthquake hazards. The amount of damage done by an earthquake depends on the geology of the area, type of building construction, magnitude of the earthquake, and duration of shaking.

- Tsunami are seismic sea waves that are produced by earthquakes, submarine landslides, and eruptions of volcanoes at sea.

- Seismic risk maps are helpful in making long-term predictions about the severity of earthquakes based on past occurrences.

- Earthquake precursors are changes preceding an earthquake that can be used to predict earthquakes. Precursors include seismic gaps, changes in surface elevations, and fluctuations of water levels in wells.

- A variety of earthquake research programs are under way in the United States, Japan, Russia, and China. Studies indicate that most people would probably not heed a short-term earthquake warning.

- Injecting fluids into locked segments of an active fault holds some promise as a possible means of earthquake control.

Important Terms

body wave (p. 250)
earthquake (p. 242)
elastic rebound theory (p. 243)
epicenter (p. 246)
focus (p. 246)
intensity (p. 254)
Love wave (L-wave) (p. 250)

magnitude (p. 255)
Modified Mercalli Intensity Scale (p. 254)
P-wave (p. 250)
Rayleigh wave (R-wave) (p. 250)
Richter Magnitude Scale (p. 255)

seismic risk map (p. 267)
seismograph (p. 246)
seismology (p. 246)
surface wave (p. 250)
S-wave (p. 250)
tsunami (p. 264)

Review Questions

1. Most earthquakes take place in the:

   a. _____ spreading-ridge zone; b. _____ Mediterranean–Asiatic belt; c. _____ rifts in continental interiors; d. _____ circum-Pacific belt; e. _____ Appalachian fault zone.

2. A P-wave is one in which:

   a. _____ movement is perpendicular to the direction of wave travel; b. _____ Earth's surface moves as a series of waves; c. _____ materials move forward and back along a line in the same direction that the wave moves; d. _____ large waves crash onto a shoreline following a submarine earthquake; e. _____ movement at the surface is similar to movement in water waves.

3. With few exceptions, the most damaging earthquakes are:

   a. _____ deep focus; b. _____ caused by volcanic eruptions; c. _____ those with Richter magnitudes of about 2; d. _____ shallow focus; e. _____ those occurring along spreading ridges.

4. A tsunami is a(n):

   a. _____ part of a fault with a seismic gap; b. _____ precursor to an earthquake; c. _____ seismic sea wave; d. _____ particularly large and destructive earthquake; e. _____ earthquake with a focal depth exceeding 300 km.

5. A qualitative assessment of the damage done by an earthquake is expressed by:

   a. _____ intensity; b. _____ dilatancy; c. _____ seismicity; d. _____ magnitude; e. _____ liquefaction.

6. It would take about _____ earthquakes with a Richter magnitude of 3 to equal the energy released in one earthquake with a magnitude of 6.

   a. _____ 9; b. _____ 30; c. _____ 27,000; d. _____ 2,000,000; e. _____ 250.

7. An earthquake's epicenter is:

   a. _____ usually in the lower part of the mantle; b. _____ a point on the surface directly above the focus; c. _____ determined by analyzing surface wave arrival times at seismic stations; d. _____ a measure of the energy released during an earthquake; e. _____ the damage corresponding to a value of IV on the Modified Mercalli Intensity Scale.

8. Earthquakes following a large earthquake are known as:

   a. _____ precursors; b. _____ tsunami; c. _____ seismic gaps; d. _____ aftershocks; e. _____ seismic moments.

9. Which of the following statements is correct?

   a. _____ tsunami are caused by especially high tides and earthquakes with epicenters on land; b. _____ monitoring precursors allows seismologists to make accurate short-range predictions of earthquakes; c. _____ the Richter Magnitude Scale is the most useful measure of an earthquake's size for insurance companies; d. _____ the time interval between arrivals of P- and S-waves at a seismic station depends on the distance from an earthquake's focus; e. _____ the safest kind of building to be in during an earthquake is one constructed of stones cemented together.

10. In which one of the following areas would you most likely experience an earthquake?
    a. _____ England; b. _____ Florida;
    c. _____ Japan; d. _____ Kansas;
    e. _____ Germany.

11. Fully describe how a tsunami is generated, how it travels, and what impact it has on shorelines.

12. What are precursors, and how can they be used to predict earthquakes?

13. What are the differences between intensity and magnitude?

14. Why are structures built on bedrock usually damaged less during an earthquake than those sited on unconsolidated material?

15. What plate tectonic settings account for earthquakes along the west coasts of North America and South America? In which of these two areas would you expect deep-focus earthquakes?

16. How does the elastic rebound theory account for energy released during an earthquake?

17. Describe the various ways earthquakes are destructive.

18. Discuss why insurance companies use the qualitative Modified Mercalli Intensity Scale instead of the quantitative Richter Magnitude Scale in classifying earthquakes.

19. From the arrival times of P- and S-waves shown in the chart and from the graph in Figure 9.10, calculate how far away from each seismograph station the earthquake occurred. How would you determine the epicenter of this earthquake?

	Arrival Time of P-Wave	Arrival Time of S-Wave
Station A	2:59:03 P.M.	3:04:03 P.M.
Station B	2:51:16 P.M.	3:01:16 P.M.
Station C	2:48:25 P.M.	2:55:55 P.M.

20. Refer to the graph in Figure 9.14. A seismograph in Berkeley, California, records the arrival time of an earthquake's P-waves at 6:59:54 P.M. and the S-waves at 7:00:02 P.M. The maximum amplitude of the S-waves as recorded on the seismogram was 75 mm. What was the magnitude of the earthquake, and how far away from Berkeley did it occur?

World Wide Web Activities

PHYSICAL
Geology ⇌ Now Assess your understanding of this chapter's topics with additional quizzing and comprehensive interactivities at

http://earthscience.brookscole.com/physgeo5e

as well as current and up-to-date weblinks, additional readings, and InfoTrac College Edition exercises.

Earth's Interior

CHAPTER 10
OUTLINE

PHYSICAL Geology⇌Now *This icon, appearing throughout the book, indicates an opportunity to explore interactive tutorials, animations, or practice problems available on the Physical GeologyNow Web site at http://earthscience.brookscole.com/physgeo5e.*

This 30-story structure on the Kola Peninsula in northwestern Russia houses a drill rig that has penetrated about 15 km into Earth's crust. A 12-km-deep drill hole is impressive, but put into perspective it would be about 0.5 mm deep if the vertical dimension of this page represented Earth's radius. Source: Cornelius Gillen

OBJECTIVES
At the end of this chapter, you will have learned that

- Geologists use seismic waves to determine Earth's internal structure.

- Earth has a central core, overlain by a thick mantle, and a thin outer layer of crust.

- Seismic tomography is yielding a more refined view of Earth's internal structure.

- Geologists determine the density, composition, and structure of the core, mantle, and crust based on studies of seismic waves, Earth's overall density, meteorites, and inclusions in volcanic rocks.

- Earth possesses considerable internal heat that continuously escapes at the surface.

- The force of gravity varies depending on several factors, thus accounting for gravity anomalies.

- The principle of isostasy holds that Earth's crust is buoyed up by a denser medium below.

- Earth possesses a magnetic field that varies through time and in intensity and direction.

Introduction

Earth's interior is so remote and inaccessible that most people think little about it. One can appreciate the stunning beauty of the aurora borealis (northern lights) and yet be completely unaware that they exist because of an interaction between Earth's internally generated magnetic field and the solar wind, a continuous stream of electrically charged particles emanating from the Sun. Much of Earth's geologic activity, such as earthquakes, volcanism, moving plates, and the origin of mountains, is caused by internal heat. In fact, the slow release of heat from Earth's interior is one major factor that makes it such a dynamic planet.

Scientists now have a good idea of Earth's overall composition and internal structure, but during most of historic time people perceived of Earth's interior as an underground world of vast caverns, heat, and sulfur gases, populated by the souls of those waiting to be judged. Many cultures have myths that explain eruptions and earthquakes as manifestations of the activities of various deities that dwell far below the surface. The Greek philosopher Aristotle (384–322 B.C.) thought that winds and fires in subterranean cavities caused earthquakes. Romans believed that Vulcan, the god of fire, had several underground workshops where his beating on anvils caused the ground to rumble and volcanoes to erupt.

These ancient Greek and Roman ideas, though interesting now, seem naïve, but some rather bizarre ideas about Earth's interior are still with us (see "Earth's Place in the Cosmos" on pages 280 and 281). For example, in 1869 Cyrus Reed Teed claimed Earth is hollow and that humans live on the inside. And in 1913 Marshall B. Gardener held that Earth is a large, hollow sphere with a 1300-km-thick outer shell surrounding a central Sun. Even today, some books and articles promote the hollow Earth idea and several Web sites are devoted to this thesis. Others hold that Earth is flat, at the center of the universe, or both.

Although making no claim to present a reliable picture of Earth's interior, Jules Verne's 1864 novel *Journey to the Center of the Earth* described the adventures of Professor Hardwigg, his nephew, and an Icelandic guide. The trio descended through a volcano in Iceland and followed a labyrinth of passageways until they arrived 140 km below the surface. Here they encountered a vast cavern illuminated by some electrical phenomenon, and they saw herds of mastodons complete with a gigantic human shepherd, Mesozoic-aged reptiles, and huge turtles dwelling in what Verne called the "central sea." Their adventure ended when they were carried back to the surface on a rising plume of water.

Even in 1864 scientists knew that Earth's overall density is about 5.5 g/cm^3 and that

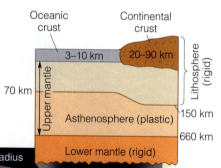

Earth's Composition and Density

	Composition	Density (g/cm^3)
Inner core	Iron with 10–20% nickel	12.6–13.0
Outer core	Iron with perhaps 12% sulfur, silicon, oxygen, nickel, and potassium	9.9–12.2
Mantle	Peridotite (composed mostly of ferromagnesian silicates)	3.3–5.7
Oceanic crust	Upper part basalt, lower part gabbro	~3.0
Continental crust	Average composition of grandiorite	~2.7

■ **Figure 10.1**

Earth's internal structure. The inset shows Earth's outer part in more detail. The asthenosphere is solid but behaves plastically and flows.

pressure and temperature increase with depth. Little else was known, although humans had probed beneath the surface with mines and wells for centuries. But even the deepest mines extend to only about 3 km below the surface, and the deepest well at about 15 km penetrates only 0.2% of the distance to Earth's center (see the chapter opening photo).

Scientists have no direct observations of Earth's interior but nevertheless know something about its internal composition and structure (■ Figure 10.1). No vast caverns exist at great depth as in Jules Verne's story, even at the modest depth of 140 km that Professor Hardwigg and his companions are supposed to have visited. At this depth rock is so hot and under such great pressure that it flows even though it remains solid. Even in deep mines rock behaves differently than it does at the surface. Recall from Chapter 5 that because of intense pressure, rock bursts and popping are constant problems.

EARTH'S SIZE, DENSITY, AND INTERNAL STRUCTURE

Among the terrestrial planets, Earth, with a diameter of 12,760 km at the equator, is slightly larger than Venus and much larger than Mercury, Mars, or Earth's Moon. Actually, Earth is an oblate spheroid because its equatorial diameter is slightly greater than its polar diameter. Scientists have known for more than 200 years that planet Earth is not homogeneous throughout. Indeed, Sir Isaac Newton (1642–1727) noted that Earth's average density—that is, its mass per unit volume—is 5.0 to 6.0 g/cm^3, and in 1797 Henry Cavendish calculated a density value very close to the 5.5 g/cm^3 now accepted.

To accurately determine Earth's density, first its size must be known. In the third century B.C. the Greek librarian Eratosthenes (276?–195? B.C.) determined that Earth's circumference is about 40,000 km—a figure only slightly different from the one accepted now. From this, one can calculate Earth's diameter and radius, and then figure out the volume of our nearly spherical planet. Density is another matter; it can be determined only indirectly by, for example, comparing the gravitational attraction between Earth and its Moon and between metal spheres of known mass. After carrying out these calculations, scientists derived an average value for the planet of 5.52 g/cm^3. But we know that most of Earth's surface and near-surface rocks have densities of only 2.5 to 3.0 g/cm^3. Accordingly, Earth's interior must be made up of materials much denser than those at or near the surface.

Seismic Waves and What They Tell Us About Earth's Interior

The behavior and travel times of P- and S-waves provide geologists with much information about Earth's internal structure. Seismic waves travel outward as wave fronts from their source areas, although it is most convenient to depict them as *wave rays*, which are lines showing the direction of movement of small parts of wave fronts (■ Figure 10.2). Any disturbance such as a passing train or construction equipment causes seismic waves, but only those generated by large earthquakes, explosive volcanism, asteroid impacts, and nuclear explosions travel completely through Earth.

As we noted in Chapter 9, P- and S-wave velocity is determined by the density and elasticity of the materials they travel through, both of which increase with depth. Wave velocity is slowed by increasing density but increases in materials with greater elasticity. Because elasticity increases with depth faster than density, a general increase in seismic wave

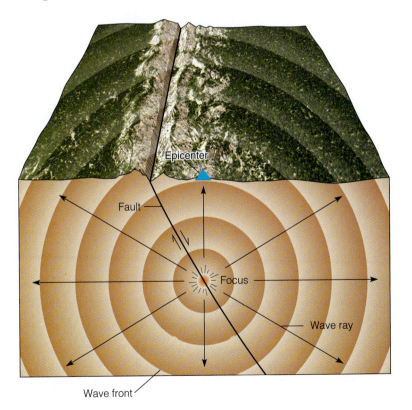

■ **Figure 10.2**

Seismic wave fronts move out in all directions from their source, the focus of an earthquake in this example. Wave rays are lines drawn perpendicular to wave fronts.

Earth's Place in the Cosmos

Various views of Earth's size, shape, interior, and its place in the cosmos. During much of historic time people viewed Earth as flat, hollow, inhabited by demons, and the center of the cosmos. We now know that Earth's internal heat is responsible for plate movements, earthquakes, and volcanism, and electrical currents in Earth's core generate the magnetic field.

Even though the ancient Greeks accepted that Earth is spherical, the idea of a flat Earth persisted for centuries. In this depiction, the flat Earth has a crystal half-sphere over it that keeps the oceans from spilling off the edges.

During the 14th century, Dante Alighieri attempted to reconcile Greek science with religious doctrine when he proposed that Earth is a stationary sphere surrounded by translucent, revolving spheres. The ninth sphere, the *Primum Mobile*, is the prime mover of all the spheres. The *Empyrean Paradise,* according to Dante, is where God and the angels reside.

Alexandria

Well at Syene

Shadow

During the third century B.C., the Greek librarian Eratosthenes knew that at the summer solstice the sun cast no shadow as it shined into a well at Syene (now Aswan), but at Alexandria it cast a shadow of $7\frac{1}{4}°$. He knew the distance between these two points and calculated Earth's circumference.

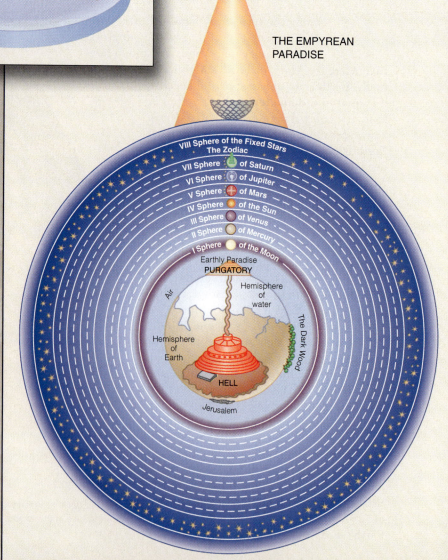

THE EMPYREAN PARADISE

VIII Sphere of the Fixed Stars
The Zodiac
VII Sphere of Saturn
VI Sphere of Jupiter
V Sphere of Mars
IV Sphere of the Sun
III Sphere of Venus
II Sphere of Mercury
I Sphere of the Moon
Earthly Paradise
PURGATORY

Air

Hemisphere of water

Hemisphere of Earth

The Dark Wood

HELL

Jerusalem

The opening to Mammoth Cave in Kentucky is one of thousands of labyrinthine cavities in Earth's crust. But contrary to popular belief even the deepest caves extend to depths of only a few hundred meters.

Jules Verne made no claims of scientific accuracy in his book *A Journey to the Center of the Earth*, but the ideas of passages leading to great depths and a hollow Earth persist even now. The figure on the right is from an 1871 translation of the book; the Central Sea with its inhabitants is shown.

In this view from the 1959 movie *Journey to the Center of the Earth*, actors James Mason and Arlene Dahl explore the remains of a lost civilization as members of an expedition to Earth's center.

A variety of ideas persist on the concept of a hollow Earth. In this view, Earth consists of a 1290-km-thick crust with vast openings at the poles, and it has a central sun about 970 km in diameter. People live on Earth's outer surface, some live in middle Earth, that is, within the crust, and some live on the crust's inner surface.

Opening at North Pole

Crust

Central sun

1290 km

970 km diameter

2256 km

Opening at South Pole

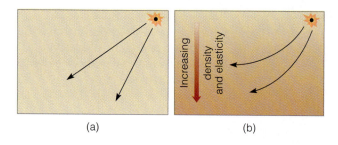

(a)   (b)

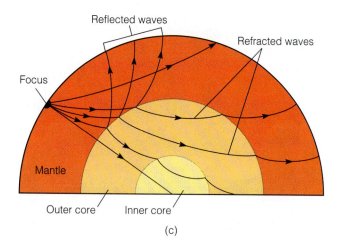

(c)

■ Figure 10.3

(a) If Earth were homogeneous throughout, seismic wave rays would follow straight paths. (b) Because density and elasticity increase with depth, wave rays are continuously refracted so that their paths are curved. (c) Refraction and reflection of P-waves as they encounter boundaries separating materials of different density or elasticity. Notice that the only wave ray not refracted is the one perpendicular to boundaries.

velocity takes place as the waves penetrate to greater depths. P-waves travel faster than S-waves under all circumstances, but unlike P-waves, S-waves are not transmitted through a liquid because liquids have no shear strength (rigidity); liquids simply flow in response to shear stress.

If Earth were a homogeneous body, P- and S-waves would travel in straight paths as shown in ■ Figure 10.3a. But as a seismic wave travels from one material into another of different density and elasticity, its velocity and direction of travel change. That is, the wave is bent, a phenomenon known as **refraction,** in much the same way as light waves are refracted as they pass from air into more dense water. Because seismic waves pass through materials of differing density and elasticity, they are continually refracted so that their paths are curved; wave rays travel in a straight line only when their direction of travel is perpendicular to a boundary (Figure 10.3b, c).

In addition to refraction, seismic waves are **reflected,** much as light is reflected from a mirror. When seismic waves encounter a boundary separating materials of different density or elasticity, some of a wave's energy is *reflected* back to the surface (Figure 10.3c). If we know the wave velocity and the time required for the wave to travel

from its source to the boundary and back to the surface, we can calculate the depth of the reflecting boundary. Such information is useful in determining not only Earth's internal structure but also the depths of sedimentary rocks that may contain petroleum. Seismic reflection is a common tool used in petroleum exploration.

Although changes in seismic wave velocity occur continuously with depth, P-wave velocity increases suddenly at the base of the crust and decreases abruptly at a depth of about 2900 km (■ Figure 10.4). These marked changes in seismic wave velocity indicate a boundary called a **discontinuity** across which a significant change in Earth materials or their properties occurs. These discontinuities are the basis for subdividing Earth's interior into concentric layers.

The contribution of seismology to the study of Earth's interior cannot be overstated. Beginning in the

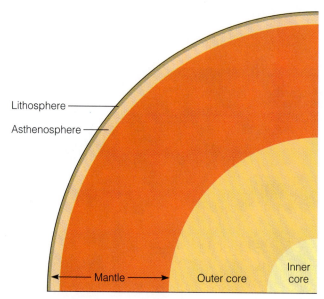

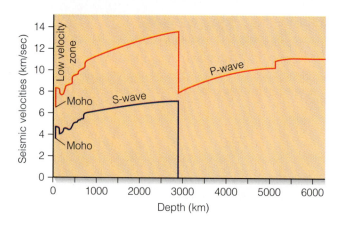

■ Figure 10.4

Profiles showing seismic wave velocities versus depth. Several discontinuities are shown across which seismic wave velocities change rapidly. Source: From G. C. Brown and A. E. Musset, *The Inaccessible Earth.* (Kluwer Academic Publishers, 1981), Figure 12.7a. Reprinted with the kind permission of Kluwer Academic Publishers.

early 1900s, scientists recognized the utility of seismic wave studies and, between 1906 and 1936, largely worked out Earth's internal structure on the basis of these studies.

Seismic Tomography and Earth's Interior

The model of Earth's interior consisting of a core and a mantle is probably accurate, but not very precise. In recent years, geophysicists have developed a technique called **seismic tomography** that allows them to develop more accurate models of Earth's interior. In seismic tomography, numerous crossing seismic waves are analyzed much as CAT (computerized axial tomography) scans are analyzed. In CAT scans, X rays penetrate the body, and a two-dimensional image of its interior is formed. Repeated CAT scans from slightly different angles are stacked to produce a three-dimensional image.

In a similar manner, geophysicists use seismic waves to probe Earth's interior. In seismic tomography, the average velocities of numerous crossing seismic waves are analyzed so that "slow" and "fast" areas of wave travel are detected. Remember that seismic wave velocity depends partly on elasticity; cold rocks have greater elasticity and thus transmit seismic waves faster than hotter rocks.

As a result of studies in seismic tomography, a much clearer picture of Earth's interior is emerging. It has already given us a better understanding of complex convection within the mantle and a clearer picture of the nature of the mantle–core boundary.

EARTH'S CORE

arth's innermost part, its **core**, lies about 2900 km below the surface. Its diameter of 6960 km is about the same as that of Mars and constitutes 16.4% of Earth's volume and nearly one third of its mass (Figure 10.1). Most of our knowledge about the core comes from studies of seismic waves, although meteorites and experiments provide additional information. One contribution made by seismic tomography is that the core–mantle boundary is not a smooth surface as depicted in Figure 10.1. The surface of the core has broad depressions and rises extending several kilometers into the mantle. And, of course, the mantle possesses the same features in reverse, and it now seems that the core's surface is continually deformed by sinking and rising masses of mantle material.

Discovery of Earth's Core

In 1906 R. D. Oldham of the Geological Survey of India realized that seismic waves arrived later than expected at seismic stations more than 130 degrees from an earthquake focus. He postulated that Earth has a core that transmits seismic waves more slowly than shallower Earth materials. We now know that P-wave velocity decreases markedly at a depth of 2900 km, which indicates an important discontinuity now recognized as the core–mantle boundary (Figure 10.4).

Because of the sudden decrease in P-wave velocity at the core–mantle boundary, P-waves are refracted in the core so that little P-wave energy reaches the surface between 103 and 143 degrees from an earthquake focus (■ Figure 10.5a). This **P-wave shadow zone,** as it is called, was discovered in 1914 by the German seismologist Beno Gutenberg. However, the P-wave shadow zone is a not a perfect shadow zone because some weak P-wave energy is recorded within it. Scientists proposed several hypotheses to account for this observation, but all were rejected by the Danish seismologist Inge Lehmann (Figure 10.5b), who in 1936 postulated that the core is not entirely liquid as previously thought. She proposed that seismic wave reflection from a solid inner core accounted for the arrival of weak P-wave energy in the P-wave shadow zone (Figure 10.5c), a proposal that was quickly accepted by seismologists.

In 1926 the British physicist Harold Jeffreys realized that S-waves were not simply slowed by the core but were completely blocked by it. So, besides a P-wave shadow zone, a much larger and more complete **S-wave shadow zone** also exists (■ Figure 10.6). At locations greater than 103 degrees from an earthquake focus, no S-waves are recorded, which indicates that S-waves cannot be transmitted through the core. S-waves will not pass through a liquid, so it seems that the outer core must be liquid or behave as a liquid. The inner core, however, is thought to be solid because P-wave velocity increases at the base of the outer core.

Density and Composition of the Core

We can estimate the core's density and composition by using seismic evidence and laboratory experiments. For instance, geologists use a diamond-anvil pressure cell in which small samples are studied while subjected to pressures and temperatures similar to those in the core. Furthermore, meteorites, which are thought to represent remnants of the material from which the solar system formed, are used to make estimates of density and composition. For example, the irons—meteorites composed of iron and nickel alloys—may represent the differentiated

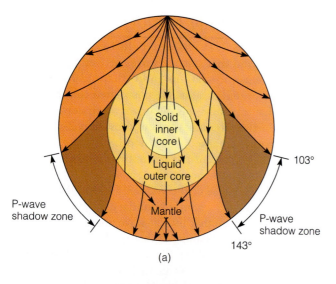

(a)

(b)

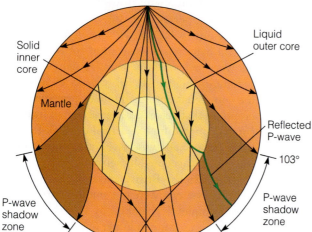

(c)

PHYSICAL
Geology⇌Now ■ **Active Figure 10.5**

(a) P-waves are refracted so that no direct P-wave energy reaches the surface in the P-wave shadow zone. (b) Inge Lehmann, a Danish seismologist who in 1936 postulated that Earth has a solid inner core. (c) According to Lehmann, reflection from an inner core could explain the arrival of weak P-wave energy in the P-wave shadow zone.

interiors of large asteroids and approximate the density and composition of Earth's core. The density of the outer core varies from 9.9 to 12.2 g/cm^3, and that of the inner core ranges from 12.6 to 13.0 g/cm^3 (Figure 10.1). At Earth's center, the pressure is equivalent to about 3.5 million times normal atmospheric pressure.

The core cannot be composed of minerals common at the surface because, even under the tremendous pressures at great depth, they would still not be dense enough to yield an average density of 5.5 g/cm^3 for Earth. Both the outer and inner cores are thought to be composed largely of iron, but pure iron is too dense to be the sole constituent of the outer core. It must be "diluted" with elements of lesser density. Laboratory experiments and comparisons with iron meteorites indicate that perhaps 12% of the outer core consists of sulfur and possibly some silicon, oxygen, nickel, and potassium (Figure 10.1).

In contrast, pure iron is not dense enough to account for the estimated density of the inner core, so perhaps 10 to 20% of the inner core consists of nickel. These metals form an iron–nickel alloy thought to be

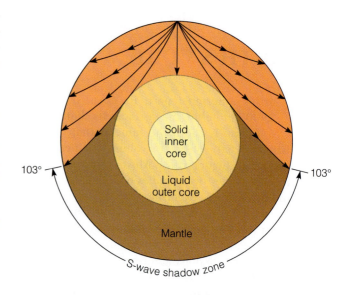

■ **Figure 10.6**

The presence of an S-wave shadow zone indicates that S-waves are being blocked within Earth.

sufficiently dense under the pressure at that depth to account for the density of the inner core.

Any model of the core's composition and physical state must explain not only variations in density but also (1) why the outer core is liquid whereas the inner core is solid and (2) how the magnetic field is generated within the core (discussed later in this chapter). When the core formed during early Earth history, it was probably entirely molten and has since cooled so that its interior has crystallized. Indeed, the inner core continues to grow as Earth slowly cools, and liquid of the outer core crystallizes as iron. Recent evidence also indicates that the inner core rotates faster than the outer core, moving about 20 km/yr relative to the outer core.

The temperature at the core–mantle boundary is estimated at 2500° to 5000°C, yet the high pressure within the inner core prevents melting. In contrast, the outer core is under less pressure, but more important than the differences in pressure are compositional differences between the inner and outer cores. The sulfur content of the outer core helps depress its melting temperature. An iron–sulfur mixture melts at a lower temperature than does pure iron, or an iron–nickel alloy, so despite the high pressure, the outer core is molten.

EARTH'S MANTLE

All of Earth's interior from the surface of the core at a depth of 2900 km to the base of the crust is the **mantle**. It makes up 83.02% of Earth's volume, but because it is less dense than the core, it comprises only 67.77% of Earth's mass. Beginning in 1958, a project called *Mohole* was funded to drill through the oceanic crust and into the mantle. Initially it received considerable support, but it was abandoned in 1966 when funding dried up because of technological problems. Although the mantle has been observed only along some seafloor fractures (see Chapter 11) geologists know some details about its composition and structure.

The Moho

Another significant discovery about Earth's interior was made in 1909 when Yugoslavian seismologist Andrija Mohorovičić detected a discontinuity at a depth of about 30 km. While studying arrival times of P-waves from Balkan earthquakes, Mohorovičić noticed that seismic stations a few hundred kilometers from an earthquake's epicenter were recording two distinct sets of P- and S-waves.

From his observations, Mohorovičić concluded that a sharp boundary separating rocks with different properties exists at a depth of about 30 km. He postulated that P-waves below this boundary travel at 8 km/sec, whereas those above the boundary travel at 6.75 km/sec. When an earthquake occurs, some waves travel directly from the focus to a seismic station, while others travel through the deeper layer and some of their energy is refracted back to the surface (■ Figure 10.7). Waves traveling through the deeper layer travel farther to a seismic station, but they do so more rapidly and arrive before those in the shallower layer. The boundary identified by Mohorovičić separates the crust from the mantle and is now called the **Mohorovičić discontinuity,** or simply the **Moho.** It is present everywhere except beneath spreading ridges, but its depth varies: Beneath the continents, it ranges from 20 to 90 km, with an average of 35 km; beneath the seafloor, it is 5 to 10 km deep.

The Mantle's Structure, Density, and Composition

Although seismic wave velocity in the mantle increases with depth, several discontinuities also exist. Between depths of 100 and 250 km, both P- and S-wave velocities decrease markedly (■ Figure 10.8). This 100- to 250-km-deep layer is the **low-velocity zone,** which corresponds closely to the **asthenosphere,** a layer in which the rocks are close to their melting point and are less elastic, accounting for the observed decrease in seismic wave velocity. The asthenosphere is an important zone because it may be where some magma is generated. Furthermore, it lacks strength, flows plastically, and is thought to be the layer over which the plates of the outer, rigid **lithosphere** move.

Even though the low-velocity zone and the asthenosphere closely correspond, they are still distinct. The asthenosphere appears to be present worldwide, but the low-velocity zone is not. In fact, the low-velocity zone appears to be poorly defined or even absent beneath the ancient shields of continents.

Other discontinuities are also present at deeper levels within the mantle. But unlike

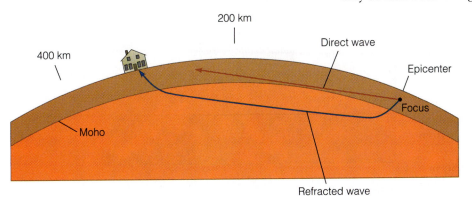

■ **Figure 10.7**

Andrija Mohorovičić studied seismic waves and detected a seismic discontinuity at a depth of about 30 km. The deeper, faster seismic waves arrive at seismic stations first, even though they travel farther. This discontinuity, now known as the Moho, is between the crust and mantle.

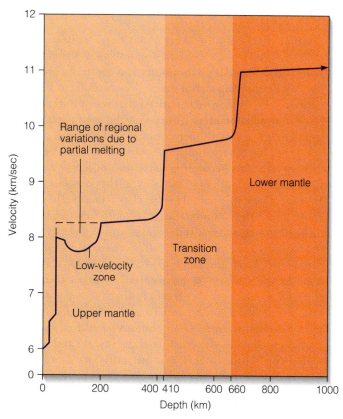

■ Figure 10.8

Variations in P-wave velocity in the upper mantle and transition zone. Source: From G. C. Brown and A. E. Musset, *The Inaccessible Earth* (Kluwer Academic Publishers, 1981), Figure 7.11. Reprinted with the kind permission of Kluwer Academic Publishers.

What Would You Do

Of course, novels such as *Journey to the Center of the Earth* are fiction, but it is surprising how many people think that vast caverns and cavities exist deep within the planet. How would you explain that even though we have no direct observations at great depth, we can still be sure that these proposed openings do not exist?

tle react with liquid from the outer core, thus forming a vertically and laterally heterogeneous layer. Some geologists think that mantle plumes originate in the D″ layer.

Although the mantle's density, which varies from 3.3 to 5.7 g/cm³, can be inferred rather accurately from seismic waves, its composition is less certain. The igneous rock *peridotite* is considered the most likely component. Peridotite contains mostly ferromagnesian silicates (60% olivine and 30% pyroxene) and about 10% feldspars (see Figures 3.9 and 3.10). Peridotite is considered the most likely candidate for three reasons. First, laboratory experiments indicate that it possesses physical properties that account for the mantle's density and observed rates of seismic wave transmissions. Second, peridotite forms the lower parts of igneous rock sequences thought to be fragments of the oceanic crust and upper mantle, called *ophiolites*, emplaced on land (see Chapter 11). And third, peridotite is found as inclusions in volcanic rock bodies such as *kimberlite pipes* that are known to have come from depths of 100 to 300 km. These inclusions appear to be pieces of the mantle (■ Figure 10.9).

those between the crust and mantle or between the mantle and core, these probably represent structural changes in minerals rather than compositional changes. In other words, geologists think the mantle is composed of the same material throughout, but the structural states of minerals such as olivine change with depth. At a depth of 410 km, seismic wave velocity increases slightly as a consequence of such changes in mineral structure (Figure 10.8). Another velocity increase occurs at about 660 km, where the minerals break down into metal oxides, such as FeO (iron oxide) and MgO (magnesium oxide), and silicon dioxide (SiO_2). These two discontinuities define the top and base of a *transition zone* separating the upper mantle from the lower mantle (Figure 10.8).

A decrease in seismic wave velocity in a zone extending 200 to 300 km above the core–mantle boundary is recognized as the D″ layer. Although commonly included within the lower mantle, the D″ layer might be considerably different in composition. Experiments indicate that silicates in the man-

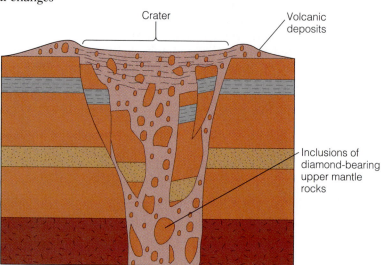

■ Figure 10.9

Generalized cross section of a kimberlite pipe. The rock in kimberlite pipes is dark gray or blue igneous rock known as kimberlite. Inclusions of probable mantle rock from depths of 100 to 300 km are found in these igneous bodies. Kimberlite is also the source of diamonds. Most kimberlite pipes measure less than 500 m across at the surface.

GEOLOGY
IN UNEXPECTED PLACES

Diamonds and Earth's Interior

Diamond—a jewel, a shape, a tool. Chances are you or someone you know owns a diamond ring, necklace, or bracelet, because diamond, which symbolizes strength and purity, is the most popular and sought-after gemstone. The value of gem-quality diamonds is determined by their color (colorless ones are most desirable), clarity (lack of flaws), and carat (1 carat = 200 milligrams). A diamond's value also depends on the cut, the way it is cleaved, cut, and polished to yield small plane surfaces known as *facets*, which enhance the quality of reflected light (■ Figure 1 and see Figure 2.1b).

Most of the world's diamonds come from stream and beach placer deposits, but the ultimate source of most gem-quality diamonds and industrial diamonds is kimberlite pipes (see Figure 10.9) composed of dark gray or blue igneous rock that originated at great depths. In fact, diamond is composed of carbon that forms at pressures found at least 100 km deep—that is, in Earth's mantle. Diamond establishes a minimum depth for the magma that cools to form kimberlite, a form of silica also in these rocks that indicates that the magma originated between 100 and 300 km below the surface. In addition to diamonds and silica, kimberlite commonly contains inclusions of peridotite (see Figure 3.10) that are most likely pieces of the mantle.

Because diamond is so hard (see Table 2.3) it is used for abrasives and for tools that cut other hard substances such as other gemstones, eyeglasses, and even computer chips. In road construction, diamonds are used to grind down old pavement before a new layer of blacktop is poured. Even petroleum companies use diamond-studded drill bits.

Peter Kaskons/Index Stock

■ **Figure 1**

A diamond's value is determined by color, clarity, carat, and cut, the four C's.

EARTH'S OUTERMOST PART

Earth's **crust** is the most accessible and best studied of its concentric layers, but it is also the most complex both chemically and physically. Whereas the core and mantle seem to vary mostly in a vertical dimension, the crust shows considerable vertical and lateral variation. (More lateral variation exists in the mantle than was once thought.) The crust along with that part of the upper mantle above the low-velocity zone constitutes the *lithosphere* of plate tectonic theory.

Continental Crust

The two types of crust—continental crust and oceanic crust—are both less dense than the underlying mantle.

Continental crust is the more complex, consisting of a wide variety of igneous, sedimentary, and metamorphic rocks (Figure 10.1). It is generally described as "granitic," meaning that its overall composition is similar to that of granitic rocks. Specifically, its overall composition corresponds closely to that of granodiorite, an igneous rock that has a chemical composition between granite and diorite (see Figure 3.9).

Continental crust varies in density depending on rock type, but with the exception of metal-rich rocks, such as iron ore deposits, most rocks have densities of 2.0 to 3.0 g/cm^3, and the overall density is about 2.70 g/cm^3. P-wave velocity in the continental crust is about 6.75 km/sec; at the base of the crust, P-wave velocity abruptly increases to about 8 km/sec.

Continental crust averages 35 km thick but is much thinner in such areas as the Rift Valleys of East Africa and a large area called the Basin and Range Province in the western United States. The crust in these areas is being stretched and thinned in what appear to be the early stages of rifting and is as thin as 20 km. In contrast, continental crust beneath mountain ranges is much thicker, as much as 90 km, and projects deep into the mantle. Crustal thickening beneath mountain ranges is an important point that will be discussed in the section on isostasy later in the chapter.

Oceanic Crust

Although variations also occur in **oceanic crust,** they are not as distinct as those for the continental crust. Oceanic crust varies from 5 to 10 km thick, being thinnest at spreading ridges (Figure 10.1). It is denser than continental crust, averaging 3.0 g/cm^3, and transmits P-waves at about 7 km/sec. Just as beneath the continental crust, P-wave velocity increases at the Moho. The P-wave velocity of oceanic crust is what one would expect if it were composed of basalt. Direct observations of oceanic crust from submersibles and deep-sea drilling confirm that its upper part is indeed composed of basalt. The lower part of the oceanic crust is composed of gabbro, the intrusive equivalent of basalt. (See Chapter 11 for a more detailed description of the oceanic crust.)

EARTH'S INTERNAL HEAT AND HEAT FLOW

During the 19th century, scientists realized that the temperature in deep mines increases with depth. Indeed, very deep mines must be air-conditioned so that the miners can survive. More recently, the same trend has been observed in deep drill holes. This temperature increase with depth, or **geothermal gradient,** near the surface is about 25°C/km. In areas of active or recently active volcanism, the geothermal gradient is greater than in adjacent nonvolcanic areas, and temperature rises faster beneath spreading ridges than elsewhere beneath the seafloor.

Most of Earth's internal heat is generated by radioactive decay, especially the decay of isotopes of uranium and thorium and to a lesser degree of potassium 40. When these isotopes decay, they emit energetic particles and gamma rays that heat surrounding rocks. Because rock is such a poor conductor of heat, it takes little radioactive decay to build up considerable heat, given enough time.

Unfortunately, the geothermal gradient is not useful for estimating temperatures at great depth. If we were simply to extrapolate from the surface downward, the temperature at 100 km would be so high that, despite the great pressure, all known rocks would melt. Yet except for pockets of magma, it appears that the mantle is solid rather than liquid because it transmits S-waves. Accordingly, the geothermal gradient must decrease markedly.

Current estimates of the temperature at the base of the crust are 800° to 1200°C. The latter figure seems to be an upper limit: If it were any higher, melting would be expected. Furthermore, fragments of mantle rock in kimberlite pipes, thought to have come from depths of 100 to 300 km, appear to have reached equilibrium at these depths at a temperature of about 1200°C. At the core–mantle boundary, the temperature is probably between 2500° and 5000°C; the wide range of values indicates the uncertainties of such estimates. If these figures are reasonably accurate, the geothermal gradient in the mantle is only about 1°C/km.

Because the core is so remote and its composition uncertain, only very general estimates of its temperature are possible. Based on various experiments, the maximum temperature at the center of the core is estimated to be 6500°C, very close to the estimated temperature for the surface of the Sun!

Even though rocks are poor conductors of heat, detectable amounts of heat from Earth's interior escape at the surface by **heat flow.** The amount of heat lost is small and can be detected only by sensitive instruments. Heavy, cylindrical probes are dropped into soft seafloor sediments, and temperatures are measured at various depths along the cylinder. On the continents, temperature measurements are made at various depths in drill holes and mines.

As one would expect, heat flow is greater in areas of active or recently active volcanism. For instance, greater heat flow occurs at spreading ridges, and lower than average values are recorded at subduction zones (■ Figure 10.10). Any area that has higher than average heat-flow values is a potential area for the development of geothermal energy (see Chapter 16).

More than 70% of the total heat lost by Earth is lost through the seafloor, but heat-flow values for both oceanic basins and continents decrease with increasing age. In the ocean basins, heat flow is higher through younger oceanic crust. This result is expected because high heat-flow values occur at spreading ridges where oceanic crust is continuously formed by igneous activity. Spreading ridges are also the sites of hydrothermal vents where considerable heat is transported upward by hydrothermal convection. Heat flow through the continental crust is not as well understood, but it too shows lower values for older crust.

It should be apparent that if heat is escaping from within Earth, the interior should be cooling unless a mechanism exists to replenish it. Radioactive decay generates heat continuously, but the quantity of radioactive isotopes (except carbon 14) decreases with time as they decay to stable daughter products. Accordingly, during its early history Earth possessed more internal heat and has been cooling continuously since then.

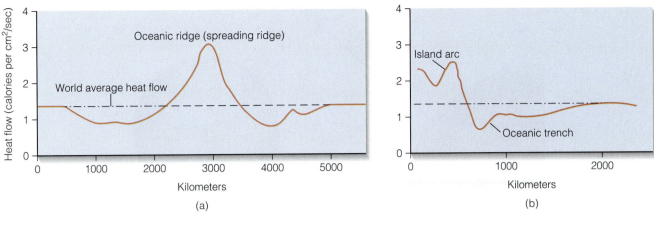

■ **Figure 10.10**

Variations in heat flow. (a) Higher than average heat flow occurs at spreading ridges and island arcs, both of which are characterized by volcanism. (b) Oceanic trenches show lower than average heat flow.

WHAT IS GRAVITY, AND HOW IS ITS FORCE DETERMINED?

Sir Isaac Newton (1642–1727) devised the *law of universal gravitation*, in which the force of gravitational attraction (*F*) between two masses (m_1 and m_2) is directly proportional to the products of the two masses and inversely proportional to the distance (*D*) between their centers of mass:

$$F = G \frac{m_1 \times m_2}{D^2}$$

G in this equation is the universal gravitational constant. Accordingly, **gravity** is the attractive force that exists between any two bodies. Our main concern here is with the attraction between any two bodies on Earth, but the gravitational attraction on Earth's surface waters by the Sun and the Moon, which is responsible for tides, is also important (see Chapter 19).

The equation tells us that the gravitational force (*F*) is greater between two massive bodies—the Moon and Earth, for instance—if the distance between the centers of mass remain the same (■ Figure 10.11a). But because (*F*) is inversely proportional to the square of distance, it decreases by a factor of 4 when the distance is doubled (see Geo-Focus 10.1). We generally refer to the gravitational force between an object and Earth as its *weight*.

Gravity

Gravitational attraction would be the same everywhere on the surface if Earth were perfectly spherical, homogeneous throughout, and not rotating. As a consequence of rotation, however, a centrifugal force is generated that partly counteracts the force of gravity (Figure 10.11b), so an object at the equator weighs slightly less than the same object would at the poles. The force of gravity also varies with distance between the centers of masses, so an object would weigh slightly less above the surface than if it were at sea level (Figure 10.11a).

Gravity Anomalies

Geologists use a sensitive instrument called a *gravimeter* to measure variations in the force of gravity. A gravimeter is simple in principle; it contains a weight suspended on a spring that responds to variations in gravity (■ Figure 10.12). Gravimeters are used extensively in exploration for hydrocarbons and mineral resources. Long ago, geologists realized that anomalous gravity values should exist over buried bodies of ore minerals and salt domes and that geologic structures such as faulted strata could be located by surface gravity surveys (Figure 10.12).

Gravity measurements are higher over an iron ore deposit than over unconsolidated sediment because of the ore's greater density (Figure 10.12a). Such departures from the expected force of gravity are **gravity anomalies.** In other words, the measurement over the body of iron ore indicates an excess of dense material, or simply a *mass excess*, between the surface and the center of Earth and is considered a **positive gravity anomaly.** A **negative gravity anomaly** indicating *mass deficiency* exists over low-density sediments because the force of gravity is less than expected (Figure 10.12b). Large negative gravity anomalies also exist over salt domes (Figure 10.12c) and at subduction zones, indicating that the crust is not in equilibrium.

Departures from Earth's expected gravitational attraction (gravity anomalies) certainly exist, but what of the tourist sites around the United States that claim

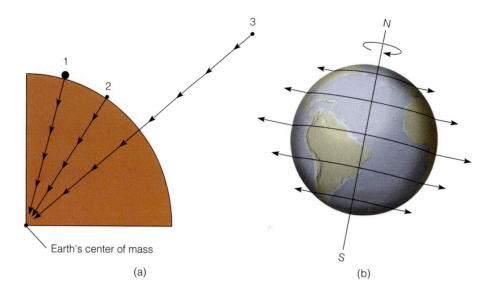

Earth's center of mass

(a)   (b)

■ Figure 10.11

(a) Earth's gravitational attraction pulls all objects toward its center of mass. Objects 1 and 2 are the same distance from Earth's center of mass, but the gravitational attraction on 1 is greater because it is more massive. Objects 2 and 3 have the same mass, but the gravitational attraction on 3 is four times less than on 2 because it is twice as far from Earth's center of mass. (b) Earth's rotation generates a centrifugal force that partly counteracts the force of gravity. Centrifugal force is zero at the poles and maximum at the equator.

gravity has somehow gone awry? All kinds of mysterious gravity-defying effects are claimed to occur in these areas, including objects rolling uphill, unsupported objects clinging to walls, and the famous plank illusion in which the heights of people on a level plank change when they switch positions. All are actually clever optical illusions that can be duplicated by anyone with the interest in doing so.

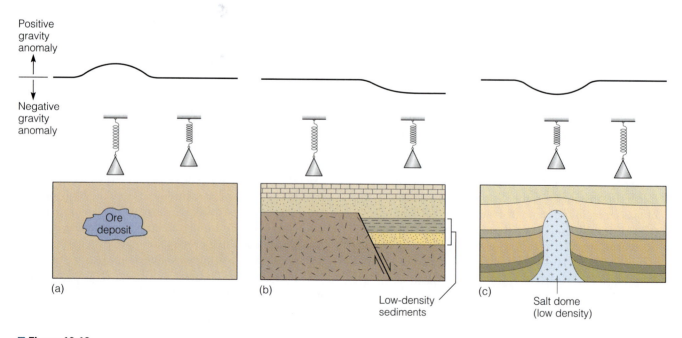

■ Figure 10.12

(a) The mass suspended from a spring in the gravimeter, shown diagrammatically, is pulled downward more over the dense body of ore than it is in adjacent areas, indicating a positive gravity anomaly. (b) A negative gravity anomaly over a buried structure. (c) Rock salt is less dense than most other types of rocks. A gravity survey over a salt dome shows a negative gravity anomaly.

FLOATING CONTINENTS?—THE PRINCIPLE OF ISOSTASY

More than 150 years ago, British surveyors in India detected a discrepancy of 177 m when they compared the results of two measurements between points 600 km apart. Even though this discrepancy was small, only about 0.03%, it was an unacceptably large error. The surveyors realized that the gravitational attraction of the nearby Himalaya Mountains probably deflected the plumb line (a cord with a suspended weight) of their surveying instruments from the vertical, thus accounting for the error. Calculations revealed, however, that if the Himalayas were simply thicker crust piled on denser material, the error should have been greater than that observed (■ Figure 10.13a).

In 1865 George Airy proposed that, in addition to projecting high above sea level, the Himalayas—and other mountains as well—project far below the surface and thus have a low-density root (Figure 10.13b). In effect, he was saying that mountains float on denser rock at depth. Their excess mass above sea level is compensated for by a mass deficiency at depth, which would account for the observed deflection of the plumb line during the British survey (Figure 10.13).

Another explanation was proposed by J. H. Pratt, who thought that the Himalayas were high because they were composed of rocks of lesser density than those in adjacent regions. Although Airy was correct with respect to the Himalayas, and mountains in general, Pratt was correct in that there are indeed places where the crust's elevation is related to its density. For example, (1) continental crust is thick and less dense than oceanic crust and thus stands high, and (2) the mid-oceanic ridges stand higher than adjacent areas because the crust there is hot and less dense than cooler oceanic crust elsewhere.

Gravity studies have revealed that mountains do indeed have a low-density "root" projecting deep into the mantle. If it were not for this low-density root, a gravity survey across a mountainous area would reveal a huge positive gravity anomaly. The fact that no such anomaly exists indicates that a mass excess is not present, so some of the dense mantle at depth must be displaced by lighter crustal rocks, as shown in Figure 10.13b. (Seismic wave studies also confirm the existence of low-density roots beneath mountains.)

Both Airy and Pratt agreed that Earth's crust is in floating equilibrium with the more dense mantle below, and now their proposal is known as the **principle of isostasy**. This phenomenon is easy to understand by analogy to an iceberg (■ Figure 10.14). Ice is slightly less dense than water, and thus it floats. According to Archimedes's principle of buoyancy, an iceberg sinks in water until it displaces a volume of water that equals its total weight. When the iceberg has sunk to an equilibrium position, only about 10% of its volume is above water level. If some of the ice above water level should

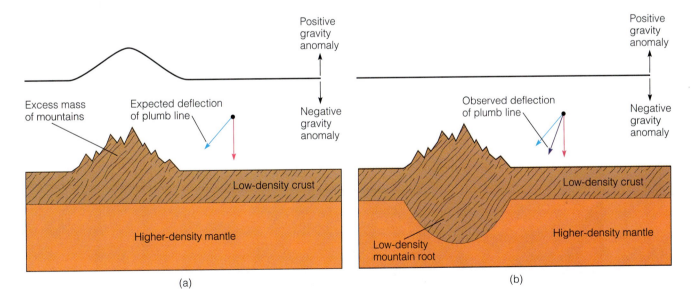

■ Figure 10.13

(a) A plumb line is normally vertical, pointing to the Earth's center of gravity. Near a mountain range, the plumb line should be deflected as shown if the mountains are simply thicker, low-density material resting on denser material, and a gravity survey across the mountains would indicate a positive gravity anomaly. (b) The actual deflection of the plumb line during the survey in India was less than expected. It was explained by postulating that the Himalayas have a low-density root. A gravity survey in this case would show no anomaly because the mass of the mountains above the surface is compensated for at depth by low-density material displacing denser material.

GEOFOCUS

10.1

Planetary Alignments, Gravity, and Catastrophes

Many of you probably recall the dire predictions for May 5, 2000, that appeared in tabloids, sensational books, at least one television show, and various radio talk shows. On and about this date the five visible planets, Mercury, Venus, Mars, Jupiter, and Saturn, plus the Sun and Moon lined up on one side of Earth in a Grand Alignment, as it was called by some. One author claimed that the planets, Sun, and Moon would be in nearly a straight line for the first time in 6000 years.

Actually, this was not the first alignment in 6000 years; 71 alignments at least as notable as the one on May 5, 2000, have taken place during the last 8200 years. Furthermore, in this most recent Grand Alignment, the planets, Sun, and Moon were not nearly in a straight line. They were only roughly aligned, being spread over about 25 degrees of space (■ Figure 1). Because of the greater gravitational attraction on Earth during this alignment, several people predicted more earthquakes, perhaps of catastrophic proportions, sea level rises of up to 90 m, huge shifts in Earth's crust, and winds as strong as 3200 km/hr. Of course, May 5, 2000, passed without any of the predicted events taking place.

This was not the first time such predictions had been made. In 1974 two astronomers published *The Jupiter Effect,* in which they predicted that increased gravitational attraction during a planetary alignment in 1982 would trigger earthquakes. According to their view, tidal forces resulting from a planetary alignment would cause solar flares, and streams of solar particles entering Earth's atmosphere would bring about weather changes, slow Earth's rate of rotation, and cause earthquakes. Actually, *The Jupiter Effect* concentrated on California's San Andreas fault while not mentioning other earthquake-prone areas, probably because ". . . in California better than any other place on Earth, one finds a fear of earthquakes combined with a proven market for sensational books."*

During the 1982 planetary alignment, the planets were not precisely aligned but rather spread over about 60 degrees of space. There were indeed earthquakes in 1982 and some areas experienced

■ **Figure 10.14**

An iceberg sinks to an equilibrium position with about 10% of its mass above water level. The larger iceberg sinks farther below and rises higher above the water surface than does the smaller one. If some of the ice above water level should melt, the icebergs will rise to maintain the same proportion of ice above and below water level. Earth's crust floating in more dense material below is analogous to this example.

unusual weather, but earthquakes and unusual weather occur somewhere on Earth every year. In any event, 1982 came and went with no noticeable change in earthquake frequency, unusual weather, or slowing of Earth's rotation.

Intuitively it might seem that planetary alignments would cause increased gravitational attraction and could have a negative effect on Earth. So why don't they? The answer is related to the equation for the law of universal gravitation. Remember the denominator in the equation tells us that the gravitational attraction between any two bodies decreases by a factor of 4 when the distance between them is doubled (see Figure 10.11a). So even when planets roughly align on one side of Earth, their combined effect is still so small that they have a minimal impact on our planet. Indeed, even Jupiter, the giant among planets, when in this configuration has only about 1/500,000 of the tidal force generated by the Moon. In short, scientists have never found any relationship between planetary alignments and changes in Earth's surface or internal processes.

When will the next alignment take place? The six inner planets align every 50 to 100 years, but a Grand Alignment similar to the

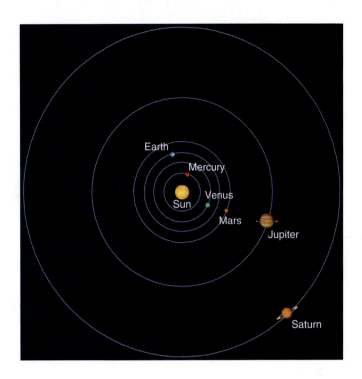

■ **Figure 1**

The Grand Alignment of May 2000 as seen from deep space far above the plane of the solar system. Earth's moon is not shown, but it too was part of this alignment. The alignment was a remarkable event, but the planets, Moon, and Sun were not in a straight line as some people claimed.

one of May 5, 2000, will not occur until 2438. Still, predictions of catastrophes are popular and will no doubt be made again prior to the next planetary alignment, be it a minor alignment or a Grand Alignment.

*J. Mosley, *Planetary Alignments in 2000.* http://www.griffithobs.org/SkyAlignments. html

---

melt, the iceberg rises in order to maintain the same proportion of ice above and below water (Figure 10.14).

Earth's crust is similar to the iceberg, or a ship, in that it sinks into the mantle to its equilibrium level. Where the crust is thickest, as beneath mountain ranges, it sinks farther down into the mantle but also rises higher above the equilibrium surface (Figure 10.13b). Continental crust being thicker and less dense than oceanic crust stands higher than the ocean basins. Earth's crust responds isostatically to widespread erosion and sediment deposition (■ Figure 10.15). And it also responds to loading when vast glaciers form and depress the crust into the mantle to maintain equilibrium (■ Figure 10.16). In Greenland and Antarctica, the crust has been depressed below sea level by the weight of glacial ice.

Unloading of the crust causes it to respond by rising upward until equilibrium is again attained. This phenomenon, known as **isostatic rebound,** takes place in areas that are deeply eroded and in areas that were formerly covered by vast glaciers. Scandinavia, which was covered by an ice sheet until about 10,000 years ago, is still rebounding at a rate of up to 1 m per century (■ Figure 10.17). Coastal cities in Scandinavia have been uplifted rapidly enough that docks constructed several centuries ago are now far from shore. Isostatic rebound has also occurred in eastern Canada, where the land has risen as much as 100 m during the last 6000 years.

If the principle of isostasy is correct, it implies that the mantle behaves like a liquid. In preceding discussions, however, we said that the mantle must be solid because it transmits S-waves, which will not move through a liquid. How can this apparent paradox be resolved? When considered in terms of the short time necessary for S-waves to pass through it, the mantle is indeed solid. But when subjected to stress over long periods, it

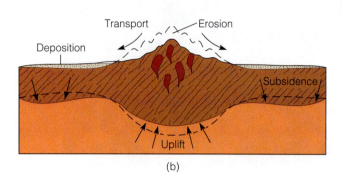

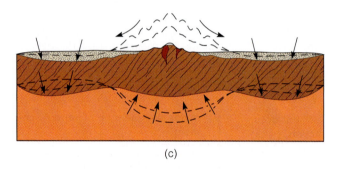

(c)

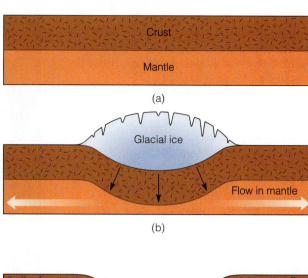

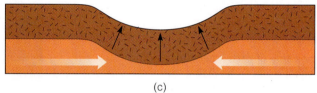

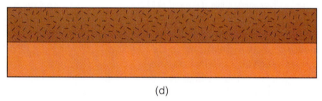

■ **Figure 10.16**

A diagrammatic representation of the response of Earth's crust to the added weight of glacial ice. (a) The crust and mantle before glaciation. (b) The weight of glacial ice depresses the crust into the mantle. (c) When the glacier melts, isostatic rebound begins, and the crust rises to its former position. (d) Isostatic rebound is complete.

PHYSICAL
Geology⇌Now ■ **Active Figure 10.15**

A diagrammatic representation showing the isostatic response of the crust to erosion (unloading) and widespread deposition (loading).

will yield by flowage and at these time scales is a viscous liquid. Silly Putty, a familiar substance that has the properties of a solid or a liquid depending on how rapidly de-

forming forces are applied, will flow under its own weight if given enough time, but shatters as a brittle solid if struck a sharp blow.

EARTH'S MAGNETIC FIELD

Although we cannot see gravity, we certainly deal with its effects, and we use the law of universal gravitation in many endeavors. Likewise, magnetism is not visible but we can detect it, measure its intensity, and even study it in ancient rocks. We already mentioned that *gravity* is the attractive force between any two masses, but how does it differ from magnetism? After all, a magnetic substance is attracted to another magnetic substance. In fact, several cen-

What Would You Do

While teaching a high school science class, you mention that Earth's crust behaves as if floating in the denser mantle below. It's obvious that your students are having difficulties grasping the concept. After all, how can a solid float in a solid? Can you think of any analogies that might help them understand? Also, what kinds of experiments might you devise to demonstrate that both crustal composition and thickness play a role in isostasy?

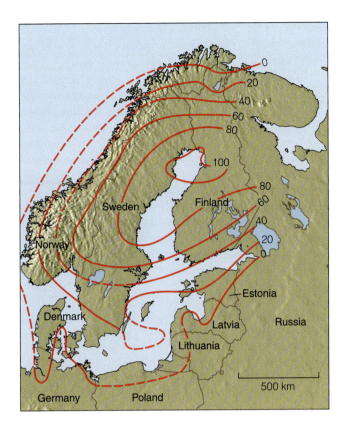

■ Figure 10.17

Isostatic rebound in Scandinavia. The lines show rates of uplift in centimeters per century. Source: From Beno Gutenburg, *Physics of the Earth's Interior* (Orlando, Florida: Academic Press, 1959), 194, Figure 9.1.

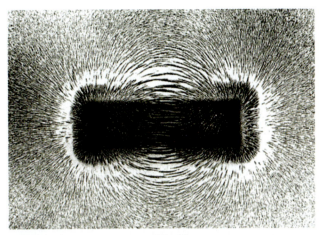

(a)

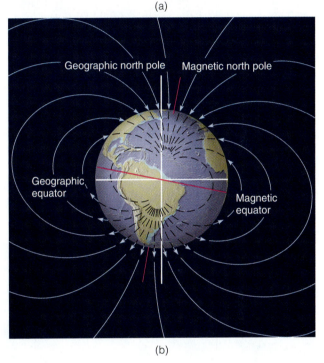

(b)

■ Figure 10.18

(a) Iron filings align along the lines of magnetic force emanating from a bar magnet. (b) Earth's magnetic field has lines of force like those of a bar magnet.

turies ago no distinction was made between these phenomena, but they are in fact different. **Magnetism** is a physical phenomenon resulting from moving electricity (it does not depend on mass as gravity does).

A simple bar magnet has a **magnetic field**, an area in which magnetic substances are affected by lines of magnetic force emanating from the magnet (■ Figure 10.18a). The magnetic field in Figure 10.18a is *dipolar*, meaning that it possesses two unlike magnetic poles referred to as the north and south poles. Earth possesses a dipolar magnetic field that resembles, on a large scale, that of a bar magnet (Figure 10.18b).

Humans have been aware of Earth's magnetic field for centuries, but only rather recently have we had any knowledge of how it is generated, even though many aspects of it are still poorly understood. We can be sure that it does not emanate from a body of deeply buried magnetic materials such as the mineral magnetite, because magnetic substances lose their magnetism when heated above a temperature known as the **Curie point.** The Curie point for magnetite is 580°C, which is far below its melting temperature. At a depth of 80 to 100 km, the temperature is high enough that magnetic substances lose their magnetism. The fact that the locations of the magnetic poles vary through time also indicates that buried magnetite is not the source of the magnetic field.

Instead, the magnetic field appears to be generated in the liquid outer core by electrical currents (an electrical current is a flow of electrons that always generates a magnetic field). Experts on magnetism do not fully understand how the magnetic field is generated, but most agree that it is continuously generated; otherwise, it would decay and Earth would have no magnetic field in as little as 20,000 years. The model most widely accepted now is that thermal and compositional convection within the liquid outer core coupled with Earth's rotation produce complex electrical currents or a *self-exciting dynamo* that in turn generates the magnetic field.

Inclination and Declination of the Magnetic Field

Notice in Figure 10.18b that the lines of magnetic force around Earth parallel its surface only near the equator. As these lines approach the poles, they are oriented at increasingly large angles with respect to the surface, and the strength of the magnetic field increases; it is weakest at the equator and strongest at the poles. Accordingly, a compass needle mounted so that it can rotate both horizontally and vertically not only points north but also is inclined with respect to the surface, except at the magnetic equator. The degree of inclination depends on the needle's location along a line of magnetic force (■ Figure 10.19).

This deviation of the magnetic field from the horizontal is **magnetic inclination.** To compensate for inclination, compasses used in the Northern Hemisphere have a small weight on the south end of the needle. This property of the magnetic field is important in determining the ancient geographic positions of tectonic plates (see Chapter 12).

Another important aspect of the magnetic field is that the magnetic poles, where the lines of force leave and enter Earth, do not coincide with the geographic (rotational) poles. At present, an $11\frac{1}{2}$-degree angle exists between the two (Figure 10.19). Studies of the magnetic field show that the locations of the magnetic poles vary slightly over time but still correspond closely, on the average, with the locations of the geographic poles.

A compass points to the north magnetic pole in the Canadian Arctic islands, some 1290 km away from the geographic pole (true north); only along the line shown in Figure 10.20 does a compass needle point to both the magnetic and geographic north poles. From any other location, an angle called **magnetic declination** exists between lines drawn from the compass position to the magnetic pole and the geographic pole (■ Figure 10.20). Magnetic declination, which in some locations is as much as 30 degrees, must be taken into account during surveying and navigation because, for most places on Earth, compass needles point east or west of true north.

Magnetic Anomalies

Variations in the normal strength of the magnetic field occur on both regional and local scales, giving rise to **magnetic anomalies.** Regional variations are most likely related to the complexities of convection within the outer core where the magnetic field is generated. Local variations result from lateral or vertical variations in rock types within the crust.

An instrument called a *magnetometer* detects slight variations in the strength of the magnetic field, and deviations from the normal are characterized as positive or negative. A **positive magnetic anomaly** exists in areas where the rocks have more iron-bearing minerals than elsewhere. In the Great Lakes region of the United States and Canada, huge iron ore deposits containing hematite and magnetite add their magnetism to that of the magnetic field, resulting in a positive magnetic anomaly (■ Figure 10.21a). Positive magnetic anomalies also exist where extensive basaltic volcanism has taken place because basalt contains appreciable quantities of iron-bearing minerals (Figure 10.21b). Areas underlain by basalt lava flows, such as the Columbia River basalts of the northwestern United States (see Figure 4.18), possess positive magnetic anomalies, whereas adjacent areas underlain by sedimentary rocks show **negative magnetic anomalies** (Figure 10.21b).

Geologists have used magnetometers for magnetic sur-

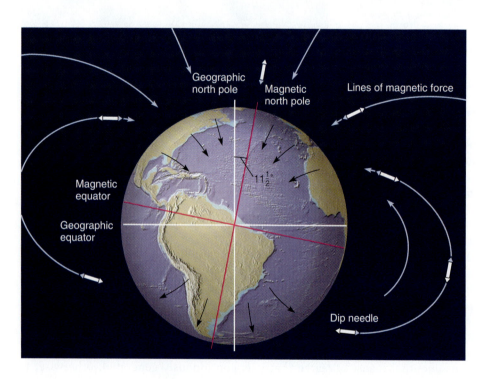

■ **Figure 10.19**

Magnetic inclination. The strength of the magnetic field changes uniformly from the magnetic equator to the magnetic poles. This change in strength causes a dip needle to parallel Earth's surface only at the magnetic equator, whereas its inclination with respect to the surface increases to 90 degrees at the magnetic poles. Notice the 11½-degree angle between the geographic and magnetic poles.

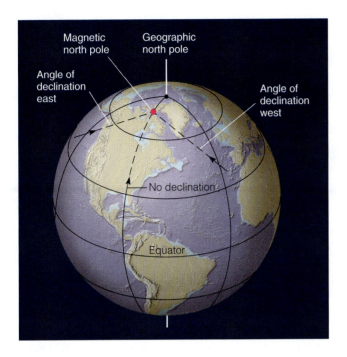

■ Figure 10.20

Magnetic declination. A compass needle points to the magnetic north pole rather than to the geographic pole (true north). The angle formed by the lines from the compass position to the two poles is the magnetic declination.

veys for decades because they can detect iron-bearing rocks by a positive magnetic anomaly even if they are deeply buried. In addition, magnetometers detect a variety of buried geologic structures, such as salt domes, which show negative magnetic anomalies (Figure 10.21c); these can be detected by gravity surveys as well.

Magnetic Reversals

As lava cools through the Curie point, its iron-bearing minerals gain their magnetism and align with the magnetic field, recording both its direction and strength. If the rock is not heated above the Curie point, it retains that magnetic record; if it is heated above the Curie point, it loses its original magnetism. Iron-bearing minerals in some sedimentary rocks, especially those deposited on the deep seafloor, also preserve a record of the magnetic field at the time of their origin. Geologists use the magnetic data preserved in lava flows and some sedimentary rocks to determine ancient directions to the magnetic poles and the latitude of the rocks when they formed.

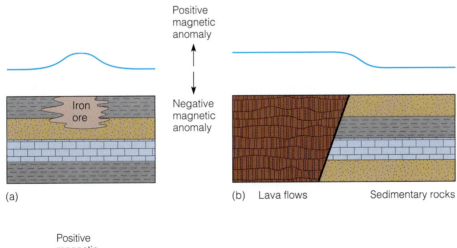

(a)   (b)   Lava flows   Sedimentary rocks

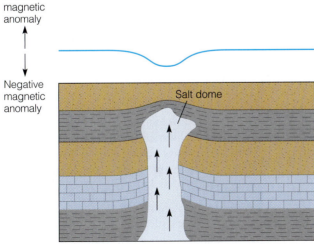

(c)

■ Figure 10.21

Positive and negative magnetic anomalies. (a) Positive magnetic anomaly over an iron ore deposit. (b) Positive magnetic anomaly over lava flows and negative magnetic anomaly over adjacent sedimentary rocks. (c) Negative magnetic anomaly over a salt dome.

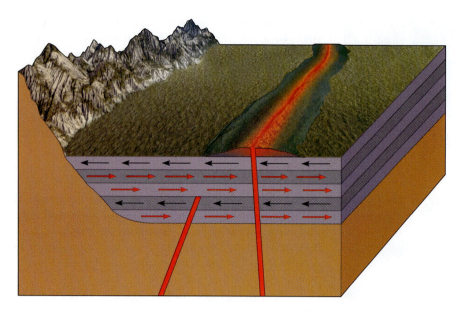

■ **Figure 10.22**

Magnetic reversals recorded in a succession of lava flows are shown diagrammatically by red arrows, and the record of normal polarity events is shown by black arrows.

The study of the ancient magnetic field, or **paleomagnetism,** relies on this remnant magnetism preserved in rocks. Geologists refer to the present magnetic field as *normal*—that is, with the north and south magnetic poles located near the north and south geographic poles. But as early as 1906, rocks were discovered showing reversed magnetism, meaning the north and south magnetic poles were switched. Studies initially conducted on lava flows on the continents clearly showed that these **magnetic reversals** have occurred many times during the past (■ Figure 10.22).

Rocks with a magnetic record the same as the present magnetic field have *normal polarity,* whereas those with opposite magnetism have *reversed polarity.* These same patterns of normal and reversed polarity first discovered in lava flows on land were subsequently discovered in oceanic crust as well (see Chapter 12). Although reversals appear to be related to changes in the intensity of the magnetic field, their cause is not completely understood. Calculations indicate that the magnetic field weakened by about 5% during the 19th century. If this trend continues, there will be a period during the next few thousand years when the magnetic field will be nonexistent and then will reverse. After a reversal, the magnetic field will rebuild itself with opposite polarity.

10   REVIEW WORKBOOK

Chapter Summary

- Earth has an outer layer of oceanic and continental crust below which lies a rocky mantle and an iron-rich core with a solid inner part and a liquid outer part.

- Studies of P- and S-waves, laboratory experiments, comparisons with meteorites, and studies of inclusions in volcanic rock provide evidence about the composition and structure of Earth's interior.

- Seismic wave velocity depends on the density and elasticity of the rocks the waves travel through. Changes in wave velocity at discontinuities within Earth establish the depth to the mantle and the core.

- Seismic waves are refracted as their direction of travel changes with depth, and they are also reflected when some of their energy returns to the surface.

- The inner core is probably made up of iron and nickel, whereas the outer core is mostly iron with 10–20% other substances.

- Peridotite is the most likely rock making up Earth's mantle.

- Oceanic crust is composed of basalt and gabbro, while continental crust has an overall composition similar to granite. The Moho is the boundary between the crust and the mantle.

- The geothermal gradient of 25°C/km cannot continue to great depth; within the mantle and core it is probably about 1°C/km.

- Detectable amounts of heat escape at the surface by heat flow. Most of Earth's internal heat comes from radioactive decay, but some comes from the molten core.

- The principle of isostasy holds that the less dense crust "floats" in the more dense mantle below. Continental crust stands high because it is thicker and less dense than oceanic crust.

- Positive and negative gravity anomalies occur where mass excesses and deficiencies occur. Gravity surveys are useful in exploration for minerals and hydrocarbons.

- The lines of magnetic force around Earth are inclined with respect to the surface, except at the magnetic equator, accounting for magnetic inclination.

- For most places on Earth, an angle called magnetic declination exists between lines drawn from a compass position and the geographic and magnetic poles.

- Scientists use a magnetometer to detect departures from the normal magnetic field, which are characterized as positive or negative.

- The cause of magnetic reversals is not understood, but we know from rocks that the magnetic field has reversed many times during the past.

Important Terms

asthenosphere (p. 285)
continental crust (p. 287)
core (p. 283)
crust (p. 287)
Curie point (p. 295)
discontinuity (p. 282)
geothermal gradient (p. 288)
gravity (p. 289)
gravity anomaly (positive and negative) (p. 289)
heat flow (p. 288)

isostatic rebound (p. 293)
lithosphere (p. 285)
low-velocity zone (p. 285)
magnetic anomaly (positive and negative) (p. 296)
magnetic declination (p. 296)
magnetic field (p. 295)
magnetic inclination (p. 296)
magnetic reversal (p. 298)
magnetism (p. 295)
mantle (p. 285)

Mohoroovičić discontinuity (Moho) (p. 285)
oceanic crust (p. 288)
paleomagnetism (p. 298)
principle of isostasy (p. 291)
P-wave shadow zone (p. 283)
reflection (p. 282)
refraction (p. 282)
seismic tomography (p. 283)
S-wave shadow zone (p. 283)

Review Questions

1. Peridotite is:
   a. _____ found mostly in Earth's core; b. _____ what causes the S-wave shadow zone; c. _____ the most likely rock type in the mantle; d. _____ probably responsible for Earth's magnetic field; e. _____ a type of sedimentary rock.

2. Iron-bearing minerals gain their magnetism as they cool through the:
   a. _____ Curie point; b. _____ isostatic curve; c. _____ Moho discontinuity; d. _____ geothermal gradient; e. _____ heat flow index.

3. Oceanic crust is:
   a. _____ composed of granite; b. _____ thickest at spreading ridges; c. _____ the source of Earth's magnetic field; d. _____ 20–90 km thick; e. _____ denser than continental crust.

4. The Moho is:
   a. _____ a type of inclusion in kimberlite pipes; b. _____ a seismic discontinuity at the base of the crust; c. _____ between 410 and 660 km below the surface; d. _____ a layer made up of iron and nickel; e. _____ a zone in which rocks yield by plastic flow.

5. According to the principle of isostasy:

a. _____ magnetism results from the attraction between masses; b. _____ Earth's interior is cooler than it was during early Earth history; c. _____ the crust "floats" in the more dense mantle; d. _____ the asthenosphere behaves like a brittle solid; e. _____ most of the lithosphere is molten.

6. Earth's temperature increase with depth is the:

a. _____ geothermal gradient; b. _____ heat-flow indicator; c. _____ seismic reflection index; d. _____ discontinuity phase change; e. _____ low-velocity zone.

7. Earth's average density is 5.5 g/cm³, but rocks near the surface average 2.5–3.0 g/cm³, so it follows that:

a. _____ the magma for kimberlite pipes came from great depth; b. _____ Earth materials at depth must be denser than those near the surface; c. _____ the inner core is solid whereas the outer core is liquid; d. _____ the mantle is composed of granite and basalt; e. _____ the magnetic field is generated by a huge body of buried magnetite.

8. Which one of the following statements is *incorrect*?

a. _____ meteorites provide some evidence for the core's composition; b. _____ the base of the crust is between 200 and 300 km deep; c. _____ the mantle is probably composed of peridotite; d. _____ the lithosphere consists of the crust and the upper part of the mantle; e. _____ seismic discontinuities are present within the mantle.

9. A change in the direction of travel of seismic waves within the planet is called:

a. _____ attenuation; b. _____ elasticity; c. _____ lithification; d. _____ refraction; e. _____ inversion.

10. The conclusion that the outer core is liquid is based on the fact that Earth has:

a. _____ an S-wave shadow zone; b. _____ magnetic reversals; c. _____ gravity anomalies; d. _____ granitic continental crust; e. _____ a geothermal gradient.

11. Earth's crust is deeply eroded in one area and loaded by widespread, thick deposits in another. Explain fully how the crust will respond in these two areas.

12. If Earth were solid and homogeneous throughout, how would P- and S-wave behave as they traveled through the planet? How do they actually behave?

13. What is the geothermal gradient and why must it decrease with depth?

14. How does the lithosphere differ from the asthenosphere?

15. Explain how the record of magnetic inclination preserved in rocks is used to determine the latitude at which ancient rocks formed.

16. Why do scientists think the inner core is solid but the outer core is liquid?

17. How is it possible to determine the overall mass and density of Earth?

18. What accounts for the two seismic discontinuities within the mantle?

19. Explain why continents stand higher than ocean basins.

20. How can positive and negative gravity anomalies be used to explore for mineral resources?

World Wide Web Activities

PHYSICAL Geology⇌Now Assess your understanding of this chapter's topics with additional quizzing and comprehensive interactivities at

http://earthscience.brookscole.com/physgeo5e

as well as current and up-to-date weblinks, additional readings, and InfoTrac College Edition exercises.

The Seafloor

CHAPTER 11

OUTLINE

PHYSICAL Geology⇌Now *This icon, appearing throughout the book, indicates an opportunity to explore interactive tutorials, animations, or practice problems available on the Physical GeologyNow Web site at http://earthscience.brookscole.com/physgeo5e.*

OBJECTIVES

At the end of this chapter, you will have learned that

- Scientists use echo sounding, seismic profiling, sampling, and observations from submersibles to study the largely hidden seafloor.

- Oceanic crust is thinner and compositionally less complex than continental crust.

- The margins of continents consist of a continental shelf and slope and in some cases a continental rise with adjacent abyssal plains. The elements making up a continental margin depend on the geologic activity that takes place in these marginal areas.

- Although the seafloor is flat and featureless in some places, it also possesses ridges, trenches, seamounts, and other features.

- Geologic activities at or near divergent and convergent plate boundaries account for distinctive seafloor features such as submarine volcanoes and deep-sea trenches.

- Most seafloor sediment comes from weathering and erosion of continents and oceanic islands, and from the shells of tiny marine organisms.

- Organisms in warm, shallow seas build wave-resistant structures known as reefs.

- Several important resources such as common salt come from seawater, and hydrocarbons are found in some seafloor sediments.

Pillow lava on the Mid-Atlantic Ridge. Source: Woods Hole Oceanographic Institute

Introduction

According to two dialogues written in about 350 B.C. by the Greek philosopher Plato, a huge continent called Atlantis existed in the Atlantic Ocean west of the Pillars of Hercules, or what we now call the Strait of Gibraltar (■ Figure 11.1). According to Plato's account, Atlantis controlled a large area extending as far east as Egypt. Yet despite its vast wealth, advanced technology, and large army and navy, Atlantis was defeated in war by Athens. And following the conquest of Atlantis:

> . . . there were violent earthquakes and floods and one terrible day and night came when . . . Atlantis . . . disappeared beneath the sea. And for this reason even now the sea there has become unnavigable and un-searchable, blocked as it is by the mud shallows which the island produced as it sank.*

No "mud shallows" exist in the Atlantic as Plato asserted, but present-day proponents of Atlantis have claimed that the Azores, Bermuda, the Bahamas, and the Mid-Atlantic Ridge are remnants of this lost continent. Actually, if a continent had sunk in the Atlantic, or anywhere else for that matter, it could be easily detected by a gravity survey. In short, no geologic evidence indicates that Atlantis ever existed. Then why has the legend persisted for so long?

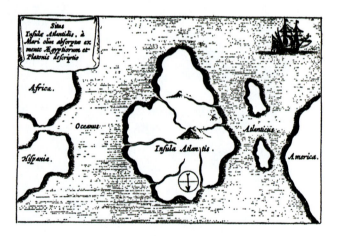

■ Figure 11.1

According to Plato, Atlantis was a continent west of the Pillars of Hercules, now called the Strait of Gibraltar. In this map from Anthanasium Kircher's *Mundus Subterraneus* (1664), north is toward the bottom of the map. The Strait of Gibraltar is the narrow area between Hispania (Spain) and Africa.

*From the *Timaeus*, Quoted in E. W. Ramage, Ed., *Atlantis: Fact or Fiction?* (Bloomington: Indiana University Press, 1978), p. 13.

One reason is that sensational stories of lost civilizations are popular, but another is that until recently no one had any real knowledge of what lies beneath the oceans. Much like Earth's interior, the seafloor is largely a hidden domain, so myths and legends about what lies below the ocean's surface were widely accepted. Indeed, the most basic observation we can make about Earth is that it has vast water-covered areas hidden from direct view and continents that at first glance might seem to be nothing more than parts of our planet not covered by water. Nevertheless, considerable differences exist between the continents and the ocean basins.

Ocean basins are of course lower than continents, but why should this be so? Recall from Chapter 10 that continental crust is thicker and less dense than oceanic crust, so according to the principle of isostasy, continental crust should stand higher. Furthermore, oceanic crust is composed of basalt and gabbro, whereas continental crust is made up of all rock types, although its overall composition compares closely to granite. Oceanic crust is continually produced at spreading ridges and consumed at subduction zones, so none of it is very old, geologically speaking. The oldest oceanic crust is about 180 million years old, but rocks on continents are as old as 3.96 billion years; remember that continental lithosphere is not dense enough to be subducted.

Why should you study the seafloor? One important reason is that it constitutes the largest part of Earth's surface (■ Figure 11.2), and despite the commonly held misconception that it is flat and featureless, it has topography as varied as that of the continents. Furthermore, many seafloor features as well as several aspects of the oceanic crust provide important evidence for plate tectonic theory (see Chapter 12). And finally, natural resources are present on the marginal parts of continents, in seawater, and on the seafloor.

As we begin our investigation of the seafloor, you should be aware that our discussion focuses on (1) the physical attributes and composition of the oceanic crust; (2) the composition and distribution of seafloor sediments; (3) seafloor topography; and (4) the origin and evolution of the continental margins. *Oceanographers* study these topics too, but in addition they study the chemistry and physics of seawater as well as oceanic circulation patterns and many aspects of marine biology. We should also point out that whereas the oceans and their marginal seas (Figure 11.2) are largely underlain by oceanic crust, the same is not true of such areas as the Dead Sea, Salton Sea, and Caspian Sea; these are actually saline lakes on the continents.

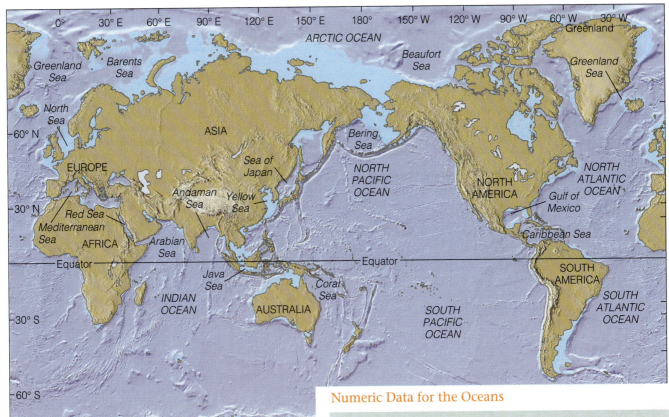

Numeric Data for the Oceans

Ocean*	Surface Area (Million KM²)	Water Volume (Million KM³)	Average Depth (FM)	Maximum Depth (KM)D
Pacific	180	700	4.0	11.0
Atlantic	93	335	3.6	9.2
Indian	77	285	3.7	7.5
Arctic	15	17	1.1	5.2

Source: P. R. Pinet, 1992. Oceanography. (St. Paul: West.)

*Excludes adjacent seas, such as the Caribbean Sea and Sea of Japan, which are marginal parts of oceans.

■ **Figure 11.2**

Map showing the four major oceans and many of the world's seas, which are marginal parts of oceans.

EXPLORING THE OCEANS

During most of historic time, people knew little of the oceans and, until fairly recently, thought that the seafloor was a vast, featureless plain. In fact, through most of this time the seafloor in one sense was more remote than the Moon's surface because it could not even be observed.

Early Exploration

The ancient Greeks had determined Earth's size and shape rather accurately, but western Europeans were not aware of the vastness of the oceans until the 1400s and 1500s, when explorers sought trade routes to the Indies. Even when Christopher Columbus set sail on August 3, 1492, in an effort to find a route to the Indies, he greatly underestimated the width of the Atlantic Ocean. Contrary to popular belief, he was not attempting to demonstrate Earth's spherical shape; its shape was well accepted by then. The controversy was over Earth's circumference and the shortest route to China; on these points, Columbus's critics were correct.

These voyages added considerably to our knowledge of the oceans, but truly scientific investigations did not begin until the late 1700s. Great Britain was the dominant maritime power, and to maintain that dominance, the British sought to increase their knowledge of the oceans. The earliest British scientific voyages were led by Captain James Cook in 1768, 1772, and 1777. In 1872 the converted British warship HMS *Challenger* began a 4-year voyage to sample seawater, collect and analyze seafloor sediment and rocks, and name and classify thousands of species of marine organisms.

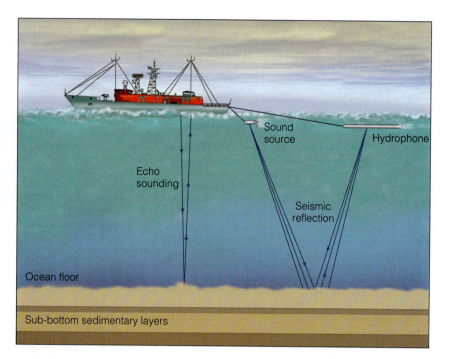

Figure 11.3

Diagram showing how echo sounding and seismic profiling are used to study the seafloor. Some of the energy generated at the energy sources is reflected from various horizons back to the surface, where it is detected by hydrophones.

Continuing exploration of the oceans revealed that the seafloor is not flat and featureless as formerly believed. Indeed, scientists discovered that the seafloor possesses varied topography including oceanic trenches, submarine ridges, broad plateaus, hills, and vast plains.

How Are Oceans Explored Today?

Measuring the length of a weighted line lowered to the seafloor was the first method for determining oceanic depths. Now scientists use an instrument called an *echo sounder*, which detects sound waves that travel from a ship to the seafloor and back (■ Figure 11.3). Depth is calculated by knowing the velocity of sound in water and the time required for the waves to reach the seafloor and return to the ship, thus yielding a continuous profile of seafloor depths along the ship's route. **Seismic profiling** is similar to echo sounding but even more useful. Strong waves from an energy source reflect from the seafloor, and some of the waves penetrate seafloor layers and reflect from various horizons back to the surface (Figure 11.3). Seismic profiling is particularly useful for mapping the structure of the oceanic crust where it is buried beneath seafloor sediments.

The Deep Sea Drilling Project, an international program sponsored by several oceanographic institutions, began in 1968. Its first research vessel, the *Glomar Challenger,* could drill in water more than 6000 m deep and recover long cores of seafloor sediment and oceanic crust. The *Glomar Challenger* drilled more than 1000 holes in

the seafloor during the 15 years of the program. The Deep Sea Drilling Project ended in 1983, but beginning in 1985 the Ocean Drilling Program with its research vessel the JOIDES* *Resolution* continued to explore the seafloor (■ Figure 11.4a). Research vessels also sample the seafloor using *clamshells samplers* and *piston corers* (■ Figure 11.5).

In addition to surface vessels, submersibles are now important vehicles for seafloor exploration. Some, such as the *Argo*, are remotely controlled and towed by a surface vessel. In 1985 the *Argo*, equipped with sonar and television systems, provided the first views of the British ocean liner HMS *Titanic* since it sank in 1912. The U.S. Geological Survey uses a towed device with sonar to produce seafloor images resembling aerial photographs. Scientists aboard submersibles such as *Alvin* (Figure 11.4b) have descended to the seafloor in many areas to make observations and collect samples.

Scientific investigations have yielded important information about the oceans for more than 200 years, but much of our current knowledge has been acquired since World War II (1939–1945). This is particularly true of the seafloor because only in recent decades has instrumentation been available to study this largely hidden domain.

OCEANIC CRUST—ITS STRUCTURE AND COMPOSITION

We have mentioned that oceanic crust is composed of basalt and gabbro and is continually generated at spreading ridges. Of course, drilling into the oceanic crust gives some details about its composition and structure, but it has never been completely penetrated and sampled. So how do we know what it is composed of and how it varies with depth? Actually, even before it was sampled and observed, these details about the oceanic crust were known.

Remember that oceanic crust is consumed at subduction zones and thus most of it is recycled, but a small amount is found in mountain ranges on land where it

---

*JOIDES is an acronym for Joint Oceanographic Institutions for Deep Earth Sampling.

Ocean Drilling Program, Texas A&M University

(a)

VU/WHOI-R. Catarach

(b)

■ **Figure 11.4**

Oceanographic research vessels. (a) The JOIDES *Resolution* is capable of drilling the deep seafloor. (b) The submersible *Alvin* is used for observing and sampling the deep seafloor.

was emplaced by moving along large fractures called thrust faults (faults are discussed more fully in Chapter 13). These preserved slivers of oceanic crust along with part of the underlying upper mantle are known as **ophiolites.** Detailed studies reveal that an ideal ophiolite consists of deep-sea sedimentary rocks underlain by rocks of the upper oceanic crust, especially pillow lava and sheet lava flows (■ Figure 11.6). Proceeding downward in an ophiolite is a sheeted dike complex, consisting of vertical basaltic dikes, then massive gabbro and layered gabbro that appear to have formed in the upper part of a magma chamber. And finally, the lowermost

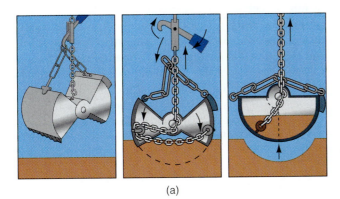

(a)

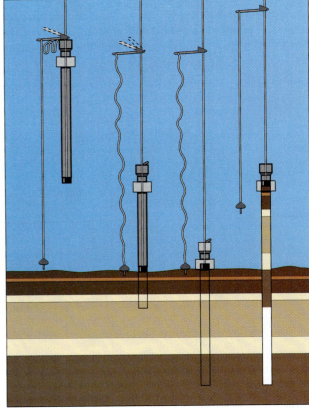

(b)

■ **Figure 11.5**

Sampling the seafloor. (a) A clamshell sampler taking a seafloor sample. (b) A piston corer falls to the seafloor, penetrates the sediment, and then is retrieved.

What Would You Do

As the only person in your community with any geologic training, you are often called on the explain local geologic features and identify fossils. Several school children on a natural history field trip picked up some rocks that you recognize as peridotite. When you visit the site where the rocks were collected, you also notice some pillow lava in the area and what appear to be dikes composed of basalt. What other rock types might you expect to find here? How would you explain (1) the association of these rocks with one another, and (2) how they came to be on land?

unit is peridotite, representing the upper mantle; this is sometimes altered by metamorphism to a greenish rock known as serpentinite. Thus a complete ophiolite consists of deep-sea sedimentary rocks underlain by rocks of the oceanic crust and upper mantle (Figure 11.6).

Sampling and drilling at oceanic ridges reveal that oceanic crust is indeed made up of pillow lava and sheet lava flows underlain by a sheeted dike complex, just as predicted from studies of ophiolites. But it was not until 1989 that a submersible carrying scientists descended to the walls of a seafloor fracture in the North Atlantic and verified what lay below the sheeted dike complex. Just as expected from ophiolites on land, the lower oceanic crust consists of gabbro and the upper mantle is made up of peridotite. In short, the composition and structure of oceanic crust and upper mantle were inferred from rocks on land and then these inferences were verified by sampling, drilling, and observing seafloor rocks.

WHAT ARE CONTINENTAL MARGINS?

In the Introduction we made the point that continents are not simply areas above sea level, although most people perceive of continents as land areas outlined by the seas. The true geologic margin of a continent—that is, where granitic continental crust changes to oceanic crust composed of basalt and gabbro—is below sea level. Accordingly, the margins of continents are submerged, and we recognize **continental margins** as separating the part of a continent above sea level from the deep seafloor.

A continental margin is made up of a gently sloping continental shelf, a more steeply inclined continental slope, and in some cases, a deeper, gently sloping continental rise (■ Figure 11.7). Seaward of the continental margin lies the deep ocean basin. Thus the continental margins extend to increasingly greater depths until they merge with the deep seafloor. The change from continental crust to oceanic crust takes places somewhere beneath the continental rise, so part of the continental slope and the continental rise actually rest on oceanic crust.

The Continental Shelf

As one proceeds seaward from the shoreline across the continental margin, the first area encountered is a gently sloping **continental shelf** lying between the shore and the more steeply dipping continental slope (Figure 11.7). The width of the continental shelf varies considerably, ranging from a few tens of meters to more than 1000 km; the shelf terminates where the inclination of the seafloor increases abruptly from 1 degree or less to several degrees. The outer margin of the continental shelf, or simply the *shelf–slope break,* is at an average depth of 135 m, so by oceanic standards the continental shelves are covered by shallow water.

At times during the Pleistocene Epoch (1.6 million to 10,000 years ago), sea level was as much as 130 m lower than it is now. As a result, the continental shelves were above sea level and were areas of stream channel and floodplain deposition. In addition, in many parts of northern Europe and North America, glaciers extended well

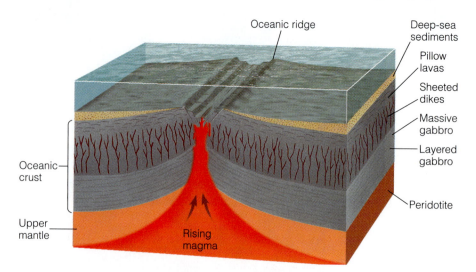

■ **Figure 11.6**

New oceanic crust consisting of the layers shown here forms as magma rises beneath oceanic ridges. The composition of the oceanic crust is known from ophiolites, sequences of rock on land consisting of deep-sea sediments, oceanic crust, and upper mantle.

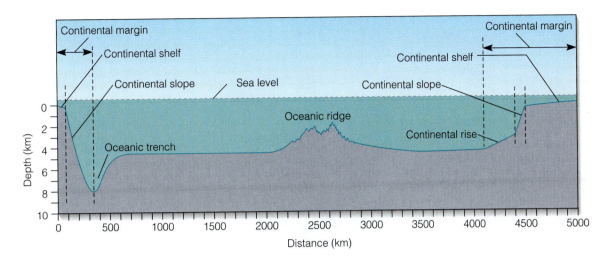

■ **Figure 11.7**

A generalized profile of the seafloor showing features of the continental margins. The vertical dimensions of the features in this profile are greatly exaggerated because the vertical and horizontal scales differ.

out onto the continental shelves and deposited gravel, sand, and mud. Since the Pleistocene ended, sea level has risen, submerging these deposits, which are now being reworked by marine processes. Evidence that these sediments were in fact deposited on land includes remains of human settlements and fossils of a variety of land-dwelling animals.

The Continental Slope and Rise

The seaward margin of the continental shelf is marked by the *shelf–slope break* (at an average depth of 135 m) where the more steeply inclined **continental slope** begins (Figure 11.7). In most areas around the margins of the Atlantic, the continental slope merges with a more gently sloping **continental rise.** This rise is absent around the margins of the Pacific where continental slopes descend directly into an oceanic trench (Figure 11.7).

The shelf–slope break, marking the boundary between the shelf and slope, is an important feature in terms of sediment transport and deposition. Landward of the break— that is, on the shelf—sediments are affected by waves and tidal currents, but these processes have no effect on sediments seaward of the break, where gravity is responsible for their transport and deposition on the slope and rise. In fact, much of the land-derived sediment crosses the shelves and is eventually deposited on the continental slopes and rises, where more than 70% of all sediments in the oceans are found. Much of this sediment is transported through submarine canyons by turbidity currents.

Submarine Canyons, Turbidity Currents, and Submarine Fans

In Chapter 6, we discussed the origin of graded bedding, most of which results from **turbidity currents,** underwater flows of sediment–water mixtures with densities greater than that of sediment-free water. As a turbidity current flows onto the relatively flat seafloor, it slows and begins depositing sediment, the largest particles first, followed by progressively smaller particles, thus forming a layer with graded bedding (see Figure 6.18). Deposition by turbidity currents results in the origin of a series of overlapping **submarine fans,** which constitute a large part of the continental rise (■ Figure 11.8). Submarine fans are distinctive features, but their outer margins are difficult to discern because they grade into deposits of the deep-ocean basin.

No one has ever observed a turbidity current in progress in the oceans, so for many years some doubted their existence. However, evidence now available removes all doubt. In 1971, for instance, abnormally turbid water was sampled just above the seafloor in the North Atlantic, indicating that a turbidity current had recently occurred. In addition, seafloor samples from many areas show a succession of layers with graded bedding and the remains of shallow-water organisms that were displaced into deeper water by turbidity currents.

Perhaps the most compelling evidence for turbidity currents is the pattern of trans-Atlantic cable breaks that took place in the North Atlantic near Newfoundland on November 18, 1929. Initially, an earthquake was assumed to have ruptured several telephone and telegraph cables. However, although the breaks on the continental shelf near the epicenter occurred when the earthquake struck, cables farther seaward were broken later and in succession. The last cable to break was 720 km from the source of the earthquake, and it did not snap until 13 hours after the first break. In 1949 geologists realized that an earthquake-generated turbidity current had moved downslope, breaking the cables in succession

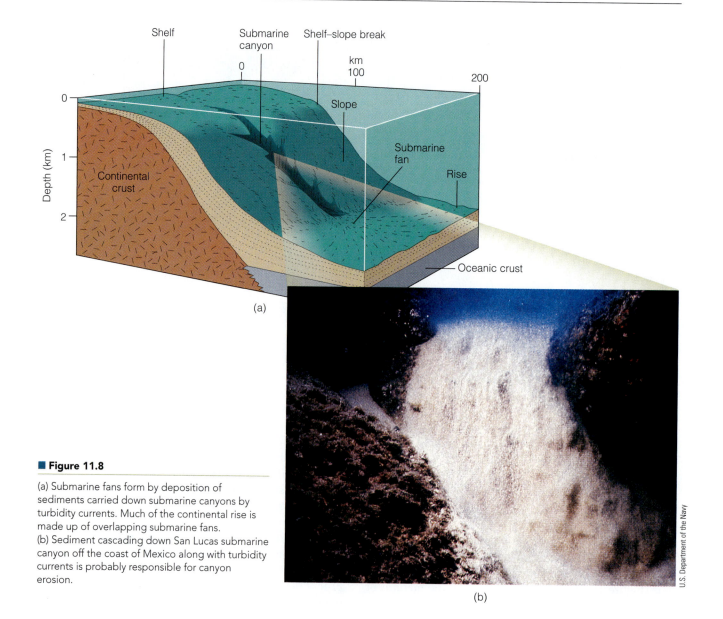

■ Figure 11.8

(a) Submarine fans form by deposition of sediments carried down submarine canyons by turbidity currents. Much of the continental rise is made up of overlapping submarine fans.
(b) Sediment cascading down San Lucas submarine canyon off the coast of Mexico along with turbidity currents is probably responsible for canyon erosion.

(a)

(b)

U.S. Department of the Navy

(■ Figure 11.9). The precise time at which each cable broke was known, so calculating the velocity of the turbidity current was simple. It moved at about 80 km/hr on the continental slope, but slowed to about 27 km/hr when it reached the continental rise.

Deep, steep-sided **submarine canyons** are present on continental shelves, but they are best developed on continental slopes (Figure 11.8). Some submarine canyons can be traced across the shelf to rivers on land; they apparently formed as river valleys when sea level was lower during the Pleistocene. However, many have no such association, and some extend far deeper than can be accounted for by river erosion during times of lower sea level. Scientists know that strong currents move through submarine canyons and perhaps play some role in their origin. Furthermore, turbidity currents periodically move through these canyons and are now thought to be the primary agent responsible for their erosion.

Types of Continental Margins

Continental margins are characterized as either *active* or *passive,* depending on their relationship to plate boundaries. An **active continental margin** develops at the leading edge of a continental plate where oceanic lithosphere is subducted (■ Figure 11.10). The western margin of South America is a good example. Here, an oceanic plate is subducted beneath the continent, resulting in seismic activity, a geologically young mountain range, and active volcanism. In addition, the continental shelf is narrow, and the continental slope descends directly into the Peru–Chile Trench, so sediment is dumped into the trench and no continental rise develops. The western margin of North America is also considered an active continental margin, although much of it is now bounded by transform faults rather than a subduction zone. However, plate convergence

■ Figure 11.9

(a) Submarine cable breaks caused by a turbidity current south of Newfoundland in 1929. The vertical dimension of this profile is greatly exaggerated. The profile labeled "No vertical exaggeration" shows what the seafloor actually looks like in this area. (b) The propeller of a submarine caused this turbidity current that flows down a slope near Jamaica. Source: (a) From Bruce E. Heezen and Charles D. Hollister, *The Face of the Deep* (New York: Oxford University Press, 1971): Figure 8.15, page 297.

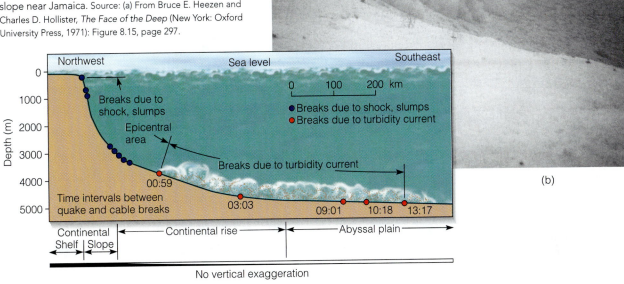

(a)

(b)

and subduction still take place in the Pacific Northwest along the continental margins of northern California, Oregon, and Washington.

The continental margins of eastern North America and South America differ considerably from their western margins. For one thing, they possess broad continental shelves as well as a continental slope and rise; also, vast, flat areas known as *abyssal plains* are present adjacent to the rises (Figure 11.10). Furthermore, these **passive continental margins** are within a plate rather than at a plate boundary, and they lack the typical volcanic and seismic activity found at active continental margins. Nevertheless, earthquakes do take place occasionally at these margins—the Charleston, South Carolina, earthquake of 1886, for instance (see Figure 9.7).

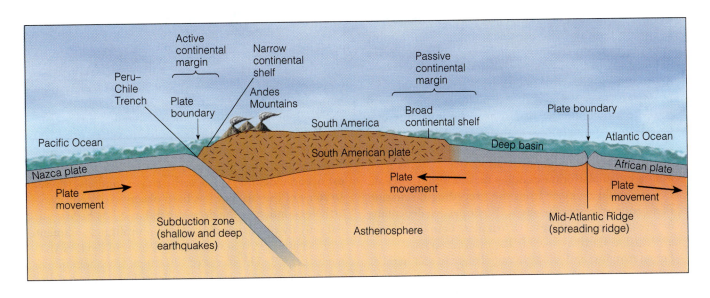

■ Figure 11.10

Active and passive continental margins along the west and east coasts of South America. Notice that passive continental margins are much wider than active margins. Seafloor sediment is not shown.

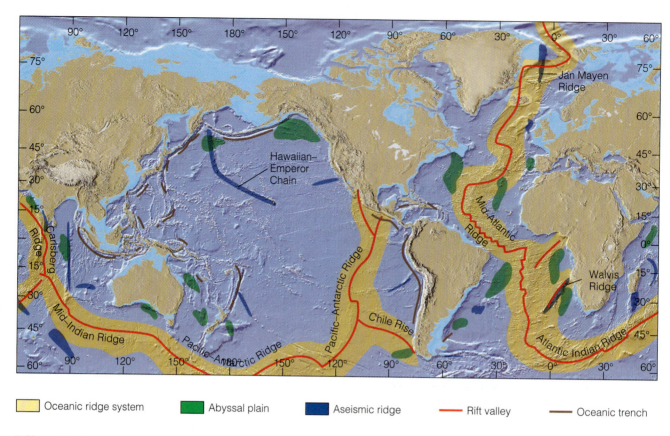

Oceanic ridge system	Abyssal plain	Aseismic ridge	—— Rift valley	—— Oceanic trench

■ **Figure 11.11**

Features found on the deep seafloor include oceanic trenches (brown), abyssal plains (green), the oceanic ridge system (yellow), rift valleys (red), and some aseismic ridges (blue). Other features such as seamounts and guyots are shown in Figure 11.16. Source: From Alyn and Alison Duxbury, *An Introduction to the World's Oceans.* Copyright © 1984 McGraw-Hill.

Active and passive continental margins share some features, but they are notably different in the widths of their continental shelves, and active margins have an oceanic trench but no continental rise. Why the differences? At both types of continental margins, turbidity currents transport sediment into deeper water. At passive margins, the sediment forms a series of overlapping submarine fans and thus develops a continental rise, whereas at an active margin, sediment is simply dumped into the trench and no rise forms. The proximity of a trench to a continent also explains why the continental shelves of active margins are so narrow. In contrast, land-derived sedimentary deposits at passive margins have built a broad platform extending far out into the ocean.

It should be clear from the preceding discussion that continental margins are active or passive depending on their location with respect to a plate boundary. However, just as Earth as a whole has evolved, so have continental margins, and one type can change to another. For instance, eastern North America now has a passive continental margin, but during the Paleozoic Era it was bounded by an active continental margin.

WHAT FEATURES ARE FOUND IN THE DEEP-OCEAN BASINS?

Considering that the oceans average more than 3.8 km deep, most of the seafloor lies far below the depth of sunlight penetration, which is rarely more than 100 m. Accordingly, most of the seafloor is completely dark, the temperature is generally just above 0°C, and the pressure varies from 200 to more than 1000 atmospheres depending on depth. Scientists in submersibles have descended to the greatest oceanic depths, the oceanic ridges, and elsewhere, so some of the seafloor has been observed directly. Nevertheless, much of the deep-ocean basin has been studied only by echo sounding, seismic profiling, sediment and oceanic crust sampling, and remote devices that have descended in excess of 11,000 m. Oceanographers are developing a more thorough understanding of the oceans and now know that the seafloor has vast plains, trenches, and ridges (■ Figure 11.11).

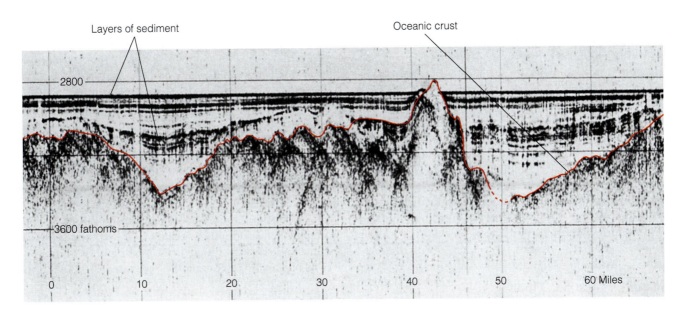

■ Figure 11.12

Seismic profile showing rugged seafloor topography covered by sediments of the Northern Madeira Abyssal Plain in the Atlantic Ocean.
Source: From Bruce E. Heezen and Charles D. Hollister, *The Face of the Deep* (New York: Oxford University Press, 1971): Figure 8.38, page 329.

Abyssal Plains

Beyond the continental rises of passive continental margins are **abyssal plains,** flat surfaces covering vast areas of the seafloor. In some places, they are interrupted by peaks rising more than 1 km, but they are nevertheless the flattest, most featureless areas on Earth (Figure 11.11). The flat topography is a result of sediment deposition; where sediment accumulates in sufficient quantities it buries rugged seafloor (■ Figure 11.12).

Seismic profiles and seafloor samples reveal that abyssal plains are covered with fine-grained sediment derived mostly from the continents and deposited by turbidity currents. Abyssal plains are invariably found adjacent to the continental rises, which are composed mostly of overlapping submarine fans that owe their origin to deposition by turbidity currents (Figure 11.9). Along active continental margins, sediments derived from the shelf and slope are trapped in an oceanic trench, and abyssal plains fail to develop. Accordingly, abyssal plains are common in the Atlantic Ocean basin but rare in the Pacific Ocean basin (Figure 11.11).

Oceanic Trenches

Oceanic trenches constitute a small percentage of the seafloor, probably less than 2%, but they are very important, for it is here that lithospheric plates are consumed by subduction (see Chapter 12). Oceanic trenches are long, narrow features restricted to active continental margins, so they are common around the margins of the Pacific Ocean basin (Figure 11.11). For instance, the Peru–Chile Trench west of South America is 5900 km long but only 100 km wide. It is more than 8000 m deep. On the landward side of oceanic trenches, the continental slope descends at angles of up to 25 degrees (Figure 11.10). Oceanic trenches are also the sites of the greatest oceanic depths; a depth of more than 11,000 m has been recorded in the Challenger Deep of the Marianas Trench in the Pacific Ocean.

Oceanic trenches show anomalously low heat flow compared with other areas of oceanic crust, indicating that the crust at trenches is cooler and slightly more dense than elsewhere. Gravity surveys across trenches reveal huge negative gravity anomalies because the crust is not in isostatic equilibrium. In fact, oceanic crust at trenches is subducted, giving rise to Benioff zones in which earthquake foci become progressively deeper in the direction the subducted plate descends (see Figure 9.5). These inclined seismic zones account for most intermediate- and deep-focus earthquakes—such as the June 1994 magnitude 8.2 earthquake in Bolivia, which had a focal depth of 640 km. Finally, subduction at oceanic trenches also results in volcanism, either as an arcuate chain of volcanic islands on oceanic crust or as a chain of volcanoes along the margin of a continent, as in western South America.

Oceanic Ridges

When the first submarine cable was laid between North America and Europe during the late 1800s, a feature called the Telegraph Plateau was discovered in the North Atlantic. Using data from the 1925–1927 voyage of the German research vessel *Meteor,* scientists proposed that the plateau was actually a continuous ridge extending the length of the Atlantic Ocean basin. Subsequent investigations revealed that this proposal was correct, and we now call this feature the Mid-Atlantic Ridge (Figure 11.11).

The Mid-Atlantic Ridge is more than 2000 km wide and rises 2 to 2.5 km above the adjacent seafloor. Furthermore, it is part of a much larger **oceanic ridge** system of mostly submarine mountainous topography. It runs from the Arctic Ocean through the middle of the Atlantic, curves around South Africa where the Indian Ridge continues into the Indian Ocean, then the Pacific–Antarctic Ridge extends eastward and a branch of this, the East Pacific Rise, trends northeast until it reaches the Gulf of California (Figure 11.11). The entire system is at least 65,000 km long, far exceeding the length of any mountain range on land. Oceanic ridges are composed almost entirely of basalt and gabbro and possess features produced by tensional forces. Mountain ranges on land, in contrast, consist of igneous, metamorphic, and sedimentary rocks, and formed when rocks were folded and fractured by compressive forces (see Chapter 13).

Oceanic ridges are mostly below sea level, but they rise above the sea in some places such as Iceland, the Azores, and Easter Island. Of course, oceanic ridges are the sites where new oceanic crust is generated and plates diverge (see Chapter 12). The rate of plate divergence is important because it determines the cross-section profile of a ridge. For example, the Mid-Atlantic Ridge has a comparatively steep profile because divergence is slow, allowing the new oceanic crust to cool, shrink, and subside closer to the ridge crest than it does in areas of faster divergence such as at the East Pacific Rise. A ridge may also have a rift along its crest that opens in response to tension (■ Figure 11.13). A rift is particularly obvious along the Mid-Atlantic Ridge but appears to be absent along parts of the East Pacific Rise. Rifts are commonly 1 to 2 km deep and several kilometers wide. They open as seafloor spreading takes place (discussed in Chapter 12) and are characterized by shallow-focus earthquakes, basaltic volcanism, and high heat flow.

Even though most oceanographic research is still done by echo sounding, seismic profiling, and seafloor sampling, scientists have been making direct observations of oceanic ridges and their rifts since 1974. As part of Project FAMOUS (French-American Mid-Ocean Undersea Study), submersibles have descended to the ridges and into their rifts in several areas. No active volcanism has been observed, but researchers did see pillow lavas (see Figure 4.9), lava tubes, and sheet lava flows, some of which seemed to have formed very recently. In fact, on return visits to a site they have seen the effects of volcanism that occurred since their last visit. And on January 25, 1998, a submarine volcano began erupting along the Juan de Fuca Ridge west of Oregon.

Submarine Hydrothermal Vents

Scientists first saw **submarine hydrothermal vents** on the seafloor in 1979 when they descended about 2500 m to the Galapagos Rift in the eastern Pacific Ocean. Since 1979 scientists have seen similar vents in several other areas in the Pacific (■ Figure 11.14a), Atlantic, and Sea of Japan. The vents are at or near spreading ridges, where cold seawater seeps through oceanic crust, is heated by the hot rocks at depth, and then rises and discharges into the seawater as plumes of hot water—temperatures of 400°C have been recorded. Many of the plumes are black because dissolved minerals give them the appearance of black smoke—hence, the name **black smoker.**

Submarine hydrothermal vents are interesting from biologic, geologic, and economic points of view. Near the vents live communities of organisms

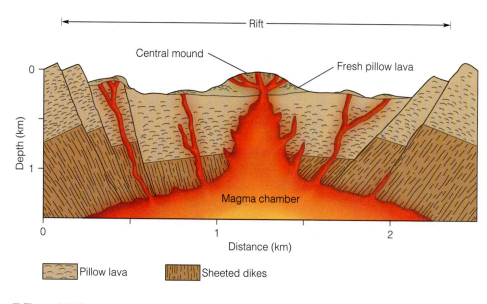

■ **Figure 11.13**

Cross section of the Mid-Atlantic Ridge showing its central rift where recent moundlike accumulations of volcanic rocks, mostly pillow lava, are found.

such as bacteria, crabs, mussels, starfish, and tube worms, many of which had never been seen before (Figure 11.14b). No sunlight is available, so the organisms in these communities depend on bacteria that oxidize sulfur compounds for their ultimate source of nutrients. The vents are also interesting because of their economic potential. The heated seawater reacts with oceanic crust to transform it into a metal-rich solution. As the hot solution discharges into seawater, it cools, precipitating iron, copper, and zinc sulfides and other minerals to form a chimneylike vent. The vents eventually collapse and form a mound of sediments rich in the elements mentioned above.

Apparently the chimneys through which black smokers discharge grow rapidly. A 10-m-high chimney accidentally knocked over in 1991 by the submersible *Alvin* grew to 6 m in just three months. Also in 1991 scientists aboard *Alvin* saw the results of a submarine eruption on the East Pacific Rise, which they missed by less than two weeks. Fresh lava and ash covered the area as well as the remains of tube worms killed during the eruption. And in a nearby area, a new fissure opened in

What Would You Do

Hydrothermal vents on the seafloor are known sites of several metals of great importance to industrialized societies. Furthermore, it appears that these metals are being deposited even now, so if we mine one area, more of the same resources form elsewhere. Given these conditions, it would seem that our problems of diminishing resources are solved. So why not simply mine the seafloor? Also, many chemical elements are present in seawater. The technology exists to extract elements such as gold, uranium, and others, so why not do so?

the seafloor; by December 1993 a new hydrothermal vent community had become well established.

In 2001 scientists announced another kind of seafloor vent in the North Atlantic responsible for massive pillars and spires up to 60 m high. Unlike the black

(a)

(b)

■ **Figure 11.14**

(a) A hydrothermal vent known as a *black smoker* at 2800 m on the East Pacific Rise. Seawater seeps down through the oceanic crust, becomes heated, and then rises and builds chimneys composed of anhydrite ($CaSO_4$) and sulfides of iron, copper, and zinc. The plume of "black smoke" is simply heated water saturated with dissolved minerals. (b) Several types of organisms including these tube worms live near black smokers.

BOB EMBLEY

Secrets Beneath the Sea

Dr. Bob Embley is a marine geologist with the Pacific Marine Environmental Laboratory, National Oceanic and Atmospheric Administration, in Newport, Oregon. After receiving a degree in geology from Rutgers University, he went to sea for a year and then began graduate school at Columbia University and Lamont–Doherty Geological (now Earth) Observatory, where he received a Ph.D. in marine geology and geophysics. He has participated in more than 50 oceanographic expeditions in the Atlantic, Pacific, and Antarctic Oceans over 37 years.

I feel extremely privileged to have participated in the scientific exploration of the oceans. Since the 1960s, when I began my career, technology has advanced such that we can now image the ocean floor literally a hundred times better. What hasn't changed is the sense of excitement that I've maintained since grade school. What I think set me on a path that led to a lifetime study of the ocean floor was the generosity of a teacher in third grade. Probably the only thing I remember clearly from that period is her placing in my hands a book about mountains. To this day, I clearly remember the mesmerizing smell of that new book, and by the tenth grade I had decided to become a geologist who specialized in studying the ocean floor.

My first ocean expedition came the summer following my graduation from college. I had begun working at the Lamont–Doherty Geological Observatory (Columbia University) the previous summer, and in 1966 I was assigned to be the geologist on Columbia University's research vessel *Vema*. The *Vema* was a 202-foot, 3-masted schooner built in 1922 as a luxury yacht. After serving as a training vessel during the war years, it was bought by Columbia University in the 1950s and refitted as a research vessel. I had been working toward this voyage for years, and it was truly exciting to leave New York on a journey that would take us north of Iceland and past the Arctic Circle. That night, as the ship steamed through Long Island Sound and out into the North Atlantic, I struggled

to get through my "sea watch" (a sea watch is traditionally 4 hours on and 8 off) in the laboratory where the echo sounders were spewing out some rather noxious fumes in addition to new information about the bottom. Fortunately, I got my "sea legs" after a couple of days. Unfortunately, about that time the

Vema sprung a leak! In order to stay afloat, we had to transfer several tons of lead ballast from one side of the ship to the other. The ship was repaired in Halifax, Nova Scotia, and we set off for the Arctic Circle. The roughest weather came not in the far north but in the nasty waters south of Iceland. One evening, after I recovered a seafloor core sample, the weather became particularly violent. I have a recurring vision of a giant wave topping the ship's bridge ready to descend on me as I was holding onto the core sample. I released it to hold on for dear life. A moment later the core was neatly returned to whence it had come, leaving me still on the ship to continue my career.

In 1984 I came to Oregon to start a new program to study hydrothermal vents and their effects on the oceans. The Mid-Ocean Ridge stretches more than 60,000 km around the earth, making it the most significant geologic structure on Earth. More than 70% of Earth's volcanic activity takes place unseen beneath the ocean, primarily along the Mid-Ocean Ridge, but also on submarine volcanoes behind many deep-sea trenches (island arcs). Most of the ridge is remote and difficult to study, lying days or weeks from the nearest port. However, a 1000-km-long section of it lies a few hundred miles west of North America and Canada. Although I didn't realize it at the time, this portion of the ridge was going to yield great scientific treasures over the next 20 years. During the next decade, we mapped it with the latest technology, and we did initial probing and sampling using cameras and human-occupied submersibles. But we still had no real idea about the volcanic processes occurring along the ridge. How big were the eruptions? How often did they occur? No one really knew.

In 1993 the U.S. Navy gave us a gift. During the Cold War, the United States had tracked Soviet submarines using supersecret hydrophone arrays (underwater microphones) cabled to shore stations. Using a "channel" in the upper ocean that traps sound waves (discovered during World War II), the Navy could "hear" a Russian submarine turn on its engine from distances of thousands of miles. In June 1993 we gained access to these underwater microphones in the northeast Pacific and began to monitor the seismic activity along the spreading centers with more than a hundred times the sensitivity available from seismic stations based in North America. Only with this capability could we "hear" earthquakes associated with the movement of magma beneath the seafloor. Amazingly, within one week, a swarm of earthquakes began on the central portion of the Juan de Fuca Ridge that indicated a possible deep-water eruption! Within a few days, a Canadian vessel in the vicinity confirmed that there was a large warm plume of water over the site. The previous year we had begun a collaboration with Canadian colleagues to use their new deep-sea robotic vehicle *ROPOS* for research on the hydrothermal vents and we were scheduled for another expedition with it on the NOAA research vessel *Discoverer*. Fortunately, we were able to divert this scheduled expedition to the site of the earthquake swarm, arriving there about two weeks after it began.

That turned out to be one of the true "highlight" cruises of my career. We lowered the vehicle to the seafloor several hundred meters east of the center of the earthquake swarm. As we drove the remotely operated vehicle west over old, sediment-covered pillow lavas, the sense of imminent discovery was palpable in the control room. At first it was difficult to see the bottom, but as the cameras auto-adjusted to the light level, it became obvious that we were looking at black, unsedimented lavas! But there was still no sign of shimmering water—a sure sign of new, cooling lava. We climbed a steep slope and began noticing more and more patches of yellow material in the depressions within the lava flow. As we crested the lava mound, the camera appeared to blur as it tried to focus on vivid white patches within the yellow zones. This was the *schlieren* effect caused by the refraction of the light through the warm water emanating from the still cooling lava! That was one of the most exciting moments I've

Figure 1

Bob Embley (left) and colleague Ed Baker discuss deployment of the CTD (conductivity, temperature, depth) instrument from R/V *T.G. Thompson* on the 2003 expedition to Mariana Arc. The device's primary function is to detect how the conductivity and temperature of the water column changes relative to depth.

ever experienced at sea. The highest temperature we recorded was about 60°C (a very warm day in Death Valley). Our investigations in the area then and later showed there had been two other eruptions along that part of the ridge between 1981 and 1993. Since 1993 that part of the ridge has been quiet, suggesting the accretion of new ocean crust at the mid-ocean ridge occurs in spurts separated by periods of quiescence. In 1996 and again in 1998 we detected and verified eruptions on other sections of the ridge off Oregon.

The ocean floor is still a frontier for scientific exploration, as it was when I began my career. Today there are many new tools available that allow us to explore the seafloor at ever greater detail. Our research team is now involved in an explorative study of the submarine volcanoes and hydrothermal systems of the island arcs of the western Pacific. During our initial survey in February 2003 we mapped 50 submarine volcanoes with stunning clarity and discovered 10 new hydrothermal systems along the arc extending north from the island of Guam. We're very excited at the prospect of returning next year with a robotic vehicle to sample and explore these sites.

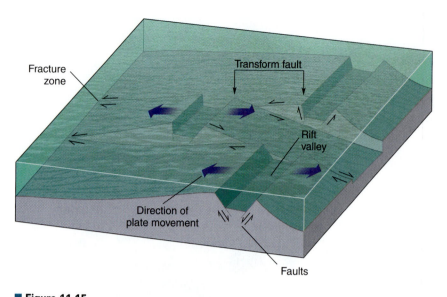

■ Figure 11.15

Diagrammatic view of an oceanic ridge offset along fractures. That part of a fracture between displaced segments of the ridge crest is known as a *transform fault* (see Chapter 12).

smokers, though, these vents are 14 or 15 km from spreading ridges, and they consist of light-colored minerals that were derived by chemical reaction between seawater and minerals in the oceanic crust.

Seafloor Fractures

Oceanic ridges are not continuous features winding without interruption around the globe. They abruptly terminate where they are offset along fractures oriented more or less at right angles to ridge axes (■ Figure 11.15). These fractures run for hundreds of kilometers, although they are difficult to trace where buried beneath seafloor sediments. Many geologists are convinced that some geologic features on continents can best be accounted for by the extension of these fractures into continents.

Where oceanic ridges are offset, they are characterized by shallow seismic activity only in the area between the displaced ridge segments (Figure 11.15). Furthermore, because ridges are higher than the seafloor adjacent to them, the offset segments yield vertical relief on the seafloor. Nearly vertical escarpments 2 or 3 km high develop, as illustrated in Figure 11.15. The reason oceanic ridges show so many fractures is that plate divergence takes place irregularly on a sphere, resulting in stresses that cause fracturing. We will have more to say about these fractures, known as *transform faults* between offset ridge segments, in Chapter 12.

Seamounts, Guyots, and Aseismic Ridges

As noted, the seafloor is not a flat, featureless plain except for the abyssal plains, and even these are underlain by rugged topography (Figure 11.12). In fact, a large number

of volcanic hills, seamounts, and guyots rise above the seafloor in all ocean basins, but they are particularly abundant in the Pacific. All are of volcanic origin and differ mostly in size. **Seamounts** rise more than 1 km above the seafloor; if they are flat topped, they are called **guyots** rather than seamounts (■ Figure 11.16). Guyots are volcanoes that originally extended above sea level. But as the plate they rested on continued to move, they were carried away from a spreading ridge, and the oceanic crust cooled and descended to greater oceanic depths. So what was once an island was eroded by waves as it slowly sank beneath the sea, giving it the typical flat-topped appearance.

Many other volcanic features smaller than seamounts are also present on the seafloor, but they probably originated in the same way. These so-called *abyssal hills*, averaging only about 250 m high, are common on the seafloor and underlie thick sediments on the abyssal plains.

Other common features in the ocean basins are long, narrow ridges and broad plateaulike features rising as much as 2 to 3 km above the surrounding seafloor. These are known as **aseismic ridges** because they lack seismic activity. A few of these ridges are probably small fragments separated from continents during rifting. These are referred to as *microcontinents* and are represented by such features as the Jan Mayen Ridge in the North Atlantic (Figure 11.11).

Most aseismic ridges form as a linear succession of hot spot volcanoes. These may develop at or near an oceanic ridge, but each volcano so formed is carried laterally with the plate on which it originated. The net result is a sequence of seamounts/guyots extending from an oceanic ridge (Figure 11.16); the Walvis Ridge in the South Atlantic is a good example (Figure 11.11). Aseismic ridges also form over hot spots unrelated to ridges. The Hawaiian–Emperor chain in the Pacific formed in such a manner (Figure 11.11).

SEDIMENTATION AND SEDIMENTS ON THE DEEP SEAFLOOR

D eep-sea sediments are mostly fine grained, consisting of silt- and clay-sized particles, because few mechanisms transport coarse-grained sediment (sand and gravel) far from land. Ice-

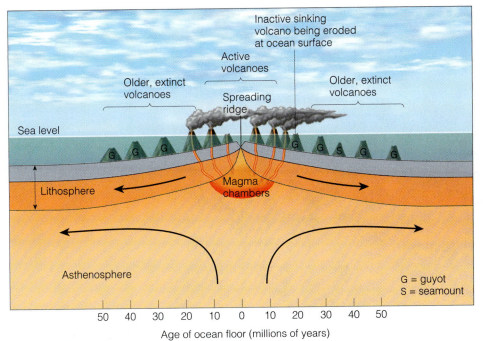

Figure 11.16

The origin of seamounts (S) and guyots (G). As a plate on which a volcano rests moves into greater depths, a volcanic island may be eroded and become a flat-topped guyot. Source: From Garrison, T., *Oceanography*, 4 ed., Figure 3.31. Wadsworth Publishing.

bergs transport coarse sediment into the ocean basins, though, and a broad band of glacial–marine sediment is present adjacent to Antarctica and Greenland (■ Figure 11.17).

Most of the fine-grained sediment in the deep sea is windblown dust and volcanic ash from the continents and oceanic islands and the shells of microscopic organisms that live in the near-surface waters. Other sources of sediment include cosmic dust and deposits resulting from chemical reactions in seawater. The manganese nodules that are fairly common in all the ocean basins are a good example of the latter (■ Figure 11.18). They are composed mostly of manganese and iron oxides but also contain copper, nickel, and cobalt. Nodules may be an important source of some metals in the future; the United States, which imports most of its manganese and cobalt, is particularly interested in this potential resource.

The contribution of cosmic dust to deep-sea sediment is negligible. Even though some researchers estimate that as much as 40,000 metric tons of cosmic dust falls to Earth each year, this is a trivial quantity compared to the volume of sediments derived from other sources.

Most deep seafloor sediments are *pelagic*, meaning that they settled from suspension far from land. Two categories of pelagic sediment are recognized: pelagic clay and ooze (■ Figure 11.19). Brown or reddish **pelagic clay**, composed of clay-sized particles derived from continents and oceanic islands, covers most of the deeper parts of the ocean basins. **Ooze**, in contrast, is composed mostly of shells of microscopic marine animals and plants. It is characterized as *calcareous ooze* if it contains mostly calcium carbonate ($CaCO_3$) skeletons of tiny marine organisms such as foraminifera. *Siliceous ooze* is composed of the silica (SiO_2) skeletons of such single-celled organisms as radiolarians (animals) and diatoms (plants) (Figure 11.19).

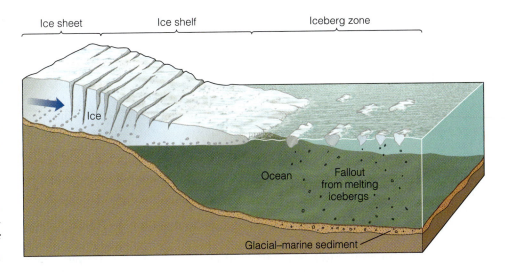

Figure 11.17

The formation of glacial–marine sediments by ice rafting, a process in which icebergs transport sediment into the ocean basins, releasing it as they melt.

Institute of Oceanographic Sciences/Nerc/SPL/Photo Researchers, Inc.

(a)

Dr. James Andrews

(b)

■ **Figure 11.18**

(a) Manganese nodules on the seafloor. (b) Sectioned manganese nodule showing its internal structure consisting of concentric layers. This nodule measures about 6 cm across.

HOW DO ORGANISMS CONSTRUCT REEFS?

The term **reef** has a variety of meanings, such as shallowly submerged rocks that pose a hazard to navigation, but here we restrict it to mean a moundlike, wave-resistant structure composed of the skeletons of marine organisms. Although commonly called *coral reefs,* they actually have a solid framework composed of skeletons of corals and various mollusks, such as clams, and encrusting organisms including sponges and algae. Reefs are restricted to shallow, tropical seas where the water is clear and the temperature does not fall below about 20°C. The depth to which reefs can grow, rarely more than 50 m, depends on sunlight penetration because many of the corals rely on symbiotic algae that depend on sunlight

for energy. (See "Reefs: Rocks Made by Organisms" on pages 322 and 323.)

Reefs are found in a variety of shapes and sizes, but most are one of three basic types: fringing, barrier, and atoll. *Fringing reefs* are solidly attached to the margins of an island or continent. They have a rough, tablelike surface, are as much as 1 km wide, and, on their seaward side, slope steeply down to the seafloor. *Barrier reefs* are similar to fringing reefs, except that a lagoon separates them from the mainland. Probably the best-known barrier reef in the world is the Great Barrier Reef of Australia, which is 2000 km long.

An *atoll* is a circular to oval reef surrounding a lagoon. Atolls form around volcanic islands that subside below sea level as the plate they rest on is carried progressively farther from an oceanic ridge. As subsidence proceeds, the reef organisms construct the reef upward so that the living part of the reef remains in shallow

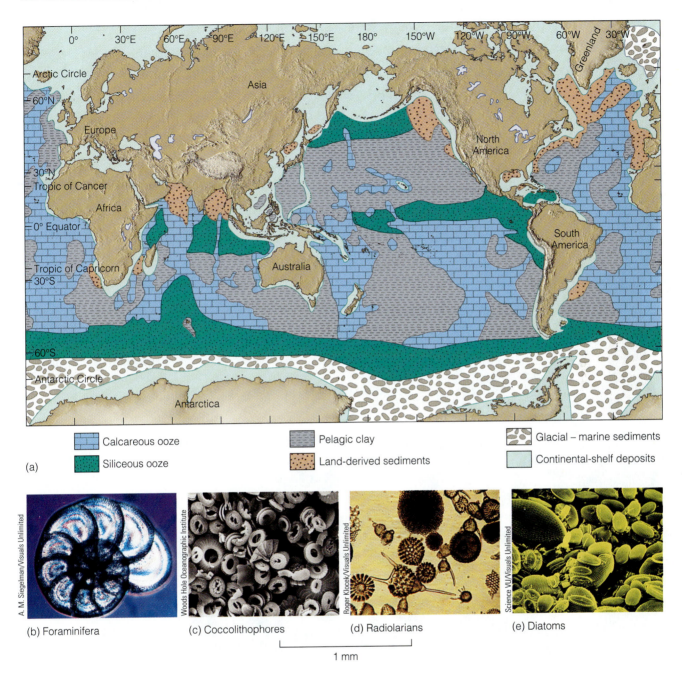

(a)

Calcareous ooze

Siliceous ooze

Pelagic clay

Land-derived sediments

Glacial – marine sediments

Continental-shelf deposits

(b) Foraminifera   (c) Coccolithophores   (d) Radiolarians   (e) Diatoms

1 mm

■ **Figure 11.19**

(a) A variety of sediments are present in the ocean basins, but most on the deep seafloor are pelagic clay and calcareous and siliceous ooze. Common constituents of calcareous ooze are skeletons of (b) foraminifera (floating single-celled animals) and (c) coccolithophores (floating single-celled plants), whereas siliceous ooze consists of skeletons of (d) radiolarians (single-celled floating animals) and (e) diatoms (single-celled floating plants).

water. The island eventually subsides below sea level, leaving a circular lagoon surrounded by a more or less continuous reef. Atolls are particularly common in the western Pacific Ocean basin. Many of these began as fringing reefs, but as the plate they were on was carried into deeper water, they evolved first to barrier reefs and finally to atolls.

This scenario for the evolution of reefs from fringing to barrier to atoll was proposed more than 150 years ago by Charles Darwin while he was serving as a naturalist

Reefs: Rocks Made by Organisms

Reefs are wave-resistant structures composed of the skeletons of corals, mollusks, sponges, and encrusting algae. Most reefs are characterized as fringing, barrier, or atolls, all of which actively grow in shallow, warm seawater where there is little or no influx of detrital sediment, especially mud. Reef rock is a type of limestone that forms directly as a solid, rather than from sediment that is later lithified. Ancient reefs are important reservoirs for hydrocarbons in some areas.

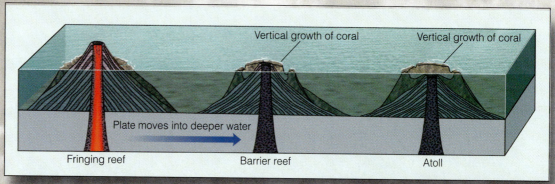

Vertical growth of coral

Vertical growth of coral

Plate moves into deeper water

Fringing reef

Barrier reef

Atoll

Three stages in the evolution of a reef. A fringing reef forms around a volcanic island, but as the island is carried into deeper water on a moving plate, the reef is separated from the island by a lagoon and becomes a barrier reef. Continued plate movement carries the island into even deeper water. The island disappears below sea level but the reef grows upward, forming an atoll.

Underwater views of reefs in the Red Sea and in Hawaii.

The white line of breaking waves marks the site of a barrier reef around Rarotonga in the Cook Islands in the Pacific Ocean. The island is only about 12 km long.

Courtesy of Carl Roessler

Courtesy of Jeff and Diana Munroe

Patrick Ward/Corbis

Sue Monroe

Talus   Reef core   Back reef

This oval reef with a central lagoon in the Pacific Ocean is an atoll. How does it differ from the reef shown at the bottom of the previous page?

An ancient reef in Australia. You can see the reef talus on the left side of the image sloping away from the reef core, which has no layering. To the right of the reef core are back-reef deposits, which show horizontal bedding.

Block diagram showing the various environments in a reef complex.

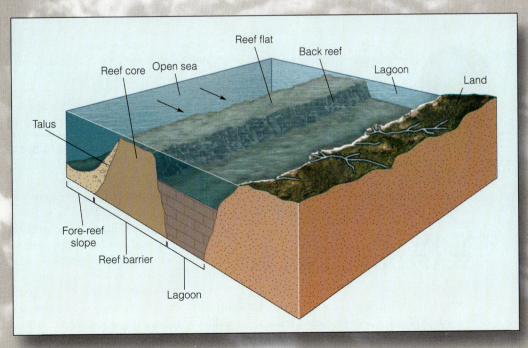

Reef core   Open sea   Reef flat   Back reef   Lagoon   Land

Talus

Fore-reef slope

Reef barrier

Lagoon

GEOLOGY
IN UNEXPECTED PLACES

Ancient Seafloor in San Francisco

At 5:13 A.M. on April 18, 1906, a huge earthquake shook San Francisco, California, killing thousands and causing tremendous property damage. San Francisco is well known for past and potential earthquakes generated by movements on the San Andreas fault (see Chapter 10). But most people, including the city's residents, are unaware that rock exposures there provide excellent examples of ancient seafloor sediments and volcanic rocks that formed at or near spreading ridges. Indeed, the city is underlain by a succession of rocks scraped off against the continental margin as an oceanic plate was subducted beneath North America during the Jurassic and Cretaceous periods (see Figure 1.17). Some of the rocks are called *mélange,* a geologic term for a chaotic mixture of rock fragments set in a matrix of clay formed by intense shearing during subduction.

Excellent exposures of some of these ancient rocks are on Alcatraz Island, which served as a maximum-security federal penitentiary from 1934 to 1963 and where such infamous criminals as Al Capone were once imprisoned. The rocks consist of thick-bedded graywacke, a type of sandstone with a large amount of rock fragments and clay (■ Figure 1). The original sediments were probably deposited

in an oceanic trench much like the present-day Peru–Chili trench along South America's west coast.

Other interesting rocks in and around San Francisco are pillow lava that originated by submarine volcanism, and thinly bedded, highly deformed, bedded chert, another type of rock that originated on the deep seafloor (see image). So in addition to outstanding scenery and its cultural atmosphere, San Francisco has an interesting geologic history.

James S. Monroe

■ Figure 1

The rocks exposed on Alcatraz Island in San Francisco Bay are thick-bedded sandstones that were deposited on the deep seafloor. The inset shows intensely folded, thinly bedded chert that also formed on the deep seafloor.

on the ship HMS *Beagle.* Drilling into atolls has revealed that they do indeed rest on a basement of volcanic rocks, confirming Darwin's hypothesis.

SEAWATER AND SEAFLOOR RESOURCES

Seawater contains many elements in solution, some of which are extracted for various industrial and domestic uses. Sodium chloride (table salt) is produced by the evaporation of seawater, and a large proportion of the world's magnesium comes from seawater. Numerous other elements and compounds are also extracted from seawater, but for many, such as gold, the cost is prohibitive.

In addition to substances in seawater, deposits on the seafloor or within seafloor sediments are becoming increasingly important. Many of these potential resources lie well beyond continental margins, so their ownership is a political and legal problem that has not yet been resolved. Most nations bordering an ocean claim those resources within their adjacent continental margin. The United States, by a presidential proclama-

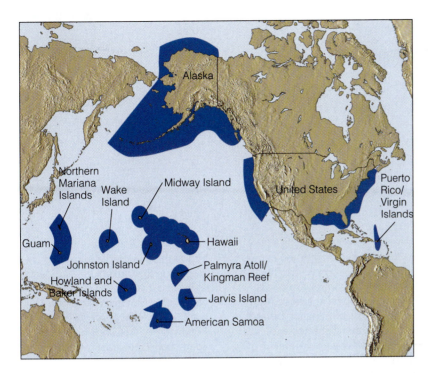

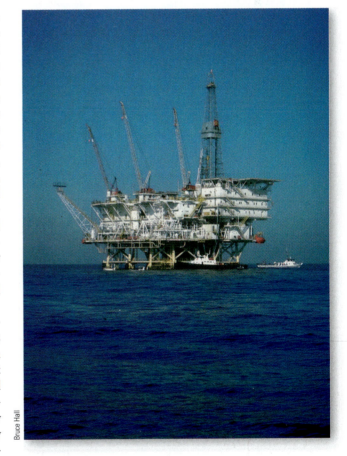

■ **Figure 11.20**

The Exclusive Economic Zone (EEZ), shown in dark blue, includes a vast area adjacent to the United States and its possessions.

tion issued on March 10, 1983, claims sovereign rights over an area designated as the **Exclusive Economic Zone (EEZ).** The EEZ extends seaward 200 nautical miles (371 km) from the coast, giving the United States jurisdiction over an area about 1.7 times larger than its land area (■ Figure 11.20). Also included within the EEZ are the areas adjacent to U.S. territories, such as Guam, American Samoa, Wake Island, and Puerto Rico (Figure 11.20). In short, the United States claims a huge area of the seafloor and any resources on or beneath it.

Numerous resources are found within the EEZ, some of which have been exploited for many years. Sand and gravel for construction are mined from the continental shelf in several areas. About 17% of U.S. oil and natural gas production comes from wells on the continental shelf (■ Figure 11.21). Some 30 sedimentary basins are known within the EEZ, several of which contain hydrocarbons, whereas others are areas of potential hydrocarbon production.

A potential resource within the EEZ is methane hydrate, which consists of single methane molecules bound up in networks formed by frozen water. Although methane hydrates have been known since the early part of the 19th century, they have only recently received much attention. Most of these deposits lie within continental margins, but so far it is not known whether they can be effectively recovered and used as an energy source. According to one estimate, the amount of carbon in methane hydrates is double that in all coal, oil, and conventional natural gas reserves.

Methane hydrates are stable at water depths exceeding 500 m and at near-freezing temperatures. Along the Atlantic continental margin of the United States, numerous submarine landslide scars are present where the

■ **Figure 11.21**

An oil-drilling platform off the coast of southern California. Although this one is in fairly shallow water, a similar platform off the Louisiana coast is anchored in water 872 m deep. About 17% of all U.S. oil production comes from wells on the continental shelves.

GEOFOCUS

11.1

Oceanic Circulation and Resources from the Sea

Earth's oceans are in constant motion. Huge quantities of water circulate in surface and deep currents as water is transferred from one part of an ocean basin to another. The Gulf Stream and South Equatorial Current carry great quantities of water toward the poles and have an important modifying effect on climate. In addition to surface and deep currents that carry water horizontally, vertical circulation takes place when *upwelling* slowly transfers cold water from depth to the surface and *downwelling* transfers warm surface water to depth.

Upwelling is of more than academic interest. It not only transfers water from depth to the surface but also carries nutrients, especially nitrates and phosphate, into the zone of sunlight penetration where they sustain high concentrations of floating organisms, which in turn support other organisms. In fact, other than the continental shelves and areas adjacent to hydrothermal vents on the seafloor, areas of upwelling are the only parts of the oceans where biological productivity is very high. Less than 1% of the ocean surfaces are areas of upwelling, yet they support more than 50% (by weight) of all fishes.

Scientists recognize three types of upwelling, but only one need concern us here—*coastal upwelling*. Most coastal upwelling takes place along the west coasts of Africa, North America, and South America, although one notable exception is present in the Indian Ocean. Coastal upwelling involves movement of water offshore, which is replaced by water rising from depth (■ Figure 1). For instance, the winds blowing south along the coast of Peru, coupled with the Coriolis effect, transport surface water seaward, and cold, nutrient-rich water from depth rises to replace it. This area, too, is a major fishery, and changes in the surface-water circulation every 3 to 7 years adjacent to South America are associated with the onset of El Niño, a weather phenomenon with far-reaching consequences.

Our interest here lies in the fact that among the nutrients in

seafloor should be stable. Geologists at the U.S. Geological Survey think that methane hydrates contribute to these slides, during which methane is released into the atmosphere, where along with carbon dioxide it contributes to global warming. Its contribution to global warming must be assessed because a volume of methane 3000 times greater than that in the atmosphere is present in seafloor sediments, and it is 10 times more effective than carbon dioxide as a factor in global warming.

Other seafloor resources of interest include massive sulfide deposits that form by submarine hydrothermal activity at spreading ridges. These deposits containing iron, copper, zinc, and other metals have been identified within the EEZ at the Gorda Ridge off the coasts of California and Oregon; similar deposits occur at the Juan de Fuca Ridge within the Canadian EEZ.

Other potential resources include the manganese nodules, discussed previously (Figure 11.18), and met-

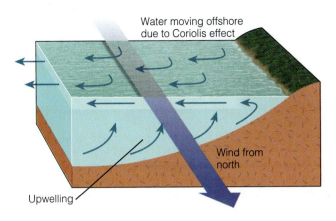

Water moving offshore due to Coriolis effect

Wind from north

Upwelling

■ Figure 1

Wind from the north along the west coast of a continent coupled with the Coriolis effect causes surface water to move offshore, resulting in upwelling of cold, nutrient-rich deep water.

upwelling oceanic waters is considerable phosphorus, an essential element for animal and plant nutrition. Although present in minute quantities in many sedimentary rocks, most commercial phosphorus is derived from *phosphorite*, a sedimentary rock containing such phosphate-rich minerals as fluorapatite ($Ca_5(PO_4)_3F$). Areas of upwelling along the outer margins of continental shelves are the sites of deposition of most of

these so-called bedded phosphorites, which are interlayered with carbonate rocks, chert, and detrital rocks such as mudrocks and sandstone. Vast deposits in the Permian-aged Phosphoria Formation of Montana, Wyoming, and Idaho formed in this manner.

Upwelling accounts for most of Earth's phosphate-rich sedimentary rocks, but some forms by other processes. In *phosphatiza-*

tion, carbonate grains such as animal skeletons and ooids are replaced by phosphate, and *guano* is made up of calcium phosphate from bird or bat excrement. Another type of phosphate deposit is essentially a placer deposit in which the skeletons of vertebrate animals are found in large numbers (vertebrate skeletons are made up mostly of hydroxyapatite ($Ca_5(PO_4)_3OH$). The 3- to 15-million-year-old Bone Valley Formation of Florida is a good example.

The United States is the world leader in production and consumption of phosphate rock, most of it coming from deposits in Florida and North Carolina, but some is also mined in Idaho and Utah. More than 90% of all phosphate rock mined in this country is used to make chemical fertilizers and animal-feed supplements. It also has other uses in metallurgy, preserved foods, ceramics, and matches.

PHYSICAL Geology⇌Now   Click Geology Interactive to work through an activity on World Currents through Waves, Tides and Currents; El Nino through Weather and Climate.

alliferous oxide crusts found on seamounts. Manganese nodules contain manganese, cobalt, nickel, and copper; the United States is heavily dependent on imports of the first three of these elements. Within the EEZ, manganese nodules are found near Johnston Island in the Pacific Ocean and on the Blake Plateau off the coast of South Carolina and Georgia. In addition, seamounts and seamount chains within the EEZ in the Pacific are known to have metalliferous oxide crusts several centimeters thick from which cobalt and manganese could be mined.

Another important resource forming in shallow seawater is phosphate-rich sedimentary rock known as *phosphorite* (see Geo-Focus 11.1).

11 REVIEW WORKBOOK

Chapter Summary

- Scientific investigations of the oceans began more than 200 years ago, but much of our knowledge comes from studies done during the last few decades.

- Present-day research vessels are equipped to investigate the seafloor by sampling and drilling, echo sounding, and seismic profiling. Scientists also use submersibles in their studies.

- Continental margins consist of a gently sloping continental shelf, a more steeply inclined continental slope, and in some cases a continental rise.

- The width of continental shelves varies considerably. They slope seaward to the shelf–slope break at a depth averaging 135 m, where the seafloor slope increases abruptly.

- Submarine canyons, mostly on continental slopes, carry huge quantities of sediment by turbidity currents into deeper water, where it is deposited as overlapping submarine fans that make up a large part of the continental rise.

- Active continental margins at the leading edge of a tectonic plate have a narrow shelf and a slope that descends directly into an oceanic trench. Volcanism and seismic activity also characterize these margins.

- Passive continental margins lie within a tectonic plate and have wide continental shelves, and the slope merges with a continental rise that grades into an abyssal plain. These margins show little seismic activity and no volcanism.

- Long, narrow oceanic trenches are found where oceanic lithosphere is subducted beneath either oceanic lithosphere or continental lithosphere. The trenches are the sites of the greatest oceanic depths, low heat flow, and negative gravity anomalies.

- Oceanic ridges are composed of volcanic rocks, and many have a central rift caused by tensional forces. Basaltic volcanism and shallow-focus earthquakes occur at ridges, which are offset by fractures that cut across them.

- Seamounts, guyots, and abyssal hills rising from the seafloor are common features that differ mostly in scale and shape. Many aseismic ridges on the seafloor consist of chains of seamounts, guyots, or both.

- Moundlike, wave-resistant structures called reefs consisting of animal skeletons are found in a variety of shapes, but most are classified as fringing reefs, barrier reefs, or atolls.

- Deep-sea drilling and observations on land and on the seafloor confirm that oceanic crust is made up, in descending order, of pillow lava/sheet lava flows, sheeted dikes, and gabbro.

- The United States claims rights to all resources within 200 nautical miles of its shorelines. Resources including sand and gravel as well as metals are found within this Exclusive Economic Zone.

Important Terms

abyssal plain (p. 313)
active continental margin (p. 310)
aseismic ridge (p. 318)
black smoker (p. 314)
continental margin (p. 308)
continental rise (p. 309)
continental shelf (p. 308)
continental slope (p. 309)

Exclusive Economic Zone (EEZ) (p. 325)
guyot (p. 318)
oceanic ridge (p. 314)
oceanic trench (p. 313)
ooze (p. 319)
ophiolite (p. 307)
passive continental margin (p. 311)

pelagic clay (p. 319)
reef (p. 320)
seamount (p. 318)
seismic profiling (p. 306)
submarine canyon (p. 310)
submarine fan (p. 309)
submarine hydrothermal vent (p. 314)
turbidity current (p. 309)

Review Questions

1. An atoll is:

a. _____ a circular to oval reef enclosing a lagoon; b. _____ a type of deep-sea sediment made up of clay; c. _____ found on the deep seafloor adjacent to a black smoker; d. _____ a deposit composed of copper and zinc sulfides; e. _____ made up of sediment that settles from suspension far from land.

2. The greatest oceanic depths are found at:

a. _____ aseismic ridges; b. _____ seamounts; c. _____ oceanic trenches; d. _____ passive continental margins; e. _____ submarine canyons.

3. A large part of a continental rise consists of overlapping:

a. _____ guyots; b. _____ hydrothermal vents; c. _____ continental margins; d. _____ abyssal plains; e. _____ submarine fans.

4. The two types of sediment most common on the deep seafloor are:

a. _____ gravel and limestone; b. _____ pelagic clay and ooze; c. _____ manganese nodules and cosmic dust; d. _____ reefs and sand; e. _____ seamounts and guyots.

5. The gently sloping part of the continental margin adjacent to the continent is the continental:

a. _____ slope; b. _____ plain; c. _____ profile; d. _____ shelf; e. _____ ridge.

6. A flat-topped seamount rising more than 1 km above the seafloor is a(n):

a. _____ guyot; b. _____ reef; c. _____ black smoker; d. _____ oceanic ridge; e. _____ aseismic plateau.

7. Which one of the following is characteristic of an active continental margin?

a. _____ wide continental shelf; b. _____ continental shelf merging with a continental rise; c. _____ seismic activity; d. _____ vast abyssal plains; e. _____ submarine hydrothermal vents.

8. Turbidity current deposits typically show:

a. _____ graded bedding; b. _____ a large component of siliceous ooze; c. _____ many skeletons of corals; d. _____ sulfide minerals; e. _____ ophiolites.

9. Which one of the following statements is *incorrect?*

a. _____ most intermediate- and deep-focus earthquakes occur at active continental margins; b. _____ submarine hydrothermal vents are found near spreading ridges; c. _____ oceanic crust is made up of granite and sandstone; d. _____ most continental margins around the Pacific are active; e. _____ pelagic clay covers much of the deep seafloor.

10. A broad, flat area adjacent to the continental rise is called:

a. _____ oceanic crust; b. _____ submarine canyon; c. _____ abyssal plain; d. _____ seamount; e. _____ passive rise.

11. Identify the types of continental margins (Pacific-type and Atlantic-type) in Figure 11.7. What are the characteristics of each?

12. Describe submarine canyons and explain how scientists think they form. Is there any evidence to support their ideas?

13. Explain how a reef evolves from fringing to barrier to an atoll.

14. How do mid-oceanic ridges form, and how do they differ from mountain ranges on land?

15. Why are abyssal plans common around the margins of the Atlantic but rare in the Pacific Ocean basin?

16. How did geologists figure out the nature of the upper mantle and oceanic crust even before they observed mantle and crust rocks in the ocean basins?

17. What is calcareous ooze and pelagic clay, and where are they found?

18. How are seismic profiling and echo sounding used to study the seafloor?

19. The most distant part of a 30-million-year-old aseismic ridge is 1000 km from an oceanic ridge. How fast, on average, did the plate with this ridge move in centimeters per year?

20. During the Pleistocene Epoch (Ice Age), sea level was as much as 130 m lower than it is today. What effect did this lower sea level have on rivers? Is there any evidence from continental shelves that might bear on this question? If so, what?

World Wide Web Activities

PHYSICAL Geology Now Assess your understanding of this chapter's topics with additional quizzing and comprehensive interactivities at

http://earthscience.brookscole.com/physgeo5e

as well as current and up-to-date weblinks, additional readings, and InfoTrac College Edition exercises.

Plate Tectonics:
A Unifying Theory

CHAPTER 12
OUTLINE

PHYSICAL
Geology⇌Now *This icon, appearing throughout the book, indicates an opportunity to explore interactive tutorials, animations, or practice problems available on the Physical GeologyNow Web site at http://earthscience.brookscole.com/physgeo5e.*

OBJECTIVES

At the end of this chapter, you will have learned that

- Plate tectonics is the unifying theory of geology and has revolutionized geology.

- The hypothesis of continental drift was based on considerable geologic, paleontologic, and climatologic evidence.

- The hypothesis of seafloor spreading accounts for continental movement and that thermal convection cells provide a mechanism for plate movement.

- The three types of plate boundaries are divergent, convergent, and transform. Along these boundaries new plates are formed, consumed, or slide past one another

- Interaction along plate boundaries accounts for most of Earth's earthquake and volcanic activity.

- The rate of movement and motion of plates can be calculated in several ways.

- Some type of convective heat system is involved in plate movement.

- Plate movements affect the distribution of natural resources.

Image of South America generated from the Shuttle Radar Topography Mission aboard the Space Shuttle Endeavour, *launched on February 11, 2000. The Andes Mountains can be clearly seen along the Pacific Coast and are the result of subduction of the Nazca plate beneath the South American plate. The Amazon River can be seen to the east of the Andes Mountains draining much of northern South America.*

Introduction

At 8.46 A.M. on January 26, 2001, an earthquake of magnitude 7.7 rocked the Gujarat region of India as well as neighboring Pakistan. Flattening villages and toppling high-rise buildings in cities, this earthquake caused an estimated $1.3 billion in damages. It is estimated that more than 20,000 people died, 167,000 were injured, and 600,000 were left homeless. It was the most powerful earthquake to strike India since 1950, when an 8.5 magnitude earthquake killed more than 1500 people.

On June 15, 1991, Mount Pinatubo in the Philippines erupted violently, discharging huge quantities of ash and gases into the atmosphere. Fortunately, warnings of an impending eruption were heeded, and 200,000 people were evacuated from areas around the volcano, yet the eruption still caused 722 deaths.

What do these two tragic events and other equally destructive volcanic eruptions and earthquakes have in common? They are part of the dynamic interactions involving Earth's plates. When two plates come together, one plate is pushed or pulled under the other plate, triggering large earthquakes such as the one that shook Turkey in 1999 and India in 2001. As the descending plate moves downward and is assimilated into Earth's interior, magma is generated. Being less dense than the surrounding material, the magma rises toward the surface, where it may erupt as a volcano such as Mount Pinatubo did in 1991 and others have since. It therefore should not be surprising that the distribution of volcanoes and earthquakes closely follows plate boundaries.

As we stated in Chapter 1, plate tectonic theory has had significant and far-reaching consequences in all fields of geology because it provides the basis for relating many seemingly unrelated phenomena (Table 12.1). The interactions between moving plates determine the location of conti-nents, ocean basins, and mountain systems, which in turn affects atmospheric and oceanic circulation patterns that ultimately determine global climates. Plate movements have also profoundly influenced the geographic distribution, evolution, and extinction of plants and animals. Furthermore, the formation and distribution of many geologic resources, such as metal ores, are related to plate tectonic processes, leading geologists to incorporate plate tectonic theory into their prospecting efforts.

If you're like most people, you probably have no idea or only a vague notion of what plate tectonic theory is. Yet plate tectonics affects all of us, whether in terms of the destruction caused by volcanoes and earthquakes, or politically and economically (see Geo-Focus 12.1). It is therefore important to understand this unifying theory, not only because it affects us as individuals and citizens of nation-states, but also because it ties together all aspects of the geology you will be studying.

WHAT WERE SOME OF THE EARLY IDEAS ABOUT CONTINENTAL DRIFT?

The idea that Earth's past geography was different from today is not new. The earliest maps showing the east coast of South America and the west coast of Africa probably provided people with the first evidence that continents may have once been joined together, then broken apart and moved to their present positions.

Table 12.1

Plate Tectonics and Earth Systems

Solid Earth	Plate tectonics is driven by convection in the mantle and in turn drives mountain-building processes and associated igneous and metamorphic activity.
Atmosphere	Arrangement of continents affects solar heating and cooling and thus winds and weather systems. Rapid plate spreading and hot spot activity may release volcanic carbon dioxide and affect global climate.
Hydrosphere	Continental arrangement affects ocean currents. Rate of spreading affects volume of mid-ocean ridges and hence sea level. Placement of continents may contribute to onset of ice ages.
Biosphere	Movement of continents creates corridors or barriers to migration and thus creates ecological niches. Habitats may be transported into more or less favorable climates.
Extraterrestrial	Arrangement of continents affects free circulation of ocean tides and influences tidal slowing of Earth's rotation.

GEO**FOCUS**

12.1

Oil, Plate Tectonics, and Politics

It is certainly not surprising that oil and politics are closely linked. The Iran–Iraq War of 1980–1989 and the Gulf War of 1990–1991 were both fought over oil (■ Figure 1). Indeed, many of the conflicts in the Middle East have had as their underlying cause, control of the vast deposits of petroleum in the region. Most people, however, are not aware of why there is so much oil in this part of the world.

Although large concentrations of petroleum occur in many areas of the world, more than 50% of all proven reserves are in the Persian Gulf region. Interestingly, however, this region did not become a significant petroleum-producing area until the economic recovery following World War II. After the war, Western Europe and Japan in particular became dependent on Persian Gulf oil and still rely heavily on this region for most of their supply. The United States is also dependent on imports from the Persian Gulf, but receives significant quantities of petroleum from other sources such as Mexico and Venezuela.

Why is so much oil in the Persian Gulf region? The answer lies in the paleogeography and plate movements of this region during the Mesozoic and Cenozoic eras. During the Mesozoic Era, and particularly the Cretaceous Period when most of the petroleum formed, the Persian Gulf area was a broad, stable marine shelf extending eastward from Africa. This passive continental margin lay near the equator where countless microorganisms lived in the surface waters. The remains of these organisms accumulated with the bottom sediments and were buried, beginning the complex process of petroleum generation and the formation of source beds.

As a consequence of rifting in the Red Sea and Gulf of Aden during the Cenozoic Era, the Arabian plate is moving northeast away from Africa and subducting beneath Iran. As the sediments of the passive continental margin were initially subducted, during the early stages of collision between Arabia and Iran, the heating broke down the organic molecules and led to the formation of petroleum. The tilting of the Arabian block to the northeast allowed the newly formed petroleum to migrate upward into the interior of the Arabian plate. The continued subduction and collision with Iran folded the rocks, creating traps for petroleum to accumulate, such that the vast area south of the collision zone (known as the Zagros suture) is a major oil-producing region.

Reuters NewMedia Inc./Corbis

■ **Figure 1**

The Kuwaiti night skies were illuminated by 700 blazing oil wells set on fire by Iraqi troops during the 1991 Gulf War. The fires continued for nine months.

Patricia G. Gensel

■ **Figure 12.1**

Glossopteris leaves from the Upper Permian Dunedoo Formation, Australia. Fossils of the *Glossopteris* flora are found on all five of the Gondwana continents, thus providing evidence that these continents were at one time connected.

During the late 19th century, Austrian geologist Edward Suess noted the similarities between the Late Paleozoic plant fossils of India, Australia, South Africa, and South America as well as evidence of glaciation in the rock sequences of these southern continents. The plant fossils make up a unique flora that occurs in the coal layers just above the glacial deposits of these southern continents. This flora is very different from the contemporaneous coal swamp flora of the northern continents and is collectively known as the *Glossopteris* **flora,** after its most conspicuous genus (■ Figure 12.1).

In his book, *The Face of the Earth,* published in 1885, Suess proposed the name *Gondwana-land* (or **Gondwana** as we will use here) for a supercontinent composed of the aforementioned southern continents. Abundant fossils of the *Glossopteris* flora are found in coal beds in Gondwana, a province in India. Suess thought these southern continents were connected by land bridges over which plants and animals migrated. Thus, in his view, the similarities of fossils on these continents were due to the appearance and disappearance of the connecting land bridges.

The American geologist Frank Taylor published a pamphlet in 1910 presenting his own theory of continental drift. He explained the formation of mountain ranges as a result of the lateral movement of continents. He also envisioned the present-day continents as parts of larger polar continents that eventually broke apart and migrated toward the equator after Earth's rotation was supposedly slowed by gigantic tidal forces. According to Taylor, these tidal forces were generated when Earth captured the Moon about 100 million years ago.

Although we now know that Taylor's mechanism is incorrect, one of his most significant contributions was his suggestion that the Mid-Atlantic Ridge, discovered by the 1872–1876 British HMS *Challenger* expeditions, might mark the site along which an ancient continent broke apart to form the present-day Atlantic Ocean.

Alfred Wegener and the Continental Drift Hypothesis

Alfred Wegener, a German meteorologist (■ Figure 12.2), is generally credited with developing the hypothesis of **continental drift.** In his monumental book, *The Origin of Continents and Oceans* (first published in 1915), Wegener proposed that all landmasses were originally united into a single supercontinent that he named **Pangaea,** from the Greek meaning "all land." Wegener portrayed his grand concept of continental movement in a series of maps showing the breakup of Pangaea and the movement of the various continents to their present-day locations. Wegener amassed a tremendous amount of geologic, paleontologic, and climatologic evidence in support of continental drift, but the initial reaction of scientists to his then-heretical ideas can best be described as mixed.

Bildarchiv Preussischer Kulterbestiz

■ **Figure 12.2**

Alfred Wegener, a German meteorologist, proposed the continental drift hypothesis in 1912 based on a tremendous amount of geologic, paleontologic, and climatologic evidence. He is shown here waiting out the Arctic winter in an expedition hut in Greenland.

Opposition to Wegener's ideas became particularly widespread in North America after 1928, when the American Association of Petroleum Geologists held an international symposium to review the hypothesis of continental drift. After each side had presented its arguments, the opponents of continental drift were clearly in the majority, even though the evidence in support of continental drift, most of which came from the Southern Hemisphere, was impressive and difficult to refute. The main problem with the hypothesis was its lack of a mechanism to explain how continents, composed of granitic rocks, could seemingly move through the denser basaltic oceanic crust.

Nevertheless, the eminent South African geologist Alexander du Toit further developed Wegener's arguments and gathered more geologic and paleontologic evidence in support of continental drift. In 1937 du Toit published *Our Wandering Continents,* in which he contrasted the glacial deposits of Gondwana with coal deposits of the same age found in the continents of the Northern Hemisphere. To resolve this apparent climatologic paradox, du Toit moved the Gondwana continents to the South Pole and brought the northern continents together such that the coal deposits were located at the equator. He named this northern landmass **Laurasia.** It consisted of present-day North America, Greenland, Europe, and Asia (except for India).

Despite what seemed to be overwhelming evidence, most geologists still refused to accept the idea that continents moved. Not until the 1960s, when oceanographic research provided convincing evidence that the continents had once been joined together and subsequently separated, did the hypothesis of continental drift finally become widely accepted.

WHAT IS THE EVIDENCE FOR CONTINENTAL DRIFT?

What, then, was the evidence Wegener, du Toit, and others used to support the hypothesis of continental drift? First, they noted that the shorelines of continents fit together to form a large supercontinent and that marine, nonmarine, and glacial rock sequences of Pennsylvanian to Jurassic age are almost identical for all five Gondwana continents, strongly indicating that they were joined at one time. Furthermore, mountain ranges and glacial deposits match up when continents are united into a single landmass. And last, many of the same extinct plant and animal groups are found today on widely separated continents, indicating the continents must have been in proximity at one time. Wegener and his supporters argued that this vast amount of evidence from a variety of sources surely indicated that the continents must have been close together in the past.

PHYSICAL
Geology ⇌ Now ■ **Active Figure 12.3**

The best fit between continents occurs along the continental slope, where erosion would be minimal.

Continental Fit

Wegener, like some before him, was impressed by the close resemblance between the coastlines of continents on opposite sides of the Atlantic Ocean, particularly South America and Africa. He cited these similarities as partial evidence that the continents were at one time joined together as a supercontinent that subsequently split apart. As his critics pointed out, though, the configuration of coastlines results from erosional and depositional processes and therefore is continuously being modified. So even if the continents had separated during the Mesozoic Era, as Wegener proposed, it is not likely that the coastlines would fit exactly.

A more realistic approach is to fit the continents together along the continental slope where erosion would be minimal. In 1965 Sir Edward Bullard, an English geophysicist, and two associates showed that the best fit between the continents occurs at a depth of about 2000 m (■ Figure 12.3). Since then, other reconstructions using the latest ocean basin data have confirmed the close fit between continents when they are reassembled to form Pangaea.

Similarity of Rock Sequences and Mountain Ranges

If the continents were at one time joined, then the rocks and mountain ranges of the same age in adjoining locations on the opposite continents should closely match.

GEOLOGY
IN UNEXPECTED PLACES

A Man's Home Is His Castle

Imagine visiting a castle in Devonshire, England, or Glasgow, Scotland, and noticing that the same rocks used in the construction of those castles can be seen in the Catskill Mountains of New York (■ Figure 1). Whereas most people who visit castles in Great Britain and elsewhere in Europe are learning about the history of the castles—when they were built, why they were built, who lived in them, and other historical facts—geologists frequently are looking at the rocks that make up the walls of a castle and trying to determine what type of rock it is, how old it is, and anything else they can learn about it.

During the Devonian Period (408 to 360 million years ago), the North American continent and what is now Europe were moving toward each other along an oceanic–continental convergent plate boundary. As movement along this boundary continued, the ocean basin separating these landmasses shrunk until the two continents collided in what is known as the Acadian Orogeny. This mountain-building episode formed a large mountain range. Just as is happening today, those mountains began weathering and their eroded sediments were carried by streams and deposited as large deltas in the shallow seas adjacent to the mountains. The sediments deposited in what is now New York are referred to by geologists as the Catskill Delta. The counterpart in Great Britain is known as the Old Red Sandstone (■ Figure 2). These sediments were deposited in the same environment and reflect the conditions at the time of deposition.

Later during the Mesozoic Era (245 to 66 million years ago), as the supercontinent Pangaea broke apart along divergent plate boundaries, the Atlantic Ocean basin formed, separating North America from Europe. Even though the Devonian rocks of the present-day Catskill Mountains in New York are separated by several thousand kilometers from the Old Red Sandstone rocks of Great Britain, they were formed at the same time and deposited in the same environment hundreds of millions of years ago. That is why they have the same red color and mineralogy and contain many of the same fossils.

So, the next time you see someone closely inspecting the rocks of a castle wall, there is a good chance that person is a geologist or someone, like yourself, who took a geology course.

■ **Figure 1**

Remains of Goodrich Castle in Herefordshire, England, dated to the period between 1160 and 1270. Goodrich Castle is a good example of a structure built from rocks quarried from the Old Red Sandstone, a Devonian-age formation that is the counterpart of the rocks found in the Catskill Mountains of New York.

■ **Figure 2**

Closeup of cross-bedding in the Old Red Sandstone. This rock comes from a building in Glasgow, Scotland, and reflects that the sediments were deposited by a stream.

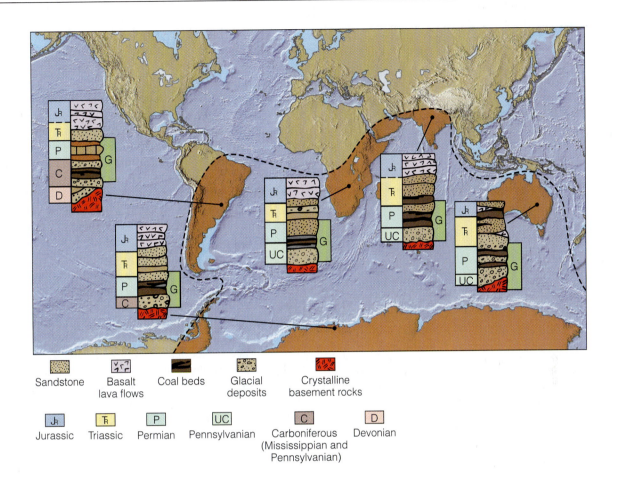

Sandstone — **Basalt lava flows** — **Coal beds** — **Glacial deposits** — **Crystalline basement rocks**

Jurassic (JR) — **Triassic** (TR) — **Permian** (P) — **Pennsylvanian** (UC) — **Carboniferous (Mississippian and Pennsylvanian)** (C) — **Devonian** (D)

■ **Figure 12.4**

Marine, nonmarine, and glacial rock sequences of Pennsylvanian to Jurassic age are nearly the same for all Gondwana continents. Such close similarity strongly suggests that they were joined at one time. The range indicated by G is that of the *Glossopteris* flora. Source: From *General Geology*, 5/e by R. J. Foster. Copyright © 1988. Reprinted by permission of Prentice-Hall, Inc., Upper Saddle River, NJ.

Such is the case for the Gondwana continents (■ Figure 12.4). Marine, nonmarine, and glacial rock sequences of Pennsylvanian to Jurassic age are almost identical for all five Gondwana continents, strongly indicating that they were joined at one time.

The trends of several major mountain ranges also support the hypothesis of continental drift. These mountain ranges seemingly end at the coastline of one continent only to apparently continue on another continent across the ocean. The folded Appalachian Mountains of North America, for example, trend northeastward through the eastern United States and Canada and terminate abruptly at the Newfoundland coastline. Mountain ranges of the same age and deformational style occur in eastern Greenland, Ireland, Great Britain, and Norway. Even though these mountain ranges are currently separated by the Atlantic Ocean, they form an essentially continuous mountain range when the continents are positioned next to each other (■ Figure 12.5).

Glacial Evidence

During the Late Paleozoic Era, massive glaciers covered large continental areas of the Southern Hemisphere.

Evidence for this glaciation includes layers of till (sediments deposited by glaciers) and striations (scratch marks) in the bedrock beneath the till. Fossils and sedimentary rocks of the same age from the Northern Hemisphere, however, give no indication of glaciation. Fossil plants found in coals indicate that the Northern Hemisphere had a tropical climate during the time that the Southern Hemisphere was glaciated.

All the Gondwana continents except Antarctica are currently located near the equator in subtropical to tropical climates. Mapping of glacial striations in bedrock in Australia, India, and South America indicates that the glaciers moved from the areas of the present-day oceans onto land. This would be highly unlikely because large continental glaciers (such as occurred on the Gondwana continents during the Late Paleozoic Era) flow outward from their central area of accumulation toward the sea.

If the continents did not move during the past, one would have to explain how glaciers moved from the oceans onto land and how large-scale continental glaciers formed near the equator. But if the continents are reassembled as a single landmass with South Africa located at the South Pole, the direction of movement of Late Paleozoic continental glaciers

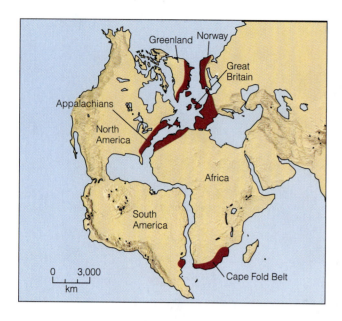

■ Figure 12.5

When continents are brought together, their mountain ranges form a single continuous range of the same age and style of deformation throughout. Such evidence indicates that the continents were at one time joined and were subsequently separated.

makes sense. Furthermore, this geographic arrangement places the northern continents nearer the tropics, which is consistent with the fossil and climatologic evidence from Laurasia (■ Figure 12.6).

Fossil Evidence

Some of the most compelling evidence for continental drift comes from the fossil record (■ Figure 12.7). Fossils of the *Glossopteris* flora are found in equivalent Pennsylvanian- and Permian-aged coal deposits on all five Gondwana continents. The *Glossopteris* flora is characterized by the seed fern *Glossopteris* (Figure 12.1) as well as by many other distinctive and easily identifiable plants. Pollen and spores of plants can be dispersed over great distances by wind, but *Glossopteris*-type plants produced seeds that are

too large to have been carried by winds. Even if the seeds had floated across the ocean, they probably would not have remained viable for any length of time in salt-water.

The present-day climates of South America, Africa, India, Australia, and Antarctica range from tropical to polar and are much too diverse to support the type of plants in the *Glossopteris* flora. Wegener therefore reasoned that these continents must once have been joined so that these widely separated localities were all in the same latitudinal climatic belt (Figure 12.7).

The fossil remains of animals also provide strong evidence for continental drift. One of the best examples is *Mesosaurus*, a freshwater reptile whose fossils are found in Permian-aged rocks in certain regions of Brazil and South Africa and nowhere else in the world (■ Figures 12.7 and 12.8). Because the physiologies of freshwater and marine animals are completely different, it is hard to imagine how a freshwater reptile could have swum across the Atlantic Ocean and found a freshwater environment nearly identical to its former habitat. Moreover, if *Mesosaurus* could have swum across the ocean, its fossil remains should be widely dispersed. It is more logical to assume that *Mesosaurus* lived in lakes in what are now adjacent areas of South America and Africa, but were then united into a single continent.

Lystrosaurus and *Cynognathus* are both land-dwelling reptiles that lived during the Triassic Period;

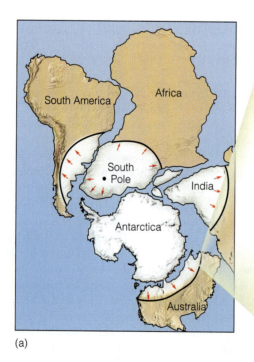

(a)

(b)

■ Figure 12.6

(a) By moving the Gondwana continents together so that South Africa is located at the South Pole, the glacial movements indicated by the striations (red arrows) makes sense. In this situation, the glacier (white area), located in a polar climate, moved radially outward from a thick central area toward its periphery.
(b) Permian-aged glacial striations in bedrock exposed at Hallet's Cove, Australia, indicate the direction of glacial movement more than 200 million years ago.

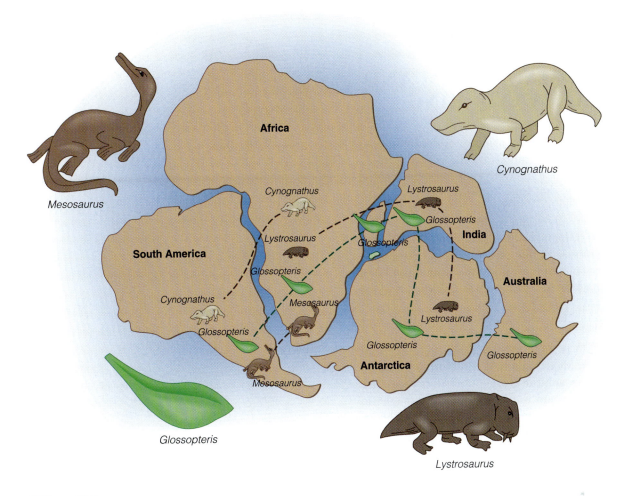

■ Figure 12.7

Some of the animals and plants whose fossils are found today on the widely separated continents of South America, Africa, India, Australia, and Antarctica. These continents were joined together during the Late Paleozoic to form Gondwana, the southern landmass of Pangaea. *Glossopteris* and similar plants are found in Pennsylvanian- and Permian-aged deposits on all five continents. *Mesosaurus* is a freshwater reptile whose fossils are found in Permian-aged rocks in Brazil and South Africa. *Cynognathus* and *Lystrosaurus* are land reptiles that lived during the Early Triassic Period. Fossils of *Cynognathus* are found in South America and Africa, and fossils of *Lystrosaurus* have been recovered from Africa, India, and Antarctica. Source: Modified form E. H. Colbert, *Wandering Lands and Animals* (1973): 72, Figure 31.

their fossils are found only on the present-day continental fragments of Gondwana (Figure 12.7). Because they are both land animals, they certainly could not have swum across the oceans currently separating the Gondwana continents. Therefore the continents must once have been connected.

Notwithstanding all of the empirical evidence presented by Wegener and later by du Toit and others, most geologists simply refused to entertain the idea that continents might have moved in the past. The geologists were not necessarily being obstinate about accepting new ideas; rather, they found the evidence for continental drift inadequate and unconvincing. In part, this was because no one could provide a suitable mechanism to explain how continents could move over Earth's surface. Not until new evidence from studies of Earth's magnetic field and oceanographic research showed that the ocean

Reed Wicander

■ Figure 12.8

Mesosaurus, a Permian-aged freshwater reptile whose fossil remains are found in Brazil and South Africa, indicating these two continents were joined at the end of the Paleozoic Era.

basins were geologically young features was interest in continental drift revived.

PALEOMAGNETISM AND POLAR WANDERING

Some of the most convincing evidence for continental drift came from the study of paleomagnetism, a relatively new discipline during the 1950s. During that time, some geologists were researching past changes of Earth's magnetic field in order to better understand the present-day magnetic field. As so often happens in science, these studies led to other discoveries. In this case, they led to the discovery that the ocean basins are geologically young and that the continents have indeed moved during the past, just as Wegener and others had proposed.

Recall from Chapter 10 that Earth's magnetic poles correspond closely to the location of the geographic poles (see Figure 10.18b). When magma cools, the iron-bearing minerals align themselves with Earth's magnetic field when they reach the Curie point, thus recording both the direction and the intensity of the magnetic field. This information can be used to determine the location of Earth's magnetic poles and the latitude of the rock when it formed.

As paleomagnetic research progressed in the 1950s, some unexpected results emerged. When geologists measured the magnetism of recent rocks, they found it was generally consistent with Earth's current magnetic field. The paleomagnetism of ancient rocks, though, showed different orientations. For example, paleomagnetic studies of Silurian lava flows in North America indicated that the north magnetic pole was located in the western Pacific Ocean at that time, whereas the paleomagnetic evidence from Permian lava flows pointed to yet another location in Asia. When plotted on a map, the paleomagnetic readings of numerous lava flows from all ages in North America trace the apparent movement of the magnetic pole through time (■ Figure 12.9). This paleomagnetic evidence from a single continent could be interpreted in three ways: The continent remained fixed and the north magnetic pole moved; the north magnetic pole stood still and the continent moved; or both the continent and the north magnetic pole moved.

On analysis, magnetic minerals from European Silurian and Permian lava flows pointed to different magnetic pole locations than those of the same age from North America (Figure 12.9). Furthermore, analysis of lava flows from all continents indicated each continent had its own series of magnetic poles. Does this mean there were different north magnetic poles for each continent? That would be highly unlikely and difficult to reconcile with the theory accounting for Earth's magnetic field.

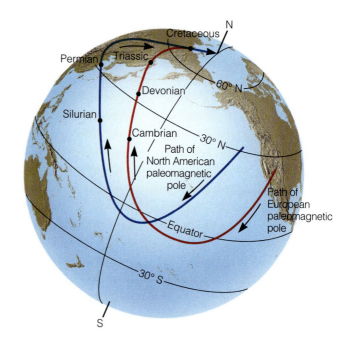

■ **Figure 12.9**

The apparent paths of polar wandering for North America and Europe. The apparent location of the north magnetic pole is shown for different periods on each continent's polar wandering path. Source: From A. Cox and R. R. Doell, "Review of Paleomagnetism," *G. S. A. Bulletin*, vol. 71, figure 33, page 758, with permission of the publisher, the Geological Society of America, Boulder, Colorado. USA. Copyright © 1955 Geological Society of America.

The best explanation for such data is that the magnetic poles have remained at their present locations near the geographic north and south poles and the continents have moved. When the continental margins are fitted together so that the paleomagnetic data point to only one magnetic pole, we find, just as Wegener did, that the rock sequences and glacial deposits match, and that the fossil evidence is consistent with the reconstructed paleogeography.

WHAT IS SEAFLOOR SPREADING?

A renewed interest in oceanographic research led to extensive mapping of the ocean basins during the 1960s. Such mapping revealed an oceanic ridge system more than 65,000 km long, constituting the most extensive mountain range in the world. Perhaps the best-known part of the ridge system is the Mid-Atlantic Ridge, which divides the Atlantic Ocean basin into two nearly equal parts (■ Figure 12.10).

As a result of the oceanographic research conducted during the 1950s, Harry Hess of Princeton University

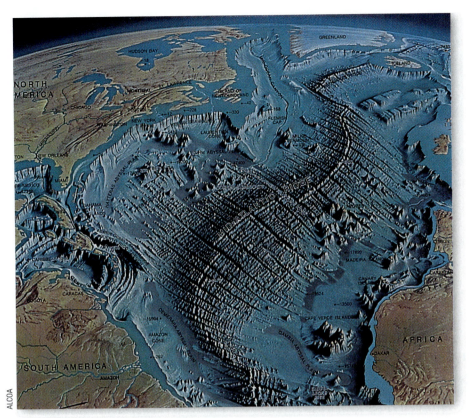

ALCOA

■ **Figure 12.10**

Artistic view of what the Atlantic Ocean basin would look like without water. The major feature is the Mid-Atlantic Ridge.

proposed the theory of **seafloor spreading** in 1962 to account for continental movement. Hess suggested that continents do not move across oceanic crust, but rather that the continents and oceanic crust move together. He suggested that the seafloor separates at oceanic ridges where new crust is formed by upwelling magma. As the magma cools, the newly formed oceanic crust moves laterally away from the ridge.

As a mechanism to drive this system, Hess revived the idea (proposed in the 1930s and 1940s by Arthur Holmes and others) of **thermal convection cells** in the mantle; that is, hot magma rises from the mantle, intrudes along fractures defining oceanic ridges, and thus forms new crust. Cold crust is subducted back into the mantle at oceanic trenches, where it is heated and recycled, thus completing a thermal convection cell (see Figure 1.11).

How could Hess's hypothesis be confirmed? If new crust is forming at oceanic ridges and Earth's magnetic field is periodically reversing itself, then these magnetic reversals should be preserved as magnetic anomalies in the rocks of the oceanic crust (■ Figure 12.11).

Around 1960 scientists from the Scripps Institution of Oceanography in California gathered magnetic data that indicated an unusual pattern of alternating positive and negative magnetic anomalies for the Pacific Ocean

seafloor off the West Coast of North America. The pattern consisted of a series of roughly north–south parallel stripes, but they were broken and offset by essentially east–west fractures. Not until 1963 did F. Vine and D. Matthews of Cambridge University and L. W. Morley, a Canadian geologist, independently arrive at a model that explained this pattern of magnetic anomalies.

These three geologists proposed that when magma intruded along the crests of oceanic ridges, it recorded the magnetic polarity at the time it cooled. As the ocean floor moved away from these oceanic ridges, repeated intrusions would form a symmetric series of magnetic stripes, recording periods of normal and reverse polarity (Figure 12.11). Shortly thereafter, the Vine, Matthews, and Morley proposal was supported by evidence from magnetic readings across the Reykjanes Ridge, part of the Mid-Atlantic Ridge south of Iceland. A group from the Lamont-Doherty Geological Observatory at Columbia University found that magnetic anomalies in this area did form stripes that were distributed parallel to and symmetric about the oceanic ridge. By the end of the 1960s, comparable magnetic anomaly patterns were found surrounding most oceanic ridges, and thus conclusively confirming Hess's theory of seafloor spreading.

Magnetic surveys of the ocean floor demonstrate that the youngest oceanic crust is adjacent to the spreading ridges and that the age of the crust increases with distance from the ridge axis, as would be expected according to the seafloor spreading hypothesis (■ Figure 12.12). Furthermore, the oldest oceanic crust is less than 180 million years old, whereas the oldest continental crust is 3.96 billion years old; this difference in age provides confirmation that the ocean basins are geologically young features whose openings and closings are partially responsible for continental movement.

Deep-Sea Drilling and the Confirmation of Seafloor Spreading

For many geologists, the paleomagnetic data amassed in support of continental drift and seafloor spreading were convincing. Results from the Deep-Sea Drilling Project (see Chapter 11) have confirmed the interpretations

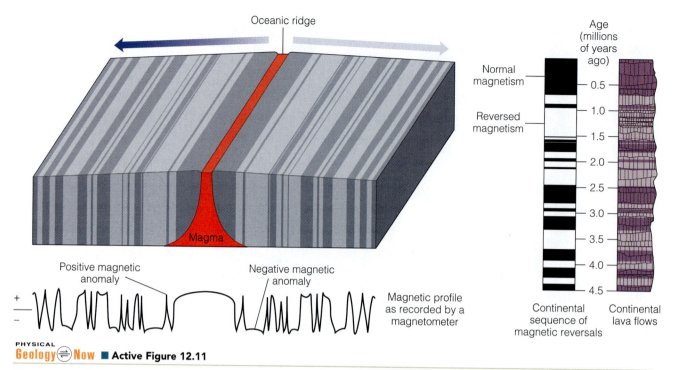

Oceanic ridge

Normal
magnetism

Reversed
magnetism

Age
(millions
of years
ago)

Magma

Positive magnetic
anomaly

Negative magnetic
anomaly

Magnetic profile
as recorded by a
magnetometer

Continental
sequence of
magnetic reversals

Continental
lava flows

Geology Now ■ Active Figure 12.11

The sequence of magnetic anomalies preserved within the oceanic crust on both sides of an oceanic ridge is identical to the sequence of magnetic reversals already known from continental lava flows. Magnetic anomalies are formed when basaltic magma intrudes into oceanic ridges; when the magma cools below the Curie point, it records Earth's magnetic polarity at the time. Seafloor spreading splits the previously formed crust in half so that it moves laterally away from the oceanic ridge. Repeated intrusions record a symmetric series of magnetic anomalies that reflect periods of normal and reversed polarity. The magnetic anomalies are recorded by a magnetometer, which measures the strength of the magnetic field. Source: Reprinted with permission from A. Cox, "Geomagnetic Reversals," *Science 163*, January 17, 1969. Copyright © 1969 American Association for the Advancement of Science.

made from earlier paleomagnetic studies. Cores of deep-sea sediments and seismic profiles obtained by the *Glomar Challenger* and other research vessels have provided much of the data that support the seafloor spreading hypothesis.

According to this hypothesis, oceanic crust is continuously forming at mid-oceanic ridges, moves away from these ridges by seafloor spreading, and is consumed at subduction zones. If this is the case, oceanic crust should be youngest at the ridges and become progressively older with increasing distance away from them. Moreover, the age of the oceanic crust should be symmetrically distributed about the ridges. As we have just noted, paleomagnetic data confirm these statements. Furthermore, fossils from sediments overlying the oceanic crust and radiometric dating of rocks found on oceanic islands both substantiate this predicted age distribution.

Sediments in the open ocean accumulate, on average, at a rate of less than 0.3 cm per 1000 years. If the ocean basins were as old as the continents, we would expect deep-sea sediments to be several kilometers thick. However, data from numerous drill holes indicate that deep-sea sediments are at most only a few hundred meters thick and are thin or absent at oceanic ridges. Their near-absence at the ridges should come as no surprise because these are the areas where new crust is continuously produced by volcanism and seafloor spreading. Ac-

cordingly, sediments have had little time to accumulate at or very close to spreading ridges where the oceanic crust is young, but their thickness increases with distance away from the ridges (■ Figure 12.13).

WHY IS PLATE TECTONICS A UNIFYING THEORY?

late tectonic theory is based on a simple model of Earth. The rigid lithosphere, consisting of both oceanic and continental crust as well as the underlying upper mantle, consists of many variable-sized pieces called **plates** (■ Figure 12.14). The plates vary in thickness; those composed of upper mantle and continental crust are as much as 250 km thick, whereas those of upper mantle and oceanic crust are up to 100 km thick.

The lithosphere overlies the hotter and weaker semiplastic asthenosphere. It is thought that movement resulting from some type of heat transfer system within the asthenosphere causes the overlying plates to move. As plates move over the asthenosphere, they separate, mostly at oceanic ridges; in other areas such as at oceanic trenches, they collide and are subducted back into the mantle.

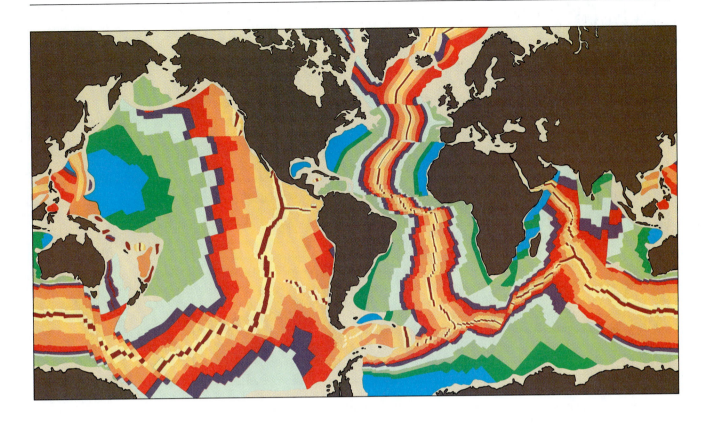

Legend:

- Pleistocene to Recent (0–1.6 M.Y.A.)
- Pliocene (1.6–5 M.Y.A.)
- Miocene (5–24 M.Y.A.)
- Oligocene (24–37 M.Y.A.)
- Eocene (37–58 M.Y.A.)
- Paleocene (58–66 M.Y.A.)
- Late Cretaceous (66–88 M.Y.A.)
- Middle Cretaceous (88–118 M.Y.A.)
- Early Cretaceous (118–144 M.Y.A.)
- Late Jurassic (144–161 M.Y.A.)

■ **Figure 12.12**

The age of the world's ocean basins established from magnetic anomalies demonstrates that the youngest oceanic crust is adjacent to the spreading ridges and that its age increases away from the ridge axis. Source: From Larson, R. L. et al. (1985). *The Bedrock Geology of the World*, W. H. Freeman and Co., New York, NY.

An easy way to visualize plate movement is to think of a conveyor belt moving luggage from an airplane's cargo hold to a baggage cart. The conveyor belt represents convection currents within the mantle, and the luggage represents Earth's lithospheric plates. The luggage is moved along by the conveyor belt until it is dumped into the baggage cart in the same way plates are moved by convection cells until they are subducted into

■ **Figure 12.13**

The total thickness of deep-sea sediments increases away from oceanic ridges. This is because oceanic crust becomes older away from oceanic ridges, and there has been more time for sediment to accumulate.

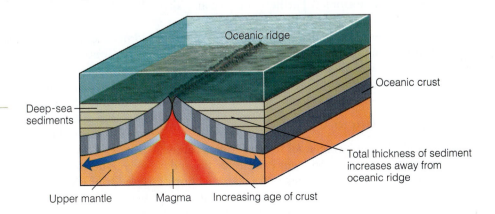

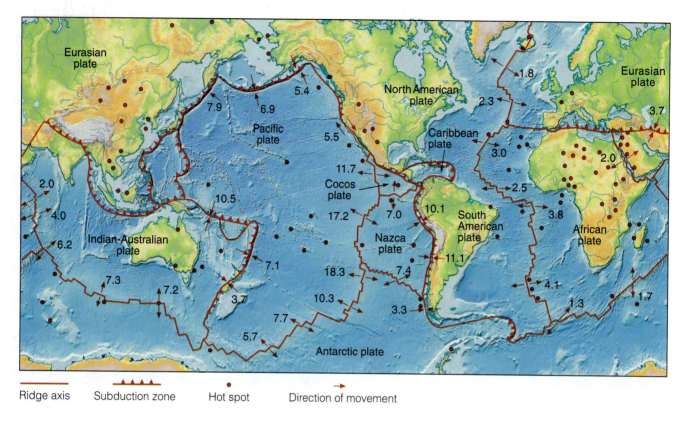

Ridge axis   Subduction zone   Hot spot   Direction of movement

■ **Figure 12.14**

A map of the world showing the plates, their boundaries, relative motion and rates of movement in centimeters per year, and hot spots.

Earth's interior. Although this analogy allows you to visualize how the mechanism of plate movement takes place, remember that this analogy is limited. The major limitation is that, unlike the luggage, plates consist of continental and oceanic crust, which have different densities, and only oceanic crust is subducted into Earth's interior. Nonetheless, this analogy does provide an easy way to visualize plate movement.

Most geologists accept plate tectonic theory, in part because the evidence for it is overwhelming and it ties together many seemingly unrelated geologic features and events and shows how they are interrelated. Consequently, geologists now view such geologic processes as mountain building, seismicity, and volcanism from the perspective of plate tectonics. Furthermore, because all inner planets have had a similar origin and early history, geologists are interested in determining whether plate tectonics is unique to Earth or operates in the same way on other planets (see "Tectonics of the Terrestrial Planets" on pages 346 and 347).

The Supercontinent Cycle

As a result of plate movement, all the continents came together to form the supercontinent Pangaea by the end of the Paleozoic Era. Pangaea began fragmenting during the Triassic Period and continues to do so, thus accounting for the present distribution of continents and ocean basins. It has been proposed that supercontinents consisting of all or most of Earth's landmasses form, break up, and reform in a cycle spanning about 500 million years.

The *supercontinent cycle hypothesis* is an expansion on the ideas of the Canadian geologist J. Tuzo Wilson. During the early 1970s, Wilson proposed a cycle (now known as the Wilson cycle) that includes continental fragmentation, the opening and closing of an ocean basin, and reassembly of the continent. According to the supercontinent cycle hypothesis, heat accumulates beneath a supercontinent because rocks of continents are poor conductors of heat. As a result of the heat accumulation, the supercontinent domes upward and fractures. Basaltic magma rising from below fills the fractures. As a basalt-filled fracture widens, it begins subsiding and forms a long, narrow ocean such as the present-day Red Sea. Continued rifting eventually forms an expansive ocean basin such as the Atlantic.

According to proponents of the supercontinent cycle, one of the most convincing arguments for their hypothesis is the "surprising regularity" of mountain building caused by compression during continental collisions. These mountain-building episodes occur about every 400 to 500 million years and are followed by an episode of rifting about 100 million years later. In other words, a supercontinent fragments and its individual plates disperse following a rifting episode, an interior

What Would You Do

You've been selected to be part of the first astronaut team to go to Mars. While your two fellow crewmembers descend to the Martian surface, you'll be staying in the command module and circling the Red Planet. As part of the geologic investigation of Mars, one of the crewmembers will be mapping the geology around the landing site and deciphering the geologic history of the area. Your job will be to observe and photograph the planet's surface and try to determine whether Mars had an active plate tectonic regime in the past and whether there is current plate movement. What features would you look for, and what evidence might reveal current or previous plate activity?

ocean forms, and then the dispersed fragments reassemble to form another supercontinent.

The supercontinent cycle is yet another example of how interrelated the various systems and subsystems of Earth are and how they operate over vast periods of geologic time.

WHAT ARE THE THREE TYPES OF PLATE BOUNDARIES?

Because it appears that plate tectonics has operated since at least the Proterozoic Eon, it is important that we understand how plates move and interact with each other and how ancient plate boundaries are recognized. After all, the move-

ment of plates has profoundly affected the geologic and biologic history of this planet.

Geologists recognize three major types of plate boundaries: *divergent, convergent,* and *transform* (Table 12.2). Along these boundaries new plates are formed, are consumed, or slide laterally past one another. Interaction of plates at their boundaries accounts for most of Earth's seismic and volcanic activity and, as will be apparent in the next chapter, the origin of mountain systems.

Divergent Boundaries

Divergent plate boundaries or *spreading ridges* occur where plates are separating and new oceanic lithosphere is forming. Divergent boundaries are places where the crust is extended, thinned, and fractured as magma, derived from the partial melting of the mantle, rises to the surface. The magma is almost entirely basaltic and intrudes into vertical fractures to form dikes and pillow lava flows (see Figure 4.9). As successive injections of magma cool and solidify, they form new oceanic crust and record the intensity and orientation of Earth's magnetic field (Figure 12.11). Divergent boundaries most commonly occur along the crests of oceanic ridges—for example, the Mid-Atlantic Ridge. Oceanic ridges are thus characterized by rugged topography with high relief resulting from displacement of rocks along large fractures, shallow-focus earthquakes, high heat flow, and basaltic flows or pillow lavas.

Divergent boundaries are also present under continents during the early stages of continental breakup (■ Figure 12.15). When magma wells up beneath a continent, the crust is initially elevated, stretched, and thinned, producing fractures and rift valleys (Figure 12.15a). During this stage, magma typically intrudes into the faults and fractures forming sills, dikes, and lava flows; the latter often cover the rift valley floor (Figure 12.15b). The East

Table 12.2

Types of Plate Boundaries

Type	Example	Landforms	Volcanism
Divergent			
Oceanic	Mid-Atlantic Ridge	Mid-oceanic ridge with axial rift valley	Basalt
Continental	East African Rift Valley	Rift valley	Basalt and rhyolite, no andesite
Convergent			
Oceanic–oceanic	Aleutian Islands	Volcanic island arc, offshore oceanic trench	Andesite
Oceanic–continental	Andes	Offshore oceanic trench, volcanic mountain chain, mountain belt	Andesite
Continental–Continental	Himalayas	Mountain belt	Minor
Transform	San Andreas fault	Fault valley	Minor

Tectonics of the Terrestrial Planets

The four inner, or terrestrial, planets—Mercury, Venus, Earth, and Mars—all had a similar early history involving accretion, differentiation into a metallic core and silicate mantle and crust, and formation of an early atmosphere by outgassing. Their early history was also marked by widespread volcanism and meteorite impacts, both of which helped modify their surfaces.

Whereas the other three terrestrial planets as well as some of the Jovian moons display internal activity, Earth appears to be unique in that its surface is broken into a series of plates.

A color-enhanced photomosaic of Mercury shows its heavily cratered surface, which has changed very little since its early history.

JPL/NASA

Courtesy of Victor Royer

Images of **Mercury** sent back by *Mariner 10* show a heavily cratered surface with the largest impact basins filled with what appear to be lava flows similar to the lava plains on Earth's moon. The lava plains are not deformed, however, indicating that there has been little or no tectonic activity.

Another feature of Mercury's surface is a large number of scarps, a feature usually associated with earthquake activity. Yet, some scientists think that these scarps formed when Mercury cooled and contracted.

Seven scarps (indicated by arrows) can clearly be seen in this image. These scarps might have formed when Mercury cooled and contracted early in its history.

Of all the planets, **Venus** is the most similar in size and mass to Earth, but it differs in most other respects. Whereas Earth is dominated by plate tectonics, volcanism seems to have been the dominant force in the evolution of the Venusian surface. Even though no active volcanism has been observed on Venus, the various-sized volcanic features and what appear to be folded mountains indicate a once-active planetary interior. All of these structures appear to be the products of rising convection currents of magma pushing up under the crust and then sinking back into the Venusian interior.

JPL/NASA

A color-enhanced photomosaic of Venus based on radar images beamed back to Earth by the *Magellan* spacecraft. This image shows impact craters and volcanic features characteristic of the planet.

JPL/NASA

Arrows point to a 600-km segment of Venus' 6800-km long Baltis Vallis, the longest known lava flow channel in our solar system.

NASA

Venus' Aine Corona, about 200 km in diameter, is ringed by concentric faults, suggesting that it was pushed up by rising magma. A network of fractures is visible in the upper right of this image as well as a recent lava flow at the center of the corona, several volcanic domes in the lower portion of the image, and a large volcanic pancake dome in the upper left of the image.

Volcano Sapas Mons contains two lava-filled calderas and is flanked by lava flows, attesting to the volcanic activity that was once common on Venus.

Mars, the Red Planet, has numerous features that indicate an extensive early period of volcanism. These include Olympus Mons, the solar system's largest volcano, lava flows, and uplifted regions thought to have resulted from mantle convection. In addition to volcanic features, Mars displays abundant evidence of tensional tectonics, including numerous faults and large fault-produced valley structures. Whereas Mars was tectonically active during the past, no evidence indicates that plate tectonics comparable to that on Earth has ever occurred there.

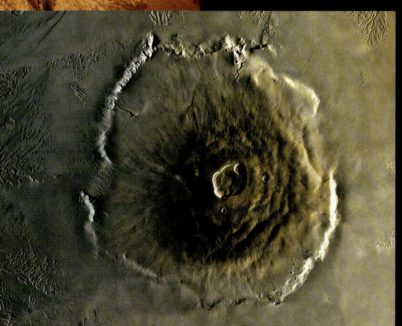

A photomosaic of Mars shows a variety of geologic structures, including the southern polar ice cap.

A vertical view of Olympus Mons, a shield volcano and the largest volcano in our solar system. The edge of the Olympus Mons caldera is marked by a cliff several kilometers high rather than a moat as in Mauna Loa, Earth's largest shield volcano.

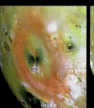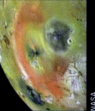

Although not a terrestrial planet, **Io**, the innermost of Jupiter's Galilean moons, must be mentioned. Images from the *Voyager* and *Galileo* spacecrafts show that Io has no impact craters. In fact, more than a hundred active volcanoes are visible on the moon's surface, and the sulfurous gas and ash erupted by these volcanoes bury any newly formed meteorite impact craters. Because of its proximity to Jupiter, the heat source of Io is probably tidal heating, in which the resulting friction is enough to at least partially melt Io's interior and drive its volcanoes.

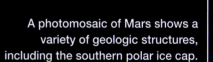

50 km

Volcanic features of Io, the innermost moon of Jupiter. As shown in these digitally enhanced color images, Io is a very volcanically active moon.

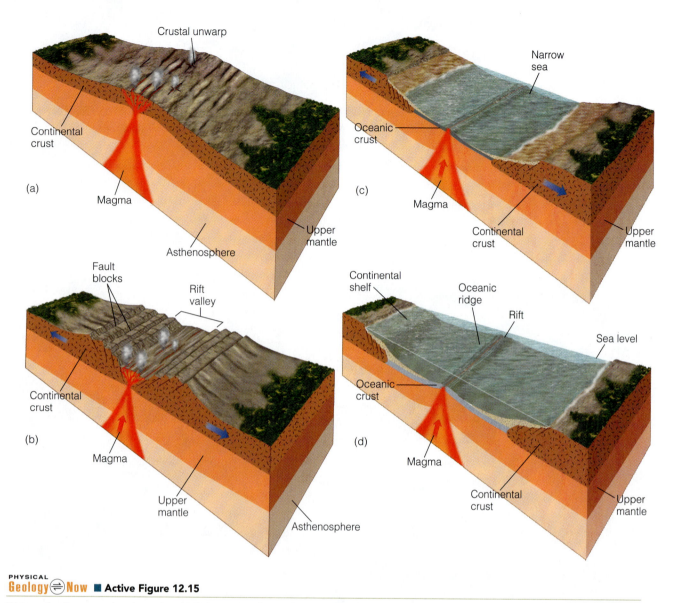

PHYSICAL
Geology⇌Now ■ Active Figure 12.15

History of a divergent plate boundary. (a) Rising magma beneath a continent pushes the crust up, producing numerous cracks and fractures. (b) As the crust is stretched and thinned, rift valleys develop and lava flows onto the valley floors. (c) Continued spreading further separates the continent until a narrow seaway develops. (d) As spreading continues, an oceanic ridge system forms, and an ocean basin develops and grows.

African Rift Valley is an excellent example of this stage of continental breakup (■ Figure 12.16).

As spreading proceeds, some rift valleys will continue to lengthen and deepen until the continental crust eventually breaks, and a narrow linear sea is formed, separating two continental blocks (Figure 12.15c). The Red Sea separating the Arabian Peninsula from Africa (Figure 12.16a) and the Gulf of California, which separates Baja California from mainland Mexico, are good examples of this more advanced stage of rifting.

As a newly created narrow sea continues to enlarge, it may eventually become an expansive ocean basin such as the Atlantic Ocean basin is today, separating North and South America from Europe and Africa by thousands of kilometers (Figure 12.15d). The Mid-Atlantic Ridge is the boundary between these diverging plates; the American plates are moving westward, and the Eurasian and African plates are moving eastward.

PHYSICAL
Geology⇌Now Click Geology Interactive to work through an activity on Plate Boundaries through Plate Tectonics.

An Example of Ancient Rifting

What features in the geologic record can geologists use to recognize ancient rifting? Associated with regions of continental rifting are faults, dikes, sills, lava flows, and thick sedimentary sequences within rift valleys. The Triassic fault basins of the eastern United States are a good example of ancient continental rifting (■ Figure

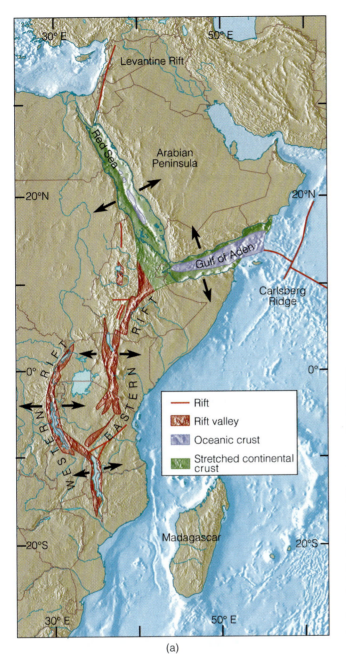

(a)

(b)

Robert Caputo/Aurora

Legend:
— Rift
▨ Rift valley
▨ Oceanic crust
▨ Stretched continental crust

■ **Figure 12.16**

(a) The East African Rift Valley is being formed by the separation of eastern Africa from the rest of the continent along a divergent plate boundary. The Red Sea represents a more advanced stage of rifting, in which two continental blocks are separated by a narrow sea. (b) View looking down the Great Rift Valley of Africa. Little Magadi, seen in the background, is one of numerous soda lakes forming in the valley. Because of high evaporation rates and lack of any drainage outlets, these lakes are very saline. The Great Rift Valley is part of the system of rift valleys resulting from stretching of the crust as plates move away from each other in eastern Africa.

12.17a). These fault basins mark the zone of rifting that occurred when North America split apart from Africa. The basins contain thousands of meters of continental sediment and are riddled with dikes and sills.

Convergent Boundaries

Whereas new crust forms at divergent plate boundaries, older crust must be destroyed and recycled in order for the entire surface area of Earth to remain the same. Otherwise, we would have an expanding Earth.

Such plate destruction occurs at **convergent plate boundaries,** where two plates collide and the leading edge of one plate is subducted beneath the margin of the other plate and eventually is incorporated into the as-

thenosphere. A dipping plane of earthquake foci, referred to as a *Benioff zone,* defines subduction zones (see Figure 9.5). Most of these planes dip from oceanic trenches beneath adjacent island arcs or continents, marking the surface of slippage between the converging plates.

Convergent boundaries are characterized by deformation, volcanism, mountain building, metamorphism, seismicity, and important mineral deposits. Three types of convergent plate boundaries are recognized: oceanic–oceanic, oceanic–continental, and continental–continental.

Oceanic–Oceanic Boundaries   When two oceanic plates converge, one is subducted beneath the other along an **oceanic–oceanic plate boundary**

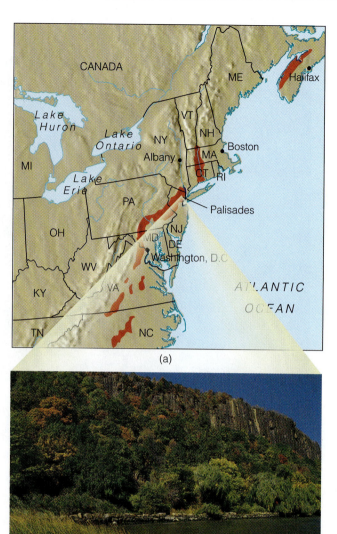

(a)

(b)

■ **Figure 12.17**

(a) Areas where Triassic fault-block basin deposits crop out in eastern North America. (b) Palisades of the Hudson River. This sill was one of many that were intruded into the fault-block basin sediments during the Late Triassic rifting that marked the separation of North America from Africa.

(■ Figure 12.18a). The subducting plate bends downward to form the outer wall of an oceanic trench. A *subduction complex*, composed of wedge-shaped slices of highly folded and faulted marine sediments and oceanic lithosphere scraped off the descending plate, forms along the inner wall. As the subducting plate descends into the mantle, it is heated and partially melted, generating magma, commonly of andesitic composition. This magma is less dense than the surrounding mantle rocks and rises to the surface of the nonsubducted plate, forming a curved chain of volcanoes called a *volcanic island arc* (any plane intersecting a sphere makes an arc). This arc is nearly parallel to the oceanic trench and is separated from it by a distance of up to several

hundred kilometers—the distance depending on the angle of dip of the subducting plate (Figure 12.18a).

In those areas where the rate of subduction is faster than the forward movement of the overriding plate, the lithosphere on the landward side of the volcanic island arc may be subjected to tensional stress and stretched and thinned, resulting in the formation of a *back-arc basin*. This back-arc basin may grow by spreading if magma breaks through the thin crust and forms new oceanic crust (Figure 12.18a). A good example of a back-arc basin associated with an oceanic–oceanic plate boundary is the Sea of Japan between the Asian continent and the islands of Japan.

Most present-day active volcanic island arcs are in the Pacific Ocean basin and include the Aleutian Islands, the Kermadec–Tonga arc, and the Japanese (Figure 12.18b) and Philippine Islands. The Scotia and Antillean (Caribbean) island arcs are present in the Atlantic Ocean basin.

Oceanic–Continental Boundaries   When a denser oceanic plate is subducted under continental crust an **oceanic–continental plate boundary** occurs (■ Figure 12.19a). Just as at oceanic–oceanic convergent plate boundaries, the descending oceanic plate forms the outer wall of an oceanic trench.

The magma generated by subduction rises beneath the continent and either crystallizes as large plutons before reaching the surface or erupts at the surface to produce a chain of volcanoes (also called a *volcanic arc*). An excellent example of an oceanic–continental plate boundary is the Pacific coast of South America, where the oceanic Nazca plate is currently being subducted beneath South America (Figure 12.19b; see also Chapter 13). The Peru–Chile Trench marks the site of subduction, and the Andes Mountains are the resulting volcanic mountain chain on the nonsubducting plate.

Continental–Continental Boundaries   Two continents approaching each other will initially be separated by an ocean floor that is being subducted under one continent. The edge of that continent will display the features characteristic of oceanic–continental convergence. As the ocean floor continues to be subducted, the two continents will come closer together until they eventually collide. Because continental lithosphere, which consists of continental crust and the upper mantle, is less dense than oceanic lithosphere (oceanic crust and upper mantle), it cannot sink into the asthenosphere. Although one continent may partly slide under the other, it cannot be pulled or pushed down into a subduction zone (■ Figure 12.20a).

When two continents collide, they are welded together along a zone marking the former site of subduction. At this **continental–continental plate boundary,** an interior mountain belt is formed consisting of deformed sedimentary rocks, igneous intrusions, metamor-

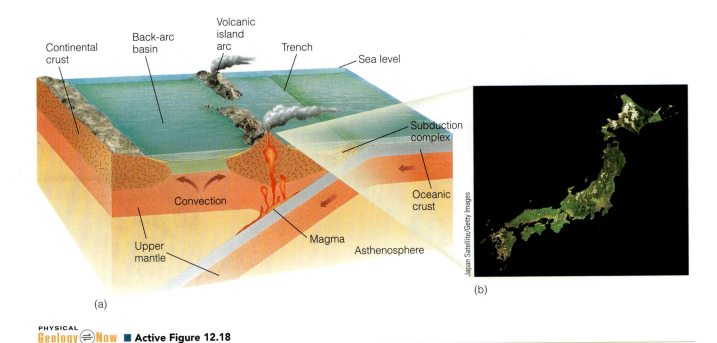

(a)

(b)

PHYSICAL
Geology ⇌ Now ■ **Active Figure 12.18**

Oceanic–oceanic plate boundary. (a) An oceanic trench forms where one oceanic plate is subducted beneath another. On the nonsubducted plate, a volcanic island arc forms from the rising magma generated from the subducting plate. (b) Satellite image of Japan. The Japanese Islands are a volcanic island arc resulting from the subduction of one oceanic plate beneath another oceanic plate.

phic rocks, and fragments of oceanic crust. In addition, the entire region is subjected to numerous earthquakes. The Himalayas in central Asia, the world's youngest and highest mountain system, resulted from the collision between India and Asia that began 40 to 50 million

years ago and is still continuing (Figure 12.20b; see Chapter 13).

PHYSICAL
Geology ⇌ Now Click Geology Interactive to work through an activity on Plate Boundaries through Plate Tectonics.

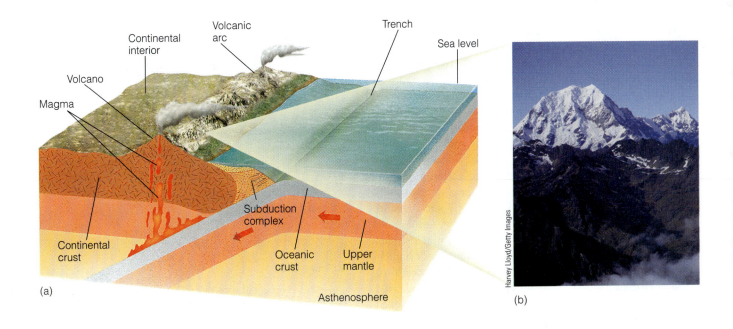

(a)

(b)

PHYSICAL
Geology ⇌ Now ■ **Active Figure 12.19**

Oceanic–continental plate boundary. (a) When an oceanic plate is subducted beneath a continental plate, an andesitic volcanic mountain range is formed on the continental plate as a result of rising magma. (b) Aerial view of the Andes Mountains in Peru. The Andes are one of the best examples of continuing mountain building at an oceanic–continental plate boundary.

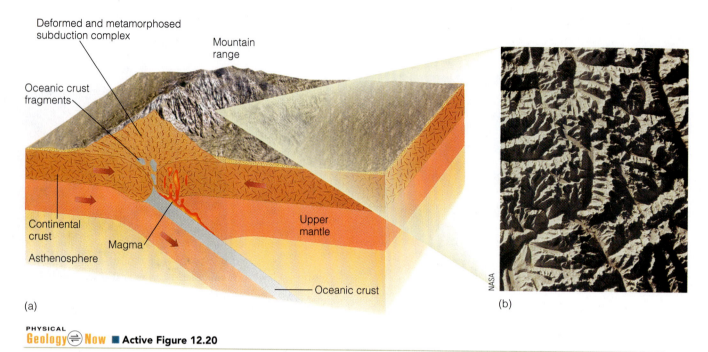

(a)

(b)

PHYSICAL
Geology⇌Now ■ Active Figure 12.20

Continental–continental plate boundary. (a) When two continental plates converge, neither is subducted because of their great thickness and low and equal densities. As the two continental plates collide, a mountain range is formed in the interior of a new and larger continent. (b) Vertical view of the Himalayas, the youngest and highest mountain system in the world. The Himalayas began to form when India collided with Asia 40 to 50 million years ago.

Recognizing Ancient Convergent Plate Boundaries

How can former subduction zones be recognized in the geologic record? Igneous rocks provide one clue. The magma erupted at the surface, forming island arc volcanoes and continental volcanoes, is of andesitic composition. Another clue can be found in the zone of intensely deformed rocks between the deep-sea trench where subduction is taking place and the area of igneous activity. Here, sediments and submarine rocks are folded, faulted, and metamorphosed into a chaotic mixture of rocks termed a *mélange*.

During subduction, pieces of oceanic lithosphere are sometimes incorporated into the mélange and accreted onto the edge of the continent. Such slices of oceanic crust and upper mantle are called *ophiolites* (■ Figure 12.21). They consist of a layer of deep-sea sediments that include graywackes (poorly sorted sandstones containing abundant feldspars and rock fragments, usually in a clay-rich matrix), black shales, and cherts. These deep-sea sediments are underlain by pillow lavas, a sheeted dike complex, massive gabbro, and layered gabbro, all of which form the oceanic crust. Beneath the gabbro is peridotite, which probably represents the upper mantle. Ophiolites are key features in recognizing plate convergence along a subduction zone.

Elongate belts of folded and faulted marine sedimentary rocks, andesites, and ophiolites are found in the Appalachians, Alps, Himalayas, and Andes Mountains.

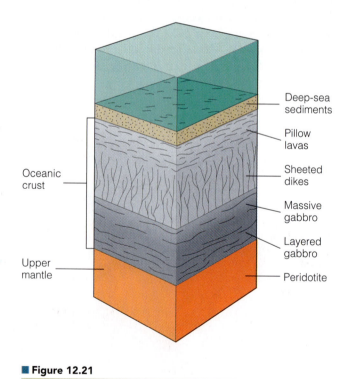

■ Figure 12.21

Ophiolites are sequences of rock on land consisting of deep-sea sediments, oceanic crust, and upper mantle.

The combination of such features is good evidence that these mountain ranges resulted from deformation along convergent plate boundaries.

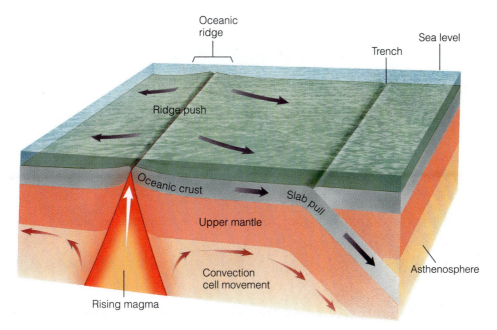

Figure 12.27

Plate movement is also thought to occur because of gravity-driven "slab-pull" or "ridge-push" mechanisms. In slap-pull, the edge of the subducting plate descends into the interior, and the rest of the plate is pulled downward. In ridge-push, rising magma pushes the oceanic ridges higher than the rest of the oceanic crust. Gravity thus pushes the oceanic lithosphere away from the ridges and toward the trenches.

still moving today has been proven beyond a doubt. And although a comprehensive theory of plate movement has not yet been developed, more and more of the pieces are falling into place as geologists learn more.

HOW DOES PLATE TECTONICS AFFECT THE DISTRIBUTION OF NATURAL RESOURCES?

Besides being responsible for the major features of Earth's crust, plate movements affect the formation and distribution of some natural resources. Consequently, geologists are using plate tectonic theory in their search for petroleum (see Geo-Focus 12.1) and mineral deposits and in explaining the occurrence of these natural resources.

It is becoming increasingly clear that if we are to keep up with the continuing demands of a global industrialized society, the application of plate tectonic theory to the origin and distribution of natural resources is essential.

Mineral Deposits

Many metallic mineral deposits such as copper, gold, lead, silver, tin, and zinc are related to igneous and associated hydrothermal activity, so it is not surprising that a close relationship exists between plate boundaries and the occurrence of these valuable deposits.

The magma generated by partial melting of a subducting plate rises toward the surface, and as it cools, it precipitates and concentrates various metallic ores. Many of the world's major metallic ore deposits are associated with convergent plate boundaries including those in the Andes of South America, the Coast Ranges and Rockies of North America, Japan, the Philippines, Russia, and a zone extending from the eastern Mediterranean region to Pakistan. In addition, the majority of the world's gold is associated with sulfide deposits located at ancient convergent plate boundaries in such areas as South Africa, Canada, California, Alaska, Venezuela, Brazil, southern India, Russia, and western Australia.

The copper deposits of western North and South America are an excellent example of the relationship between convergent plate boundaries and the distribution, concentration, and exploitation of valuable metallic ores (■ Figure 12.28a). The world's largest copper deposits are found along this belt. The majority of the copper deposits in the Andes and the southwestern United States were formed less than 60 million years ago when oceanic plates were subducted under the North and South American plates. The rising magma and associated hydrothermal fluids carried minute amounts of copper, which was originally widely disseminated but eventually became concentrated in the cracks and fractures of the surrounding andesites. These low-grade copper deposits contain from 0.2 to 2% copper and are extracted from large open-pit mines (Figure 12.28b).

Divergent plate boundaries also yield valuable resources. The island of Cyprus in the Mediterranean is rich in copper and has been supplying all or part of the world's needs for the last 3000 years. The concentration of copper on Cyprus formed as a result of precipitation adjacent to hydrothermal vents along a divergent plate

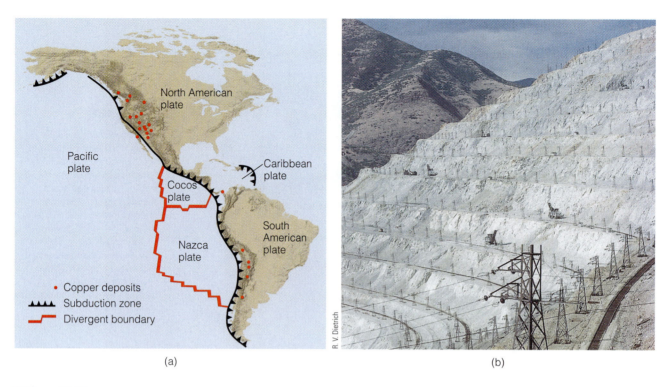

(a)   (b)

R. V. Dietrich

■ **Figure 12.28**

(a) Important copper deposits are located along the west coasts of North and South America. (b) Bingham Mine in Utah is a huge open-pit copper mine with reserves estimated at 1.7 billion tons. More than 400,000 tons of rock are removed each day.

What Would You Do

You are part of a mining exploration team that is exploring a promising and remote area of central Asia. You know that former convergent and divergent plate boundaries are frequently sites of ore deposits. What evidence would you look for to determine whether the area you're exploring might be an ancient convergent or divergent plate boundary? Is there anything you can do before visiting the area that might help you to determine the geology of the area?

boundary. This deposit was brought to the surface when the copper-rich seafloor collided with the European plate, warping the seafloor and forming Cyprus.

Studies indicate that minerals of such metals as copper, gold, iron, lead, silver, and zinc are currently forming as sulfides in the Red Sea. The Red Sea is opening as a result of plate divergence and represents the earliest stage in the growth of an ocean basin (Figures 12.15c and 12.16a).

12 REVIEW WORKBOOK

Chapter Summary

- The concept of continental movement is not new. The earliest maps showing the similarity between the east coast of South America and the west coast of Africa provided the first evidence that continents might once have been united and subsequently separated from each other.

- Alfred Wegener is generally credited with developing the hypothesis of continental drift. He provided abundant geologic and paleontologic evidence to show that the continents were once united into one supercontinent he named Pangaea. Unfortunately, Wegener could not explain how the continents moved, and most geologists ignored his ideas.

- The hypothesis of continental drift was revived during the 1950s when paleomagnetic studies of rocks indicated the presence of multiple magnetic north poles instead of just one as there is today. This paradox was resolved by constructing a hypothetical map and moving the continents into different positions, making the paleomagnetic data consistent with a single magnetic north pole.

- Magnetic surveys of the oceanic crust revealed magnetic anomalies in the rocks, indicating that Earth's magnetic field has reversed itself in the past. Because the anomalies are parallel and form symmetric belts adjacent to the oceanic ridges, new oceanic crust must have formed as the seafloor was spreading.

- Seafloor spreading has been confirmed by dating the sediments overlying the oceanic crust and by radiometric dating of rocks on oceanic islands. Such dating reveals that the oceanic crust becomes older with distance from spreading ridges.

- Plate tectonic theory became widely accepted by the 1970s because of the overwhelming evidence supporting it and because it provides geologists with a powerful theory for explaining such phenomena as volcanism, seismicity, mountain building, global climatic changes, the distribution of the world's biota, and the distribution of some mineral resources.

- The supercontinent cycle indicates that all or most of Earth's landmasses form, break up, and re-form in cycles of about 500 million years.

- Three types of plate boundaries are recognized: divergent boundaries, where plates move away from each other; convergent boundaries, where two plates collide; and transform boundaries, where two plates slide past each other.

- Ancient plate boundaries can be recognized by their associated rock assemblages and geologic structures. For divergent boundaries, these may include rift valleys with thick sedimentary sequences and numerous dikes and sills. For convergent boundaries, ophiolites and andesitic rocks are two characteristic features. Transform faults generally do not leave any characteristic or diagnostic features in the geologic record.

- The average rate of movement and relative motion of plates can be calculated in several ways. The results of these different methods all agree and indicate that the plates move at different average velocities.

- Absolute motion of plates can be determined by the movement of plates over mantle plumes. A mantle plume is an apparently stationary column of magma that rises to the surface where it becomes a hot spot and forms a volcano.

- Although a comprehensive theory of plate movement has yet to be developed, geologists think that some type of convective heat system is the major driving force.

- A close relationship exists between the formation of some mineral deposits and plate boundaries. Furthermore, the formation and distribution of some natural resources are related to plate movements.

Important Terms

continental–continental plate
 boundary (p. 350)
continental drift (p. 334)
convergent plate boundary
 (p. 349)
divergent plate boundary (p. 345)
Glossopteris flora (p. 334)
Gondwana (p. 334)

hot spot (p. 353)
Laurasia (p. 335)
oceanic–continental plate
 boundary (p. 350)
oceanic–oceanic plate boundary
 (p. 349)
Pangaea (p. 334)
plate (p. 342)

plate tectonic theory (p. 342)
seafloor spreading (p. 341)
thermal convection cell (p. 341)
transform fault (p. 353)
transform plate boundary
 (p. 353)

Review Questions

1. The man credited with developing the continental drift hypothesis is:
 a. _____ Wilson; b. _____ Wegener;
 c. _____ Hess; d. _____ du Toit;
 e. _____ Vine.

2. The name of the supercontinent that formed
 at the end of the Paleozoic Era is:
 a. _____ Laurasia; b. _____ Gondwana;
 c. _____ Panthalassa; d. _____ Atlantis;
 e. _____ Pangaea.

3. Hot spots and aseismic ridges can be used to
 determine:
 a. _____ the location of divergent plate
 boundaries; b. _____ the absolute motion of
 plates; c. _____ the location of magnetic
 anomalies in oceanic crust; d. _____ the relative motion of plates; e. _____ the location of
 convergent plate boundaries.

4. Subduction occurs along what type of plate
 boundary?
 a. _____ divergent; b. _____ transform;
 c. _____ convergent; d. _____ answers a
 and b; e. _____ answers a and c.

5. Magnetic surveys of the ocean basins indicate
 that:
 a. _____ the oceanic crust is youngest adjacent to mid-oceanic ridges; b. _____ the
 oceanic crust is oldest adjacent to mid-
 oceanic ridges; c. _____ the oceanic crust is
 youngest adjacent to the continents; d. _____
 the oceanic crust is the same age everywhere;
 e. _____ answers b and c.

6. The driving mechanism of plate movement is
 thought to be:
 a. _____ isostasy; b. _____ Earth's rotation;
 c. _____ thermal convection cells; d. _____
 magnetism; e. _____ polar wandering.

7. Convergent plate boundaries are zones
 where:
 a. _____ new continental lithosphere is forming; b. _____ new oceanic lithosphere is
 forming; c. _____ two plates come together;
 d. _____ two plates slide past each other;
 e. _____ two plates move away from each
 other.

8. A distinctive assemblage of deep-sea
 sediments, oceanic crustal rocks, and upper
 mantle constitutes a(n):
 a. _____ back-arc basin; b. _____ volcanic island arc; c. _____ ophiolite; d. _____ subduction complex; e. _____ none of the preceding
 answers.

9. The Northern Hemisphere landmass of Pangaea is called:
 a. _____ Laurentia; b. _____ Gondwana;
 c. _____ Laurasia; d. _____ America;
 e. _____ Euramerica.

10. The San Andreas fault is an example of what
 type of plate boundary?
 a. _____ divergent; b. _____ convergent;
 c. _____ transform; d. _____ oceanic–
 continental; e. _____ continental–
 continental.

11. Which of the following allows geologists to
 determine absolute plate motion?
 a. _____ hot spots; b. _____ the age of the
 sediment directly above any portion of the
 oceanic crust; c. _____ magnetic reversals in
 the oceanic crust; d. _____ satellite–laser
 ranging techniques; e. _____ all of the preceding answers.

12. The East African Rift Valleys is a good example of what type of plate boundary?

a. _____ continental–continental; b. _____ oceanic–oceanic; c. _____ oceanic–continental; d. _____ divergent; e. _____ transform.

13. Iron-bearing minerals in magma gain their magnetism and align themselves with the magnetic field when they cool through the:

a. _____ Curie point; b. _____ magnetic anomaly point; c. _____ thermal convection point; d. _____ hot spot point; e. _____ isostatic point.

14. Using the age for each of the Hawaiian Islands in Figure 12.24, calculate the average rate of movement per year for the Pacific plate since each island formed. Is the average rate of movement the same for each island? Would you expect it to be? Explain why it may not be.

15. What is the supercontinent cycle? Who proposed this concept, and what kinds of geologic data were needed to support such a concept?

16. What evidence convinced Wegener that the continents were once joined and subsequently broke apart?

17. Estimate the age of seafloor crust, and the age and thickness of the oldest sediment off the eastern coast of the United States (e.g., Virginia). In so doing, refer to Figure 12.12 for ages and the deep-sea sediment accumulation rate stated in this chapter.

18. How have plate tectonic processes affected the formation and distribution of natural resources?

19. If the movement along the San Andreas fault, which separates the Pacific plate from the North American plate, averages 5.5 cm per year, how long will it take before Los Angeles is opposite San Francisco?

20. Why is plate tectonics the unifying theory of geology?

World Wide Web Activities

PHYSICAL Geology⇌Now Assess your understanding of this chapter's topics with additional quizzing and comprehensive interactivities at

http://earthscience.brookscole.com/physgeo5e

as well as current and up-to-date weblinks, additional readings, and InfoTrac College Edition exercises.

Deformation, Mountain Building, and the Evolution of Continents

CHAPTER 13 OUTLINE

OBJECTIVES

At the end of this chapter, you will have learned that

- Rock deformation involves changes in the shape or volume or both of rocks in response to applied forces.

- Geologists use several criteria to differentiate among geologic structures such as folds, joints, and faults.

- Correctly interpreting geologic structures is important in human endeavors such as constructing highways and dams, choosing sites for power plants, and finding and extracting some resources.

- Deformation and the origin of geologic structures are important in the origin and evolution of mountains.

- Most of Earth's large mountain systems formed, and in some cases continue to form, at or near the three types of convergent plate boundaries.

- Terranes have special significance in mountain building.

- Specific processes account for the origin and continuing evolution of continents.

PHYSICAL Geology⇌Now *This icon, appearing throughout the book, indicates an opportunity to explore interactive tutorials, animations, or practice problems available on the Physical GeologyNow Web site at http://earthscience.brookscole.com/physgeo5e.*

This polished specimen of the Moine Schist on display in the Royal Museum in Edinburgh, Scotland, shows intense deformation. Notice that many layers within the rock are intricately folded, and some layers have been displaced along small fractures. Source: Sue Monroe

Introduction

"**S**olid as rock" implies permanence and durability, but you know from earlier chapters that physical and chemical processes disaggregate and decompose rocks, and rocks behave very differently at great depth than they do at or near the surface. Indeed, under the tremendous pressures and high temperature at several kilometers below the surface, rock layers actually crumple or fold yet remain solid, and at shallower depths, they yield by fracturing or a combination of folding and fracturing (chapter opening photo and ■ Figure 13.1). In either case dynamic forces within Earth cause **deformation,** a general term encompassing all changes in the shape or volume (or both) of rocks.

The fact that dynamic forces are active within Earth is obvious from ongoing seismic activity, volcanism, plate movements, and the continuing evolution of mountains in several areas including South America and Asia. In short, Earth is an active planet with a variety of processes driven by internal heat, particularly plate movements; most of Earth's seismic activity, volcanism, and rock deformation take place at divergent, convergent, and transform plate boundaries.

Mountains, areas that stand significantly higher than adjacent areas, may form by processes unrelated to deformation, but the origin of Earth's truly large mountain ranges on the continents involves tremendous deformation, usually accompanied by emplacement of plutons, volcanism, and metamorphism, at convergent plate boundaries. The Appalachians of North America, the Alps in Europe, the Himalayas of Asia, and the Andes in South America all owe their existence to deformation at convergent plate boundaries. And in some cases this activity continues even now. Thus deformation and mountain building are closely related topics and accordingly we will consider both in this chapter.

The past and continuing evolution of continents also involves deformation at continental margins, where new material is added to an existing continent, a phenomenon known as *continental accretion.* North America, for instance, has not always had its present shape and area. Indeed, it began evolving during the Archean Eon (4.0–2.5 billion years ago) as new material was added to the continent at deformation belts along its margins. The other continents have had similar histories, but in this chapter we will concentrate on the geologic evolution of North America.

Much of this chapter is devoted to a review of the various types of *geologic structures,* such as folded and fractured rock layers resulting from deformation, their descriptive terminology, and the forces responsible for them. Even so, there are several practical reasons to study deformation and mountain building. For one thing, crumpled and fractured rock layers provide a record of the kinds and intensities of forces that operated during the past. Thus interpretations of these structures allow us to satisfy our curiosity about Earth history, and in addition such studies are essential in engineering endeavors such as choosing sites for dams, bridges, and nuclear power plants, especially if they are in areas of ongoing deformation. Also, many aspects of mining and exploration for petroleum and natural gas rely on correctly identifying geologic structures.

David J. Matty

(a)

Reed Wicander

(b)

■ **Figure 13.1**

Many rocks show the effect of deformation. (a) Small-scale folds in sedimentary rocks. The pen is 13.5 cm long. (b) These rocks have been deformed by folding and fracturing. Notice the light pole for scale. The nearly vertical fracture where light-colored rocks were displaced is a fault, a fracture along which rocks on opposite sides of the fracture have moved parallel with the fracture surface.

ROCK DEFORMATION— HOW DOES IT OCCUR?

We have defined *deformation* as a general term referring to changes in the shape or volume (or both) of rocks. That is,

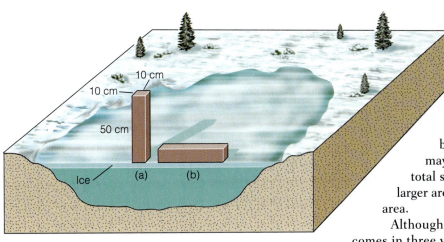

■ Figure 13.2

Stress and strain exerted on an ice-covered pond. The vertical object (a) has a density of 1 g/cm^3 and a volume of 5000 cm^3. The area of the object on the ice is 100 cm^2, so the stress exerted on the ice is 50 g/cm^2. The object on its side (b) has an area of 500 cm^2 in contact with the ice. Accordingly, the stress exerted on the ice is only 10 g/cm^2. The total stress at (a) and (b) is exactly the same, but in (b) it is spread over a larger area.

rocks may be crumpled into folds or fractured as a result of **stress,** which results from force applied to a given area of rock. If the intensity of the stress is greater than the rock's internal strength, the rock undergoes **strain,** which is simply deformation caused by stress. The terminology is a little confusing at first, but keep in mind that *deformation* and *strain* are synonyms, and stress is the force that causes deformation or strain. The following discussion and reference to ■ Figure 13.2 will help clarify the meaning of stress and the distinction between stress and strain.

Stress and Strain

Remember that stress is the force applied to a given area of rock, usually expressed in kilograms per square centimeter (kg/cm^2). For example, the stress, or force, exerted by a person walking on an ice-covered pond is a function of the person's weight and the area beneath her or his feet. The ice's internal strength resists the stress unless the

stress is too great, in which case the ice may bend or crack as it is strained (deformed). In our simplified example in Figure 13.2, we use a vertical, rectangular object rather than a person to simplify the calculations. In any case, to avoid breaking through the ice, the person may lie down; this does not reduce the total stress but it does distribute it over a larger area, thus reducing the stress per unit area.

Although stress is force per unit area, it comes in three varieties, *compression, tension,* and *shear,* depending on the direction of the applied forces. In **compression,** rocks, or any other object, are squeezed or compressed by forces directed toward one another along the same line, as when you squeeze a rubber ball in your hand. Rock layers in compression tend to be shortened in the direction of stress by either folding or fracturing (■ Figure 13.3a). **Tension,** on the other hand, results from forces acting along the same line but in opposite directions, and tends to lengthen rocks or pull them apart (Figure 13.3b). Incidentally, rocks are much stronger in compression than they are in tension. In **shear stress,** forces act parallel to one another but in opposite directions, resulting in deformation by displacement along closely spaced planes (Figure 13.3c).

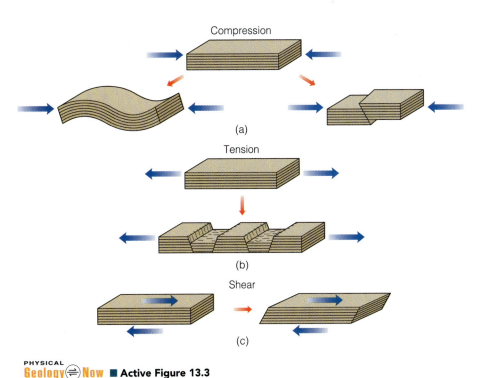

**PHYSICAL
Geology ⇌ Now ■ Active Figure 13.3**

Stress and possible types of resulting deformation. (a) Compression causes shortening of rock layers by folding or faulting. (b) Tension lengthens rock layers and causes faulting. (c) Shear stress causes deformation by displacement along closely spaced planes.

Types of Strain

Geologists characterize strain as **elastic strain** if deformed rocks return to their original shape when the deforming forces are relaxed. In Figure 13.2, for example, the ice on the pond may bend under a person's weight but return to its original shape once the person leaves. As you might expect, rocks are not very elastic. Nevertheless, Earth's crust behaves elastically when loaded by glacial ice and is depressed into the mantle. Recall from Chapter 10 that Earth's crust has responded elastically by isostatic rebound in large parts of Scandinavia and Canada following the last Ice Age (see Figure 10.17).

As stress is applied to rocks, rocks respond first by elastic strain, but when strained beyond their elastic limit, they undergo **plastic strain** as when they yield by folding, or they behave like brittle solids and **fracture** (Figure 13.3). In either folding or fracturing, the strain is permanent; that is, the rocks do not recover their original shape or volume even if the stress is removed.

Whether strain is elastic, plastic, or fracture depends on the kind of stress applied, pressure and temperature, rock type, and the length of time rocks are subjected to stress. A small stress applied over a long period, as on a mantelpiece supported only at its ends, will cause the rock to sag; that is, the rock deforms plastically (■ Figure 13.4). By contrast, a large stress applied rapidly to the same object, as when struck by a hammer, results in fracture. Rock type is important because not all rocks have the same internal strength and thus respond to stress differently. Some rocks are *ductile* whereas others are *brittle,* depending on the amount of

plastic strain they exhibit. Brittle rocks show little or no plastic strain before they fracture, but ductile rocks exhibit a great deal (Figure 13.4).

Many rocks show the effects of plastic deformation that must have taken place deep within the crust. In Chapters 7 and 10 we noted that the behavior of rocks under great pressure and at high temperature is very different from their behavior at or near the surface. At or near the surface, rocks commonly behave like brittle solids and fracture, but at depth they more often yield by plastic deformation; they become more ductile with increasing pressure and temperature. Most earthquake foci are at depths of less than 30 km, indicating that deformation by fracturing becomes increasingly difficult with depth, and no fracturing is known deeper than about 700 km.

STRIKE AND DIP— DETERMINING THE ORIENTATION OF ROCK LAYERS

During the 1660s, Nicolas Steno, a Danish anatomist, proposed several principles essential for deciphering Earth history from the record preserved in rocks. One is the *principle of original horizontality,* meaning that sediments accumulate in horizontal or nearly horizontal layers. Thus, if we observe steeply inclined sedimentary rocks, we are justified in inferring that they were deposited horizontally, lithified, and then tilted into their present position (■ Figure 13.5). If rock layers have been deformed by folding or faulting or both, geologists use the concepts of *strike* and *dip* to describe their orientation with respect to a horizontal plane.

By definition, **strike** is the direction of a line formed by the intersection of a horizontal plane and an inclined plane. For example, the surfaces of the rock layers in

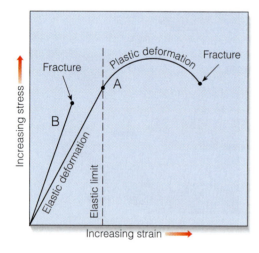

■ Figure 13.4

Rocks initially respond to stress by elastic deformation and return to their original shape when the stress is released. If the elastic limit is exceeded, as in curve A, rocks deform plastically, which is permanent deformation. The amount of plastic deformation rocks exhibit before fracturing depends on their ductility. If they are ductile, they show considerable plastic deformation (curve A), but if they are brittle, they show little or no plastic deformation before failing by fracture (curve B).

GEOLOGY
IN UNEXPECTED PLACES

Ancient Ruins and Geology

Ancient Roman and Greek ruins might not seem like a good place to study geology, but they do tell us something about the use of stone in construction and about stress and strain. Stone is incredibly strong, but its strength varies depending on how it is used. Remember, it is stronger in compression than in tension. Thus, when stones are used for a chimney, for example, all stresses act vertically, with the lower stones supporting the weight of those above. In other words, the stones are in compression.

Given that stone is so strong, why did ancient Greek and Roman builders erect buildings in which horizontal beams were supported by closely spaced vertical columns (■ Figure 1)? Wouldn't it have been more efficient to eliminate every other column and save on both materials and labor? The problem relates to the strength of stone in compression versus tension. If the horizontal beams spanned great distances, they would simply collapse under their own weight because a stone beam is compressed on its upper side but subjected to tension on its underside. The stress on its top is about the same as on the bottom, but the top contributes little to the beam's overall strength. So if used this way,

tension fractures would form on the bottom and travel upward to the top, causing failure.

Ancient builders were aware of stone's strength, probably from trial and error, and accordingly used closely spaced columns to support horizontal beams. Builders today are similarly restricted. Can you figure out why simply doubling our hypothetical beam's dimensions, other than its length, would not solve the problem of spanning great distances between columns?

■ **Figure 1**

These Greek ruins show closely spaced columns supporting horizontal beams.

Jim Winkley/Ecoscene/Corbis

■ Figure 13.6 are inclined planes, whereas the water surface is a horizontal plane. The direction of the line formed at the intersection of the water surface and the inclined surface of a rock layer is the strike. The strike line's orientation is determined by using a compass to measure its angle with respect to north. **Dip** is a measure of an inclined plane's deviation from horizontal, so it must be measured at right angles to the strike direction (Figure 13.6).

Geologic maps showing the age, distribution, and geologic structures of rocks in an area employ a special symbol to indicate strike and dip. A long line oriented in the appropriate direction indicates strike, and a short line perpendicular to the strike line shows the direction of dip (Figure 13.6). Adjacent to the strike and dip symbol is a number corresponding to the dip angle. The usefulness of strike and dip symbols will become apparent in the following sections on folds and faults.

DEFORMATION AND GEOLOGIC STRUCTURES

Recall that *deformation* and its synonym *strain* refer to changes in the shape or volume of rocks. During deformation rocks might be crumpled into folds, or they might be fractured, or perhaps folded and fractured. Any of these features resulting from deformation is referred to as a **geologic**

(a)   (b)

■ **Figure 13.5**

(a) These sedimentary rock layers in Valley of the Gods, Utah, are horizontal as when they were deposited and lithified. (b) We can infer that these sandstone beds in Colorado were deposited horizontally, lithified, and then tilted into their present position.

■ **Figure 13.6**

Strike and dip. (a) The intersection of a horizontal plane (the water surface) and an inclined plane (the surface of the rock layer) forms a line known as the *strike*. The *dip* is the maximum angular deviation of the inclined layer from horizontal. Notice the strike and dip symbol with 50 adjacent to it indicating the angle of dip. (b) Natural example of strike and dip. The strike of the dipping rock surface is marked by its intersection with the water surface.

structure. Various geologic structures, or simply *structures*, are present almost everywhere that rock exposures can be observed, and many are detected far below the surface by drilling and several geophysical techniques.

Folded Rock Layers

Geologic structures known as **folds,** in which planar features such as bedding are bent, are quite common. Most folding takes place in response to compression, as when you place your hands on a tablecloth and move them toward one another, thereby producing a series of up- and down-arched folds in the fabric. Rock layers in Earth's crust respond similarly to compression, but unlike the tablecloth, folding in rock layers is permanent. That is, plastic strain has taken place—so once folded, the rocks stay folded. Most folding probably takes place deep in the crust where rocks are more ductile than they are at or near the surface. The configuration of folds and the intensity of folding vary considerably, but only three basic types of folds are recognized: *monoclines, anticlines,* and *synclines.*

Monoclines   A simple bend or flexure in otherwise horizontal or uniformly dipping rock layers is a **monocline** (■ Figure 13.7a). The large monocline in Figure 13.7b formed when the Bighorn Mountains in Wyoming were uplifted along a fault. The fault did not penetrate to the surface, so as uplift proceeded the near-surface rock layers were bent so that they now appear draped over the margin of the uplifted block.

Anticlines and Synclines   Monoclines are not rare, but they are not nearly as common as anticlines and synclines. An **anticline** is an up-arched or convex upward fold with the oldest rock layers in its core, whereas a **syncline** is a down-arched or concave upward

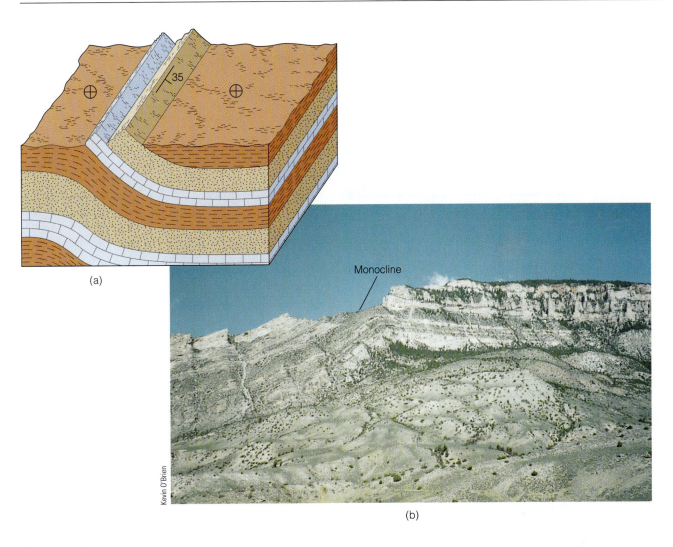

(a)

Monocline

Kevin O'Brien

(b)

■ **Figure 13.7**

(a) A monocline. Notice the strike and dip symbol and the circled cross, which is the symbol for horizontal layers. (b) A monocline in the Bighorn Mountains in Wyoming.

fold in which the youngest rock layers are in its core (■ Figure 13.8).

Anticlines and synclines also possess an axial plane connecting the points of maximum curvature in each folded layer (■ Figure 13.9); the axial plane divides folds into halves, each half being a *limb*. Because folds most often are found as a series of anticlines alternating with synclines, an anticline and an adjacent syncline share a limb. It is important to remember that anticlines and synclines are simply folded rock layers and do not necessarily correspond to high and low areas at Earth's surface (■ Figure 13.10).

Folds are commonly exposed to view in areas of deep erosion, but even where eroded, strike and dip and the relative ages of the folded rock layers easily distinguish anticlines and synclines from one another. Notice in ■ Figure 13.11 that in the surface view of the anticline, each limb dips outward or away from the center of the fold where the oldest exposed rocks are. In an eroded syncline, though, each limb dips inward

to the fold's center, where the youngest exposed rocks are found.

Thus far, we have described upright folds in which the axial plane is vertical and each fold limb dips at the same angle (Figure 13.11). If the axial plane is inclined, however, the limbs dip at different angles, and the fold is characterized as *inclined* (■ Figure 13.12a). In an *overturned fold*, both limbs dip in the same direction. In other words, one fold limb has been rotated more than 90 degrees from its original position so that it is now upside down (Figure 13.12b). Folds in which the axial plane is horizontal are referred to as *recumbent* (Figure 13.12c). Overturned and recumbent folds are particularly common in mountain ranges that formed by compression at convergent plate boundaries (discussed later in this chapter).

Our definitions of anticlines and synclines as up- and down-arched folds are not sufficient for folds tipped on their sides or upside down. In Figure 13.12c, for instance, one cannot determine which fold is an anticline

Reed Wicander

■ **Figure 13.8**

Folded rocks in the Calico Mountains of southeastern California. Three folds are visible from left to right: a syncline, an anticline, and another syncline. We can infer that compression was responsible for these folds.

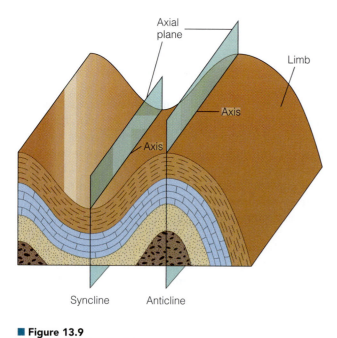

■ **Figure 13.9**

Syncline and anticline showing the axial plane, axis, and fold limbs.

or a syncline with the information available. Even strike and dip will not help in this case, but the relative ages of the folded rock layers will resolve the problem. Recall that an anticline has the oldest rocks in its core, so the fold nearest the surface is an anticline and the lower fold is a syncline.

Plunging Folds   Geologists further characterize folds as *nonplunging* or *plunging*. In the former, the fold *axis*, a line formed by the intersection of the axial plane with the folded beds, is horizontal (Figure 13.9). But it is much more common for the axis to be inclined so that it appears to plunge beneath the surrounding rocks; folds possessing an inclined axis are **plunging folds** (■ Figure 13.13a). To differentiate plunging anticlines from plunging synclines, geologists use exactly the same criteria used for nonplunging folds: All rocks dip away from the fold axis in plunging anticlines; in plunging synclines, all rocks dip inward toward the axis. The oldest exposed rocks are in the center of an eroded plunging anticline, whereas the youngest exposed rocks are in the center of an eroded plunging syncline (Figure 13.13b).

(a)

(b)

■ **Figure 13.10**

Folds and their relationship to topography. (a) Cross section illustrating that anticlines and synclines do not necessarily correspond to high and low areas of the surface. Notice that folds even underlie the rather flat area. (b) A syncline is at the peak of this mountain in Kootenay National Park, British Columbia, Canada. Lower on the left flank of the mountain, an anticline and another syncline are also visible.

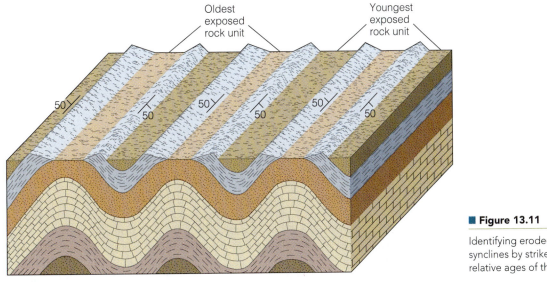

■ **Figure 13.11**

Identifying eroded anticlines and synclines by strike and dip and the relative ages of the folded rock layers.

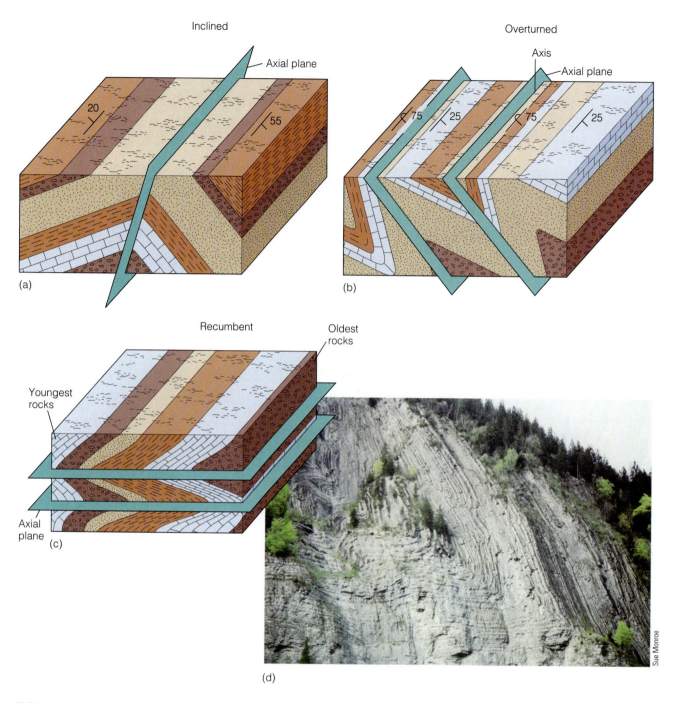

■ Figure 13.12

(a) An inclined fold. The axial plane is not vertical, and the fold limbs dip at different angles. (b) Overturned folds. Both fold limbs dip in the same direction, but one limb is inverted. Notice the special strike and dip symbol to indicate overturned beds. (c) Recumbent folds. (d) Recumbent fold in Switzerland.

In Chapter 6 we noted that anticlines form one type of structural trap for petroleum and natural gas (see Figure 6.24). As a matter of fact, most of the world's petroleum production comes from anticlinal traps, although several other types are also important. Accordingly, geologists are particularly interested in correctly identifying the geologic structures in areas of potential petroleum and natural gas production.

Domes and Basins Domes and basins are the circular and oval equivalents of anticlines and synclines, respectively (■ Figure 13.14). They tend to be rather equidimensional, whereas anticlines and synclines are elongate structures. Essentially the same criteria used to recognize anticlines and synclines are used to differentiate domes and basins. In an eroded dome the oldest exposed rocks are in the center, whereas in a basin the

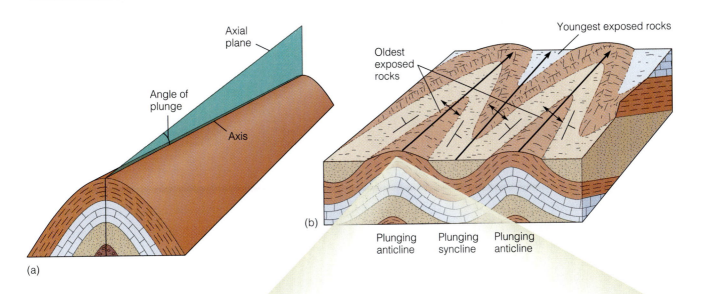

■ Figure 13.13

(a) A plunging fold. (b) Surface and cross-sectional views of plunging folds. The long arrow at the center of each fold is the standard geologic symbol used to depict plunging anticlines and synclines. The arrow at the end of the line shows the direction of plunge. (c) View of the eroded, plunging Sheep Mountain anticline in Wyoming. Notice the smaller fold on the right flank of the larger one. Can you tell from the strike and dip symbols that this is a plunging anticline?

opposite is true. All rocks in a dome dip away from a central point (as opposed to dipping away from a fold axis, which is a line). By contrast, all rocks in a basin dip inward toward a central point (Figure 13.14).

The terms *dome* and *basin* are also used to designate high and low areas on Earth's surface, but as with anticlines and synclines, domes and basins resulting from deformation do not necessarily correspond to mountains and valleys. In several instances in the following discus-

sions and chapters, we will mention basins in other contexts, and we have already used the word *dome*, as in "exfoliation dome," so here we will modify these two terms with *structural*. Accordingly, a structural dome or structural basin is an area of deformation corresponding to our previous definition, with no reference to surface elevations.

Some structural domes and basins are small structures that are easily recognized by their surface exposure

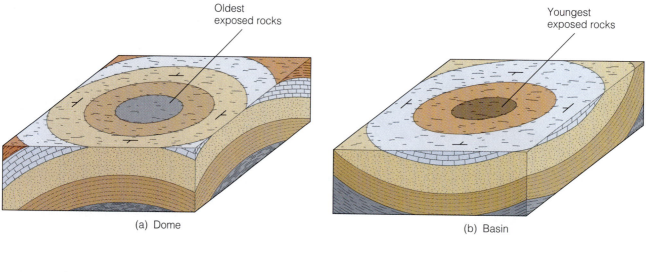

(a) Dome

(b) Basin

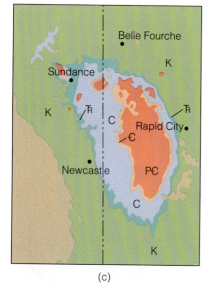

(c)

K	Cretaceous	Younger
Ŧ	Triassic	
C	Carboniferous	
Є	Cambrian	
PЄ	Precambrian	

■ **Figure 13.14**

A dome (a) and a basin (b). Notice that in a dome the oldest exposed rocks are in the center and all rocks dip outward from a central point, whereas in a basin the youngest exposed rocks are in the center and all rocks dip inward toward a central point. (c) This geologic map, a surface view only, uses colors and symbols to depict rocks and geologic structures in the Black Hills of South Dakota. From the information provided can you determine whether this map depicts a dome or a basin?

patterns, but many are so large that they can be visualized only on geologic maps or aerial photographs. Many of these large-scale structures formed in the continental interior, not by compression but as a result of vertical uplift of parts of the crust with little additional folding and faulting.

The Black Hills of South Dakota are a large eroded structural dome with a core of ancient rocks surrounded by progressively younger rocks (Figure 13.14c). The rocks dip outward rather uniformly from the Black Hills, which have been uplifted so that they now stand more than 2000 m above the adjacent plains. One of the best-known structural basins in the United States is the Michigan Basin. Most of the Michigan Basin is buried beneath younger rocks, so it is not directly observable at the surface. Nevertheless, strike and dip of exposed rocks near the basin margin and thousands of drill holes for oil and gas clearly show that the rock layers beneath the surface are deformed into a large structural basin.

Joints

Besides folding, rocks are also permanently deformed by fracturing. **Joints** are fractures along which no movement has taken place parallel with the fracture surface (■ Figure 13.15), although they may open up; that is, they show movement perpendicular to the fracture. Coal miners used the term *joint* long ago for cracks in rocks that they thought were surfaces where the adjacent blocks were "joined" together.

Remember that rocks near the surface are brittle and therefore commonly fail by fracturing when subjected to stress. Hence, nearly all near-surface rocks possess joints that form in response to compression, tension, and shearing. They vary from minute fractures to those extending for many kilometers and are often arranged in two or perhaps three prominent sets. Regional mapping reveals that joints and sets of joints are usually related to other geologic structures such as large folds and faults.

Galen Rowell/Peter Arnold, Inc.

(a)

C. G. Tillman

(b)

■ Figure 13.15

(a) Erosion along parallel joints in Arches National Park, Utah. (b) Joints intersecting at right angles yield this rectangular pattern in Wales.

It may seem odd that joints, which indicate brittle behavior, are so common in folded rocks that have been deformed plastically. The crest of an anticline provides a good example of how this might occur. Although anticlines are produced by compression, the rock layers are arched so that tension occurs perpendicular to fold crests, and joints form parallel to the long axis of the fold in the upper part of a folded layer.

We discussed two other types of joints in earlier chapters: columnar joints and sheet joints. Columnar joints form in some lava flows and in some plutons as cooling magma contracts and tensional stresses form polygonal fracture patterns (see Figure 4.7). Sheet joints form in response to pressure release (see Figure 5.4).

Faults

Another type of fracture known as a **fault** is one along which blocks of rock on opposite sides of the fracture have moved parallel with the fracture surface. The surface along which movement takes place is a **fault plane** (■ Figure 13.16a). Not all faults penetrate to the surface, but those that do might show a *fault scarp,* a bluff or cliff formed by vertical movement (Figure 13.16b). Fault scarps are usually quickly eroded and obscured. When movement takes place on a fault plane, the rocks on opposite sides may be scratched and polished (Figure 13.16b), or crushed and shattered into angular blocks forming *fault breccia* (Figure 13.16c).

Refer to Figure 13.16a and notice the designations *hanging wall block* and *footwall block*. The **hanging wall block** consists of the rock overlying a fault, whereas the **footwall block** lies beneath a fault plane. You can recognize these two blocks on any fault except one—that is, one that dips at 90 degrees. To identify some kinds of faults you must not only correctly identify these two blocks but also determine which one moved relatively up or down. We use the phrase *relative movement* because you cannot usually tell which block actually moved or if both moved. In Figure 13.16a the footwall block may have moved up, the hanging wall block may have moved down, or both could have moved. Nevertheless, the hanging wall block appears to have moved down relative to the footwall block.

Remember our discussion of strike and dip of rock layers. Fault planes are also inclined planes, and they too are characterized by strike and dip (Figure 13.16a). In fact, the two basic varieties of faults are defined by whether the blocks on opposite sides of the fault plane moved parallel to the direction of dip (dip-slip faults) or along the direction of strike (strike-slip faults). See "Types of Faults" on pages 378 and 379.

Dip-Slip Faults All movement on **dip-slip faults,** takes place parallel with the fault's dip; that is, movement is vertical, either up or down the fault plane. In

■ **Figure 13.16**

(a) Fault terminology. (b) Polished, scratched fault plane and fault scarp near Klamath Falls, Oregon. (c) Fault breccia, the zone of rubble along a fault in the Bighorn Mountains, Wyoming. If arrows were not shown could you determine which side of the fault moved relatively up?

■ Figure 13.17a, for instance, the hanging wall block moved down relative to the footwall block, giving rise to a **normal fault.** In contrast, in a **reverse fault** the hanging wall block moves up relative to the footwall block (Figure 13.17b). In Figure 13.17c, the hanging wall block also moved up relative to the footwall block, but the fault has a dip of less than 45 degrees and is a special variety of reverse fault known as a **thrust fault.**

Reference to Figure 13.3b shows that normal faults result from tension. Numerous normal faults are present along one or both sides of mountain ranges in the Basin and Range Province of the western United States where the crust is being stretched and thinned. The Sierra Nevada at the western margin of the Basin and Range is bounded by normal faults, and the range has risen along these faults so that it now stands more than 3000 m above the lowlands to the east. Also, an active normal

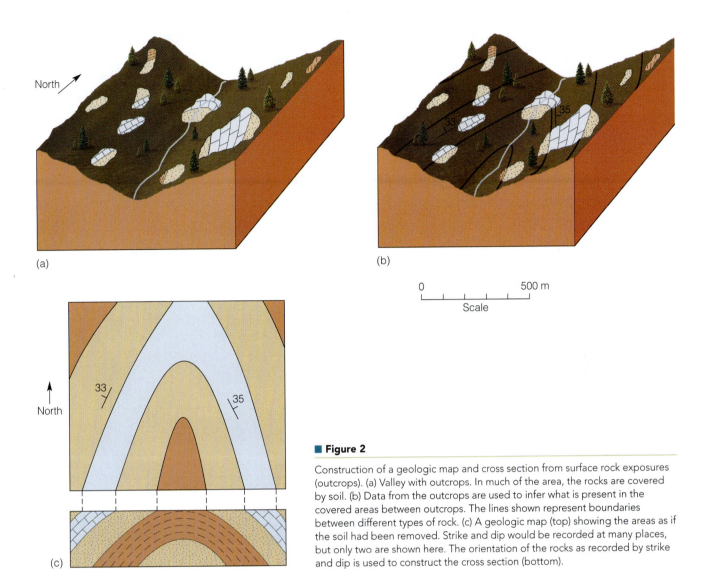

North

(a)

(b)

33   35

0   500 m
Scale

North

33   35

(c)

■ **Figure 2**

Construction of a geologic map and cross section from surface rock exposures (outcrops). (a) Valley with outcrops. In much of the area, the rocks are covered by soil. (b) Data from the outcrops are used to infer what is present in the covered areas between outcrops. The lines shown represent boundaries between different types of rock. (c) A geologic map (top) showing the areas as if the soil had been removed. Strike and dip would be recorded at many places, but only two are shown here. The orientation of the rocks as recorded by strike and dip is used to construct the cross section (bottom).

maps. You, for example, may one day be a member of a safety planning board of a city with known active faults. Geologic maps might be useful in developing zoning regulations and construction codes. Given the position of the active fault in Figure 1, zoning this area for a housing development would not seem prudent, but it might be perfectly satisfactory for agriculture or some other land use. Perhaps you will become a land-use planner or member of a county commission charged with selecting a site for a sanitary landfill or securing an adequate supply of groundwater for your community. Geologic maps would almost certainly be used in these endeavors.

Geologic maps are used extensively in planning and constructing dams, choosing the best route for a highway through a mountainous region, and selecting the sites for nuclear reactors. It would certainly be unwise to build a dam in a valley with a known active fault or with rocks too weak to support such a large structure. And in areas of continuing volcanic activity, geologic maps provide essential information about the kinds of volcanic processes that might occur in the future, such as lava flows, ashfalls, or mudflows. Recall from Chapter 4 that one way in which volcanic hazards are assessed is by mapping and determining the geologic history of an area.

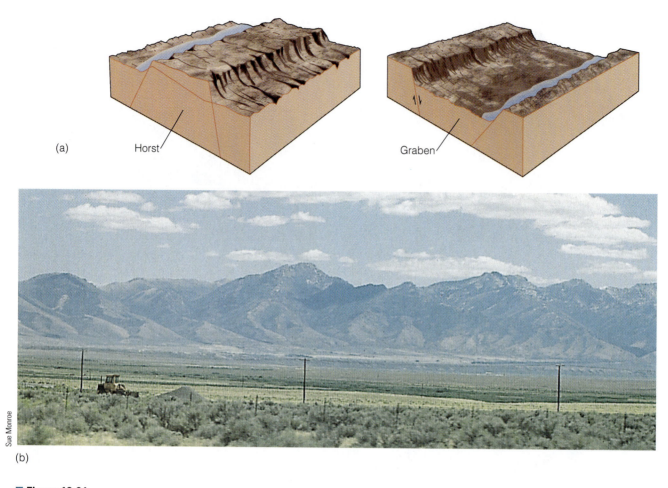

■ **Figure 13.21**

(a) Block-faulting and the origin of horsts and grabens. Many of the mountain ranges in the Basin and Range Province of the western United States and northern Mexico formed in this manner. (b) View of the Humboldt Range in Nevada, which is a horst bounded by normal faults.

consist of resistant plutonic rocks exposed following up-lift and erosion of the softer overlying sedimentary rocks.

Block-faulting is yet another way to form mountains, but it involves considerable deformation (■ Figure 13.21). Block-faulting entails movement on normal faults so that one or more blocks are elevated relative to adjacent areas. A classic example is the large-scale active block-faulting in the Basin and Range Province of the west-ern United States, a large area centered on Nevada but extending into several adjacent states and northern Mexico. In the Basin and Range Province, the crust is being stretched in an east–west direction; thus ten-sional stresses produce north–south oriented, range-bounding faults. Differential movement on these faults has produced uplifted blocks called *horsts* and down-dropped blocks called *grabens* (Figure 13.21) bounded by parallel normal faults. Erosion of the horsts has yielded the mountainous topography now present, and the grabens have filled with sediments eroded from the horsts (Figure 13.21).

The processes discussed in this section can certainly yield mountains. However, the truly large mountain sys-tems of the continents, such as the Alps of Europe and the Appalachians in North America, formed by compres-sion along convergent plate boundaries.

Plate Tectonics and Mountain Building

Geologists define the term **orogeny** as an episode of mountain building during which intense deformation takes place, generally accompanied by metamorphism, the emplacement of plutons, especially batholiths, and thickening of Earth's crust. The processes responsible for an orogeny are still not fully understood, but it is known that mountain building is related to plate move-ments. In fact, the advent of plate tectonic theory has completely changed the way geologists view the origin of mountain systems.

Any theory accounting for mountain building must adequately explain the characteristics of mountains, such as their geometry and location; they tend to be long and narrow and at or near plate margins. Mountains also show intense deformation, especially compression-induced overturned and recumbent folds as well as reverse

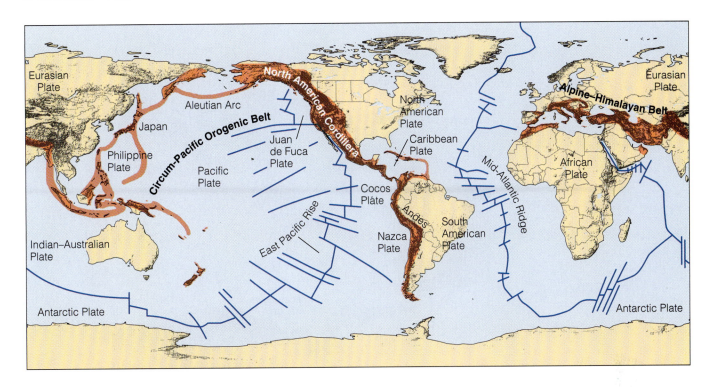

■ **Figure 13.22**

Most of Earth's geologically recent and present-day orogenic activity is concentrated in the circum-Pacific and Alpine–Himalayan orogenic belts.

and thrust faults. Furthermore, granitic plutons and regional metamorphism characterize the interiors or cores of mountain ranges. Another feature is sedimentary rocks now far above sea level that were clearly deposited in shallow and deep marine environments.

Deformation and associated activities at convergent plate boundaries are certainly important processes in mountain building. They account for a mountain range's location and geometry as well as complex geologic structures, plutons, and metamorphism. Yet the present-day topographic expression of mountains is also related to several surface processes such as mass wasting (gravity-driven processes including landslides), glaciers, and running water. In other words, erosion also plays an important role in the evolution of mountains.

Most of Earth's geologically recent and present-day orogenic activity is concentrated in two major zones or belts: the *Alpine–Himalayan orogenic belt* and the *circum-Pacific orogenic belt* (■ Figure 13.22). Both belts are composed of a number of smaller segments known as *orogens*, each a zone of deformed rocks, many of which have been metamorphosed and intruded by plutons.

In fact, we can explain most of Earth's past and present orogenies in terms of the geologic activity at convergent plate boundaries. Recall from Chapter 12 that convergent plate boundaries might be oceanic–oceanic, oceanic–continental, or continental–continental.

Orogenies at Oceanic–Oceanic Plate Boundaries

Orogenies that occur where oceanic lithosphere is subducted beneath oceanic lithosphere are characterized by the formation of a volcanic island arc and by deformation and igneous activity. Deformation takes place when sediments from the volcanic island arc are deposited on the adjacent seafloor and in the oceanic trench, and then deformed and scraped off against the landward side of the trench (■ Figure 13.23). These form a subduction complex, or *accretionary wedge,* of intricately folded rocks cut by numerous thrust faults. In addition, orogenies generated by plate convergence result in low-temperature, high-pressure metamorphism characteristic of the blueschist facies (see Figure 7.17).

Deformation caused largely by the emplacement of plutons also takes place in the island arc system where many rocks show evidence of high-temperature, low-pressure metamorphism. The overall effect of island arc orogenesis is the origin of two more or less parallel orogenic belts consisting of a landward volcanic island arc underlain by batholiths and a seaward belt of deformed trench rocks (Figure 13.23). The Japanese Islands are a good example of this type of deformation.

In the area between an island arc and its nearby continent, the back-arc basin, volcanic rocks and sediments derived from the island arc and the adjacent continent are also deformed as the plates continue to converge. The

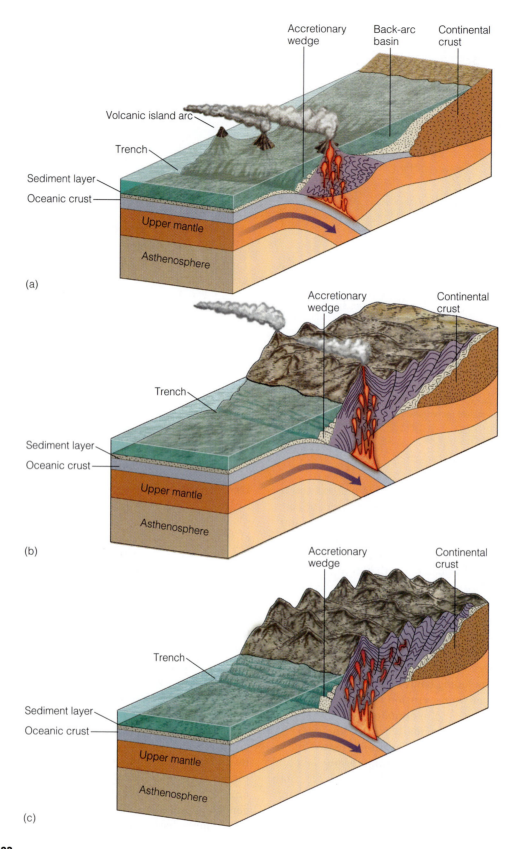

■ **Figure 13.23**

Orogeny and the origin of a volcanic island arc at an oceanic–oceanic plate boundary. (a) Subduction of an oceanic plate beneath an island arc. (b) Continued subduction, emplacement of plutons, and beginning of deformation by thrusting and folding of back-arc basin sediments. (c) Thrusting of back-arc basin sediments onto the adjacent continent and suturing of the island arc to the continent.

sediments are intensely folded and displaced toward the continent along low-angle thrust faults. Eventually, the entire island arc complex is fused to the edge of the continent, and the back-arc basin sediments are thrust onto the continent, forming a thick stack of thrust sheets (Figure 13.23).

Orogenies at Oceanic–Continental Plate Boundaries

Several mountain systems such as the Alps of Europe and the Andes of South America formed at oceanic–continental plate boundaries where oceanic lithosphere is subducted. The Andes are perhaps the best example of such continuing orogeny. Among the ranges of the Andes are the highest mountain peaks in the Americas and many active volcanoes. Furthermore, the west coast of South America is an extremely active segment of the circum-Pacific earthquake belt; and one of Earth's great oceanic trench systems, the Peru–Chile Trench, lies just off the west coast of South America.

Prior to 200 million years ago, the western margin of South America was a passive continental margin where sediments accumulated much as they do now along the east coast of North America. However, when Pangaea split apart along what is now the Mid-Atlantic Ridge, the South American plate moved westward. As a consequence, the oceanic lithosphere west of South America began subducting beneath the continent (■ Figure 13.24). Subduction resulted in partial melting of the descending plate, which produced the andesitic volcano arc of composite volcanoes, and the west coast became an active continental margin. Felsic magmas, mostly of granitic composition, were emplaced as large plutons beneath the arc (Figure 13.24).

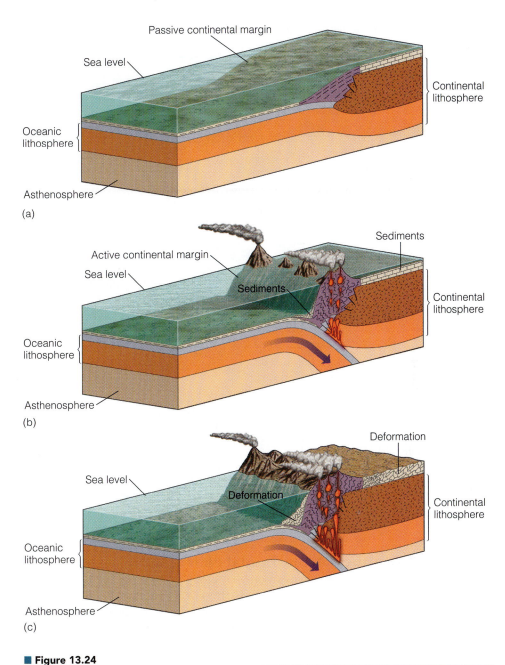

■ Figure 13.24

Generalized diagrams showing three stages in the development of the Andes of South America at an oceanic–continental plate boundary. (a) Prior to 200 million years ago, the west coast of South America was a passive continental margin. (b) An orogeny began when the west coast of South America became an active continental margin. (c) Continued deformation, volcanism, and plutonism.

As a result of the events just described, the Andes Mountains consist of a central core of granitic rocks capped by andesitic volcanoes. To the west of this central core along the coast are the deformed rocks of the accretionary wedge. And to the east of the central core are intensely folded sedimentary rocks that were thrust eastward onto the continent (Figure 13.24). Present-day subduction, volcanism, and seismicity along South America's west coast indicate that the Andes Mountains are still actively forming.

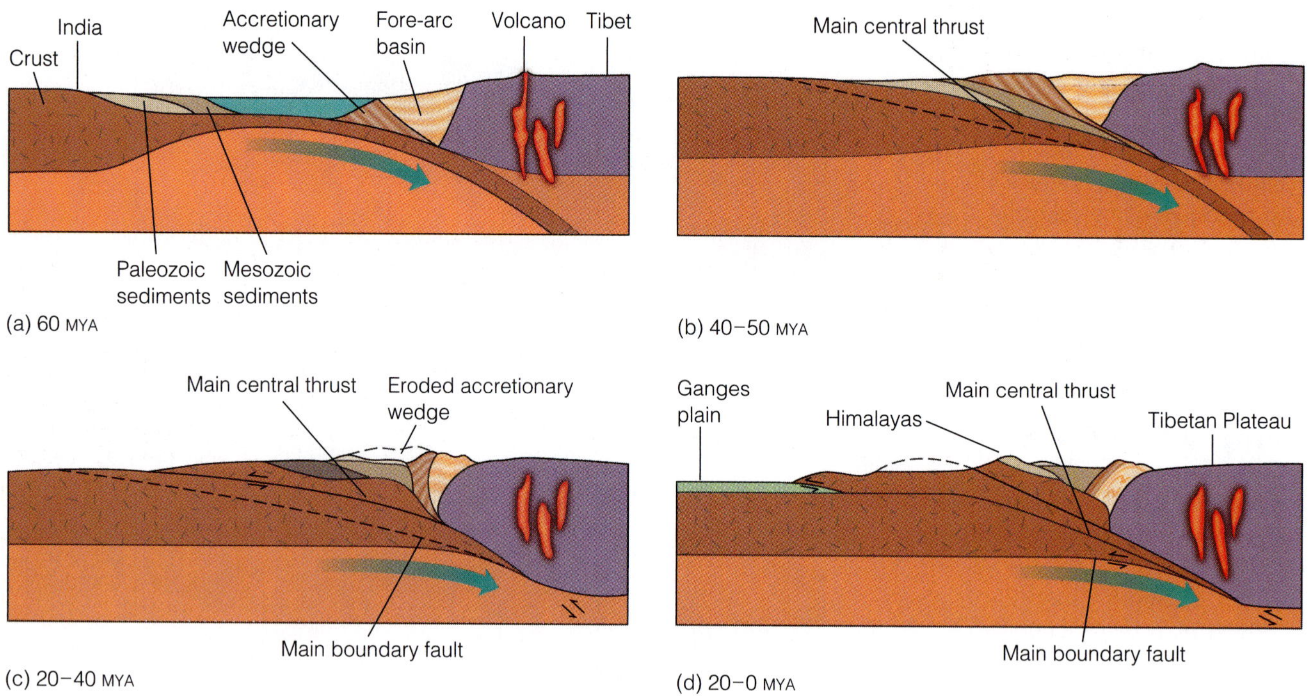

■ Figure 13.25

Simplified cross sections showing the collision of India with Asia and the origin of the Himalayas. (a) The northern margin of India before its collision with Asia. Subduction of oceanic lithosphere beneath southern Tibet as India approached Asia. (b) About 40 to 50 MYA, India collided with Asia, but because India was too light to be subducted, it was underthrust beneath Asia. (c) Continued convergence accompanied by thrusting of rocks of Asian origin onto the Indian subcontinent. (d) Since about 10 MYA, India has moved beneath Asia along the main boundary fault. Shallow marine sedimentary rocks that were deposited along India's northern margin now form the higher parts of the Himalayas. Sediment eroded from the Himalayas has been deposited on the Ganges Plain. Source: From Peter Molnar, "The Geological History and the Structure of the Himalayas." *American Scientist 74*: 148–149, Figure 4, Journal of Sigma Xi, The Scientific Research Group.

Orogenies at Continental–Continental Plate Boundaries The best example of an orogeny along a continental–continental plate boundary is the Himalayas of Asia. The Himalayas began forming when India collided with Asia about 40 to 50 million years ago. Prior to that time, India was far south of Asia and separated from it by an ocean basin (■ Figure 13.25a). As the Indian plate moved northward, a subduction zone formed along the southern margin of Asia where oceanic lithosphere was consumed. Partial melting generated magma, which rose to form a volcanic arc, and large granite plutons were emplaced into what is now Tibet. At this stage, the activity along Asia's southern margin was similar to what is now occurring along the west coast of South America.

The ocean separating India from Asia continued to close, and India eventually collided with Asia (Figure 13.25b). As a result, two continental plates became welded, or sutured, together. Thus the Himalayas are within a continent rather than along a continental margin. The exact time of India's collision with Asia is uncertain, but between 40 and 50 million years ago, India's rate of northward drift decreased abruptly from 15 to 20 cm per year to about 5 cm per year. Because continental lithosphere is not dense enough to be subducted, this decrease seems to mark the time of collision and India's resistance to subduction. Consequently, the leading margin of India was thrust beneath Asia, causing crustal thickening, thrusting, and uplift. Sedimentary rocks that had been deposited in the sea south of Asia were thrust northward, and two major thrust faults carried rocks of Asian origin onto the Indian plate (Figure 10.25c, d). Rocks deposited in the shallow seas along India's northern margin now form the higher parts of the Himalayas. Since its collision with Asia, India has been underthrust about 2000 km beneath Asia, and now moves north at about 5 cm per year.

A number of other mountain systems also formed as a result of collisions between two continental plates. The Urals in Russia and the Appalachians of North America formed by such collisions. In addition, the Arabian plate is now colliding with Asia along the Zagros Mountains of Iran.

Terranes and the Origin of Mountains

In the preceding sections, we discussed orogenies along convergent plate boundaries resulting in the addition of new material to a continent, a process termed **continental accretion**. Much of the material accreted to continents during these events is simply eroded older continental crust, but a significant amount of new material is added to continents as well—igneous rocks that formed during subduction and partial melting and the suturing of an island arc to a continent, for example. Although subduction is the predominant influence on tectonic history, in many regions of orogenies, other processes are also involved in mountain building and continental accretion, especially the accretion of blocks of rock known as *terranes*.

During the late 1970s and 1980s, geologists discovered that parts of many mountain systems are composed of small accreted lithospheric blocks that are clearly of foreign origin. These **terranes*** differ completely in their fossil content, structural trends, and paleomagnetic properties from the rocks of the surrounding mountain system. In fact, they are so different from adjacent rocks that most geologists think they formed elsewhere and were carried great distances as parts of other plates until they collided with continents (■ Figure 13.26).

Geologic evidence indicates that much of the Pacific coast from Alaska to Baja California consists of accreted terranes or igneous intrusions. They are composed of volcanic island arcs, oceanic ridges, seamounts, and small fragments of continents that were scraped off and accreted to the continent's margin as the oceanic plate upon which they were carried was subducted. Numerous ridges, plateaus, and seamounts in today's oceans are potential terranes, especially in the Pacific Ocean (see Figure 11.11). According to one estimate, more than 100 different-sized terranes have been added to the western margin of North America during the last 200 million years (■ Figure 13.27).

The basic plate tectonic reconstruction of orogenies and continental accretion remains unchanged, but the details of these reconstructions are decidedly different in view of terrane accretion. For one thing, these accreted terranes are often new additions to a continent rather than reworked older continental material.

Most of the terranes identified so far are in mountains of the North American Pacific coast region, but a number of such plates are suspected to be present in other mountain systems as well. They are more difficult to recognize in older mountain systems, such as the Appalachians. Nevertheless, about a dozen terranes have been identified, although their boundaries are difficult to discern.

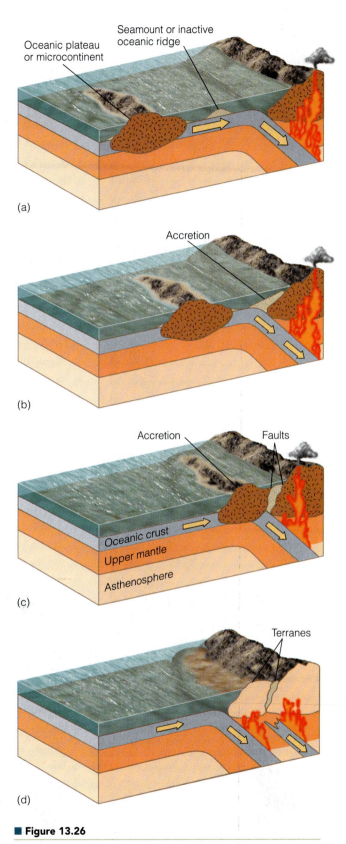

(a)

(b)

(c)

(d)

■ **Figure 13.26**

(a)–(d) The origin of terranes and their incorporation into a continent. Most terranes are made up of inactive oceanic ridges, oceanic plateaus, seamounts, and guyots, but oceanic lithosphere is also sometimes "scraped off" onto continental lithosphere, thus forming rock complexes geologists call ophiolites (see Chapter 11).

*Some geologists prefer the term *suspect terrane, exotic terrane,* or *displaced terrane.* Notice also the spelling of *terrane* as opposed to the more familiar *terrain,* the later a geographic term indicating a particular area of land.

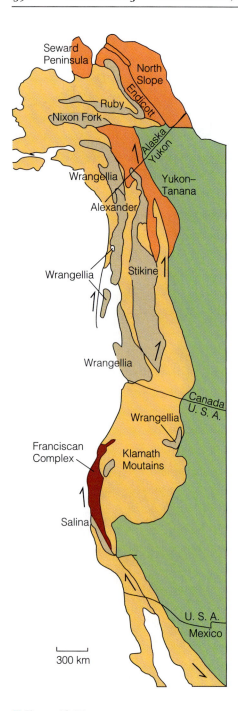

■ Figure 13.27

Some of the accreted lithospheric blocks called *terranes* that form the western margin of North America. The light brown blocks probably originated as parts of continents other than North America. The reddish brown blocks are possibly displaced parts of North America. The Franciscan Complex consists of a variety of seafloor rocks. Source: From Zvi Ben-Avraham, "The Movement of Continents." *American Scientist 69:* 291–299, Figure 9, p. 298, Journal of Sigma Xi, The Scientific Research Group.

HOW DID THE CONTINENTS FORM AND EVOLVE?

I f we could somehow go back and visit Earth shortly after it formed 4.6 billion years ago, we would see a hot, barren, waterless planet bombarded by comets and meteorites, no continents, intense cosmic radiation, and ubiquitous volcanism. Judging from the oldest known rocks, the 3.96-billion-year-old Acasta Gneiss in Canada and rocks in Montana, some continental crust existed by that time. And sedimentary rocks in Australia contain detrital zircons ($ZrSiO_4$) dated at 4.2 billion years, so source rocks at least this old existed.

According to one model for the origin of continents, the earliest crust was thin, unstable, and composed of ultramafic rock. Upwelling basaltic magmas disrupted this ultramafic crust, which was destroyed as it was consumed at subduction zones. A second stage in crustal evolution took place when basaltic crust, which formed when ultramafic crust was disrupted, formed island arcs and partially melted to form crust of a more intermediate (andesitic) composition. Partial melting of andesite yielded granitic magmas that were emplaced in the island arcs, which became more like granitic continental crust. And by 3.96 billion years ago, plate movements along with subduction and collisions of island arcs formed several granitic continental nuclei (■ Figure 13.28).

Shields, Platforms, and Cratons

Continents are more than simply land areas above sea level. You already know that they have an overall composition similar to granite and that continental crust is thicker and less dense than oceanic crust, which is made up of basalt and gabbro. In addition, a **shield** consisting of a vast area or areas of exposed ancient (Precambrian) rocks is found on all continents (see Figure 7.1). Continuing outward from shields are broad **platforms** of buried ancient rocks that underlie much of each continent. Collectively, a shield and platform made up a **craton,** which we can think of as a continent's ancient nucleus.

The cratons are the foundations of the continents, and along their margins more continental crust was added as they evolved to their present sizes and shapes, a phenomenon called *continental accretion.* Many of the rocks along the margins of the cratons show evidence of deformation accompanied by metamorphism, igneous activity, and mountain building. In North America the exposed part of the craton is the **Canadian shield,** which occupies most of northeastern Canada, a large part of Greenland, the Adirondak Mountains of New

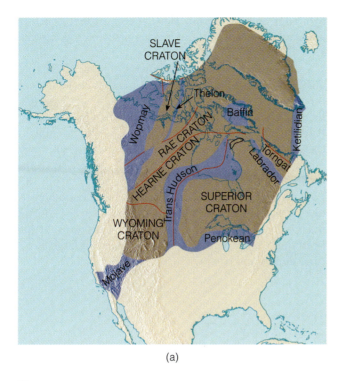

(a)

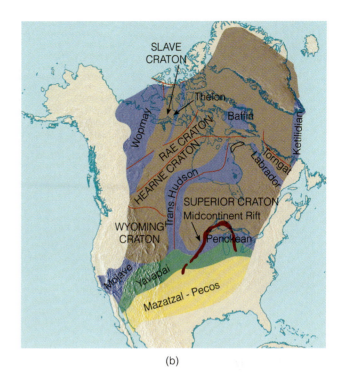

(b)

Years ago:

- 900 million–1.2 billion
- 1.6 billion–1.75 billion
- 1.75 billion–1.8 billion
- 1.8 billion–2.0 billion
- 2.5 billion–3.0 billion

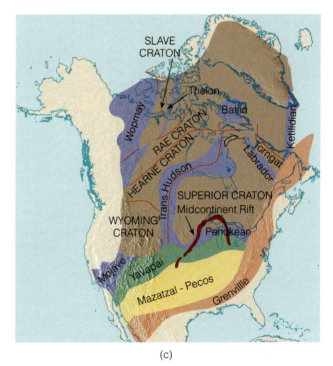

(c)

■ **Figure 13.28**

Three stages in the early evolution of North America. (a) By 1.8 BYA, the continent consisted of the elements shown here. The several orogens formed when older cratons collided to form a larger craton. (b), (c) Continental accretion took place along the southern and eastern margins of North America. By about 1.0 BYA, North America had the approximate size and shape shown in (c). The rest of North America evolved during the Phanerozoic Eon—that is, during the last 545 million years. Source: Reprinted with permission from Kent C. Condie, *Plate Tectonics and Crustal Evolution*, 4d, p. 65 (Fig. 2.26), © 1997, Butterworth-Heinemann.

York, and parts of the Lake Superior region of Minnesota, Wisconsin, and Michigan.

The Geologic Evolution of North America

To the extent that all continents evolved by accretion along their margins they have similar histories, although the details for each differ. Here we concentrate on the evolution of North America. Figure 13.28a shows North America's approximate size and shape at 1.8 billion years ago. But even this early continent was made up of smaller units or cratons, including the Superior and Wyoming cratons that had collided along deformation belts such as the Trans-Hudson orogen (remember an orogeny is an episode of mountain building). By 1.6 billion years ago more accretion had taken place along the southern margin of the continent (Figure 13.28b), and by about 900 million years

What Would You Do ?

As a member of a planning commission, you are charged with developing zoning regulations and building codes for an area with known active faults, steep hills, and deep soils. A number of contractors as well as developers and citizens in your community are demanding action because they want to begin several badly needed housing developments. How might geologic maps and an appreciation of geologic structures influence you in this endeavor? Do you think, considering the probable economic gains for your community, that the regulations you draft should be rather lenient or very strict? If the latter, how would you explain why you favored regulations that would involve additional cost for houses?

ago, additional accretion along the continent's eastern margin gave rise to a landmass approximating the size and shape of that in Figure 13.28c.

The sequence of events shown in Figure 13.28 took place during that part of geologic time that geologists call the Precambrian, which ended 545 million years ago (see Figure 1.17). Since the Precambrian—that is, during the Paleozoic, Mesozoic, and Cenozoic eras—continental accretion has occurred mostly along the eastern, southern, and western margins of the craton, giving rise to the present configuration of North America. During this time, though, the craton itself has been remarkably stable, but it has been invaded six times by the seas during marine transgressions followed by regressions. It has been only mildly deformed into a number of large structural domes and basins.

13 REVIEW WORKBOOK

Chapter Summary

- Folded and fractured rocks have been deformed or strained by applied stresses.

- Stress is compression, tension, or shear. Elastic strain is not permanent, but plastic strain and fracture are, meaning that rocks do return to their original shape or volume when the deforming forces are removed.

- Strike and dip are used to define the orientation of deformed rock layers. This same concept applies to other planar features such as fault planes.

- Anticlines and synclines are up- and down-arched folds, respectively. They are identified by strike and dip of the folded rocks and the relative ages of rocks in the cores of these folds.

- Structural domes and basins are the circular to oval equivalents of anticlines and synclines but are commonly much larger structures.

- The two structures resulting from fracture are joints and faults. Joints may open up but they show no movement parallel with the fracture surface, whereas faults do show movement parallel with the fracture surface.

- Joints are very common and form in response to compression, tension, and shear.

- On dip-slip faults, all movement is up or down the dip of the fault; if the hanging wall moves relatively down it is a normal fault, but if the hanging wall moves relatively up it is a reverse fault. Normal faults result from tension; reverse faults from compression.

- In strike-slip faults, all movement is along the strike of the fault. These faults are either right-lateral or left-lateral, depending on the apparent direction of offset of one block relative to the other.

- Oblique-slip faults show components of both dip-slip and strike-slip movement.

- A variety of processes account for the origin of mountains, some involving little or no deformation, but the large mountain systems on the continents resulted from deformation at convergent boundaries.

- A volcanic island arc, deformation, igneous activity, and metamorphism characterize orogenies at oceanic-oceanic plate boundaries, whereas subduction at an oceanic-continental plate boundary also results in orogeny.

- Some mountain systems are within continents far from a present-day plate boundary. These mountains formed when two continental plates collided and became sutured.

- Geologists now realize that orogenies also involve collisions of terranes with continents.

- A craton made up of an exposed shield and buried platform of ancient rocks forms the nucleus of a continent.

- Accretion along the margins of cratons accounts for the evolution of the continents.

- In North America, the craton evolved during Precambrian time by collisions of smaller cratons along deformation belts. Since then, accretion has taken place along the craton's margin.

Important Terms

anticline (p. 368)
basin (p. 372)
Canadian shield (p. 390)
compression (p. 365)
continental accretion (p. 389)
craton (p. 390)
deformation (p. 364)
dip (p. 367)
dip-slip fault (p. 375)
dome (p. 372)
elastic strain (p. 366)
fault (p. 375)
fault plane (p. 375)

fold (p. 368)
footwall block (p. 375)
fracture (p. 366)
geologic structure (p. 367)
hanging wall block (p. 375)
joint (p. 374)
monocline (p. 368)
normal fault (p. 376)
oblique-slip fault (p. 380)
orogeny (p. 384)
plastic strain (p. 366)
platform (p. 390)

plunging fold (p. 370)
reverse fault (p. 376)
shear stress (p. 365)
shield (p. 390)
strain (p. 365)
stress (p. 365)
strike (p. 366)
strike-slip fault (p. 380)
syncline (p. 368)
tension (p. 365)
terrane (p. 389)
thrust fault (p. 376)

Review Questions

1. Rocks characterized as ductile:
   a. _____ fracture easily when in compression;
   b. _____ show a great amount of plastic strain; c. _____ are found along the crests of anticlines; d. _____ are common along dip-slip faults; e. _____ are the main rocks in shields.

2. A structural basin is a(n):
   a. _____ fault on which the hanging wall block moved down; b. _____ elongate fold with all strata dipping away from the fold axis; c. _____ kind of fracturing that occurs in lava as it cools; d. _____ oval fold with the youngest exposed rocks at its center; e. _____ type of deformation found adjacent to strike-slip faults.

3. The process by which new material is added to the margins of cratons is:
   a. _____ continental accretion; b. _____ shear stress; c. _____ thrust faulting; d. _____ elastic strain; e. _____ platform evolution.

4. The line formed by the intersection of a horizontal plane and an inclined plane is the definition of:
   a. _____ stress; b. _____ brittle behavior; c. _____ jointing; d. _____ uplift; e. _____ strike.

5. The fault illustrated in Figure 13.17e shows both _____ and _____ faulting.
   a. _____ hanging wall block uplift/footwall block subsidence; b. _____ low-angle thrust/normal; c. _____ thrust/reverse; d. _____ reverse dip-slip/left-lateral strike-slip; e. _____ right-lateral strike-slip/normal.

6. Orogeny is the geologic term used for:
   a. _____ deformation with little or no plastic strain; b. _____ the origin of large circular folds; c. _____ an episode of deformation and the origin of mountains; d. _____ rocks that show both elastic and plastic deformation; e. _____ a type of fold with its axis inclined.

7. The ancient stable core of a continent is known as its:

    a. _____ craton; b. _____ active interior;
    c. _____ thrust belt; d. _____ structural dome;
    e. _____ dome.

8. An oceanic plate colliding with a continental plate accounts for the ongoing mountain building:

    a. _____ in the Rocky Mountains;
    b. _____ along the west coast of South America; c. _____ where the Pacific plate collides with Japan; d. _____ along North America's eastern margin; e. _____ in a large region in Africa.

9. Most folding of rock layers results from:

    a. _____ tension; b. _____ shear stress;
    c. _____ convection; d. _____ compression;
    e. _____ fracturing.

10. A fault along which the hanging wall block moved down relative to the footwall block is a _____ fault.

    a. _____ normal; b. _____ strike-slip;
    c. _____ thrust; d. _____ oblique;
    e. _____ reverse.

11. What kinds of evidence indicate that mountain building took place in the Canadian shield where mountains are no longer present?

12. Rocks are displaced 200 km along a strike-slip fault during a period of 5 million years. What was the average rate of movement per year? Is the average likely to represent the actual rate of displacement on this fault? Explain.

13. How would you explain stress and strain to someone unfamiliar with the concepts?

14. Discuss the features of mountains formed at an oceanic-continental plate boundary, and give an example of where such activity is presently taking place.

15. Explain what is meant by the term *terrane* and how terranes are incorporated into continents.

16. Describe how time, rock type, pressure, and temperature influence rock deformation.

17. What is continental accretion and how does it take place?

18. Illustrate a recumbent anticline and explain what criteria are necessary to distinguish it from a recumbent syncline.

19. Notice that the monocline in Figure 13.7b is deeply eroded where flexure of the rock layers is greatest. Why is this so?

20. What is meant by the elastic limit of rocks, and what happens when rocks are strained beyond their elastic limit?

World Wide Web Activities

PHYSICAL Geology⇌Now Assess your understanding of this chapter's topics with additional quizzing and comprehensive interactivities at

http://earthscience.brookscole.com/physgeo5e

as well as current and up-to-date weblinks, additional readings, and InfoTrac College Edition exercises.

Mass Wasting

CHAPTER 14
OUTLINE

PHYSICAL
Geology⇌Now *This icon, appearing throughout the book, indicates an opportunity to explore interactive tutorials, animations, or practice problems available on the Physical GeologyNow Web site at http://earthscience.brookscole.com/physgeo5e.*

OBJECTIVES
At the end of this chapter, you will have learned that

- It is important to understand the different types of mass wasting because mass wasting affects us all and causes economically significant destruction.

- Factors such as slope angle, weathering and climate, water content, vegetation, and overloading are interrelated, and all contribute to mass wasting.

- Mass movements can be triggered by such factors as overloading, soil saturation, and ground shaking.

- Mass wasting can be categorized as resulting from either rapid mass movements or slow mass movements.

- The different types of rapid mass movements are rockfalls, slumps, rock slides, mudflows, debris flows, and quick clays; each type has recognizable characteristics.

- The different types of slow mass movements are earthflows, solifluction, and creep; each type has recognizable characteristics.

- People can minimize the effects of mass wasting by conducting geologic investigations of an area and stabilizing slopes to prevent and ameliorate movement.

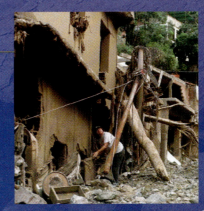

Residents of Caracas, Venezuela, clean up the debris from massive flooding and mudslides that devastated large areas of Venezuela during December 1999. Source: AP/Wide World Photos

Introduction

Triggered by relentless torrential rains that began on December 15, 1999, the floods and mudslides that devastated Venezuela were some of the worst ever to strike that country. Although a reliable death toll is impossible to determine, it is estimated that between 10,000 and 30,000 people were killed, 100,000 to 150,000 were left homeless, 35,000 to 40,000 homes were destroyed or buried by mudslides, and between $10 billion and $20 billion in damage was done before the rains and slides abated (see the chapter opening photo). It is easy to mention numbers of dead and homeless, but the human side of the disaster was most vividly brought home by a mother who described standing helplessly by and watching her four small children buried alive in the family car as a raging mudslide carried it away.

Mudslides engulfed and buried not only homes, buildings, and roads, but even entire communities. Some areas were covered with as much as 7 m of mud. In addition, flooding and the accompanying mudslides swept away large parts of many of Venezuela's northern coastal communities, leaving huge areas uninhabitable.

This terrible tragedy illustrates the close link between geology and individuals, governments, and society in general, a theme we stressed in Chapter 1. The underlying causes of the mudslides are not unique to Venezuela; they can be found anywhere in the world. By being able to recognize and understand these causes and what the result may be, we can find ways to reduce hazards and minimize damage, in terms of both human suffering and misery as well as property damage. This tragedy shows how geology affects all our lives and how interconnected the various systems and subsystems of Earth are.

The topography of land areas is the result of the interaction among Earth's internal processes, type of rocks exposed at the surface, effects of weathering, and the erosional agents of water, ice, and wind. The specific type of landscape developed depends, in part, on which agent of erosion is dominant. *Landslides* (a general term for mass movements), which can be very destructive, are part of the normal adjustments of slopes to changing surface conditions.

Mass wasting (also called *mass movement*) is defined as the downslope movement of material under the direct influence of gravity. Most types of mass wasting are aided by weathering and usually involve surficial material. The material moves at rates ranging from almost imperceptible, as in the case of creep, to extremely fast, as in a rockfall or slide.

Table 14.1

Selected Landslides, Their Cause, and the Number of People Killed

Date	Location	Type	Deaths
218 B.C.	Alps (European)	Avalanche—destroyed Hannibal's army	18,000
1556	China (Hsian)	Landslides—earthquake triggered	1,000,000
1806	Switzerland (Goldau)	Rock slide	457
1903	Canada (Frank, Alberta)	Rock slide	70
1920	China (Kansu)	Landslides—earthquake triggered	~200,000
1941	Peru (Huaraz)	Avalanche and mudflow	7,000
1962	Peru (Mt. Hauscarán)	Ice avalanche and mudflow	~4,000
1963	Italy (Vaiont Dam)	Landslide—subsequent flood	~2,000
1966	United Kingdom (Aberfan, South Wales)	Debris flow—collapse of mining-waste tip	144
1970	Peru (Mt. Hauscarán)	Rockfall and debris avalanche—earthquake triggered	25,000
1981	Indonesia (West Irian)	Landslide—earthquake triggered	261
1987	El Salvador (San Salvador)	Landslide	1,000
1989	Tadzhikistan	Mudflow—earthquake triggered	274
1994	Colombia (Paez River Valley)	Avalanche—earthquake triggered	>300
1999	Venezuela	Mudflow	>10,000
2003	Southern California (U.S.A.)	Mudflow	10

Source: Data from J. Whittow, *Disasters: The Anatomy of Environmental Hazards* (Athens: University of Georgia Press, 1979); *Geotimes;* and *Earth.*

Although water can play an important role, the relentless pull of gravity is the major force behind mass wasting.

Mass wasting is an important geologic process that can occur at any time and almost any place. It is thus important to study this phenomenon because it affects all of us, no matter where we live (Table 14.1). In the United States alone, mass wasting occurs in all 50 states and causes economically significant destruction in more than 25 states (■ Figure 14.1). Furthermore, between 25 and 50 people, on average, are killed each year by landslides in the United States, and the annual cost in damages exceeds $1.5 billion! Although all major landslides have natural causes, many smaller ones are the result of human activity and could have been prevented or their damage minimized. In this chapter we will examine the factors leading to mass wasting, and discuss ways to prevent or minimize the damage they cause.

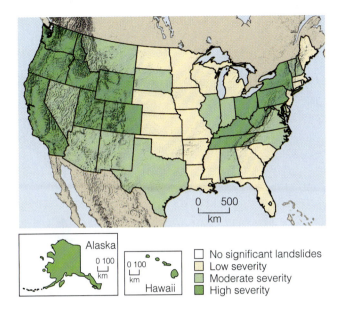

■ **Figure 14.1**

Severity of landslides in the United States. Note that areas of greatest severity occur in the coastal mountain ranges of the West Coast, the Rocky Mountains, and Appalachian Mountains.

WHAT FACTORS INFLUENCE MASS WASTING?

When the gravitational force acting on a slope exceeds its resisting force, slope failure (mass wasting) occurs. The resisting forces that help to maintain slope stability include the slope material's strength and cohesion, the amount of internal friction between grains, and any external support of the slope (■ Figure 14.2). These factors collectively define a slope's **shear strength.**

Opposing a slope's shear strength is the force of gravity. Gravity operates vertically but has a component acting parallel to the slope, thereby causing instability (Figure 14.2). The steeper a slope's angle, the greater the component of force acting parallel to the slope, and the greater the chance for mass wasting. The steepest angle that a slope can maintain without collapsing is its *angle of repose*. At this angle, the shear strength of the slope's material exactly counterbalances the force of gravity. For unconsolidated material, the angle of repose normally ranges from 25 to 40 degrees. Slopes steeper than 40 degrees usually consist of unweathered rock.

All slopes are in a state of *dynamic equilibrium*, which means that they are constantly adjusting to new conditions. Although we tend to view mass wasting as a disruptive and usually de-

structive event, it is one of the ways that a slope adjusts to new conditions. Whenever a building or road is constructed on a hillside, the equilibrium of that slope is affected. The slope must then adjust, perhaps by mass wasting, to this new set of conditions.

Many factors can cause mass wasting: a change in slope angle, weakening of material by weathering, increased water content, changes in the vegetation cover, and overloading. Although most of these are interrelated,

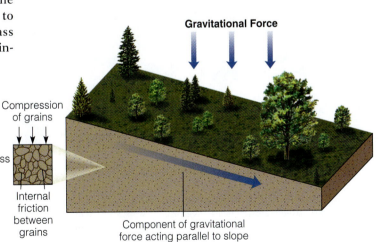

■ **Figure 14.2**

A slope's shear strength depends on the slope material's strength and cohesion, the amount of internal friction between grains, and any external support of the slope. These factors promote slope stability. The force of gravity operates vertically but has a component acting parallel to the slope. When this force, which promotes instability, exceeds a slope's shear strength, slope failure occurs.

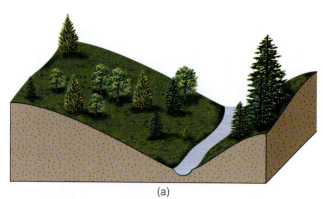

(a)

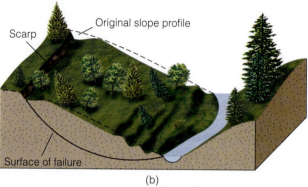

Original slope profile

Scarp

Surface of failure

(b)

Reed Wicander

(c)

■ **Figure 14.3**

Undercutting by stream erosion (a) removes a slope's base, which increases the slope angle and (b) can lead to slope failure. (c) Undercutting by stream erosion caused slumping along this stream near Weidman, Michigan.

we will examine them separately for ease of discussion, but we will also show how they individually and collectively affect a slope's equilibrium.

Slope Angle

Slope angle is probably the major cause of mass wasting. Generally speaking, the steeper the slope, the less stable it is. Therefore steep slopes are more likely to experience mass wasting than gentle ones.

A number of processes can oversteepen a slope. One of the most common is undercutting by stream or wave action (■ Figure 14.3). This removes the slope's base, increases the slope angle, and thereby increases the gravitational force acting parallel to the slope. Wave action, especially during storms, often results in mass movements along the shores of oceans or large lakes (■ Figure 14.4).

Excavations for road cuts and hillside building sites are another major cause of slope failure (■ Figure 14.5). Grading the slope too steeply, or cutting into its side increases the stress in the rock or soil until it is no longer strong enough to remain at the steeper angle and mass movement ensues. Such action is analogous to undercutting by streams or waves and has the same result, thus explaining why so many mountain roads are plagued by frequent mass movements.

Weathering and Climate

Mass wasting is more likely to occur in loose or poorly consolidated slope material than in bedrock. As soon as rock is exposed at Earth's surface, weathering begins to disintegrate and decompose it, reducing its shear strength and increasing its susceptibility to mass wasting. The deeper the weathering zone extends, the greater the likelihood of some type of mass movement.

Recall that some rocks are more susceptible to weathering than others and that climate plays an important role in the rate and type of weathering. In the tropics, where temperatures are high and considerable rainfall occurs, the effects of weathering extend to depths of several tens of meters, and mass movements most commonly occur in the deep weathering zone. In arid and semiarid regions, the weathering zone is usually considerably shallower. Nevertheless, intense, localized cloudbursts can drop large quantities of water on an area in a short time. With little vegetation to absorb this water, runoff is rapid and frequently results in mudflows.

■ **Figure 14.4**

This sea cliff north of Bodega Bay, California, was undercut by waves during the winter of 1997–1998. As a result, part of the land slid into the ocean, damaging several houses.

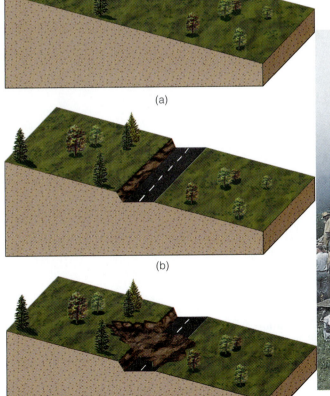

(a)

(b)

(c)

■ **Figure 14.5**

(a) Highway excavations disturb the equilibrium of a slope by (b) removing a portion of its support as well as oversteepening it at the point of excavation. (c) Such action can result in landslides. (d) Cutting into the hillside to construct this portion of the Pan-American Highway in Mexico resulted in a rockfall that completely blocked the road.

(d)

GEOLOGY
IN UNEXPECTED PLACES

New Hampshire Says Good-Bye to the "Old Man"

Imagine waking up one morning and discovering that your state symbol had suddenly disappeared during the night. That's exactly what happened to the residents of New Hampshire when they discovered that "The Old Man of the Mountain," a landmark symbolizing their state's independence and stubbornness, had collapsed from natural causes sometime during the night of May 5, 2003.

The Old Man of the Mountain, located in Franconia Notch State Park, was composed of a series of five horizontal granite ledges that had weathered over millions of years to form a man's profile. Overlooking Profile Lake 400 m below, the Old Man measured about 13 m from chin to forehead and jutted out about 8 m from the main part of the mountain.

The natural forces that shaped the Old Man also brought about its demise. Freezing temperatures, high winds, and heavy rains all contributed to finally overcoming the forces that held the rock together. Despite nearly 100 years of effort to protect the landmark from destruction, the forces of nature eventually won. All that's left of the Old Man now is a pile of rubble at the mountain's base and some stabilizing cables and epoxy where the Old Man once looked out on the state it symbolized for so long.

White Mountains Attractions Association

■ **Figure 1**

The Old Man of the Mountain before it collapsed.

Water Content

The amount of water in rock or soil influences slope stability. Large quantities of water from melting snow or heavy storms greatly increase the likelihood of slope failure. The additional weight that water adds to a slope can be enough to cause mass movement. Furthermore, water percolating through a slope's material helps decrease friction between grains, contributing to a loss of cohesion. For example, slopes composed of dry clay are usually quite stable, but when wetted they quickly lose cohesiveness and internal friction and become an unstable slurry. This occurs because clay, which can hold large quantities of water, consists of platy particles that easily slide over each other when wet. For this reason, clay beds are frequently the slippery layer along which overlying rock units slide downslope (see Geo-Focus 14.1 on page 406).

Vegetation

Vegetation affects slope stability in several ways. By absorbing the water from a rainstorm, vegetation decreases water saturation of a slope's material that would otherwise lead to a loss of shear strength. Vegetation's root system also helps stabilize a slope by binding soil particles together and holding the soil to bedrock.

The removal of vegetation by either natural or human activity is a major cause of many mass move-

T. Spencer/Colorific

■ **Figure 1**

Location map and aerial view of the Aberfan tip disaster in which 144 people died.

Tonya Paul/Oroville Mercury Register

■ **Figure 14.10**

A huge rockfall closed both lanes of traffic on Highway 70 near Rogers Flat, California, on July 25, 2003. Rocks the size of large dump trucks had to be blasted into smaller pieces to clear the highway. Despite the pavement cracking caused by the falling boulders, geologists determined that the roadbase was undamaged and the road would be safe following cleanup operations.

WILLIAM M. GOODMAN

Applied Hydrogeology: Understanding Water As an Agent of Mass Wasting

William M. Goodman is Senior Hydrogeologist for an engineering and architectural firm based in Rochester, New York. He graduated with a B.A. and M.A. in geology from Temple University in 1982 and 1985, respectively, and a Ph.D. in geology from the University of Rochester in 1993. He is currently a licensed Professional Geologist in Pennsylvania and Kansas.

Over the course of my college education and professional career in geology, I have come to appreciate the awesome power of water as an agent of geological change. Sometimes the effects of water are subtle, gradual, and perceptible only by comparison of topographic maps and/or photographs of a locale that span decades. In some situations, however, water-driven surficial processes are both episodic and catastrophic. It is one role of the professional hydrogeologist to understand both the gradual and episodic effects of water-induced surficial processes, so that we can predict what can happen in the future and advise the public accordingly.

As a hydrogeologist, I am responsible for studying the interaction between water and geologic materials. Accordingly, I fully appreciate the role water plays in the triggering of landslides, rock falls, slope failures, and other types of mass wasting processes. I work with highway design engineers, waste disposal companies, and mining interests to provide an understanding of groundwater flow conditions that may affect the stability of soil and rock on road cuts, on landfill sites, and above deep subsurface salt mines.

Up here in the seasonally frigid Northeast, the role of frozen water in producing rock falls on road cuts and natural rock outcrops cannot be ignored. We have many steep rock escarpments and river gorges in this part of the country. During the winter, spectacular ice flows develop as groundwater seeping through fractures and along bedding planes reaches the cold air near the ground surface. The ice is often tinted red or other colors due to the presence of rock and mineral debris plucked from cliff faces as the ice is forced outward from the rock face. Icy fingers penetrate bedrock joints, forcing them open to form the boundaries of loose, massive rock blocks. When the spring melt comes, it can be extremely dangerous at the base of the exposures. The conversion of ice to water unbinds winter's grip on those loosened masses of soil and rock and allows them to crash to the roads and river beds below!

Steep topographic relief is common in New York State. Construction on steep side-slopes can be a challenge given the nature of our low permeability glacial till soils. Because our silt-rich soils don't allow water to efficiently percolate to appreciable depths, the upper few meters of soil become water-logged in the spring and during strong storm events. Over the 10,000 years since glaciers retreated from this area, these seasonally saturated soils have weathered to a brown color that is different than their dark gray or deep maroon original colors. The innumerable

Sear Brown

freeze–thaw cycles and the leaching of minerals by water have altered the structure of this upper "oxidized" zone so that the soil is less cohesive. Consequently, when it becomes water-saturated, the weathered soil profile can potentially be prone to failure when landfills or earthen berms with steep side-slopes that create unnatural loading are constructed on them. To avoid soil failure and landslides, careful description of the groundwater flow patterns by a hydrogeologist provides critical information to a geotechnical engineer who can design a pore-water suppression system that effectively depressurizes the soil profile while a load is imposed on it, thereby making it stronger and less resistant to failure. This is a critical task in the construction of landfills in upstate New York. If a landfill side-slope were to fail, leachate (i.e., water tainted with dissolved constituents from the disposed wastes) would be free to run downslope and potentially contaminate both groundwater and surface-water bodies.

Although it is not commonly cited as an example of mass wasting because of its rarity, catastrophic ground failure due to mine flooding is a phenomenon that has happened here in upstate New York in the past decade. We are blessed with bedded rock salt that has been commercially mined for over a century at subsurface depths of roughly 1000 to 2000 feet. The long history of mining at one of the oldest, and certainly the largest, salt mines in New York was abruptly terminated in 1994 when the mine advanced under an anomalously deep subsurface pocket of saline groundwater. The weight and pressure of this brine pocket created a localized failure in the mine roof. The water flowed rapidly into the mine, and because it was not fully saturated with salt, the inflowing water dissolved pillars that were holding up the roof

of the mine. As a consequence, the roof of the mine rapidly sagged to meet the floor of the mine, a vertical distance of about 10 feet. This rapid mine closure resulted in immediate ground subsidence at the surface over the mine, damaging a bridge and several homes. In less than 24 months, a mine roughly the size of Manhattan was completely flooded with water! Once the rate of ground subsidence began to wane, I spearheaded a group of geologists and geophysicists to collect seismic images of subsurface conditions near the collapsed bridge. Because our data suggested that all the void space had appeared to be closed, the state department of transportation proceeded to rebuild the bridge.

So, in conclusion, the concept of mass wasting has very real meaning to us in this part of the country. As a professional hydrogeologist, I see and appreciate the dynamic role that water plays as an agent of geologic change. The effects of water on surficial geologic materials can be dangerous, even deadly. Professional geologists and hydrogeologists need to work with engineers, state transportation departments, environmental regulatory agencies, and the public to inform them of ways in which water weakens soil and bedrock on steep slopes, so that they can design and implement systems and policies to protect human health and the environment. Many states already recognize the valuable contribution that geologists and hydrogeologists make to protecting the welfare of their citizenry through professional licensure. We are a bit behind in New York State; our licensure bill for professional geologists stalled in the legislature. It shouldn't be long, however, before our state legislators come to appreciate how important it is that applied geoscience is conducted by licensed professionals!

Sue Monroe

James S. Monroe

(a)

(b)

■ **Figure 14.11**

Minimizing damage from rock falls.
(a) Wire mesh has been used to cover this steep slope in Hawaii. This is a common practice in mountainous areas to prevent rocks from falling on the road. (b) A wire mesh fence along the base of this hillside of Highway 44 in California has caught many boulders and prevented them from rolling onto the highway.

A **slump** involves the downward movement of material along a curved surface of rupture and is characterized by the backward rotation of the slump block (■ Figure 14.12). Slumps usually occur in unconsolidated or weakly consolidated material and range in size from small individual sets, such as occur along stream banks, to massive, multiple sets that affect large areas and cause considerable damage.

Slumps can be caused by a variety of factors, but the most common is erosion along the base of a slope, which removes support for the overlying material. This local steepening may be caused naturally by stream erosion along its banks (Figure 14.12) or by wave action at the base of a coastal cliff (■ Figure 14.13). Slope oversteepening can also be caused by human activity, such as the construction of highways and housing developments. Slumps are particularly prevalent along highway cuts, where they are generally the most frequent type of slope failure observed.

Although many slumps are merely a nuisance, large-scale slumps involving populated areas and highways can cause extensive damage. Such is the case in coastal southern California where slumping and sliding have been a constant problem. Many areas along the coast are underlain by poorly to weakly consolidated silts, sands, and gravels interbedded with clay layers, some of which are weathered ashfalls. In addition, southern California is tectonically active so that many of these deposits are cut by faults and joints, which allow the infrequent rains to percolate downward rapidly, wetting and lubricating the clay layers.

Southern California lies in a semiarid climate and is dry most of the year. When it does rain, typically between November and March, large amounts of rain can fall in a short time. Thus the ground quickly becomes saturated, leading to landslides along steep canyon walls as well as along coastal cliffs (Figure 14.13). Most of the slope failures along the southern California coast are the result of slumping. These slumps have destroyed many expensive homes and forced many roads to be closed and relocated.

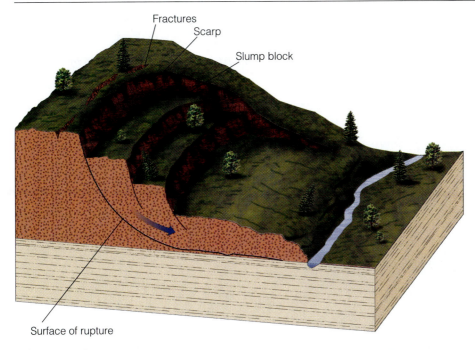

Fractures
Scarp
Slump block

Surface of rupture

■ **Figure 14.12**

In a slump, material moves downward along the curved surface of a rupture, causing the slump block to rotate backward. Most slumps involve unconsolidated or weakly consolidated material and are typically caused by erosion along the slope's base.

Pacific Palisades
Los Angeles
Santa Monica
Pacific Ocean
Palos Verdes
Long Beach

■ **Figure 14.13**

Undercutting of steep sea cliffs by wave action resulted in massive slumping in the Pacific Palisades area of southern California on March 31 and April 3, 1958. Highway 1 was completely blocked. Note the heavy earth-moving equipment for scale.

Point Fermin—Slip Sliding Away

Dubbed the "sunken city" by residents of the area, Point Fermin in southern California is famous for its numerous examples of mass wasting. The area is underlain by fine-grained sedimentary rocks interbedded with diatomite layers and volcanic ash. When these layers get wet, they get slippery and tend to slide easily. The rocks also dip slightly toward the ocean and form steep coastal bluffs that are being undercut by constant wave action at their base. This wave action results in oversteeping of the cliffs, which causes slumping.

Mass wasting began in 1929 with minor slumping in the area. In the early 1940s, water mains in the region were broken and several individual blocks began slumping. Movement largely ceased following this main phase of slumping, but it has continued intermittently until the present and residents have paid the price for living in an unstable coastal area.

Eleanora Robbins / USGS

A view of one portion of the Point Fermin slide area showing the fine-grained sedimentary rocks dipping slightly toward the ocean and the oversteepened cliffs resulting from slumping and sliding in the foreground.

A map of southern California showing the location of Point Fermin and an aerial view at low tide of the sliding that has taken place. Note the numerous slump blocks and oversteepened cliffs. The continuous pounding of waves and surf along the base of the cliffs further erodes and undercuts them, leading to even more slumping and sliding.

Reed Wicander

A view of one slump block shows remnants of a former road and a palm tree still growing as if nothing has happened

Reed Wicander

An abandoned house on the edge of an oversteepened coastal bluff. Note the concrete blocks placed along the beach to absorb the erosive energy of the incoming waves and slow down the erosion to the cliffs.

Reed Wicander

Creep and minor slumping are evident in this photo. Note the two small slump scarps. The smaller one in the background is mostly grass-covered, whereas the one in the foreground has bare spots and is apparently moving at a slightly faster rate. Notice the effect of creep on the right-hand wall of the house. The bottom part of the wall is moving toward the right of the photo as a result of creep, producing a bend of the wall that can be seen clearly near its base.

Reed Wicander

Slumping along the oversteepened cliffs at Point Fermin. The abandoned house in the photo above is just to the right of this view at the top of the cliff.

Reed Wicander

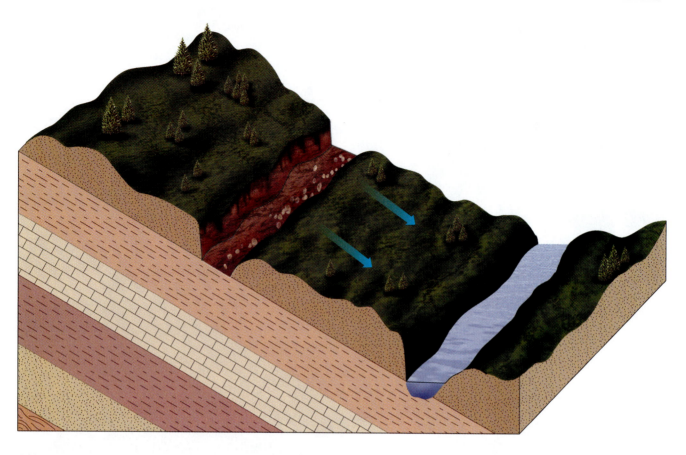

■ **Figure 14.14**

Rock slides occur when material moves downslope along a generally planar surface.

A **rock** or *block* **slide** occurs when rocks move downslope along a more or less planar surface. Most rock slides take place because the local slopes and rock layers dip in the same direction (■ Figure 14.14), although they can also occur along fractures parallel to a slope. In addition to slumping, rock slides are common occurrences along the southern California coast. At Point Fermin, seaward-dipping rocks with interbedded slippery clay layers are undercut by waves, causing numerous slides. See "Point Fermin—Slip Sliding Away" on the preceding pages.

Farther south, in the town of Laguna Beach, startled residents watched as a rock slide destroyed or damaged 50 homes on October 2, 1978 (■ Figure 14.15). Just as at Point Fermin, the rocks at Laguna Beach dip about 25 degrees in the same direction as the slope of the canyon walls and contain clay beds that "lubricate" the overlying rock layers, causing the rocks and the houses built on them to slide. In addition, percolating water from the previous winter's heavy rains wet a subsurface clayey siltstone, thus reducing its shear strength and helping activate the slide. Although the 1978 slide covered only about 5 acres, it was part of a larger ancient slide complex.

Not all rock slides are the result of rocks dipping in the same direction as a hill's slope. The rock slide at Frank, Alberta, Canada, on April 29, 1903, illustrates how nature and human activity can combine to create a situation with tragic results (■ Figure 14.16).

It would appear at first glance that the coal-mining town of Frank, lying at the base of Turtle Mountain, was in no danger from a landslide (Figure 14.16). After all, many of the rocks dipped away from the mining valley, unlike the situations at Point Fermin and Laguna Beach. The joints in the massive limestone composing Turtle Mountain, however, dip steeply toward the valley and are essentially parallel with the slope of the mountain itself. Furthermore, Turtle Mountain is supported by weak limestones, shales, and coal layers that underwent slow plastic deformation from the weight of the overlying massive limestone. Coal mining along the base of the valley also contributed to the stress on the rocks by removing some of the underlying support. All these factors, as well as frost action and chemical weathering that widened the joints, finally resulted in a massive rock slide. Almost 40 million m^3 of rock slid down Turtle Mountain along joint planes, killing 70 people and partially burying the town of Frank.

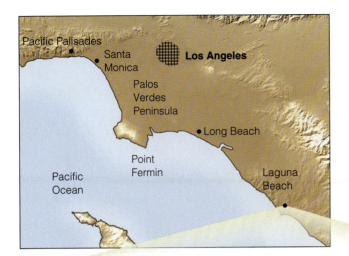

■ **Figure 14.15**

A combination of interbedded clay beds that become slippery when wet, rocks dipping in the same direction as the slope of the sea cliffs, and undercutting of the sea cliffs by wave action activated a rock slide at Laguna Beach, California, that destroyed numerous homes and cars on October 2, 1978.

Steven R. Lower, GeoPhoto Publishing Co.

Flows

Mass movements in which material flows as a viscous fluid or displays plastic movement are termed *flows*. Their rate of movement ranges from extremely slow to extremely rapid (Table 14.2). In many cases, mass movements begin as falls, slumps, or slides and change into flows farther downslope.

Of the major mass movement types, **mudflows** are the most fluid and move most rapidly (at speeds of up to 80 km/hr). They consist of at least 50% silt- and clay-sized material combined with a significant amount of water (up to 30%). Mudflows are common in arid and semiarid environments where they are triggered by heavy rainstorms that quickly saturate the regolith, turning it into a raging flow of mud that engulfs everything in its path. Mudflows can also occur in mountain regions (■ Figure 14.17) and in areas covered by volcanic ash

where they can be particularly destructive (see Chapter 4). Because mudflows are so fluid, they generally follow preexisting channels until the slope decreases or the channel widens, at which point they fan out.

As urban areas in arid and semiarid climates continue to expand, mudflows and the damage they create are becoming problems. Mudflows are common, for example, in the steep hillsides around Los Angeles, where they have damaged or destroyed many homes.

Debris flows are composed of larger-sized particles than those in mudflows and do not contain as much water. Consequently, they are usually more viscous than mudflows, typically do not move as rapidly, and rarely are confined to preexisting channels. Debris flows can be just as damaging, though, because they can transport large objects (■ Figure 14.18).

Earthflows move more slowly than either mudflows or debris flows. An earthflow slumps from the upper part

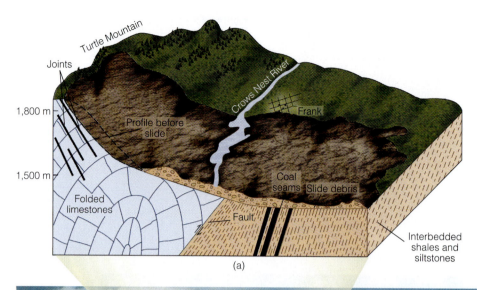

(a)

(b)

B. Bradley, Geology Dept. of the University of Colorado

■ **Figure 14.16**

(a) The tragic Turtle Mountain rock slide that killed 70 people and partially buried the town of Frank, Alberta, Canada, on April 29, 1903, was caused by a combination of factors. These included joints that dipped in the same direction as the slope of Turtle Mountain, a fault partway down the mountain, weak shale and siltstone beds underlying the base of the mountain, and mined-out coal seams. (b) Results of the 1903 rock slide at Frank.

of a hillside, leaving a scarp, and flows slowly downslope as a thick, viscous, tongue-shaped mass of wet regolith (■ Figure 14.19). Like mudflows and debris flows, earthflows can be of any size and are frequently destructive. They occur most commonly in humid climates on grassy, soil-covered slopes, following heavy rains.

Some clays spontaneously liquefy and flow like water when they are disturbed. Such **quick clays** have caused serious damage and loss of lives in Sweden, Norway, eastern Canada (■ Figure 14.20), and Alaska (Table 14.1). Quick clays are composed of fine silt and clay particles made by the grinding action of glaciers. Geologists think these fine sediments were originally deposited in a marine environment where their pore space was filled with saltwater. The ions in saltwater helped establish strong bonds between the clay particles, thus stabilizing and strengthening the clay. When the clays were subsequently uplifted above sea level, the saltwater was flushed out by fresh groundwater, reducing the effectiveness of the ionic bonds between the clay particles and thereby reducing the overall strength and cohesiveness of the clay. Consequently, when the clay is disturbed by a sudden shock or shaking, it essentially turns to a liquid and flows.

An example of the damage that can be done by quick clays occurred in the Turnagain Heights area of Anchorage, Alaska, in 1964 (■ Figure 14.21). Underlying most of the Anchorage area is the Bootlegger Cove Clay, a massive clay unit of poor permeability. Because the Bootlegger Cove Clay forms a barrier preventing groundwater from flowing through the adjacent glacial deposits to the sea, considerable hydraulic pressure builds up behind the clay. Some of this water has flushed out the saltwater in the clay and has saturated the lenses of sand and silt associated with the clay beds. When the magnitude 8.6 Good Friday earthquake struck on March 27, 1964, the shaking turned parts of the Bootlegger Cove Clay into a quick clay and precipitated a series of massive slides in the coastal bluffs that destroyed most of the homes in the Turnagain Heights subdivision (Figure 14.21).

James S. Monroe

■ **Figure 14.17**

A mudflow near Estes Park, Colorado.

B. Pipkin, University of Southern California

■ **Figure 14.18**

A debris flow and damaged house in lower Ophir Creek, western Nevada. Note the many large boulders that are part of the debris flow.

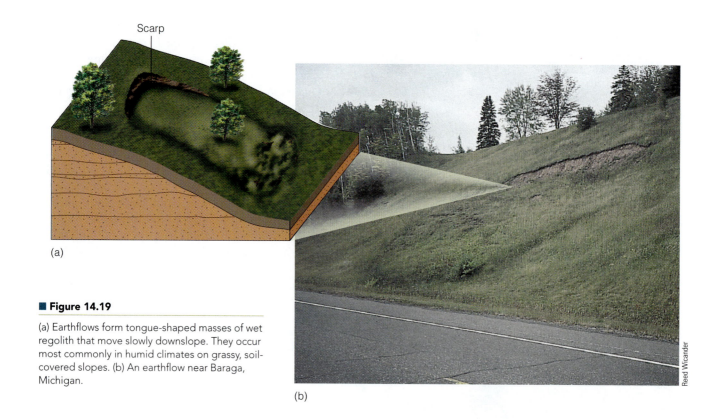

Scarp

(a)

(b)

■ Figure 14.19

(a) Earthflows form tongue-shaped masses of wet regolith that move slowly downslope. They occur most commonly in humid climates on grassy, soil-covered slopes. (b) An earthflow near Baraga, Michigan.

Reed Wicander

Canadian Air Force

■ Figure 14.20

Quick-clay slide at Nicolet, Quebec, Canada. The house on the slide (to the right of the bridge) traveled several hundred feet with relatively little damage.

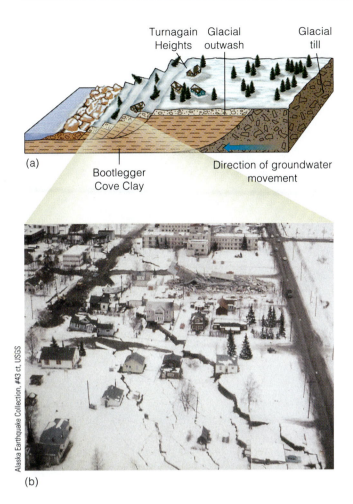

Turnagain Heights   Glacial outwash   Glacial till

(a)

Bootlegger Cove Clay   Direction of groundwater movement

Alaska Earthquake Collection, #43 ct, USGS

(b)

■ **Figure 14.21**

(a) Groundshaking by the 1964 Alaska earthquake turned parts of the Bootlegger Cove Clay into a quick clay, causing numerous slides. (b) Low-altitude photograph of the Turnagain Heights subdivision of Anchorage shows some of the numerous landslide fissures that developed as well as the extensive damage to buildings in the area. The remains of the Four Seasons apartment building can be seen in the background.

Solifluction is the slow downslope movement of water-saturated surface sediment. Solifluction can occur in any climate where the ground becomes saturated with water, but is most common in areas of permafrost.

Permafrost, ground that remains permanently frozen, covers nearly 20% of the world's land surface (■ Figure 14.22a). During the warmer season when the upper portion of the permafrost thaws, water and surface sediment form a soggy mass that flows by solifluction and produces a characteristic lobate topography (Figure 14.22b).

As might be expected, many problems are associated with construction in a permafrost environment. A good example is what happens when an uninsulated building is constructed directly on permafrost. Heat escapes through the floor, thaws the ground below, and turns it into a soggy, unstable mush. Because the ground is no

longer solid, the building settles unevenly into the ground, and numerous structural problems result (■ Figure 14.23).

Construction of the Alaska pipeline from the oil fields in Prudhoe Bay to the ice-free port of Valdez raised numerous concerns about the effect it might have on the permafrost and the potential for solifluction. Some thought that oil flowing through the pipeline would be warm enough to melt the permafrost, causing the pipeline to sink farther into the ground and possibly rupture. After numerous studies were conducted, scientists concluded that the pipeline, completed in 1977, could safely be buried for more than half of its 1280-km length; where melting of the permafrost might cause structural problems to the pipe, it was insulated and installed above ground.

Creep, the slowest type of flow, is the most widespread and significant mass-wasting process in terms of the total amount of material moved downslope and the monetary damage it does annually. Creep involves extremely slow downhill movement of soil or rock. Although it can occur anywhere and in any climate, it is most effective and significant as a geologic agent in humid regions. In fact, it is the most common form of mass wasting in the southeastern United States and the southern Appalachian Mountains.

Because the rate of movement is essentially imperceptible, we are frequently unaware of creep's existence until we notice its effects: tilted trees and power poles, broken streets and sidewalks, or cracked retaining walls or foundations (■ Figure 14.24). Creep usually involves the whole hillside and probably occurs, to some extent, on any weathered or soil-covered, sloping surface.

Creep is difficult not only to recognize but also to control. Although engineers can sometimes slow or stabilize creep, many times the only course of action is to simply avoid the area if at all possible or, if the zone of

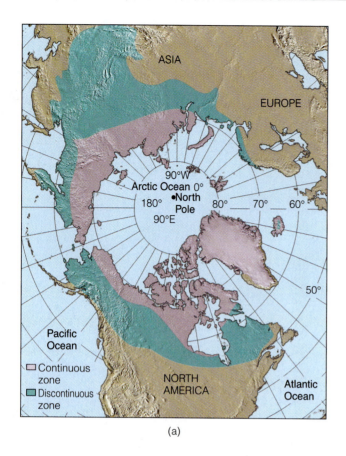

(a)

(b)

■ **Figure 14.22**

(a) Distribution of permafrost areas in the Northern Hemisphere.
(b) Solifluction flows near Suslositna Creek, Alaska, show the typical lobate topography that is characteristic of solifluction conditions.

B. Bradley, Geology Dept. of the University of Colorado

O. J. Ferrains, Jr./USGS

■ **Figure 14.23**

This house, south of Fairbanks, Alaska, has settled unevenly because the underlying permafrost in fine-grained silts and sands has thawed.

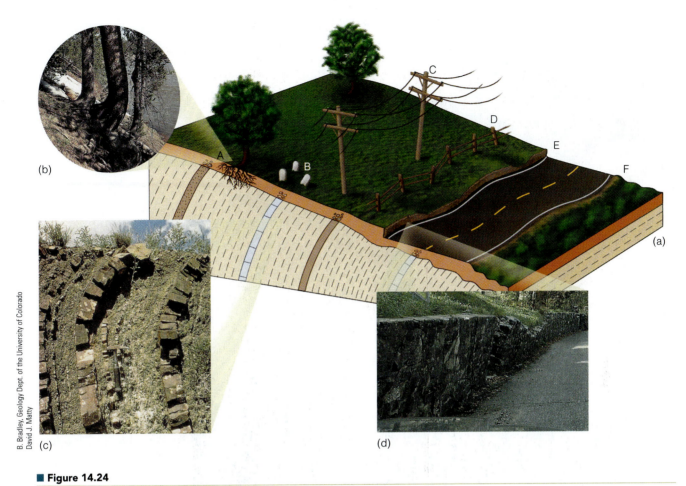

(b)

(c)

(d)

(a)

■ **Figure 14.24**

(a) Some evidence of creep: (A) curved tree trunks; (B) displaced monuments; (C) tilted power poles; (D) displaced and tilted fences; (E) roadways moved out of alignment; (F) hummocky surface. (b) Trees bent by creep, Wyoming. (c) Creep has bent these sandstone and shale beds of the Haymond Formation near Marathon, Texas. (d) Stone wall tilted due to creep, Champion, Michigan.

creep is relatively thin, design structures that can be anchored into the solid bedrock.

Complex Movements

Recall that many mass movements are combinations of different movement types. When one type is dominant, the movement can be classified as one of those described thus far. If several types are more or less equally involved, it is called a **complex movement.**

The most common type of complex movement is the slide-flow, in which there is sliding at the head and then some type of flowage farther along its course. Most slide-flow landslides involve well-defined slumping at the head,

followed by a debris flow or earth-flow (■ Figure 14.25). Any combination of different mass movement types can, however, be classified as a complex movement.

A *debris avalanche* is a complex movement that often occurs in very steep mountain ranges. Debris avalanches typically start out as rockfalls when large quantities of rock, ice, and snow are dislodged from a mountainside, frequently as a result of an earthquake. The material then slides or flows down the mountainside, picking up additional surface material and increasing in speed. The 1970 Peru earthquake (Table 14.1) set in motion the debris avalanche that destroyed the towns of Yungay and Ranrahirca, Peru, and killed more than 25,000 people (■ Figure 14.26).

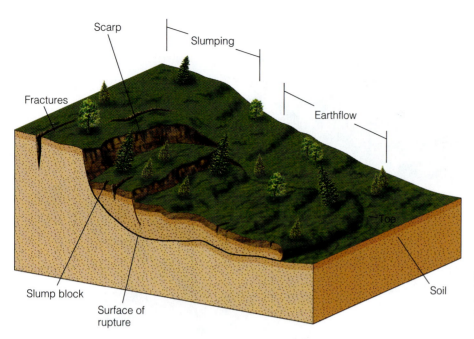

Figure 14.25

A complex movement in which slumping occurs at the head, followed by an earthflow.

What Would You Do?

You've found your dream parcel of land in the hills of northern Baja California, where you plan to retire someday. Because you want to make sure the area is safe to build a house, you decide to do your own geologic investigation of the area to make sure there aren't any obvious geologic hazards. What specific things would you look for that might indicate mass wasting in the past? Even if there is no obvious evidence of rapid mass wasting, what features would you look for that might indicate a problem with slow types of mass wasting such as creep?

HOW CAN WE RECOGNIZE AND MINIMIZE THE EFFECTS OF MASS MOVEMENTS?

The most important factor in eliminating or minimizing the damaging effects of mass wasting is a thorough geologic investigation of the region in question. In this way, former landslides and areas susceptible to mass movements can be identified and perhaps avoided. By assessing the risks of possible mass wasting before construction begins, engineers can take steps to eliminate or minimize the effects of such events.

Figure 14.26

An earthquake 65 km away triggered a landslide on Nevado Huascarán, Peru, that destroyed the towns of Yungay and Ranrahirca and killed more than 25,000 people.

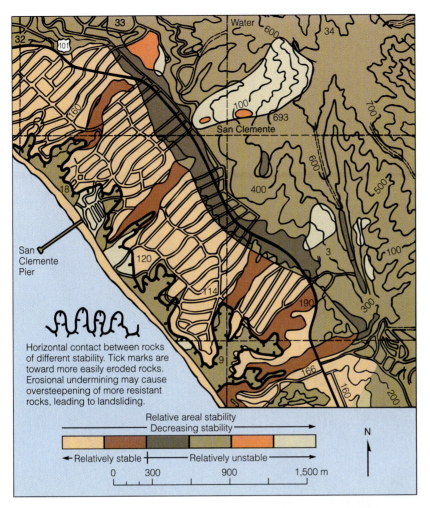

■ **Figure 14.27**

Relative slope-stability map of part of San Clemente, California, showing areas delineated according to relative stability.

Identifying areas with a high potential for slope failure is important in any hazard-assessment study; these studies include identifying former landslides as well as sites of potential mass movement. Scarps, open fissures, displaced or tilted objects, a hummocky surface, and sudden changes in vegetation are some of the features indicating former landslides or an area susceptible to slope failure. The effects of weathering, erosion, and vegetation may, however, obscure the evidence of previous mass wasting.

Soil and bedrock samples are also studied, in both the field and laboratory, to assess such characteristics as composition, susceptibility to weathering, cohesiveness, and ability to transmit fluids. These studies help geologists and engineers predict slope stability under a variety of conditions.

The information derived from a hazard-assessment study can be used to produce *slope-stability maps* of the area (■ Figure 14.27). These maps allow planners and developers to make decisions about where to site roads, utility lines, and housing or industrial developments based on the relative stability or instability of a particular location. In addition, the maps indicate the extent of an area's landslide problem and the type of mass movement that may occur. This information is important for grading slopes or building structures, to prevent or minimize slope-failure damage.

Although most large mass movements usually cannot be prevented, geologists and engineers can employ various methods to minimize the danger and damage resulting from them. Because water plays such an important role in many landslides, one of the most effective and inexpensive ways to reduce the potential for slope failure or to increase existing slope stability is through surface and subsurface drainage of a hillside. Drainage serves two purposes. It reduces the weight of the material likely to slide and increases the shear strength of the slope material by lowering pore pressure.

Surface waters can be drained and diverted by ditches, gutters, or culverts designed to direct water away from slopes. Drainpipes perforated along one surface and driven into a hillside can help remove subsurface water (■ Figure 14.28). Finally, planting vegetation on hillsides helps stabilize slopes by holding the soil together and reducing the amount of water in the soil.

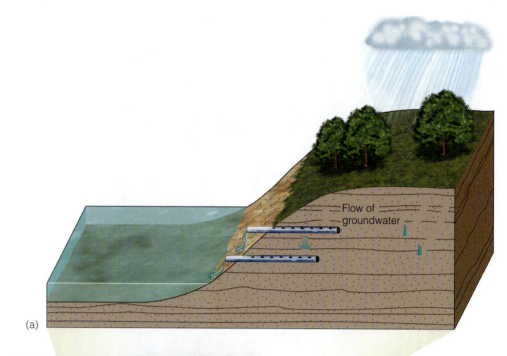

(a)

(b)

Reed Wicander

■ **Figure 14.28**

(a) Driving drainpipes that are perforated on one side into a hillside, with the perforated side up, can remove some subsurface water and help stabilize the hillside. (b) A drainpipe driven into the hillside at Point Fermin, California.

Another way to help stabilize a hillside is to reduce its slope. Recall that overloading and oversteepening by grading are common causes of slope failure. By reducing the angle of a hillside, the potential for slope failure is decreased. Two methods are usually employed to reduce a slope's angle. In the *cut-and-fill* method, material is removed from the upper part of the slope and used as fill at the base, thus providing a flat surface for construction and reducing the slope (■ Figure 14.29). The second method, which is called *benching,* involves cutting a series of benches or steps into a hillside (■ Figure 14.30). This process reduces the overall average slope, and the benches serve as collecting sites for small landslides or rockfalls that might occur. Benching is most commonly used on steep hillsides in conjunction with a system of surface drains to divert runoff.

In some situations, retaining walls are constructed to provide support for the base of the slope (■ Figure 14.31). These are usually anchored well into bedrock, backfilled with crushed rock, and provided with drain holes to prevent the buildup of water pressure in the hillside.

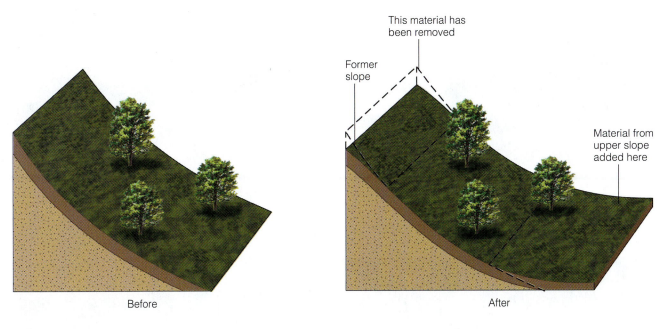

Before

After

This material has been removed

Former slope

Material from upper slope added here

■ **Figure 14.29**

One common method used to help stabilize a hillside and reduce its slope is the cut-and-fill method. Here, material from the steeper upper part of the hillside is removed, thereby reducing the slope angle, and is used to fill in the base. This provides some additional support at the base of the slope.

Before

Former slope

(a)　　After

John D. Cunningham/Visuals Unlimited

(b)

■ **Figure 14.30**

(a) Another common method used to stabilize a hillside and reduce its slope is benching. This process involves making several cuts along a hillside to reduce the overall slope. Furthermore, individual slope failures are now limited in size, and the material collects on the benches. (b) Benching is used in nearly all road cuts.

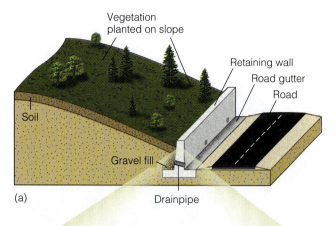

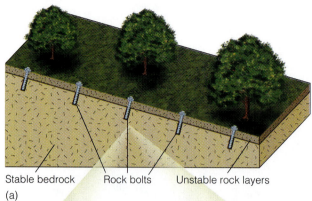

■ **Figure 14.31**

(a) Retaining walls anchored into bedrock, backfilled with gravel, and provided with drainpipes can support a slope's base and reduce landslides. (b) Steel retaining wall built to stabilize the slope and keep falling and sliding rocks off the highway.

■ **Figure 14.32**

(a) Rock bolts secured in bedrock can help stabilize a slope and reduce landslides. (b) Rock bolts and wire mesh are used to help secure rock on a steep hillside in Brisbane, Australia.

Rock bolts, similar to those employed in tunneling and mining, can sometimes be used to fasten potentially unstable rock masses into the underlying stable bedrock (■ Figure 14.32). This technique has been used successfully on the hillsides of Rio de Janeiro, Brazil, and to help secure the slopes at the Glen Canyon Dam on the Colorado River.

Recognition, prevention, and control of landslide-prone areas are expensive, but not nearly as expensive as the damage can be when such warning signs are ignored or not recognized. The collapse of Tip No. 7 at Aberfan, Wales (see Geo-Focus 14.1), and other landfill and dam collapses are tragic examples in which the warning signs of impending disaster were ignored.

14 REVIEW WORKBOOK

Chapter Summary

- Mass wasting is the downslope movement of material under the influence of gravity. It occurs when the gravitational force acting parallel to a slope exceeds the slope's strength.

- Mass wasting frequently results in loss of life as well as millions of dollars in damages annually.

- Mass wasting can be caused by many factors, including slope angle, weathering of slope material, water content, overloading, and removal of vegetation. Usually several of these factors in combination contribute to slope failure.

- Mass movements are generally classified on the basis of their rate of movement (rapid versus slow), type of movement (falling, sliding, or flowing), and type of material (rock, soil, or debris).

- Rockfalls are a common mass movement in which rocks free-fall.

- Two types of slides are recognized. Slumps are rotational slides involving movement along a curved surface; they are most common in poorly consolidated or unconsolidated material. Rock slides occur when movement takes place along a more or less planar surface; they usually involve solid pieces of rock.

- Several types of flows are recognized on the basis of their rate of movement (rapid versus slow), type of material (rock, sediment, soil), and amount of water.

- Mudflows consist of mostly clay- and silt-sized particles and contain more than 30% water. They are most common in semiarid and arid environments and generally follow preexisting channels.

- Debris flows are composed of larger particles and contain less water than mudflows. They are more viscous and do not flow as rapidly as mudflows.

- Earthflows move more slowly than either debris flows or mudflows; they move downslope as thick, viscous, tongue-shaped masses of wet regolith.

- Quick clays are clays that spontaneously liquefy and flow like water when they are disturbed.

- Solifluction is the slow downslope movement of water-saturated surface material and is most common in areas of permafrost.

- Creep, the slowest type of flow, is the imperceptible downslope movement of soil or rock. It is the most widespread of all types of mass wasting.

- Complex movements are combinations of different types of mass movements in which one type is not dominant. Most complex movements involve sliding and flowing.

- The most important factor in reducing or eliminating the damaging effects of mass wasting is a thorough geologic investigation to outline areas susceptible to mass movements.

- Slopes can be stabilized by building retaining walls, draining excess water, regrading slopes, and planting vegetation.

Important Terms

complex movement (p. 421)	mudflow (p. 415)	shear strength (p. 399)
creep (p. 419)	quick clay (p. 417)	slide (p. 405)
debris flow (p. 415)	rapid mass movement (p. 404)	slow mass movement (p. 404)
earthflow (p. 415)	rockfall (p. 404)	slump (p. 410)
mass wasting (p. 398)	rock slide (p. 414)	solifluction (p. 419)

Review Questions

1. The force opposing a slope's shear strength is:
   a. _____ external support; b. _____ gravity;
   c. _____ internal support; d. _____ cohesion;
   e. _____ internal friction.

2. The downslope movement of material along a curved surface of rupture is a(n):
   a. _____ rockfall; b. _____ slump; c. _____ mudflow; d. _____ earthflow; e. _____ rock slide.

3. The most widespread and costly of all mass-wasting processes is:
   a. _____ slumps; b. _____ creep; c. _____ mudflows; d. _____ rockfalls; e. _____ quick clays.

4. Where can mass wasting occur?
   a. _____ only on steep slopes; b. _____ only in temperate climates; c. _____ anywhere;
   d. _____ only on gentle slopes; e. _____ only where bedrock is exposed.

5. Where in the United States are the areas of greatest landslide severity *least* likely to be found?
   a. _____ Sierra Nevada; b. _____ Appalachian Mountains; c. _____ Cascade Range;
   d. _____ Rocky Mountains; e. _____ Midwest.

6. The movement of material along a surface or surfaces of failure is a:
   a. _____ flow; b. _____ fall; c. _____ slide;
   d. _____ landslide; e. _____ none of the preceding answers.

7. Which of the following is a factor influencing mass wasting?
   a. _____ slope angle; b. _____ vegetation;
   c. _____ weathering; d. _____ water content;
   e. _____ all of the preceding answers.

8. Solifluction occurs most commonly in which areas?
   a. _____ beaches; b. _____ deserts;
   c. _____ permafrost; d. _____ tropical forests;
   e. _____ none of the preceding answers.

9. Which of the following helps reduce the slope angle, or provides support at the base, of a hillside?
   a. _____ cut and fill; b. _____ retaining walls;
   c. _____ benching; d. _____ all of the preceding answers; e. _____ none of the preceding answers.

10. Former landslides and areas currently susceptible to slope failure can be identified by which of the following features?
    a. _____ tilted objects; b. _____ open fissures; c. _____ scarps;

d. _____ hummocky surfaces; e. _____ all of the preceding answers.

11. Discuss some of the ways slope stability can be maintained so as to reduce the likelihood of mass movements.

12. What potential value would a slope-stability map be to a person seeking to purchase property for a new home? Using the slope-stability map in Figure 14.27, locate the line that shows a horizontal contact between rocks of different stability. What is the potential for mass wasting along this line and why?

13. Discuss how topography and the underlying geology contribute to slope failure.

14. Why are slumps such a problem along highways and railroad tracks in areas with relief?

15. Using Figure 14.14 as an example, explain how the geologic planes of weakness in this slope plus water from rainfall influenced development of the depicted slide.

16. Discuss how the different factors that influence mass wasting are interconnected.

17. How could mass wasting be recognized on other planets or moons? What would that tell us about the geology and perhaps the atmosphere of the planet or moon on which it occurred?

18. If an area has a documented history of mass wasting that has endangered or taken human life, how should people and governments keep such events from happening again? Are most large mass-wasting events preventable or predictable?

19. Why is it important to know about the different types of mass wasting?

20. How would removing preexisting vegetation tend to affect most slopes in humid regions with respect to mass wasting?

World Wide Web Activities

PHYSICAL Geology⇌Now Assess your understanding of this chapter's topics with additional quizzing and comprehensive interactivities at

http://earthscience.brookscole.com/physgeo5e

as well as current and up-to-date weblinks, additional readings, and InfoTrac College Edition exercises.

Running Water

PHYSICAL GeologyNow *This icon, appearing throughout the book, indicates an opportunity to explore interactive tutorials, animations, or practice problems available on the Physical GeologyNow Web site at http://earthscience.brookscole.com/physgeo5e.*

OBJECTIVES
At the end of this chapter, you will have learned that

- Running water, one part of the hydrologic cycle, does considerable geologic work.

- Water is continually cycled from the oceans to land and back to the oceans.

- Running water transports large quantities of sediment and deposits sediment in or adjacent to braided and meandering rivers and streams.

- Alluvial fans (on land) and deltas (in a standing body of water) are deposited when a stream's capacity to transport sediment decreases.

- Flooding is a natural part of stream activity that takes place when a channel receives more water than it can handle.

- The several types of structures to control floods are only partly effective.

- Rivers and streams continuously adjust to changes.

- The concept of a graded stream is an ideal, although many rivers and streams approach the graded condition.

- Most valleys form and change in response to erosion by running water coupled with other geologic processes such as mass wasting.

The South Arm of Rice Creek in northern California cascades about 7 m over volcanic rocks. Source: Sue Monroe

 Introduction

Earth is unique among the terrestrial planets in having abundant liquid water. Both Mercury and Earth's Moon are too small to retain any water, and Venus, because of its high temperature, is too hot for surface water to exist. At present, Mars has only some frozen water and trace amounts of water in its atmosphere, but studies of *Mariner* and *Voyager* spacecraft images and other data reveal areas with winding valleys that were apparently carved by running water during Mars's early history. In contrast, 71% of Earth's surface is covered by oceans and seas (see Figure 11.2), and small but important quantities of water are present in the atmosphere and on land.

When we consider the interactions among Earth's systems, certainly the hydrosphere has a tremendous impact on the surface. Of course, the hydrosphere consists of several elements such as water vapor in the atmosphere, groundwater (see Chapter 16), water frozen in glaciers (see Chapter 17), water in the oceans (see Chapters 11 and 19), and a small but very important amount of water on land. Water on land consists of all water in the groundwater system, lakes, swamps, and bogs and, our main concern here, the small percentage of running water confined to channels.

To appreciate the power of running water you need only read some of the vivid accounts of floods. For example, at 4:07 P.M. on May 31, 1889, the residents of Johnstown, Pennsylvania, heard "a roar like thunder," and within 10 minutes a catastrophic flood destroyed the town, leaving at least 2200 people dead, many of whom were never recovered. An 18-m-high wall of water roared through the town at more than 60 km/hr, sweeping up houses, debris, and hundreds of people (■ Figure 15.1). According to one account, "Thousands of people desperately tried to escape the wave. Those caught by the wave found themselves swept up in a torrent of oily, muddy water, surrounded by tons of grinding debris. . . . Many became hopelessly entangled in miles of barbed wire from the destroyed wire works."*

The fact that Johnstown was built on a floodplain in a narrow valley contributed to the tragedy, but the failure of the South Fork Dam about 22 km upstream on the Conemaugh River was the main cause of the disaster. Unfortunately, the dam was poorly maintained and unable to withstand the added stress from the 20–25 cm of rain that fell in 24 hours. This was not the last dam failure in the United States, but it certainly was the most tragic one.

Floods caused by purely natural causes as well as those resulting from human carelessness are a continuing threat. Indeed, floods are so common that reports of damage, injuries, and fatalities appear in the news regularly. Floods

---

*National Park Service—U.S. Department of Interior, Johnstown Information Service Online.

(a)

(b)

■ **Figure 15.1**

Johnstown, Pennsylvania, before (a) and after (b) the May 31, 1889, flood.

■ Figure 15.2

(a) Oroville Dam on the Feather River in California helps control floods, provides water for irrigation, and is a recreation area. Electricity is also generated at the Hyatt Power Plant seen at the base of the dam. This 235-m-high dam is the highest in the United States. (b) Inland waterways are important avenues of commerce. These freight barges are traveling north on the Tennessee River. (c) A popular recreation area on the McCloud River in California. The small waterfall plunges over an escarpment developed on a lava flow.

occur several times every year in North America, but the last flood of truly vast proportions was the Flood of 1993 that inundated large parts of the central United States, especially Iowa, Wisconsin, Illinois, Missouri, Minnesota, North and South Dakota, Kansas, and parts of some adjacent states. At last 50 people died, 70,000 were homeless, and property damage was estimated at $15 to $20 billion. See "The Flood of '93" on pages 434 and 435.

Nevertheless, there are also many benefits from running water, even from some floods. Running water in channels—that is, streams and rivers—is one important source of

freshwater for industry, agriculture, domestic use, and recreation, and about 8% of all electricity used in North America is generated by falling water at hydroelectric power plants (■ Figure 15.2; see Geo-Focus 15.1). Large waterways throughout the world are major avenues of commerce (Figure 15.2b). When Europeans explored the interior of North America, they followed the St. Lawrence, Mississippi, Missouri, Columbia, and Ohio Rivers. And even though rivers periodically flood, some floods are truly beneficial. For instance, the agricultural lands along the Nile River in Egypt depend on annual flood-derived deposits to remain fertile.

The Flood of '93

Although several floods take place each year that cause damage, injuries, and fatalities, the last truly vast flooding in North America occurred during June and July of 1993. Now called the Flood of '93, it was responsible for 50 deaths and 70,000 were left homeless. Extensive property damage occured in several states, but particularly hard hit were Missouri and Iowa (see chart). Unusual behavior of the jet stream and the convergence of air masses over the Midwest were responsible for numerous thunderstorms that caused the flooding.

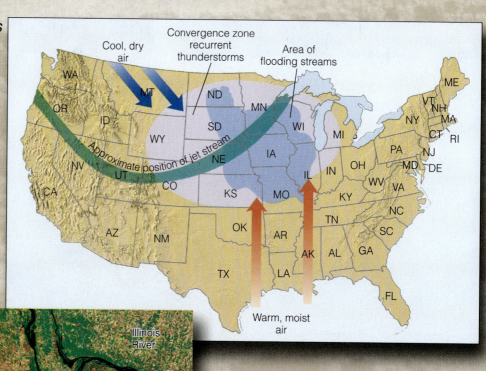

The dominant weather pattern for June and July 1993. The jet stream remained over the Midwest during the summer rather than shifting north over Canada as it usually does. Thunderstorms developed in the convergence zone where warm, moist air and cool, dry air met.

Satellite images of the Mississippi, Missouri, and Illinois rivers near the juncture of three rivers during the drought of 1988 (left), and during the flood of 1993 (below). The "x" marks the site of Portage des Sioux, Missouri (see facing page).

Portage des Sioux, St. Charles County, Missouri, on July 16, 1993. The channel of the Mississippi River is on the far right.

Floodwaters in Portage des Sioux covered the 5.5-m-high pedestal of this statue on the bank of the Mississippi River.

Breached levee on the Mississippi River near Davenport, Iowa. This is one of 800 levees that failed or was overtopped during the flood.

Damage caused by the Flood of '93. Compiled by the U. S. Army Corps of Engineers, figures are rounded to the nearest $1000.

State	Residential	Agricultural	Other	Total
Illinois	$176,833,000	$166,502,000	$409,020,000	$752,355,000
Iowa	57,827,000	1,030,030,000	334,835,000	1,422,692,000
Kansas	35,829,000	855,849,000	176,162,000	1,067,840,000
Minnesota	16,940,000	694,041,000	286,540,000	997,521,000
Missouri	405,175,000	540,666,000	1,255,191,000	2,201,032,000
Nebraska	18,584,000	120,521,000	67,899,000	207,004,000
North Dakota	10,138,000	35,039,000	25,806,000	70,983,000
South Dakota	21,919,000	276,218,000	195,408,000	493,545,000
Wisconsin	17,747,000	133,835,000	135,917,000	287,499,000
Total—All States	760,992,000	3,852,701,000	2,886,778,000	7,500,471,000

Sue Monroe

Chris Stewart/ Black Star

GEOFOCUS

15.1

Dams, Reservoirs, and Hydroelectric Power Plants

Flip a switch and we illuminate our homes and workplaces; turn a dial and we heat our homes or perhaps cook our food or wash our clothes; and some of our public transportation relies on an unseen but important energy source—electricity. In fact, about 40% of the total energy used in industrialized nations is converted to electricity. Most electricity is generated at power plants that burn fossil fuels (oil, natural gas, and especially coal), but nuclear power plants are common in some areas. Geothermal energy (see Chapter 16), wind, and tidal power (see Chapter 19) account for only a small percentage of all electricity.

Electricity generated at *hydroelectric power plants*—that is, power plants that use moving water—accounts for only 8% of the total electricity generated in North America, but its importance varies from area to area. For instance, less than 2% of the electricity generated in Florida, Ohio, and Texas comes from hydroelectric plants, whereas Washington, Oregon, and Idaho derive more than 80% of their electricity from this source.

At all power plants a spinning turbine connected to a generator containing a coil of wire produces electricity. However, the energy used to turn turbines differs. In fossil-fuel-burning plants, for example, coal, oil, or natural gas is burned to heat water and generate steam, which in turn serves as the energy source to turn turbines. Steam is also generated in nuclear power plants.

In hydroelectric power plants, moving water rather than steam is used to turn turbines. To provide the necessary water, a dam is built impounding a reservoir where the water is higher than the power plant. Water moves from the reservoir through a large pipe called a *penstock* and spins a turbine (■ Figure 1). In other words, the potential energy of the water in the reservoir is converted to electrical energy at the power plant. Regardless of the power source—fossil fuels, moving water, or uranium—once electricity is generated, it is transmitted to areas of use by power lines.

Falling water was first used to generate electricity in 1892 at Appleton, Wisconsin, and since then hydroelectric power plants have become common features on many of the world's waterways. Currently the largest in terms of generating capacity is on the Parana River on the Paraguay–Brazil border, and an even larger one was recently completed in the People's Republic of China; the reservoir behind the Three Gorges Dam on the Yangtze River began filling in April 2003. Many similar facilities are also present in North America. A large region of the Pacific Northwest depends on electricity generated at the Grand Coulee Dam in Washington, and hydroelectric plants on the Niagara River on the U.S.–Canadian border provide electricity to a large area.

One might be curious about why agencies and governments do not simply increase their hydroelectric output. After all, hydroelectric power generation has several appealing aspects, not the least of which is that it is a renewable resource. However, not all areas have this potential; suitable sites for dams and reservoirs might not be present, for instance. In addition, dams are very expensive to build, reservoirs fill with sediment, and during droughts enough water might not be available to keep reservoirs sufficiently full. And, of course, people must be relocated from areas where reservoirs are im-

WATER ON EARTH

According to one estimate, 1.36 billion km³ of water is present on Earth; most (97.2%) is in the oceans (■ Figure 15.3). About 2.15% is frozen in glaciers on land, especially in Antarctica and Greenland, with most of the remaining 0.65% constituting water in the atmosphere, groundwater, lakes, swamps, and bogs. Only about 0.0001% of all water in the hydrosphere is in stream and river channels at any one time (Figure 15.3). Nevertheless, running water, with its resultant erosion, transport, and deposition, is with few exceptions the most important geologic agent bringing about changes

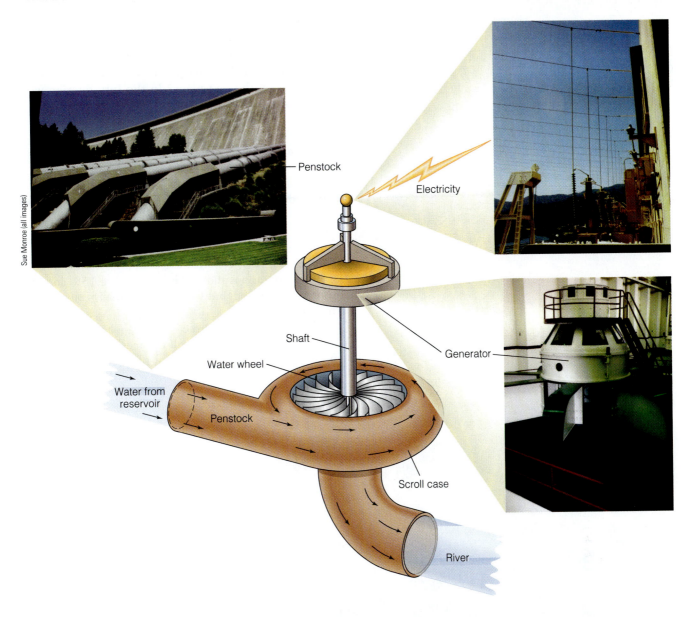

Labels on figure:
Penstock
Electricity
Shaft
Water wheel
Generator
Water from reservoir
Penstock
Scroll case
River
Sue Monroe (all images)

■ **Figure 1**

Electricity is produced at a generator when water rushes through the penstock, where it spins a water wheel connected by a shaft to an electromagnet within a coil of wire. Each penstock shown here measures about 4 m in diameter and carries water at 16 to 22 km/hr. The dam from which water enters the penstocks is visible in the background. Once electricity is generated, it is transmitted to areas of use by power lines.

pounded and from the discharge areas downstream from dams. So

although hydroelectric dams remain important, we cannot realistically

look forward to very much additional use of this energy source.

to Earth's land surface. Only in areas covered by vast glaciers or parts of some deserts are other geologic agents more important than running water. Even in most deserts, though, the effects of running water are conspicuous, although channels are dry most of the time.

Much of the following discussion of running water is necessarily descriptive, but always be aware that streams and rivers are dynamic systems that must continuously respond to change. For example, paving in urban areas increases surface runoff to waterways, while other human activities such as building dams and impounding reservoirs also alter the dynamics of stream and river systems. Natural changes, too, affect the complex interacting parts of stream and river systems.

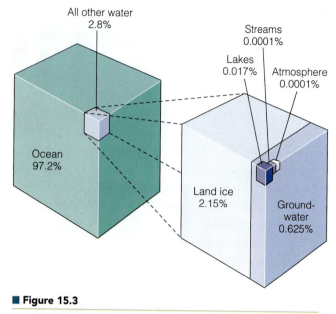

■ Figure 15.3

The relative amounts of water on Earth. Of the estimated 1.36 billion km³ of water, 97.2% is in the oceans; most of the rest, 2.15%, is frozen in glaciers on land.

The Hydrologic Cycle

The connection between precipitation and clouds is obvious, but where does the moisture for rain and snow come from in the first place? In the Introduction we noted that 97.2% of all water on Earth is in the oceans, so one might immediately suspect that the oceans are the ultimate source of precipitation. In fact, water is continually recycled from the oceans, through the atmosphere, to the continents, and back to the oceans. This **hydrologic cycle,** as it is called, is powered by solar radiation and is possible because water changes easily from liquid to gas (water vapor) under surface conditions (■ Figure 15.4). A volume of water corresponding to a layer about 1 m thick evaporates from the oceans each year, constituting about 85% of all water entering the atmosphere. The remaining 15% comes from water on land, but this water originally came from the oceans as well.

Regardless of its source, water vapor rises into the atmosphere, where the complex processes of cloud formation and condensation take place. Much of the

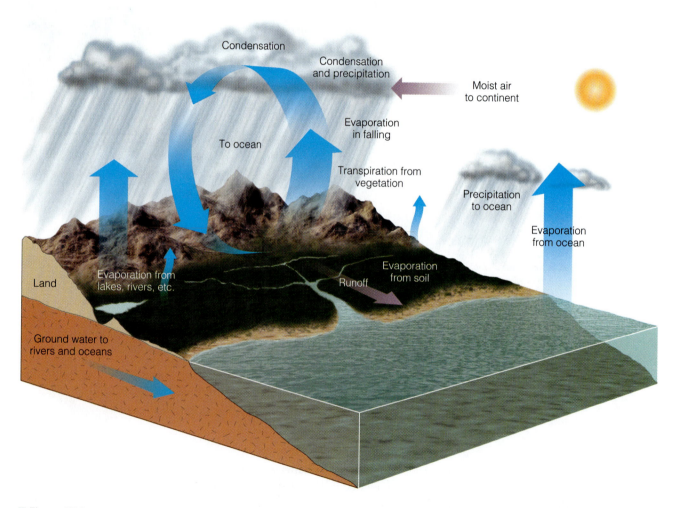

■ Figure 15.4

During the hydrologic cycle, water evaporates from the oceans and rises as water vapor to form clouds that release their precipitation over oceans or over land. Much of the precipitation falling on land returns to the oceans by surface runoff, thus completing the cycle.

world's precipitation, about 80%, falls directly back into the oceans, in which case the hydrologic cycle is limited to a three-step process of evaporation, condensation, and precipitation. For the 20% of all precipitation falling on land, the hydrologic cycle is more complex, involving evaporation, condensation, movement of water vapor from the oceans to land, precipitation, and runoff. Although some precipitation evaporates as it falls and reenters the cycle, about 36,000 km³ of the precipitation falling on land returns to the oceans by **runoff,** the surface flow in streams and rivers.

Not all precipitation returns directly to the oceans, though. Some is temporarily stored in lakes and swamps, snowfields and glaciers, or seeps below the surface where it enters the groundwater system (see Chapter 16). Water might remain effectively stored in some of these reservoirs for thousands of years, but eventually glaciers melt, lakes and groundwater feed streams and rivers, and this water returns to the oceans. Even the water used by plants evaporates, a process known as *transpiration,* and returns to the atmosphere. In short, all water derived from the oceans eventually makes it back to the oceans and can thus begin the hydrologic cycle again (Figure 15.4). Our concern here is with the comparatively small quantity of water that returns to the oceans by surface runoff.

Fluid Flow

Solids are rigid substances that retain their shapes unless deformed by a force, but fluids—that is, liquids and gases—have no strength so they flow in response to any force no matter how slight. Liquid water, which is our concern here, flows downslope in response to gravity, but its flow may be *laminar* or *turbulent*. In laminar flow, lines of flow called streamlines parallel one another with little or no mixing between adjacent layers (■ Figure 15.5). All flow is in one direction only and it remains unchanged through time. In turbulent flow, streamlines intertwine, causing complex mixing within the moving fluid (Figure 15.5). If we could trace a single water molecule in turbulent flow, it may move in any direction at a particular time although its overall movement would be in the direction of flow.

You can easily observe laminar flow in viscous fluids such as cold motor oil or syrup, both of which flow slowly and with difficulty. When heated, however, both flow much more readily. Viscosity is also a consideration in running water, but its viscosity is so low that it most often moves by turbulent flow. Temperature exerts some control on water's viscosity, but the primary controls are velocity and the roughness of the surface over which flow occurs.

Although some mudflows and lava flows undoubtedly move by laminar flow, examples of laminar flow in

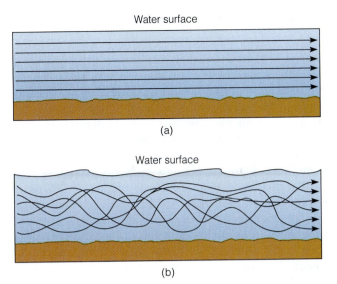

Water surface

(a)

Water surface

(b)

■ **Figure 15.5**

(a) In laminar flow, streamlines are parallel to one another, and little or no mixing takes place between adjacent layers in the fluid. (b) In turbulent flow, streamlines are complexly intertwined, indicating mixing between layers. Most flow in streams is turbulent.

running water are difficult to find. Laminar flow takes place as groundwater moves slowly through the tiny pores in soil, sediment, and rocks (see Chapter 13), but velocity and roughness in almost all surface flow ensure that it is fully turbulent. Even if laminar flow should occur at the surface, it is generally too slow and too shallow to cause any erosion. Turbulent flow, on the other hand, is much more energetic and therefore capable of considerable erosion and sediment transport.

Runoff during a rainstorm depends on **infiltration capacity,** the rate at which surface materials absorb water. Several factors control infiltration capacity, including intensity and duration of rainfall. If rain is absorbed as fast as it falls, no surface runoff takes place. For instance, loosely packed dry soil absorbs water faster than tightly packed wet soil, and thus more rain must fall on loose dry soil before runoff begins. Regardless of the initial condition of surface materials, once they are saturated, excess water collects on the surface, and if on a slope, it moves downhill.

RUNNING WATER

We have mentioned that moving water is the most important geologic agent modifying Earth's land surface. The only exceptions are where vast glaciers are present and in parts of some deserts, but otherwise the effects of running water are ubiquitous.

Sheet Flow and Channel Flow

Even on steep slopes, flow is initially slow and hence causes little or no erosion. As water moves downslope, though, it accelerates and may move by *sheet flow*, a more or less continuous film of water flowing over the surface. Sheet flow is not confined to depressions, and it accounts for *sheet erosion*, a particular problem on some agricultural lands (see Chapter 5).

In *channel flow*, surface runoff is confined to troughlike depressions that vary in size from tiny rills with a trickling stream of water to the Amazon River in South America, which is 6450 km long and at one place 2.4 km wide and 90 m deep. We describe flow in channels with terms such as *rill, brook, creek, stream*, and *river*, most of which are distinguished by size and volume. Here we use the terms *stream* and *river* more or less interchangeably, although the latter usually refers to a larger body of running water.

Streams and rivers receive water from several sources, including sheet flow and rain falling directly into their channels. Far more important, though, is water supplied by soil moisture and groundwater, both of which flow downslope and discharge into waterways (Figure 15.4). In areas where groundwater is plentiful, streams and rivers maintain a fairly stable flow year-round because their water supply is continuous. In contrast, the amount of water in arid and semiarid streams and rivers fluctuates widely because they depend more on infrequent rainstorms and surface runoff for their water.

Stream Gradient

Water in any channel flows downhill over a slope known as its **gradient.** For example, suppose a river has its headwaters (source) 1000 m above sea level and it flows 500 km to the sea, so it drops vertically 1000 m over a horizontal distance of 500 km. Its gradient is found by dividing the vertical drop by the horizontal distance, which in this example is 1000 m/500 km = 2 m/km (■ Figure 15.6). We can say that on the average this river drops vertically 2 m for every kilometer along its course.

In the preceding example we calculated the average gradient for a hypothetical river. Gradients vary not only among channels, however, but even along the course of a single channel. Rivers and streams are steeper in their upper reaches (near their headwaters) where they may have gradients of several tens of meters per kilometer, but they have gradients of only a few centimeters per kilometer where they discharge into the sea.

Velocity and Discharge

The **velocity** of running water is simply a measure of the downstream distance water travels in a given time. It is usually expressed in meters per second (m/sec) or feet per second (ft/sec), and it varies across a channel's

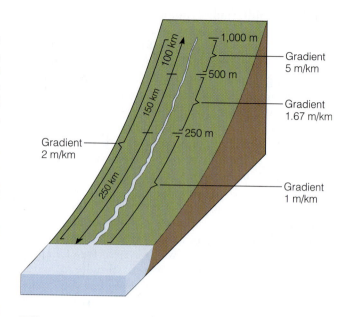

■ **Figure 15.6**

The average gradient of this stream is 2 m/km, but gradient can be calculated for any segment of a stream, as shown in this example. Notice that the gradient is steepest in the headwaters area and decreases in a downstream direction.

width as well as along its length. Water moves more slowly and with greater turbulence near a channel's bed and banks because friction is greater there than it is some distance from these boundaries (■ Figure 15.7a). Channel shape and roughness also influence flow velocity. Broad, shallow channels and narrow, deep channels have proportionately more water in contact with their perimeters than channels with semicircular cross sections (Figure 15.7b). So if other variables are the same, water flows faster in a semicircular channel because of less frictional resistance. As one would expect, rough channels, such as those strewn with boulders, offer more frictional resistance to flow than do channels with a bed and banks composed of sand or mud.

Intuitively you might suspect that gradient is the greatest control on velocity—the steeper the gradient, the greater the velocity. In fact, a channel's average velocity actually increases downstream even though its gradient decreases! Keep in mind that we are talking about average velocity for a long segment of a channel, not velocity at a single point. Three factors account for this general downstream increase in velocity. First, velocity increases even with decreasing gradient in response to the acceleration of gravity unless other factors retard flow. Second, the upstream reaches of channels tend to be boulder-strewn, broad, and shallow, so frictional resistance to flow is high, whereas downstream segments of the same channels are generally more semicircular and have banks composed of finer materials. And finally, the number of smaller tributaries joining a larger channel increases downstream. Thus the total vol-

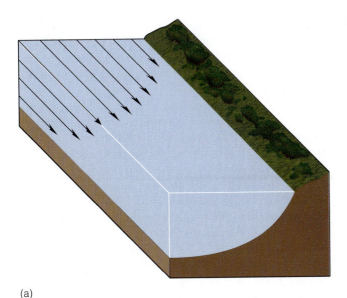

(a)

■ Figure 15.7

Flow velocity in rivers and streams varies as a result of friction with their banks and beds. (a) The maximum flow velocity is near the center and top of a straight channel where friction is least. The arrows are proportional to velocity. (b) These three differently shaped channels have the same cross-sectional area, but the semicircular one has less water in contact with its perimeter and thus less frictional resistance to flow.

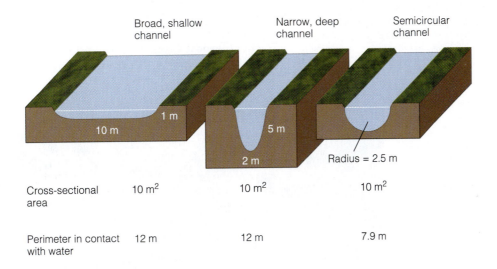

(b)

ume of water (discharge) increases, and increasing discharge results in greater velocity.

We mentioned discharge in the preceding paragraph but noted only that it refers to the volume of water. More specifically, **discharge,** the volume of water passing a particular point in a given period of time, is found by knowing the dimensions of a water-filled channel—that is, its cross-sectional area (A) and its flow velocity (V). Discharge (Q) is then calculated with the formula $Q = VA$ and is expressed in cubic meters per second (m³/sec) or cubic feet per second (ft³/sec) (■ Figure 15.8). The Mississippi River has an average discharge of 18,000 m³/sec, and the average discharge for the Amazon River in South America is 200,000 m³/sec.

In most rivers and streams, discharge increases downstream as more and more water enters a channel. However, there are a few exceptions. Because of high evaporation rates and infiltration, the flow in some desert waterways actually decreases downstream until

they disappear. And even in perennial rivers and streams, discharge is obviously greatest during times of heavy rainfall and at a minimum during the dry season.

HOW DOES RUNNING WATER ERODE AND TRANSPORT SEDIMENT?

Streams and rivers possess two kinds of energy: potential and kinetic. *Potential energy* is the energy of position, such as the energy of water behind a dam or at a high elevation. In stream flow, potential energy is converted to *kinetic energy,* the energy of motion. Most of this kinetic energy is dissipated as heat within streams by fluid turbulence, but a small amount, perhaps 5%, is available to erode and transport

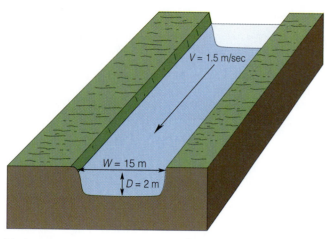

Discharge (Q) = VA = 1.5 m/sec × 30 m² = 45 m³/sec

■ **Figure 15.8**

Discharge (Q) is found by multiplying a river or stream's velocity (V) by its cross-sectional area (A). Cross-sectional area (A) is the product of depth (D) times width (W) of the water-filled part of the channel. The channel in this example is rectangular to make the calculation easier, but natural channels are irregular, so determining (A) is more difficult.

sediment. Erosion involves the physical removal of dissolved substances, minerals, and loose particles of soil and rock from a source area. Thus the sediment transported in a stream consists of both dissolved materials and solid particles.

Because the *dissolved load* of a stream is invisible, it is commonly overlooked, but it is an important part of the total sediment load. Some of it is acquired from the streambed and banks where soluble rocks such as limestone and dolostone are exposed, but much of it is carried into streams by sheet flow and by groundwater.

The solid sediment ranges from clay-sized particles to large boulders, much of it supplied by mass wasting, but some is derived directly from the streambed and banks (■ Figure 15.9). The power of running water, called **hydraulic action,** is sufficient to set particles in motion. Everyone has seen the results of hydraulic action, although perhaps not in streams. For example, if the flow from a garden hose is directed onto loose soil, hydraulic action soon gouges out a hole.

Another process of erosion in streams is **abrasion,** in which exposed rock is worn and scraped by the impact of solid particles. If running water contains no sed-

(a)

■ **Figure 15.9**

(a) This stream acquires some of its sediment load by undercutting its banks. (b) Some of the sediment in the Snake River of Idaho comes from these talus cones that accumulated as a result of mass wasting.

(b)

James S. Monroe

(a)

James S. Monroe

(b)

Sue Monroe

(c)

■ **Figure 15.10**

(a) Close-up view of a 2-m-diameter pothole in Lucerne, Switzerland. Notice the two large boulders in the pothole. (b) These potholes in the bed of the Chippewa River in Ontario, Canada, measure about 1 m across. Two potholes at the top center have merged to form a larger, composite pothole. (c) These stones measuring 7 to 8 cm from a pothole are remarkably spherical and smooth because of abrasion.

iment, no abrasion of rock surfaces results, but if the water is transporting sand and gravel, the impact of these particles abrades exposed rock surfaces. *Potholes* in the beds of streams are one obvious manifestation of abrasion (■ Figure 15.10). These circular to oval holes form where eddying currents containing sand and gravel swirl around and erode depressions into rock.

Running water transports a **dissolved load** consisting of materials taken into solution during chemical weathering, and it also transports sedimentary particles. The smallest of these particles, mostly silt and clay, are kept suspended by fluid turbulence and transported as a **suspended load** (■ Figure 15.11). The Mississippi River transports nearly 200 million metric tons of suspended load past Vicksburg, Mississippi, each year, and the Yellow River of China carries almost four times as much. Particles transported in suspension are deposited only where turbulence is minimal, as in lakes and lagoons.

Streams and rivers also transport a **bed load**, consisting of larger particles of sand and gravel (Figure 15.11). Because fluid turbulence is insufficient to keep sand and gravel suspended, they move along the streambed. However, part of the bed load can be suspended at least temporarily, as when an eddying current swirls across a streambed and lifts sand grains into the water. These particles move forward at approximately the flow velocity, but at the same time they settle toward the streambed where they come to rest, to be moved again later by the same process. This process of intermittent bouncing and skipping along the streambed is called *saltation* (Figure 15.11).

Particles too large to be suspended even temporarily are transported by rolling or sliding (Figure 15.11). Obviously, greater flow velocity is required to move particles of these sizes. The maximum-sized particles that a stream can carry define its *competence*, a factor related to flow velocity. Figure 15.11b shows the velocities required to erode, transport, and deposit particles of various sizes. As expected, high velocity is necessary to erode and transport gravel-sized particles, whereas sand is eroded and transported at lower velocities. Notice, though, that high velocity is also needed to erode clay because clay deposits are very cohesive; the tiny clay particles adhere to one another and only energetic flow can disrupt them. Once eroded, however, very little energy is needed to keep clay particles in motion.

Capacity is a measure of the total load a stream can carry. It varies as a function of discharge; with greater

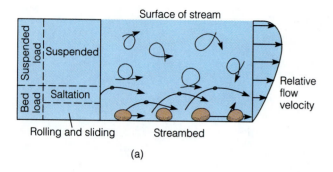

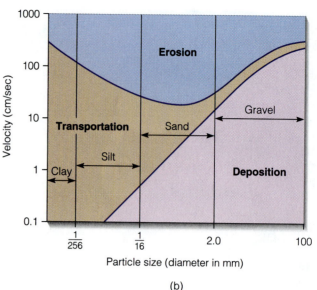

■ **Figure 15.11**

(a) Sediment transported as bed load and suspended load. The arrows in the velocity profile on the right are proportional to flow velocity, which is greatest near the surface. (b) Sediment erosion, transport, and deposition by running water are related to particle size and flow velocity.

discharge, more sediment can be carried. Capacity and competence may seem similar, but they are actually related to different aspects of stream transport. For instance, a small, swiftly flowing stream may have the competence to move gravel-sized particles but not to transport a large volume of sediment, so it has a low capacity. A large, slow-flowing stream, on the other hand, has a low competence but may have a very large suspended load and hence a large capacity.

DEPOSITION BY RUNNING WATER

Some of the sediment now being deposited in the Gulf of Mexico by the Mississippi River came from such distant sources as Pennsylvania, Minnesota, and Alberta, Canada. In short, transport might be lengthy but deposition eventually takes place. Some deposits accumulate along the way in channels, on adjacent floodplains, or where rivers and streams discharge from mountains onto nearby lowlands or where they flow into lakes or seas.

Rivers and streams constantly erode, transport, and deposit sediment, but most of their geologic work takes place when they flood. Consequently, their deposits, collectively called **alluvium,** do not represent the day-to-day activities of running water, but rather the periodic, large-scale events of sedimentation that take place during floods. Remember from Chapter 5 that sediments accumulate in *depositional environments* characterized as continental, transitional, and marine. Deposits of rivers and streams are found in the first two of these settings; however, much of the detrital sediment found on continental margins is derived from the land and transported to the oceans by running water.

The Deposits of Braided Streams

Braided streams possess an intricate network of dividing and rejoining channels separated from one another by sand or gravel bars (■ Figure 15.12). Seen from above, the channels resemble the complex strands of a braid. Braided channels develop when sediment exceeds transport capacity, resulting in the deposition of sand and gravel bars. During high-water stages, the bars are submerged, but when the water is low, they are exposed and divide a single channel into multiple channels. Braided streams have broad, shallow channels and are characterized as bed-load transport streams because they transport and deposit mostly sand and gravel (Figure 15.12).

Braided channels are common in arid and semiarid regions with little vegetation and high erosion rates. Streams with easily eroded banks are also likely to be braided. In fact, a stream that is braided where its banks are easily eroded may have a single sinuous or meandering channel when it flows into an area of more resistant materials. Streams fed by melting glaciers are also commonly braided because melting glacial ice yields so much sediment (see Chapter 17).

Meandering Streams and Their Deposits

Meandering streams have a single, sinuous channel with broadly looping curves known as *meanders* (■ Figure 15.13). These channels are semicircular in cross section along straight reaches, but at meanders they are markedly asymmetric, being deepest near the outer bank, which commonly descends vertically into the channel (■ Figure 15.14). The outer bank is called the *cut bank* because greater velocity and turbulence on that side of the channel erode it. Despite the known dynamics of erosion along the outer banks of meanders, it is remarkable that houses and other structures are built

(a)

(b)

■ **Figure 15.12**

Braided streams and their deposits. (a) A braided stream in Alaska. The deposits of this stream are composed mostly of sand. (b) A braided stream with gravel bars near Chester, California.

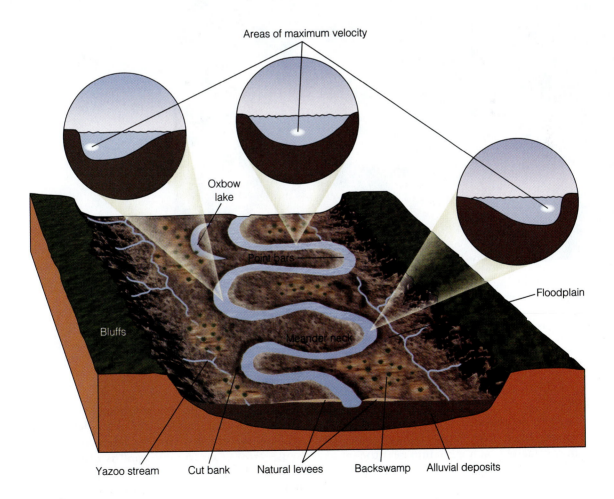

■ **Figure 15.13**

Diagrammatic view of a meandering river.

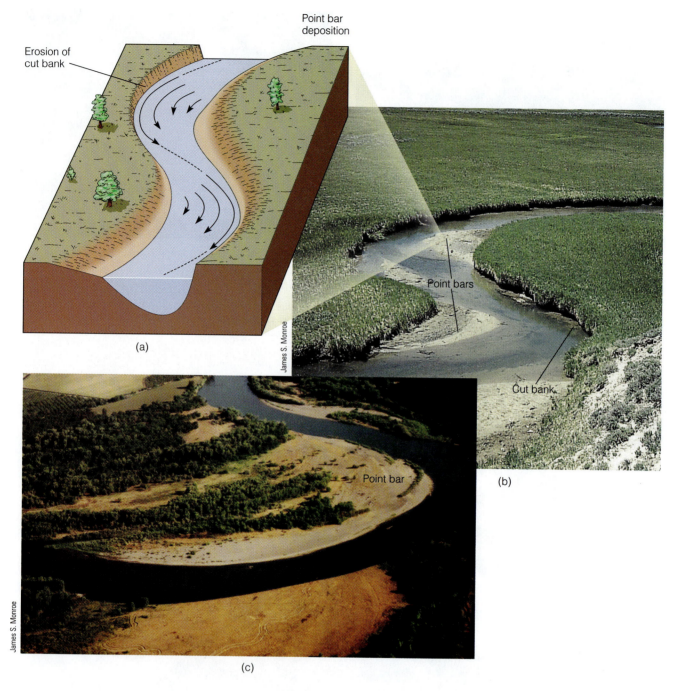

Erosion of cut bank

Point bar deposition

James S. Monroe

(a)

Point bars

Cut bank

(b)

Point bar

James S. Monroe

(c)

■ **Figure 15.14**

(a) In a meandering channel, flow velocity is greatest near the outer bank. The dashed line follows the path of maximum velocity, and the solid arrows are proportional to velocity. Because of varying velocity across the channel, the outer bank or cut bank is eroded but a point bar is deposited on the opposite side of the channel. (b) Two small point bars of sand in a meandering stream. Notice how they are inclined into the deeper part of the channel. Also note the cut bank. (c) This point bar measures several hundred meters across.

overlooking cut banks. The life expectancy of these structures is short because during a single flood, several meters of erosion might take place along a cut bank. In contrast, flow velocity is at a minimum near the inner bank, which slopes gently into the channel (Figure 15.14).

As a result of the unequal distribution of flow velocity across meanders, the cut bank erodes, and deposition takes place along the opposite side of the channel. The net effect is that a meander migrates laterally, and the channel maintains a more or less constant width because erosion on the cut bank is offset by an equal amount of deposition on the opposite side of the channel. The deposit formed in this manner is a **point bar** consisting of cross-bedded sand or, in some cases, gravel (Figure 15.14).

(a)

(b)

Deposits of silt and clay

Abandoned channel

(c)

Oxbow lake

(d)

(e)

James S. Monroe

Four stages in the origin of an oxbow lake. In (a) and (b), the meander neck becomes narrower. (c) The meander neck is cut off, and part of the channel is abandoned. (d) When it is completely isolated from the main channel, the abandoned meander is an oxbow lake. (e) This oxbow lake in Wyoming formed recently.

It is not uncommon for meanders to become so sinuous that the thin neck of land separating adjacent meanders is eventually cut off during a flood. The valley floors of meandering streams are commonly marked by crescent-shaped **oxbow lakes,** which are actually cutoff meanders (Figures 15.13 and ■ 15.15). These oxbow lakes may persist as lakes for some time, but they eventually fill with organic matter and fine-grained sediment carried by floods. Even after they are filled, they remain visible on floodplains.

One immediate effect of meander cutoff is an increase in flow velocity; following the cutoff, the stream abandons part of its old course and flows a shorter distance, thereby increasing its gradient. Numerous cutoffs would, of course, significantly shorten a meandering stream, but streams usually establish new meanders elsewhere when old ones are cut off.

Floodplain Deposits

Rivers and streams periodically receive more water than their channels can carry, so they spread across low-lying, relatively flat **floodplains** adjacent to their channels (Figure 15.13). Even small streams commonly have

What Would You Do ?

Given what is known about the dynamics of running water in channels, it is remarkable that houses are still built on the cut banks of meandering rivers. No doubt property owners think these locations provide good views because they sit high above the adjacent channel. Explain why you would or would not build a house in such a location. What recommendations would you make to a planning commission on land use for areas as described here? Are there any specific zoning regulations or building codes you might favor?

a floodplain, but this feature is usually proportional to the size of the stream; thus small streams have narrow floodplains, whereas the lower Mississippi and other large rivers have floodplains many kilometers wide. Streams restricted to deep, narrow valleys have little or no floodplain.

Some floodplains are composed mostly of sand and gravel that were deposited as point bars. When a meandering stream erodes its cut bank and deposits on the opposite bank, it migrates laterally across its floodplain. As lateral migration occurs, a succession of point bars develops by *lateral accretion* (■ Figure 15.16). That is, the deposits build laterally as a result of repeated episodes of sedimentation on the inner banks of meanders.

When a stream overflows its banks and floods, the velocity of the water spilling onto the floodplain diminishes rapidly because of greater frictional resistance to flow as the water spreads out as a broad, shallow sheet. In response to the diminished velocity, ridges of sandy alluvium known as **natural levees** are deposited along the margins of the channel (■ Figure 15.17). Natural levees build up by repeated deposition of sediment during numerous floods. These natural levees separate most of the floodplain from the stream channel, so floodplains are commonly poorly drained and swampy. In fact, tributary streams may parallel

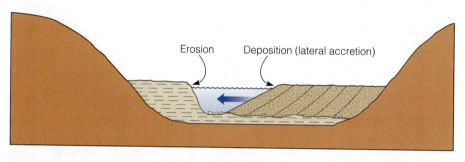

■ **Figure 15.16**

Floodplain deposits forming by lateral accretion of point bars.

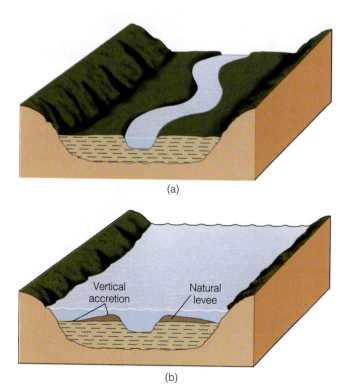

(a)

(b)

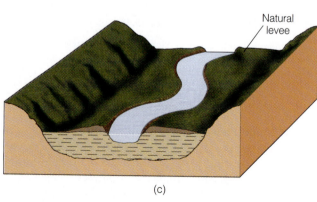

(c)

■ **Figure 15.17**

Three stages in the formation of vertical accretion deposits on a floodplain. (a) Stream at low-water stage. (b) Flooding stream and deposition of natural levees. The levees form after many such episodes of flooding. (c) After flooding. Notice the tributary stream, which parallels the main stream until it finds a way through the natural levees.

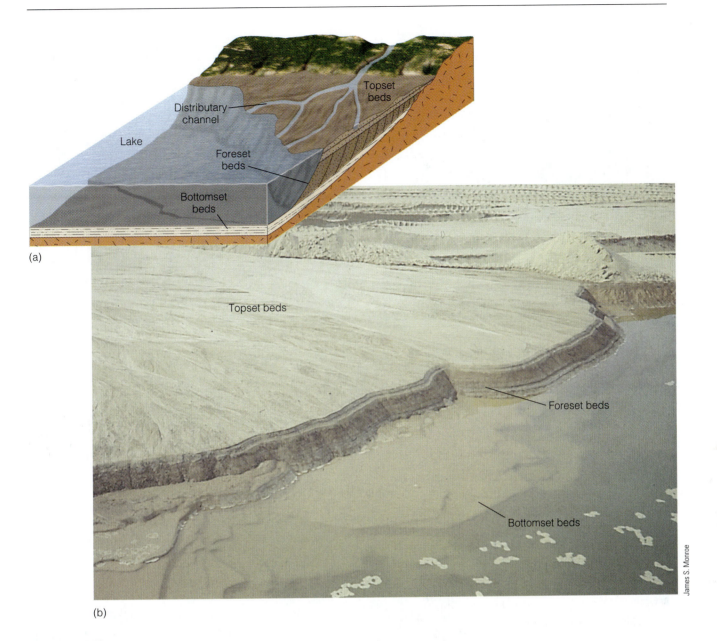

(a)

(b)

James S. Monroe

■ Figure 15.18

(a) Internal structure of the simplest type of prograding delta. (b) A small delta, measuring about 20 m across, in which bottomset, foreset, and topset beds are visible.

the main stream for many kilometers until they find a way through the natural levee system (Figure 15.17).

Floodwaters spilling from a main channel carry large quantities of silt- and clay-sized sediment beyond the natural levees and onto the floodplain. During the waning stages of a flood, floodwaters may flow very slowly or not at all, and the suspended silt and clay eventually settle as layers of mud that build upward by deposition during successive floods, a process known as *vertical accretion* (Figure 15.17).

Deltas

The process of delta formation is rather simple: Where running water flows into another body of water, its flow velocity decreases rapidly and it deposits sediment. As a result, a **delta** forms, causing the local shoreline to build out, or *prograde* (■ Figure 15.18), unless the stream-deposited sediment is swept along the shoreline or into deeper water by marine currents.

The simplest prograding deltas exhibit a characteristic vertical sequence in which *bottomset beds* are successively overlain by *foreset beds* and *topset beds* (Figure 15.18a). This sequence develops when a stream enters another body of water and the finest sediments are carried some distance beyond the stream's mouth, where they settle from suspension and form bottomset beds. Nearer the stream's mouth, foreset beds form as sand and silt are deposited in gently inclined layers. The topset beds consist of coarse-grained

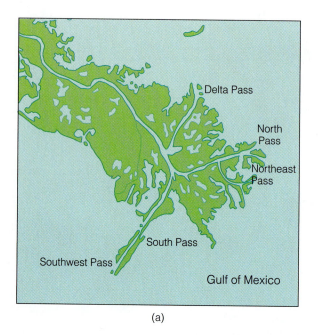

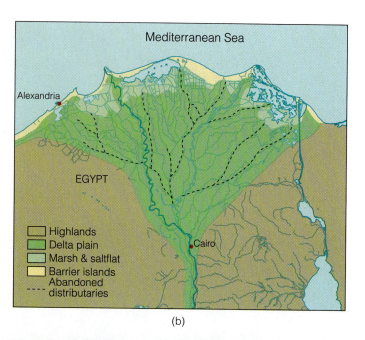

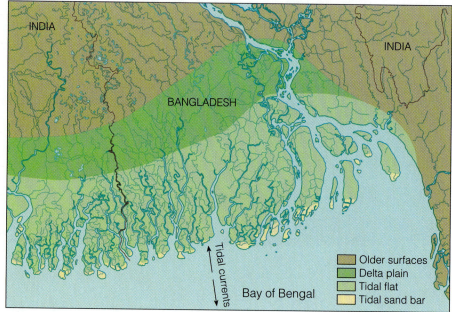

■ Figure 15.19

(a) The Mississippi River delta of the U.S. Gulf Coast is stream-dominated. (b) The Nile delta of Egypt is wave-dominated. (c) The Ganges–Brahmaputra delta of Bangladesh is tide-dominated.

sediments deposited in a network of *distributary channels* traversing the top of the delta.

Many small deltas in lakes have the three-part division described, but large marine deltas are usually much more complex. Depending on the relative importance of stream, wave, and tidal processes, geologists identify three major types of marine deltas (■ Figure 15.19). *Stream-dominated deltas,* such as the Mississippi River delta, consist of long fingerlike sand bodies, each deposited in a distributary channel that progrades far seaward. These deltas are commonly called *bird's-foot deltas* because the projections resemble the toes of a bird. In contrast, the Nile delta of Egypt is *wave-dominated,* although it also possesses distributary channels; the sea-

ward margin of the delta consists of a series of barrier islands formed by reworking of sediments by waves, and the entire margin of the delta progrades seaward. *Tide-dominated deltas,* such as the Ganges–Brahmaputra of Bangladesh, are continually modified into tidal sand bodies that parallel the direction of tidal flow.

Coal forms in a variety of environments, including the freshwater marshes between distributary channels of deltas (Figure 15.18a). These marshes are dominated by nonwoody plants whose remains accumulate to form peat, the first stage in the origin of coal (see Chapter 6).

Delta progradation is one way that potential reservoirs for oil and gas form. Because the sediments of distributary sand bodies are porous and in proximity to

GEOLOGY
IN UNEXPECTED PLACES

Floating Burial Chambers

Have you ever thought about how geologic conditions affect where and how cemeteries are constructed? Did you know that architects design cemeteries and must consider the geology of an area when planning a site? Let's begin our investigation into this topic by considering the cemetery in Leesville, Louisiana, where stone burial chambers rest on the ground surface because of the area's unusually high water table. Many of the burial chambers are now nearly afloat as the land subsides, resulting in a concurrent rise in sea level, and of course the water table rises as well.

The subsidence-rising water table problem is not confined to Leesville but affects much of southern Louisiana, especially the southern part of the Mississippi River delta (■ Figure 1). Flood-control projects on the Mississippi River prevent most sediment from reaching the seaward margin of the delta, which contributes to loss of land due to wave erosion. Even more important is the natural propensity for delta deposits, especially mud, to compact under their own weight and subside. As a result many areas on the delta, including New Orleans, are now below sea level and must be protected by systems of levees.

Each year about 65 km^2 of Louisiana's coastline is lost to these processes and, ironically, human efforts to prevent floods have only worsened the situation. And what about global warming, which causes thermal expansion of the ocean's surface waters and melting of glaciers? This will definitely contribute to the region's problems as sea level rises and much of the delta continues to subside. What do you think the future holds for the cemetery at Leesville, Louisiana?

JPL/NASA

■ **Figure 1**

The Mississippi River delta from space.

---

organic-rich marine sediments, they commonly contain oil and gas that can be extracted in economic quantities. Much of the oil and gas production of the Gulf Coast of Texas comes from buried delta deposits. Some of the older deposits of the Niger River delta of Africa and the Mississippi River delta also contain vast reserves of oil and gas.

Alluvial Fans

Lobate deposits of alluvium on land known as **alluvial fans** form best on lowlands with adjacent highlands in arid and semiarid regions where little vegetation exists to stabilize surface materials (■ Figure 15.20). During periodic rainstorms, surface materials are quickly saturated and surface runoff is funneled into a mountain canyon leading to adjacent lowlands. In the mountain canyon, the runoff is confined so it cannot spread laterally, but when it discharges onto the lowlands it quickly spreads out, its velocity diminishes, and deposition ensues. Repeated episodes of this kind of sedimentation result in the accumulation of a fan-shaped body of alluvium.

Deposition by running water in the manner just described is responsible for many alluvial fans. In this case they are composed mostly of sand and gravel, both of which contain a variety of sedimentary structures. In some cases, though, the water flowing through a canyon picks up so much sediment that it becomes a viscous debris flow. Consequently, some alluvial fans consist mostly of debris flow deposits showing little or no layering. Of course, the dominant type of deposition can change through time, so a particular fan might have both types of deposits.

(a)

(b)

John S. Shelton

■ **Figure 15.20**

(a) Alluvial fans form where streams discharge from mountain canyons onto adjacent lowlands. (b) Alluvial fans adjacent to the Panamint Range on the margin of Death Valley, California.

Alluvial fan

Channels are dry most of the time

Alluvial fans

What Would You Do?

The largest part of many states' mineral revenues comes from sand and gravel, most of which is used in construction. You become aware of a sand and gravel deposit that you can acquire for a small investment. How would the proximity of the deposit to potential markets influence your decision? Assuming the deposit was stream deposited, would it be important to know whether deposition took place in braided or meandering channels? How could you tell one from the other?

CAN FLOODS BE PREDICTED AND CONTROLLED?

When a river or stream receives more water than its channel can handle, it floods, occupying part or all of its floodplain. Indeed, floods are so common that, unless they cause considerable property damage or fatalities, they rarely rate more than a passing note in the news. The most extensive recent flooding in the United States was the Flood of '93 in the Midwest (see pp. 434 and 435), but since then several other areas have experienced serious flooding. For instance, a series of storms from late December 1996 to early January 1997 caused widespread flooding in several western states and resulted in at least 28 deaths and billions of dollars in property damage. And in April 1997, the spring thaw of a record snowfall caused the Red River to flood, leading to the evacuation of all 50,000 residents of Grand Forks, North Dakota, and an additional 28,000 people from areas adjacent to the river in Manitoba, Canada. Floods and mudslides triggered by heavy rain killed tens of thousands during December 1999 in Venezuela (see Chapter 14).

Annual property damage from flooding in the United States exceeds $100 million. Despite the completion of more and more flood-control projects, the amount of property damage is not decreasing. In fact, the combination of fertile soils, level surfaces for construction, and proximity to water for agriculture, industry, and domestic uses makes floodplains popular sites for settlement. Unfortunately, urbanization greatly increases surface runoff because surface materials are compacted or covered by concrete and asphalt, reducing their infiltration capacity. Storm drains in urban areas quickly carry water to nearby waterways, many of which flood much more commonly than they did in the past.

To monitor stream behavior, the U.S. Geological Survey maintains more than 11,000 stream-gauging stations, and various state agencies also monitor streams. Data collected at a gauging station are used to

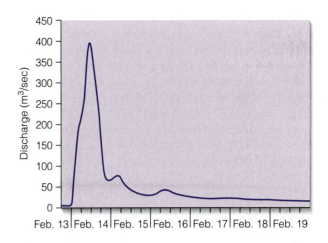

■ **Figure 15.21**

Hydrograph for Sycamore Creek near Ashland City, Tennessee, for the February 1989 flood.

construct a *hydrograph* showing how a stream's discharge varies over time (■ Figure 15.21). Hydrographs are useful in planning irrigation and water-supply projects, and they give planners a better idea of what to expect during floods.

Stream gauge data are also used to construct *flood-frequency curves* (■ Figure 15.22). To make such a curve, the peak discharges for a stream are first arranged in order of volume; the flood with the greatest discharge

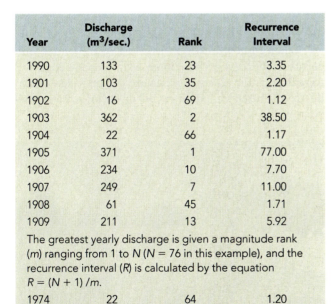

Year	Discharge (m³/sec.)	Rank	Recurrence Interval
1990	133	23	3.35
1901	103	35	2.20
1902	16	69	1.12
1903	362	2	38.50
1904	22	66	1.17
1905	371	1	77.00
1906	234	10	7.70
1907	249	7	11.00
1908	61	45	1.71
1909	211	13	5.92

The greatest yearly discharge is given a magnitude rank (*m*) ranging from 1 to *N* (*N* = 76 in this example), and the recurrence interval (*R*) is calculated by the equation $R = (N + 1)/m$.

1974	22	64	1.20
1975	68	43	1.79

Source: U.S. Geological Survey Open-File Report 79–681.

(a)

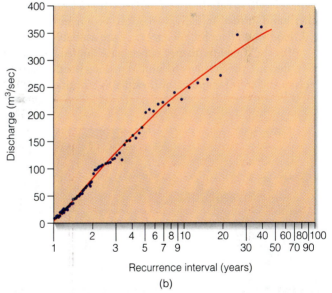

(b)

■ **Figure 15.22**

(a) Selected data and recurrence intervals for the Rio Grande near Lobatos, Colorado. (b) The data in (a) were used to construct this flood-frequency curve for the Rio Grande near Lobatos, Colorado.

has a magnitude rank of 1, the second largest is 2, and so on (Figure 15.22). The *recurrence interval*—that is, the time period during which a flood of a given magnitude or larger can be expected over an average of many years—is determined by the equation shown in Figure 15.22. For this stream, floods with magnitude ranks of 1 and 23 have recurrence intervals of 77.00 and 3.35 years, respectively. Once the recurrence interval has been calculated, it is plotted against discharge, and a line is drawn through the data points.

According to Figure 15.22, the 10-year flood for the Rio Grande near Lobatos, Colorado, has a discharge of 245 m³/sec. This means that, on average, we can expect one flood of this size or greater to occur within a 10-year interval. One cannot, however, predict that a flood of this size will take place in any particular year, only that it has a probability of 1 in 10 (1/10) of occurring in any year. Furthermore, 10-year floods are not necessarily separated by 10 years. That is, two such floods could occur in the same year or in successive years, but over a period of centuries their average occurrence would be once every 10 years.

Unfortunately, stream gauge data in North America have been available for only a few decades and rarely for more than a century. Accordingly, we have a good idea of stream behavior over short periods—the 2- and 5-year floods, for example—but our knowledge of long-term behavior is limited by the short recordkeeping period. Accordingly, predictions of 50- or 100-year floods from Figure 15.22 are unreliable. In fact, the largest magnitude flood shown in Figure 15.22 may have been a unique event for this stream that will never be repeated. On the other hand, it may actually turn out to be a magnitude 2 or 3 flood when data for a longer time are available.

Although flood-frequency curves have limited applicability, they are nevertheless helpful in making flood-control decisions. Careful mapping of floodplains identifies areas at risk for floods of a given magnitude. For a particular stream, planners must decide what magnitude of flood to protect against because the cost goes up faster than the increasing sizes of floods would indicate.

Federal, state, and local agencies and land-use planners use flood-frequency analyses to develop recommendations and regulations concerning construction on and use of floodplains. Geologists and engineers are interested in these analyses for planning appropriate flood-control projects. They must decide, for example, where dams and basins should be constructed to contain the excess water of floods.

Flood control has been practiced for thousands of years. Common practices are to construct dams that impound reservoirs (see Figure 15.2a) and to build levees along stream banks. Levees raise the banks of a stream, thereby restricting flow during floods (■ Figure 15.23a). Unfortunately, deposition within the channel results in raising the streambed, making the levees useless unless they too are raised. Levees along the banks of the Huang He in China caused the streambed to rise more than 20 m above its surrounding floodplain in 4000 years. When the Huang He breached its levees in 1887, more than 1 million people were killed.

Dams and levees alone are insufficient to control large floods, so in many areas floodways are also used (Figure 15.23b). These usually consist of a channel constructed to divert part of the excess water in a stream around populated areas or areas of economic importance. Reforestation of cleared land also helps reduce the potential for flooding because vegetated soil helps prevent runoff by absorbing more water.

When flood-control projects are well planned and constructed, they are functional. What many people fail to realize is that these projects are designed to contain floods of a given size; if larger floods occur, streams spill

(a)

(b)

■ **Figure 15.23**

Flood control. (a) This levee, an artificial embankment along a waterway, helps protect nearby areas from flooding. (b) In addition to dams, engineers construct floodways to carry excess water from a river (not visible) around a small community.

onto floodplains anyway. Furthermore, dams occasionally collapse, and reservoirs eventually fill with sediment unless dredged. In short, flood-control projects are not only initially expensive but also require constant, costly maintenance. Such costs must be weighed against the cost of damage if no control projects were undertaken.

DRAINAGE BASINS AND DRAINAGE PATTERNS

Thousands of waterways, which are parts of larger drainage systems, flow directly or indirectly into the oceans. The only exceptions are some rivers and streams that flow into desert basins surrounded by higher areas. But even these are parts of larger systems consisting of a main channel with all its tributaries—that is, streams that contribute water to another stream. The Mississippi River and its tributaries, such as the Ohio, Missouri, Arkansas, and Red Rivers and thousands of smaller ones, or any other drainage system for that matter, carries runoff from an area known as a **drainage basin.** A topographically high area called a **divide** separates a drainage basin from adjoining ones (■ Figure 15.24). The continental divide along the crest of the Rocky Mountains in North America, for instance, separates drainage in opposite directions; drainage to the west goes to the Pacific, whereas that to the east eventually reaches the Gulf of Mexico.

Various arrangements of channels within an area are classified as different types of **drainage patterns.** *Dendritic drainage,* consisting of a network of channels resembling tree branching, is the most common (■ Figure 15.25a). It develops on gently sloping surfaces composed of materials that respond more or less homogeneously to erosion. Areas underlain by nearly horizontal sedimentary rocks and some areas of igneous and metamorphic rock commonly display dendritic drainage.

In dendritic drainage, tributaries join larger channels at various angles, but *rectangular drainage* is characterized by right-angle bends and tributaries joining larger channels at right angles (Figure 15.25b). Such regularity in channels is strongly controlled by geologic structures, particularly regional joint systems that intersect at right angles.

Trellis drainage, consisting of a network of nearly parallel main streams with tributaries joining them at right angles, is common in some parts of the eastern United States. In Virginia and Pennsylvania, for example, erosion of folded sedimentary rocks developed a landscape of alternating ridges on resistant rocks and valleys underlain by easily eroded rocks. Main waterways follow the valleys, and short tributaries flowing from the nearby ridges join the main channels at nearly right angles (Figure 15.25c).

In *radial drainage,* streams flow outward in all directions from a central high point such as a large volcano

(Figure 15.25d). Many of the volcanoes in the Cascade Range of western North America have radial drainage patterns. In all the types of drainage mentioned so far, some kind of pattern is easily recognized. *Deranged drainage,* however, is characterized by irregularity, with streams flowing into and out of swamps and lakes, streams with only a few short tributaries, and vast swampy areas between channels (Figure 15.25e). The presence of this kind of drainage indicates that it developed recently and has not yet formed a fully organized drainage system. In parts of Minnesota, Wisconsin, and Michigan, where glaciers obliterated the previous drainage, only 10,000 years has elapsed since the glaciers melted. As a result, drainage systems have not fully developed and large areas still remain undrained.

The Significance of Base Level

Just how deeply can a stream or river erode? The Grand Canyon in Arizona is more than 1.6 km deep, but the bottom of the canyon is still far above sea level. Obviously channels must maintain some gradient, so they are restricted by **base level,** the lowest limit to which running water can erode. Sea level is called *ultimate base level,* and theoretically a channel could erode deeply enough so that its gradient rose ever so gently inland from the sea (■ Figure 15.26). Ultimate base level applies to an entire stream or river system, but channels may also have *local* or *temporary base levels.* For instance, a local base level may be a lake or another stream, or where a stream or river flows across particularly resistant rocks and a waterfall develops (Figure 15.26).

Ultimate base level is sea level, but suppose sea level should drop or rise with respect to the land, or suppose that the land should rise or subside? In any of these cases, base level would change, bringing about changes in stream and river systems. For example, during the Pleistocene Epoch (Ice Age) sea level was about 130 m lower than it is now, and streams adjusted by eroding deeper valleys (they had steeper gradients) and extended well out onto the continental shelves. Rising sea level at the end of the Ice Age accounted for rising base level, decreased stream gradients, and deposition within channels.

Natural changes, such as fluctuations in sea level during the Pleistocene, alter the dynamics of rivers and streams, but so does human intervention. Geologists and engineers are well aware that building a dam to impound a reservoir creates a local base level (■ Figure 15.27a). A stream entering a reservoir slows down and deposits sediment, so unless dredged, reservoirs eventually fill. In addition, the water discharged at a dam is largely sediment-free but still possesses energy to carry a sediment load. As a result it is not uncommon for streams to erode vigorously downstream from a dam to acquire a sediment load.

Draining a lake may seem like a small change and well worth the time and expense to expose dry land for

(a)

James S. Monroe

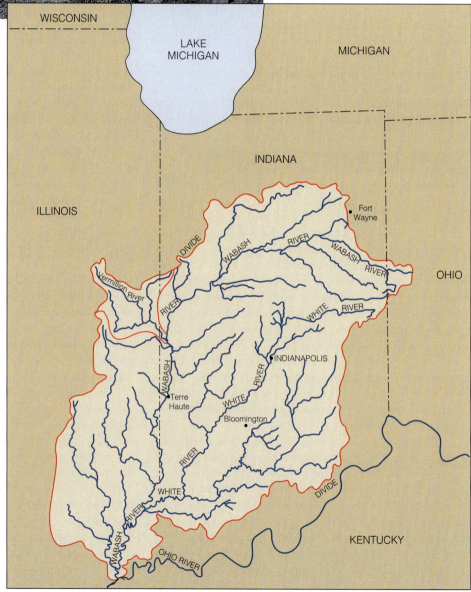

(b)

■ **Figure 15.24**

(a) Small drainage basins separated from one another by divides (dashed lines), which are along the crests of the ridges between channels (solid lines). (b) The drainage basin of the Wabash River, which is one of the tributaries of the Ohio River. All tributary streams within the drainage basin, such as the Vermillion River, have their own smaller drainage basins. Divides are shown by red lines.

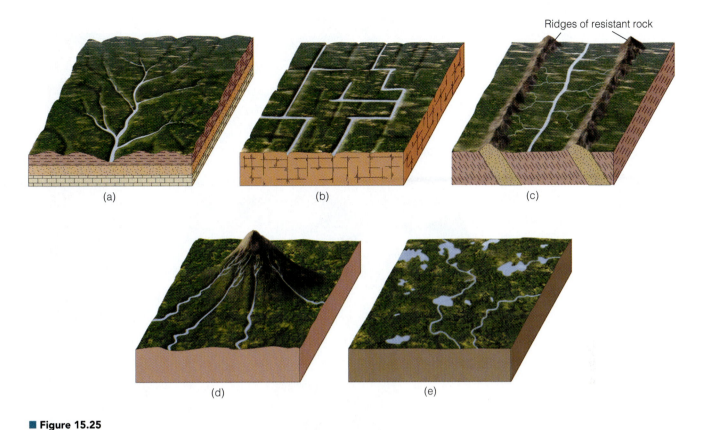

Ridges of resistant rock

(a)   (b)   (c)

(d)   (e)

■ **Figure 15.25**

Examples of drainage patterns: (a) dendritic drainage, (b) rectangular drainage, (c) trellis drainage, (d) radial drainage, and (e) deranged drainage.

agriculture or commercial development. But in doing so a local base level has been eliminated, and a stream that originally flowed into the lake responds by rapidly eroding a deeper valley as it adjusts to a new base level (Figure 15.27b).

What Is a Graded Stream?

The *longitudinal profile* of any waterway shows the elevations of a channel along its length as viewed in cross section (■ Figure 15.28). For some rivers and streams, the longitudinal profile is smooth, but for others, a number of irregularities such as lakes and waterfalls are present, all of which are local base levels (Figure 15.26). Over time these irregularities tend to be eliminated because deposition takes place where the gradient is insufficient to maintain sediment transport, and erosion decreases the gradient where it is steep. So given enough time, rivers and streams develop a smooth, concave longitudinal profile or equilibrium, meaning that all parts of the system dynamically adjust to one another.

A **graded stream** is one with an equilibrium profile in which a delicate balance exists between gradient, discharge, flow velocity, channel shape, and sediment load so that neither significant erosion nor deposition takes place within its channel. Such a delicate balance is rarely attained, so the concept of a graded stream is an ideal. Nevertheless, the graded condition is closely approached in many streams, although only temporarily and not necessarily along their entire lengths.

Even though the concept of a graded stream is an ideal, we can anticipate the response of a graded stream to changes that alter its equilibrium. For instance, a change in base level would cause a stream to adjust as previously discussed. Increased rainfall in a stream's drainage basin would result in greater discharge and flow velocity. In short, the stream would now possess greater energy—energy that must be dissipated within the stream system by, for example, a change from a semicircular to a broad, shallow channel that would dissipate more energy by friction. On the other hand, the stream may respond by eroding a deeper valley and effectively reduce its gradient until it is once again graded.

Vegetation inhibits erosion by having a stabilizing effect on soil and other loose surface materials. So a decrease in vegetation in a drainage basin might lead to higher erosion rates, causing more sediment to be washed into a stream than it can effectively carry. Accordingly, the stream may respond by deposition within its channel, which increases its gradient until it is sufficiently steep to transport the greater sediment load.

■ Figure 15.26

(a) Sea level is ultimate base level, whereas a resistant rock layer forms a local base level. (b) Cumberland Falls on the Big South Fork River in Cumberland Falls State Resort Park, Kentucky, plunges 18 m over a local base level. This is the second highest waterfall east of the Rocky Mountains. (c) Ultimate base level and a local base level where a stream flows into a lake. (d) The Au Train River in Michigan (right) flows into Lake Superior, its local base level.

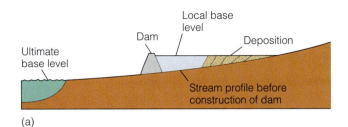

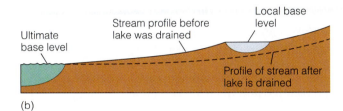

(b)

■ **Figure 15.27**

(a) Constructing a dam and impounding a reservoir create a local base level. A stream deposits much of its sediment load where it flows into a reservoir. (b) A stream adjusts to a lower base level when a lake is drained.

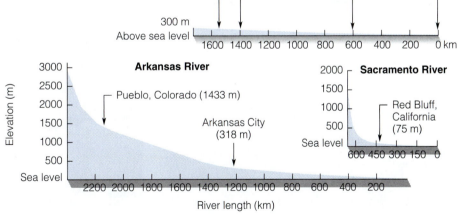

■ **Figure 15.28**

(a) An ungraded stream has irregularities in its longitudinal profile. (b) Erosion and deposition along the course of a stream tend to eliminate irregularities and yield the smooth, concave profile of a graded stream. (c) The actual measured profiles of three rivers. All profiles are concave but differ in the degree of concavity because each stream has a different slope. The vertical scale is exaggerated 275 times in this illustration.

HOW DO VALLEYS FORM AND EVOLVE?

Low areas on land known as **valleys** are bounded by higher land, and most of them have a river or stream running their length, with tributaries draining the nearby high areas. Valleys are common landforms and with few exceptions they form and evolve in response to erosion by running water, although other processes, especially mass wasting, contribute. The shapes and sizes of valleys vary from small, steep-sided *gullies* to those that are broad with gently sloping valley walls. Steep-walled, deep valleys of vast size are *canyons,* and particularly narrow and deep ones are *gorges* (■ Figure 15.29).

Erosion of a valley might be initiated where runoff has sufficient energy to dislodge surface materials and excavate a small rill. Once formed, a rill collects more runoff and becomes deeper and wider and continues to do so until a full-fledged valley develops. Processes related to running water that contribute to valley formation include downcutting, lateral erosion, headward erosion, and sheet wash. A variety of mass wasting processes are also important.

Downcutting takes place when a river or stream has more energy than it needs to transport sediment, so some of its excess energy is used to deepen its valley. If downcutting were the only process operating, valleys would be narrow and steep sided. In most cases, though, the valley walls are undercut by stream action, a process called *lateral erosion,* creating unstable slopes that may fail by one or more mass wasting processes. Furthermore, erosion by sheet wash and erosion by tributary streams carry materials from the valley walls into the main stream in the valley.

Valleys not only become deeper and wider but also become longer by *headward erosion,* a phenomenon involving erosion by entering runoff at the upstream end of a valley (■ Figure 15.30). Continued headward erosion commonly results in *stream piracy,* the breaching of a drainage divide and diversion of part of the drainage of another stream (Figure 15.30). Once stream piracy takes place, both drainage systems must adjust to these new conditions; one system now has greater discharge and the potential to do more erosion and sediment transport, whereas the other is diminished in its ability to accomplish these tasks.

Stream Terraces

Adjacent to many streams are erosional remnants of floodplains formed when the streams were flowing at a higher level. These **stream terraces** consist of a fairly flat upper surface and a steep slope descending to the level of the lower, present-day floodplain (■ Figure 15.31). In some cases, a stream has several steplike sur-

(a)

(b)

■ **Figure 15.29**

Valleys of various sizes and shapes form mostly by running water and mass wasting, but they may be modified by other processes such as glaciers (see Chapter 17). (a) Many valleys have gently sloping walls much like this one. The river at the bottom of this valley is not visible. (b) The valley of the Colorado River near Moab, Utah, has steep to nearly vertical walls.

(a)   (b)

■ **Figure 15.30**

Two stages in the evolution of a valley. (a) The stream widens its valley by lateral erosion and mass wasting while simultaneously extending its valley by headward erosion. (b) As the larger stream continues to erode headward, stream piracy takes place when it captures some of the drainage of the smaller stream. Notice also that the larger valley is wider in (b) than it was in (a).

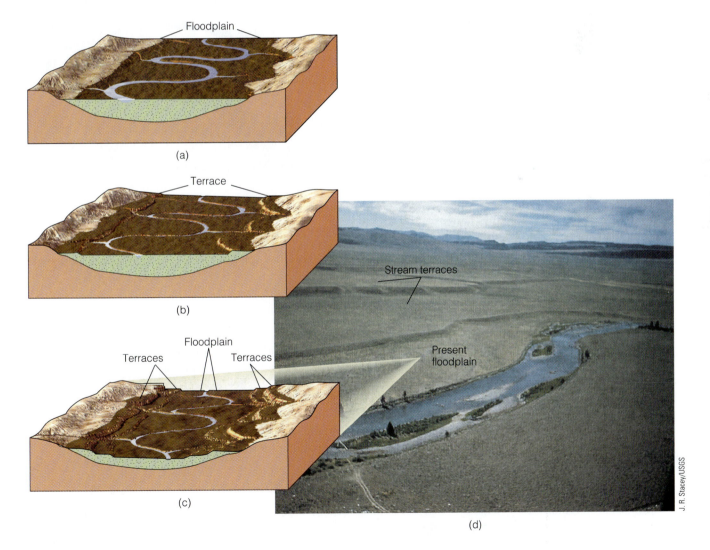

■ **Figure 15.31**

Origin of stream terraces. (a) A stream has a broad floodplain. (b) The stream erodes downward and establishes a new floodplain at a lower level. Remnants of its old, higher floodplain are stream terraces. (c) Another level of stream terraces forms as the stream erodes downward again. (d) Stream terraces along the Madison River in Montana.

(a)

(b)

(c)

(d)

(e)

■ **Figure 15.32**

(a) A meandering river flows across a gently sloping surface and becomes incised (b) as it erodes downward, thus occupying a meandering canyon. (c) In some cases the thin wall of rock between adjacent incised meanders is cut off, forming a natural bridge (e). (d) The Colorado River at Dead Horse State Park, Utah, is incised to a depth of 600 m. (e) Sipapu Bridge in Natural Bridges National Monument, Utah, formed as shown in (c). It stands 67 m above the canyon floor and has a span of 81.5 m.

faces above its present-day floodplain, indicating that stream terraces formed several times.

Although all stream terraces result from erosion, they are preceded by an episode of floodplain formation and sediment deposition. Subsequent erosion causes the stream to cut downward until it is once again graded (Figure 15.31). Once the stream again becomes graded, it begins eroding laterally and establishes a new floodplain at a lower level. Several such episodes account for the multiple terrace levels adjacent to some streams.

Renewed erosion and the formation of stream terraces are usually attributed to a change in base level. Either uplift of the land a stream flows over or lowering of

sea level yields a steeper gradient and increased flow velocity, thus initiating an episode of downcutting. When the stream reaches a level at which it is once again graded, downcutting ceases. Although changes in base level no doubt account for many stream terraces, greater runoff in a stream's drainage basin can also result in the formation of terraces.

Incised Meanders

Some streams are restricted to deep, meandering canyons cut into bedrock, where they form features called **incised meanders** (■ Figure 15.32). For example, the San Juan River in Utah occupies a meandering canyon more than

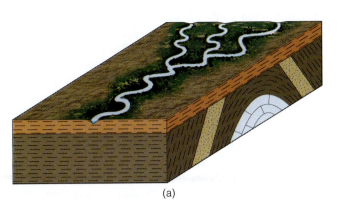

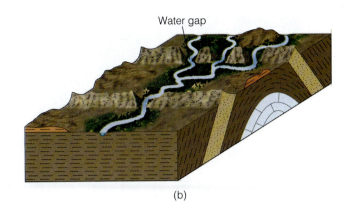

(a)

(b)

■ **Figure 15.33**

The origin of a superposed stream. (a) A stream begins cutting down into horizontal strata. (b) The horizontal layer is removed by erosion, exposing the underlying structure. The stream cuts narrow valleys in resistant rocks that form the ridges, forming water gaps.

390 m deep. Streams restricted by rock walls are generally ineffective in eroding laterally; thus they lack a floodplain and occupy the entire width of the canyon floor.

It is not difficult to understand how a stream can cut downward into solid rock, but forming a meandering pattern in bedrock is another matter. Because lateral erosion is inhibited once downcutting begins, one must infer that the meandering course was established when the stream flowed across an area covered by alluvium. For example, suppose that a stream near base level has established a meandering pattern. If the land the stream flows over is uplifted, erosion is initiated, and the meanders become incised into the underlying bedrock.

Superposed Streams

Streams flow downhill in response to gravity, so their courses are determined by preexisting topography. Yet a number of streams seem, at first glance, to have defied this fundamental control. For example, the Delaware, Potomac, and Susquehanna Rivers in the eastern United States have valleys that cut directly through ridges lying in their paths. The Madison River in Montana meanders northward through a broad valley, then enters a narrow canyon cut into bedrock that leads to the next valley where the river resumes meandering.

All of these examples are of **superposed streams.** In order to understand superposition, it is necessary to know the geologic histories of these streams. In the case of the Madison River, the valleys it now occupies were once filled with sedimentary rocks so that the river flowed on a surface at a higher level (■ Figure 15.33). As the river eroded downward, it was superposed directly upon a preexisting knob of more resistant rock, and instead of changing its course, it cut a narrow, steep-walled canyon called a *water gap*.

Superposition also accounts for the fact that the Delaware, Potomac, and Susquehanna Rivers flow through water gaps. During the Mesozoic Era, the Appalachian Mountain region was eroded to a sediment-covered plain across which numerous streams flowed generally eastward. During the Cenozoic Era, regional uplift commenced, and as a result of the uplift, the streams began eroding downward and were superposed on resistant strata, thus forming water gaps (Figure 15.33).

VALLEY DEVELOPMENT— A SUMMARY

According to one concept, stream erosion of an area uplifted above sea level yields a distinctive series of landscapes. When erosion begins, streams erode downward; their valleys are deep, narrow, and V-shaped in cross section; and their profiles show a number of irregularities (■ Figure 15.34a). As streams cease eroding downward, they start eroding laterally, establishing a meandering pattern and a broad floodplain (Figure 15.34b). Finally, with continued erosion, a vast, rather featureless plain develops (Figure 15.34c).

Many streams do indeed show an association of features typical of these stages. For instance, the Colorado River flows through the Grand Canyon and closely matches the features in the initial stage shown in Figure 15.34a. Streams in many areas approximate the second stage of development, and certainly the lower Mississippi closely resembles the last stage. Nevertheless, the idea of a sequential development of

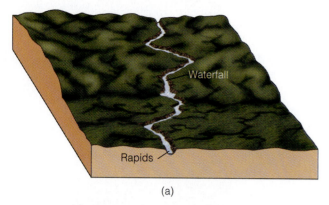

(a)

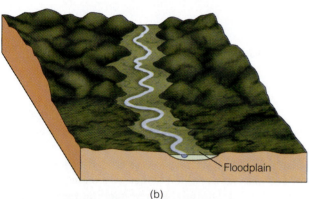

(b)

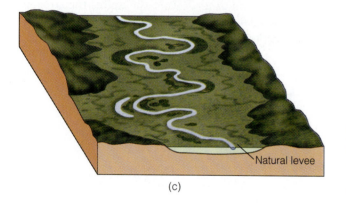

(c)

stream-eroded landscapes has been largely abandoned because there is no reason to think that streams necessarily follow this idealized cycle. Indeed, a stream on a gently sloping surface near sea level could develop features of the last stage very early in its history. In addition, as long as the rate of uplift exceeds the rate of downcutting, a stream will continue to erode downward and be confined to a narrow canyon.

Although it is doubtful that streams go through the three-stage sequence just outlined, they do indeed evolve in response to changing conditions in the variables that influence them. Of course, human intervention can change a stream's capacity for erosion and sediment transport, and increase the frequency of flooding. However, natural processes such as uplift or subsidence, climatic change, cutting a deeper outlet and draining a lake, or encountering a particularly resistant layer of rock can also influence stream activity. In short, they are dynamic systems continuously adjusting to changes.

PHYSICAL
Geology ⇌ Now ■ Active Figure 15.34

Idealized stages in the development of a stream and its associated landforms. According to this idea, an uplifted area begins eroding as in (a), and with time a landscape evolves as illustrated in (b) and finally in (c).

REVIEW
WORKBOOK

Chapter Summary

- Water continuously evaporates from the oceans, rises as water vapor, condenses, and falls as precipitation, about 20% of which falls on land and eventually returns to the oceans, mostly by surface runoff.

- Running water moves by laminar flow, in which streamlines parallel one another, and by turbulent flow, in which streamlines complexly intertwine. Almost all flow in channels is turbulent.

- Runoff takes place by sheet flow, a thin, more-or-less continuous sheet of water, and by channel flow confined to long troughlike depressions.

- The vertical drop in a given distance, or the gradient, for a channel varies from steep in its upper reaches to more gentle in its lower reaches.

- Flow velocity and discharge are related, so that if either changes, the other changes as well.

- Rivers and streams carry runoff from their drainage basins, which are separated from one another by divides.

- Erosion by running water takes place by hydraulic action, abrasion, and dissolution of soluble substances.

- Bedload in channels is made up of sand and gravel, whereas suspended load consists of silt- and clay-sized particles. Running water also transports a dissolved load.

- The competence of a stream or river is a measure of the maximum-sized particles it can carry and is related to velocity, whereas capacity, a measure of total sediment load, is related to discharge.

- Braided waterways have a complex of dividing and rejoining channels, and their deposits are mostly sheets of sand and gravel.

- A single sinuous channel is typical of meandering streams that deposit mostly mud with subordinate point bar deposits of sand.

- The broad, flat floodplains adjacent to channels are commonly the sites of oxbow lakes, which are simply abandoned meanders.

- An alluvial deposit at a river's mouth is a delta. Some deltas conform to the three-part division of bottomset, foreset, and topset beds, but large marine deltas are much more complex and characterized as stream/river-, wave-, or tide-dominated.

- Alluvial fans are lobate deposits of sand and gravel on land that form best in semiarid regions. They form mostly by deposition from running water, but debris flows are also important.

- Sea level is ultimate base level, the lowest level to which streams or rivers can erode. However, they also have local base levels such as lakes or where they flow across resistant rocks.

- Graded streams tend to eliminate irregularities in their channels so they develop a smooth, concave profile of equilibrium. In these channels all parameters are in balance so that little or no erosion or deposition takes place.

- A combination of processes, including downcutting, lateral erosion, sheet wash, mass wasting, and headward erosion, are responsible for the origin and evolution of valleys.

- Stream terraces and incised meanders usually form when a stream or river that was formerly in equilibrium begins a new episode of downcutting.

Important Terms

abrasion (p. 442)
alluvial fan (p. 451)
alluvium (p. 444)
base level (p. 455)
bed load (p. 443)
braided stream (p. 444)
delta (p. 449)
discharge (p. 441)
dissolved load (p. 443)
divide (p. 455)

drainage basin (p. 455)
drainage pattern (p. 455)
floodplain (p. 447)
graded stream (p. 457)
gradient (p. 440)
hydraulic action (p. 442)
hydrologic cycle (p. 438)
incised meander (p. 462)
infiltration capacity (p. 439)
meandering stream (p. 444)

natural levee (p. 448)
oxbow lake (p. 447)
point bar (p. 446)
runoff (p. 439)
stream terrace (p. 460)
superposed stream (p. 463)
suspended load (p. 443)
valley (p. 460)
velocity (p. 440)

Review Questions

1. The discharge of a stream or river is:
    a. _____ the quantity of water moving past a specific place in a given amount of time; b. _____ how fast water moves around the outside of a meander; c. _____ the distance water flows from its source to the ocean; d. _____ the amount of water lost to evaporation and infiltration; e. _____ a measure of its total load of sand, gravel, and dissolved materials.

2. Running water carrying sand and gravel effectively erodes by:
    a. _____ solution; b. _____ piracy;
    c. _____ abrasion; d. _____ deposition;
    e. _____ sheetwash.

3. A stream characterized as braided is one that has:
    a. _____ a single sinuous channel; b. _____ a deep, narrow valley; c. _____ numerous point bar deposits; d. _____ a delta dominated by wave action; e. _____ multiple interconnected channels.

4. In which one of the following areas would radial drainage likely develop?
    a. _____ composite volcano; b. _____ alluvial fan; c. _____ point bar; d. _____ oxbow lake;
    e. _____ terrace.

5. The bedload of a stream is made up of:
    a. _____ dissolved materials and organic matter; b. _____ clay, silt, and runoff; c. _____ alluvial fans and deltas; d. _____ sand and gravel; e. _____ all materials in solution.

6. The lowest level to which a stream or river can erode is known as:
    a. _____ base level; b. _____ incised piracy;
    c. _____ dendritic drainage; d. _____ natural levee; e. _____ gradient.

7. Erosion by the direct impact of water is:
    a. _____ point bar collapse; b. _____ capacity;
    c. _____ hydraulic action; d. _____ superposition; e. _____ alluvial disruption.

8. The topographically high area between adjacent drainage basins is a:
   a. _____ profile of equilibrium; b. _____ divide; c. _____ floodplain; d. _____ delta; e. _____ discharge.

9. An erosional remnant of a floodplain higher than a stream's current floodplain is a(n):
   a. _____ lateral accretion deposit; b. _____ terrace; c. _____ incised meander; d. _____ point bar; e. _____ divide.

10. A narrow, deep valley is known as a(n):
    a. _____ canyon; b. _____ gully; c. _____ gorge; d. _____ alluvium; e. _____ graded stream.

11. Why is Earth the only planet in the solar system with abundant liquid water?

12. Calculate the *daily discharge* for a river 148 m wide and 2.6 m deep with a flow velocity of 0.3 m/sec.

13. Describe how a point bar and a natural levee form.

14. How is it possible for a meandering stream to erode laterally and yet maintain a more or less constant channel width? A diagram would be helpful.

15. A river 2000 m above sea level flows 1500 km to the ocean. What is its gradient? Do you think your calculated gradient is valid for all segments of this river? Explain.

16. How do alluvial fans and deltas compare and differ?

17. Explain how a stream can lengthen its channel at both its upper and lower ends.

18. About 10.75 km³ of sediment erodes from the continents annually, and the volume of the continents above sea level is 93,000,000 km³. Thus the continents should erode to sea level in just a little more than 8,600,000 years. The erosion rate and volume of the continents are reasonably accurate, but something is seriously wrong with the assumption that the continents would be leveled in so short a time. Explain.

19. Explain the concept of a graded stream, and describe conditions that might upset the graded condition.

20. The steeper a stream or river's gradient, the greater the flow velocity. Right? Explain.

World Wide Web Activities

Groundwater

16

CHAPTER 16
OUTLINE

PHYSICAL
Geology⇌Now *This icon, appearing throughout the book, indicates an opportunity to explore interactive tutorials, animations, or practice problems available on the Physical GeologyNow Web site at http://earthscience.brookscole.com/physgeo5e.*

OBJECTIVES

At the end of this chapter, you will have learned that

- Groundwater is one reservoir of the hydrologic cycle and represents approximately 22% of the world's supply of freshwater.

- Porosity and permeability are largely responsible for the amount, availability, and movement of groundwater.

- The water table separates the zone of aeration from the underlying zone of saturation and is a subdued replica of the overlying land surface.

- Groundwater moves downward due to the force of gravity.

- An artesian system is one in which groundwater is confined and builds up high hydrostatic pressure.

- Groundwater is an important agent of both erosion and deposition and is responsible for karst topography and a variety of cave features.

- Modifications of the groundwater system may result in lowering of the water table, saltwater incursion, subsidence, and contamination.

- Hot springs and geysers result when groundwater is heated, typically in regions of recent volcanic activity.

- Geothermal energy is a desirable and relatively nonpolluting alternative form of energy.

A variety of cave deposits are present in Mammoth Cave, Kentucky.
Source: David Muench/Corbis

Introduction

Within the limestone region of western Kentucky lies the largest cave system in the world. In 1941 approximately 51,000 acres were set aside and designated as Mammoth Cave National Park. In 1981 it became a World Heritage Site.

From ground level, the topography of the area is unimposing with gently rolling hills. Beneath the surface, however, are more than 540 km of interconnected passageways whose spectacular geologic features have been enjoyed by numerous cave explorers and tourists.

During the War of 1812, approximately 180 metric tons of saltpeter, used in the manufacture of gunpowder, were mined from Mammoth Cave. At the end of the war, the saltpeter market collapsed and Mammoth Cave was developed as a tourist attraction, easily overshadowing the other caves in the area. During the next 150 years, the discovery of new passageways and caverns helped establish Mammoth Cave as the world's premier cave and the standard against which all others were measured.

The formation of the caves themselves began about 3 million years ago when groundwater began dissolving the region's underlying St. Genevieve Limestone to produce a complex network of openings, passageways, and huge chambers that constitute present-day Mammoth Cave. Flowing through the various caverns is the Echo River, a system of streams that eventually joins the Green River at the surface.

The colorful cave deposits are the primary reason millions of tourists have visited Mammoth Cave over the years. Hanging down from the ceiling and growing up from the floor are spectacular icicle-like structures as well as columns and curtains in a variety of colors (see chapter opening photo). In addition, intricate passageways connect various-sized rooms. The cave is also home to more than 200 species of insects and other animals, including about 45 blind species.

In addition, to the beautiful caves, caverns, and cave deposits produced by groundwater movement, groundwater is an important source of freshwater for agriculture, industry, and domestic users. More than 65% of the groundwater used in the United States each year goes for irrigation, with industrial use second, followed by domestic needs. These demands have severely depleted the groundwater supply in many areas and led to such problems as ground subsidence and saltwater contamination. In other areas, pollution from landfills, toxic waste, and agriculture has rendered the groundwater supply unsafe.

As the world's population and industrial development expand, the demand for water, particularly groundwater, will increase. Not only must new groundwater sources be located, but once found, these sources must be protected from pollution and managed properly to ensure that users do not withdraw more water than can be replenished. It is therefore important that people become aware of what a valuable resource groundwater is, so that they can assure future generations of a clean and adequate supply of this water source.

GROUNDWATER AND THE HYDROLOGIC CYCLE

Groundwater, water that fills open spaces in rocks, sediment, and soil beneath the surface, is one reservoir in the hydrologic cycle, representing approximately 22% (8.4 million km^3) of the world's supply of freshwater (see Figure 15.4). Like all other water in the hydrologic cycle, the ultimate source of groundwater is the oceans, but its more immediate source is the precipitation that infiltrates the ground and seeps down through the voids in soil, sediment, and rocks. Groundwater may also come from water infiltrating from streams, lakes, swamps, artificial recharge ponds, and water treatment systems.

Regardless of its source, groundwater moving through the tiny openings between soil and sediment particles and the spaces in rocks filters out many impurities such as disease-causing microorganisms and many pollutants. Not all soils and rocks are good filters, though, and sometimes so much undesirable material may be present that it contaminates the groundwater. Groundwater movement and its recovery at wells depend on two critical aspects of the materials it moves through: *porosity* and *permeability*.

HOW DO EARTH MATERIALS ABSORB WATER?

Porosity and permeability are important physical properties of Earth materials and are largely responsible for the amount, availability, and movement of groundwater. Water soaks into the ground because soil, sediment, and rock have open spaces or pores. **Porosity** is simply the percentage of a material's total volume that is pore space. Porosity most often consists of the spaces between particles in soil, sediments, and sedimentary rocks, but other types of porosity include cracks, fractures, faults, and vesicles in volcanic rocks (■ Figure 16.1).

Porosity varies among different rock types and depends on the size, shape, and arrangement of the mate-

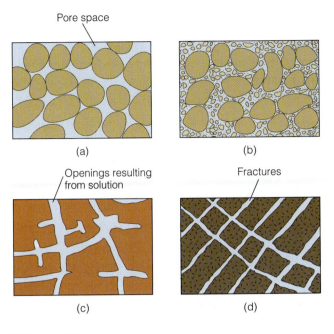

Pore space

(a)

(b)

Openings resulting from solution

Fractures

(c)

(d)

■ **Figure 16.1**

A rock's porosity depends on the size, shape, and arrangement of the material composing the rock. (a) A well-sorted sedimentary rock has high porosity, whereas (b) a poorly sorted one has lower porosity. (c) In soluble rocks such as limestone, porosity can be increased by solution, whereas (d) crystalline metamorphic and igneous rocks can be rendered porous by fracturing. Source: Modified from *U.S. News and World Report* (8 March 1991): 72–73.

rial composing the rock (Table 16.1). Most igneous and metamorphic rocks, as well as many limestones and dolostones, have very low porosity because they consist of tightly interlocking crystals. Their porosity can be increased, however, if they have been fractured or weathered by groundwater. This is particularly true for massive limestone and dolostone whose fractures can be enlarged by acidic groundwater.

By contrast, detrital sedimentary rocks composed of well-sorted and well-rounded grains can have high porosity because any two grains touch at only a single point, leaving relatively large open spaces between the grains (Figure 16.1a). Poorly sorted sedimentary rocks, on the other hand, typically have low porosity because smaller grains fill in the spaces between the larger grains, further reducing porosity (Figure 16.1b). In addition, the amount of cement between grains can decrease porosity.

Porosity determines the amount of groundwater Earth materials can hold, but it does not guarantee that the water can be easily extracted as well. So in addition to being porous, Earth materials must have the capacity to transmit fluids, a property known as **permeability**. Thus both porosity and permeability play important roles in groundwater movement and recovery. Permeability is dependent not only on porosity but also on the size of the pores or fractures and their interconnections. For example, deposits of silt or clay are typically more porous than sand or gravel, but they have low permeability because pores between the particles are very small, and molecular attraction between the particles and water is great, thereby preventing movement of the water. In contrast, pore spaces between grains in sandstone and conglomerate are much larger, and molecular attraction on the water is therefore low. Chemical and biochemical sedimentary rocks, such as limestone and dolostone, and many igneous and metamorphic rocks that are highly fractured can also be very permeable if the fractures are interconnected.

The contrasting porosity and permeability of familiar substances are well demonstrated by sand versus clay. Pour some water on sand and it rapidly sinks in, whereas water poured on clay simply remains on the surface. Furthermore, wet sand dries quickly, but once clay absorbs water, it may take days to dry out because of its low permeability. Neither sand nor clay makes a good substance in which to grow crops or gardens, but a mixture of the two plus some organic matter in the form of humus makes an excellent soil for farming and gardening (see Chapter 5).

A permeable layer transporting groundwater is an *aquifer*, from the Latin *aqua*, "water." The most effective aquifers are deposits of well-sorted and well-rounded sand and gravel. Limestones in which fractures and bedding planes have been enlarged by solution are also good aquifers. Shales and many igneous and metamorphic rocks make poor aquifers because they are typically impermeable unless fractured. Rocks such as these and any other materials that prevent the movement of groundwater are *aquicludes*.

Table 16.1

Porosity Values for Different Materials

Material	Percentage Porosity
Unconsolidated sediment	
Soil	55
Gravel	20–40
Sand	25–50
Silt	35–50
Clay	50–70
Rocks	
Sandstone	5–30
Shale	0–10
Solution activity in limestone, dolostone	10–30
Fractured basalt	5–40
Fractured granite	10

Source: U.S. Geological Survey, Water Supply Paper 2220 (1983) and others.

WHAT IS THE WATER TABLE?

Some of the precipitation on land evaporates, and some enters streams and returns to the oceans by surface runoff; the remainder seeps into the ground. As this water moves down from the surface, a small amount adheres to the material it moves through and halts its downward progress. With the exception of this *suspended water,* however, the rest seeps farther downward and collects until it fills all the available pore spaces. Thus two zones are defined by whether their pore spaces contain mostly air, the **zone of aeration,** and the underlying **zone of saturation.** The surface separating these two zones is the **water table** (■ Figure 16.2). The base of the zone of saturation varies from place to place but usually extends to a depth where an impermeable layer is encountered or to a depth where confining pressure closes all open space. Extending irregularly upward a few centimeters to several meters from the zone of saturation is the *capillary fringe.* Water moves upward in this region because of surface tension, much as water moves upward through a paper towel.

In general, the configuration of the water table is a subdued replica of the overlying land surface; that is, it rises beneath hills and has its lowest elevations beneath valleys. Several factors contribute to the surface configuration of a region's water table. These include regional differences in amount of rainfall, permeability, and rate of groundwater movement. During periods of high rain-

fall, groundwater tends to rise beneath hills because it cannot flow fast enough into adjacent valleys to maintain a level surface. During droughts, the water table falls and tends to flatten out because it is not being replenished. In arid and semiarid regions, the water table is usually quite flat regardless of the overlying land surface.

PHYSICAL Geology⇌Now   Click Geology Interactive to work through an activity on Groundwater Basics through Groundwater.

HOW DOES GROUNDWATER MOVE?

Gravity provides the energy for the downward movement of groundwater. Water entering the ground moves through the zone of aeration to the zone of saturation (■ Figure 16.3). When water reaches the water table, it continues to move through the zone of saturation from areas where the water table is high toward areas where it is lower, such as streams, lakes, or swamps. Only some of the water follows the direct route along the slope of the water table. Most of it takes longer curving paths down and then enters a stream, lake, or swamp from below, because it moves from areas of high pressure toward areas of lower pressure within the saturated zone.

Groundwater velocity varies greatly and depends on many factors. Velocities range from 250 m per day in

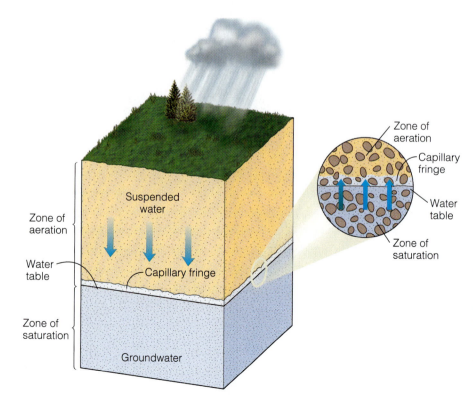

■ Figure 16.2

The zone of aeration contains both air and water within its open spaces, whereas all open spaces in the zone of saturation are filled with groundwater. The water table is the surface separating the zones of aeration and saturation. Within the capillary fringe, water rises by surface tension from the zone of saturation into the zone of aeration.

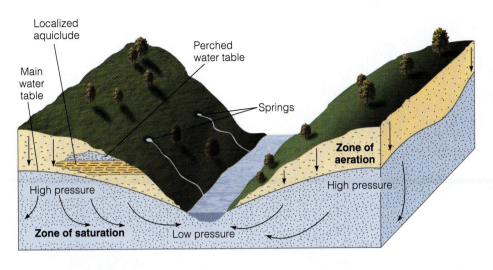

Localized
aquiclude

Perched
water table

Main
water
table

Springs

**Zone of
aeration**

High pressure

High pressure

Zone of saturation

Low pressure

■ **Figure 16.3**

Groundwater moves down through the zone of aeration to the zone of saturation. Then some of it moves along the slope of the water table, and the rest moves through the zone of saturation from areas of high pressure toward areas of low pressure. Some water might collect over a local aquiclude, such as a shale layer, thus forming a perched water table.

some extremely permeable material to less than a few centimeters per year in nearly impermeable material. In most ordinary aquifers, the average velocity of groundwater is a few centimeters per day.

WHAT ARE SPRINGS, WATER WELLS, AND ARTESIAN SYSTEMS?

You can think of the water in the zone of saturation much like a reservoir whose surface rises or falls depending on additions as opposed to natural and artificial withdrawals. *Recharge*—that is, additions to the zone of saturation—comes from rainfall or melting snow, or water might be added artificially at wastewater-treatment plants or recharge ponds constructed for just this purpose. But if groundwater is discharged naturally or withdrawn at wells without sufficient recharge, the water table drops, just as a savings account diminishes if withdrawals exceed deposits. Withdrawals from the groundwater system take place where groundwater flows laterally into streams, lakes, or swamps, where it discharges at the surface as *springs,* and where it is withdrawn from the system at water wells.

Springs

Places where groundwater flows or seeps out of the ground as **springs** have always fascinated people. The water flows out of the ground for no apparent reason and from no readily identifiable source. So it is not sur-

prising that springs have long been regarded with superstition and revered for their supposed medicinal value and healing powers. Nevertheless, there is nothing mystical or mysterious about springs.

Although springs can occur under a wide variety of geologic conditions, they all form in basically the same way (■ Figure 16.4a). When percolating water reaches the water table or an impermeable layer, it flows laterally, and if this flow intersects the surface, the water discharges as a spring (Figure 16.4b). The Mammoth Cave area in Kentucky is underlain by fractured limestones whose fractures have been enlarged into caves by solution activity (see the chapter opening photo). In this geologic environment, springs occur where the fractures and caves intersect the ground surface, allowing groundwater to exit onto the surface. Most springs are along valley walls where streams have cut valleys below the regional water table.

Springs can also develop wherever a perched water table intersects the surface (Figure 16.3). A *perched water table* may occur wherever a local aquiclude is present within a larger aquifer, such as a lens of shale within sandstone. As water migrates through the zone of aeration, it is stopped by the local aquiclude, and a localized zone of saturation "perched" above the main water table forms. Water moving laterally along the perched water table may intersect the surface to produce a spring.

Water Wells

Water wells are simply openings made by drilling or digging down into the zone of saturation. Once the zone of saturation has been penetrated, water percolates into the well, filling it to the level of the water table. A few wells are free flowing (see the next section), but for most the water must be brought to the surface by pumping. In some parts of the world, water is raised to the surface with nothing more than a bucket on a rope or a hand-operated pump. In many parts of the United States and Canada, one can see windmills from times past that employed wind power to pump water. Most of these are no longer in use, having been replaced by more efficient electric pumps (■ Figure 16.5).

When groundwater is pumped from a well, the water table in the area around the well is lowered, forming a

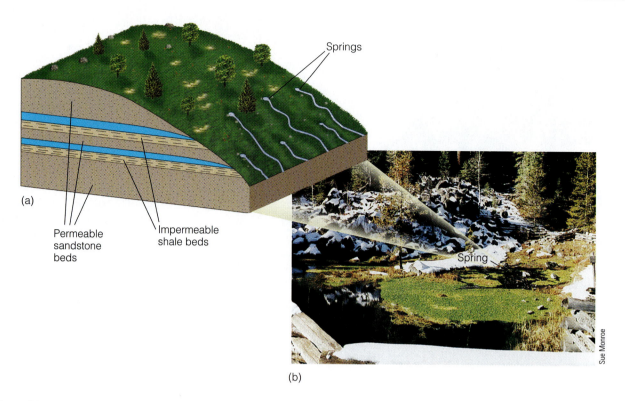

(a)

Springs

Permeable
sandstone
beds

Impermeable
shale beds

Spring

Sue Monroe

(b)

■ **Figure 16.4**

Springs form wherever laterally moving groundwater intersects Earth's surface. (a) Most commonly, springs form when percolating water reaches an impermeable layer and migrates laterally until it seeps out at the surface. (b) This spring issues from rocks at the base of the mountain in the background.

(a)

(b)

■ **Figure 16.5**

(a) During the past, windmills like this one were used extensively to pump water to the surface. This one still functions and supplies the water for the small cemetery in the foreground. (b) Electric pumps such as this one in a farmer's field have largely replaced windmills.

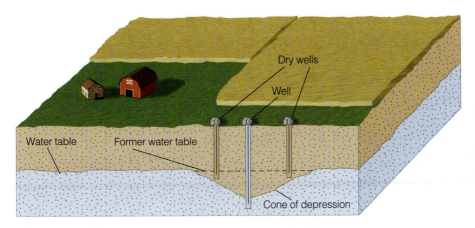

■ Figure 16.6

A cone of depression forms whenever water is withdrawn from a well. If water is withdrawn faster than it can be replenished, the cone of depression will grow in depth and circumference, lowering the water table in the area and causing nearby shallow wells to go dry.

cone of depression. A cone of depression forms because the rate of water withdrawal from the well exceeds the rate of water inflow to the well, lowering the water table around the well. This lowering normally does not pose a problem for the average domestic well, provided that the well is drilled deep enough into the zone of saturation. The tremendous amounts of water used by industry and for irrigation, however, may create a large cone of depression that lowers the water table sufficiently to cause shallow wells in the immediate area to go dry (■ Figure 16.6). This situation is not uncommon and frequently results in lawsuits by the owners of the shallow dry wells. Furthermore, lowering of the regional water table is becoming a serious problem in many areas, particularly in the southwestern United States where rapid growth has placed tremendous demands on the groundwater system. Unrestricted withdrawal of groundwater cannot continue indefinitely, and the rising costs and decreasing supply of groundwater should soon limit the growth of this region of the United States.

PHYSICAL Geology ⇌ Now Click Geology Interactive to work through an activity on Tapping the Ground plus Potentiometric Surface through Groundwater.

Artesian Systems

The word *artesian* comes from the French town and province of Artois (called Artesium during Roman times) near Calais, where the first European artesian well was drilled in 1126 and is still flowing today. The term **artesian system** can be applied to any system in which groundwater is confined and builds up high hydrostatic (fluid) pressure. Water in such a system is able to rise above the level of the aquifer if a well is drilled through the confining layer, thereby reducing the pressure and forcing the water upward. For an artesian system to develop, three geologic conditions must be present (■ Figure 16.7): (1) The aquifer must be confined above and below by aquicludes to prevent water from escaping; (2) the rock sequence is usually tilted and exposed at the surface, enabling the aquifer to be

recharged; and (3) there is sufficient precipitation in the recharge area to keep the aquifer filled.

The elevation of the water table in the recharge area and the distance of the well from the recharge area determine the height to which artesian water rises in a well. The surface defined by the water table in the recharge area, called the *artesian-pressure surface*, is indicated by the sloping dashed line in Figure 16.7. If there were no friction in the aquifer, well water from an artesian aquifer would rise exactly to the elevation of the artesian-pressure surface. Friction, however, slightly reduces the pressure of the aquifer water and consequently the level to which artesian water rises. This is why the pressure surface slopes.

An artesian well will flow freely at the ground surface only if the wellhead is at an elevation below the artesian-pressure surface. In this situation, the water flows out of the well because it rises toward the artesian-pressure surface, which is at a higher elevation than the wellhead. In a nonflowing artesian well, the wellhead is above the artesian-pressure surface, and the water will rise in the well only as high as the artesian-pressure surface.

In addition to artesian wells, many artesian springs exist. Such springs form if a fault or fracture intersects the confined aquifer, allowing water to rise above the aquifer. Oases in deserts are commonly artesian springs.

Because the geologic conditions necessary for artesian water can occur in a variety of ways, artesian systems are common in many areas of the world underlain by sedimentary rocks. One of the best-known artesian systems in the United States underlies South Dakota and extends southward to central Texas. The majority of the artesian water from this system is used for irrigation. The aquifer of this artesian system, the Dakota Sandstone, is recharged where it is exposed along the margins of the Black Hills of South Dakota. The hydrostatic pressure in this system was originally great enough to produce free-flowing wells and to operate waterwheels. The extensive use of water for irrigation over the years has reduced the pressure in many of the wells so that they are no longer free-flowing and the water must be pumped.

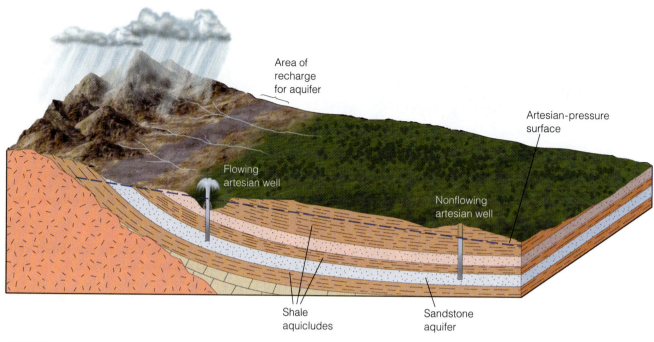

PHYSICAL
Geology⇌Now ■ **Active Figure 16.7**

An artesian system must have an aquifer confined above and below by aquicludes, the aquifer must be exposed at the surface, and there must be sufficient precipitation in the recharge area to keep the aquifer filled. The elevation of the water table in the recharge area, which is indicated by a sloping dashed line (the artesian-pressure surface), defines the highest level to which well water can rise. If the elevation of a wellhead is below the elevation of the artesian-pressure surface, the well will be free-flowing because the water will rise toward the artesian-pressure surface, which is at a higher elevation than the wellhead. If the elevation of a wellhead is at or above that of the artesian-pressure surface, the well will be nonflowing.

Another example of an important artesian system is the Floridan aquifer system. Here Tertiary-aged carbonate rocks are riddled with fractures, caves, and other openings that have been enlarged and interconnected by solution activity. These carbonate rocks are exposed at the surface in the northwestern and central parts of the state where they are recharged, and they dip toward both the Atlantic and Gulf coasts where they are covered by younger sediments. The carbonates are interbedded with shales, forming a series of confined aquifers and aquicludes. This artesian system is tapped in the southern part of the state where it is an important source of freshwater and one that is being rapidly depleted.

One final comment on artesian wells/springs and the water derived from them. It is not unusual for advertisers to tout the quality of artesian water as somehow superior to other groundwater. Some artesian water might in fact be of excellent quality, but its quality is not dependent on the fact that water rises above the surface of an aquifer. Rather, its quality is a function of dissolved minerals and any introduced substances, so it really is no different from any other groundwater. The myth of its superiority probably arises from the fact that people have always been fascinated by water that flows freely from the ground.

What Would You Do

Your well periodically goes dry and never produces the amount of water you would like or need. Is your well too shallow, is it drilled into Earth materials with low permeability, or are nearby pumps used for irrigation causing your problems? What would you do to resolve this problem?

HOW DOES GROUNDWATER ERODE AND DEPOSIT MATERIAL?

When rainwater begins seeping into the ground, it immediately starts to react with the minerals it contacts, weathering them chemically. In an area underlain by soluble

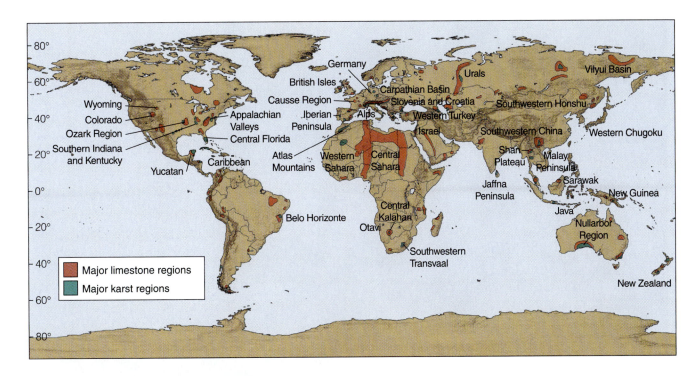

■ Figure 16.8

The distribution of the major limestone and karst areas of the world.

rock, groundwater is the principal agent of erosion and is responsible for the formation of many important features of the landscape.

Limestone, a common sedimentary rock composed primarily of the mineral calcite ($CaCO_3$), underlies large areas of Earth's surface (■ Figure 16.8). Although limestone is practically insoluble in pure water, it readily dissolves if a small amount of acid is present. Carbonic acid (H_2CO_3) is a weak acid that forms when carbon dioxide combines with water ($H_2O + CO_2 \rightarrow H_2CO_3$) (see Chapter 5). Because the atmosphere contains a small amount of carbon dioxide (0.03%) and carbon dioxide is also produced in soil by the decay of organic matter, most groundwater is slightly acidic. When groundwater percolates through the various openings in limestone, the slightly acidic water readily reacts with the calcite to dissolve the rock by forming soluble calcium bicarbonate, which is carried away in solution (see Chapter 5).

Sinkholes and Karst Topography

In regions underlain by soluble rock, the ground surface may be pitted with numerous depressions that vary in size and shape. These depressions, called **sinkholes** or merely *sinks,* mark areas with underlying soluble rock (■ Figure 16.9). Most sinkholes form in one of two ways. The first is when soluble rock below the soil is dissolved by seeping water, and openings in the rock are enlarged and filled in by the overlying soil. As the groundwater continues to dissolve the rock, the soil is eventually removed, leaving shallow depressions with gently sloping sides. When adjacent sinkholes merge, they form a network of larger, irregular, closed depressions called *solution valleys.*

Sinkholes also form when a cave's roof collapses, usually producing a steep-sided crater. Sinkholes formed in this way are a serious hazard, particularly in populated areas. In regions prone to sinkhole formation, the depth and extent of underlying cave systems must be mapped before any development to ensure that the underlying rocks are thick enough to support planned structures.

Karst topography, or simply *karst,* develops largely by groundwater erosion in many areas underlain by soluble rocks (■ Figure 16.10). The name *karst* is derived from the plateau region of the border area of Slovenia, Croatia, and northeastern Italy where this type of topography is well developed. In the United States, regions of karst topography include large areas of southwestern Illinois, southern Indiana, Kentucky, Tennessee, northern Missouri, Alabama, and central and northern Florida (Figure 16.8).

Karst topography is characterized by numerous caves, springs, sinkholes, solution valleys, and disappearing streams (Figure 16.10). *Disappearing streams* are so named because they typically flow only a short distance at the surface and then disappear into a sinkhole. The water continues flowing underground through various fractures or caves until it surfaces again at a spring or other stream.

Karst topography varies from the spectacular high-relief landscapes of China to the subdued and pockmarked

Frank Kujawa/University of Central Florida. GeoPhoto

(a)

James S. Monroe

(b)

■ **Figure 16.9**

(a) This sinkhole formed on May 8 and 9, 1981, in Winter Park, Florida. It formed in previously dissolved limestone following a drop in the water table. The 100-m-wide, 35-m-deep sinkhole destroyed a house, numerous cars, and a municipal swimming pool. (b) A small sinkhole in Montana now occupied by a lake. The water enters the lake from a hot spring so it remains warm year round. In fact, tropical fish have been introduced into the lake and thus live in an otherwise inhospitable climate.

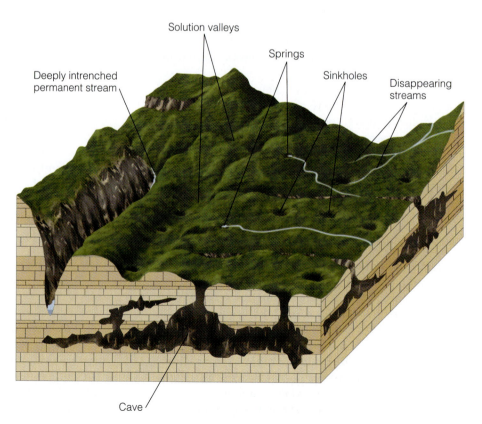

Solution valleys

Springs

Sinkholes

Disappearing streams

Deeply intrenched permanent stream

Cave

■ **Figure 16.10**

Some of the features of karst topography.

(a)

(b)

■ Figure 16.11

(a) The Stone Forest, 125 km southeast of Kunming, China, is a high-relief karst landscape formed by the dissolution of carbonate rocks.
(b) Solution valleys, sinkholes, and sinkhole lakes dominate the subdued karst topography east of Bowling Green, Kentucky.

landforms in Kentucky (■ Figure 16.11). Common to all karst topography, though, is thick-bedded, readily soluble rock at the surface or just below the soil, and enough water for solution activity to occur (see "The Burren Area of Ireland" on pages 480 and 481). Karst topography is therefore typically restricted to humid and temperate climates.

Caves and Cave Deposits

Caves are perhaps the most spectacular examples of the combined effects of weathering and erosion by groundwater. As groundwater percolates through carbonate rocks, it dissolves and enlarges original fractures and

openings to form a complex interconnecting system of crevices, caves, caverns, and underground streams. A **cave** is usually defined as a naturally formed subsurface opening that is generally connected to the surface and is large enough for a person to enter. A *cavern* is a very large cave or a system of interconnected caves.

More than 17,000 caves are known in the United States. Most of them are small, but some are large and spectacular. Some of the more famous ones are Mammoth Cave, Kentucky (see the Introduction); Carlsbad Caverns, New Mexico; Lewis and Clark Caverns, Montana; Wind Cave and Jewel Cave, South Dakota; Lehman Cave, Nevada; and Meramec Caverns, Missouri, which Jesse James and his outlaw band used as a

The Burren Area of Ireland

The Burren region in northwest county Clare, Ireland, covers more than 100 square miles and is one of the finest examples of karst topography in Europe. Although the Burren landscape is frequently referred to as lunarlike because it is seemingly barren and lifeless, it is actually teeming with life and is world-famous for its variety of vegetation. In fact, because of the heat-retention capacity of the massive limestones and the soil trapped in the vertical joints of bare limestone pavement, an extremely diverse community of plants abounds.

A map of Ireland showing the Burren region of county Clare.

The present landscape is best described as glaciated karst. Like most of Ireland, the Burren was covered by a warm, shallow sea some 340 million years ago. As much as 780 m of interbedded marine limestones and shales were deposited at this time. These marine limestones and shales were then covered by nearly 330 m of sandstones, siltstones, and shales. During the Pleistocene Epoch, glaciers stripped off most of the detrital rocks, thus exposing the underlying limestones to weathering and erosion. The main erosional agents in the Burren are rain water, which is slightly acidic, and the humic acids produced by localized vegetation. Together they have been the driving force producing the distinctive karst topography we find today.

Bare limestone pavement with a small wedge tomb from the Neolithic period—approximately 6000 years ago—in the background.

Bare limestone pavements typically display a blocky appearance. The network of vertical cracks is the result of weathering of the joint pattern produced in the limestone during uplift of the region.

A closeup shows the characteristic karren weathering pattern produced by solution of the limestone. *Karren* is used to describe the various microsolutional features of limestone pavement.

Reed Wicander

Despite a lack of significant soil cover, a profusion of plants live in the soil trapped in the cracks, joints, and solution cavities of the limestone beds in the Burren.

Perhaps the most famous Irish Neolithic dolmen is the Poulnabrone portal tomb, which dates from about 5800 years ago. The name Poulnabrone literally means "the hole of the sorrows." The capstone of the Poulnabrone dolmen rests on two 1.8-m-high portal stones to create a chamber in a low circular cairn, 9 m in diameter.

A typical small wedge tomb. The Burren is famous for its monuments, characteristic of every period from the Neolithic to the present. The legacy of early settlers is the many Neolithic tombs as well as stone structures and walls.

Reed Wicander

Excavation of the Poulnabrone dolmen revealed that the tomb contained the remains of up to 22 people, buried over a period of six centuries, as well as a polished stone axe, two stone disc beads, flint and chert arrowheads and scrapers, and more than 60 shards of pottery.

Reed Wicander

Reed Wicander

A Burren limestone wall. Burren limestone has been used for centuries in the construction of tombs, forts, houses, and walls. Today it is in demand for limestone-finished houses, walls, and garden features.

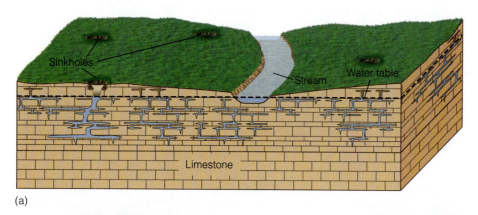

(a)

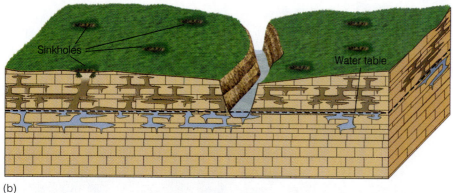

(b)

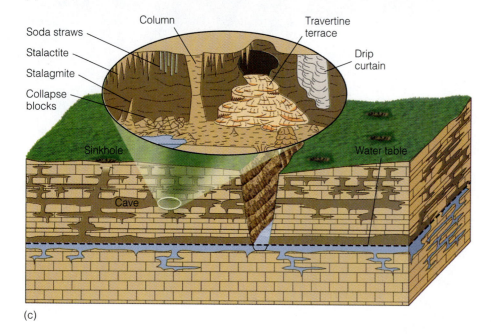

(c)

■ Figure 16.12

The formation of caves. (a) As groundwater percolates through the zone of aeration and flows through the zone of saturation, it dissolves the carbonate rocks and gradually forms a system of passageways. (b) Groundwater moves along the surface of the water table, forming a system of horizontal passageways through which dissolved rock is carried to the surface streams, thus enlarging the passageways. (c) As the surface streams erode deeper valleys, the water table drops, and the abandoned channelways form an interconnecting system of caves and caverns.

hideout. The United States has many famous caves, but Canada also has many caves, including 536-m-deep Arc-

tomys Cave in Mount Robson Provincial Park, British Columbia, the deepest known cave in North America.

Caves and caverns form as a result of the dissolution of carbonate rocks by weakly acidic groundwater (■ Figure 16.12). Groundwater percolating through the zone of aeration slowly dissolves the carbonate rock and enlarges its fractures and bedding planes. On reaching the water table, groundwater migrates toward the region's surface streams. As the groundwater moves through the zone of saturation, it continues to dissolve the rock and gradually forms a system of horizontal passageways through which the dissolved rock is carried to the streams. As the surface streams erode deeper valleys, the water table drops in response to the lower elevation of the streams. The water that flowed through the system of horizontal passageways now percolates to the lower water table where a new system of passageways begins to form. The abandoned channelways form an interconnecting system of caves and caverns. Caves eventually become unstable and collapse, littering the floor with fallen debris.

When most people think of caves, they think of the seemingly endless variety of colorful and bizarre-shaped deposits found in them. Although a great many different types of cave deposits exist, most form in essentially the same manner and are collectively known as *dripstone*. As water seeps into a cave, some of the dissolved carbon dioxide in the water escapes, and a small amount of calcite is precipitated. In this manner, the various dripstone deposits are formed.

Stalactites are icicle-shaped structures hanging from cave ceilings that form as a result of precipitation from

Stone/Getty Images

■ **Figure 16.13**

Stalactites are the icicle-shaped structures hanging from the cave's ceiling, whereas the upward-pointing structures on the floor are stalagmites. Columns result when stalactites and stalagmites meet. All three structures are visible in Buchan Cave, Victoria, Australia.

Lowering the Water Table

Withdrawing groundwater at a significantly greater rate than it is replaced by either natural or artificial recharge can have serious effects. For example, the High Plains aquifer is one of the most important aquifers in the United States. It underlies more than 450,000 km², including most of Nebraska, large parts of Colorado and Kansas, portions of South Dakota, Wyoming, and New Mexico, as well as the panhandle regions of Oklahoma and Texas, and accounts for approximately 30% of the groundwater used for irrigation in the United States (■ Figure 16.14). Irrigation from the High Plains aquifer is largely responsible for the region's agricultural productivity, which includes a significant percentage of the nation's corn, cotton, and wheat, and half of U.S. beef cattle. Large areas of land (about 18 million acres) are currently irrigated with water pumped from the High Plains aquifer. Irrigation is popular because yields from irrigated lands can be triple what they would be without irrigation.

dripping water (■ Figure 16.13). With each drop of water, a thin layer of calcite is deposited over the previous layer, forming a cone-shaped projection that grows down from the ceiling.

The water that drips from a cave's ceiling also precipitates a small amount of calcite when it hits the floor. As additional calcite is deposited, an upward-growing projection called a *stalagmite* forms (see Geo-Focus 8.1; Figure 16.13). If a stalactite and stalagmite meet, they form a *column*. Groundwater seeping from a crack in a cave's ceiling may form a vertical sheet of rock called a *drip curtain*, while water flowing across a cave's floor may produce *travertine terraces* (Figure 16.12).

HOW DO HUMANS IMPACT THE GROUNDWATER SYSTEM?

Groundwater is a valuable natural resource that is rapidly being exploited with little regard to the effects of overuse and misuse. Currently about 20% of all water used in the United States is groundwater. This percentage is rapidly increasing, and unless this resource is used more wisely, sufficient amounts of clean groundwater will not be available in the future. Modifications of the groundwater system may have many consequences, including (1) lowering of the water table, causing wells to dry up; (2) loss of hydrostatic pressure, causing once free-flowing wells to require pumping; (3) saltwater incursion; (4) subsidence; and (5) contamination.

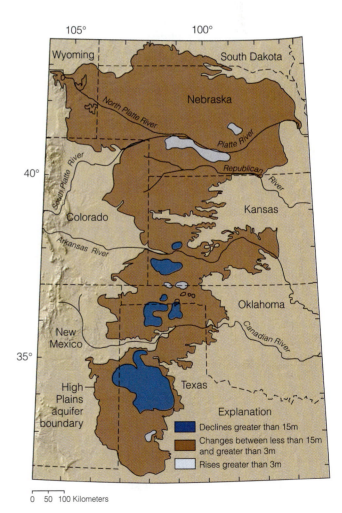

Explanation
■ Declines greater than 15m
■ Changes between less than 15m and greater than 3m
■ Rises greater than 3m

0 50 100 Kilometers

■ **Figure 16.14**

Geographic extent of the High Plains aquifer and changes in water level from predevelopment through 1993. Source: From J. B. Weeks et al., *U.S. Geological Survey Professional Paper* 1400-A, 1988.

Although the High Plains aquifer has contributed to the high productivity of the region, it cannot continue to provide the quantities of water that it has in the past. In some parts of the High Plains, from 2 to 100 times more water is being pumped annually than is being recharged. Consequently, water is being removed from the aquifer faster than it is being replenished, causing the water table to drop significantly in many areas (Figure 16.14).

What will happen to this region's economy if long-term withdrawal of water from the High Plains aquifer greatly exceeds its recharge rate so that it can no longer supply the quantities of water necessary for irrigation? Solutions range from going back to farming without irrigation to diverting water from other regions such as the Great Lakes. Farming without irrigation would result in greatly decreased yields and higher costs and prices for agricultural products. The diversion of water from elsewhere would cost billions of dollars and the price of agricultural products would still rise.

Another excellent example of what we might call deficit spending with regard to groundwater took place in California during the drought of 1987 to 1992. During that time, the state's aquifers were overdrawn at the rate of 10 million acre/feet per year (an acre/foot is the amount of water that covers 1 acre 1 foot deep). In short, during each year of the drought California was withdrawing 12 km^3 of groundwater more than was being replaced. Unfortunately, excessive depletion of the groundwater reservoir has other consequences, such as subsidence, which involves sinking or settling of the ground surface (discussed in a later section).

Water supply problems certainly exist in many areas, but on the positive side, water use in the United States actually declined during the five years following 1980 and has remained nearly constant since then, even though the population has increased. This downturn in demand resulted largely from improved techniques in irrigation, more efficient industrial water use, and a general public awareness of water problems coupled with conservation practices. Nevertheless, the rates of withdrawal of groundwater from some aquifers still exceed their rate of recharge, and population growth in the arid to semiarid Southwest is putting greater demands on an already limited water supply.

Saltwater Incursion

The excessive pumping of groundwater in coastal areas can result in *saltwater incursion* such as occurred on Long Island, New York, during the 1960s. Along coastlines where permeable rocks or sediments are in contact with the ocean, the fresh groundwater, being less dense than seawater, forms a lens-shaped body above the underlying saltwater (■ Figure 16.15a). The weight of the freshwater exerts pressure on the underlying saltwater. As long as rates of recharge equal rates of with-

drawal, the contact between the fresh groundwater and the seawater will remain the same. If excessive pumping occurs, however, a deep cone of depression forms in the fresh groundwater (Figure 16.15b). Because some of the pressure from the overlying freshwater has been removed, saltwater forms a *cone of ascension* as it rises to fill the pore space that formerly contained freshwater. When this occurs, wells become contaminated with saltwater and remain contaminated until recharge by freshwater restores the former level of the fresh-groundwater water table.

Saltwater incursion is a major problem in many rapidly growing coastal communities. As the population

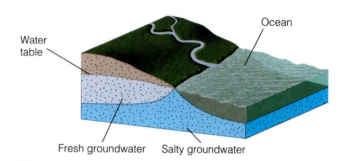
(a)

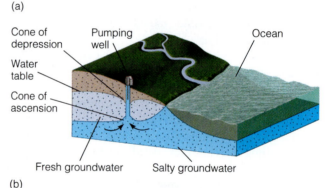
(b)

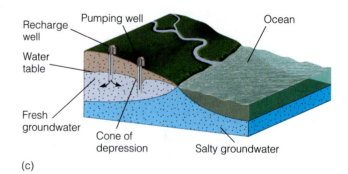
(c)

■ **Figure 16.15**

Saltwater incursion. (a) Because freshwater is not as dense as saltwater, it forms a lens-shaped body above the underlying saltwater. (b) If excessive pumping occurs, a cone of depression develops in the fresh groundwater, and a cone of ascension forms in the underlying salty groundwater that may result in saltwater contamination of the well. (c) Pumping water back into the groundwater system through recharge wells can help lower the interface between the fresh groundwater and the salty groundwater and reduce saltwater incursion.

in these areas grows, greater demand for groundwater creates an even greater imbalance between recharge and withdrawal.

Not only is saltwater incursion a major concern for some coastal communities, it is also becoming a problem in the Salinas Valley, California, which produces fruits and vegetables valued at about $1.7 billion annually. Here, in an area encompassing about 160,000 acres of rich farmland 160 km south of San Francisco, saltwater incursion caused by overpumping of several shallow-water aquifers is threatening the groundwater supply used for irrigation. Because of droughts in recent years and increased domestic needs caused by a burgeoning population, overpumping has resulted in increased seepage of saltwater into the groundwater system such that large portions of some of the aquifers are now too salty even for irrigation. At some locations in the Salinas Valley, seawater has migrated more than 11 km inland during the past 13 years. If this incursion is left unchecked, the farmlands of the valley could become too salty to support most agriculture. Uncontaminated deep aquifers could be tapped to replace the contaminated shallow groundwater, but doing so is too expensive for most farmers and small towns.

To counteract the effects of saltwater incursion, recharge wells are often drilled to pump water back into the groundwater system (Figure 16.15c). Recharge ponds that allow large quantities of fresh surface water to infiltrate the groundwater supply may also be constructed. Both of these methods are successfully used on Long Island, New York, which has had a saltwater incursion problem for several decades.

Subsidence

As excessive amounts of groundwater are withdrawn from poorly consolidated sediments and sedimentary rocks, the water pressure between grains is reduced, and the weight of the overlying materials causes the grains to pack closer together, resulting in *subsidence* of the ground. As more and more groundwater is pumped to meet the increasing needs of agriculture, industry, and population growth, subsidence is becoming more prevalent.

The San Joaquin Valley of California is a major agricultural region that relies largely on groundwater for irrigation. Between 1925 and 1977, groundwater withdrawals in parts of the valley caused subsidence of nearly 9 m (■ Figure 16.16). Other areas in the United States that have experienced subsidence are New Orleans, Louisiana, and Houston, Texas, both of which have subsided more than 2 m, and Las Vegas, Nevada, where 8.5 m of subsidence has taken place (Table 16.2).

Elsewhere in the world, the tilt of the Leaning Tower of Pisa is partly due to groundwater withdrawal (■ Figure 16.17). The tower started tilting soon after construction began in 1173 because of differential com-

■ Figure 16.16

The dates on this power pole dramatically illustrate the amount of subsidence in the San Joaquin Valley, California. Because of groundwater withdrawals and subsequent sediment compaction, the ground subsided nearly 9 m between 1925 and 1977. For a time, surface water use reduced subsidence, but during the drought of 1987 to 1992 it started again as more groundwater was withdrawn.

paction of the foundation. During the 1960s, the city of Pisa withdrew ever larger amounts of groundwater, causing the ground to subside further; as a result, the tilt of the tower increased until it was considered in danger of falling over. Strict control of groundwater withdrawal and stabilization of the foundation have reduced the amount of tilting to about 1 mm per year, thus ensuring that the tower should stand for several more centuries.

A spectacular example of continuing subsidence is taking place in Mexico City, which was built on a former lakebed. As groundwater is removed for the increasing needs of the city's 15.6 million people, the water table has been lowered up to 10 m. As a result, the fine-grained lake deposits are compacting, and Mexico City is slowly and unevenly subsiding. Its opera house has settled more than 3 m, and half of the first floor is now below ground level. Other parts of the city have subsided

Table 16.2

Subsidence of Cities and Regions Due Primarily to Groundwater Removal

Location	Maximum Subsidence (m)	Area Affected (km2)
Mexico City, Mexico	8.0	25
Long Beach and Los Angeles, California	9.0	50
Taipei Basin, Taiwan	1.0	100
Shanghai, China	2.6	121
Venice, Italy	0.2	150
New Orleans, Louisiana	2.0	175
London, England	0.3	295
Las Vegas, Nevada	8.5	500
Santa Clara Valley, California	4.0	600
Bangkok, Thailand	1.0	800
Osaka and Tokyo, Japan	4.0	3,000
San Joaquin Valley, California	9.0	9,000
Houston, Texas	2.7	12,100

Source: Data from R. Dolan and H. G. Goodell, "Sinking Cities," *American Scientist 74* (1986): 38–47; and J. Whittow, *Disasters: The Anatomy of Environmental Hazards* (Athens: University of Georgia Press, 1979).

Sarah Stone (Stone/Getty Images)

as much as 7.5 m, creating similar problems for other structures (■ Figure 16.18). The fact that 72% of the city's water comes from the aquifer beneath the metropolitan area ensures that problems of subsidence will continue.

The extraction of oil can also cause subsidence. Long Beach, California, has subsided 9 m as a result of 34 years of oil production. More than $100 million of damage was done to the pumping, transportation, and harbor facilities in this area because of subsidence and encroachment of the sea (■ Figure 16.19). Once water was pumped back into the oil reservoir, thus stabilizing it, subsidence virtually stopped.

Groundwater Contamination

A major problem facing our society is the safe disposal of the numerous pollutant by-products of an industrialized economy. We are becoming increasingly aware that streams, lakes, and oceans are not unlimited reservoirs for waste, and that we must find new safe ways to dispose of pollutants.

The most common sources of groundwater contamination are sewage, landfills, toxic-waste-disposal sites, and agriculture. Once pollutants get into the groundwa-

■ **Figure 16.17**

The Leaning Tower of Pisa, Italy. The tilting is partly the result of subsidence due to the removal of groundwater.

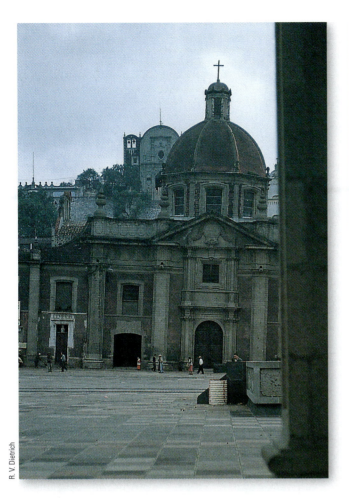

R. V. Dietrich

■ **Figure 16.18**

Excessive withdrawal of groundwater from beneath Mexico City has resulted in subsidence and uneven settling of buildings. The right side of this church (Our Lady of Guadalupe) has settled slightly more than a meter.

What Would You Do

Americans generate tremendous amounts of waste. Some of this waste, such as battery acid, paint, cleaning agents, insecticides, and pesticides, can easily contaminate the groundwater system. Your community is planning to construct a city dump to contain waste products, but it simply wants to dig a hole, dump waste in, and then bury it. What do you think of this plan? Are you skeptical of this plan's merits, and if so, what would you suggest to remedy any potential problems?

ter system, they spread wherever groundwater travels, which can make their containment difficult (see Geo-Focus 16.1). Furthermore, because groundwater moves so slowly, it takes a long time to cleanse a groundwater reservoir once it has become contaminated.

In many areas, septic tanks are the most common way of disposing of sewage. A septic tank slowly releases sewage into the ground, where it is decomposed by oxidation and microorganisms and filtered by the sediment as it percolates through the zone of aeration. In most situations, by the time the water from the sewage reaches the zone of saturation, it has been cleansed of any impurities and is safe to use (■ Figure 16.20a). If the water table is close to the surface or if the rocks are very permeable, water entering the zone of saturation may still be contaminated and unfit to use.

Landfills are also potential sources of groundwater contamination (Figure 16.20b). Not only does liquid

City of Long Beach, Dept. of Oil Properties

TOTAL SUBSIDENCE
1928 TO 1968

■ **Figure 16.19**

The withdrawal of petroleum from the oil field in Long Beach, California, resulted in up to 9 m of ground subsidence because of sediment compaction. It was not until water was pumped back into the reservoir to replace the petroleum that ground subsidence essentially ceased.

GEOFOCUS

16.1

Arsenic and Old Lace

Many people probably learned that arsenic is a poison from either reading or seeing the play *Arsenic and Old Lace*, written by Joseph Kesselring. In the play, the elderly Brewster sisters poison lonely old men by adding a small amount of arsenic to their homemade elderberry wine.

Arsenic is a naturally occurring toxic element found in the environment, and several types of cancer have been linked to arsenic in water. Arsenic also harms the central and peripheral nervous systems and may cause birth defects and reproductive problems. In fact, because of arsenic's prevalence in the environment and its adverse health

effects, Congress included it in the amendments to the Safe Drinking Water Act in 1996. Arsenic gets into the groundwater system mainly as a result of arsenic-bearing minerals dissolving in the natural weathering process of rocks and soils.

A map published in 2001 by the U.S. Geological Survey (USGS) shows the extent and concentration

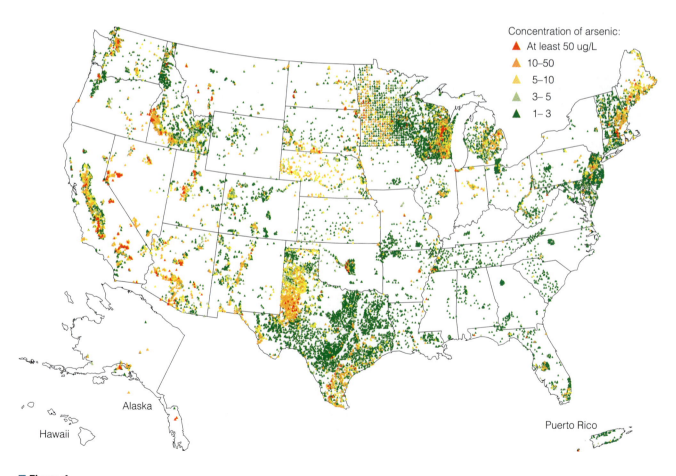

Concentration of arsenic:
▲ At least 50 ug/L
▲ 10–50
▲ 5–10
▲ 3–5
▲ 1–3

Alaska

Hawaii

Puerto Rico

■ **Figure 1**

Arsenic concentrations for 31,350 groundwater samples collected from 1973 to 2001.

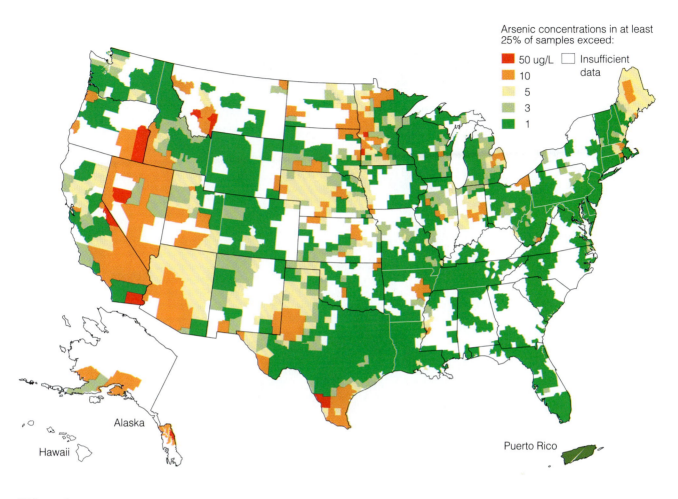

Arsenic concentrations in at least 25% of samples exceed:

50 ug/L   Insufficient data
10
5
3
1

Alaska

Hawaii

Puerto Rico

■ **Figure 2**

County map showing arsenic concentrations found in at least 25% of groundwater samples per county. The map is based on 31,350 groundwater samples collected between 1973 and 2001.

of arsenic in the nation's groundwater supply (■ Figure 1). The highest concentrations of groundwater arsenic were found throughout the West and in parts of the Midwest and Northeast. Although the map is not intended to provide specific information for individual wells or even a locality within a county, it helps researchers and policymakers identify areas of high concentration so that they can make informed decisions about water use. We should point out, however, that a high degree of local variability in the amount of arsenic in the groundwater can be caused by local geology,

type of aquifer, depth of well, and other factors. The only way to learn the arsenic concentration in any well is to have it tested.

What is considered a safe level of arsenic in drinking water? In 2001 the U.S. Environmental Protection Agency (USEPA, or EPA) lowered the maximum level of arsenic permitted in drinking water from 50 micrograms of arsenic per liter to 10 micrograms of arsenic per liter. This is the same figure used by the World Health Organization. From the data in Figure 1, additional maps were created. ■ Figure 2 shows arsenic concentrations found in at least 25% of groundwater

samples per county. Based on these data, approximately 10% of the samples in the USGS study exceed the new standard of 10 micrograms of arsenic per liter of drinking water.

Public water supply systems that exceed the existing EPA arsenic standard are required to either treat the water to remove the arsenic or find an alternative supply. Although reducing the acceptable level of arsenic in drinking water will surely increase the cost of water to consumers, it will also decrease their exposure to arsenic and the possible adverse health effects associated with this toxic element.

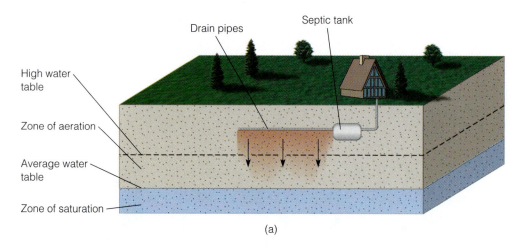

High water table

Zone of aeration

Average water table

Zone of saturation

Drain pipes

Septic tank

(a)

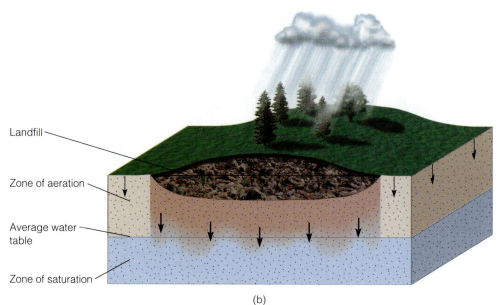

Landfill

Zone of aeration

Average water table

Zone of saturation

(b)

■ **Figure 16.20**

(a) A septic system slowly releases sewage into the zone of aeration. Oxidation, bacterial degradation, and filtering usually remove impurities before they reach the water table. However, if the rocks are very permeable or the water table is too close to the septic system, contamination of the groundwater can result. (b) Unless there is an impermeable barrier between a landfill and the water table, pollutants can be carried into the zone of saturation and contaminate the groundwater supply.

waste seep into the ground, but rainwater also carries dissolved chemicals and other pollutants down into the groundwater reservoir. Unless the landfill is carefully designed and lined with an impermeable layer such as clay, many toxic compounds such as paints, solvents, cleansers, pesticides, and battery acid will find their way into the groundwater system.

Toxic-waste sites where dangerous chemicals are either buried or pumped underground are an increasing source of groundwater contamination. The United States alone must dispose of several thousand metric tons of hazardous chemical waste per year. Unfortunately, much of this waste has been and still is being improperly dumped and is contaminating the surface water, soil, and groundwater.

Examples of indiscriminate dumping of dangerous and toxic chemicals can be found in every state. Perhaps the most famous is Love Canal, near Niagara Falls, New York. During the 1940s, the Hooker Chemical Company dumped approximately 19,000 tons of chemical waste into Love Canal. In 1953 it covered one of the dump sites with dirt and sold it for $1 to the Niagara Falls Board of Education, which built an elementary school and playground on the site. Heavy rains and snow during the winter of 1976–1977 raised the water table and turned the area into a muddy swamp in the spring of 1977. Mixed with the mud were thousands of toxic, noxious chemicals that formed puddles in the playground, oozed into people's basements, and covered gardens and lawns. Trees, lawns, and gardens began to die, and many of the residents of the area suffered from serious illnesses. The cost of cleaning up the Love Canal site and relocating its residents exceeded $100 million, and the site and neighborhood are now vacant.

PHYSICAL Geology⇌Now Click Geology Interactive to work through an activity on Contamination through Groundwater.

GEOLOGY
IN UNEXPECTED PLACES

Water Treatment Plants

Your local or regional water treatment plant may not be the first place you think of when it comes to geology, but surprisingly a fair amount of geology is associated with it. To start off, there is the source of water, which frequently comes from wells drilled below the local water table. The quality of this water depends on the local geology. The water frequently contains high concentrations of iron, calcium, and magnesium; the latter two contribute to the hardness of the water. These elements are usually removed to soften the water.

The first step in softening the water is to aerate it with oxygen to help remove iron, carbon dioxide, and other gases from the water. Removing the carbon dioxide in the water reduces the amount of chemicals needed in subsequent stages of the softening process.

The next step in softening the water is adding lime and sodium hydroxide to precipitate the dissolved calcium and magnesium. This precipitate settles out to form a layer of sludge that is pumped to holding ponds; it can be used later as an agricultural soil conditioner. Following the removal of calcium and magnesium, carbon dioxide is again added to the water to reduce its pH to about 9, which means it is slightly alkaline. This step stops the softening process and stabilizes the water's chemical composition.

The final process is to pass the treated water through filters to remove final particles and then add chemicals to ensure that the water is safe for drinking. The water is now ready to be pumped to the plant's storage reservoirs, from which it is distributed to the final customers.

■ **Figure 1**

One of the sludge ponds of the Mt. Pleasant, Michigan, Water Treatment Plant. The sludge produced by the softening process is diverted into a sludge pond. The sludge will be reused later as an agricultural soil conditioner.

Groundwater Quality

Finding groundwater is rather easy because it is present beneath the land surface nearly everywhere, although the depth to the water table varies considerably. But just finding water is not enough. Sufficient amounts in porous and permeable materials must be located if it is to be withdrawn for agricultural, industrial, or domestic use. The availability of groundwater was important in the westward expansion in both Canada and the United States, and now more than one third of all water for irrigation comes from the groundwater system. More than 90% of the water used for domestic purposes in rural America and the water for a number of large cities comes from groundwater, and, as you would expect, quality is more important here than it is for most other purposes.

Discounting contamination by humans from landfills, septic systems, toxic-waste sites, and industrial effluents, groundwater quality is mostly a function of (1) the kinds of materials making up an aquifer, (2) the residence time of water in an aquifer, and (3) the solubility of rocks and minerals. These factors account for the amount of dissolved materials in groundwater, such as calcium, iron, fluoride, and several others. Most pose no health problems, but some have undesirable effects such as deposition of minerals in water pipes and water heaters and offensive taste or smell, or they may stain clothing and fixtures, or inhibit the effectiveness of detergents.

Everyone has heard of *hard water* and knows that *soft water* is more desirable for most purposes. But just what is hard water, and how does it become hard? Recall from Chapter 5 that some rocks, especially limestone, composed of calcite ($CaCO_3$), and dolostone, composed of dolomite [$CaMg(CO_3)_2$], are readily soluble in water that contains a small amount of carbonic acid. In fact, the amount of dissolved calcium (Ca^{+2}) and magnesium (Mg^{+2}) is what determines whether water is hard or soft. Water with less than 60 milligrams per liter (mg/L) is soft, whereas 61–120 mg/L indicate moderately hard water, values from 121 to 180 mg/L characterize hard water, and any water with more than 180 mg/L is very hard. One negative aspect of hard water is the precipitation of scale (Ca and Mg salts) in water pipes, water heaters, dishwashers, and even on glasses and dinnerware. The gray, hard, scaly deposits in the bottoms of teakettles result from the same process. Furthermore, soaps and detergents do not lather well in hard water, and dirt and soap combine to form an unpleasant scum.

Hard water is a problem in many areas, especially those underlain by limestone and dolostone. To remedy this problem, many households have a water softener, a device in which sodium (Na^+) replaces calcium and magnesium using an ion exchanger or a mineral sieve. In either case, the amount of calcium and magnesium is reduced and the water is more desirable for most domestic purposes. However, people on low-sodium diets, such as those with hypertension (high blood pressure), are cautioned not to drink softened water because it contains more sodium.

Iron in water is common in some areas, especially places underlain by rocks with abundant iron-bearing minerals. If present, it imparts a disagreeable taste and might leave red stains in toilets, sinks, and other plumbing fixtures. Light-colored clothing washed in iron-rich water takes on a rusty color. Even though iron poses no health threat, it is so disagreeable that most people remove it. Water softeners will remove some of it, but in many cases some kind of iron filter must be placed on the incoming water line before it reaches the softener. In any case, the problem is usually easily solved, but of course it adds another expense to household maintenance.

Not all dissolved materials in groundwater are undesirable, at least in small quantities. Fluoride (F^-), for instance, if present in amounts of 1.0 to 1.5 parts per million (ppm), combines with the calcium phosphate in teeth and makes them more resistant to decay. Too much fluoride—more than 4.0 ppm—though, gives children's teeth a dark, blotchy appearance. Fluoride in natural waters is rare, so not many communities benefit from its presence, and artificially adding fluorine to water has met with considerable opposition because of possible health risks.

HYDROTHERMAL ACTIVITY—WHAT IS IT, AND WHERE DOES IT OCCUR?

Hydrothermal is a term referring to hot water. Some geologists restrict the meaning to encompass only water heated by magma, but here we use it to mean any hot subsurface water and surface activity resulting from its discharge. One manifestation of hydrothermal activity in areas of active or recently active volcanism is the discharge of gases, much of it as steam, at vents known as *fumeroles* (see Figure 4.3). Of more immediate concern here, however, is the groundwater that rises to the surface as *hot springs* or *geysers*. It may be heated by its proximity to magma or by Earth's geothermal gradient, because it circulates deeply.

Hot Springs

A **hot spring** (also called a *thermal spring* or *warm spring*) is any spring in which the water temperature is higher than 37°C, the temperature of the human body (■ Figure 16.21a). Some hot springs are much hotter, with temperatures up to the boiling point in many instances (Figure 16.21b). Another type of hot spring, called a *mud pot*, results when chemically altered rocks yield clays that bubble as hot water and steam rise through them (Figure 16.21c). Of the approximately 1100 known hot springs in the United States, more than 1000 are in the far West, with the others in the Black Hills of South Dakota, Georgia, the Appalachian region, and the Ouachita region of Arkansas. At Hot Springs National Park in Arkansas, 47 hot springs with temperatures of 62°C are found on the slopes of Hot Springs Mountain. Hot springs are also common in other parts of the world. One of the most famous is at Bath, England, where shortly after the Roman conquest of Britain in A.D. 43, numerous bathhouses and a temple were built around the hot springs (■ Figure 16.22).

The heat for most hot springs comes from magma or cooling igneous rocks. The geologically recent igneous activity in the western United States accounts for the large number of hot springs in that region. The water in some hot springs, however, circulates deep into Earth, where it is warmed by the normal increase in temperature, the geothermal gradient. For example, the spring

■ Figure 16.21

Hot springs. (a) Hot spring in the West Thumb Geyser Basin, Yellowstone National Park, Wyoming. (b) The water in this hot spring at Bumpass Hell in Lassen Volcanic National Park, California, is boiling. (c) Mud pot at the Sulfur Works, also in Lassen Volcanic National Park. (d) The Park Service warns of the dangers in hydrothermal areas, but some people ignore the warnings and are injured or killed.

British Tourist Authority

■ **Figure 16.22**

One of the many bathhouses in Bath, England, that were built around hot springs shortly after the Roman conquest in A.D. 43.

water of Warm Springs, Georgia, is heated in this manner. This hot spring was a health and bathing resort long before the Civil War; later, with the establishment of the Georgia Warm Springs Foundation, it was used to help treat polio victims.

Geysers

Hot springs that intermittently eject hot water and steam with tremendous force are known as **geysers**. The word comes from the Icelandic *geysir*, "to gush" or "to rush forth." One of the most famous geysers in the world is Old Faithful in Yellowstone National Park in Wyoming (■ Figure 16.23a). With a thunderous roar, it erupts a column of hot water and steam every 30 to 90 minutes. Other well-known geyser areas are found in Iceland and New Zealand.

Geysers are the surface expression of an extensive underground system of interconnected fractures within hot igneous rocks (■ Figure 16.24). Groundwater percolating down into the network of fractures is heated as it comes into contact with the hot rocks. Because the water near the bottom of the fracture system is under greater pressure than that near the top, it must be heated to a higher temperature before it will boil. Thus, when the deeper water is heated to near the boiling point, a slight rise in temperature or a drop in pressure, such as from escaping gas, will instantly change it to steam. The expanding steam quickly pushes the water above it out of the ground and into the air, producing a geyser eruption. After the eruption, relatively cool

James S. Monroe

(a)

(b)

■ **Figure 16.23**

Geysers in Yellowstone National Park, Wyoming. (a) Old Faithful Geyser erupts every 30 to 90 minutes, spewing water from 32 to 56 m high. (b) A small geyser erupting in Norris Geyser Basin.

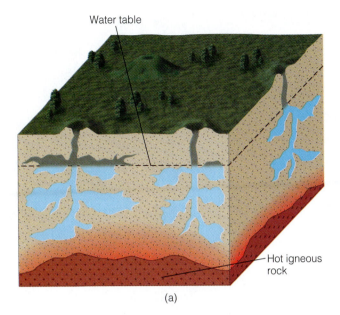

(a)

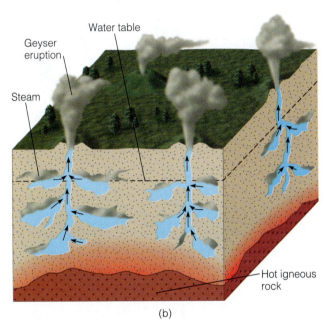

(b)

■ **Figure 16.24**

The eruption of a geyser. (a) Groundwater percolates down into a network of interconnected openings and is heated by the hot igneous rocks. The water near the bottom of the fracture system is under greater pressure than that near the top and consequently must be heated to a higher temperature before it will boil. (b) Any rise in temperature of the water above its boiling point or a drop in pressure will cause the water to change to steam, which quickly pushes the water above it up and out of the ground, producing a geyser eruption.

groundwater starts to seep back into the fracture system, where it heats to near its boiling temperature and the eruption cycle begins again. Such a process explains how geysers can erupt with some regularity.

Hot-spring and geyser water typically contains large quantities of dissolved minerals because most minerals dissolve more rapidly in warm water than in cold water.

Because of this high mineral content, some believe the waters of many hot springs have medicinal properties. Numerous spas and bathhouses have been built at hot springs throughout the world to take advantage of these supposed healing properties.

When the highly mineralized water of hot springs or geysers cools at the surface, some of the material in solution precipitates, forming various types of deposits. The amount and type of precipitated minerals depend on the solubility and composition of the material that the groundwater flows through. If the groundwater contains dissolved calcium carbonate ($CaCO_3$), then *travertine* or *calcareous tufa* (both varieties of limestone) are precipitated. Spectacular examples of hot-spring travertine deposits are found at Pamukhale in Turkey and at Mammoth Hot Springs in Yellowstone National Park (■ Figure 16.25a). Groundwater containing dissolved silica will, upon reaching the surface, precipitate a soft, white, hydrated mineral called *siliceous sinter* or *geyserite*, which can accumulate around a geyser's opening (Figure 16.25b).

Geothermal Energy

Geothermal energy is defined as any energy produced from Earth's internal heat. In fact, the term *geothermal* comes from *geo*, "Earth," and *thermal*, "heat." Several forms of internal heat are known, such as hot dry rocks and magma, but so far only hot water and steam are used. Approximately 1 to 2% of the world's current energy needs could be met by geothermal energy. In those areas where it is plentiful, geothermal energy can supply most, if not all, of the energy needs, sometimes at a fraction of the cost of other types of energy. Some of the countries currently using geothermal energy in one form or another are Iceland, the United States, Mexico, Italy, New Zealand, Japan, the Philippines, and Indonesia.

The city of Rotorua, New Zealand, is world famous for its volcanoes, hot springs, geysers, and geothermal fields. Since the first well was sunk in the 1930s, more than 800 wells have been drilled to tap the hot water and steam. Geothermal energy in Rotorua is used in a variety of ways, including heating homes, commercial buildings, and greenhouses.

In the United States, the first commercial geothermal electrical generating plant was built in 1960 at The Geysers, about 120 km north of San Francisco, California. Here, wells were drilled into the numerous nearvertical fractures underlying the region. As pressure on the rising groundwater decreases, the water changes to steam, which is piped directly to electricity-generating turbines and generators (■ Figure 16.26).

As oil reserves decline, geothermal energy is becoming an attractive alternative, particularly in parts of the western United States, such as the Salton Sea area of southern California, where geothermal exploration and development have begun.

(a)

(b)

■ **Figure 16.25**

Hot-spring deposits in Yellowstone National Park, Wyoming.
(a) Minerva Terrace formed when calcium-carbonate-rich hot-spring
water cooled, precipitating travertine. (b) Liberty Cap is a geyserite
mound formed by numerous geyser eruptions of silicon-dioxide-
rich hot-spring water.

Julie Donnelly-Nolan/USGS

■ **Figure 16.26**

Steam rising from geothermal power plants at The Geysers in Sonoma County, California.

16 REVIEW WORKBOOK

Chapter Summary

- Groundwater consists of all subsurface water trapped in the pores and other open spaces in rocks, sediment, and soil.

- About 22% of the world's supply of freshwater is groundwater, which constitutes one reservoir in the hydrologic cycle.

- For groundwater to move through materials, they must be porous and permeable. Any material that transmits groundwater is an aquifer, whereas materials that prevent groundwater movement are aquicludes.

- The zone of saturation (in which pores are filled with water) is separated from the zone of aeration (in which pores are filled with air and water) by the water table. The water table is a subdued replica of the land surface in most places.

- Groundwater moves slowly through the pore spaces in the zone of aeration and moves through the zone of saturation to outlets such as streams, lakes, and swamps.

- Springs are found wherever the water table intersects the surface. Some springs are the result of a perched water table—that is, a localized aquiclude within an aquifer and above the regional water table.

- Water wells are made by digging or drilling into the zone of saturation. When water is pumped out of a well, a cone of depression forms.

- In an artesian system, confined groundwater builds up high hydrostatic pressure. Three conditions must generally be met for an artesian system to form: The aquifer must be confined above and below by aquicludes, the aquifer is usually tilted and exposed at the surface so it can be recharged, and precipitation must be sufficient to keep the aquifer filled.

- Karst topography results from groundwater weathering and erosion and is characterized by sinkholes, caves, solution valleys, and disappearing streams.

- Caves form when groundwater in the zone of saturation weathers and erodes soluble rock such as limestone. Cave deposits, called dripstone, result from the precipitation of calcite.

- Modifications of the groundwater system can cause serious problems. Excessive withdrawal of groundwater may result in dry wells, loss of hydrostatic pressure, saltwater incursion, and ground subsidence.

■ Groundwater contamination is becoming a serious problem and can result from sewage, landfills, and toxic waste.

■ Groundwater may be heated by magma or by the geothermal gradient as it circulates deeply. In either case, the water commonly rises to the surface, thus accounting for hydrothermal activity in the form of hot springs, geysers, and other features.

■ Geothermal energy comes from the steam and hot water trapped within Earth's crust. It is a relatively nonpolluting form of energy that is used as a source of heat and to generate electricity.

Important Terms

artesian system (p. 475)
cave (p. 479)
cone of depression (p. 475)
geothermal energy (p. 495)
geyser (p. 494)
groundwater (p. 470)

hot spring (p. 492)
hydrothermal (p. 492)
karst topography (p. 477)
permeability (p. 471)
porosity (p. 470)
sinkhole (p. 477)

spring (p. 473)
water table (p. 472)
water well (p. 473)
zone of aeration (p. 472)
zone of saturation (p. 472)

Review Questions

1. Two features typical of areas of karst topography are:

    a. _____ geysers and hot springs; b. _____ hydrothermal activity and springs; c. _____ saltwater incursion and pollution; d. _____ sinkholes and disappearing streams; e. _____ dripstone and a cone of depression.

2. Which of the following is a cave deposit?

    a. _____ aquiclude; b. _____ chamber; c. _____ artesian spring; d. _____ stalagmite; e. _____ sinkhole.

3. What is the correct order, from highest to lowest, of groundwater usage in the United States?

    a. _____ industrial, agricultural, domestic; b. _____ agricultural, industrial, domestic; c. _____ agricultural, domestic, industrial; d. _____ domestic, agricultural, industrial; e. _____ industrial, domestic, agricultural.

4. The porosity of Earth materials is defined as:

    a. _____ their ability to transmit fluids; b. _____ the depth of the zone of saturation; c. _____ the percentage of void spaces; d. _____ their solubility in the presence of weak acids; e. _____ the temperature of groundwater.

5. Hydrothermal is a term referring to:

    a. _____ calcareous tufa; b. _____ groundwater contamination; c. _____ hot water; d. _____ sinkhole formation; e. _____ artesian wells.

6. A cone of depression forms when:

    a. _____ a stream flows into a sinkhole; b. _____ water in the zone of aeration is replaced by water from the zone of saturation; c. _____ a spring forms where a perched water table intersects the surface; d. _____ water is withdrawn faster than it can be replaced; e. _____ the ceiling of a cave collapses, forming a steep-sided crater.

7. Which of the following conditions must exist for an artesian system to form?

    a. _____ an aquifer must be confined above and below by aquicludes; b. _____ groundwater must circulate near magma; c. _____ water must rise very high in the capillary fringe; d. _____ the rocks below the surface must be especially resistant to solution; e. _____ the water table must be at or very near the surface.

8. A hot spring that periodically erupts is known as a:

    a. _____ mud pot; b. _____ geyser; c. _____ cone of ascension; d. _____ travertine terrace; e. _____ stalactite.

9. When water is pumped from wells in some coastal areas, a problem arises known as:

    a. _____ artesian recharge; b. _____ dripstone deposition; c. _____ geothermal depression; d. _____ saltwater incursion; e. _____ permeability decrease.

10. Why isn't geothermal energy a virtually unlimited source of energy?

11. Why should we be concerned about how fast the groundwater supply is being depleted in some areas?

12. Describe three features you might see in an active hydrothermal area. Where in the United States would you go to see such activity?

13. Describe the configurations of the water table beneath a humid area and an arid region. Why do they differ?

14. Discuss the role of groundwater in the hydrologic cycle.

15. Explain how some Earth materials can be porous, yet not permeable. Give an example.

16. Why does groundwater move so much slower than surface water?

17. Explain how groundwater weathers and erodes Earth materials.

18. Describe some ways to quantitatively measure the rate of groundwater movement.

19. Explain how saltwater incursion takes place and why it is a problem in coastal areas.

20. One concern geologists have about burying nuclear waste in present-day arid regions such as Nevada is that the climate may change during the next several thousand years and become more humid, thus allowing more water to percolate through the zone of aeration. Why is this a concern? What would the average rate of groundwater movement have to be during the next 5000 years to reach canisters containing radioactive waste buried at a depth of 400 m?

World Wide Web Activities

PHYSICAL Geology⇌Now Assess your understanding of this chapter's topics with additional quizzing and comprehensive interactivities at

http://earthscience.brookscole.com/physgeo5e

as well as current and up-to-date weblinks, additional readings, and InfoTrac College Edition exercises.

Glaciers and Glaciation

CHAPTER 17
OUTLINE

PHYSICAL Geology⇌Now *This icon, appearing throughout the book, indicates an opportunity to explore interactive tutorials, animations, or practice problems available on the Physical GeologyNow Web site at http://earthscience.brookscole.com/physgeo5e.*

OBJECTIVES

At the end of this chapter, you will have learned that

- Moving bodies of ice on land known as glaciers cover about 10% of Earth's land surface.

- During the Pleistocene Epoch (Ice Age) glaciers were much more widespread than they are now.

- Water frozen in glaciers constitutes one reservoir in the hydrologic cycle.

- In any area with a yearly net accumulation of snow, the snow is first converted to granular ice known as firn and eventually into glacial ice.

- The concept of the glacial budget is important when considering the dynamics of any glacier.

- Glaciers move by a combination of basal slip and plastic flow, but several factors determine their rates of movement, and under some conditions they may move rapidly.

- Glaciers effectively erode, transport, and deposit sediment, thus accounting for the origin of several distinctive landforms.

- As a result of glaciation during the Ice Age, Earth's crust was depressed into the mantle and has since risen, sea level fluctuated widely, and lakes were present in areas now quite arid.

- A current widely accepted theory explaining the onset of ice ages relies on irregularities in Earth's rotation and orbit.

Harvard Glacier in College Fiord, Alaska. The dark streaks on the glacier consist of sediment transported by the glacier. Source: Sue Monroe

Introduction

Scientists know that Earth's surface temperatures have increased during the last few decades, although they disagree about how much humans have contributed to climatic change. Is the increase part of a normal climatic fluctuation, or is the introduction of greenhouse gases into the atmosphere actually having an adverse effect on climate? These questions are not fully answered, but we do know from the geologic record that an Ice Age took place between 1.6 million and 10,000 years ago, and since the end of the Ice Age Earth has experienced several climatic fluctuations. About 6000 years ago, during the Holocene Maximum, temperatures on average were slightly warmer than they are now, and some of today's arid regions such as the Sahara Desert of north Africa had enough precipitation to support lush vegetation, swamps, and lakes.

Following the Holocene Maximum was a time of cooler temperatures, but from about A.D. 1000 to 1300, Europe experienced what is known as the Medieval Warm Period during which wine grapes grew 480 km farther north than they do now. However, a cooling trend beginning in about A.D.

1300 led to the *Little Ice Age*, from about 1500 to the middle or late 1800s. During this time, glaciers expanded (■ Figure 17.1), winters were colder, summers were cooler and wetter, sea ice at high latitudes persisted for long periods, and several widespread famines took place.

Conditions varied considerably during the Little Ice Age. Some winters were mild and the growing seasons long enough to support the largely agrarian society of Europe. But at other times, cool, wet summers and colder winters contributed to poor harvests and widespread famine. During the coldest part of the Little Ice Age, from about 1680 to 1730, the growing season in England was about five weeks shorter than it was during the 1900s, and in 1695 Iceland was surrounded by sea ice for much of the year. To the surprise of people living in the Orkney Islands, off Scotland's north coast, Eskimos were sighted offshore on several occasions during the late 1600s and early 1700s.

Most of you have some idea of what a glacier is and have probably heard of the Ice Age, although it is doubtful that you know much of the Little Ice Age. In any case, geolo-

Reprinted with permission from T. H. Van Andel, *New Views on an Old Planet.*

(a)

■ **Figure 17.1**

(a) During the Little Ice Age, many glaciers in Europe extended much farther down their valleys than they do now. Samuel Birmann (1793–1847) painted this view, titled *The Unterer Grindlewald*, in 1826. (b) The same glacier today, its terminus now hidden behind a rock projection at the lower end of its valley.

Sue Monroe

(b)

gists define a **glacier** as a mass of ice on land consisting of compacted snow that flows either downslope or outward from a central area. We will address exactly how glaciers flow in a later section. Our definition excludes sea ice as in the North Polar region and frozen seawater adjacent to Antarctica. Drifting icebergs are not glaciers either, although they may have come from glaciers that flowed into the sea.

Among the various surface processes modifying Earth, glaciers are particularly effective at erosion, transport, and deposition. They deeply scour the land they move over, producing a number of easily recognized landforms, and they eventually deposit huge quantities of sediment. Indeed, in many northern states and Canada, glacial deposits are important sources of sand and gravel for construction. Glaciers covered far more area during the Pleistocene Epoch (Ice Age), but many are still present and now cover about 10% of Earth's land surface.

Unfortunately, our period of recordkeeping is too short to resolve the question of whether the last Ice Age and the Little Ice Age are truly events of the past or simply parts of long-term climatic events and likely to occur again. Nevertheless, studying glaciers and their possible causes may help clarify some aspects of long-term climatic changes and possibly tell us something about the debate on global warming. Glaciers are very sensitive even to short-term climatic changes, so they are closely monitored to see if they advance, remain stationary, or retreat.

GLACIERS AND GLACIATION

Presently glaciers cover nearly 1.5 million km², or about 10% of Earth's land surface (Table 17.1). As a matter of fact, if all the glacial ice on Earth were in the United States and Canada, these countries would be covered by ice about 1.5 km thick! Small glaciers are common in the high mountains of the western United States, especially Alaska, and western Canada, as well as the Andes in South America, the Alps of Europe, and the Himalayas in Asia. Even some of the highest peaks in Africa, though near the equator, have glaciers. Australia is the only continent with no glaciers. The small glaciers in mountains are picturesque, but the truly vast glaciers are in Antarctica and Greenland, which contain 99% of all glacial ice (Table 17.1).

At first glance glaciers appear static. Even briefly visiting a glacier may not dispel this impression, because although glaciers move, they usually do so slowly. Nevertheless, they do move and just like other geologic agents such as running water, glaciers are dynamic systems that continuously adjust to changes. For example, a stream's capacity to erode and transport varies depending on velocity and discharge. Likewise, the amount of ice in a glacier or the amount of water present determines rates of movement and thus a glacier's ability to erode and transport sediment. So glaciers also respond to changes; they just do so more slowly.

Glaciers—Part of the Hydrologic Cycle

Recall from Chapter 1 that one of Earth's systems is the hydrosphere, consisting of all surface water in the oceans, lakes, streams, the atmosphere, and frozen in glaciers (see Table 1.1). It is the comparatively small amount of water in glaciers, about 2.15% of the total but more than 75% of Earth's freshwater, that interests us here.

Glaciers constitute one reservoir in the hydrologic cycle where water is stored for long periods, but even this water eventually returns to its original source, the oceans (see Figure 15.4). Many glaciers at high latitudes, as in Alaska and northern Canada, flow directly into the seas where they melt, or icebergs break off (a process known as *calving*) and drift out to sea where they eventually melt. At lower elevations or areas remote from the sea, glaciers flow to lower elevations where melting takes place and the water yielded either enters the groundwater system (another reservoir in the hydrologic cycle) or returns to the seas by surface runoff. In some parts of the western United States and Canada, glaciers are important freshwater reservoirs that release water to streams during the dry season.

Melting is the most important process in returning glacial ice to the hydrologic cycle, but glaciers also lose water by *sublimation,* a process in which ice changes to water vapor without an intermediate liquid phase. Sublimation is easy to understand if one thinks of ice cubes stored in a container in a freezer where no melting can take place. Because of sublimation, the older ice cubes at the bottom of the container are much smaller than the more recently formed ones. In any event, the water vapor thus derived from glaciers enters the atmosphere where it may condense and fall again as rain or snow, but in the long run that water too returns to the oceans.

How Do Glaciers Form and Move?

In Chapter 2 we briefly mentioned that ice is crystalline and possesses characteristic physical and chemical properties, and thus is a mineral. Accordingly, glacial ice is a type of rock, but one that is easily deformed. Glacial ice forms in a very straightforward manner (■ Figure 17.2). In any area where more snow falls than melts during the warmer seasons, a net accumulation occurs. Freshly fallen snow has about 80% air-filled pore space and 20% solids, but it compacts as it accumulates, partly thaws, and refreezes, converting to a granular type of snow known as **firn.** As more snow accumulates, firn is buried and further compacted and recrystallizes until it becomes **glacial ice,** consisting of about 90% solids (Figure 17.2). As we mentioned in the Introduction, a *glacier* is a moving mass of recrystallized snow on land, but just how is movement accomplished?

Table 17.1

Present-Day Ice-Covered Areas

	Area (km²)	Volume (km³)	Percent of Total
Antarctica	12,588,000	30,109,800	91.49
Greenland	1,802,600	2,620,000	7.96
Canadian Arctic islands	153,169		
Iceland	12,173		
Svalbard	58,016		
Russian Arctic islands	55,541		
Alaska	51,476		
United States (other than Alaska)	513		
Western Canada	24,880		
South America	26,500		
Europe	7,410		
Asia	116,854		
Africa	12		
New Zealand	1,000		
Others	176	180,000[a]	0.55
	14,898,276	32,909,800	100.00

[a]Total volume of glacial ice outside Antarctica and Greenland.

Source: U.S. Geological Survey Professional Paper 1386-A.

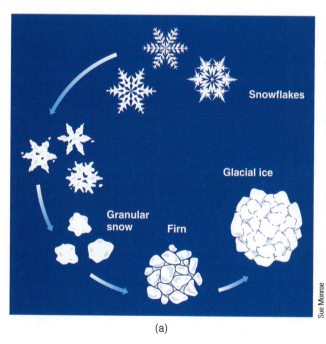

(a)

Sue Monroe

(b)

■ **Figure 17.2**

(a) The conversion of freshly fallen snow to firn and glacial ice. (b) This iceberg in Portage Lake, Alaska, shows the typical blue color of glacial ice. The longer wavelengths of white light are absorbed by the ice, but blue (short wavelength) is transmitted into the ice and scattered, accounting for the blue color.

At this time it is useful to recall some of the terms from Chapter 13 on deformation. Remember that *stress* is force per unit area and *strain* is a change in the shape or volume or both of solids. When snow and ice reach a critical thickness of about 40 m, the stress on the ice at dept is great enough to induce **plastic flow,** a type of permanent deformation involving no fracturing. Glaciers move primarily by plastic flow, but they may also slide over their underlying surface by **basal slip** (■ Figure 17.3). Basal slip is facilitated by the presence of water, which reduces friction between the underlying surface and a glacier. The total movement of a glacier, then, results from a combination of plastic flow and basal slip, although the former occurs continuously whereas the latter varies depending on the season. Indeed, if a glacier is solidly frozen to the surface below, it moves only by plastic flow.

You now have some idea of how glaciers form in the first place, but what controls their distribution? As you probably suspect, temperature and the amount of snowfall are important factors. Of course temperature varies with elevation and latitude, so we would expect to find glaciers in high mountains and at high latitudes, if the areas receive enough snow. Many small glaciers are present in the Sierra Nevada of California but only at eleva-

tions exceeding 3900 m. In fact, the high mountains in California, Oregon, and Washington all possess glaciers because in addition to their elevations they receive huge amounts of snow. Indeed, Mount Baker in Washington had almost 29 m of snow during the winter of 1998–1999, and average snowfalls of 10 m or more are common in many parts of these mountains. Glaciers are also present in the mountains along the Pacific Coast of Canada, which also receive considerable snow and in addition are farther north. Some of the higher peaks in the Rocky Mountains of both the United States and Canada also support glaciers.

WHAT KINDS OF GLACIERS ARE THERE?

All glaciers share some characteristics, but they also vary in several ways. Some are confined to mountain valleys or bowl-shaped depressions on mountainsides, and flow from higher to lower elevations. Others are of much greater thickness and extent, and flow outward from centers of accumulation and are completely unconfined by topography. Thus we recognize two basic types of glaciers: *valley* and *continental,* and some variations on these basic types.

Valley Glaciers

A **valley glacier,** as its name implies, is confined to a mountain valley through which it flows from higher to lower elevations (■ Figure 17.4). We use the term *valley glacier* here, but *alpine glacier* and *mountain glacier* are synonyms. Many valley glaciers have smaller tributary glaciers entering them, just as rivers and streams have tributaries, thus forming a network of glaciers in

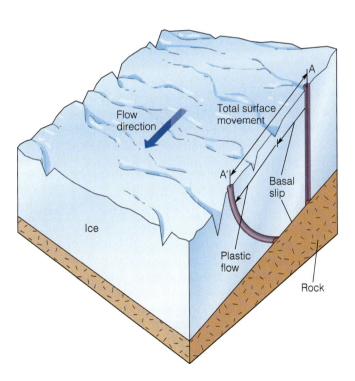

■ Figure 17.3

Part of a glacier showing movement by a combination of plastic flow and basal slip. Plastic flow involves internal deformation within the ice, whereas basal slip is sliding over the underlying surface. If a glacier is solidly frozen to its bed, it moves only by plastic flow. Notice that the top of the glacier moves farther in a given time than the bottom does.

Engineering Mechanics, Virginia Polytechnic Institute and State University

(a)

Sue Monroe

(b)

■ **Figure 17.4**

(a) View of a valley glacier in Alaska. Notice the tributaries to the large glacier. (b) This view is from the upper end of a valley glacier in Switzerland. This glacier's terminus is about 22 km in the distance.

a system of interconnected valleys. A valley glacier's shape is obviously controlled by the shape of the valley it occupies, so these glaciers are long, narrow tongues of moving ice. Where a valley glacier flows from a valley onto a wider plain and spreads out, or where two or more valley glaciers coalesce at the base of a mountain range, they form a more extensive ice cover called a *piedmont glacier.*

Valley glaciers are invariably small compared to continental glaciers, but even so they may be several kilometers across, 200 km long, and hundreds of meters thick. For instance, the Bering Glacier in Alaska is about 200 km long, and the Saskatchewan Glacier in Canada is 555 m thick. Erosion and deposition by valley glaciers were responsible for much of the spectacular scenery in such places as Grand Teton National Park, Wyoming; Glacier National Park, Montana; and Waterton, Banff, and Jasper National Parks in Canada.

Continental Glaciers and Ice Caps

Continental glaciers, also known as *ice sheets,* cover at least 50,000 km^2 and are unconfined by topography (■ Figure 17.5). That is, their shape and movement are

not controlled by the underlying landscape. Valley glaciers flow downhill within the confines of a valley, but continental glaciers flow outward in all directions from central areas of accumulation in response to variations in ice thickness.

Continental glaciers are currently present in only Greenland and Antarctica. In both areas, the ice is more than 3000 m thick in their central areas, becomes thinner toward the margins, and covers all but the highest mountains (Figure 17.5b). The aerial extent of the continental glacier in Greenland is about 1,800,000 km^2, and in Antarctica the East and West Antarctic Glaciers merge to form a continuous ice sheet covering more than 12,650,000 km^2. During the Pleistocene Epoch, continental glaciers covered large parts of the Northern Hemisphere continents. They account for many erosional and depositional landforms in Canada and the states from Washington to Maine.

Although valley and continental glaciers are easily differentiated by size and location, an intermediate variety called an **ice cap** is also recognized (Figure 17.5c). Ice caps are similar to, but smaller than, continental glaciers, covering less than 50,000 km^2. The 6000 km^2 Penny Ice Cap on Baffin Island, Canada, and the Juneau

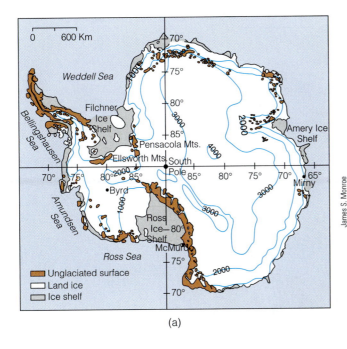

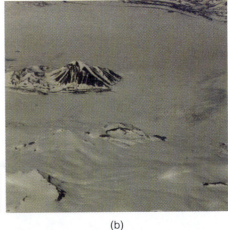

■ Figure 17.5

(a) The West Antarctic and much larger East Antarctic ice sheets merge to form a nearly continuous ice cover averaging 2160 m thick and reaching a maximum thickness of about 4000 m. (b) View of part of the ice sheet that covers most of Greenland. (c) An ice cap on Baffin Island, Canada.

Icefield in Alaska and Canada at about 3900 km² are good examples. Some ice caps form when valley glaciers grow and overtop the divides and pass between adjacent valleys and coalesce to form a continuous ice cap. They also form on fairly flat terrain in Iceland and some of the islands of the Canadian Arctic.

THE GLACIAL BUDGET— ACCUMULATION AND WASTAGE

Just as a savings account grows and shrinks as funds are deposited and withdrawn, glaciers expand and contract in response to accumulation and wastage. We describe their behavior in terms of a **glacial budget,** which is essentially a balance sheet of accumulation and wastage. The upper part of a valley glacier is a **zone of accumulation,** where additions exceed losses and the glacier's surface is perennially cov-

ered with snow. In contrast, the lower part of the same glacier is a **zone of wastage,** where losses from melting, sublimation, and calving of icebergs exceed the rate of accumulation (■ Figure 17.6).

At the end of winter, a glacier's surface is completely covered with the accumulated seasonal snowfall. During spring and summer, the snow begins to melt, first at lower elevations and then progressively higher up the glacier. The elevation to which snow recedes during a wastage season is called the *firn limit* (Figure 17.6). One can easily identify the zones of accumulation and wastage by noting the position of the firn limit.

Observations of a single glacier reveal that the position of the firn limit usually changes from year to year. If it does not change or shows only minor fluctuations, the glacier is said to have a balanced budget; that is, additions in the zone of accumulation are exactly balanced by losses in the zone of wastage, and the distal end, or *terminus,* of the glacier remains stationary (Figure 17.6a). When the firn limit moves down the glacier, the glacier has a positive budget; its additions exceed its losses, and its terminus advances (Figure 17.6b). If the budget is negative, the glacier recedes, and its terminus

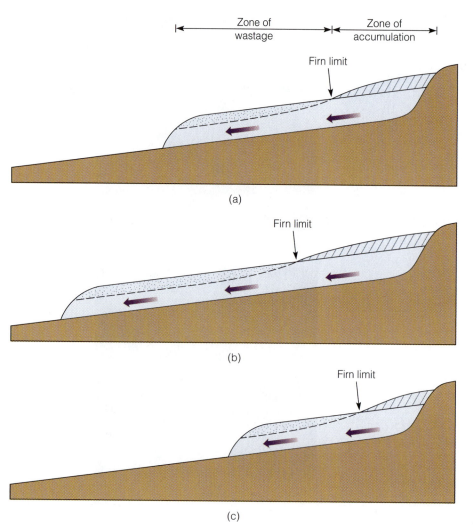

Zone of wastage   Zone of accumulation

(a)

(b)

(c)

Firn limit

■ Figure 17.6

Response of a hypothetical glacier to changes in its budget. (a) If losses in the zone of wastage, shown by stippling, equal additions in the zone of accumulation, shown by cross-hatching, the terminus of the glacier remains stationary. (b) Gains exceed losses, and the glacier's terminus advances. (c) Losses exceed gains, and the glacier's terminus retreats, although the glacier continues to flow.

retreats up the glacial valley (Figure 17.6c). But even though a glacier's terminus may recede, the glacial ice continues to move toward the terminus by plastic flow and basal slip. If a negative budget persists long enough, though, a glacier recedes and thins until it no longer flows and becomes a *stagnant glacier.*

Although we used a valley glacier as an example, the same budget considerations control the flow of continental glaciers as well. The entire Greenland ice sheet, for example, is in the zone of accumulation, but it flows into the ocean where wastage occurs (■ Figure 17.7).

How Fast Do Glaciers Move?

In general, valley glaciers move more rapidly than continental glaciers, but the rates for both vary, ranging from centimeters to tens of meters per day. Valley glaciers moving down steep slopes flow more rapidly than glaciers of comparable size on gentle slopes, assuming that all other variables are the same. The main glacier in a valley glacier system contains a greater volume of ice and thus has a greater discharge and flow velocity

than its tributaries (Figure 17.4a). Temperature exerts a seasonal control on valley glaciers because, although plastic flow remains rather constant year-round, basal slip is more important during warmer months when meltwater is more abundant.

Flow rates also vary within the ice itself. For example, flow velocity increases in the zone of accumulation until the firn limit is reached; from that point, the velocity becomes progressively slower toward the glacier's terminus. Valley glaciers are similar to streams, in that the valley walls and floor cause frictional resistance to flow, so the ice in contact with the walls and floor moves more slowly than the ice some distance away (■ Figure 17.8).

Notice in Figure 17.8 that flow velocity increases upward until the top few tens of meters of ice are reached, but little or no additional increase occurs after that point. This upper ice constitutes the rigid part of the glacier that is moving as a result of basal slip and plastic flow below. The fact that this upper 40 m or so of ice behaves like a brittle solid is clearly demonstrated by large fractures called *crevasses* that develop when a valley glacier flows over a step in its valley floor where the

James S. Monroe

■ Figure 17.7

All of the ice sheet in Greenland is in the zone of accumulation, but as shown here it flows into the ocean where icebergs break off and float out to sea.

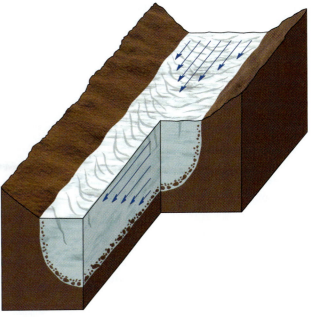

■ Figure 17.8

Flow velocity in a valley glacier varies both horizontally and vertically. Velocity is greatest at the top center of the glacier because friction with the walls and floor of the trough causes flow to be slower adjacent to these boundaries. The lengths of the arrows in the figure are proportional to velocity.

slope increases or where it flows around a corner (■ Figure 17.9a). In either case, the glacial ice is stretched (subjected to tension), and large crevasses develop, but they extend downward only to the zone of plastic flow. In some cases, a valley glacier descends over such a steep precipice that crevasses break up the ice into a jumble of blocks and spires, and an *ice fall* develops (Figure 17.9b).

Continental glaciers ordinarily flow at a rate of centimeters to meters per day. Nevertheless, even a rather modest rate of a meter or so per day has a great cumulative effect after several decades. One reason continental glaciers move comparatively slowly is that they exist at higher latitudes and are frozen to the underlying surface most of the time, which limits the amount of basal slip.

Some basal slip does occur even beneath the Antarctic ice sheet, but most of its movement is by plastic flow. Nevertheless, some parts of continental glaciers manage to achieve extremely high flow rates. Near the margins of the Greenland ice sheet, the ice is forced between mountains in what are called *outlet glaciers*. In some of these outlets, flow velocities exceed 100 m per day.

R. V. Dietrich

(a)

Sue Monroe

(b)

■ Figure 17.9

(a) Crevasses on glaciers in Alaska. (b) The area of chaotic ice in the middle of this image is an ice fall on a small glacier in Switzerland.

In parts of the continental glacier covering West Antarctica, ice streams have been identified in which flow rates are considerably greater than in adjacent glacial ice. Drilling has revealed a 5-m-thick layer of water-saturated sediment beneath these ice streams, which apparently acts to facilitate movement of the ice above. Some geologists think that geothermal heat from active volcanism melts the underside of the ice, thus accounting for the layer of water-saturated sediment.

Glacial Surges

A **glacial surge** is a short-lived episode of accelerated flow in a glacier during which the glacier's surface breaks into a maze of crevasses and its terminus advances markedly. Although surges are best documented in valley glaciers, they also take place in ice caps and perhaps continental glaciers. During a surge a glacier may advance at rates of several tens of meters per day for weeks or months and then return to its normal flow rate. Surging glaciers constitute only a tiny proportion of all glaciers, and none are present in the United States outside Alaska. Even in Canada they are found only in the Yukon Territory and the Queen Elizabeth Islands.

The fastest glacial surge ever recorded took place in 1953 in the Kutiah Glacier in Pakistan; the glacier advanced 12 km in three months. In 1986 the terminus of Hubbard Glacier in Alaska began advancing at about 10 m per day (■ Figure 17.10) and, more recently, in 1993 Alaska's Bering Glacier advanced more than 1.5 km in just three weeks. During the following year, two *glacial outburst floods* took place when water was suddenly released from cavities within or beneath the glacier.

The onset of a glacial surge is heralded by a thickened bulge in the upper part of a glacier that begins to move at several times the glacier's normal velocity toward the terminus. When the bulge reaches the terminus, it causes rapid movement and displacement of the terminus by as much as 20 km. Surges are probably related to accelerated rates of basal slip rather than more rapid plastic flow. One theory holds that thickening in the zone of accumulation with concurrent thinning in the zone of wastage increases the glacier's slope and accounts for accelerated flow. But another theory holds that pressure on soft sediment beneath a glacier forces fluids through the sediment, thereby allowing the overlying glacier to slide more effectively.

AP Wide World

■ Figure 17.10

During a 1986 surge, the terminus of Hubbard Glacier in Alaska advanced across Russell Fiord, the shallow embayment at the right back. Environmentalists saved some of the marine mammals trapped in the former bay but hundreds of seals and porpoises died.

GLACIAL EROSION AND TRANSPORT

Glaciers are now limited in areal extent, but during the Pleistocene Epoch, they covered much larger areas and were more important than their present distribution indicates. Glaciers as moving solids erode and transport huge quantities of materials, especially unconsolidated sediment and soil. Important erosional processes include bulldozing, plucking, and abrasion. *Bulldozing*, although not a formal geologic term, is fairly self-explanatory: A glacier simply shoves or pushes unconsolidated materials in its path. This effective process was aptly described in 1744 during the Little Ice Age by an observer in Norway:

> When at times [the glacier] pushes forward a great sound is heard, like that of an organ and it pushes in front of it unmeasurable masses of soil, grit and

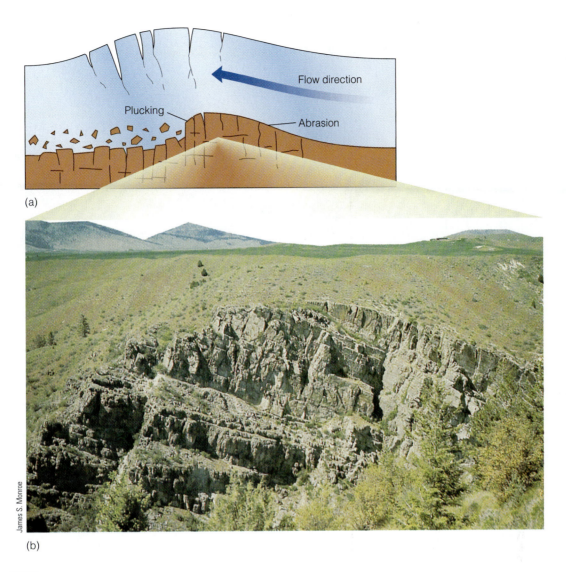

(a)

(b)

James S. Monroe

■ **Figure 17.11**

(a) Origin of a roche moutonnée. As the ice moves over a hill, it smooths the "upstream" side by abrasion and shapes the "downstream" side by plucking. (b) A roche moutonnée in Montana.

rocks bigger than any house could be, which it then crushes small like sand.*

Plucking, also called *quarrying,* results when glacial ice freezes in the cracks and crevices of a bedrock projection and eventually pulls it loose. One manifestation of plucking is a landform called a *roche moutonnée,* a French term for "rock sheep." As shown in ■ Figure 17.11, a glacier smooths the "upstream" side of an obstacle, such as a small hill, and plucks pieces of rock from the "downstream" side by repeatedly freezing and pulling away from the obstacle.

Bedrock over which sediment-laden glacial ice moves is effectively eroded by **abrasion** and develops a **glacial polish,** a smooth surface that glistens in re-

flected light (■ Figure 17.12a). Abrasion also yields **glacial striations,** consisting of rather straight scratches rarely more than a few millimeters deep on rock surfaces (Figure 17.12b). Abrasion also thoroughly pulverizes rocks, yielding an aggregate of clay- and silt-sized particles that have the consistency of flour—hence, the name *rock flour.* Rock flour is so common in streams discharging from glaciers that the water has a milky appearance (Figure 17.12c).

Continental glaciers derive sediment from mountains projecting through them, and windblown dust settles on their surfaces. Otherwise, they obtain most of their sediment from the surface they move over which is transported in the lower part of the ice sheet. In contrast, valley glaciers carry sediment in all parts of the ice, but it is concentrated at the base and along the margins (■ Figure 17.13). Some of the marginal sediment is derived by abrasion and plucking, but much of it is

*Quoted in C. Officer and J. Page, *Tales of the Earth* (New York: Oxford University Press, 1993), p. 99.

(a)

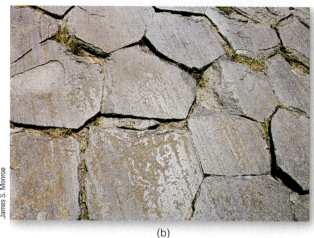

(b)

(c)

■ **Figure 17.12**

When sediment-laden ice moves over rocks, it abrades them and imparts a sheen known as glacial polish (a) as on this gneiss in Michigan. Glacial polish is also visible in (b) and so are straight scratches called glacial striations. The rock is basalt at Devil's Postpile National Monument, California. (c) The water in this stream in Switzerland is discolored by rock flour, small particles yielded by glacial abrasion.

supplied by mass wasting processes, as when soil, sediment, or rock falls or slides onto the glacier's surface.

Erosion by Valley Glaciers

Erosion by valley glaciers has yielded some of the world's most inspiring scenery. Many mountain ranges are scenic to begin with, but when modified by valley glaciers they take on a unique appearance of angular ridges and peaks in the midst of broad valleys. Several of the national parks and monuments in the western United States and Canada owe their scenic appeal to erosion by valley glaciers (see Geo-Focus 17.1). The erosional landforms that result from valley glaciation are easily recognized and enable us to appreciate the tremendous erosive power of moving ice (■ Figure 17.14). See "Valley Glaciers and Erosion" on pages 514 and 515.

■ **Figure 17.13**

Debris on the surface of the Mendenhall Glacier in Alaska. The largest boulder measures about 2 m across. Notice the ice fall in the background. To get some idea of scale, notice the person at the left center of the image.

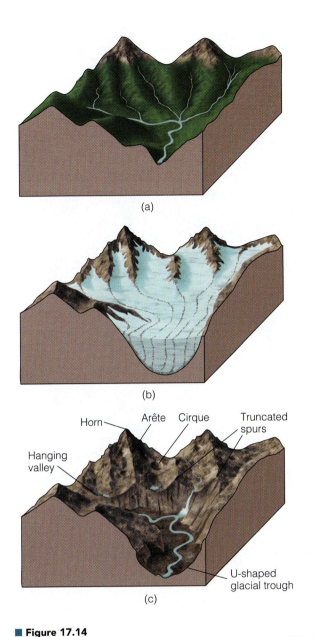

(a)

(b)

Horn   Arête   Cirque   Truncated spurs

Hanging valley

U-shaped glacial trough

(c)

■ **Figure 17.14**

Erosional landforms produced by valley glaciers. (a) A mountain area before glaciation. (b) The same area during the maximum extent of valley glaciers. (c) After glaciation.

U-Shaped Glacial Troughs A **U-shaped glacial trough** is one of the most distinctive features of valley glaciation (Figure 17.14c). Mountain valleys eroded by running water are typically V-shaped in cross section; that is, they have valley walls that descend steeply to a narrow valley bottom (Figure 17.14a). In contrast, valleys scoured by glaciers are deepened, widened, and straightened so that they have very steep or vertical walls but broad, rather flat valley floors; thus they exhibit a U-shaped profile.

Many glacial troughs contain triangular-shaped *truncated spurs,* which are cutoff or truncated ridges that extend into the preglacial valley (Figure 17.14c). Another common feature is a series of steps or rock basins in the valley floor where the glacier eroded rocks of varying resistance; many of the basins now contain small lakes.

During the Pleistocene, when glaciers were extensive, sea level was about 130 m lower than at present, so glaciers flowing into the sea eroded their valleys to much greater depths than they do now. When the glaciers melted at the end of the Pleistocene, sea level rose, and the ocean filled the lower ends of the glacial troughs so that now they are long, steep-walled embayments called **fiords.**

Fiords are restricted to high latitudes where glaciers exist even at low elevations, such as Alaska, western Canada, Scandinavia, Greenland, southern New Zealand, and southern Chile. Lower sea level during the Pleistocene was not entirely responsible for the formation of all fiords. Unlike running water, glaciers can erode a considerable distance below sea level. In fact, a glacier 500 m thick can stay in contact with the seafloor and effectively erode it to a depth of about 450 m before the buoyant effects of water cause the glacial ice to float! The depth of some fiords is impressive; some in Norway and southern Chile are about 1300 m deep.

Hanging Valleys Waterfalls form in several ways, but some of the world's highest and most spectacular are found in recently glaciated areas. Bridalveil Falls in Yosemite National Park, California, plunge from a **hanging valley,** which is a tributary valley whose floor is at a higher level than that of the main valley. Where the two valleys meet, the mouth of the hanging valley is perched far above the main valley's floor (Figure 17.14c). Accordingly, streams flowing through hanging valleys plunge over vertical or steep precipices.

Although not all hanging valleys form by glacial erosion, many do. As Figure 17.14 shows, the large glacier in the main valley vigorously erodes, whereas the smaller glaciers in tributary valleys are less capable of large-scale erosion. When the glaciers disappear, the smaller tributary valleys remain as hanging valleys.

Cirques, Arétes, and Horns Perhaps the most spectacular erosional landforms in areas of valley glaciation are at the upper ends of glacial troughs and along the divides that separate adjacent glacial troughs. Valley glaciers form and move out from steep-walled, bowl-shaped depressions called **cirques** at the upper end of their troughs (Figure 17.14c). Cirques are typically steep-walled on three sides, but one side is open and leads into the glacial trough. Some cirques slope continuously into the glacial trough, but many have a lip or threshold at their lower end (Figure 17.14c).

The details of cirque origin are not fully understood, but they apparently form by erosion of a preexisting depression on a mountain side. As snow and ice accumulate in the depression, frost wedging and plucking enlarge it until it takes on the typical cirque shape. In cirques with a lip or threshold, the glacial ice not only moves out but rotates as well, scouring out a depression rimmed by rock. These depressions commonly contain a small lake known as a *tarn.*

Valley Glaciers and Erosion

Valley glaciers effectively erode and produce several easily recognized landforms. Where glaciers move through mountain valleys, the valleys are deepened and widened giving them a distinctive U-shaped profile. The peaks and ridges rising above valley glaciers are also eroded and they become jagged and angular. Much of the spectacular scenery in Grand Teton National Park, Wyoming, Yosemite National Park, California, and Glacier National Park, Montana, resulted from erosion by valley glaciers. In fact, valley glaciers remain active in some of the mountains of western North America, especially in Alaska and Canada.

Part of southwestern Greenland (right) where valley glaciers merged to form an ice cap. Should these glaciers melt, several landforms like those in Figure 17.14c would be present. The Teton Range in Wyoming (above) acquired its angular peaks and ridges and broadly rounded valleys as a result of erosion by valley glaciers.

U-shaped glacial troughs. The glacial trough above is in northern Montana, whereas the one above right is in southern Germany. The lake is impounded behind a glacial deposit known as an end moraine. The steep-walled glacial trough in Norway (right) extends below sea level, so it is a fiord.

The bowl-shaped depression on Mount Wheeler in Great Basin National Park, Nevada, is a cirque. It has steep walls on three sides and opens out into a glacial trough.

James S. Monroe

Lake Helen on Lassen Peak in Lassen Volcanic National Park, California, is a tarn, that is, a lake in a cirque.

James S. Monroe

Sue Monroe

Bridalveil Falls in Yosemite National Park, California, plunges 190 m from a hanging valley. The valley in the foreground is a huge U-shaped glacial trough.

James S. Monroe

Swiss National Tourist Office

The Matterhorn (above) in Switzerland is a well-known horn. This view on the Jungfrau (left) in Switzerland shows two small Glaciers, a cirque headwall, and an arête.

GEOFOCUS

17.1

Glaciers in the U.S. and Canadian National Parks

Glaciers are present in several national parks in the United States and Canada, but here we concentrate on only two: Waterton Lakes National Park in Alberta and Glacier National Park in Montana. In 1932 both parks were designated an international peace park, the first of its kind.

Both parks have spectacular scenery; interesting wildlife such as mountain goats, bighorn sheep, and grizzly bears; and an impressive geologic history. The present-day landscapes resulted from deformation and uplift from Cretaceous to Eocene times, followed by deep erosion by glaciers and running water. Park visitors can see the results of the phenomenal forces at work during deformation by visiting sites where a large fault is visible, and glacial landforms such as U-shaped glacial troughs, arétes, cirques, and horns are some of the finest in North America (■ Figure 1).

Most of the rocks exposed in Glacier National Park belong to the Late Proterozoic-aged Belt Super-group,* whereas those in Waterton Lakes National Park are assigned to the Purcell Supergroup. The names differ north and south of the border, but the rocks are the same. These Belt-Purcell rocks are nearly 4000 m thick and were deposited between 1.45 billion and 850 million years ago. The rocks themselves are interesting, and some are attractive, especially red and green rocks consisting mostly of mud and thick limestone formations. In addition, many of the rocks contain a variety of sedimentary structures such as mud cracks, ripple marks, and cross-bedding that help geologists interpret how they were deposited in the first place.

Dark-colored mudstones and sandstones were deposited during the Cretaceous Period, when a marine transgression took place that covered a large part of North America.

(a)

(b)

■ **Figure 1**

The sharp angular peaks and ridges and rounded valleys are typical of areas eroded by valley glaciers such as (a) Glacier National Park, Montana, and (b) Waterton Lakes National Park, Alberta.

Supergroup is a geologic term for two or more groups that in turn are composed of two or more formations.

Grinnell Glacier 1850–1981

1850

1968   1937

1981

Carl H. Key/USGS

■ **Figure 2**

Grinnell Glacier in Glacier National Park, Montana. In 1850, at the end of the Little Ice Age, the glacier extended much farther and covered about 2.33 km². By 1981 its terminus had retreated to the position shown, and in 1993 it covered only about 0.88 km².

buried the entire area. In fact, several episodes of Pleistocene glaciation took place, but the evidence for the most recent one is most obvious. These glaciers flowed outward in all directions, and in the east they merged with the continental glacier covering most of Canada and the northern states. Much of the parks' landscapes developed during these glacial episodes as valleys were gouged deeper and widened, and cirques, arétes, and horns developed (Figure 1).

Today only a few dozen small glaciers remain active in the parks. But just like earlier glaciers, they continue to erode, transport, and deposit sediment, only at a considerably reduced rate. In fact, many of the 150 or so glaciers present in Glacier National Park in 1850 are now gone or remain only as patches of stagnant ice. And even among the others it is difficult to determine exactly how many are active because they are so small and move so slowly, only a few meters per year. They did, however, expand markedly during the Little Ice Age, but have since retreated (■ Figure 2). For example, Grinnell Glacier covered only 0.88 km² in 1993 as opposed to 2.33 km² in 1850 (Figure 2), and during the same time Sperry Glacier was down to 0.87 km² from 3.76 km².

It seems that these small glaciers are shrinking as a result of the 1°C increase in average summer temperatures in this region since 1900. According to one U.S. Geological Survey report, expected increased warming will eliminate the glaciers by 2030, and certainly by 2100 even if no additional warming takes place.

None of the active glaciers in either park can be reached by road, but several are visible from a distance. Nevertheless, Pleistocene glaciers and the remaining active ones were responsible for much of the striking scenery. Now weathering, mass wasting, and streams are modifying the glacial landscape.

The most impressive geologic structure in the parks is the Lewis overthrust,* a large fault along which Belt-Purcell rocks have moved at least 75 km eastward so that they now rest on much younger Cretaceous-aged rocks. (See "Types of Faults" on pages 378–379).

During the Pleistocene Epoch, glaciers formed and grew, overtopping the divides between valleys and thus forming an ice cap that nearly

---

*An overthrust fault is simply a very low angle thrust fault along which movement is usually measured in kilometers.

Cirques become wider and are cut deeper into mountainsides by headward erosion as a result of abrasion, plucking, and several mass wasting processes. For example, part of a steep cirque wall may collapse while frost wedging continues to pry loose rocks that tumble downslope, so a combination of processes erode a small mountainside depression into a large cirque; the largest one known is the Walcott Cirque in Antarctica, which is 16 km wide and 3 km deep.

The fact that cirques expand laterally and by headward erosion accounts for the origin of two other distinctive erosional features, arêtes and horns. **Arêtes**—narrow, serrated ridges—form in two ways. In many cases, cirques form on opposite sides of a ridge, and headward erosion reduces the ridge until only a thin partition of rock remains (Figure 17.14c). The same effect occurs when erosion in two parallel glacial troughs reduces the intervening ridge to a thin spine of rock.

The most majestic of all mountain peaks are **horns,** steep-walled, pyramidal peaks formed by headward erosion of cirques. For a horn to form, a mountain peak must have at least three cirques on its flanks, all of which erode headward (Figure 17.14c). Excellent examples of horns are Mount Assiniboine in the Canadian Rockies, the Grand Teton in Wyoming, and the most famous of all, the Matterhorn in Switzerland.

Continental Glaciers and Erosional Landforms

Areas eroded by continental glaciers tend to be smooth and rounded because these glaciers bevel and abrade high areas that projected into the ice. Rather than yielding the sharp, angular landforms typical of valley glaciation, they produce a landscape of rather flat topography interrupted by rounded hills.

In a large part of Canada, particularly the vast Canadian Shield region, continental glaciation has stripped off the soil and unconsolidated surface sediment, revealing extensive exposures of striated and polished bedrock. These areas have deranged drainage (see Figure 15.25e), numerous lakes and swamps, low relief, extensive bedrock exposures, and little or no soil. They are referred to as *ice-scoured plains* (■ Figure 17.15). Similar though smaller bedrock exposures are also widespread in the northern United States from Maine through Minnesota.

GLACIAL DEPOSITS

Both valley and continental glaciers effectively erode and transport, but eventually they deposit their sediment load as **glacial drift,** a general term for all deposits resulting from glacial activity. A vast sheet of Pleistocene glacial drift is present in the northern tier of the United States and adjacent parts of Canada. Smaller but similar deposits are found where valley glaciers existed or remain active. The appearance of these deposits may not be inspiring as are some landforms resulting from glacial erosion, but they are important as reservoirs of groundwater and in many areas they are exploited for their sand and gravel. As a matter of fact, sand and gravel constitute a large part of the mineral extraction economies of several states and provinces.

All glacial drift has been carried far from its source, but one conspicuous element of drift is boulders scattered around that were obviously not derived from the area in which they now rest. These **glacial erratics,** as they are called, have simply been eroded and transported from some distant region and then deposited (■ Figure 17.16). A good example is the popular decorative stone (called pudding stone) in Michigan consisting of quartzite containing conspicuous pieces of red jasper that was eroded from surface exposures in Ontario, Canada.

As noted, *glacial drift* is a general term, and geologists define two types of drift: till and stratified drift. **Till** consists of sediments deposited directly by glacial ice. They are not sorted by particle size or density, and they exhibit no layering or stratification. The till of both valley and continental glaciers is similar, but that of continental glaciers is much more extensive and usually has been transported much farther.

As opposed to till, **stratified drift** is layered—that is,

■ **Figure 17.15**

An ice-scoured plain in the Northwest Territories of Canada.

(a)

(b)

R. V. Dietrich

James S. Monroe

■ **Figure 17.16**

Glacial erratics at (a) Hammond, New York, and (b) Beaver Island in Lake Michigan. Both boulders have been transported far from their sources and are unlike the rocks at their present locations.

stratified—and it invariably exhibits some degree of sorting by particle size. As a matter of fact, most stratified drift is actually layers of sand and gravel or mixtures thereof that accumulated in braided stream channels. In Chapter 15 we mentioned that streams issuing from melting glaciers are commonly braided because they receive more sediment than they can effectively transport.

Landforms Composed of Till

Landforms composed of till include several types of *moraines* and elongated hills known as *drumlins*.

End Moraines   The terminus of either a valley or a continental glacier may become stabilized in one position for some period of time, perhaps a few years or even decades. Stabilization of the ice front does not mean that the glacier has ceased flowing, only that it has a balanced budget (Figure 17.6). When an ice front is stationary, flow within the glacier continues, and any sediment transported within or upon the ice is dumped as a pile of rubble at the glacier's terminus (■ Figures 17.17 and 17.18). These deposits are **end moraines,** which continue to grow as long as the ice front remains stationary. End moraines of valley glaciers are crescent-shaped

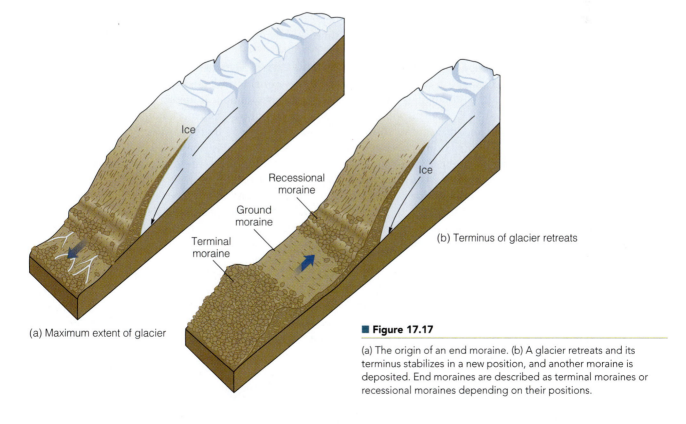

Ice

Recessional moraine

Ground moraine

Terminal moraine

(a) Maximum extent of glacier

(b) Terminus of glacier retreats

Ice

■ **Figure 17.17**

(a) The origin of an end moraine. (b) A glacier retreats and its terminus stabilizes in a new position, and another moraine is deposited. End moraines are described as terminal moraines or recessional moraines depending on their positions.

End moraine

James S. Monroe

(a)

(b)

James S. Monroe

■ **Figure 17.18**

(a) An end moraine deposited by a valley glacier. This particular end moraine is also a terminal moraine because it is the one most distant from the glacier's source. (b) Closeup of an end moraine. Notice that the deposit is not sorted by particle size, and it shows no layering or stratification.

ridges of till spanning the valley occupied by the glacier. Those of continental glaciers similarly parallel the ice front but are much more extensive.

Following a period of stabilization, a glacier may advance or retreat, depending on changes in its budget. If it advances, the ice front overrides and modifies its former moraine. Should it have a negative budget, though, the ice front retreats toward the zone of accumulation. As the ice front recedes, till is deposited as it is liberated from the melting ice and forms a layer of **ground moraine** (Figure 17.17b). Ground moraine has an irregular, rolling topography, whereas end moraine consists of long ridgelike accumulations of sediment.

After a glacier has retreated for some time, its terminus may once again stabilize, and it deposits another end moraine. Because the ice front has receded, such moraines are called **recessional moraines** (Figure 17.17b). During the Pleistocene Epoch, continental glaciers in the midcontinent region extended as far south as southern Ohio, Indiana, and Illinois. Their outermost end moraines, marking the greatest extent of the glaciers, go by the special name **terminal moraine** (valley glaciers also deposit terminal moraines). As the glaciers retreated from the positions where their terminal moraines were

deposited, they temporarily ceased retreating numerous times and deposited dozens of recessional moraines.

Lateral and Medial Moraines   Valley glaciers transport considerable sediment along their margins. Much of this sediment is abraded and plucked from the valley walls, but a significant amount falls or slides onto the glacier's surface by mass wasting processes. In any case, this sediment is deposited as long ridges of till called **lateral moraines** along the margin of the glacier (■ Figure 17.19).

Where two lateral moraines merge, as when a tributary glacier flows into a larger glacier, a **medial moraine** forms (Figure 17.19). A large glacier will often have several dark stripes of sediment on its surface, each of which is a medial moraine (see chapter opening photo). Although medial moraines are identified by their position on a valley glacier, they are, in fact, formed from the coalescence of two lateral moraines. One can generally determine how many tributaries a valley glacier has by the number of its medial moraines (Figure 17.19).

Drumlins   In many areas where continental glaciers deposited till, the till has been reshaped into elongated

Engineering Mechanics, Virginia Polytechnic Institute and State University

(a)

Peter Kresan

(b)

■ Figure 17.19

(a) Lateral and medial moraines on a glacier in Alaska. Notice that where the two large tributary glaciers converge, two lateral moraines merge to form a medial moraine. (b) The two parallel ridges extending from this mountain valley are lateral moraines.

hills known as **drumlins** (■ Figures 17.20 and 17.21). Some drumlins are as much as 50 m high and 1 km long, but most are much smaller. From the side, a drumlin looks like an inverted spoon, with the steep end on the side from which the glacial ice advanced and the gently sloping end pointing in the direction of ice movement (■ Figure 17.21). Drumlins are rarely found as single, isolated hills; instead, they occur in *drumlin fields* that contain hundreds or thousands of drumlins. Drumlin fields are found in several states and Ontario, Canada, but perhaps the finest example is near Palmyra, New York.

No one fully understands how drumlins originate. According to one hypothesis, they form when till beneath a glacier is reshaped into streamlined hills as the ice moves over it by plastic flow. Another hypothesis holds that huge floods of glacial meltwater modify till into drumlins.

Landforms Composed of Stratified Drift

As already noted, stratified drift is a type of glacial deposit that exhibits sorting and layering, both indications that it was deposited by running water. Stratified drift is deposited by streams discharging from both valley and continental glaciers, but as one would expect, it is more extensive in areas of continental glaciation.

Outwash Plains and Valley Trains Glaciers discharge meltwater laden with sediment most of the time, except perhaps during the coldest months. This meltwater forms a series of braided streams that radiate out from the front of continental glaciers over a wide region. So much sediment is supplied to these streams that much of it is deposited within the channels as sand and gravel bars. The vast blanket of sediments so formed is an **outwash plain** (■ Figures 17.20 and 17.22).

Occasionally, huge quantities of meltwater are discharged from glaciers when, for instance, a volcano erupts beneath them. A recent example is an eruption beneath Iceland's largest ice cap, which melted the base of the glacier. Meltwater accumulated in a caldera, also beneath the ice cap, and on November 5, 1996, suddenly burst forth in Iceland's largest flood in 60 years. Huge blocks of ice estimated to weigh 900 metric tons were ripped from the glacier and deposited nearly 5 km away. Damage to roads, utility lines, and bridges amounted to about $15 million, but no lives were lost.

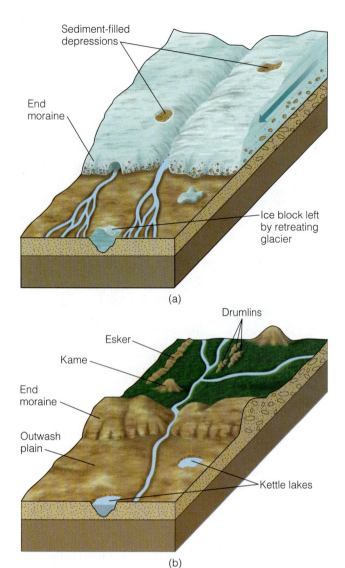

Sediment-filled depressions

End moraine

Ice block left by retreating glacier

(a)

Drumlins

Esker

Kame

End moraine

Outwash plain

Kettle lakes

(b)

Valley glaciers also discharge large amounts of melt-water and, like continental glaciers, have braided streams extending from them. However, these streams are confined to the lower parts of glacial troughs, and their long, narrow deposits of stratified drift are known as **valley trains** (Figure 17.22b).

Outwash plains, valley trains, and some moraines commonly contain numerous circular to oval depressions, many of which contain small lakes. These depressions are *kettles*; they form when a retreating ice sheet or valley glacier leaves a block of ice that is subsequently partly or wholly buried (■ Figures 17.20 and 17.23a). When the ice block eventually melts, it leaves a depression; if the depression extends below the water table, it becomes the site of a small lake. Some outwash plains have so many kettles that they are called *pitted outwash plains*.

Kames and Eskers **Kames** are conical hills as high as 50 m composed of stratified drift (Figures 17.20 and 17.23b). Many kames form when a stream deposits sediment in a depression on a glacier's surface; as the ice melts, the deposit is lowered to the surface. They also form in cavities within or beneath stagnant ice.

Long sinuous ridges of stratified drift, many of which meander and have tributaries, are **eskers** (Figures 17.20 and 17.23c). Most eskers have sharp crests and sides that slope at about 30 degrees. Some are quite high, as much as 100 m, and can be traced for more than 500 km. Eskers are usually found in areas once covered by continental

■ **Figure 17.20**

Two stages in the origin of kettles, kames, eskers, drumlins, and outwash plains: (a) during glaciation and (b) after glaciation.

Carl Guell Slide Collection

■ **Figure 17.21**

These streamlined hills are drumlins.

(a)

(b)

■ Figure 17.22

(a) Outwash plain deposits in Michigan. (b) Deposit of this valley train in Alaska consists of stratified drift.

(a)

(b)

(c)

■ Figure 17.23

(a) A kettle in a moraine in Alaska. (b) This small hill in Wisconsin is a kame. (c) This sinuous ridge near Dahlen, North Dakota, is an esker. The origin of kettles, kames, and eskers is shown in Figure 17.20.

GEOLOGY
IN UNEXPECTED PLACES

Evidence of Glaciation in New York City

Our discussion of glaciers thus far has concentrated on Antarctica, Greenland, the Canadian Arctic islands, and high mountains around the world. However, you can see the effects of continental glaciers in some urban areas. For instance, Central Park in New York City is a good place to start. As a matter of fact, the area now occupied by the city as well as Long Island to the east is an excellent place to see a variety of erosional and depositional landforms created by a continental glacier as recently as 20,000 years ago. Indeed, Long Island is made up of end moraines and outwash.

The area now comprising Central Park was an irregular region with rock outcroppings, bluffs, and marshes that was deemed unsuitable for development. Nevertheless, when the city acquired the initial 700 acres of land for the park during the mid-1800s it was necessary to displace about 1600 poor residents. In 1863 an additional 143 acres of land was acquired, bringing the park to its present size.

In Central Park many of the rocks resting on the surface are glacial erratics—that is, boulders unlike the underlying rock that were carried far from their sources similar to those in Figure 17.16. Glacial polish, striations, and grooves are

easily seen on many of the erratics as well as the exposed bedrock. In fact, the rock exposure in Figure 1 is a large roche moutonnée that formed as the "upstream" side of a bedrock projection was abraded and polished, whereas plucking accounts for the irregular "downstream" side. This roche moutonnée formed just like the one shown in Figure 17.11. The orientation of the roche moutonnée and the striations and grooves on this and other rocks indicate that the glacier moved from northwest to southeast across what is now Central Park.

■ **Figure 1**

A roche moutonnée in Central Park, New York City.

glaciers, but they are also associated with large valley glaciers. The sorting and stratification of the sediments within eskers clearly indicate deposition by running water. The properties of ancient eskers and observations of present-day glaciers indicate that they form in tunnels beneath stagnant ice (Figure 17.23c). Excellent examples of eskers can be seen at Kettle Moraine State Park in Wisconsin and in several other states, but the most extensive eskers in the world are in northern Canada.

Deposits in Glacial Lakes

Numerous lakes exist in areas of glaciation. Some formed as a result of glaciers scouring out depressions;

others occur where a stream's drainage was blocked; and others are the result of water accumulating behind moraines or in kettles. Regardless of how they formed, glacial lakes, like all lakes, are areas of deposition. Sediment may be carried into them and deposited as small deltas, but of special interest are the fine-grained deposits. Mud deposits in glacial lakes are commonly finely laminated (having layers less than 1 cm thick) and consist of alternating light and dark layers. Each light–dark couplet is a *varve* (■ Figure 17.24a), which represents an annual episode of deposition; the light layer formed during the spring and summer and consists of silt and clay; the dark layer formed during the winter when the smallest particles of clay and organic

(a)

Varves

Till

(b)

■ **Figure 17.24**

(a) Glacial varves with a dropstone. Each varve, a dark–light couplet, is an annual deposit. (b) This exposure in Alaska shows varves that formed in a glacial lake overlying glacial till.

matter settled from suspension as the lake froze over. The number of varves indicates how many years a glacial lake has existed.

Another distinctive feature of glacial lakes containing varved deposits is the presence of *dropstones* (Figure 17.24). These are pieces of gravel, some of boulder size, in otherwise very fine-grained deposits. The presence of varves indicates that currents and turbulence in such lakes was minimal; otherwise, clay and organic matter would not have settled from suspension. How then can we account for dropstones in a low-energy environment? Most of them were probably carried into the lakes by icebergs that eventually melted and released sediment contained in the ice.

THE ICE AGE—WHAT WAS IT, AND WHEN DID IT TAKE PLACE?

In hindsight, it is hard to believe that so many competent naturalists of the 1800s were skeptical that widespread glaciers existed on the northern continents during the not too distant past. Many naturalists invoked the biblical flood to account for the large boulders throughout Europe that occur far from their sources. Others believed that the boulders were rafted to their present positions by icebergs floating in floodwaters. It was not until 1837 that the Swiss naturalist Louis Agassiz argued convincingly that the displaced boulders, many sedimentary deposits, polished and striated bedrock, and many valleys in Europe resulted from huge ice masses moving over the land.

What Would You Do ?

During a visit to the Pacific Northwest, you notice that a stream has cut a gorge in which you observe the following. In the lower part of the gorge is a bedrock projection with features similar to those shown in Figure 17.11b. Overlying this bedrock is a sequence of thin (2–4 mm thick) alternating layers of dark and light very fine-grained materials (silt and clay) with a few boulders measuring 10 to 15 cm across. And in the uppermost part of the gorge is a deposit of mud, sand, and gravel showing no layering or sorting. How would you decipher the geologic history of these deposits?

We know today that the Pleistocene Ice Age began about 1.6 million years ago (MYA) and consisted of several intervals of glacial expansion separated by warmer interglacial periods. At least four major episodes of Pleistocene glaciation have been recognized in North America, and six or seven major glacial advances and retreats are recognized in Europe. And it now appears that evidence for at least 20 warm–cold cycles is present in deep-sea cores. Based on the best available evidence, the Pleistocene ended about 10,000 years ago (YA). But geologists do not know if the present interglacial period will persist indefinitely, or whether we will enter another glacial interval.

The onset of glacial conditions really began about 40 MYA when surface ocean waters at high southern latitudes suddenly cooled. By about 38 MYA, glaciers had

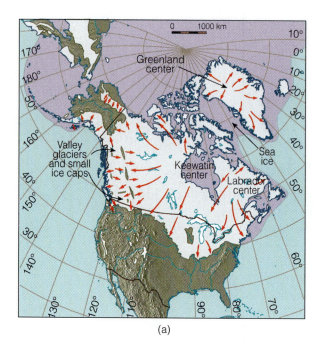

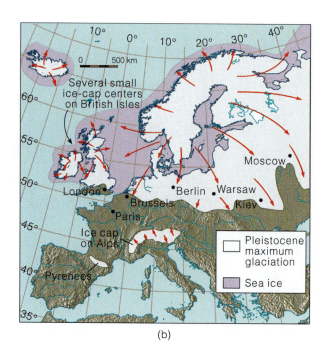

(a)   (b)

■ **Figure 17.25**

(a) Centers of ice accumulation and maximum extent of Pleistocene glaciation in North America. (b) Centers of ice accumulation and directions of ice movement in Europe during the maximum extent of Pleistocene glaciation.

formed in Antarctica, but a continuous ice sheet did not develop there until 15 MYA. Following a brief warming trend during the Late Tertiary Period, ice sheets began forming in the Northern Hemisphere, and by 1.6 MYA the Pleistocene Ice Age was under way. At their greatest extent, Pleistocene glaciers covered about three times as much of Earth's surface as they do now (■ Figure 17.25). Large areas of North America were covered by glacial ice, as were Greenland, Scandinavia, Great Britain, Ireland, and a large part of northern Russia. Mountainous areas also experienced an expansion of valley glaciers and the development of ice caps.

Pleistocene Climates

As one would expect, the climatic effects responsible for Pleistocene glaciation were worldwide. Contrary to popular belief, though, the world was not as frigid as it is commonly portrayed in cartoons and movies. During times of glacier growth, those areas in the immediate vicinity of the glaciers experienced short summers and long, wet winters.

Areas outside the glaciated regions experienced varied climates. During times of glacial growth, lower ocean temperatures reduced evaporation so that most of the world was drier than it is today, but some areas that are arid now were much wetter. For example, as the cold belts at high latitudes expanded, the temperate, subtropical, and tropical zones were compressed toward the equator, and the rain that now falls on the Mediterranean shifted so that it fell on the Sahara of North Africa, enabling lush forests to grow in what is now desert. Califor-

nia and the arid southwestern United States were also wetter because a high-pressure zone over the northern ice sheet deflected Pacific winter storms south.

Following the Pleistocene, mild temperatures prevailed between 8000 and 6000 YA. After this warm period, conditions gradually became cooler and moister, favoring the growth of valley glaciers on the Northern Hemisphere continents. Careful studies of the deposits at the margins of present-day glaciers reveal that during the last 6000 years (a time called the *Neoglaciation*), glaciers expanded several times. The last expansion, which took place between 1500 and the mid-to-late 1800s, was the Little Ice Age (see the Introduction).

Pluvial and Proglacial Lakes

During the Pleistocene, many of the basins in the western United States contained large lakes that formed as a result of more precipitation and overall cooler temperatures (especially during the summer), which lowered the evaporation rate (■ Figure 17.26). The largest of these *pluvial lakes*, as they are called, was Lake Bonneville, which attained a maximum size of 50,000 km² and a depth of at least 335 m (Figure 17.26). The vast salt deposits of the Bonneville Salt Flats west of Salt Lake City in Utah formed as parts of this ancient lake dried up: Great Salt Lake is simply the remnant of this once much larger lake.

Another large pluvial lake existed in Death Valley, California, which is now the hottest, driest place in North America. During the Pleistocene, Death Valley received enough rainfall to maintain a lake 145 km long

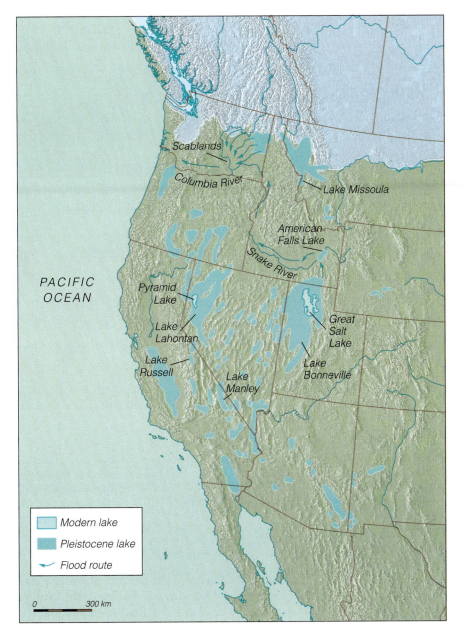

■ **Figure 17.26**

Pleistocene lakes in the western United States. Lake Missoula was a proglacial lake but all the others were pluvial lakes. The Great Salt Lake in Utah and Pyramid Lake in Nevada are shrunken remnants of once much more extensive lakes.

until the glacial ice along its northern margin melted, at which time the lake drained northward into Hudson Bay.

Before the Pleistocene, no large lakes existed in the Great Lakes region, which was then an area of lowlands with broad stream valleys draining to the north. As the glaciers advanced southward, they eroded the valleys more deeply, forming what were to become the Great Lakes basins; four of the five basins were eroded below sea level.

At their greatest extent, the glaciers covered the entire Great Lakes region and extended far to the south. As the ice sheet retreated north during the late Pleistocene, the ice front periodically stabilized, and numerous recessional moraines were deposited (Figure 17.17b). By about 14,000 YA, parts of the Lake Michigan and Lake Erie basins were ice free, and glacial meltwater began to form proglacial lakes (■ Figure 17.27). As the ice sheet continued to retreat—although periodically interrupted by minor readvances of the ice front—the Great Lakes basins were uncovered, and the lakes expanded until they eventually reached their present size and configuration. Currently, the Great Lakes contain nearly 23,000 km³ of water, about 18% of the water in all freshwater lakes.

Another proglacial lake of interest is glacial Lake Missoula in Montana, which formed when ice dammed the drainage of the Clark Fork River in Idaho (■ Figure 17.28). The shorelines of this 7800-km² lake are clearly visible on the mountainsides around Missoula, Montana. On more than one occasion the dam collapsed, resulting in huge floods responsible for gravel ridges, called giant ripple marks, as well as the *scablands* of eastern Washington where surface deposits were scoured, thus exposing the underlying bedrock (Figure 17.28).

Glaciation and Changes in Sea Level

More than 70 million km³ of snow and ice covered the continents during the maximum extent of Pleistocene glaciers. The storage of ocean waters in glaciers lowered sea level 130 m and exposed vast areas of the continen-

and 178 m deep. When the lake evaporated, the dissolved salts were precipitated on the valley floor; some of these evaporite deposits, especially borax, are important mineral resources (see Chapter 18).

In contrast to pluvial lakes, which form far from glaciers, *proglacial lakes* form by meltwater accumulating along the margins of glaciers. In fact, one shoreline of proglacial lakes is the ice front itself. Lake Agassiz, named in honor of the naturalist Louis Agassiz, was a large proglacial lake covering about 250,000 km², mostly in Manitoba, Saskatchewan, and Ontario, Canada, but extending into North Dakota and Minnesota. It persisted

13,000 years ago

11,500 years ago

9,500 years ago

6,000 years ago

■ **Figure 17.27**

Four stages in the evolution of the Great Lakes. As the glacial ice retreated northward, the lake basins began filling with meltwater. The dotted lines indicate the present-day shorelines of the lakes. Source: From V. K. Prest, *Economic Geology Report 1*, 5d, 1970, p. 90–91 (fig. 7-6), Geological Survey of Canada. Department of Energy, Mines, and Resources, reproduced with permission of Minister of Supply and Services.

tal shelves, which were quickly covered with vegetation. Indeed, a land bridge existed across the Bering Strait from Alaska to Siberia. Native Americans crossed the Bering land bridge, and various animals migrated between the continents; the American bison, for example, migrated from Asia. The British Isles were connected to Europe during the glacial intervals because the shallow floor of the North Sea was above sea level. When the glaciers disappeared, these areas were again flooded, drowning the plants and forcing the animals to migrate.

Lower sea level during the Pleistocene also affected the base level of most rivers and streams. When sea level dropped, streams eroded deeper valleys as they sought to adjust to a new lower base level (see Chapter 15). Stream channels in coastal areas were extended and deepened along the emergent continental shelves. When sea level rose at the end of the Pleistocene, the lower ends of the valleys along the East Coast of North America were flooded and are now important harbors (see Chapter 19).

A tremendous quantity of water is still stored on land in present-day glaciers (see Figure 15.3). If these glaciers should melt, sea level would rise about 70 m, flooding all coastal areas of the world where many of the large population centers are located.

Glaciers and Isostasy

In Chapter 10 we discussed the concept of isostasy and noted that the crust responds to an increased load by subsiding, and it rises when a load on it is reduced. When the Pleistocene ice sheets formed and increased in size, their weight caused the crust to respond by slowly subsiding deeper into the mantle. In some places, the surface was depressed as much as 300 m below preglacial elevations. As the ice sheets disappeared, the downwarped areas gradually rebounded to their former positions. As noted in Chapter 10, parts of Scandinavia are still rebounding at a rate of about 1 m per century (see Figure 10.17). And more than 100 m of isostatic rebound has taken place in northeastern Canada during the last 6000 years.

We noted previously that the Great Lakes evolved as the glaciers retreated to the north. As one would expect, isostatic rebound began as the ice front retreated north. Rebound began first in the southern part of the region because that area was free of ice first. Furthermore, the greatest loading by glaciers, and hence the greatest crustal depression, occurred farther north in Canada in the zones of accumulation. For these reasons, rebound

(a)

(b)

■ **Figure 17.28**

(a) The horizontal lines on Sentinel Mountain at Missoula, Montana, are wave-cut shorelines of glacial Lake Missoula. (b) These gravel ridges are the so-called giant ripple marks that formed when glacial Lake Missoula drained across this area near Camas Hot Springs, Montana.

has not been evenly distributed over the entire glaciated area: It increases in magnitude from south to north. As a result of this uneven isostatic rebound, coastal features in the Great Lakes region, such as old shorelines, are now elevated higher above their former levels in the north and thus slope to the south.

WHAT CAUSES ICE AGES?

So far we have examined the effects of glaciation but have not addressed the central questions of what causes large-scale glaciation and why there have been so few episodes of widespread glaciation. For more than a century, scientists have attempted to develop a comprehensive theory explaining all aspects of ice ages, but they have not yet been completely successful. One reason for their lack of success is that the climatic changes responsible for glaciation, the cyclic occurrence of glacial–interglacial episodes, and short-term events such as the Little Ice Age operate on vastly different time scales.

Only a few periods of glaciation are recognized in the geologic record, each separated from the others by long intervals of mild climate. Such long-term climatic changes probably result from slow geographic changes related to plate tectonic activity. Moving plates carry continents to high latitudes where glaciers exist, provided that they receive enough precipitation as snow. Plate collisions, the subsequent uplift of vast areas far above sea level, and the changing atmospheric and oceanic circulation patterns caused by the changing shapes and positions of plates also contribute to long-term climatic change.

The Milankovitch Theory

Changes in Earth's orbit as a cause of intermediate-term climatic events were first proposed during the mid-1800s, but the idea was made popular during the 1920s by the Serbian astronomer Milutin Milankovitch. He proposed that minor irregularities in Earth's rotation and orbit are sufficient to alter the amount of solar radiation received at any given latitude and hence bring about climate changes. Now called the **Milankovitch theory,** it was initially ignored but has received renewed interest since the 1970s and is widely accepted.

Milankovitch attributed the onset of the Pleistocene Ice Age to variations in three aspects of Earth's orbit. The first is *orbital eccentricity,* which is the degree to which Earth's orbit departs from a circle (■ Figure 17.29a). When the orbit is nearly circular, both the Northern and Southern Hemispheres have similar contrasts between the seasons. However, if the orbit is more elliptic, hot summers and cold winters will occur in one hemisphere, whereas warm summers and cool winters will take place in the other hemisphere. Calculations indicate a roughly 100,000-year cycle between times of maximum eccentricity, which corresponds closely to the 20 warm–cold climatic cycles that took place during the Pleistocene.

Milankovitch also pointed out that the angle between Earth's axis and a line perpendicular to the plane of the ecliptic shifts about 1.5 degrees from its current value of 23.5 degrees during a 41,000-year cycle (Figure 17.29b). Although changes in *axial tilt* have little effect on equatorial latitudes, they strongly affect the amount of solar radiation received at high latitudes and the duration of the dark period at and near Earth's poles. Coupled with the third aspect of Earth's orbit, precession of the equinoxes, high latitudes might receive as much as 15% less solar radiation, certainly enough to affect glacial growth and melting.

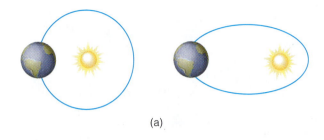

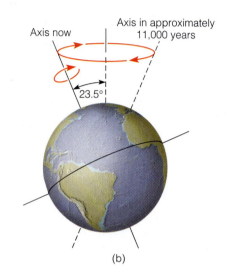

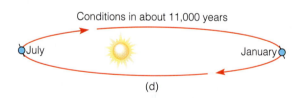

■ **Figure 17.29**

Minor irregularities in Earth's rotation and orbit may affect climatic changes. (a) Earth's orbit varies from nearly a circle (left) to an ellipse (right) and back again in about 100,000 years. (b) Earth moves around its orbit while spinning about its axis, which is tilted to the plane of its orbit around the Sun at 23.5 degrees and points toward the North Star. Earth's axis of rotation slowly moves and traces out the path of a cone in space. (c) At present, Earth is closest to the Sun in January, when the Northern Hemisphere experiences winter. (d) In about 11,000 years, as a result of precession, Earth will be closer to the Sun in July, when summer occurs in the Northern Hemisphere.

The last aspect of Earth's orbit that Milankovitch cited is *precession of the equinoxes,* which refers to a change in the time of the equinoxes. At present, the equinoxes take place on about March 21 and September 21 when the Sun is directly over the equator. But as Earth rotates on its axis, it also wobbles as its axial tilt varies 1.5 degrees from its current value, thus changing the time of the equinoxes. Taken alone, the time of the equinoxes has little climatic effect, but changes in Earth's axial tilt also change the time of *aphelion* and *perihelion,* which are, respectively, when Earth is farthest from and closest to the Sun during its orbit (Figure 17.29c, d). Earth is now at perihelion, closest to the Sun, during Northern Hemisphere winters, but in about 11,000 years perihelion will be in July. Accordingly, Earth will be at aphelion, farthest from the Sun, in January and have colder winters.

Continuous variations in Earth's orbit and axial tilt cause the amount of solar heat received at any latitude to vary slightly through time. The total heat received by the planet changes little, but according to Milankovitch, and now many scientists agree, these changes cause complex climatic variations and provided the triggering mechanism for the glacial–interglacial episodes of the Pleistocene.

Short-Term Climatic Events

Climatic events with durations of several centuries, such as the Little Ice Age, are too short to be accounted for by plate tectonics or Milankovitch cycles. Several hypotheses have been proposed, including variations in solar energy and volcanism.

Variations in solar energy could result from changes within the Sun itself or from anything that would reduce the amount of energy Earth receives from the Sun. The latter could result from the solar system passing through clouds of interstellar dust and gas or from substances in the atmosphere reflecting solar radiation back into space. Records kept over the past 85 years indicate that during this time the amount of solar radiation has varied only slightly. Although variations in solar energy may influence short-term climatic events, such a correlation has not been demonstrated.

During large volcanic eruptions, tremendous amounts of ash and gases are spewed into the atmosphere where they reflect incoming solar radiation and thus reduce atmospheric temperatures. Small droplets of sulfur gases remain in the atmosphere for years and can have a significant effect on climate. Several large-scale volcanic events have occurred, such as the 1815 eruption of Tambora, and are known to have had climatic effects. However, no relationship between periods of volcanic activity and periods of glaciation has yet been established.

REVIEW WORKBOOK

Chapter Summary

- Glaciers are masses of ice on land that move by plastic flow and basal slip. Glaciers currently cover about 10% of the land surface and contain 2.15% of all water on Earth.

- Valley glaciers are confined to mountain valleys and flow from higher to lower elevations, whereas continental glaciers cover vast areas and flow out in all directions from a zone of accumulation.

- A glacier forms when winter snowfall exceeds summer melt and therefore accumulates year after year. Snow is compacted and converted to glacial ice, and when the ice is about 40 m thick, pressure causes it to flow.

- The behavior of a glacier depends on its budget, which is the relationship between accumulation and wastage. If a glacier possesses a balanced budget, its terminus remains stationary; a positive or negative budget results in advance or retreat of the terminus, respectively.

- Glaciers move at varying rates depending on slope, discharge, and season. Valley glaciers tend to flow more rapidly than continental glaciers.

- Glaciers effectively erode and transport because they are solids in motion. They are particularly effective at eroding soil and unconsolidated sediment, and they can transport any size sediment supplied to them.

- Continental glaciers transport most of their sediment in the lower part of the ice, whereas valley glaciers carry sediment in all parts of the ice.

- Erosion of mountains by valley glaciers produces several sharp, angular landforms, including cirques, arêtes, and horns. U-shaped glacial troughs, fiords, and hanging valleys are also products of valley glaciation.

- Continental glaciers abrade and bevel high areas, forming a smooth, rounded landscape, known as an ice-scoured plain.

- Depositional landforms include moraines, which are ridgelike accumulations of till. The several types of moraines include terminal, recessional, lateral, and medial moraines.

- Drumlins are composed of till that was apparently reshaped into streamlined hills by continental glaciers or floods of glacial meltwater.

- Stratified drift consists of sediments deposited by meltwater streams issuing from glaciers; it is found in outwash plains and valley trains. Ridges called eskers and conical hills called kames are also composed of stratified drift.

- During the Pleistocene Epoch, glaciers covered about 30% of the land surface, especially in the Northern Hemisphere and Antarctica. Several intervals of widespread glaciation, separated by interglacial periods, occurred in North America.

- Areas far beyond the ice were affected by Pleistocene glaciation; climate belts were compressed toward the equator, large pluvial lakes existed in what are now arid regions, and sea level was as much as 130 m lower than at present.

- Loading of Earth's crust by Pleistocene glaciers caused isostatic subsidence. When the glaciers melted, isostatic rebound began and continues in some areas.

- Major glacial intervals separated by tens or hundreds of millions of years probably occur as a result of the changing positions of plates, which in turn cause changes in oceanic and atmospheric circulation patterns.

- Currently, the Milankovitch theory is widely accepted as the explanation for glacial–interglacial intervals.

- The reasons for short-term climatic changes, such as the Little Ice Age, are not understood. Two proposed causes for these events are changes in the amount of solar energy received by Earth and volcanism.

Important Terms

abrasion (p. 511)
arête (p. 518)
basal slip (p. 505)
cirque (p. 513)
continental glacier (p. 506)
drumlin (p. 521)
end moraine (p. 519)
esker (p. 522)
fiord (p. 513)
firn (p. 503)
glacial budget (p. 507)
glacial drift (p. 518)
glacial erratic (p. 518)

glacial ice (p. 503)
glacial polish (p. 511)
glacial striation (p. 511)
glacial surge (p. 510)
glacier (p. 503)
ground moraine (p. 520)
hanging valley (p. 513)
horn (p. 518)
ice cap (p. 506)
kame (p. 522)
lateral moraine (p. 520)
medial moraine (p. 520)

Milankovitch theory (p. 529)
outwash plain (p. 521)
plastic flow (p. 505)
recessional moraine (p. 520)
stratified drift (p. 518)
terminal moraine (p. 520)
till (p. 518)
U-shaped glacial trough (p. 513)
valley glacier (p. 505)
valley train (p. 522)
zone of accumulation (p. 507)
zone of wastage (p. 507)

Review Questions

1. Which one of the following statements is correct?

   a. _____ a cirque forms where two valley glaciers merge; b. _____ Ice Age glacial deposits are found as far south as Alabama; c. _____ most of northern Europe was covered by glacial ice during the Little Ice Age; d. _____ many of the valley glaciers in the United States are in the Appalachian Mountains; e. _____ less than 1% of all glacial ice is found outside Antarctica and Greenland.

2. If a glacier deposits a terminal moraine and then retreats and deposits another moraine, the latter is known as a(n) _____ moraine:

   a. _____ lateral; b. _____ medial;
   c. _____ recessional; d. _____ optimal;
   e. _____ outwash.

3. An arête is a(n):

   a. _____ knifelike ridge between glacial troughs; b. _____ deposit of unsorted sand and gravel; c. _____ pyramid-shaped peak eroded by continental glaciers; d. _____ type of glacier found in mountain valleys; e. _____ outwash plain with numerous kettles.

4. When freshly fallen snow compacts and partly melts and refreezes, it forms granular ice known as:

   a. _____ kame; b. _____ till; c. _____ firn;
   d. _____ drift; e. _____ cirque.

5. Glaciers move mostly by:

   a. _____ surging; b. _____ basal slip;
   c. _____ lateral compression;
   d. _____ plastic flow; e. _____ abrasion.

6. A bowl-shaped depression on a mountainside at the upper end of a glacial trough is a(n):

   a. valley train; b. _____ cirque; c. _____ esker;
   d. _____ drumlin; e. _____ erratic.

7. During the Ice Age or _____ Epoch, glaciers covered about _____ of the land surface.

   a. _____ Cretaceous/75%; b. _____ Proterozoic/10%; c. _____ Paleozoic/50%; d. _____ Pleistocene/30%; e. _____ Mesozoic/15%.

8. If a glacier has a balanced budget:

   a. _____ it stops moving; b. _____ its terminus remains stationary; c. _____ its rate of wastage exceeds its rate of accumulation;
   d. _____ the glacier's length decreases;
   e. _____ crevasses no longer form.

9. An ice-scoured plain is a(n):

   a. _____ subdued landscape resulting from erosion by a continental glacier; b. _____ area with many cirques, horns, and arêtes;
   c. _____ vast area covered by outwash deposits; d. _____ region in which drumlins and eskers are commonly found;
   e. _____ type of deposit consisting of alternating dark- and light-colored layers of clay.

10. The two present-day areas with continental glaciers are:

   a. _____ Glacier National Park, Montana, and Waterton National Park, Canada; b. _____ Mount Baker, Washington, and the Sierra Nevada of California; c. _____ Antarctica and Greenland; d. _____ Scandinavia and Canada; e. _____ Iceland and Baffin Island.

11. When at least three glaciers erode a single mountain peak, it yields a pyramid-shaped peak known as a(n):

    a. _____ horn; b. _____ drift; c. _____ esker; d. _____ striation; e. _____ fiord.

12. A glacially transported boulder now lying far from its source is a(n):

    a. _____ firn; b. _____ erratic; c. _____ moraine; d. _____ varve; e. _____ dropstone.

13. How is it possible for glaciers to erode below sea level whereas steams cannot?

14. A valley glacier has a cross-sectional area of 400,000 m² and a flow velocity of 2 m/day. How long will it take for 1 km³ of ice to move past a given point?

15. In your travels you encounter a roadside rock exposure consisting of alternating layers of dark- and light-colored laminated mud containing a few boulders measuring 20 to 40 cm across. Explain the sequence of events responsible for deposition.

16. Several of the Cascade Range volcanoes in California, Oregon, and Washington have glaciers on them. What two factors account for glaciers on these mountains?

17. Explain in terms of the glacial budget how a once-active glacier becomes stagnant.

18. What kinds of evidence would indicate that a now ice-free area was once covered by a continental glacier?

19. What are basal slip and plastic flow, what causes each, and how do they vary seasonally?

20. How does the Milankovitch theory explain the onset of Pleistocene episodes of glaciation?

21. What is the firn limit on a glacier, and how does its position relate to a glacier's budget?

22. What are hanging valleys and how do they originate?

23. How do terminal and recessional moraines form?

24. How does glacial ice originate and why is it considered a rock?

World Wide Web Activities

PHYSICAL Geology⇌Now Assess your understanding of this chapter's topics with additional quizzing and comprehensive interactivities at

http://earthscience.brookscole.com/physgeo5e

as well as current and up-to-date weblinks, additional readings, and InfoTrac College Edition exercises.

The Work of Wind
and Deserts

CHAPTER 18
OUTLINE

PHYSICAL
Geology⇌Now *This icon, appearing throughout the book, indicates an opportunity to explore interactive tutorials, animations, or practice problems available on the Physical GeologyNow Web site at http://earthscience.brookscole.com/physgeo5e.*

OBJECTIVES
At the end of this chapter, you will have learned that

- Wind transports sediment and modifies the landscape through the processes of abrasion and deflation.

- Dunes and loess are the result of wind depositing material.

- Dunes form when wind flows over and around an obstruction.

- The four major dune types are barchan, longitudinal, transverse, and parabolic.

- Loess is formed from windblown silt and clay and is derived from three main sources: deserts, Pleistocene glacial outwash deposits, and floodplains of rivers in semiarid regions.

- The global pattern of air-pressure belts and winds are responsible for Earth's atmospheric circulation patterns.

- Deserts are dry and receive less than 25 cm of rain per year, have high evaporation rates, typically have poorly developed soils, and are mostly or completely devoid of vegetation.

- The majority of deserts are found in the dry climates of the low and middle latitudes.

- Deserts have many distinctive landforms, produced by both wind and running water.

The Saharan community of El Gedida in western Egypt is slowly being overwhelmed by advancing sand. Source: George Gerster/The National Audubon Society/Photo Researchers

535

Introduction

During the past several decades, deserts have been advancing across millions of acres of productive land, destroying rangeland, croplands, and even villages (see the chapter opening photo). Such expansion, estimated at 70,000 km² per year, has exacted a terrible toll in human suffering. Because of the relentless advance of deserts, hundreds of thousands of people have died of starvation or been forced to migrate as, "environmental refugees" from their homelands to camps where the majority are severely malnourished. This expansion of deserts into formerly productive lands is called *desertification* and is a major problem in many countries.

Most regions undergoing desertification lie along the margins of existing deserts where a delicately balanced ecosystem serves as a buffer between the desert on one side and a more humid environment on the other. Their potential to adjust to increasing environmental pressures from natural causes as well as human activity is limited. Ordinarily, desert regions expand and contract gradually in response to natural processes such as climatic change, but much recent desertification has been greatly accelerated by human activities.

In many areas, the natural vegetation has been cleared as crop cultivation has expanded into increasingly drier fringes to support the growing population. Because grasses are the dominant natural vegetation in most fringe areas, raising livestock is a common economic activity. However, increasing numbers of livestock in many areas have greatly exceeded the land's ability to support them. Consequently, the vegetation cover that protects the soil has diminished, causing the soil to crumble and be stripped away by wind and water, which results in increased desertification.

One particularly hard-hit area of desertification is the Sahel of Africa (a belt 300–1100 km wide, lying south of the Sahara). Because drought is common in the Sahel, the region can support only a limited population of livestock and humans. Unfortunately, expanding human and animal populations and more intensive agriculture have increased the demands on the lands. Plagued with periodic droughts, this region has suffered tremendously as crops have failed and livestock has overgrazed the natural vegetation, resulting in thousands of deaths, displaced people, and the encroachment of the Sahara.

The tragedy of the Sahel and prolonged droughts in other desert fringe areas remind us of the delicate equilibrium of ecosystems in such regions. Once the fragile soil cover has been removed by erosion, it takes centuries for new soil to form (see Chapter 5).

There are many important reasons to study deserts and the processes that are responsible for their formation. One reason is that deserts cover large regions of Earth's surface. More than 40% of Australia is desert, and the Sahara occupies a vast part of northern Africa. Although deserts are generally sparsely populated, some desert regions are experiencing an influx of people, such as Las Vegas, Nevada, the high desert area of southern California, and various locations in Arizona. Many of these places already have problems with increased population growth.

Furthermore, with the current debate about global warming, it is important to understand how desert processes operate and how global climate changes affect various Earth systems and subsystems. By understanding how desertification operates, people can take steps to eliminate or reduce the destruction done, particularly in terms of human suffering. Understanding the underlying causes of climate change by examining ancient desert regions may provide insight into the possible duration and severity of future climatic changes. This can have important ramifications in deciding whether burying nuclear waste in a desert, such as the Yucca Mountain, Nevada, is as safe as some claim and in our best interests as a society (see Geo-Focus 18.1).

More than 6000 years ago, the Sahara was a fertile savannah supporting diverse fauna and flora, including humans. Then the climate changed, and the area became a desert. How did this happen? Will this region change back again in the future? These are some of the questions geoscientists hope to answer by studying deserts.

Finally, many agents and processes that have shaped deserts do not appear to be limited to our planet. Features found on Mars, especially as seen in images transmitted by the *Spirit* rover, are apparently the result of the same wind-driven processes operating on Earth.

HOW DOES WIND TRANSPORT SEDIMENT?

Wind is a turbulent fluid and therefore transports sediment in much the same way as running water. Although wind typically flows at a greater velocity than water, it has a lower density and thus can carry only clay- and silt-sized particles as *suspended load*. Sand and larger particles are moved along the ground as *bed load*.

Bed Load

Sediments too large or heavy to be carried in suspension by water or wind are moved as bed load either by *saltation* or by rolling and sliding. As we discussed in Chapter 15, saltation is the process by which a portion of the bed load moves by intermittent bouncing along a streambed. Saltation also occurs on land. Wind starts sand grains rolling and lifts and carries some grains short distances before they fall back to the surface. As the descending sand grains hit the surface, they strike other grains, causing them to bounce along by saltation (■ Figure 18.1). Wind-tunnel experiments have shown that once sand grains begin moving, they will continue to move, even if the wind drops below the speed necessary to start them moving! This happens because once saltation begins, it sets off a chain reaction of collisions between sand grains that keeps the grains in constant motion.

Saltating sand usually moves near the surface, and even when winds are strong, grains are rarely lifted higher than about a meter. If the winds are very strong, these wind-whipped grains can cause extensive abrasion. A car's paint can be removed by sandblasting in a short time, and its windshield will become completely frosted and translucent from pitting.

Suspended Load

Silt- and clay-sized particles constitute most of a wind's suspended load. Even though these particles are much smaller and lighter than sand-sized particles, wind usually starts the latter moving first. The reason for this phenomenon is that a very thin layer of motionless air lies next to the ground where the small silt and clay particles remain undisturbed. The larger sand grains, however, stick up into the turbulent air zone, where they can be moved. Unless the stationary air layer is disrupted, the silt and clay particles remain on the ground, providing a smooth surface. This phenomenon can be observed on a dirt road on a windy day. Unless a vehicle travels over the road, little dust is raised even though it is windy. When a vehicle moves over the road, it breaks the calm boundary layer of air and disturbs the smooth layer of dust, which is picked up by the wind and forms a dust cloud in the vehicle's wake.

In a similar manner, when a sediment layer is disturbed, silt- and clay-sized particles are easily picked up and carried in suspension by the wind, creating clouds of dust or even dust storms. Once these fine particles are lifted into the atmosphere, they may be carried thousands of kilometers from their source. For example, large

quantities of fine dust from the southwestern United States were blown east and fell on New England during the Dust Bowl of the 1930s (see Figure 5.21).

HOW DOES WIND ERODE?

Although wind action produces many distinctive erosional features and is an extremely efficient sorting agent, running water is still responsible for most erosional landforms in arid regions, even though stream channels are typically dry. Wind erodes material in two ways: abrasion and deflation.

Abrasion

Abrasion involves the impact of saltating sand grains on an object and is analogous to sandblasting. The effects of abrasion are usually minor because sand, the most common agent of abrasion, is rarely carried more than 1 m above the surface. Rather than creating major erosional features, wind abrasion typically modifies existing features by etching, pitting, smoothing, or polishing. Nonetheless, wind abrasion can produce many strange-looking and bizarre-shaped features (■ Figure 18.2).

Ventifacts are a common product of wind abrasion; these are stones whose surfaces have been polished, pitted, grooved, or faceted by the wind (■ Figures 18.3 and 18.6c). If the wind blows from different directions or if the stone is moved, the ventifact will have multiple facets. Ventifacts are most common in deserts, yet they can also form wherever stones are exposed to saltating sand grains, as on beaches in humid regions and some outwash plains in New England.

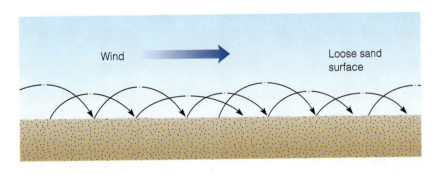

■ Figure 18.1

Most sand is moved near the ground surface by saltation. Sand grains are picked up by the wind and carried a short distance before falling back to the ground, where they usually hit other grains, causing them to bounce and move in the direction of the wind.

GEOFOCUS

18.1

Radioactive Waste Disposal— Safe or Sorry?

One problem of the nuclear age is finding safe storage sites for radioactive waste from nuclear power plants, the manufacture of nuclear weapons, and the radioactive by-products of nuclear medicine. Radioactive waste can be grouped into two categories: low-level and high-level waste. Low-level wastes are low enough in radioactivity that, when properly handled, they do not pose a significant environmental threat. Most low-level wastes can be safely buried in controlled dump sites where the geology and groundwater system are well known and careful monitoring is provided.

High-level radioactive waste, such as the spent uranium-fuel assemblies used in nuclear reactors and the material used in nuclear weapons, is extremely dangerous because of large amounts of radioactivity; it

therefore presents a major environmental problem. Currently some 77,000 tons of spent nuclear fuel are temporarily stored at 131 sites in 39 states, while awaiting shipment to a permanent site in Yucca Mountain, Nevada (■ Figure 1).

In 1987 Congress amended the Nuclear Waste Policy Act and directed the Department of Energy (DOE) to study only Yucca Mountain as a possible nuclear waste repository. With the passage and signing by President Bush of House Joint Resolution 87 in 2002, the way was cleared for the DOE to prepare an application for a Nuclear Regulatory Commission license to begin constructing a nuclear waste repository at Yucca Mountain. When completed, the repository could begin receiving shipments of spent nuclear fuel by 2010.

Why Yucca Mountain? After more than 20 years of study and $8 billion

spent on researching the region, many scientists think Yucca Mountain has the features necessary to isolate high-level nuclear waste from the environment for at least 10,000 years, which is the minimum time the waste will remain dangerous. What makes Yucca Mountain so appealing is its remote location and long distance from a large population center—in this case, Las Vegas, which is about 166 km southeast of Yucca Mountain. It has a very dry climate, with less than 15 cm of rain, and a deep water table that is 500 to 720 m below the repository. In fact, the radioactive waste will be buried in volcanic tuff at a depth of about 300 m in canisters designed to remain leakproof for at least 300 years.

What then are the concerns and why the opposition to Yucca Mountain? One of the main concerns is whether the climate will change dur-

■ Figure 18.2

Wind abrasion has formed these structures by eroding the exposed limestone in Desierto Libico, Egypt.

O. Alamany & W. Vicens/Corbis

U.S. Department of Energy

■ **Figure 1**

The location and aerial view of Nevada's Yucca Mountain.

ing the next 10,000 years. If the region should become more humid, more water will percolate through the zone of aeration. This will increase the corrosion rate of the canisters and could cause the water table to rise, thereby decreasing the travel time between the repository and the zone of saturation. This area of the country was much more humid during the Ice Age, 1.6 million to 10,000 years ago (see Chapter 17).

Another concern is the seismic activity in the area. It is in fact riddled with faults and has experienced numerous earthquakes during historic time. Nevertheless, based on underground inspections at Yucca Mountain as well as the tunnels at the Nevada Test site nearby, the DOE is convinced that earthquakes pose little danger to the underground repository itself because the disruptive effects of an earthquake are usually confined to the surface. Furthermore, it is required that the facilities, both above and below ground, be designed to withstand any severe earthquake likely to strike the area.

Finally, some people worry about the possibility of sabotage to the facility as well as the problems of transporting high-level nuclear waste to the facility. Being buried 300 m below ground renders it virtually impenetrable to acts of terrorism or sabotage, but the possibility of an accident or terrorist attack on the way to the repository is still a concern to many.

Although it appears that Yucca Mountain meets all the requirements for a safe high-level radioactive waste repository, the site is still controversial and at the time of this writing still embroiled in lawsuits seeking to block its construction and funding battles in Congress.

Yardangs are larger features than ventifacts and also result from wind erosion (■ Figure 18.4). They are elongated, streamlined ridges that look like an overturned ship's hull. Yardangs are typically found grouped in clusters aligned parallel to the prevailing winds. They probably form by differential erosion in which depressions, parallel to the direction of wind, are carved out of a rock body, leaving sharp, elongated ridges. These ridges may then be further modified by wind abrasion into their characteristic shape. Although yardangs are fairly common desert features, interest in them has renewed with images radioed back from Mars showing that they are also widespread features on the Martian surface.

What Would You Do

As an expert in desert processes, you have been assigned the job of teaching the first astronaut crew that will explore Mars all about deserts and their landforms. The reason is that many Martian features display evidence of having formed as a result of wind processes, and many landforms are the same as those found in deserts on Earth. Describe how you would teach the astronauts to recognize wind-formed features and where on Earth you would take the astronauts to show them the types of landforms they may find on Mars.

GEOLOGY
IN UNEXPECTED PLACES

Blowing in the Wind

On a hot summer day, a light breeze can make us feel cooler and more comfortable. That same breeze on a cold day in winter can make us feel even colder and more miserable. Usually we don't pay much attention to the wind and most people don't associate it with geology. Yet wind is an important geologic agent, not only in deserts but almost everywhere.

Wind can sculpt unusual shapes in rocks (see Figure 18.2), and ventifacts can be found not only in the desert (see Figure 18.3) but also on beaches and even out of this world (■ Figure 1)! Anyone who has ever been caught in a strong windstorm can attest to the erosive power of wind. Many a car's paint and windows have been ruined by sandblasting when caught in a sudden sandstorm. In areas with lots of sand, wooden telephone and telegraph poles need a metal skirting from their base up to about 6 to 8 feet to prevent them from being "cut down" by the erosive effects of wind-blown sand.

Look at the base of many cemetery head-stones and you'll see that they typically are more weathered than the rest of the headstone (■ Figure 2). This is because wind can move larger particles along the ground by saltation, resulting in more erosion near the base of headstones than farther up, where only very small particles can be carried. The next time you hear or feel the wind blowing, remember that it is also a geologic agent at work.

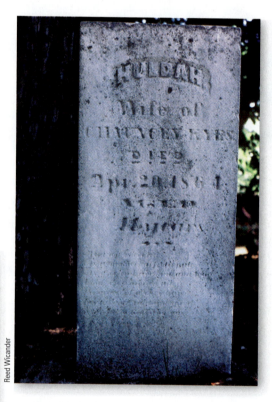

Reed Wicander

■ Figure 2

An 1864 headstone in a cemetery in Mount Pleasant, Michigan, shows the differential effects of blowing wind. Notice how the lower portion of the headstone is more weathered than the rest of it. This is partly because wind can move larger particles by saltation, and these particles in turn have greater erosive power than smaller particles that can be carried higher by the wind.

NASA

■ Figure 1

Wind has not only sandblasted the surface of these boulders strewn on the Martian surface but also probably helped shape them.

■ **Figure 18.3**

(a) A ventifact forms when wind-borne particles (1) abrade the surface of a rock (2) forming a flat surface. If the rock is moved, (3) additional flat surfaces are formed. (b) Large ventifacts lying on desert pavement in Death Valley National Monument, California.

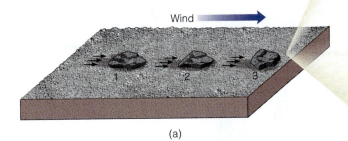

Wind

(a)

Martin G. Miller/Visuals Unlimited

(b)

Marion A. Whitney

■ **Figure 18.4**

Profile view of a streamlined yardang in the Roman playa deposits of the Kharga Depression, Egypt.

Deflation

Another important mechanism of wind erosion is **deflation,** which is the removal of loose surface sediment by the wind. Among the characteristic features of deflation in many arid and semiarid regions are *deflation hollows,* or *blowouts* (■ Figure 18.5). These shallow depressions of variable dimensions result from differential erosion of surface materials. Ranging in size from several kilometers in diameter and tens of meters deep to small depressions only a few meters wide and less than a meter deep, deflation hollows are common in the southern Great Plains region of the United States.

In many dry regions, the removal of sand-sized and smaller particles by wind leaves a surface of pebbles, cobbles, and boulders. As the wind removes the fine-grained material from the surface, the effects of gravity and occasional floodwaters rearrange the remaining coarse particles into a mosaic of close-fitting rocks called **desert pavement** (Figures 18.3b and ■ 18.6). Once desert pavement forms, it protects the underlying material from further deflation.

WHAT ARE THE DIFFERENT TYPES OF WIND DEPOSITS?

Although wind is of minor importance as an erosional agent, it is responsible for impressive deposits, which are primarily of two types. The first, dunes, occur in several distinctive types, all of which consist of sand-sized particles that are usually deposited near their source. The second is

Martin G. Miller/Visuals Unlimited

■ **Figure 18.5**

A deflation hollow in Death Valley, California.

loess, which consists of layers of windblown silt and clay deposited over large areas downwind and commonly far from their source.

The Formation and Migration of Dunes

The most characteristic features associated with sand-covered regions are **dunes,** which are mounds or ridges of wind-deposited sand (■ Figure 18.7). Dunes form when wind flows over and around an obstruction, resulting in deposition of sand grains, which accumulate and build up deposits of sand. As they grow, these sand deposits become self-generating in that they form ever-larger wind barriers that further reduce the wind's velocity, resulting in more sand deposition and growth of the dune.

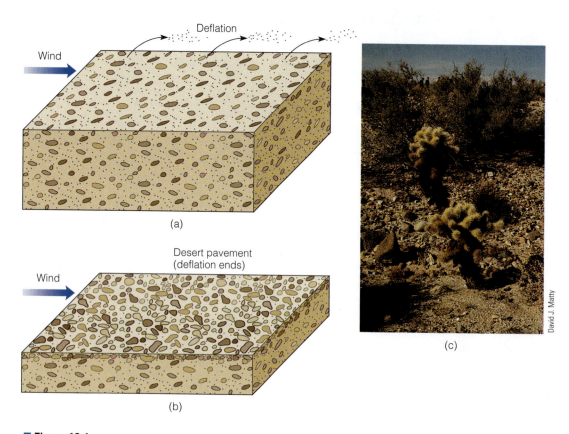

David J. Matty

■ **Figure 18.6**

Deflation and the origin of desert pavement. (a) Fine-grained material is removed by wind, (b) leaving a concentration of larger particles that form desert pavement. (c) Desert pavement in the Mojave Desert, California. Several ventifacts can be seen in the lower left of the photo.

■ Figure 18.7

Large sand dunes in Death Valley, California. Well-developed ripple marks can be seen on the surface of the dunes. The prevailing wind direction is from left to right.

Most dunes have an asymmetric profile, with a gentle windward slope and a steeper downwind, or leeward, slope that is inclined in the direction of the prevailing wind (■ Figure 18.8a). Sand grains move up the gentle windward slope by saltation and accumulate on the leeward side, forming an angle between 30 and 34 degrees from the horizontal, which is the angle of repose of dry sand. When this angle is exceeded by accumulating sand, the slope collapses and sand slides down the leeward slope, coming to rest at its base. As sand moves from a dune's windward side and periodically slides down its leeward slope, the dune slowly migrates in the direction of the prevailing wind (Figure 18.8b). When preserved in the geologic record, dunes help geologists determine the prevailing direction of ancient winds (■ Figure 18.9).

Dune Types

Geologists recognize four major dune types (barchan, longitudinal, transverse, and parabolic), although intermediate forms also exist. The size, shape, and arrangement of dunes result from the interaction of such factors as sand supply, the direction and velocity of the prevailing wind, and the amount of vegetation. Although dunes are usually found in deserts, they can also develop wherever sand is abundant, such as along the upper parts of many beaches.

Barchan dunes are crescent-shaped dunes whose tips point downwind (■ Figure 18.10). They form in areas that have a generally flat, dry surface with little vegetation, a limited supply of sand, and a nearly constant wind direction. Most barchans are small, with the largest reaching about 30 m high. Barchans are the most mobile of the major dune types, moving at rates that can exceed 10 m per year.

Longitudinal dunes (also called *seif dunes*) are long, parallel ridges of sand aligned generally parallel to the direction of the prevailing winds; they form where the sand supply is somewhat limited (■ Figure 18.11). Longitudinal dunes result when winds converge from slightly different directions to produce the prevailing wind. They range in size from about 3 m to more than 100 m high, and some stretch for more than 100 km. These dunes are especially well developed in central Australia, where they cover nearly one fourth of the

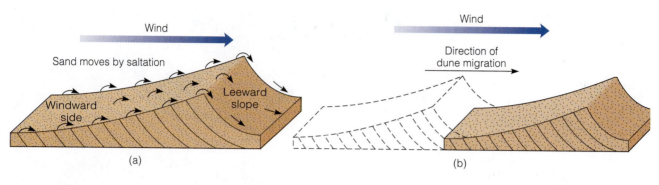

■ **Figure 18.8**

(a) Profile view of a sand dune. (b) Dunes migrate when sand moves up the windward side and slides down the leeward slope. Such movement of the sand grains produces a series of cross-beds that slope in the direction of wind movement.

■ **Figure 18.9**

Cross-bedding in this sandstone in Zion National Park, Utah, helps geologists determine the prevailing direction of the wind that formed these ancient sand dunes.

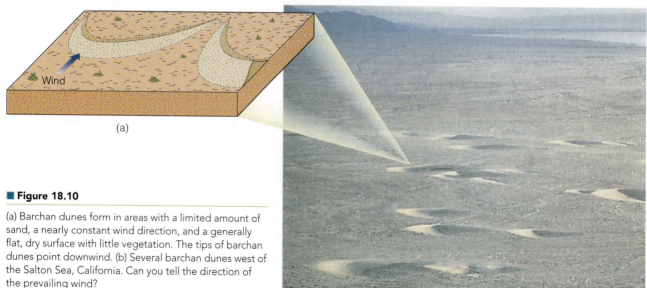

■ **Figure 18.10**

(a) Barchan dunes form in areas with a limited amount of sand, a nearly constant wind direction, and a generally flat, dry surface with little vegetation. The tips of barchan dunes point downwind. (b) Several barchan dunes west of the Salton Sea, California. Can you tell the direction of the prevailing wind?

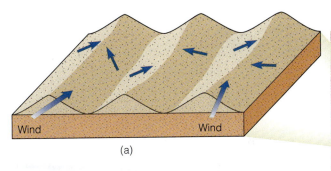

■ Figure 18.11

(a) Longitudinal dunes form long, parallel ridges of sand aligned roughly parallel to the prevailing wind direction. They typically form where sand supplies are limited. (b) Longitudinal dunes, 15 m high, in the Gibson Desert, west central Australia. The bright blue areas between the dunes are shallow pools of rainwater, and the darkest patches are areas where the Aborigines have set fires to encourage the growth of spring grasses.

■ Figure 18.12

(a) Transverse dunes form long ridges of sand that are perpendicular to the prevailing wind direction in areas of little or no vegetation and abundant sand. (b) Transverse dunes, Great Sand Dunes National Monument, Colorado. The prevailing wind direction is from lower left to upper right.

continent. They also cover extensive areas in Saudi Arabia, Egypt, and Iran.

Transverse dunes form long ridges perpendicular to the prevailing wind direction in areas with abundant sand and little or no vegetation (■ Figure 18.12). When viewed from the air, transverse dunes have a wavelike appearance and are therefore sometimes called *sand seas*. The crests of transverse dunes can be as high as 200 m, and the dunes may be as wide as 3 km. Some transverse dunes develop a clearly distinguishable barchan form and may separate into individual barchan dunes along the edges of the dune field where there is less sand. Such intermediate-form dunes are known as *barchanoid dunes*.

Parabolic dunes are most common in coastal areas with abundant sand, strong onshore winds, and a partial cover of vegetation (■ Figure 18.13). Although parabolic dunes have a crescent shape like barchan dunes, their tips point upwind. Parabolic dunes form when the vegetation cover is broken and deflation produces a deflation hollow or blowout. As the wind transports the sand out of the depression, it builds up on the convex downwind dune crest. The central part of the dune is excavated by the wind, while vegetation holds the ends and sides fairly well in place.

Another type of dune commonly found in the deserts of North Africa and Saudi Arabia is the *star dune*, so named because of its resemblance to a multipointed

Wind

(a)

Reed Wicander

(b)

■ **Figure 18.13**

(a) Parabolic dunes typically form in coastal areas with a partial cover of vegetation, a strong onshore wind, and abundant sand. (b) Parabolic dune developed along the Lake Michigan shoreline west of St. Ignace, Michigan.

representation of a star (■ Figure 18.14). Star dunes are among the tallest in the world, rising, in some cases, more than 100 m above the surrounding desert plain. They consist of pyramidal hills of sand, from which radiate several ridges of sand, and they develop where the wind direction is variable. Star dunes can remain stationary for centuries at a time and have served as desert landmarks for many nomadic peoples.

Loess

Windblown silt and clay deposits composed of angular quartz grains, feldspar, micas, and calcite are known as **loess.** The distribution of loess shows that it is derived from three main sources: deserts, Pleistocene glacial outwash deposits, and the floodplains of rivers in semi-arid regions. Loess must be stabilized by moisture and vegetation in order to accumulate. Consequently, loess is not found in deserts, even though they provide much of its material. Because of its unconsolidated nature,

loess is easily eroded, and as a result, eroded loess areas are characterized by steep cliffs and rapid lateral and headward stream erosion (■ Figure 18.15).

At present, loess deposits cover approximately 10% of Earth's land surface and 30% of the United States. The most extensive and thickest loess deposits occur in northeast China, where accumulations greater than 30 m are common. The extensive deserts in central Asia are the source for this loess. Other important loess deposits are on the North European Plain from Belgium eastward to Ukraine, in Central Asia, and in the Pampas of Argentina. In the United States, they occur in the Great Plains, the Midwest, the Mississippi River Valley, and eastern Washington.

Loess-derived soils are some of the world's most fertile (Figure 18.15). It is therefore not surprising that the world's major grain-producing regions correspond to the distribution of large loess deposits such as the North European Plain, Ukraine, and the Great Plains of North America.

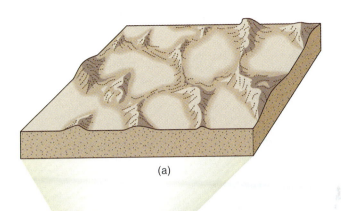

(a)

(b)

Nigel J. Dennis/Photo Researchers, Inc.

■ **Figure 18.14**

(a) Star dunes are pyramidal hills of sand that develop where the wind direction is variable. (b) Ground-level view of star dunes in Namib-Naukluft Park, Namibia.

Lowell Georgia/Corbis

■ **Figure 18.15**

Terraced wheat fields in the loess soil at Tangwa Village, China. Because of its unconsolidated nature, many farmers live in hillside caves they carved from the loess.

HOW ARE AIR-PRESSURE BELTS AND GLOBAL WIND PATTERNS DISTRIBUTED?

To understand the work of wind and the distribution of deserts, we need to consider the global pattern of air-pressure belts and winds, which are responsible for Earth's atmospheric circulation patterns. Air pressure is the density of air exerted on its surroundings (that is, its weight). When air is heated, it expands and rises, reducing its mass for a given volume and causing a decrease in air pressure. Conversely, when air is cooled, it contracts and air pressure increases. Therefore those areas of Earth's surface that receive the most solar radiation, such as the equatorial regions, have low air pressure, whereas the colder areas, such as the polar regions, have high air pressure.

Air flows from high-pressure zones to low-pressure zones. If Earth did not rotate, winds would move in a straight line from one zone to another. Because Earth rotates, however, winds are deflected to the right of their direction of motion (clockwise) in the Northern Hemisphere and to the left of their direction of motion (counterclockwise) in the Southern Hemisphere. Such a deflection of air between latitudinal zones resulting from Earth's rotation is known as the **Coriolis effect**. The combination of latitudinal pressure differences and the Coriolis effect produces a worldwide pattern of wind belts oriented east–west (■ Figure 18.16).

Earth's equatorial zone receives the most solar energy, which heats the surface air, causing it to rise. As the air rises, it cools and releases moisture that falls as rain in the equatorial region (Figure 18.16). The rising air is now much drier as it moves northward and southward toward each pole. By the time it reaches 20 to 30 degrees north and south latitudes, the air has become cooler and denser and begins to descend. Compression of the atmosphere warms the descending air mass and produces a warm, dry, high-pressure area, providing the perfect conditions for the formation of the low-latitude deserts of the Northern and Southern Hemispheres (■ Figure 18.17).

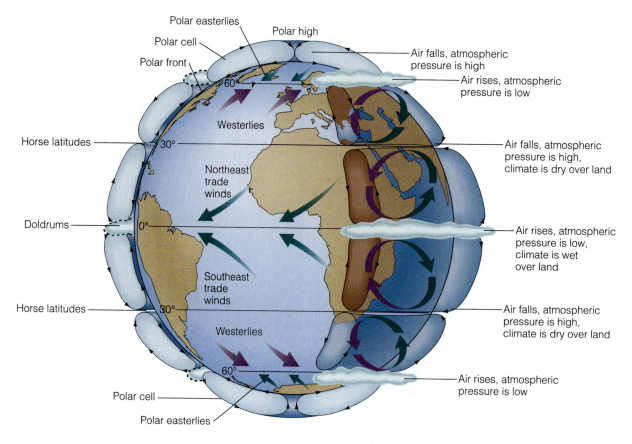

■ **Figure 18.16**

The general circulation of Earth's atmosphere.

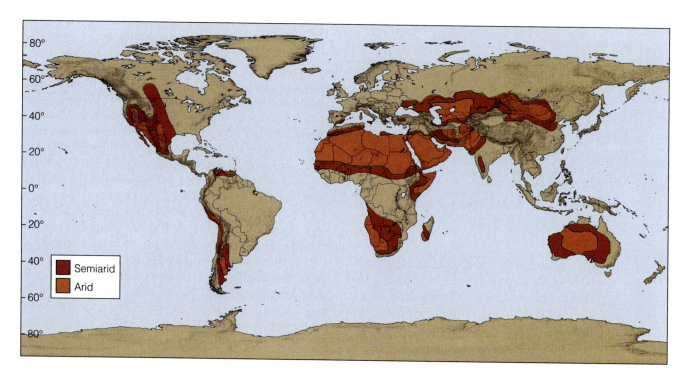

■ **Figure 18.17**

The distribution of Earth's arid and semiarid regions.

WHERE DO DESERTS OCCUR?

Dry climates occur in the low and middle latitudes where the potential loss of water by evaporation exceeds the yearly precipitation (Figure 18.17). Dry climates cover 30% of Earth's land surface and are subdivided into semiarid and arid regions. *Semiarid regions* receive more precipitation than arid regions, yet are moderately dry. Their soils are usually well developed and fertile and support a natural grass cover. *Arid regions,* generally described as **deserts,** are dry; on average, they receive less than 25 cm of rain per year, have high evaporation rates, typically have poorly developed soils, and are mostly or completely devoid of vegetation.

The majority of the world's deserts are in the dry climates of the low and middle latitudes (Figure 18.17). In North America most of the southwestern United States and northern Mexico are characterized by this hot, dry climate, whereas in South America this climate is primarily restricted to the Atacama Desert of coastal Chile and Peru. The Sahara in northern Africa, the Arabian Desert in the Middle East, and the majority of Pakistan and western India form the largest essentially unbroken desert environment in the Northern Hemisphere. More than 40% of Australia is desert, and most of the rest of it is semiarid.

The remaining dry climates of the world are found in the middle and high latitudes, mostly within continental interiors in the Northern Hemisphere (Figure 18.17). Many of these areas are dry because of their remoteness from moist maritime air and the presence of mountain ranges that produce a **rainshadow desert** (■ Figure 18.18). When moist marine air moves inland and meets a mountain range, it is forced upward. As it rises, it cools, forming clouds and producing precipitation that falls on the windward side of the mountains. The air that descends on the leeward side of the mountain range is much warmer and drier, producing a rainshadow desert.

Three widely separated areas are included within the mid-latitude dry-climate zone (Figure 18.17). The largest is the central part of Eurasia extending from just north of the Black Sea eastward to north-central China. The Gobi Desert in China is the largest desert in this region. The Great Basin area of North America is the second largest mid-latitude dry-climate zone and results from the rainshadow produced by the Sierra Nevada. This region adjoins the southwestern deserts of the United States that formed as a result of the low-latitude subtropical high-pressure zone. The smallest of the mid-latitude dry-climate areas is the Patagonian region of southern and western Argentina. Its dryness results from the rainshadow effect of the Andes. The remainder of the world's deserts are found in the cold but dry high latitudes, such as Antarctica.

WHAT ARE THE CHARACTERISTICS OF DESERTS?

To people who live in humid regions, deserts may seem stark and inhospitable. Instead of a landscape of rolling hills and gentle slopes with an almost continuous cover of vegetation, deserts are dry, have little vegetation, and consist of nearly continuous rock exposures, desert pavement, or sand dunes. And yet despite the great contrast between deserts and more humid areas, the same geologic processes are at work, only operating under different climatic conditions.

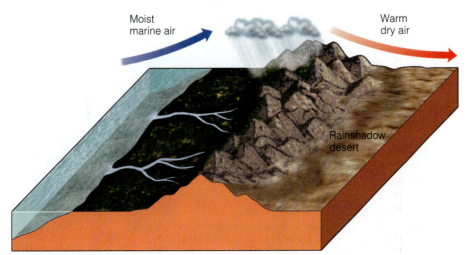

Moist marine air

Warm dry air

Rainshadow desert

■ **Figure 18.18**

Many deserts in the middle and high latitudes are rainshadow deserts, so named because they form on the leeward side of mountain ranges. When moist marine air moving inland meets a mountain range, it is forced upward where it cools and forms clouds that produce rain. This rain falls on the windward side of the mountains. The air descending on the leeward side is much warmer and drier, producing a rainshadow desert.

Temperature, Precipitation, and Vegetation

The heat and dryness of deserts are well known. Many deserts of the low latitudes have average summer temperatures that range between 32° and 38°C. It is not uncommon for some low-elevation inland deserts to record daytime highs of 46° to 50°C for weeks at a time. The highest temperature ever recorded was 58°C in El Azizia, Libya, on September 13, 1922.

During the winter months when the Sun's angle is lower and there are fewer daylight hours, daytime temperatures average between 10° and 18°C. Winter nighttime lows can be quite cold, with frost and freezing temperatures common in the more poleward deserts. Winter daily temperature fluctuations in low-latitude deserts are among the greatest in the world, ranging between 18° and 35°C. Temperatures have been known to fluctuate from below 0°C to more than 38°C in a single day!

The dryness of the low-latitude deserts results primarily from the year-round dominance of the subtropical high-pressure belt, while the dryness of the mid-latitude deserts is due to their isolation from moist marine winds and the rainshadow effect created by mountain ranges. The dryness of both is further accentuated by their high temperatures.

Although deserts are defined as regions that receive, on average, less than 25 cm of rain per year, the amount of rain that falls each year is unpredictable and unreliable. It is not uncommon for an area to receive more than an entire year's average rainfall in one cloudburst and then to receive very little rain for several years. Thus yearly rainfall averages can be misleading.

Deserts display a wide variety of vegetation (■ Figure 18.19). Although the driest deserts, or those with large areas of shifting sand, are almost devoid of vegetation, most deserts support at least a sparse plant cover. Compared to humid areas, desert vegetation may appear monotonous. A closer examination, however, reveals an amazing diversity of plants that have evolved the ability to live in the near absence of water.

Desert plants are widely spaced, typically small, and grow slowly. Their stems and leaves are usually hard and waxy to minimize water loss by evaporation and protect the plant from sand erosion. Most plants have a widespread, shallow root system to absorb the dew that forms each morning in all but the driest deserts and to help anchor the plant in what little soil there may be. In extreme cases, many plants lie dormant during particularly dry years and spring to life after the first rain shower, with a beautiful profusion of flowers.

Weathering and Soils

Mechanical weathering is dominant in desert regions. Daily temperature fluctuations and frost wedging are the primary forms of mechanical weathering (see Chapter 5). The breakdown of rocks by roots and from salt crystal growth is of minor importance. Some chemical weathering does occur, but its rate is greatly reduced by aridity and the scarcity of organic acids produced by the sparse vegetation. Most chemical weathering takes place during the winter months when more precipitation occurs, particularly in the mid-latitude deserts.

An interesting feature seen in many deserts is a thin, red, brown, or black shiny coating on the surface of many rocks (see "Rock Art for the Ages" on pages 552 and 553).

■ Figure 18.19

Desert vegetation is typically sparse, widely spaced, and characterized by slow growth rates. The vegetation shown here in Organ Pipe National Monument, Arizona, includes saguaro and cholla cacti, paloverde trees, and jojoba bushes and is characteristic of the vegetation found in the Sonoran Desert of North America.

Charlie Ott, The National Audubon Society Collection/Photo Researchers, Inc.

NICHOLAS LANCASTER

Controlling Dust Storms Through Revegetation

Nicholas Lancaster, the world's foremost expert on desert sand dunes, is a research professor with the Quaternary Sciences Center at the Desert Research Institute, part of the University and Community College System of Nevada. Scientists at the Desert Research Institute study air quality, water resources, surface processes, climate history and dynamics, and human adaptations in arid lands.

Transport of sand and dust by the wind is an important geologic process in many desert regions of the world. These areas are affected by dust storms that reduce visibility, close airports, and cause traffic accidents. Fine particles in the air can be hazardous to human health, causing respiratory diseases and possibly spreading fungal infections. Blowing sand can block roads and fill irrigation ditches, while migrating sand dunes can bury buildings, pipelines, and scarce agricultural land.

Dealing with these geologic hazards of living in desert regions requires a good understanding of the physical principles of sand and dust transport by wind and the formation and movement of sand dunes. I have traveled and conducted research in many desert regions, studying dune formation and movement as well as the ways in which sand and dust are moved by wind. My research, and that of many others, has shown that sand dunes form and move in very similar ways in widely separated desert regions. We can therefore take the results of studies conducted in one area and apply them to other locations to assist people in controlling blowing sand and dust.

One such area is Owens Lake in eastern California. Once the site of a shallow lake, the area is now a saline lake bed, or *playa,* from which dust storms arise on frequent occasions, affecting air quality throughout much of southern California. In fact, Owens Lake is regarded as the largest single source of windblown dust in the Americas. The drying up of Owens Lake is the result of diversion of water since the 1920s from the Owens River and its tributaries to the Los Angeles aqueduct, which now supplies a major part of the water used by the city of Los Angeles. Many of the dust storms at Owens Lake

occur because windblown sand grains—lifted from the sandy sediments of the former delta of the Owens River and from old beach deposits—abrade the salt-encrusted playa sediments. The fine particles released by this natural sand blasting are raised into the air by the turbulence of the wind and fill the Owens Valley with dust.

The dust problem is so severe that California state agencies were instructed to find a way to control dust emissions from Owens Lake by the end of the century. One way being considered to reduce dust storms is revegetation of parts of the dry lake bed with natural salt-tolerant grasses. Even sparse desert vegetation can have an important effect in protecting the sediments from wind (and water) erosion. Vegetation helps protect the surface by slowing the wind speed close to the ground and reducing or preventing entirely wind transport of surface sediments. The question is: How much grass cover is needed to stop sand transport and dust emissions?

In the past two years, I have been conducting field studies of sand transport in areas of different natural- and planted-vegetation cover on sand sheets at Owens Lake. The goal is to find the ground cover of salt-tolerant grass that reduces sediment transport to a minimal level and to develop a model that can be used to predict the amount of grass required to stabilize sand surfaces. Our field studies show that vegetation increases the threshold wind velocity required for sand transport so that there is an exponential decrease in the rate of sand transport with increasing vegetation cover. The result is that sand transport is effectively eliminated when the cover of salt grass is greater than 15%. We can now use these findings to predict how much vegetation is required to stabilize sandy areas at Owens Lake, as well as sand dune areas worldwide.

Rock Art for the Ages

Rock art includes rock paintings (where paints made from natural pigments are applied to a rock surface) and petroglyphs (from the Greek petro, meaning "rock," and glyph, meaning "carving or engraving"), which are the abraded, pecked, incised, or scratched marks made by humans on boulders, cliffs, and cave walls.

Rock art has been found on every continent except Antarctica and is a valuable archaeological resource that provides graphic evidence of the cultural, social, and religious relationships and practices of ancient peoples. The oldest known rock art was made by hunters in western Europe and dates back to the Pleistocene Epoch. Africa has more rock art sites than any other continent. The oldest known African rock art, found in the southern part of the continent, is estimated to be 27,000 years old.

Petroglyphs are a fragile and nonrenewable cultural resource that cannot be replaced if they are damaged or destroyed. A commitment to their preservation is essential so that future generations can study them as well as enjoy their beauty and mystery.

In the arid Southwest and Great Basin area of North America where rock art is plentiful, rock paintings and petroglyphs extend back to about 2000 B.C. Here, rock art can be divided into two categories. *Representational art* deals with life-forms such as humans, birds, snakes, and humanlike supernatural beings. Rarely exact replicas, they are more or less stylized versions of the beings depicted. *Abstract art,* in contrast, bears no resemblance to any real-life images.

Reed Wicander

Reed Wicander

Various petroglyphs exposed at an outcrop along Cub Creek Road in Dinosaur National Monument, Utah.

A humanlike petroglyph, an example of representational art, exposed at an outcrop along Cub Creek Road in Dinosaur National Monument, Utah. Note the contrast between the fresh exposure of the rock where the upper part of the petroglyph's head has been removed, the weathered brown surface of the rest of the petroglyph, and the black rock varnish coating the rock surface.

Petroglyphs are the most common form of rock art in North America and are made by pecking, incising, or scratching the rock surface with a tool harder than the rock itself. In arid regions, many rock surfaces display a patination, or thin brown or black coating, known as rock varnish (see Figure 18.20). When this coating is broken by pecking, incising, or scratching, the underlying lighter colored natural-rock surface provides an excellent contrast for the petroglyphs.

Petroglyphs are especially abundant in the Southwest and Great Basin area, where they occur by the thousands, having been made by Native Americans from many cultures during the past several thousand years. Petroglyphs can be seen in many of the U.S. national parks and monuments, such as Petrified Forest National Park, Arizona; Dinosaur National Monument, Utah; Canyonlands National Park, Utah; and Petroglyph National Monument, New Mexico—to name a few.

Examples of both representational and abstract art are displayed in these petroglyphs from Arizona.

Several petroglyphs on basalt boulders in Rinconada Canyon, Petroglyph National Monument, Albuquerque, New Mexico.

Examples of rock art from Tassili-n-Ajjer,

Rock art found in a rock shelter open to visitors in Uluru, Australia. These rock paintings are the work of Anagu artists, local Aboriginal people, and are created for religious and ceremonial expression as well as for teaching and storytelling. Each abstract symbol can have many levels of meaning and can help illustrate a story.

The paints used in this rock art are made from natural mineral substances mixed with water and sometimes with animal fat. Red, yellow, and orange pigments come from iron-stained clays, while the white pigments come from either ash or the mineral calcite. The black colors come from charcoal.

James S. Monroe

■ **Figure 18.20**

The shiny black coating on this rock exposed at Castle Valley, Utah, is rock varnish. It is composed of iron and manganese oxides.

This coating, called *rock varnish,* is composed of iron and manganese oxides (■ Figure 18.20). Because many of the varnished rocks contain little or no iron and manganese oxides, the varnish is thought to result either from wind-blown iron and manganese dust that settles on the ground or from the precipitated waste of microorganisms.

Desert soils, if developed, are usually thin and patchy because the limited rainfall and the resultant scarcity of vegetation reduce the efficiency of chemical weathering and hence soil formation. Furthermore, the sparseness of the vegetative cover enhances wind and water erosion of what little soil actually forms.

Mass Wasting, Streams, and Groundwater

When traveling through a desert, most people are impressed by such wind-formed features as moving sand, sand dunes, and sand and dust storms. They may also notice the dry washes and dry streambeds. Because of the lack of running water, most people would conclude that wind is the most important erosional agent in deserts. They would be wrong. Running water, even though it occurs infrequently, causes most of the erosion in deserts. The dry conditions and sparse vegetation characteristic of deserts enhance water erosion. If you look closely, you will see the evidence of erosion and transportation by running water nearly everywhere except in areas covered by sand dunes.

Most of a desert's average annual rainfall of 25 cm or less comes in brief, heavy, localized cloudbursts. During these times, considerable erosion takes place because the ground cannot absorb all the rainwater. With so little vegetation to hinder its flow, runoff is rapid, especially on moderately to steeply sloping surfaces, resulting in flash floods and sheetflows. Dry stream channels quickly fill with raging torrents of muddy water and mudflows, which carve out steep-sided gullies and overflow their banks. During these times, a tremendous amount of sediment is rapidly transported and deposited far downstream.

Although water is the major erosive agent in deserts today, it was even more important during the Pleistocene Epoch when these regions were more humid (see Chapter 17). During that time, many of the major topographic features of deserts were forming. Today that topography is being modified by wind and infrequently flowing streams.

Most desert streams are poorly integrated and flow only intermittently. Many of them never reach the sea because the water table is usually far deeper than the channels of most streams, so they cannot draw upon groundwater to replace water lost to evaporation and absorption into the ground. This type of drainage in which a stream's load is deposited within the desert is called *internal drainage* and is common in most arid regions.

Although most deserts have internal drainage, some deserts have permanent through-flowing streams such as the Nile and Niger Rivers in Africa, the Rio Grande and Colorado River in the southwestern United States, and the Indus River in Asia. These streams can flow through desert regions because their headwaters are well outside the desert and water is plentiful enough to offset losses resulting from evaporation and infiltration. Demands for greater amounts of water for agriculture and domestic use from the Colorado River, however, are leading to increased salt concentrations in its lower reaches and causing political problems between the United States and Mexico.

Wind

Although running water does most of the erosional work in deserts, wind can also be an effective geologic agent capable of producing a variety of distinctive erosional and depositional features (Figure 18.2). It is effective in transporting and depositing unconsolidated sand-, silt-, and dust-sized particles. Contrary to popular belief, most deserts are not sand-covered wastelands but rather vast areas of rock exposures and desert pavement. Sand-covered regions, or sandy deserts, constitute less than 25% of the world's deserts. The sand in these areas has accumulated primarily by the action of wind.

WHAT TYPES OF LANDFORMS ARE FOUND IN DESERTS?

Because of differences in temperature, precipitation, and wind, as well as the underlying rocks and recent tectonic events, landforms in arid regions vary considerably. Although wind is an important geologic agent in deserts, many distinctive landforms are produced and modified by running water.

John S. Shelton

(a)

John S. Shelton

(b)

■ **Figure 18.21**

(a) Playa lake formed after a rainstorm filled Croneis Dry Lake, Mojave Desert, California. (b) Racetrack Playa, Death Valley, California. The Inyo Mountains can be seen in the background.

After an infrequent and particularly intense rainstorm, excess water not absorbed by the ground may accumulate in low areas and form *playa lakes* (■ Figure 18.21a). These lakes are temporary, lasting from a few hours to several months. Most of them are shallow and have rapidly shifting boundaries as water flows in or leaves by evaporation and seepage into the ground. The water is often very saline.

When a playa lake evaporates, the dry lakebed is called a **playa**, or *salt pan*, and is characterized by mud cracks and precipitated salt crystals (Figure 18.21b). Salts in some playas are thick enough to be mined commercially. For example, borates have been mined in Death Valley, California, for more than 100 years.

Other common features of deserts, particularly in the Basin and Range region of the United States, are alluvial fans and bajadas. **Alluvial fans** form when sediment-laden streams flowing out from the generally straight, steep mountain fronts deposit their load on the relatively flat desert floor. Once beyond the mountain front where no valley walls contain streams, the sediment spreads out laterally, forming a gently sloping and poorly sorted fan-shaped sedimentary deposit (■ Figure 18.22). Alluvial fans are similar in origin and shape to deltas (see Chapter 15) but are formed entirely on land. Alluvial fans may

coalesce to form a *bajada*, a broad alluvial apron that typically has an undulating surface resulting from the overlap of adjacent fans (■ Figure 18.23).

Large alluvial fans and bajadas are frequently important sources of groundwater for domestic and agricultural use. Their outer portions are typically composed of fine-grained sediments suitable for cultivation, and their gentle slopes allow good drainage of water. Many alluvial fans and bajadas are also the sites of large towns and cities, such as San Bernardino, California, Salt Lake City, Utah, and Teheran, Iran.

Most mountains in desert regions, including those of the Basin and Range region, rise abruptly from gently sloping surfaces called pediments. **Pediments** are erosional bedrock surfaces of low relief that slope gently away from mountain bases (■ Figure 18.24). Most pediments are covered by a thin layer of debris, alluvial fans, or bajadas.

The origin of pediments has been the subject of much controversy. Most geologists agree that they are erosional features developed on bedrock in association with the erosion and retreat of a mountain front (Figure 18.24a). The disagreement concerns how the erosion has occurred. Although not all geologists would agree, it appears that pediments are produced by the combined activities of

Martin G. Miller/Visuals Unlimited

■ **Figure 18.22**

Aerial view of an alluvial fan, Death Valley, California

lateral erosion by streams, sheet flooding, and various weathering processes along the retreating mountain front. Thus pediments grow at the expense of the mountains, and they will continue to expand as the mountains are eroded away or partially buried.

Rising conspicuously above the flat plains of many deserts are isolated steep-sided erosional remnants called **inselbergs,** a German word meaning "island mountain." Inselbergs have survived for a longer period of time than other mountains because of their greater resistance to weathering.

Other easily recognized erosional remnants common to arid and semiarid regions are mesas and buttes

(■ Figure 18.25). A **mesa** is a broad, flat-topped erosional remnant bounded on all sides by steep slopes. Continued weathering and stream erosion form isolated pillarlike structures known as **buttes.** Buttes and mesas consist of relatively easily weathered sedimentary rocks capped by nearly horizontal, resistant rocks such as sandstone, limestone, or basalt. They form when the resistant rock layer is breached, allowing rapid erosion of the less resistant underlying sediment. One of the best-known areas of mesas and buttes in the United States is Monument Valley on the Arizona–Utah border (Figure 18.25).

Alan L. and Linda D. Mayo, GeoPhoto Publishing Co.

■ **Figure 18.23**

Coalescing alluvial fans forming a bajada at the base of the Black Mountains, Death Valley, California.

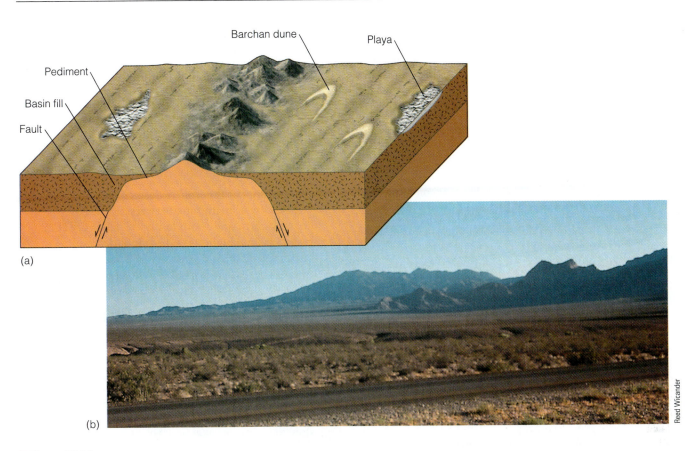

■ Figure 18.24

(a) Pediments are erosional bedrock surfaces formed by erosion along a mountain front. (b) Pediment north of Mesquite, Nevada.

■ Figure 18.25

(a) Several mesas as well as buttes can be seen in this aerial view of Monument Valley Navajo Tribal Park, Navajo Indian Reservation. The mesas are the broad, flat-topped structures, one of which is prominent in the foreground, whereas the buttes are more pilarlike structures. (b) Left Mitten Butte and Right Mitten Butte in Monument Valley on the border of Arizona and Utah.

18 REVIEW WORKBOOK

Chapter Summary

- Wind transports sediment in suspension or as bed load, which involves saltation and surface creep.

- Wind erodes material by either abrasion or deflation. Abrasion is a near-surface effect caused by the impact of saltating sand grains. Ventifacts are common wind-abraded features.

- Deflation is the removal of loose surface material by wind. Deflation hollows resulting from differential erosion of surface material are common features of many deserts, as is desert pavement, which effectively protects the underlying surface from additional deflation.

- The two major wind deposits are dunes and loess. Dunes are mounds or ridges of wind-deposited sand, whereas loess is wind-deposited silt and clay.

- The four major dune types are barchan, longitudinal, transverse, and parabolic. The amount of sand available, the prevailing wind direction, the wind velocity, and the amount of vegetation present determine which type will form.

- Loess is derived from deserts, Pleistocene glacial outwash deposits, and river floodplains in semiarid regions. Loess covers approximately 10% of Earth's land surface and weathers to a rich and productive soil.

- Deserts are very dry (averaging less than 25 cm of rain per year), have poorly developed soils, and are mostly or completely devoid of vegetation.

- The winds of the major air-pressure belts, oriented east–west, resulting from rising and cooling air, are deflected by the Coriolis effect. These belts help control the world's climate.

- Dry climates are located in the low and middle latitudes where the potential loss of water by evaporation exceeds the yearly precipitation. Dry climates cover 30% of Earth's surface and are subdivided into semiarid and arid regions.

- The majority of the world's deserts are in the low-latitude dry-climate zone between 20 and 30 degrees north and south latitudes. Their dry climate results from a high-pressure belt of descending dry air. The remaining deserts are in the middle latitudes, where their distribution is related to the rain-shadow effect and in the dry polar regions.

- Deserts are characterized by little precipitation and high evaporation rates. Furthermore, rainfall is unpredictable and, when it does occur, tends to be intense and of short duration. As a consequence of such aridity, desert vegetation and animals are scarce.

- Mechanical weathering is the dominant form of weathering in deserts. The sparse precipitation and slow rates of chemical weathering result in poorly developed soils.

- Running water is the dominant agent of erosion in deserts and was even more important during the Pleistocene, when wetter climates resulted in humid conditions.

- Wind is an erosional agent in deserts and is very effective in transporting and depositing unconsolidated fine-grained sediments.

- Important desert landforms include playas, which are dry lakebeds; when temporarily filled with water, they form playa lakes. Alluvial fans are fan-shaped sedimentary deposits that may coalesce to form bajadas.

- Pediments are erosional bedrock surfaces of low relief gently sloping away from mountain bases.

- Inselbergs are isolated, steep-sided erosional remnants that rise above the surrounding desert plains. Buttes and mesas are, respectively, pinnacle-like and flat-topped erosional remnants with steep sides.

Important Terms

abrasion (p. 537)	Coriolis effect (p. 547)	dune (p. 542)
alluvial fan (p. 555)	deflation (p. 541)	inselberg (p. 556)
barchan dune (p. 543)	desert (p. 549)	loess (p. 546)
butte (p. 556)	desert pavement (p. 541)	longitudinal dune (p. 543)

mesa (p. 556)   playa (p. 555)   transverse dune (p. 545)
parabolic dune (p. 545)   rainshadow desert (p. 549)   ventifact (p. 537)
pediment (p. 555)

Review Questions

1. Bed load is primarily transported by which of the following processes?

   a. _____ precipitation; b. _____ deflation; c. _____ saltation; d. _____ abrasion; e. _____ suspension.

2. The Coriolis effect causes wind to be deflected:

   a. _____ only to the right for both hemispheres; b. _____ to the right in the Northern Hemisphere and to the left in the Southern Hemisphere; c. _____ only to the left for both hemispheres; d. _____ to the left in the Northern Hemisphere and to the right in the Southern Hemisphere; e. _____ not at all.

3. Which of the following dunes are among the highest in the world, consist of pyramidal hills of sand, and develop where the wind direction is variable?

   a. _____ barchan; b. _____ longitudinal; c. _____ transverse; d. _____ star; e. _____ parabolic.

4. The major agent of erosion in deserts today is:

   a. _____ glaciers; b. _____ wind; c. _____ abrasion; d. _____ running water; e. _____ none of the preceding answers.

5. Deserts:

   a. _____ can be found in the low, middle, and high latitudes; b. _____ receive more than 25 cm of rain per year; c. _____ are mostly or completely devoid of vegetation; d. _____ answers a and c; e. _____ answers b and c.

6. The driest deserts in the world are found between _____ and _____ latitude in both hemispheres.

   a. _____ 60°, 80°; b. _____ 40°, 60°; c. _____ 30°, 60°; d. _____ 20°, 30°; e. _____ 10°, 20°.

7. A crescent-shaped dune whose tips point downwind is a _____ dune.

   a. _____ star; b. _____ longitudinal; c. _____ parabolic; d. _____ barchan; e. _____ transverse.

8. Airborne particles abrading the surface of a rock produce:

   a. _____ inselbergs; b. _____ dunes; c. _____ ventifacts; d. _____ playas; e. _____ loess.

9. What area of the world has the thickest and most extensive loess deposits?

   a. _____ United States; b. _____ Ukraine; c. _____ Argentina; d. _____ China; e. _____ Belgium.

10. The primary cause of dryness in low-latitude deserts is:

    a. _____ isolation from moist marine winds; b. _____ dominance of the subtropical high-pressure belt; c. _____ the Coriolis effect; d. _____ the rainshadow effect; e. _____ all of the preceding answers.

11. If deserts are dry regions in which mechanical weathering predominates, why are so many of their distinctive landforms the result of running water and not of wind?

12. Considering what you now know about deserts, their location, how they form, and the various landforms associated with them, how can you use this information to determine where deserts may have existed in the past?

13. Much of the recent desertification has been greatly accelerated by human activity. Can anything be done to slow the process?

14. Why are so many desert rock formations red?

15. Why is desert pavement important in a desert environment?

16. Is it possible to get the same type of sand dunes on Mars as on Earth? What does that tell us about the climate and geology of Mars?

17. How do dunes form and migrate? Why is dune migration a problem in some areas?

18. Why are low-latitude deserts so common?

19. What is loess, and why is it important?

20. How are pediments formed?

World Wide Web Activities

PHYSICAL
Geology⇌Now Assess your understanding of this chapter's topics with additional quizzing and comprehensive interactivities at

http://earthscience.brookscole.com/physgeo5e

as well as current and up-to-date weblinks, additional readings, and InfoTrac College Edition exercises.

Shorelines and Shoreline Processes

CHAPTER 19
OUTLINE

PHYSICAL Geology⇌Now *This icon, appearing throughout the book, indicates an opportunity to explore interactive tutorials, animations, or practice problems available on the Physical GeologyNow Web site at http://earthscience.brookscole.com/physgeo5e.*

OBJECTIVES
At the end of this chapter, you will have learned that

- Wind-generated waves and their associated nearshore currents effectively modify shorelines by erosion and deposition.

- The gravitational attraction by the Moon and Sun and Earth's rotation are responsible for the rhythmic rise and fall of sea level known as tides.

- Seacoasts and lakeshores are both modified by waves and nearshore currents, but seacoasts also experience tides, that are insignificant in even the largest lakes.

- The concept of a nearshore sediment budget considers equilibrium, losses, and gains in the amount of sediment in a coastal area.

- Distinctive erosional and depositional landforms such as wave cut platforms, spits, and barrier islands are found along shorelines.

- Shoreline management is complicated by rising sea level, and structures originally far from the shoreline are now threatened or have already been destroyed.

- Several types of coasts are recognized based on criteria such as deposition and erosion and fluctuations in sea level.

Waves pounding the shoreline near Bodega Bay, California. Source: Sue Monroe

Introduction

Intuitively we know that a **shoreline** is the area of land in contact with the sea or a lake, but we can expand this definition by noting that shorelines include the land between low tide and the highest level on land affected by storm waves. Thus a shoreline is a long, narrow zone where marine processes—particularly waves, nearshore currents, and tides—are active. In short, shorelines, just like running water in channels and glaciers, are dynamic systems where energy is expended, erosion takes place, and sediment is transported and deposited. As a matter of fact, shorelines continuously adjust to any changes that take place, such as increased or diminished sediment supply or increased wave activity.

Our main concern in this chapter is with ocean shorelines, or seashores, but waves and nearshore currents are also quite effective in large lakes, where shorelines exhibit many of the same features seen on seashores (■ Figure 19.1). Along the Great Lakes shorelines, for example, many depositional and erosional features typical of seacoasts are well represented. The most notable differences between lakeshores and seashores are that waves and nearshore currents are much more energetic along the latter, and even the largest lakes have insignificant tides.

You already know that the hydrosphere is a vast system consisting of all water on Earth, the bulk of which is in the oceans. In this enormous body of water, energy is transferred through the water to shorelines, so understanding the geologic processes operating on shorelines is important to many people. Indeed, many of the world's major centers of commerce and much of Earth's population are concentrated in a narrow band at or near shorelines. Furthermore, a number of coastal communities such as Myrtle Beach, South Carolina,

(a)

(b)

■ **Figure 19.1**

Seashores (a) and lakeshores (b) show many similar features. Both are modified by waves and nearshore currents, although these processes are more vigorous along seacoasts, and even the largest lakes have no tides of any significance. (a) The Pacific coast of the United States. (b) This erosional feature called Miner's Castle in Lake Superior in Michigan forms just as similar features do along seashores.

Fort Lauderdale, Florida, and Padre Island, Texas, depend heavily on tourists visiting and enjoying their beaches.

Geologists, oceanographers, marine biologists as well as coastal engineers, among others, are interested in the dynamic nature of shorelines. Elected officials and city planners of coastal communities must be familiar with shoreline processes so they can develop policies and zoning regulations that serve the public as well as protect the fragile shoreline environment. Such understanding is especially important now because in many parts of the world sea level is rising, so buildings that were far inland are now in peril or have been destroyed, and slumps and slides are more common along steep shorelines eroded by waves. Furthermore, hurricanes expend much of their energy on shorelines, resulting in extensive coastal flooding, numerous fatalities, and considerable property damage.

The geologic processes operating on shorelines provide another excellent example of systems interactions—in this case between part of the hydrosphere and the solid earth. But the atmosphere is also involved in transferring energy from wind to water, thus causing waves, which in turn generate nearshore currents. And, of course, the gravitational attraction of the Moon and Sun on oceanic waters is responsible for the rhythmic rise and fall of tides. Accordingly, a huge quantity of water is involved in these interactions, but much of the energy of waves and tides is expended on shorelines.

The continents have more than 400,000 km of shoreline, some of it rocky and steep as along North America's West Coast and its northeast coast in Maine and the Maritime Provinces of Canada. Other areas, including much of the East Coast of the United States as well as the Gulf Coast, have broad sandy beaches or long, narrow islands of sand just offshore. Whatever the type of shoreline, though, waves, nearshore currents, and tides continuously bring about changes, some of which are not very desirable, at least from the human perspective.

SHORELINES AND SHORELINE PROCESSES

In contrast to other geologic agents such as running water, wind, and glaciers that operate over vast areas, shoreline processes are restricted to a narrow zone at any particular time. But shorelines might migrate landward or seaward depending on changing sea level or uplift or subsidence of coastal regions. During a rise in sea level, for instance, the shoreline migrates landward, and as it does so, wave, tide, and nearshore current activity shifts landward as well; during times when sea level falls, just the opposite takes place. Recall from Chapter 6 that during marine transgressions and regressions, beach and nearshore sediments can be deposited over vast regions.

(a)

(b)

■ **Figure 19.2**

Low tide (a) and high tide (b) in Turnagain Arm, part of Cook Inlet in Alaska. The tidal range here is about 10 m. Turnagain Arm is a huge fiord now being filled in by sediment carried in by rivers. Notice the mudflats in (a).

In the marine realm, several biological, chemical, and physical processes are operating continuously. Organisms change the local chemistry of seawater and contribute their skeletons to nearshore sediments, temperature and salinity change, and internal waves occur in the oceans. However, the processes most important for modifying shorelines are purely physical ones, especially waves, tides, and nearshore currents. We cannot totally discount some of these other processes, though; offshore reefs, for instance, may protect a shoreline area from most of the energy of waves.

Tides

The surface of the oceans rises and falls twice daily in response to the gravitational attraction of the Moon and Sun. These regular fluctuations in the ocean's surface, or **tides,** result in most shoreline areas having two daily high tides and two low tides as sea level rises and falls anywhere from a few centimeters to more than 15 m (■ Figure 19.2). A complete tidal cycle includes *flood tide* that progressively covers more and more of the nearshore area until high tide is reached, followed by *ebb tide* during which the nearshore area is once again exposed.

Both the Moon and the Sun have sufficient gravitational attraction to exert tide-generating forces strong enough to deform the solid body of Earth, but they have a much greater influence on the oceans. The Sun is 27 million times more massive than the Moon, but it is 390 times as far away so its tide-generating force is only 46% as strong as that of the Moon. Accordingly, the tides are dominated by the Moon, but the Sun plays an important role as well.

If we consider only the Moon acting on a spherical, water-covered Earth, the tide-generating forces produce two bulges on the ocean surface (■ Figure 19.3a). One bulge extends toward the Moon because it is on the side of Earth where the Moon's gravitational attraction is greatest. The other bulge is on the opposite side of Earth, where the Moon's gravitational attraction is least. These two bulges always point toward and away from the Moon (Figure 19.3a), so as Earth rotates and the Moon's position changes, an observer at a particular shoreline location experiences the rhythmic rise and fall of tides twice daily. But the heights of two successive high tides may vary depending on the Moon's inclination with respect to the equator.

The Moon revolves around Earth every 28 days, so its position with respect to any latitude changes slightly each day. That is, as the Moon moves in its orbit and Earth rotates on its axis, it takes the Moon 50 minutes longer each day to return to the same position it was in the previous day. Thus an observer would experience a high tide at 1:00 P.M. on one day, for example, and at 1:50 P.M. on the following day.

Tides are also complicated by the combined effects of the Moon and the Sun. Even though the Sun's tide-generating force is weaker than the Moon's, when the Moon and Sun are aligned every two weeks, their forces added together generate *spring tides,* which are about 20% higher than average tides (Figure 19.3b). When the Moon and Sun are at right angles to each other, also at two-week intervals, the Sun's tide-generating force cancels some of the Moon's, and *neap tides* about 20% lower than average occur (Figure 19.3c).

Tidal ranges are also affected by shoreline configuration. Broad, gently sloping continental shelves as in the Gulf of Mexico have low tidal ranges, whereas steep, irregular shorelines experience much greater rise and fall of tides. Tidal ranges are greatest in some narrow, funnel-shaped bays and inlets. For example, the tidal range in the Bay of Fundy in Nova Scotia is 16.5 m, and ranges greater than 10 m occur in several other areas.

Tides have an important impact on shorelines because the area of wave attack constantly shifts onshore and offshore as the tides rise and fall. Tidal currents themselves have little modifying effect on shorelines, except in narrow passages where tidal current velocity is great enough to erode and transport sediment. Indeed, if it were not for strong tidal currents, some passageways would be blocked by sediments deposited by nearshore currents. Tides are one potential source of energy (see Geo-Focus 19.1).

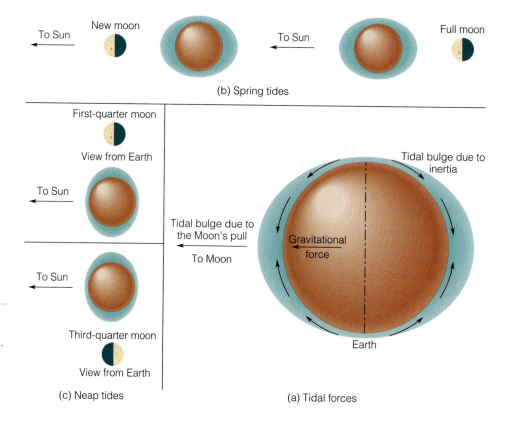

■ **Figure 19.3**

(a) Tides are caused by the gravitational pull of the Moon and, to a lesser degree, the Sun. The Earth–Moon–Sun alignments at the times of the (b) spring and (c) neap tides are shown.

PHYSICAL
Geology⇌Now Click Geology Interactive to work through an activity on High/low Tides, Tidal Forces and Tidal Shifts through Waves, Tides and Currents.

Waves

Waves, or oscillations of a water surface, are seen on all bodies of water but are most significant along seashores where they are responsible for most of the erosion, transport, and deposition of sediment. ■ Figure 19.4 shows a typical series of waves in deep water and the terminology applied to them. The highest part of a wave is its **crest,** and the low point between crests is the **trough. Wavelength** is the distance between successive wave crests (or troughs), and **wave height** is the vertical distance from trough to crest. You can calculate the speed at which a wave advances, called *celerity* (C), if you know the wavelength (L) and the **wave period** (T), which is the time required for two successive wave crests (or troughs) to pass a given point:

$$C = L/T$$

The speed of wave advance (C) is actually a measure of the velocity of the waveform rather than a measure of the speed of the molecules of water. In fact, the water in

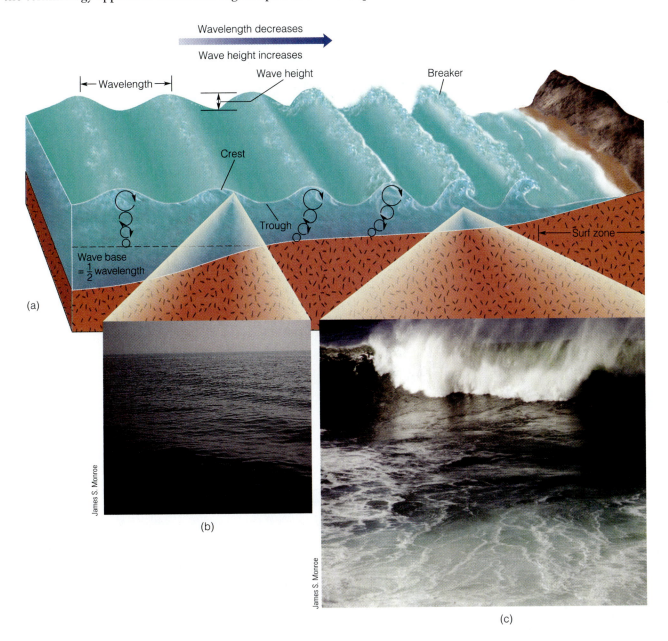

Wavelength decreases

Wave height increases

Wavelength

Wave height

Breaker

Crest

Trough

Surf zone

Wave base = ½ wavelength

(a)

James S. Monroe

(b)

James S. Monroe

(c)

■ **Figure 19.4**

(a) Waves and the terminology applied to them. As swells (b) in deep water move toward shore, the orbital motion of water within them is disrupted when they enter water shallower than wave base. Wavelength decreases while wave height increases, causing the waves to oversteepen and plunge forward as breakers (c).

GEO**FOCUS**

19.1

Energy from the Oceans

Complex interactions among systems have been a recurring theme throughout this book, and here too we emphasize these interactions as we discuss the potential to extract energy from the oceans. Solar energy is responsible for differential heating of the atmosphere and thus air circulation, which is responsible for waves and oceanic currents, although Earth's rotation also plays an important role in currents. And gravitational attraction between Earth and the Sun and Moon coupled with Earth's rotation accounts for the twice-daily rise and fall of tides seen in most coastal areas.

If we could harness the energy of waves, oceanic currents, temperature differences in oceanic waters, and tides, an almost limitless, largely nonpolluting energy supply would be ensured. Unfortunately, ocean energy is diffuse, meaning that the energy for a given volume of water is small and thus difficult to concentrate and use. Of the several sources of ocean energy, only tides show much promise for the near future.

Ocean thermal energy conversion (OTEC) exploits the temperature difference between surface waters and those at depth to run turbines and generate electricity. The amount of energy from this source is enormous, but a number of practical problems must be solved. For one thing, enormous quantities of water must be circulated through a power plant, which requires large surface areas devoted to this purpose. Despite several decades of research, no commercial OTEC plants are oper-

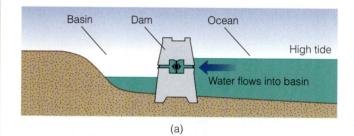

(a)

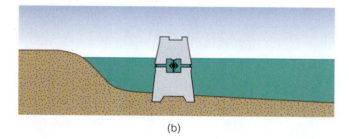

(b)

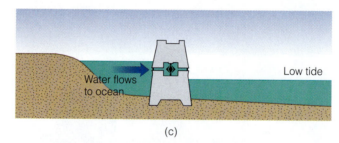

(c)

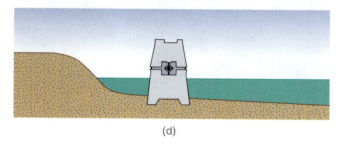

(d)

■ **Figure 1**

Rising and falling tides produce electricity by spinning turbines connected to generators, just as at hydroelectric plants. (a) Water flows from ocean to basin during flood tide. (b) Basin full. (c) Water flows from basin to ocean during ebb tide. (d) Basin water is depleted, and the cycle begins once more.

■ **Figure 2**

The La Rance River estuary tidal power-generating plant in western France. This 850-m-long dam was built where the maximum tidal range is 13.4 m. Electricity is generated during both flood and ebb tides.

ating or under construction, although small experimental ones have been tested in Hawaii and Japan.

Oceanic currents such as the Gulf Stream also possess energy that might be tapped to generate electricity. Unfortunately, these currents flow at only a few kilometers per hour at most, whereas hydroelectric power plants on land rely on water moving rapidly from higher to lower elevations. And unlike streams on land, oceanic currents cannot be dammed, their energy is diffuse, and any power plant would have to contend with unpredictable changes in flow direction.

Harnessing wave energy to generate electricity is not a new idea, and in fact it is used on an extremely limited scale. Any facility using this form of energy would obviously have to be designed to withstand the effects of storms and saltwater corrosion. No large-scale wave-energy plants are operating, but the Japanese have developed

devices to power lighthouses and buoys, and a facility with a capacity to provide power to about 300 homes began operating in Scotland during September 2000.

Tidal power has been used for centuries in some coastal areas to run mills, but its use at present for electrical generation is limited. Most coastal areas experience a twice-daily rise and fall of tides, but only a few areas are suitable for exploiting this energy source. One limitation is that the tidal range must be at least 5 m, and there must also be a coastal region where water can be stored following high tide.

Suitable sites for tidal power plants are limited not only by tidal range but also by location. Many areas along the U.S. Gulf Coast would certainly benefit from a tidal power plant, but the tidal range of generally less than 1 m precludes the possibility of development. However, an area with an appropriate tidal range in some remote area such as southern Chile or the Arctic

islands of Canada offers little potential because of the great distance from population centers. Accordingly, in North America only a few areas show much potential for developing tidal energy: Cook Inlet, Alaska; the Bay of Fundy on the U.S.–Canadian border; and some areas along the New England coast, for instance.

The idea behind tidal power is simple, although putting it into practice is not easy. First, a dam with sluice gates to regulate water flow must be built across the entrance to a bay or estuary. When the water level has risen sufficiently high during flood tide, the sluice gates are closed. Water held on the landward side of the dam is then released and electricity is generated just as it is at a hydroelectric dam. Actually, a tidal power plant can operate during both flood and ebb tides (■ Figure 1).

The first tidal power-generating facility was constructed at the La Rance River estuary in France (■ Figure 2). This 240-megawatt (MW) plant began operations in 1966 and has been quite successful. In North America, a much smaller 20-MW tidal power plant has been operating in the Bay of Fundy, Nova Scotia, where the tidal range, the greatest in the world, exceeds 16 m. The Bay of Fundy is part of the more extensive Gulf of Maine, where the United States and Canada have been considering a much larger tidal power-generating project for several decades.

Although tidal power shows some promise, it will not solve our energy needs even if developed to its fullest potential. Most analysts think that only 100 to 150 sites worldwide have sufficiently high tidal ranges and the appropriate coastal configuration to exploit this energy resource. This, coupled with the facts that construction costs are high and tidal energy systems can have disastrous effects on the ecology (biosphere) of estuaries, makes it unlikely that tidal energy will ever contribute more than a small percentage of all energy production.

James S. Monroe

(a)

James S. Monroe

(b)

■ **Figure 19.5**

(a) The waves in this lake have a wavelength of about 2 m, so we can infer that wave base is 1 m deep. Where the waves encounter wave base, they stir up the bottom sediment, thus accounting for the nearshore turbid water. (b) These 2-m-high waves at Jenner, California, and the nearshore water are brown because of sediment. If you look closely, you can see the blue water of the open ocean just beyond where the small wave is breaking in the middle distance, which is the position of wave base.

waves moves forward and back as a wave passes but has little or no net forward movement. As waves move across a water surface, the water "particles" rotate in circular orbits and transfer energy in the direction of wave advance (Figure 19.4).

The diameters of the orbits followed by water particles in waves diminish rapidly with depth, and at a depth of about one-half wavelength ($L/2$), called **wave base,** they are essentially zero (Figure 19.4). At depths exceeding wave base, the water and seafloor, or lake floor, are unaffected by surface waves (■ Figure 19.5). We will explore the significance of wave base more fully in later sections.

Wave Generation   Several processes including displacement of water by landslides, displacement of the seafloor by faulting, and volcanic explosions generate waves. Some waves so generated are huge and may be devastating to coastal areas, but most of the geologic work done on shorelines is accomplished by wind-generated waves. When wind blows over water, some of its energy is transferred to the water, causing the water surface to oscillate. The mechanism that transfers energy from wind to water is related to the frictional drag that results from one fluid (air) moving over another (water).

In an area of wave generation, as beneath a storm center at sea, sharp-crested, irregular waves called *seas* develop. Seas are an aggregate of waves of various

heights and lengths, and one wave cannot be clearly distinguished from another. As seas move out from the area of wave generation, though, they are sorted into broad *swells* that have rounded, long crests and are all about the same size (Figure 19.4b).

As one would expect, the harder and longer the wind blows, the larger are the waves. Wind velocity and duration, however, are not the only factors controlling wave size. High-velocity wind blowing over a small pond will never generate large waves regardless of how long it blows. In fact, waves are present on ponds and most lakes only while the wind is blowing; once the wind stops, the water quickly smooths out. In contrast, the surface of the ocean is always in motion, and waves in the open seas with heights of 34 m have been recorded during storms.

The reason for the disparity between wave sizes on ponds and lakes and on the oceans is the **fetch,** which is the distance the wind blows over a continuous water surface. Fetch is limited by the available water surface, so on ponds and lakes it corresponds to their length or width, depending on wind direction. To produce waves of greater length and height, more energy must be transferred from wind to water; hence, large waves form beneath large storms at sea.

In the areas of storm wave generation, waves with different lengths, heights, and periods might merge, making them smaller or larger. Occasionally, two wave crests merge to form *rogue waves* that may be three or four times higher than the average. Indeed, these rogue waves can rise unexpectedly out of an otherwise comparatively calm sea and threaten even the largest ships. In 1942 HMS *Queen Mary,* while carrying 15,000 American soldiers, was hit broadside by a rogue wave and nearly capsized.

Shallow-Water Waves and Breakers

Waves moving out from an area of generation form swells and lose only a small amount of energy as they travel across the ocean. In deep-water swells, the water surface oscillates and water particles move in orbital paths, with little net displacement of water taking place in the direction of wave advance. When these waves enter progressively shallower water, the water is displaced in the direction of wave advance (Figure 19.4).

As deep-water waves enter shallow water, they are transformed from broad, undulating swells into sharp-crested waves. This transformation begins at a water depth of wave base; that is, it begins where wave base intersects the seafloor. At this point, the waves "feel" the bottom, and the orbital motions of water particles within waves are disrupted (Figure 19.4). As they move farther shoreward, the speed of wave advance and wavelength decrease, and wave height increases. In effect, as waves enter shallower water, they become oversteepened; the wave crest advances faster than the wave form, until eventually the crest plunges forward as a **breaker** (Figure 19.4c). Breakers are commonly several times higher than deep-water waves, and when they plunge forward their kinetic energy is expended on the shoreline.

The waves just described are the classic plunging breakers that pound the shorelines of areas with steep offshore slopes, such as on the north shore of Oahu in the Hawaiian Islands (■ Figure 19.6a). In contrast, shorelines with gentle offshore slopes are characterized by spilling breakers, where the waves build up slowly and the wave's crest spills down the front of the wave (Figure 19.6b). Whether the breakers spill or plunge, the water rushes onto the shore, then returns seaward to become part of another breaking wave.

The size of breakers varies enormously. Waves tend to be larger and more energetic in winter because storms are more common during that season. In addition, waves of various sizes merging in the area where they form not only create rogue waves but also account for variations in the size of waves breaking on the shore. For instance, when waves with different lengths merge, smaller waves result, whereas merging of waves with the same length forms larger waves. As a result, several small waves might break on a beach followed by a series of larger ones. Surfers commonly take advantage of this phenomenon by swimming out to sea during a relative calm where they wait for a large set of waves to ride toward shore.

PHYSICAL Geology ⇌ Now   Click Geology Interactive to work through an activity on Wave Properties, Wave Progression and Wind-Wave Relationships through Waves, Tides and Currents.

Nearshore Currents and Sediment Transport

We identify the *nearshore zone* as the area extending seaward from the upper limits of the shoreline to just beyond where waves break. It includes a *breaker zone* and a *surf zone,* which is where breaking waves rush forward onto the shore followed by seaward movement of the water as backwash (Figure 19.4). The width of the nearshore zone varies depending on the wavelength of approaching waves because long waves break at a greater depth, and thus farther offshore, than do short

(a)   (b)

■ **Figure 19.6**

(a) A plunging breaker on the north shore of Oahu, Hawaii. (b) A spilling breaker.

Sue Monroe

■ Figure 19.7

Wave refraction (wave crests are indicated by dashed lines). These waves are refracted as they enter shallow water and more nearly parallel the shoreline. The waves generate a longshore current that flows in the direction of wave approach, from upper left to lower right (arrow) in this example.

waves. Two types of currents are important in the nearshore zone: longshore currents and rip currents.

Wave Refraction and Longshore Currents

Deep-water waves have long, continuous crests, but rarely are their crests parallel with the shoreline (■ Figure 19.7). One part of a wave enters shallow water where it encounters wave base and begins breaking before other parts of the same wave. As a wave begins breaking, its velocity diminishes, but the part of the wave still in deep water races ahead until it too encounters wave base. The net effect of this oblique approach is that waves bend so that they more nearly parallel the shoreline (Figure 19.7). This phenomenon is known as **wave refraction.**

Even though waves are refracted, they still usually strike the shoreline at some angle, causing the water between the breaker zone and the beach to flow parallel to the shoreline. These **longshore currents,** as they are called, are long and narrow and flow in the same general direction as the approaching waves (Figure 19.7). These currents are particularly important because they transport and deposit large quantities of sediment in the nearshore zone.

Rip Currents
Waves carry water into the nearshore zone, so there must be a mechanism for mass transfer of water back out to sea. One way that water moves sea-

What Would You Do ?

While swimming at your favorite beach, you suddenly notice that you are far down the beach from the point where you entered the water. Furthermore, the size of the waves you were swimming in has diminished considerably. You decide to swim back to shore and then walk back to your original starting place, but no matter how hard you swim, you are carried farther and farther away from shore! Assuming that you survive this incident, explain what happened and what you did to remedy the situation.

ward from the nearshore zone is in **rip currents,** which are narrow surface currents that flow out to sea through the breaker zone (■ Figure 19.8). Surfers commonly take advantage of rip currents for an easy ride out beyond the breaker zone, but these currents pose a danger to inexperienced swimmers. Rip currents may flow at several kilometers per hour, so if a swimmer is caught in one, it is useless to try to swim directly back to shore. Instead, because rip currents are narrow and usually nearly perpendicular to the shore, one can swim paral-

Breaker zone

High wave crests

Low wave crests

Rip current

(a)

(b)

(c)

James S. Monroe

James S. Monroe

■ **Figure 19.8**

(a) Rip currents are fed on each side by currents moving parallel to the shoreline. (b) Suspended sediment, indicated by discolored water, is being carried seaward in this rip current. (c) Sign along the California coast warning swimmers of dangerous rip currents, higher than normal (sleeper) waves, and backwash.

lel to the shoreline for a short distance and then turn shoreward with no difficulty.

Rip currents are circulating cells fed by longshore currents. When waves approach a shoreline, the amount of water builds up until the excess moves out to sea through the breaker zone. Rip currents are fed by nearshore currents that increase in velocity from midway between each rip current (Figure 19.8).

Seafloor configuration also plays an important role in determining the location of rip currents. They commonly develop where wave heights are lower than in adjacent areas. Differences in wave height are controlled by variations in water depth. For instance, if waves move over a depression, their heights over the depression tend to be less than in adjacent areas.

PHYSICAL
Geology⇌Now Click Geology Interactive to work through an activity on Dissect a Current through Waves, Tides and Currents.

DEPOSITION ALONG SHORELINES

Depositional features of shorelines include beaches, spits, baymouth bars, tombolos, and barrier islands. The characteristics of beaches are determined by wave energy and shoreline materials, and they are continuously modified by waves and longshore currents. Spits and baymouth bars both result from deposition by longshore currents, but the origin of barrier islands is controversial. Rip currents play only a minor role in the configuration of shorelines, but they do transport fine-grained sediment seaward through the breaker zone.

Beaches

Beaches are the most familiar of all coastal landforms, attracting millions of visitors each year and providing the economic base for many communities. They consist of a long, narrow strip of unconsolidated sediment, commonly sand, and are constantly changing. Depending on shoreline materials and wave intensity, beaches may be discontinuous, existing only as *pocket beaches* in protected areas such as embayments, or they may be continuous for long distances. See "Shoreline Processes and Beaches" on pages 574 and 575.

By definition, a **beach** is a deposit of unconsolidated sediment extending landward from low tide to a change in topography such as a line of sand dunes, a sea cliff, or the point where permanent vegetation begins. Typically, a beach has several component parts, including a *backshore* that is usually dry, being covered by water only during storms or exceptionally high tides. The backshore consists of one or more **berms**, platforms composed of sediment deposited by waves. Berms are nearly horizontal or slope gently landward. The sloping area below a berm that is exposed to wave swash is the *beach face*. The beach face is part of the *foreshore,* an area covered by water during high tide but exposed during low tide.

Some sediment on beaches is derived from weathering and wave erosion of the shoreline, but most of it is transported to the coast by streams and redistributed along the shoreline by longshore currents. **Longshore drift** is the phenomenon involving sand transport along a shoreline by longshore currents. As noted, waves usually strike beaches at some angle, causing the sand grains to move up the beach face at a similar angle, but as backwash carries the sand grains seaward, they move perpendicular to the long axis of the beach. So individual sand grains move in a zigzag pattern in the direction of longshore currents. This movement is not restricted to the beach but extends seaward to the outer edge of the breaker zone.

In an attempt to widen a beach or prevent erosion, shoreline residents build structures known as *groins* that project seaward at right angles from the shoreline. Groins interrupt the flow of longshore currents, causing sand deposition on their upcurrent side, widening the beach at that location. However, erosion inevitably occurs on the downcurrent side of a groin.

Many beaches are sandy, but in areas of particularly vigorous wave activity, they might be gravel covered.

Most beach sand is composed of quartz, but other minerals and rock fragments may be present as well. One of the most common accessory minerals in beach sands is magnetite; because of its high specific gravity, magnetite is commonly separated from the other minerals and is visible as thin, black layers.

Although quartz is the most common mineral in most beach sands, there are some notable exceptions. The black sand beaches of Hawaii are composed of sand-sized basalt rock fragments, and some beaches are made up of fragmented calcium carbonate shells of marine organisms. In short, beaches are composed of whatever material is available; quartz is most abundant simply because it is abundant in most areas and is the most durable and stable of the common rock-forming minerals (see Chapter 5).

Seasonal Changes in Beaches

Wave energy is dissipated on beaches, so the loose grains composing it are constantly moved by waves. But the overall configuration of a beach remains unchanged as long as equilibrium conditions persist. We can think of the beach profile as consisting of a berm or berms and a beach face shown in ■ Figure 19.9a as a profile of equilibrium; that is, all parts of the beach are adjusted to the prevailing conditions of wave intensity and nearshore currents.

Tides and longshore currents affect the configuration of beaches to some degree, but storm waves are by far the most important agent modifying their equilibrium profile. In many areas, beach profiles change with the seasons; so we recognize *summer beaches* and *winter beaches,* which have adjusted to the conditions prevailing at these times (Figure 19.9). Summer beaches are sand covered and possess a wide berm, a steep beach face, and a smooth offshore profile. During winter, however, when waves more vigorously attack the beach, a winter beach develops, which has little or no berm, a gentle slope, and offshore sand bars paralleling the shoreline (Figure 19.9). Seasonal changes can be so profound that sand moving onshore and offshore yields sand-covered beaches during summer and gravel-covered beaches during winter.

Seasonal changes in beach profiles are related to changing wave intensity. During the winter, energetic storm waves erode the sand from the beach and transport it offshore where it is stored in sand bars (Figure 19.9). The same sand that was eroded from the beach during the winter returns the next summer when it is driven onshore by the more gentle swells that occur during that season. The volume of sand in the system remains more or less constant; it simply moves offshore or onshore depending on the energy of waves.

The terms *winter beach* and *summer beach,* though widely used, are somewhat misleading. A "winter beach" profile can develop at any time of the year if a large

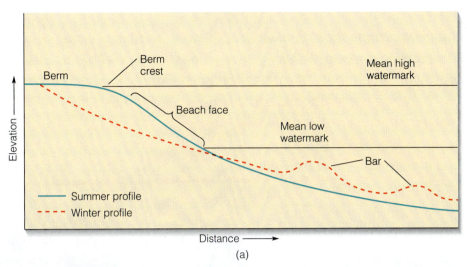

(a)

(b)

(c)

■ Figure 19.9

(a) Seasonal changes in beach profiles. (b) and (c) San Gregorio State Beach, California. These photographs were taken from nearly the same place, but (c) was taken two years after (b). Much of the change from (b) to (c) can be accounted for by beach erosion during 1997–1998 winter storms.

storm occurs, and a "summer beach" profile can develop during a prolonged calm period in the winter.

Spits, Baymouth Bars, and Tombolos

Other than beaches, the most common depositional landforms on shorelines are spits, baymouth bars, and tombolos, all of which are variations of the same feature. A **spit** is simply a continuation of a beach forming a point, or "free end," that projects into a body of water, commonly a bay. A **baymouth bar** is a spit that has grown until it completely closes off a bay from the open sea (■ Figure 19.10).

Both spits and baymouth bars form and grow as a result of deposition by longshore currents. Where currents are weak, as in the deeper water at the opening to a bay, longshore current velocity diminishes, and sedi-

ment is deposited, forming a sand bar. The free ends of many spits are curved by wave refraction or waves approaching from a different direction. Such spits are called *hooks* or *recurved spits* (Figure 19.10).

A rarer type of spit, a **tombolo,** extends out into the sea and connects an island to the mainland. Tombolos develop on the shoreward sides of islands, as shown in ■ Figure 19.11. Wave refraction around an island causes converging currents that turn seaward and deposit a sand bar connecting the shore with the island. A similar feature may form when an artificial breakwater is constructed offshore (Figure 19.11b).

Although spits, baymouth bars, and tombolos are most commonly found on irregular seacoasts, many examples of the same features are present in large lakes. Whether along seacoasts or lakeshores, these sand deposits present a continuing problem where bays must be

Shoreline Processes and Beaches

Beaches are the most familiar depositional landforms along shorelines. They are found in a variety of sizes and shapes, with long sandy beaches typical of the East and Gulf Coasts, and smaller, mostly protected beaches along the West Coast. The sand on most beaches is primarily quartz, but there are some notable exceptions: shell sand beaches in Florida and black sand beaches in Hawaii. Beaches are dynamic systems where waves, tides, and marine currents constantly bring about change.

Michael Slear

Sue Monroe

The Grand Strand of South Carolina (left), shown here at Myrtle Beach, is 100 km of nearly continuous beach. Small pocket beach (above) at Julia Pfeiffer Burns State Park, California.

Diagram of a beach showing its component parts.

Beach

Foreshore   Backshore   Dunes

Beach face   Berms

High tide level

Low tide level

The backshore area of a pocket beach along the Pacific coast. Note that the berm ends at the rocks on the right and the beach face slopes steeply seaward.

Berm

Beach face

James S. Monroe

James S. Monroe

Sand dunes are present at the upper part of this beach at Oregon Dunes National Recreation Area.

Longshore currents transport sediment along the shoreline, a phenomenon called longshore drift, between the breaker zone and the upper limit of wave action. These groins (right) at Cape May, New Jersey, interrupt the flow of longshore currents. Sand is trapped on their upcurrent sides, whereas erosion takes place on their downcurrent sides. Can you tell the direction of longshore drift in this image?

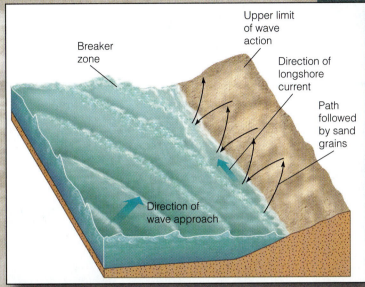

Breaker zone

Upper limit of wave action

Direction of longshore current

Path followed by sand grains

Direction of wave approach

John S. Shelton

Although most beaches are made up dominantly of quartz sand, there are exceptions. This black sand beach on Maui in Hawaii is made up of small basalt rock fragments and particles of obsidian that formed when lava flowed into the sea.

Sue Monroe

The beach shown on Oahu in Hawaii is composed mostly of fragmented shells of marine organisms, although some minerals from basalt are also included.

Sue Monroe

The California beach is made up of quartz and sand- and gravel-sized rock fragments.

Sue Monroe

James S. Monroe

James S. Monroe

(a)

(b)

(c)

■ **Figure 19.10**

(a) Spits form where longshore currents deposit sand in deeper water, as at the entrance to a bay. A baymouth bar is simply a spit that has grown until it extends across the mouth of a bay. (b) A spit at the mouth of the Russian River near Jenner, California. (c) A baymouth bar.

kept open for pleasure boating or commercial shipping. The entrances to these bays must be either regularly dredged or protected. The most common way to protect entrances to bays is to build *jetties*, which are structures extending seaward (or lakeward) that protect the bay from deposition by longshore currents.

Barrier Islands

Long, narrow islands composed of sand and separated from the mainland by a lagoon are **barrier islands** (■ Figure 19.12). On their seaward margins, barrier islands are smoothed by waves, but their lagoon sides are irregular. During large storms, waves overtop these islands and deposit lobes of sand in the lagoon. Once deposited, these lobes are modified only slightly because they are protected from further wave action. Windblown sand dunes are common and are generally the highest part of barrier islands.

The origin of barrier islands has been long debated and is still not completely resolved. It is known that they form on gently sloping continental shelves with abun-

dant sand in areas where both tidal fluctuations and wave-energy levels are low. Although barrier islands are found in many areas, most of them are along the East Coast of the United States from New York to Florida and along the U.S. Gulf Coast. Barrier islands are not nearly as common in Canada, but some are found along the coasts of New Brunswick and Prince Edward Island. Ac-

Tombolo

Wave crest

(a)

James S. Monroe

(b)

John S. Shelton

Breakwater

(c)

■ **Figure 19.11**

(a) Wave refraction around an island and the origin of a tombolo. (b) A small tombolo along the Pacific coast of the United States. (c) Soon after this breakwater, an offshore barrier, was constructed at Santa Monica, California, a bulge similar to a tombolo appeared in the beach.

cording to one model, barrier islands formed as spits that became detached from the land, whereas another model proposes that they formed as beach ridges on coasts that subsequently subsided (■ Figure 19.13).

Because sea level is currently rising, most barrier islands are migrating landward. Such migration is a natural consequence of evolution of these islands, but it is a problem for the island residents and communities. Barrier islands migrate rather slowly, but the rates for many are rapid enough to cause shoreline problems (discussed in a later section).

HOW ARE SHORELINES ERODED?

Along seacoasts, where erosion rather than deposition predominates, beaches are lacking or restricted to protected areas, and sea cliffs are present. Sea cliffs are pounded by waves, especially during storms, and the cliff face retreats landward as a result of corrosion, hydraulic action, and abrasion. *Corrosion* is an erosional process involving the

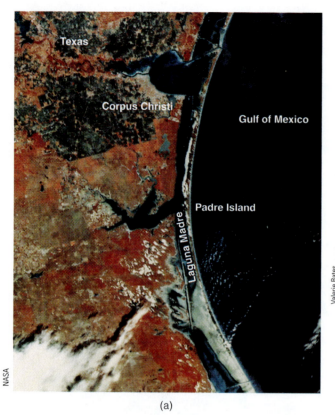

(b)

■ Figure 19.12

(a) View from space of the barrier islands along the Gulf Coast of Texas. Notice that a lagoon up to 20 km wide separates the long, narrow barrier islands from the mainland. (b) Aerial view of Padre Island on the Texas Gulf Coast. Laguna Madre is on the left, and the Gulf of Mexico is on the right.

(a)

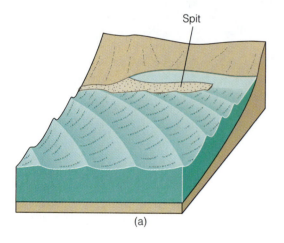

(a)

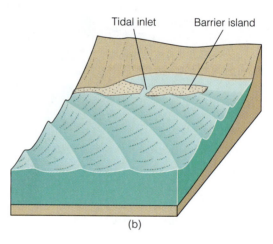

(b)

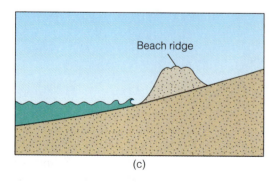

(c)

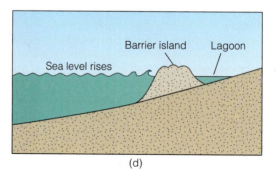

(d)

■ Figure 19.13

Two models for the origin of barrier islands. (a) Longshore currents extend a spit along a coast. (b) During a storm, the spit is breached, forming a tidal inlet and a barrier island. (c) A beach ridge forms on land. (d) Sea level rises, partly submerging the beach ridge.

James S. Monroe

(a)

Sue Monroe

(b)

■ **Figure 19.14**

(a) Rocks in the lower part of this image on a small island in the Irish Sea have been smoothed by abrasion, but the rocks higher up are out of reach of waves. (b) A combination of processes accounts for erosion of these sea cliffs near Bodega Bay, California. Although erosion was already in progress, it was particularly intense during storms of February 1998. Attempts to stabilize the shoreline have failed and some of the houses have been abandoned.

wearing away of rock by chemical processes, especially the solvent action of seawater. The force of the water itself, called *hydraulic action*, is a particularly effective erosional process. Waves exert tremendous pressure on shorelines by direct impact but are most effective on sea cliffs composed of sediment or highly fractured rocks. *Abrasion* involves the grinding action of rocks and sand carried by waves (■ Figure 19.14). Remember from Chapter 15 that running water also erodes by hydraulic action and abrasion.

Wave-Cut Platforms

The rate at which sea cliffs erode and retreat landward depends on wave intensity and the resistance of the coastal rocks or sediments. Most sea cliff retreat takes place during storms and, as one would expect, occurs most rapidly in sea cliffs composed of sediment. For example, some parts of the White Cliffs of Dover in Great Britain retreat at a rate of more than 100 m per century. By comparison, sea cliffs consisting of dense igneous or metamorphic rocks retreat much more slowly.

Sea cliffs erode mostly as a result of hydraulic action and abrasion at their bases. As a sea cliff is undercut by erosion, the upper part is left unsupported and susceptible to mass wasting processes. Thus sea cliffs retreat little by little, and as they do, they leave a beveled surface known as a **wave-cut platform** that slopes gently seaward (■ Figure 19.15). Broad wave-cut platforms exist in many areas, but invariably the water over them is shallow because the abrasive planing action of waves is effective to a depth of only about 10 m. Sediment eroded from sea cliffs is transported seaward until it reaches deeper water at the edge of the wave-cut platform. There it is deposited and forms a seaward extension of the wave-cut platform called a *wave-built platform* (Figure 19.15). Wave-cut platforms now above sea level are known as **marine terraces** (Figure 19.15c).

Sea Caves, Arches, and Stacks

Sea cliffs do not retreat uniformly because some of the materials of which they are composed are more resistant to erosion than others. Those parts of a shoreline that consist of resistant materials might form seaward-projecting headlands, whereas less resistant materials erode more rapidly, yielding embayments where pocket beaches may be present. Wave refraction around headlands causes them to erode on both sides so that *sea caves* form; if caves on the opposite

(a)

John S. Shelton

(b)

James S. Monroe

(c)

■ **Figure 19.15**

(a) Wave erosion causes a sea cliff to migrate landward, leaving a gently sloping surface known as a wave-cut platform. A wave-built platform originates by deposition at the seaward margin of the wave-cut platform. (b) Sea cliffs and a wave-cut platform. (c) This gently sloping surface along the Pacific Coast of California is a marine terrace. Notice the sea stacks rising above the terrace and others formed by erosion of the sea cliff.

sides of a headland join, they form a *sea arch* (■ Figure 19.16). Continued erosion causes the span of an arch to collapse, yielding isolated *sea stacks* on wave-cut platforms (Figure 19.16).

In the long run, shoreline processes tend to straighten an initially irregular shoreline. They do so because wave refraction causes more wave energy to be expended on headlands and less on embayments. Headlands erode, and some of the sediment yielded by erosion is deposited in the embayments, straightening the shoreline (■ Figure 19.17).

THE NEARSHORE SEDIMENT BUDGET

We can think of the gains and losses of sediment in the nearshore zone in terms of a **nearshore sediment budget** (■ Figure 19.18). If a nearshore system has a balanced budget, sediment is supplied as fast as it is removed, and the volume of sediment remains more or less constant, although sand may shift offshore and onshore with the

Sea cave

Sea arch

Sea stack

(a)

Sue Monroe

(c)

Nick Harvey

(b)

■ **Figure 19.16**

(a) Erosion of a headland and the origin of sea caves, sea arches, and sea stacks. (b) This sea stack in Australia has an arch developed in it. (c) Sea stacks on California's central coast.

changing seasons. A positive budget means gains exceed losses, whereas a negative budget means losses exceed gains. If a negative budget prevails long enough, the nearshore system is depleted and beaches may disappear (Figure 19.18).

Erosion of sea cliffs provides some sediment to beaches, but in most areas probably no more than 5 to 10% of the total sediment supply comes from this source. There are exceptions, though; almost all the sediment on the beaches of Maine is derived from the erosion of shoreline rocks. Most sediment on typical beaches is transported to the shoreline by streams and then redistributed along the shoreline by longshore drift. Thus longshore drift also plays a role in the nearshore sediment budget because it continually moves sediment into and away from beach systems (Figure 19.18).

The primary ways that a nearshore system loses sediment are longshore drift, offshore transport, wind, and deposition in submarine canyons. Offshore transport mostly involves fine-grained sediment that is carried seaward, where it eventually settles in deeper water. Wind is an important process because it removes sand from beaches and blows it inland, where it piles up as sand dunes.

If the heads of submarine canyons are nearshore, huge quantities of sand are funneled into them and deposited in deeper water. La Jolla and Scripps submarine canyons off the coast of southern California funnel off an estimated 2 million m^3 of sand each year. In most areas, however, submarine canyons are too far offshore to interrupt the flow of sand in the nearshore zone.

It should be apparent from this discussion that if a nearshore system is in equilibrium, its incoming supply of sediment exactly offsets its losses. Such a delicate balance tends to continue unless the system is somehow disrupted. One common change that affects this balance

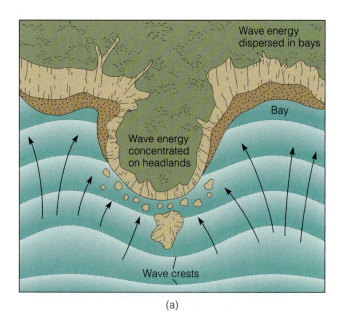

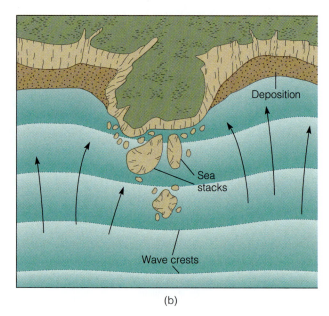

■ Figure 19.17

(a) Wave refraction acts to straighten shorelines by concentrating wave energy on headlands. (b) The same shoreline after extensive erosion of the headlands and deposition in the bays.

is the construction of dams across the streams that supply sand. Once dams have been built, all sediment from the upper reaches of the drainage system is trapped in reservoirs and thus cannot reach the shoreline.

HOW ARE COASTAL AREAS MANAGED AS SEA LEVEL RISES?

During the last century, sea level rose about 12 cm worldwide, and all indications are that it will continue to rise. The absolute rate of sea-level rise in a particular shoreline region depends on two factors. The first is the volume of water in the ocean basins, which is increasing as a result of the melting of glacial ice and the thermal expansion of near-surface seawater. Many scientists think that sea level will continue to rise because of global warming, resulting from concentrations of greenhouse gases in the atmosphere.

The second factor controlling sea level is the rate of uplift or subsidence of a coastal area. In some areas, uplift is occurring so fast that sea level is actually falling with respect to the land. In other areas, sea level is rising while the coastal region is simultaneously subsiding, resulting in a net change in sea level of as much as 30 cm per century. Perhaps such a "slow" rate of sea-level change seems insignificant; after all it amounts to only a few millimeters per year. But in gently sloping coastal areas, as in the eastern United States from New Jersey southward, even a slight rise in sea level would eventually have widespread effects.

Many of the nearly 300 barrier islands along the East and Gulf Coasts of the United States are migrating landward as sea level rises. During storms, large waves erode sand from the seaward sides of barrier islands and deposit it on their landward sides, resulting in a gradual landward shift of the entire island complex (■ Figure 19.19). During the last 120 years, Hatteras Island, North Carolina, has migrated nearly 500 m landward. At its current rate of landward migration, Assateague Island, Maryland, will eventually merge with the mainland (Figure 19.19d).

Landward migration of barrier islands would pose few problems if it were not for the communities, resorts, and vacation homes on them. Moreover, barrier islands

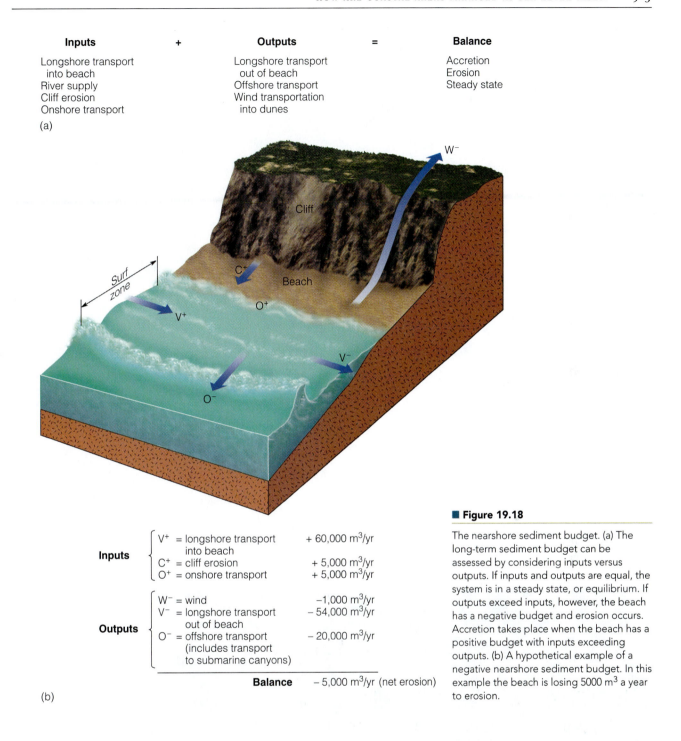

Inputs	+	Outputs	=	Balance
Longshore transport into beach		Longshore transport out of beach		Accretion
River supply		Offshore transport		Erosion
Cliff erosion		Wind transportation into dunes		Steady state
Onshore transport				

(a)

Inputs
V⁺ = longshore transport into beach   + 60,000 m³/yr
C⁺ = cliff erosion   + 5,000 m³/yr
O⁺ = onshore transport   + 5,000 m³/yr

Outputs
W⁻ = wind   −1,000 m³/yr
V⁻ = longshore transport out of beach   − 54,000 m³/yr
O⁻ = offshore transport (includes transport to submarine canyons)   − 20,000 m³/yr

Balance   − 5,000 m³/yr (net erosion)

(b)

■ Figure 19.18

The nearshore sediment budget. (a) The long-term sediment budget can be assessed by considering inputs versus outputs. If inputs and outputs are equal, the system is in a steady state, or equilibrium. If outputs exceed inputs, however, the beach has a negative budget and erosion occurs. Accretion takes place when the beach has a positive budget with inputs exceeding outputs. (b) A hypothetical example of a negative nearshore sediment budget. In this example the beach is losing 5000 m³ a year to erosion.

are not the only threatened areas. For example, Louisiana's coastal wetlands, an important wildlife habitat and seafood-producing area, are currently being lost at a rate of about 90 km² per year. Much of this loss results from sediment compaction, but sea-level rise exacerbates the problem.

Rising sea level also directly threatens many beaches that communities depend on for revenue. The beach at Miami Beach, Florida, for instance, was disappearing at an alarming rate until the Army Corps of Engineers began replacing the eroded beach sand (■ Figure 19.20).

The problem is even more serious in other countries. A rise in sea level of only 2 m would inundate large areas of the East and Gulf Coasts but would cover 20% of the entire country of Bangladesh. Other problems associated with sea-level rise include increased coastal flooding during storms and saltwater incursions that may threaten groundwater supplies (see Chapter 16).

Because we can do nothing to prevent sea level from rising, engineers, scientists, planners, and political leaders must examine what they can do to prevent or minimize the effects of shoreline erosion. At present, only a

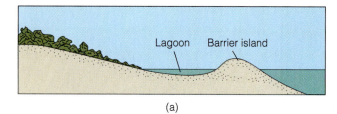

(a)

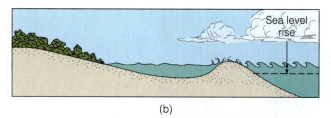

(b)

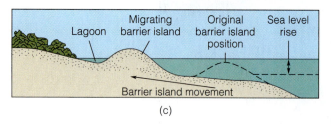

(c)

■ **Figure 19.19**

(a) Barrier island before migrating landward. (b) Sea level rises and the island migrates as storm waves carry sand from its seaward side to its lagoon. (c) Over time, the entire island complex migrates landward. (d) The seaward-projecting black lines are jetties constructed in the 1930s to protect the inlet at Ocean City, Maryland. The jetties disrupted the net southerly longshore drift and Assateague Island, starved of sediment, has migrated 500 m landward.

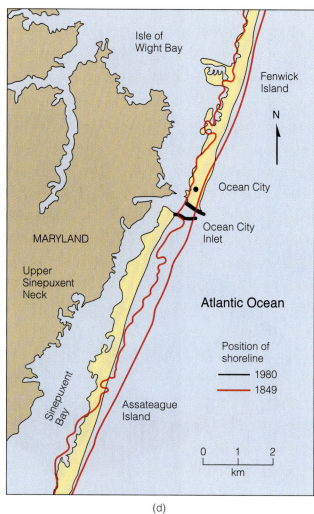

(d)

(a)

(b)

■ **Figure 19.20**

The beach at Miami Beach, Florida, (a) before and (b) after the U.S. Army Corps of Engineers' beach nourishment project.

Rosenberg Library, Galveston, Texas

■ **Figure 19.21**

Construction of this seawall to protect Galveston, Texas, from storm waves began in 1902. Notice that the wall is curved to deflect waves upward.

few viable options exist. One is to put strict controls on coastal development. North Carolina, for example, permits large structures to be sited no closer to the shoreline than 60 times the annual erosion rate. Although a growing awareness of shoreline processes has resulted in similar legislation elsewhere, some states have virtually no restrictions on coastal development.

Regulating coastal development is commendable, but it has no impact on existing structures and coastal communities. A general retreat from the shoreline may be possible but expensive for individual dwellings and small communities, but it is impractical for large population centers. Such communities as Atlantic City, New Jersey, Miami Beach, Florida, and Galveston, Texas, have adopted one of two strategies to combat coastal erosion.

One option is to build protective barriers such as seawalls. Seawalls, like the one at Galveston, can be effective, but they are tremendously expensive to construct and maintain (■ Figure 19.21). During a five-year period, more than $50 million was spent to replenish the beach sand and build a protective seawall at Ocean City, Maryland. Unfortunately, seawalls retard erosion only in the area directly behind them; Galveston Island west of the seawall has been eroded back about 45 m.

Armoring shorelines with seawalls and similar structures has had an unanticipated effect on Oahu in the Hawaiian Islands. Erosion of coastal rocks is the primary supply of sand to many Hawaiian beaches, so beaches maintain their width even as sea level rises.

However, seawalls trap sediment on their landward sides, thus diminishing the amount of sand on the beaches. In four areas studied on Oahu, 24% of the beaches either have been lost or are much narrower than they were several decades ago. Seawalls protect beachfront homes, hotels, and apartment buildings effectively, but they are expensive to build and during large storms are commonly damaged, requiring repairs. Furthermore, they commonly lead to loss of sand on beaches. North Carolina, South Carolina, Rhode Island, Oregon, and Maine no longer allow construction of seawalls.

Another option, adopted by both Atlantic City and Miami Beach, is to pump sand onto the beaches to replace sand lost to erosion (Figure 19.20). This, too, is expensive as the sand must be replenished periodically because erosion is a continuing process. In many areas, groins are constructed to preserve beaches, but unless additional sand is artificially supplied to the beaches, longshore currents invariably erode sand from the downcurrent sides of the groins.

The futility of artificially maintaining beaches is aptly demonstrated by the efforts to protect homes on the northeast end of the Isle of Palms, a barrier island off the coast of South Carolina. Following each spring tide, heavy equipment excavates sand from a deposit known as an *ebb tide delta* and constructs a sand berm to protect the houses from the next spring tide (■ Figure 19.22). Two weeks later, the process must be repeated in a never-ending cycle of erosion and artificial replenishment of the beach.

GEOLOGY
IN UNEXPECTED PLACES

Erosion and the Cape Hatteras Lighthouse

The Outer Banks of North Carolina are a series of barrier islands separated from the mainland by a lagoon up to 48 km wide. Actually, the Outer Banks are made up of several barrier islands and spits extending about 240 km from the Virginia state line to Shackleford Banks near Beaufort, North Carolina (■ Figure 1a). On September 18, 2003, Hurricane Isabel roared across the Outer Banks, causing wind damage and flooding there and in several adjacent states, particularly Virginia.

Since 1526, more than 600 ships have been lost on the dangerous sand bars adjacent to the Outer Banks, such as Diamond Shoals, an area of very shallow seawater. To reduce ship losses, several lighthouses were built on the islands; the most famous is the Cape Hatteras Lighthouse. It was constructed in 1802, although the current lighthouse was built in 1870. In 1999, however, the National Park Service, which administers Cape Hatteras National Seashore, had the lighthouse moved 500 m inland and 760 m to the southwest (see image). The reason? The Outer Banks, just like other barrier islands, are migrating toward the mainland as storm waves carry sand from the seaward margins into the lagoon. Indeed, Hurricane Isabel cut a new channel through Hatteras Island, thereby isolating communities to the north and south.

When the Cape Hatteras Lighthouse was built in 1870, it was 457 m from the shoreline, but when it was moved in 1999, it stood only 36.5 m inland (Figure 1b). Given that the annual erosion rate is about 3 m per year and all previous attempts to stabilize the shoreline had failed, the Park Service found it prudent to act when it did. The Cape Hatteras Lighthouse, the tallest brick lighthouse in the world, should now be safe for several centuries.

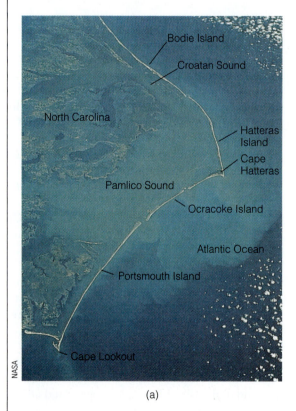

(a)

■ Figure 1

(a) These barrier islands make up the Outer Banks of North Carolina. (b) Cape Hatteras Lighthouse is shown during the 1999 move of the 60-m-high, 4381-metric-ton structure. Its previous position was at the top of the image.

(b)

Dr. Stanley R. Riggs

■ **Figure 19.22**

Heavy equipment builds a berm, an embankment of sand, on the seaward side of beach homes on the Isle of Palms, South Carolina, to protect them from erosion. During the next spring tide, the berms disappear and must be rebuilt.

STORM WAVES AND COASTAL FLOODING

Coastal flooding caused by storm-generated waves and intense rainfall is of considerable concern to shoreline residents. In fact, coastal flooding during large storms causes more damage, injuries, and fatalities than high wind does. The most recent large-scale event of this type was Hurricane Isabel that hit the U.S. East Coast along the Outer Banks of North Carolina on September 19, 2003.

Flooding during hurricanes results when large waves are driven onshore and by as much as 60 cm of rain in as little as 24 hours. In addition, as a hurricane moves over the ocean, low atmospheric pressure beneath the eye of the storm causes the ocean surface to bulge upward as much as 0.5 m. When the eye reaches the shoreline, the bulge coupled with wind-driven waves piles up in a *storm surge* that may rise several meters above normal high tide and inundate areas far inland.

In 1900 Galveston, Texas, was hit by a hurricane and storm waves surged inland, eventually covering the entire barrier island the city was built on. Buildings and other structures near the shoreline were battered to pieces, and "great beams and railway ties were lifted by the [waves] and driven like battering rams into dwellings and business houses"* farther inland. Between 6000 and 8000 people died, and in an effort to protect the city from future storms a huge seawall was

constructed (Figure 19.21) and the entire city was elevated to the level of the top of the seawall.

More recently, in 1989 Charleston, South Carolina, and nearby areas were flooded by the 2.5- to 6-m-high storm surge generated by Hurricane Hugo, which caused 21 deaths and more than $7 billion in property damage. Bangladesh is even more susceptible to storm surges; in 1970 300,000 people drowned and in 1991 another 139,000 were lost.

The problem of coastal flooding is, of course, exacerbated by rising sea level. To gain some perspective on the magnitude of the problem, consider this: Of the nearly $2 billion paid out by the federal government's National Flood Insurance Program since 1974, exclusive of Hurricane Isabel, most has gone to owners of beachfront homes.

TYPES OF COASTS

Coasts are difficult to classify because of variations in the factors that control their development and differences in their composition and configuration. Rather than attempt to categorize all coasts, we will simply note that two types of coasts have already been discussed, those dominated by deposition and those dominated by erosion, and we will look further at the changing relationships between coasts and sea level.

Depositional and Erosional Coasts

Depositional coasts, such as the U.S. Gulf Coast, are characterized by an abundance of detrital sediment and

---

*L. W. Bates, Jr., "Galveston—A City Built upon Sand," *Scientific American* 95 (1906): 64.

such depositional landforms as wide, sandy beaches, deltas, and barrier islands. In contrast, erosional coasts are steep and irregular and typically lack well-developed beaches except in protected areas (see "Shoreline Processes and Beaches" on pages 574–575). They are further characterized by sea cliffs, wave-cut platforms, and sea stacks. Many of the beaches along the West Coast of North America fall into this category.

The following section will examine coasts in terms of their changing relationships to sea level. But note that although some coasts, such as those in southern California, are described as emergent (uplifted), these same coasts may be erosional as well. In other words, coasts commonly possess features that allow them to be classified in more than one way.

Submergent and Emergent Coasts

If sea level rises with respect to the land or the land subsides, coastal regions are flooded and said to be **submer-**gent or *drowned* (■ Figure 19.23). Much of the East Coast of North America from Maine southward through South Carolina was flooded during the rise in sea level following the Pleistocene Epoch, so it is extremely irregular. Recall that during the expansion of glaciers during the Pleistocene, sea level was as much as 130 m lower than at present and that streams eroded their valleys more deeply and extended across continental shelves. When sea level rose, the lower ends of these valleys were drowned, forming *estuaries* such as Delaware and Chesapeake Bays (Figure 19.23). Estuaries are the seaward ends of river valleys where seawater and freshwater mix.

Submerged coasts are also present at higher latitudes where Pleistocene glaciers flowed into the sea. When sea level rose, the lower ends of the glacial troughs were drowned, forming fiords.

Emergent coasts are found where the land has risen with respect to sea level (■ Figure 19.24). Emergence takes place when water is withdrawn from the oceans, as occurred during the Pleistocene expansion of

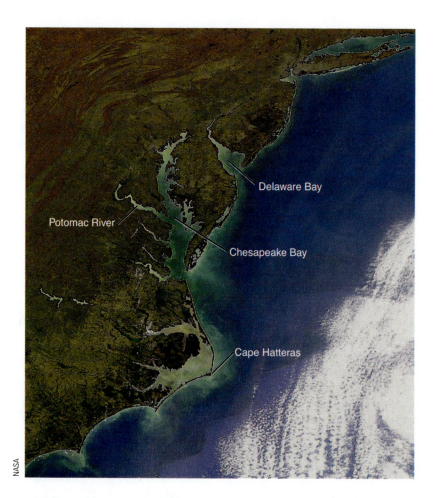

■ Figure 19.23

Submergent coasts tend to be extremely irregular with estuaries such as Chesapeake Bay. It formed when the East Coast of the United States was flooded as sea level rose following the Pleistocene Epoch.

glaciers. Presently coasts are emerging as a result of isostasy or tectonism. In northeastern Canada and the Scandinavian countries, the coasts are irregular because isostatic rebound is elevating formerly glaciated terrain from beneath the sea.

Coasts that rise in response to tectonism, in contrast, tend to be rather straight because the seafloor topography being exposed as uplift proceeds is smooth. The west coasts of North and South America are rising as a result of plate tectonics. Distinctive features of these coasts include *marine terraces* (Figure 19.15c), which are old wave-cut platforms now elevated above sea level. Uplift in such areas appears to be episodic rather than continuous, as indicated by the multiple levels of terraces in some areas. In southern California, for example, several terrace levels are present; each probably represents a period of tectonic stability followed by uplift. The highest of these terraces is now about 425 m above sea level.

Marine terrace

■ **Figure 19.24**

An emergent coast in California. Emergent coasts tend to be steep and straighter than submergent coasts. Notice the several sea stacks and the sea arch. Also, a marine terrace is visible in the distance.

19 REVIEW WORKBOOK

Chapter Summary

- Shorelines are continuously modified by the energy of waves and longshore currents and, to a lesser degree, by tidal currents.

- The gravitational attraction of the Moon and Sun causes the ocean surface to rise and fall as tides twice daily in most shoreline areas. Tidal currents usually have little effect on shorelines.

- Waves are oscillations on water surfaces that transmit energy in the direction of wave movement. Surface waves affect the water and seafloor only to wave base, which is equal to half the wavelength.

- Little or no net forward motion of water occurs in waves in the open sea. When waves enter shallow water, they are transformed into waves in which water does move in the direction of wave advance.

- Wind-generated waves, especially storm waves, are responsible for most geologic work on shorelines, but waves can also be generated by faulting, volcanic explosions, and rockfalls.

- Breakers form where waves enter shallow water and the orbital motion of water particles is disrupted.

Waves plunge or spill onto the shoreline, expending their kinetic energy.

- Waves approaching a shoreline at an angle generate a longshore current, which is capable of considerable erosion, transport, and deposition.

- Narrow surface currents called rip currents carry water from the nearshore zone seaward through the breaker zone.

- Beaches, the most common shoreline depositional features, are continuously modified by nearshore processes, and their profiles generally exhibit seasonal changes.

- Spits, baymouth bars, and tombolos all form and grow as a result of longshore current transport and deposition.

- Barrier islands are nearshore sediment deposits of uncertain origin. They parallel the mainland but are separated from it by a lagoon.

- The volume of sediment in a nearshore system remains rather constant unless the system is somehow disrupted, as when dams are built across the streams supplying sand to the system.

- Many shorelines are characterized by erosion rather than deposition. Such shorelines have sea cliffs and wave-cut platforms. Other features commonly present include sea caves, sea arches, and sea stacks.

- Depositional coasts are characterized by long sandy beaches, deltas, and barrier islands.

- Submergent and emergent coasts are defined on the basis of their relationships to changes in sea level.

Important Terms

barrier island (p. 576)
baymouth bar (p. 575)
beach (p. 574)
berm (p. 574)
breaker (p. 569)
crest (p. 565)
emergent coast (p. 588)
fetch (p. 568)
longshore current (p. 570)

longshore drift (p. 575)
marine terrace (p. 579)
nearshore sediment budget (p. 580)
rip current (p. 570)
shoreline (p. 562)
spit (p. 575)
submergent coast (p. 588)
tide (p. 563)

tombolo (p. 575)
trough (p. 565)
wave (p. 565)
wave base (p. 568)
wave-cut platform (p. 579)
wave height (p. 565)
wavelength (p. 565)
wave period (p. 565)
wave refraction (p. 570)

Review Questions

1. A berm is:

   a. _____ a gently landward-sloping part of the backshore zone; b. _____ the highest level reached during flood tide; c. _____ the lowest part of a wave between crests; d. _____ the distance that wind blows over a continuous water surface; e. _____ the angle that a wave hits the shoreline.

2. A wave-cut platform now above sea level is a:

   a. _____ wave trough; b. _____ submergent coast; d. _____ longshore current; d. _____ marine terrace; e. _____ baymouth bar.

3. A sand deposit extending into the mouth of a bay is a:

   a. _____ tombolo; b. _____ spit; c. _____ berm; d. _____ crest; e. _____ beach face.

4. A longshore current is generated by:

   a. _____ a rapidly rising tide following by ebb tide; b. _____ erosion of headlands and deposition in embayments; c. _____ a diminished amount of sand reaching a beach; d. _____ erosion of shoreline rocks; e. _____ waves approaching a shoreline at an angle.

5. A long, narrow sand deposit separated from the mainland by a lagoon is a:

   a. _____ baymouth bar; b. _____ tidal delta; c. _____ barrier island; d. _____ beach cusp; e. _____ wave-built platform.

6. The time it takes for two successive wave crests (or troughs) to pass a given point is known as the:

   a. _____ wave period; b. _____ wave celerity; c. _____ wave form; d. _____ wave height; e. _____ wave fetch.

7. Fetch is defined as the:

   a. _____ excess water in the nearshore zone that moves seaward in rip currents; b. _____ distance wind blows over a continuous water surface; c. _____ time necessary for two waves to pass a given point; d. _____ part of a wave that encounters wave base first; e. _____ effective depth to which waves can erode.

8. A spit that connects an island with the shoreline is a(n):

   a. _____ tombolo; b. _____ breaker; c. _____ emergent coast; d. _____ drift; e. _____ sea stack.

9. Although there are some exceptions, most beaches receive most of their sediment from:

   a. _____ coastal submergence; b. _____ erosion of shoreline rocks; c. _____ erosion of reefs; d. _____ streams and rivers; e. _____ breakers.

10. Erosion by water carrying sand and gravel is known as:

    a. _____ collusion; b. _____ hydraulic action; c. _____ longshore drift; d. _____ abrasion; e. _____ wave base.

11. A hypothetical shoreline has a balanced budget, but a dam is built on the river supplying sediment and a wall is built to protect sea cliffs from erosion. What will likely happen to beaches in this area? Explain.

12. Why are long sandy beaches common along North America's East Coast but found mostly in protected areas in the west?

13. How do barrier islands migrate landward? Is there any evidence indicating that U.S. barrier islands are migrating? Give some specific examples.

14. Why does an observer at a shoreline location experience two high and two low tides daily?

15. What is wave base and how does it affect waves as they enter progressively shallower water?

16. While driving along North America's West Coast, you notice a rather flat surface near the seashore with several isolated masses or rock rising above it. What is this surface and how did it form?

17. What are rip currents and how would you recognize one? If caught in a rip current, how would you get out?

18. Name and explain how three depositional landforms originate along coasts.

19. How are waves responsible for transporting sediment great distances along shorelines?

20. How and why do summer and winter beaches differ?

World Wide Web Activities

PHYSICAL Geology⬚Now Assess your understanding of this chapter's topics with additional quizzing and comprehensive interactivities at

http://earthscience.brookscole.com/physgeo5e

as well as current and up-to-date weblinks, additional readings, and InfoTrac College Edition exercises.

Physical Geology in Perspective

CHAPTER 20

Broadly speaking, we have investigated three aspects of our dynamic planet. The first was Earth materials—that is, minerals and rocks as well as the processes by which they form and change (the rock cycle; see Figure 1.14). Second, we considered internal processes such as the magnetic field and Earth's internal heat, which is responsible for volcanism, moving plates, mountain building, and seismicity. And third, we discussed various surface processes, including mass wasting, running water, glaciers, and groundwater, all of which yield distinctive landscapes. But few of you will become geologists, so how will you benefit from the information and insights you gained on these topics?

Quite simply, many of you will enter professions in which at least part of your work will involve geologic considerations. You may be a city planner, a city council person, a county commissioner, or a member of a planning board for siting a sanitary landfill, or for ensuring an adequate supply of groundwater for your community, or for developing zoning regulations for construction on floodplains or in a seismically active area. Perhaps you will become a developer or a contractor and have to contend with slope stability, expansive soils, and areas prone to flooding. Or maybe you will be an engineer involved in planning and constructing large-scale structures such as bridges, dams, power plants, and highways in tectonically active regions (■ Figure 20.1).

PHYSICAL GeologyNow *This icon, appearing throughout the book, indicates an opportunity to explore interactive tutorials, animations, or practice problems available on the Physical GeologyNow Web site at http://earthscience.brookscole.com/physgeo5e.*

About 60 million years ago, basalt lava flows covered a landscape of forests where rivers flowed over chalk (limestone) bedrock in what is now Northern Ireland. Then for 2 million years, weathering yielded laterite, a deep red soil, seen about halfway up the slope. Once again lava flows covered the area, and as before they flowed into valleys where they cooled slowly and formed columns (see Figure 4.7). Since then, glaciers, running water, waves, and mass wasting have modified the area. In short, the present-day landscape resulted from interactions among Earth's systems during the last several tens of millions of years. Source: James S. Monroe

(a)

■ **Figure 20.1**

(a) This freeway crosses a small valley a short distance from the San Andreas fault in California. Engineers had to take into account the near certainty that this structure will be badly shaken during an earthquake. (b) A bridge crossing part of a reservoir in a mountainous region. Rock type and geologic structures were important considerations when planning and erecting this structure.

(b)

Various federal and state agencies plan for and mitigate the effects of *geologic hazards* such as volcanism, earthquakes, floods, and landslides. These are the more spectacular, and sometimes catastrophic, kinds of geologic hazards, but others are more subtle—soil creep and radon gas, for example. As a matter of fact, soil creep and expansive soils account for more property damage than any other geologic hazard, but, of course, they do not cause injuries or fatalities. Radon gas, however, poses a long-term health risk in some areas.

Geololgic hazards account for thousands of fatalities and billions of dollars in damages each year, and whereas the incidence of hazardous events has not increased, fatalities and damages have because more and more people live in disaster-prone areas. As we noted in Chapter 1, most scientists agree that the single greatest environmental problem is overpopulation (see Figure 1.6).

Of course we cannot eliminate geologic hazards or other natural hazards such as hurricanes, but we can better understand the phenomena, enact zoning and land-use regulations, and at the very least decrease the amount of human suffering and damage. Unfortunately, geologic information that is readily available is often ig-

nored. A case in point is the Turnagain Heights subdivision in Anchorage, Alaska, that was so heavily damaged when the soil beneath it liquefied during the 1964 earthquake (see Figure 14.21). Not only were reports on soil stability ignored or overlooked before homes were built there, but since 1964 new homes have been built on part of the same site!

In Chapter 1 we defined a *system* as a combination of related parts that operate in an organized fashion (see Figure 1.2 and Table 1.1), and we gave examples of interacting systems throughout the text as we discussed volcanism, plate tectonics, running water, mass wasting, glaciers, and other topics. Our emphasis has, of course, been on the present-day aspects of these interacting systems and their effects on our lives as well as on planet Earth. However, we have mentioned systems in the broader context of geologic time only briefly (see the chapter opening photo).

When Earth formed 4.6 billions years ago, its various systems became operative, but not all at the same time and some not precisely in their present forms. For instance, Earth did not become differentiated into a core, mantle, and crust for millions of years after it initially formed. Once it did, though, internal heat was responsible for interactions among lithospheric plates as they began to diverge, converge, and slide past one another along transform plate boundaries. Indeed, once the nuclei of continents formed, they grew, and continue to do so, along convergent plate boundaries as new material was added to them (see Figure 13.26).

During its earliest history, Earth's surface was hot and dry, volcanism was ubiquitous, and the atmosphere was likely composed mostly of carbon dioxide. Meteorites and comets flashed through this primitive atmosphere, and because no ozone layer existed, cosmic radiation was intense. During this time, gases derived from Earth's interior formed the surface waters, but not until organisms were present—perhaps 3.6 billion years ago—did the atmosphere begin to accumulate oxygen as a waste product of photosynthesis. Once oxygen was present in sufficient quantities, an ozone layer formed and then oxygen-dependent organisms evolved. In short, as Earth's systems became established, our planet began to evolve and gradually became more like it is today.

The Moon, Mercury, and Mars show much less internal activity and surface processes are minimal compared to those on Earth. Why, then, is Earth so active? And will it eventually suffer the same fate as these other celestial bodies? Earth is, of course, much larger than the other bodies mentioned, and its greater mass accounts for greater gravitational attraction, thus allowing it to retain an atmosphere and liquid surface water (hydrosphere) (its distance from the Sun is also important). And among the planets and moons of the solar system, Earth is the only body known to support life (biosphere). Earth remains internally active because it possesses sufficient heat for volcanism, seismicity, and plate movements, but scientists are convinced that Earth's internally generated heat will gradually diminish and in the far distant future these activities will eventually cease. In the meantime, though, Earth remains a dynamic planet (■ Figure 20.2).

We mentioned in Chapter 8 that the concept of time measured in millions and billions of years sets geology apart from most of the other sciences, although astronomy certainly is concerned with time of this magnitude. Indeed, some refer to geology as a four-dimensional science, alluding to the fact that vast amounts of time (geologic time) are an essential part of the science. Geologic time has figured importantly in some of our discussions, such as plate tectonics and glaciation, but for the most part we have dealt with events of the present or not-too-distant past. The second course in geology, historical geology, as the name implies, considers the physical and biological evolution of Earth and the concept of geologic time in much greater detail. Many colleges and universities require completion of a course in physical geology as a prerequisite for historical geology.

Perhaps your chosen profession may not rely on geologic considerations, but as a concerned citizen you cannot ignore important environmental issues. Certainly global warming, acid runoff and acid rain, ozone depletion, and waste disposal have an impact on our continued existence on this planet. What are the positive and negative aspects of storing nuclear wastes at Yucca Mountain, Nevada (see Geo-Focus 18.1)? Perhaps you favor or disapprove of constructing a dam for flood control and generating electricity. In either case your appeal is better based on an understanding of the science involved than on only emotion.

In several chapters we emphasized the importance of various natural resources, such as iron ore, hydrocarbons, phosphate, and many others. These resources and reserves are important because they affect our economy and standard of living, and they figure importantly in decisions made by business, industry, and government. Indeed, as we noted in Chapter 1, the distribution of resources and our access to them are factors that shape U.S. foreign policy. Furthermore, dependence on politically unstable nations for essential commodities is risky, and the fact that we must import so much is not good for our balance of payments. As of April 2003, the United States imported almost 57% of the petroleum it needs and all or large proportions of many other mineral commodities (see Figure 2.22).

On another level, geology may enrich your appreciation of the scenic wonders in our state and national parks and monuments, allow you to understand more fully the magnitude of geologic time, and enhance your understanding of geologic processes and how they operate through time. Ten or twenty years from now you may not remember such terms as *anticline, arête, barchan,* or *tombolo,* but chances are if you see these features you will have a good idea of how they formed.

Much of this book has necessarily been descriptive; we simply presented the data and terminology on glaciers, for instance. But one objective of this book and much of your secondary education is to develop your skill as a critical thinker. As opposed to simple disagreement, critical thinking involves evaluating the support or evidence for a particular point of view. Although your exposure to geology at this time is limited, you do have the fundamentals needed to appraise why scientists accept plate tectonic theory and to logically evaluate an opposing hypothesis should one be formulated and presented.

In conclusion, the most important lesson you can learn from the study of physical geology is that Earth is an extremely complex planet in which interactions among its systems have resulted in changes in the atmosphere, hydrosphere, biosphere, and the planet itself. We tend to view our planet from the perspectives of our own lives and overlook the fact that it has changed and continues to do so but not always on a time scale we can readily appreciate. Physical geology therefore is not a static science, but one, like the dynamic Earth, that constantly evolves as new information and methods of investigation become available.

■ Figure 20.2

Some of Earth's interacting systems. (a) Plate divergence in the Red Sea (lower right) results in northward movement of the Arabian plate (upper right) against the Eurasian plate. (b) Although several processes account for this spectacular scenery in Yosemite National Park, California, erosion by glaciers and running water, both parts of the hydrosphere, was responsible for most of it. (c) This huge windblown sand dune near Fallon, Nevada, is 180 m high and about 4 km long. (d) Layering in these ancient rocks in Michigan resulted from the activities of cyanobacteria or blue-green algae. Not only have these organisms left their mark in the rocks, but they also added oxygen to early Earth's atmosphere during photosynthesis.

APPENDIX A
English–Metric Conversion Chart

	English Unit	Conversion Factor	Metric Unit	Conversion Factor	English Unit
Length	Inches (in.)	2.54	Centimeters (cm)	0.39	Inches (in.)
	Feet (ft)	0.305	Meters (m)	3.28	Feet (ft)
	Miles (mi)	1.61	Kilometers (km)	0.62	Miles (mi)
Area	Square inches (in.^{2})	6.45	Square centimeters (cm^2)	0.16	Square inches (in.^{2})
	Square feet (ft^2)	0.093	Square meters (m^2)	10.8	Square feet (ft^2)
	Square miles (mi^2)	2.59	Square kilometers (km^2)	0.39	Square miles (mi^2)
Volume	Cubic inches (in.^{3})	16.4	Cubic centimeters (cm^3)	0.061	Cubic inches (in.^{3})
	Cubic feet (ft^3)	0.028	Cubic meters (m^3)	35.3	Cubic feet (ft^3)
	Cubic miles (mi^3)	4.17	Cubic kilometers (km^3)	0.24	Cubic miles (mi^3)
Weight	Ounces (oz)	28.3	Grams (g)	0.035	Ounces (oz)
	Pounds (lb)	0.45	Kilograms (kg)	2.20	Pounds (lb)
	Short tons (st)	0.91	Metric tons (t)	1.10	Short tons (st)
Temperature	Degrees Fahrenheit (°F)	$-32° \times 0.56$	Degrees centigrade (Celsius)(°C)	$\times 1.80 + 32°$	Degrees Fahrenheit (°F)

Examples:

10 inches = 25.4 centimeters; 10 centimeters = 3.9 inches

100 square feet = 9.3 square meters; 100 square meters = 1080 square feet

50°F = 10.1°C; 50°C = 122°F

APPENDIX B
Periodic Table of the Elements

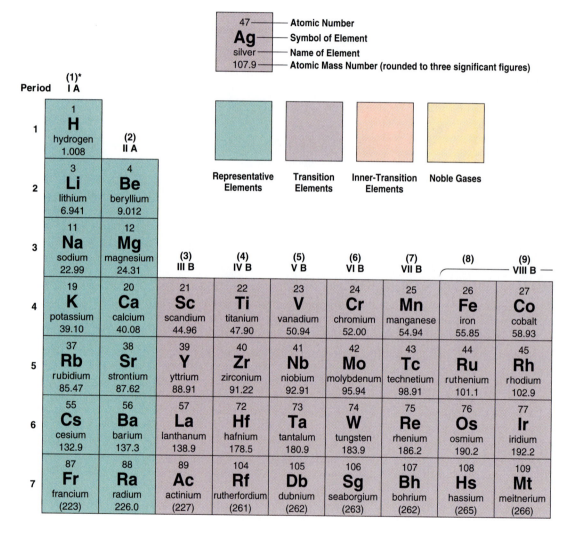

Atomic Number — 47
Symbol of Element — **Ag**
Name of Element — silver
Atomic Mass Number (rounded to three significant figures) — 107.9

| Representative Elements | Transition Elements | Inner-Transition Elements | Noble Gases |

Period

	(1)* I A	(2) II A	(3) III B	(4) IV B	(5) V B	(6) VI B	(7) VII B	(8)	(9) VIII B
1	1 **H** hydrogen 1.008								
2	3 **Li** lithium 6.941	4 **Be** beryllium 9.012							
3	11 **Na** sodium 22.99	12 **Mg** magnesium 24.31							
4	19 **K** potassium 39.10	20 **Ca** calcium 40.08	21 **Sc** scandium 44.96	22 **Ti** titanium 47.90	23 **V** vanadium 50.94	24 **Cr** chromium 52.00	25 **Mn** manganese 54.94	26 **Fe** iron 55.85	27 **Co** cobalt 58.93
5	37 **Rb** rubidium 85.47	38 **Sr** strontium 87.62	39 **Y** yttrium 88.91	40 **Zr** zirconium 91.22	41 **Nb** niobium 92.91	42 **Mo** molybdenum 95.94	43 **Tc** technetium 98.91	44 **Ru** ruthenium 101.1	45 **Rh** rhodium 102.9
6	55 **Cs** cesium 132.9	56 **Ba** barium 137.3	57 **La** lanthanum 138.9	72 **Hf** hafnium 178.5	73 **Ta** tantalum 180.9	74 **W** tungsten 183.9	75 **Re** rhenium 186.2	76 **Os** osmium 190.2	77 **Ir** iridium 192.2
7	87 **Fr** francium (223)	88 **Ra** radium 226.0	89 **Ac** actinium (227)	104 **Rf** rutherfordium (261)	105 **Db** dubnium (262)	106 **Sg** seaborgium (263)	107 **Bh** bohrium (262)	108 **Hs** hassium (265)	109 **Mt** meitnerium (266)

Lanthanides

58 **Ce** cerium 140.1	59 **Pr** praseodymium 140.9	60 **Nd** neodymium 144.2	61 **Pm** promethium (147)	62 **Sm** samarium 150.4

Actinides

90 **Th** thorium 232.0	91 **Pa** protactinium 231.0	92 **U** uranium 238.0	93 **Np** neptunium 237.0	94 **Pu** plutonium (244)

( ) Indicates mass number of isotope with longest known half-life.

* Number in ( ) heading each column represents the group designation recommended by the American Chemical Society Committee on Nomenclature.

			(13) III A	**(14)** IV A	**(15)** V A	**(16)** VI A	**(17)** VII A	**(18)** Noble Gases
								2 **He** helium 4.003
			5 **B** boron 10.81	6 **C** carbon 12.01	7 **N** nitrogen 14.01	8 **O** oxygen 16.00	9 **F** fluorine 19.00	10 **Ne** neon 20.18
(10)	**(11)** I B	**(12)** II B	13 **Al** aluminum 26.98	14 **Si** silicon 28.09	15 **P** phosphorus 30.97	16 **S** sulfur 32.06	17 **Cl** chlorine 35.45	18 **Ar** argon 39.95
28 **Ni** nickel 58.71	29 **Cu** copper 63.55	30 **Zn** zinc 65.37	31 **Ga** gallium 69.72	32 **Ge** germanium 72.59	33 **As** arsenic 74.92	34 **Se** selenium 78.96	35 **Br** bromine 79.90	36 **Kr** krypton 83.80
46 **Pd** palladium 106.4	47 **Ag** silver 107.9	48 **Cd** cadmium 112.4	49 **In** indium 114.8	50 **Sn** tin 118.7	51 **Sb** antimony 121.8	52 **Te** tellurium 127.6	53 **I** iodine 126.9	54 **Xe** xenon 131.3
78 **Pt** platinum 195.1	79 **Au** gold 197.0	80 **Hg** mercury 200.6	81 **Tl** thallium 204.4	82 **Pb** lead 207.2	83 **Bi** bismuth 209.0	84 **Po** polonium (210)	85 **At** astatine (210)	86 **Rn** radon (222)
110 **Uun** ununnilium (269)	111 **Uuu** unununium (272)	112 **Uub** ununbium (277)	113	114 **Uuq** ununquadium (289)	115	116 **Uuh** ununhexium (289)	117	118 **Uuo** ununoctium (293)

63 **Eu** europium 152.0	64 **Gd** gadolinium 157.3	65 **Tb** terbium 158.9	66 **Dy** dysprosium 162.5	67 **Ho** holmium 164.9	68 **Er** erbium 167.3	69 **Tm** thulium 168.9	70 **Yb** ytterbium 173.0	71 **Lu** lutetium 175.0
95 **Am** americium (243)	96 **Cm** curium (247)	97 **Bk** berkelium (247)	98 **Cf** californium (251)	99 **Es** einsteinium (254)	100 **Fm** fermium (257)	101 **Md** mendelevium (258)	102 **No** nobelium (255)	103 **Lr** lawrencium (256)

APPENDIX C
Mineral Identification Tables

Metallic Luster

Mineral	Chemical Composition	Color	Hardness / Specific Gravity	Other Features	Comments
Chalcopyrite	$CuFeS_2$	Brassy yellow	3.5–4 / 4.1–4.3	Usually massive; greenish black streak; iridescent tarnish	Important source of copper. Mostly in hydrothermal rocks.
Galena	PbS	Lead gray	2.5 / 7.6	Cubic crystals; 3 cleavages at right angles	The ore of lead. Mostly in hydrothermal rocks.
Graphite	C	Black	1–2 / 2.09–2.33	Greasy feel; writes on paper; 1 direction of cleavage	Used for pencil "leads" and dry lubricant. Mostly in metamorphic rocks.
Hematite	Fe_2O_3	Red brown	6 / 4.8–5.3	Usually granular or massive; reddish brown streak	Most important ore of iron. An accessory mineral in many rocks.
Magnetite	Fe_3O_4	Black	5.5–6.5 / 5.2	Strong magnetism	An important ore of iron. An accessory mineral in many rocks.
Pyrite	FeS_2	Brassy yellow	6.5 / 5.0	Cubic and octahedral crystals	Found in some igneous and hydrothermal rocks and in sedimentary rocks associated with coal.

Nonmetallic Luster

Mineral	Chemical Composition	Color	Hardness / Specific Gravity	Other Features	Comments
Anhydrite	$CaSO_4$	White, gray	3.5 / 2.9–3.0	Crystals with 2 cleavages; usually in granular masses	Found in limestones and evaporite deposits. Used as a soil conditioner.
Apatite	$Ca_5(PO_4)_3F$	Blue, green, brown, yellow, white	5 / 3.1–3.2	6-sided crystals; in massive or granular masses	The main constituent of bone and dentine. A source of phosphorus for fertilizer.

Nonmetallic Luster

Mineral	Chemical Composition	Color	Hardness		Other Features	Comments
			Specific Gravity			
Augite	$Ca(Mg,Fe,Al)(Al,Si)_2O_6$	Black, dark green	6	3.25–3.55	Short 8-sided crystals; 2 cleavages; cleavages nearly at right angles	The most common pyroxene mineral. Found mostly in mafic igneous rocks.
Barite	$BaSO_4$	Colorless, white, gray	3	4.5	Tabular crystals; high specific gravity for a nonmetallic mineral	Commonly found with metal ores and in limestones and hot spring deposits. A source of barium.
Biotite (mica)	$K(Mg,Fe)_3AlSi_3O_{10}(OH)_2$	Black, brown	2.5	2.9–3.4	1 cleavage direction; cleaves into thin sheets	Found in felsic and mafic igneous rocks, metamorphic rocks, and some sedimentary rocks.
Calcite	$CaCO_3$	Colorless, white	3	2.71	3 cleavages at oblique angles; cleaves into rhombs; reacts with dilute hydrochloric acid	The most common carbonate mineral. Main component of limestone and marble.
Cassiterite	SnO_2	Brown to black	6.5	7.0	High specific gravity for a nonmetallic mineral	The main ore of tin.
Chlorite	$(Mg,Fe)_3(Si,Al)_4O_{10}$ $(Mg,Fe)_3(OH)_6$	Green	2	2.6–3.4	1 cleavage; occurs in scaly masses	Common in low-grade metamorphic rocks such as slate.
Corundum	Al_2O_3	Gray, blue, pink, brown	9	4.0	6-sided crystals and great hardness are distinctive	An accessory mineral in some igneous and metamorphic rocks. Used as a gemstone and for abrasives.
Dolomite	$CaMg(CO_3)_2$	White, yellow, gray, pink	3.5–4	2.85	Cleavage as in calcite; reacts with dilute hydrochloric acid when powdered	The main constituent of dolostone.
Fluorite	CaF_2	Colorless, purple, green, brown	4	3.18	4 cleavage directions; cubic and octahedral crystals	Occurs mostly in hydrothermal rocks and in some limestones and dolostones. Used in the manufacture of steel and the preparation of hydrofluoric acid.
Garnet	$Fe_3Al_2(SiO_4)_3$	Dark red	7–7.5	4.32	12-sided crystals common; uneven fracture	Found mostly in gneiss and schist. Used as a semiprecious gemstone and for abrasives.
Gypsum	$CaSO_4 \cdot 2H_2O$	Colorless, white	2	2.32	Elongate crystals; fibrous and earthy masses	The most common sulfate mineral. Found mostly in evaporite deposits. Used to manufacture plaster of Paris and cements.

Nonmetallic Luster

Mineral	Chemical Composition	Color	Hardness / Specific Gravity		Other Features	Comments
Halite	$NaCl$	Colorless, white	3–4	2.2	3 cleavages at right angles; cleaves into cubes; cubic crystals; salty taste	Occurs in evaporite deposits. Used as a source of chlorine and in the manufacture of hydrochloric acid, many sodium compounds, and food seasoning.
Hornblende	$NaCa_2(Mg,Fe,Al)_5(Si,Al)_8O_{22}(OH)_2$	Green, black	6	3.0–3.4	Elongate, 6-sided crystals; 2 cleavages intersecting at 56 and 124 degrees	A common rock-forming amphibole mineral in igneous and metamorphic rocks.
Illite	$(Ca,Na,K)(Al,Fe^{+3},Fe^{+2},Mg)_2(Si,Al)_4O_{10}(OH)_2$	White, light gray, buff	1–2	2.6–2.9	Earthy masses; particles too small to observe properties	A clay mineral common in soils and clay-rich sedimentary rocks.
Kaolinite	$Al_2Si_4O_{10}(OH)_8$	White	2	2.6	Massive; earthy odor; particles too small to observe properties	A common clay mineral. The main ingredient of kaolin clay used for the manufacture of ceramics.
Muscovite (mica)	$KAl_2Si_3O_{10}(OH)_2$	Colorless	2–2.5	2.7–2.9	1 direction of cleavage; cleaves into thin sheets	Common in felsic igneous rocks, metamorphic rocks, and some sedimentary rocks. Used as an insulator in electrical appliances.
Olivine	$(Fe,Mg)_2SiO_4$	Olive green	6.5	3.3–3.6	Small mineral grains in granular masses; conchoidal fracture	Common in mafic igneous rocks.
Plagioclase feldspars	Varies from $CaAl_2Si_2O_8$ to $NaAlSi_3O_8$	White, gray, brown	6	2.56	2 cleavages at right angles	Common in igneous rocks and a variety of metamorphic rocks. Also in some arkoses.
Potassium feldspar — Microcline	$KAlSi_3O_8$	White, pink, green				Common in felsic igneous rocks, some metamorphic rocks, and arkoses. Used in the manufacture of porcelain.
Potassium feldspar — Orthoclase	$KAlSi_3O_8$	White, pink	6	2.56	2 cleavages at right angles	
Quartz	SiO_2	Colorless, white, gray, pink, green	7	2.67	6-sided crystals; no cleavage; conchoidal fracture	A common rock-forming mineral in all rock groups. Also occurs in varieties known as chert, flint, agate, and chalcedony.
Siderite	$FeCO_3$	Yellow, brown	4	3.8–4.0	3 cleavages at oblique angles; cleaves into rhombs	Found mostly in concretions and sedimentary rocks associated with coal.

Nonmetallic Luster

Mineral	Chemical Composition	Color	Hardness		Other Features	Comments
			Specific Gravity			
Smectite	$(Al,Mg)_8(Si_4O_{10})_3(OH)_{10}$ $\cdot 12H_2O$	Gray, buff, white	1–1.5 2.5		Earthy masses; particles too small to observe properties	A clay mineral with the property of swelling and contracting as it absorbs and releases water.
Sphalerite	ZnS	Yellow, brown, black	3.5–4 4.0–4.1		6 cleavages; cleaves into dodecahedra	The most important ore of zinc. Commonly found in hydrothermal rocks.
Talc	$Mg_3Si_4O_{10}(OH)_2$	White, green	1 2.82		1 cleavage direction; usually in compact masses	Formed by the alteration of magnesium silicates. Mostly in metamorphic rocks. Used in ceramics and cosmetics and as a filler in paints.
Topaz	$Al_2SiO_4(OH,F)$	Colorless, white, yellow, blue	8 3.5–3.6		High specific gravity; 1 cleavage direction	Found in pegmatites, granites, and hydrothermal rocks. An important gemstone.
Zircon	Zr_2SiO_4	Brown, gray	7.5 3.9–4.7		4-sided, elongate crystals	A common accessory in granitic rocks. An ore of zirconium and used as a gemstone.

APPENDIX D
Topographic Maps

Nearly everyone has used a map of one kind or another and is probably aware that a map is a scaled-down version of the area depicted. For a map to be of any use, however, one must understand what is shown on a map and how to read it. A particularly useful type of map for geologists, and people in many other professions, is a *topographic map,* which shows the three-dimensional configuration of Earth's surface on a two-dimensional sheet of paper.

Maps showing relief—differences in elevation in adjacent areas—are actually models of Earth's surface. Such maps are available for some areas, but they are expensive, difficult to carry, and impossible to record data on. Thus paper sheets that show relief by using lines of equal elevation known as *contours* are most commonly used. Topographic maps depict (1) relief, which includes hills, mountains, valleys, canyons, and plains; (2) bodies of water such as rivers, lakes, and swamps; (3) natural features such as forests, grasslands, and glaciers; and (4) various cultural features, including communities, highways, railroads, land boundaries, canals, and power transmission lines.

Topographic maps known as *quadrangles* are published by the U.S. Geological Survey (USGS). The area depicted on a topographic map is identified by referring to the map's name in the upper right and lower right corners, which is usually derived from some prominent geographic feature (Lincoln Creek Quadrangle, Idaho) or community (Mt. Pleasant Quadrangle, Michigan). In addition, most maps have a state outline map along the bottom margin, and shown within the outline is a small black rectangle indicating the part of the state represented by the map.

CONTOURS

Contour lines, or simply contours, are lines of equal elevation used to show topography. Think of contours as the lines formed where imaginary horizontal planes intersect Earth's surface at specific elevations. On maps, contours are brown, and every fifth contour, called an *index contour,* is darker than adjacent ones and labeled with its elevation (■ Figure D1). Elevations on most USGS topographic maps are in feet, although a few use meters; in either case, the specified elevation is above or below mean sea level. Because contours are defined as lines of equal elevation, they cannot divide or cross one another, although they will converge and appear to join in areas with vertical or overhanging cliffs. Notice in Figure D1 that where contours cross a stream they form a V that points upstream toward higher elevations.

The vertical distance between contours is the *contour interval.* If an area has considerable relief, a large contour interval is used, perhaps 80 or 100 feet, whereas a small interval such as 5, 10, or 20 feet is used in areas with little relief. The values recorded on index contours are always multiples of the map's contour interval, shown at the bottom of the map. For instance, if a map has a contour interval of 10 feet, index contour values such as 3600, 3650, and 3700 feet might be shown (Figure D1). In addition to contours, specific elevations are shown at some places on maps and may be indicated by a small ✕, next to which is a number. A specific elevation might also be shown adjacent to the designation *BM* (benchmark), a place where the elevation and location are precisely known.

Contour spacing depends on slope, so in areas with steep slopes, contours are closely spaced because there is a considerable increase in elevation in a short distance. In contrast, if slopes are gentle, contours are widely spaced (Figure D1). Furthermore, if contour spacing is uniform, the slope angle remains constant, but if spacing changes, the slope angle changes. However, one must be careful in comparing slopes on maps with different contour intervals or different scales.

Topographic features such as hills, valleys, plains, and so on are easily shown by contours. For instance, a hill is shown by a concentric pattern of contours with the highest elevation in the central part of the pattern. All contours must close on themselves, but they may do so beyond the confines of a particular map. A concentric

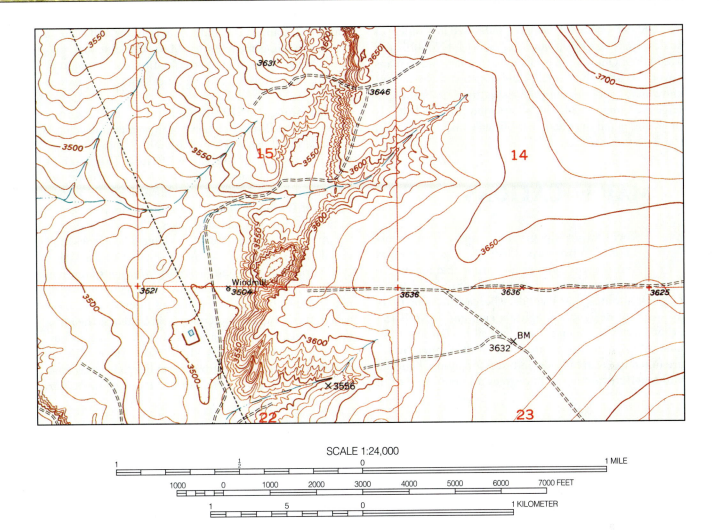

■ Figure D1

Part of the Bottomless Lakes Quadrangle, New Mexico, which has a contour interval of 10 feet; every fifth contour is darker and labeled with its elevation. Notice that contours are widely spaced where slopes are gentle and more closely spaced where they are steeper, as in the central part of the map. Hills are shown by contours that close on themselves, whereas depressions are indicated by contours with hachure marks pointing toward the center of the depression. The dashed blue lines on the map represent intermittent streams; notice that where contours cross a stream's channel they form a V that points upstream.

contour pattern also might show a closed depression, but in this case special contours with short bars perpendicular to the contour pointing toward the central part of the depression are used (Figure D1).

MAP SCALES

All maps are scaled-down versions of the areas shown, so to be of any use they must have a scale. Highway maps, for example, commonly have a scale such as "1 inch equals 10 miles," by which one can readily determine distances. Two types of scales are used on topographic maps. The first and most easily understood is a graphic scale, which is simply a bar subdivided into appropriate units

of length (Figure D1). This scale appears at the bottom center of the map and may show miles, feet, kilometers, or meters. Indeed, graphic scales on USGS topographic maps generally show both English and metric distance units.

A ratio or fractional scale, which represents the degree of reduction of the area depicted, appears above the graphic scale. On a map with a ratio scale of 1:24,000, for instance, the area shown is 1/24,000th the size of the actual land area (Figure D1). Another way to express this relationship is to say that any unit of length on the map equals 24,000 of the same units on the ground. Thus 1 inch on the map equals 24,000 inches on the ground, which is more meaningful if one converts inches to feet, making 1 inch equal to 2000 feet. A few maps have scales of 1:63,360, which converts to 1 inch equals 5280 feet, or 1 inch equals 1 mile.

USGS topographic maps are published in a variety of scales such as 1:50,000, 1:62,500, 1:125,000, and 1:250,000. One should also realize that large-scale maps cover less area than small-scale maps, and the former show much more detail than the latter. For example, a large-scale map (1:24,000) shows more surface features in greater detail than does a small-scale map (1:125,000) for the same area.

MAP LOCATIONS

Location on topographic maps can be determined in two ways. First, the borders of maps correspond to lines of latitude and longitude. Latitude is measured north and south of the equator in degrees, minutes, and seconds, whereas the same units are used to designate longitude east and west of the prime meridian, which passes through Greenwich, England. Maps depicting all areas within the United States are noted in north latitude and west longitude. Latitude and longitude are noted in degrees and minutes at the corners of maps, but usually only minutes and seconds are shown along the margins. Many USGS topographic maps cover $7\frac{1}{2}$ or 15 minutes of latitude and longitude and are thus referred to as $7\frac{1}{2}$- and 15-minute quadrangles.

Beginning in 1812, the General Land Office (now known as the Bureau of Land Management) developed a standardized method for accurately defining the location of property in the United States. This method, known as the General Land Office Grid System, has been used for all states except those along the eastern seaboard (except Florida), parts of Ohio, Tennessee, Kentucky, West Virginia, and Texas.

As new land acquired by the United States was surveyed, the surveyors laid out north–south lines they called *principal meridians* and east–west lines known as *base lines*. These intersecting lines form a set of coordinates for locating specific pieces of property. The basic unit in the General Land Office Grid System is the *township*, an area measuring 6 miles on a side and thus covering 36 square miles (■ Figure D2). Townships are numbered north and south of base lines and are designated as T.1N., T.1S., and so on. Rows of townships

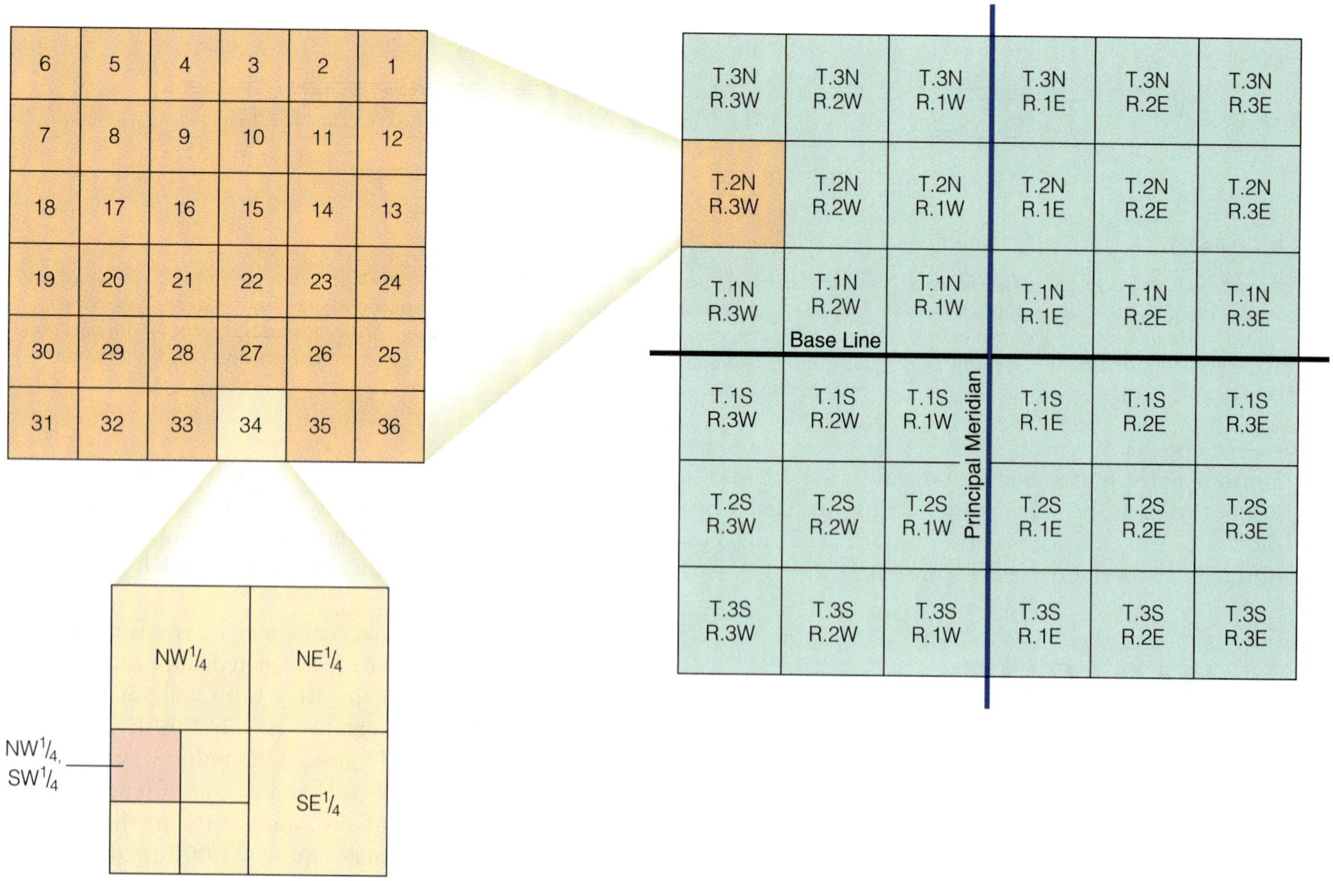

■ **Figure D2**

The General Land Office Grid System. Each 36-square-mile township is designated by township and range numbers. Townships are subdivided into sections, which can be further subdivided into quarter sections and quarter-quarter sections.

known as *ranges* are numbered east and west of principal meridians—R.2W and R.4E, for example. Note in Figure D2 that each township has a unique designation of township and range numbers.

Townships are subdivided into 36 1-square-mile (640-acre) *sections* numbered from 1 to 36. Because of surveying errors and the adjustments necessary to make a grid system conform to Earth's curved surface, not all sections are exactly 1 mile square. Nevertheless, each section can be further subdivided into half sections and quarter sections designated NE$\frac{1}{4}$, NW$\frac{1}{4}$, SE$\frac{1}{4}$, and SW$\frac{1}{4}$, and each quarter section can be further divided into quarter-quarter sections. To show the complete designation for an area, the smallest unit is noted first (quarter-quarter section) followed by quarter section, section number, township, and range. For example, the area shown in Figure D2 is the NW$\frac{1}{4}$, SW$\frac{1}{4}$, Sec. 34, T.2N., R.3W.

Because only a few principal meridians and base lines were established, they do not appear on most topographic maps. Nevertheless, township and range numbers are printed along the margins of 7$\frac{1}{2}$- and 15-minute quadrangles, and a grid consisting of red land boundaries depicts sections. In addition, each section number is shown in red within the map. However, small-scale maps show only township and range.

WHERE TO OBTAIN TOPOGRAPHIC MAPS

Many people find topographic maps useful. Land use planners, personnel in various local, state, and federal agencies, as well as engineers and real estate developers might use these maps for a variety of reasons. In addition, hikers, backpackers, and others interested in exploring undeveloped areas commonly use topographic maps because trails are shown by black dashed lines. Furthermore, map users can readily determine their location by interpreting the topographic features depicted by contours, and they can anticipate the type of terrain they will encounter during off-road excursions.

Topographic maps for local areas are available at some sporting goods stores, at National Park Visitor Centers, and from some state geologic surveys. Free index maps showing the names and locations of all quadrangles for each state are available from the USGS to anyone uncertain of which specific map is needed. Any published topographic map can be purchased from two main sources. For maps of areas east of the Mississippi River, write to

Branch of Distribution
U.S. Geological Survey
1200 S. Eads Street
Arlington, Virginia 22202

Maps for areas west of the Mississippi River can be obtained from

Branch of Distribution
U.S. Geological Survey
Box 25286 Federal Center
Denver, Colorado 80225

ANSWERS
Multiple-Choice Review Questions

Chapter 1
1. c; 2. e; 3. b; 4. d; 5. b; 6. d; 7. c; 8. d; 9. c; 10. a; 11. b; 12. e

Chapter 2
1. b; 2. e; 3. a; 4. c; 5. b; 6. a; 7. d; 8. e; 9. c; 10. b

Chapter 3
1. c; 2. e; 3. a; 4. b; 5. a; 6. d; 7. c; 8. b; 9. b; 10. a

Chapter 4
1. c; 2. e; 3. a; 4. b; 5. a; 6. d; 7. c; 8. b; 9. a; 10. a

Chapter 5
1. a; 2. c; 3. e; 4. c; 5. a; 6. b; 7. b; 8. b; 9. e; 10. d

Chapter 6
1. b; 2. e; 3. b; 4. e; 5. c; 6. a; 7. b; 8. a; 9. a; 10. e

Chapter 7
1. e; 2. b; 3. e; 4. c; 5. c; 6. d; 7. e; 8. c; 9. d; 10. d

Chapter 8
1. d; 2. e; 3. c; 4. a; 5. b; 6. c; 7. c; 8. b; 9. c; 10. a; 11. b

Chapter 9
1. d; 2. c; 3. d; 4. c; 5. a; 6. c; 7. b; 8. d; 9. d; 10. c

Chapter 10
1. c; 2. a; 3. e; 4. b; 5. c; 6. a; 7. b; 8. b; 9. d; 10. a

Chapter 11
1. a; 2. c; 3. e; 4. b; 5. d; 6. a; 7. c; 8. a; 9. c; 10. c

Chapter 12
1. b; 2. e; 3. b; 4. c; 5. a; 6. c; 7. c; 8. c; 9. c; 10. c; 11. a; 12. d

Chapter 13
1. b; 2. d; 3. a; 4. e; 5. d; 6. c; 7. a; 8. b; 9. a; 10. a

Chapter 14
1. b; 2. b; 3. b; 4. c; 5. e; 6. c; 7. e; 8. c; 9. d; 10. e

Chapter 15
1. a; 2. c; 3. e; 4. a; 5. d; 6. a; 7. c; 8. b; 9. b; 10. c

Chapter 16
1. d; 2. d; 3. b; 4. c; 5. c; 6. d; 7. a; 8. b; 9. d

Chapter 17
1. e; 2. c; 3. a; 4. c; 5. d; 6. b; 7. d; 8. b; 9. a; 10. c

Chapter 18
1. c; 2. b; 3. d; 4. d; 5. d; 6. d; 7. d; 8. c; 9. d; 10. b

Chapter 19
1. a; 2. d; 3. b; 4. e; 5. c; 6. a; 7. b; 8. a; 9. d; 10. d

Glossary

A

aa   A lava flow with a surface of rough, angular blocks and fragments.

abrasion   The process whereby exposed rock is worn and scraped by the impact of solid particles.

absolute dating   The process of assigning ages in years before the present to geologic events. Various radioactive decay-dating techniques yield absolute ages. See also *relative dating*.

abyssal plain   A vast flat area on the seafloor adjacent to the continental rises of passive continental margins.

active continental margin   A continental margin characterized by volcanism and seismicity at the leading edge of a continental plate where oceanic lithosphere is subducted. See also *passive continental margin*.

alluvial fan   A cone-shaped alluvial deposit formed where a stream flows from mountains onto an adjacent lowland.

alluvium   A general term for all detrital material transported and deposited by running water.

alpha decay   A type of radioactive decay involving the emission of a particle consisting of two protons and two neutrons from the nucleus of an atom; decreases the atomic number by 2 and the atomic mass number by 4.

angular unconformity   An unconformity below which older strata dip at a different angle (usually steeper) than the overlying strata. See also *disconformity* and *nonconformity*.

anticline   An up-arched fold in which the oldest exposed rocks coincide with the fold axis and all strata dip away from the axis.

aphanitic texture   An igneous texture in which individual minerals are too small to be seen without magnification; results from rapid cooling and generally indicates an extrusive origin.

arête   A narrow, serrated ridge separating two glacial valleys or adjacent cirques.

artesian system   A confined groundwater system in which high hydrostatic (fluid) pressure builds up causing water to rise above the level of the aquifer.

aseismic ridge   A ridge or broad, plateaulike feature rising as much as 2 to 3 km above the surrounding seafloor and lacking seismic activity.

ash   Pyroclastic material measuring less than 2 mm.

assimilation   A process in which magma changes composition as it reacts with country rock with which it comes in contact.

asthenosphere   The part of the mantle that lies below the lithosphere; behaves plastically and flows.

atom   The smallest unit of matter that retains the characteristics of an element.

atomic mass number   The total number of protons and neutrons in the nucleus of an atom.

atomic number   The number of protons in the nucleus of an atom.

aureole   A zone surrounding a pluton in which contact metamorphism has taken place.

B

barchan dune   A crescent-shaped dune with the tips of the crescent pointing downwind.

barrier island   A long, narrow island composed of sand oriented parallel to a shoreline but separated from the mainland by a lagoon.

basal slip   A type of glacial movement in which a glacier slides over its underlying surface.

basalt plateau   A large area built up by numerous flat-lying lava flows erupted from fissures.

base level   The lowest limit to which a stream can erode.

basin   The circular equivalent of a syncline. All strata in a basin dip toward a central point, and the youngest exposed rocks are in the center.

batholith   A discordant, irregularly shaped pluton composed chiefly of granitic rocks with surface area of at least 100 km^2.

baymouth bar   A spit that has grown until it cuts off a bay from the open sea.

beach A deposit of unconsolidated sediment extending landward from low tide to a change in topography or where permanent vegetation begins.

bed (bedding) A bed is an individual layer of rock, especially sedimentary rock, whereas bedding is the layered arrangement of rocks. See also *stratification.*

bed load The part of a stream's sediment load transported along its bed; consists of sand and gravel.

berm The backshore area of a beach, consisting of a platform composed of sediment deposited by waves. Berms are nearly horizontal or slope gently landward.

beta decay A type of radioactive decay during which a fast-moving electron emitted from a neutron is converted to a proton; results in an increase of one atomic number but does not change atomic mass number.

Big Bang A model for the evolution of the universe in which a dense, hot state was followed by expansion, cooling, and a less dense state.

biochemical sedimentary rock A sedimentary rock resulting from the chemical processes of organisms.

black smoker A type of submarine hydrothermal vent that emits a plume of black water colored by dissolved minerals.

body wave An earthquake wave that travels through Earth. Both P- and S-waves are body waves.

bonding The process whereby atoms are joined to other atoms.

Bowen's reaction series A mechanism accounting for the derivation of intermediate and felsic magmas from mafic magma. It has a discontinuous branch of ferromagnesian minerals that change from one to another over specific temperature ranges and a continuous branch of plagioclase feldspars whose composition changes as the temperature decreases.

braided stream A stream possessing an intricate network of dividing and rejoining channels. Braiding occurs when sand and gravel bars are deposited within channels.

breaker A wave that steepens as it enters shallow water until its crest plunges forward.

butte An isolated, steep-sided, pinnacle-like erosional feature formed by the breaching of a resistant cap rock, which allows rapid erosion of the less resistant underlying rocks.

C

caldera A large, steep-sided circular to oval volcanic depression usually formed by summit collapse resulting from partly draining of the underlying magma chamber.

Canadian shield The exposed part of the North American craton consisting of various ancient rocks, many of them metamorphic; exposed mostly in Canada but also seen in Minnesota, Wisconsin, Michigan, and New York.

carbon 14 dating technique An absolute dating method that relies on determining the ratio of C^{14} to C^{12} in a sample; useful back to about 70,000 years ago; can be applied only to organic substances.

carbonate mineral A mineral containing the negatively charged carbonate ion $(CO_3)^{-2}$, e.g., calcite $(CaCO_3)$ and dolomite $[CaMg(CO_3)_2]$.

carbonate rock A rock containing mostly carbonate minerals (such as limestone and dolostone).

Cascade Range A mountain range stretching from northern California through Oregon and Washington and into southern British Columbia, Canada. Made up of about a dozen large composite volcanoes and thousands of smaller volcanic vents.

cave A naturally formed subsurface opening that is generally connected to the surface and is large enough for a person to enter.

cementation The precipitation of minerals as binding material in and around sediment grains, thus converting sediment to sedimentary rock.

chemical sedimentary rock Rock formed of minerals derived from materials dissolved during chemical weathering.

chemical weathering The decomposition of rocks by chemical alteration of parent material.

cinder cone A small, steep-sided volcano composed of pyroclastic materials that accumulate around a vent.

circum-Pacific belt A zone of seismic and volcanic activity that nearly encircles the margins of the Pacific Ocean basin.

cirque A steep-walled, bowl-shaped depression formed by erosion at the upper end of a glacial valley.

cleavage The breaking or splitting of mineral crystals along planes of weakness. Cleavage is determined by the strength of the bonds within minerals.

columnar joint Joints in some igneous rocks consisting of six-sided columns that form as a result of shrinkage during cooling.

compaction A method of lithification whereby the pressure exerted by the weight of overlying sediment reduces the amount of pore space and thus the volume of a deposit.

complex movement A combination of different types of mass movements in which one type is not dominant. Most complex movements involve sliding and flowing.

composite volcano A volcano composed of pyroclastic layers, lava flows typically of intermediate composition, and mudflows; also called *stratovolcano.*

compound A substance resulting from the bonding of two or more different elements, such as water (H_2O) and quartz (SiO_2).

compression Stress resulting when rocks are squeezed by external forces directed toward one another.

concordant pluton Pluton whose boundaries are parallel to the layering in the country rock. See also *discordant pluton*.

cone of depression A cone-shaped depression in the water table around a well, resulting from pumping water from an aquifer faster than it is replenished.

contact metamorphism Metamorphism in which a body of magma alters the surrounding country rock.

continental accretion The phenomenon whereby continents grow by additions of new material along their margins.

continental–continental plate boundary A convergent plate boundary along which two continental lithospheric plates collide, such as the collision of India with Asia.

continental crust The rocks of continents overlying the upper mantle and consisting of a wide variety of igneous, sedimentary, and metamorphic rocks. It has an overall granitic composition and an average density of about 2.70 g/cm^3.

continental drift The theory that the continents were once joined into a single landmass that broke apart with the various fragments (continents) moving with respect to one another.

continental glacier A glacier covering a vast area (at least $50,000 \text{ km}^2$) and unconfined by topography. Also called an *ice sheet*.

continental margin The area separating the part of a continent above sea level from the deep seafloor.

continental rise The gently sloping area of the seafloor beyond the base of the continental slope.

continental shelf The area between the shoreline and the continental slope where the seafloor slopes gently seaward.

continental slope The relatively steep area between the shelf–slope break (at an average depth of 135 m) and the more gently sloping continental rise or an oceanic trench.

convergent plate boundary The boundary between two plates that are moving toward one another; three types of convergent plate boundaries are recognized. See also *continental–continental plate boundary, oceanic–continental plate boundary,* and *oceanic–oceanic plate boundary.*

core The interior part of Earth, beginning at a depth of about 2900 km; probably composed mostly of iron and nickel; divided into an outer liquid core and an inner solid core.

Coriolis effect The deflection of winds to the right of their direction of motion (clockwise) in the Northern Hemisphere and to the left of their direction of motion (counterclockwise) in the Southern Hemisphere, due to Earth's rotation.

correlation The demonstration of the physical continuity or time equivalency of rock units in different areas.

country rock Any rock that is invaded by and surrounds a pluton.

covalent bond A bond formed by the sharing of electrons between atoms.

crater A circular or oval depression at the summit of a volcano resulting from the extrusion of gases, pyroclastic materials, and lava; connected by a conduit to a magma chamber below Earth's surface.

craton The relatively stable part of a continent; consists of a shield and a platform, a buried extension of a shield; the ancient nucleus of a continent.

creep A type of mass wasting in which soil or rock moves slowly downslope.

crest The highest part of a wave.

cross-bedding Layers in sedimentary rocks deposited at an angle to the surface on which they accumulated.

crust Earth's outermost layer; the upper part of the lithosphere, which is separated from the mantle by the Moho; divided into continental and oceanic crust.

crystal A naturally occurring solid of an element or a compound with a specific internal structure that is manifested externally by planar faces, sharp corners, and straight edges.

crystal settling The physical separation and concentration of minerals in the lower part of a magma chamber or pluton by crystallization and gravitational settling.

crystalline solid A solid in which the constituent atoms are arranged in a regular, three-dimensional framework.

Curie point The temperature at which iron-bearing minerals in a cooling magma attain their magnetism.

D

debris flow A mass wasting process involving the flow of a viscous mixture of water, soil, and rocks; much like a mudflow, but at least half the particles are larger than sand.

deflation The removal of loose surface sediment by wind.

deformation Any change in shape or volume, or both, of rocks in response to stress. Deformation involves folding and fracturing.

delta An alluvial deposit formed where a stream flows into a lake or the sea.

density The mass of an object per unit of volume; usually expressed in grams per cubic centimeter (g/cm^3).

depositional environment Any area where sediment is deposited, such as a floodplain or a beach.

desert Any area that receives less than 25 cm of rain per year and has a high evaporation rate.

desert pavement   A surface mosaic of close-fitting pebbles, cobbles, and boulders found in many dry regions; formed by the removal of sand-sized and smaller particles by wind.

detrital sedimentary rock   Rock consisting of the solid particles (detritus) of preexisting rocks.

differential pressure   Pressure that is not applied equally to all sides of a rock body.

differential weathering   Weathering of rock at different rates, producing an uneven surface.

dike   A tabular or sheetlike discordant pluton.

dip   A measure of the maximum angular deviation of an inclined plane from horizontal.

dip-slip fault   A fault on which all movement is parallel with the dip of the fault plane. See also *normal fault* and *reverse fault.*

discharge   The volume of water in a stream moving past a particular point in a given period of time.

disconformity   An unconformity above and below which the strata are parallel. See also *angular unconformity* and *nonconformity.*

discontinuity   A boundary across which seismic wave velocity or direction changes abruptly, such as the mantle–core boundary.

discordant pluton   Pluton whose boundaries cut across the layering in the country rock. See also *concordant pluton.*

dissolved load   The part of a stream's load consisting of ions in solution.

divergent plate boundary   The boundary between two plates that are moving apart.

divide   A topographically high area that separates adjacent drainage basins.

dome   A circular equivalent of an anticline. All strata in a dome dip away from a central point, and the oldest exposed rocks are at the dome's center.

drainage basin   The surface area drained by a stream and its tributaries.

drainage pattern   The regional arrangement of channels in a drainage system.

drumlin   An elongate hill of till formed by the movement of a continental glacier or floods.

dune   A mound or ridge of wind-deposited sand.

dynamic metamorphism   Metamorphism occurring in fault zones where rocks are subjected to high differential pressure.

E

earthflow   A mass wasting process involving downslope flow of water-saturated soil.

earthquake   Vibrations caused by the sudden release of energy, usually as a result of the displacement of rocks along faults.

elastic rebound theory   A theory that explains how energy is suddenly released during earthquakes: When rocks are deformed, they store energy and bend; when the inherent strength of the rocks is exceeded, they rupture and release energy, causing earthquakes.

elastic strain   A type of deformation in which the material returns to its original shape when stress is relaxed.

electron   A negatively charged particle of very little mass that encircles the nucleus of an atom.

electron capture   A type of radioactive decay in which an electron is captured by a proton and converted to a neutron; results in a loss of one atomic number but no change in atomic mass number.

electron shell   Electrons orbit rapidly around the nuclei of atoms at specific distances known as electron shells.

element   A substance composed of atoms all having the same properties; atoms of one element can change to atoms of another element by radioactive decay, but otherwise they cannot be changed by ordinary chemical means.

emergent coast   A coast where the land has risen with respect to sea level.

end moraine   A pile of rubble deposited at the terminus of a glacier. See also *recessional moraine* and *terminal moraine.*

epicenter   The point on Earth's surface vertically above the focus of an earthquake.

erosion   The removal of weathered materials from their source area.

esker   A long, sinuous ridge of stratified drift formed by deposition by running water in tunnels beneath stagnant ice.

evaporite   A sedimentary rock that formed by inorganic chemical precipitation of minerals from an evaporating solution (such as rock salt and rock gypsum).

Exclusive Economic Zone   An area extending 371 km seaward from the coast of the United States and its territories in which the United States claims all sovereign rights.

exfoliation   The process whereby slabs of rock bounded by sheet joints slip or slide off the host rock.

exfoliation dome   A large rounded dome of rock resulting from the process of exfoliation.

expansive soil   A soil in which the volume increases when water is present.

F

fault   A fracture along which movement has occurred parallel to the fracture surface.

fault plane   A fracture surface along which blocks of rock on opposite sides have moved relative to one another.

felsic magma   Magma containing more than 65% silica and considerable sodium, potassium, and aluminum but little calcium, iron, and magnesium. See also *intermediate magma* and *mafic magma.*

ferromagnesian silicate   A silicate mineral containing iron or magnesium or both.

fetch   The distance the wind blows over a continuous water surface.

fiord   An arm of the sea extending into a U-shaped glacial trough eroded below sea level.

firn   Granular snow formed by partial melting and refreezing of snow.

fission track dating   The process of dating samples by counting the number of small linear tracks (fission tracks) that result when a mineral crystal is damaged by rapidly moving alpha particles generated by radioactive decay of uranium.

fissure eruption   An eruption in which lava or pyroclastic material is emitted from a long, narrow fissure or group of fissures.

floodplain   A low-lying, relatively flat area adjacent to a river or stream, which is partly or completely covered with water when the stream overflows its banks.

fluid activity   An agent of metamorphism in which water and carbon dioxide promote metamorphism by increasing the rate of chemical reactions.

focus   The place within Earth where an earthquake originates and energy is released.

fold   A type of geologic structure in which planar features in rock layers such as bedding and foliation have been bent.

foliated texture   A texture of metamorphic rocks in which platy and elongate minerals are arranged in a parallel fashion.

footwall block   The block of rock that lies beneath a fault plane.

fossil   Remains or traces of prehistoric organisms preserved in rocks.

fracture   A break in a rock resulting from intense applied pressure.

frost action   The disaggregation of rocks by repeated freezing and thawing of water in cracks and crevices.

frost heaving   The process whereby a mass of sediment or soil undergoes freezing, expansion, and actual lifting, followed by thawing, contraction, and lowering of the mass.

frost wedging   The opening and widening of cracks by the repeated freezing and thawing of water.

G

geologic structure   Any feature in rocks resulting from deformation, such as folds, joints, and faults.

geologic time scale   A chart with the designation for the earliest interval of geologic time at the bottom, followed upward by designations for progressively more recent time intervals.

geology   The science concerned with the study of Earth; includes studies of Earth materials (minerals and rocks), surface and internal processes, and Earth history.

geothermal energy   Energy that comes from the steam and hot water trapped within the crust.

geothermal gradient   The temperature increase with depth; it averages 25°C/km near the surface but varies from area to area.

geyser   A hot spring that intermittently ejects hot water and steam.

glacial budget   The balance between expansion and contraction of a glacier in response to accumulation and wastage.

glacial drift   A collective term for all sediment deposited by glaciers or associated processes; includes till deposited by ice and outwash deposited by streams derived from melting ice.

glacial erratic   A rock fragment carried some distance from its source by a glacier and usually deposited on bedrock of a different composition.

glacial ice   Water in the solid state within a glacier; forms as snow partially melts and refreezes and is compacted so that it is transformed first into firn and then into glacial ice.

glacial polish   A smooth, glistening rock surface formed by the movement of a sediment-laden glacier over it.

glacial striation   A straight scratch rarely more than a few millimeters deep on a rock caused by the movement of sediment-laden glacial ice.

glacial surge   A time of greatly accelerated flow in a glacier. Commonly results in displacement of the glacier's terminus by several kilometers.

glacier   A mass of ice on land that moves by plastic flow and basal slip.

***Glossopteris* flora**   A Late Paleozoic association of plants found only on the Southern Hemisphere continents and India.

Gondwana   One of six major Paleozoic continents; composed of the present-day continents of South America, Africa, Antarctica, Australia, and India and parts of other continents such as southern Europe, Arabia, and Florida.

graded bedding   A type of sedimentary bedding in which an individual bed is characterized by a decrease in grain size from bottom to top.

graded stream   A stream possessing an equilibrium profile in which a delicate balance exists between gradient, discharge, flow velocity, channel characteristics, and sediment load such that neither significant erosion nor deposition occurs within its channel.

gradient   The slope over which a stream flows; expressed in m/km or ft/mi.

gravity   The force exerted between any two bodies in the universe as a function of their mass and the distance between their centers of mass.

gravity anomaly (positive and negative)   A departure from the expected force of gravity; a gravity anomaly might be positive, indicating a mass excess, or negative, indicating a mass deficiency.

ground moraine   The layer of sediment liberated from melting ice as a glacier's terminus retreats.

groundwater   Underground water stored in the pore spaces of rock, sediment, or soil.

guide fossil   Any fossil that can be used to determine the relative geologic ages of rocks and to correlate rocks of the same relative age in different areas.

guyot   A flat-topped seamount of volcanic origin rising more than 1 km above the seafloor.

H

half-life   The time required for one-half of the original number of atoms of a radioactive element to decay to a stable daughter product (e.g., the half-life of potassium 40 is 1.3 billion years).

hanging valley   A tributary glacial valley whose floor is at a higher level than that of the main glacial valley.

hanging wall block   The block of rock that overlies a fault plane.

hardness   A term used to express the resistance of a mineral to abrasion.

heat   An agent of metamorphism. Heat comes from increasing depth, magma, and applied pressure.

heat flow   The flow of heat from Earth's interior to its surface.

horn   A steep-walled, pyramidal peak formed by the headward erosion of at least three cirques.

hot spot   A localized zone of melting below the lithosphere.

hot spring   A spring in which the water temperature is warmer than the temperature of the human body (37°C).

humus   The material in soils derived by bacterial decay of organic matter.

hydraulic action   The power of moving water.

hydrologic cycle   The continuous recycling of water from the oceans, through the atmosphere, to the continents, and back to the oceans.

hydrolysis   The chemical reaction between the hydrogen (H^+) ions and hydroxyl (OH^-) ions of water and a mineral's ions.

hydrothermal   A term referring to hot water as in hot springs or geysers.

hypothesis   A provisional explanation for observations; subject to continual testing and modification. If well supported by evidence, hypotheses are then generally called *theories*.

I

ice cap   A dome-shaped mass of glacial ice covering less than 50,000 km^2.

ice sheet   See *continental glacier*.

igneous rock   Any rock formed by cooling and crystallization of magma or by the accumulation and consolidation of pyroclastic materials such as ash.

incised meander   A deep, meandering canyon cut into bedrock by a stream.

index mineral   A mineral that forms within specific temperature and pressure ranges during metamorphism.

infiltration capacity   The maximum rate at which soil or sediment absorbs water.

inselberg   An isolated steep-sided erosional remnant rising above a surrounding desert plain.

intensity   The subjective measure of the kind of damage done by an earthquake, as well as people's reaction to it.

intermediate magma   Magma having a silica content between 53% and 65% and an overall composition intermediate between felsic and mafic magmas. See also *felsic magma* and *mafic magma*.

intrusive igneous rock   See *plutonic rock*.

ion   An electrically charged atom produced by adding or removing electrons from the outermost electron shell.

ionic bond   A bond that results from the attraction of positively and negatively charged ions.

isostasy   See *principle of isostasy*.

isostatic rebound   The phenomenon in which unloading of Earth's crust causes it to rise upward until equilibrium is again attained. See also *principle of isostasy*.

J

joint   A fracture along which no movement has occurred or where movement has been perpendicular to the fracture surface.

Jovian planet   Any of the four planets (Jupiter, Saturn, Uranus, and Neptune) that resemble Jupiter. All are large and have low mean densities, indicating they are composed mostly of lightweight gases, such as hydrogen and helium, and frozen compounds, such as ammonia and methane. See also *terrestrial planet*.

K

kame Conical hill of stratified drift originally deposited in a depression on a glacier's surface.

karst topography Topography consisting of numerous caves, sinkholes, and solution valleys developed by groundwater solution of rocks such as limestone and dolostone; also characterized by springs, disappearing streams, and underground drainage.

L

laccolith A concordant pluton with a mushroomlike geometry.

lahar A mudflow composed of volcanic materials such as ash.

lateral moraine The sediment deposited as a long ridge of till along the margin of a valley glacier.

laterite A red soil, rich in iron or aluminum or both, that forms in the tropics by intense chemical weathering.

Laurasia A Late Paleozoic, Northern Hemisphere continent composed of the present-day continents of North America, Greenland, Europe, and Asia.

lava Magma at Earth's surface.

lava dome A bulbous, steep-sided structure formed by viscous magma moving upward through a volcanic conduit.

lava flow A stream of magma flowing over Earth's surface.

lava tube A tunnel beneath the solidified surface of a lava flow through which a molten flow continues to move. Also, the hollow space left when the lava within the tube drains away.

lithification The process of converting sediment into sedimentary rock.

lithosphere Earth's outer, rigid part consisting of the upper mantle, oceanic crust, and continental crust.

lithostatic pressure Pressure exerted on rock by the weight of overlying rocks; it is applied equally in all directions.

loess Windblown silt and clay deposits.

longitudinal dune A long ridge of sand generally parallel to the direction of the prevailing wind.

longshore current A current between the breaker zone and the beach that flows parallel to the shoreline and is produced by wave refraction.

longshore drift The movement of sediment along a shoreline by longshore currents.

Love wave (L-wave) A surface wave in which the individual particles of material move only back and forth in a horizontal plane perpendicular to the direction of wave travel.

low-velocity zone The zone within the mantle between 100 and 250 km deep where the velocity of both P- and S-waves decreases markedly. It corresponds closely to the asthenosphere.

luster The appearance of a mineral in reflected light. The two basic types of luster are metallic and nonmetallic, although the latter has several subcategories.

M

mafic magma Silica-poor magma containing between 45% and 52% silica and proportionately more calcium, iron, and magnesium than intermediate and felsic magmas. See also *felsic magma* and *intermediate magma*.

magma Molten rock material generated within Earth.

magma chamber A reservoir of magma within the upper mantle or lower crust.

magma mixing The process of mixing magmas of different composition, thereby producing a modified version of the parent magmas.

magnetic anomaly Any change, such as a change in average strength, of Earth's magnetic field.

magnetic declination The angle between lines drawn from a compass position to the north magnetic and geographic poles.

magnetic field The area in which magnetic substances are affected by lines of magnetic force emanating from Earth.

magnetic inclination The deviation from horizontal of the magnetic lines of force around Earth.

magnetic reversal The phenomenon in which the north and south magnetic poles are completely reversed.

magnetism A physical phenomenon resulting from moving electricity in which magnetic substances are attracted toward one another.

magnitude The total amount of energy released by an earthquake at its source. See also *Richter Magnitude Scale*.

mantle The thick layer between Earth's crust and core.

marine regression The withdrawal of the sea from a continent or coastal area, resulting in the emergence of the land as sea level falls or the land rises with respect to sea level.

marine terrace A wave-cut platform now elevated above sea level.

marine transgression The invasion of coastal areas or much of a continent by the sea, resulting from a rise in sea level or subsidence of the land.

mass wasting The downslope movement of material under the influence of gravity.

meandering stream A stream possessing a single, sinuous channel with broadly looping curves.

mechanical weathering Disaggregation of rocks by physical processes that yield smaller pieces retaining the same composition as the parent material.

medial moraine   A moraine formed where two lateral moraines merge.

Mediterranean belt   A zone of seismic and volcanic activity extending westerly from Indonesia through the Himalayas, across Iran and Turkey, and through the Mediterranean region of Europe; about 20% of all active volcanoes and 15% of all earthquakes occur in this belt.

mesa   A broad, flat-topped erosional remnant bounded on all sides by steep slopes; forms when resistant cap rock is breached, allowing rapid erosion of the less resistant underlying sedimentary rock.

metamorphic facies   A group of metamorphic rocks characterized by particular mineral assemblages formed under the same broad temperature–pressure conditions.

metamorphic rock   Any rock altered by high temperature and pressure and the chemical activities of fluids is said to have been metamorphosed (such as slate, gneiss, or marble).

metamorphic zone   The region between lines of equal metamorphic intensity known as isograds.

Milankovitch theory   A theory that explains cyclic variations in climate and the onset of ice ages as a result of irregularities in Earth's rotation and orbit.

mineral   A naturally occurring, inorganic, crystalline solid having characteristic physical properties and a narrowly defined chemical composition.

Modified Mercalli Intensity Scale   A scale having values ranging from I to XII used to characterize earthquake intensity based on damage.

Moho   See *Mohorovičić discontinuity*.

Mohorovičić discontinuity   The boundary between the crust and mantle; also called the Moho.

monocline   A simple bend or flexure in otherwise horizontal or uniformly dipping rock layers.

mud crack   A sedimentary structure found in clay-rich sediment that has dried out. When such sediment dries, it shrinks and forms intersecting fractures.

mudflow   A flow consisting of mostly clay- and silt-sized particles and more than 30% water; most common in semi-arid and arid environments.

N

native element   A mineral composed of a single element (such as gold).

natural levee   A ridge of sandy alluvium deposited along the margins of a channel during floods.

nearshore sediment budget   The balance between additions and losses of sediment in the nearshore zone.

neutron   An electrically neutral particle found in the nucleus of an atom.

nonconformity   An unconformity in which stratified sedimentary rocks above an erosion surface overlie igneous or metamorphic rocks. See also *angular unconformity* and *disconformity*.

nonferromagnesian silicate   A silicate mineral that has no iron or magnesium.

nonfoliated texture   A metamorphic texture in which there is no discernable preferred orientation of mineral grains.

normal fault   A dip-slip fault on which the hanging wall block has moved downward relative to the footwall block. See also *reverse fault*.

nucleus   The central part of an atom consisting of one or more protons and neutrons.

nuée ardente   A mobile dense cloud of hot pyroclastic materials and gases ejected from a volcano.

O

oblique-slip fault   A fault having both dip-slip and strike-slip movement.

oceanic–continental plate boundary   A type of convergent plate boundary along which oceanic lithosphere and continental lithosphere collide; characterized by subduction of the oceanic plate beneath the continental plate and by volcanism and seismicity.

oceanic crust   The crust underlying the ocean basins. It ranges from 5 to 10 km thick, is composed of gabbro and basalt, and has an average density of 3.0 g/cm^3.

oceanic–oceanic plate boundary   A type of convergent plate boundary along which two oceanic lithospheric plates collide and one is subducted beneath the other.

oceanic ridge   A submarine mountain system found in all the oceans. It is composed of volcanic rock (mostly basalt) and displays features produced by tension.

oceanic trench   A long, narrow feature restricted to active continental margins and along which subduction occurs.

ooze   Deep-sea sediment composed mostly of shells of marine animals and plants.

ophiolite   A sequence of igneous rocks representing a fragment of oceanic lithosphere; composed of peridotite overlain successively by gabbro, sheeted basalt dikes, and pillow lavas.

orogeny   The process of forming mountains, especially by folding and thrust faulting; an episode of mountain building.

outwash plain   The sediment deposited by the meltwater discharging from a continental glacier's terminus.

oxbow lake   A cutoff meander filled with water.

oxidation   The reaction of oxygen with other atoms to form oxides or, if water is present, hydroxides.

P

pahoehoe   A type of lava flow with a smooth, ropy surface.

paleocurrent   The direction of an ancient current as indicated by sedimentary structures such as cross-bedding.

paleomagnetism   Remnant magnetism in rocks, studied to determine the intensity and direction of Earth's past magnetic field.

Pangaea   The name Alfred Wegener proposed for a supercontinent consisting of all Earth's landmasses that existed at the end of the Paleozoic Era.

parabolic dune   A crescent-shaped dune in which the tips point upwind.

parent material   The material that is chemically and mechanically weathered to yield sediment and soil.

passive continental margin   The trailing edge of a continental plate consisting of a broad continental shelf and a continental slope and rise. A vast, flat abyssal plain is commonly present adjacent to the rise. See also *active continental margin*.

pedalfer   A soil formed in humid regions with an organic-rich A horizon and aluminum-rich clays and iron oxides in horizon B.

pediment   An erosion surface of low relief gently sloping away from a mountain base.

pedocal   A soil characteristic of arid and semiarid regions with a thin A horizon and a calcium carbonate–rich B horizon.

pelagic clay   Generally brown or reddish deep-sea sediment composed of clay-sized particles derived from the continents and oceanic islands.

permeability   A material's capacity for transmitting fluids.

phaneritic texture   A texture in igneous rocks in which minerals are easily visible without magnification; results from slow cooling and generally indicates an intrusive origin.

pillow lava   Bulbous masses of basalt, resembling pillows, formed when lava is rapidly chilled under water.

plastic flow   The flow that occurs in response to pressure and causes permanent deformation.

plastic strain   The result of stress in which a material cannot recover its original shape and retains the configuration produced by the stress such as folding of rocks.

plate   An individual segment of lithosphere that moves over the asthenosphere.

plate tectonic theory   The theory that large segments of the outer part of Earth (lithospheric plates) move relative to one another.

platform   That part of a craton that lies buried beneath flat-lying or only mildly deformed sedimentary rocks. A platform and shield constitute a craton.

playa   A dry lakebed found in deserts.

plunging fold   A fold with an inclined axis.

pluton   An intrusive igneous body that forms when magma cools and crystallizes within the crust (such as a batholith and sill).

plutonic (intrusive igneous) rock   Igneous rock that crystallizes from magma intruded into or formed in place within Earth's crust.

point bar   The sediment body deposited on the gently sloping side of a meander loop.

porosity   The percentage of a material's total volume that is pore space.

porphyritic texture   An igneous texture with minerals of markedly different sizes.

pressure release   A mechanical weathering process in which rocks that formed under pressure expand on being exposed at the surface.

primary wave   See *P-wave*.

principle of cross-cutting relationships   A principle used to determine the relative ages of events; holds that an igneous intrusion or fault must be younger than the rocks it intrudes into or cuts.

principle of fossil succession   A principle holding that fossils, and especially assemblages of fossils, succeed one another through time in a regular and determinable order.

principle of inclusions   A principle holding that inclusions, or fragments, in a rock unit are older than the rock unit itself (for example, granite fragments in a sandstone are older than the sandstone).

principle of isostasy   The theoretical concept of Earth's crust "floating" on a denser underlying layer.

principle of lateral continuity   A principle holding that sediment layers extend outward in all directions until they terminate.

principle of original horizontality   A principle holding that sediment layers are deposited horizontally or very nearly so.

principle of superposition   A principle holding that younger rocks are deposited on top of older rocks.

principle of uniformitarianism   A principle holding that we can interpret past events by understanding present-day processes; based on the assumption that natural laws have not changed through time.

proton   A positively charged particle found in the nucleus of an atom.

P-wave   A compressional, or push–pull, wave; the fastest seismic wave and one that can travel through solids, liquids, and gases; also known as a primary wave.

P-wave shadow zone   The area between 103 and 143 degrees from an earthquake focus where little P-wave energy is recorded by seismographs.

pyroclastic materials   Fragmental substances, such as ash, explosively ejected from a volcano.

pyroclastic (fragmental) texture   A fragmental texture characteristic of igneous rocks composed of pyroclastic materials.

pyroclastic sheet deposit   Vast, sheetlike deposit of felsic pyroclastic materials erupted from fissures.

Q

quick clay   A clay that spontaneously liquefies and flows like water when disturbed.

R

radioactive decay   The spontaneous change of an atom to an atom of a different element by emission of a particle from its nucleus (alpha and beta decay) or by electron capture.

rainshadow desert   A desert found on the lee side of a mountain range; forms because moist marine air moving inland forms clouds and produces precipitation on the windward side of the mountain range so that the air descending on the leeward side is much warmer and drier.

rapid mass movement   Any kind of mass movement involving a visible downslope displacement of material.

Rayleigh wave (R-wave)   A surface wave in which the individual particles of material move in an elliptic path within a vertical plane oriented in the direction of wave movement.

recessional moraine   A type of end moraine formed when a glacier's terminus retreats, then stabilizes and till is deposited. See also *end moraine* and *terminal moraine*.

reef   A moundlike, wave-resistant structure composed of the skeletons of organisms.

reflection   The return to the surface of some of a seismic wave's energy when it encounters a boundary separating materials of different density or elasticity.

refraction   The change in direction and velocity of a seismic wave when it travels from one material into another of different density and elasticity.

regional metamorphism   Metamorphism that occurs over a large area, resulting from high temperatures, tremendous pressures, and the chemical activity of fluids within the crust.

regolith   The layer of unconsolidated rock and mineral fragments and soil that covers most of the land surface.

relative dating   The process of determining the age of an event relative to other events; involves placing geologic events in their correct chronologic order but involves no consideration of when the events occurred in terms of number of years ago. See also *absolute dating*.

reserve   The part of the resource base that can be extracted economically.

resource   A concentration of naturally occurring solid, liquid, or gaseous material in or on Earth's crust in such form and amount that economic extraction of a commodity from the concentration is currently or potentially feasible.

reverse fault   A dip-slip fault in which the hanging wall block has moved upward relative to the footwall block. See also *normal fault*.

Richter Magnitude Scale   An open-ended scale that measures the amount of energy released during an earthquake.

rill erosion   Erosion by running water that scours small channels in the ground.

rip current   A narrow surface current that flows out to sea through the breaker zone.

ripple mark   Wavelike (undulating) structure produced in granular sediment such as sand by unidirectional wind and water currents or by oscillating wave currents.

rock   An aggregate of one or more minerals, as in limestone or granite, or a consolidated aggregate of rock fragments, as in conglomerate; includes rocklike materials such as coal and natural glass.

rock cycle   A group of processes through which Earth materials may pass as they are transformed from one rock type to another.

rockfall   A common type of extremely rapid mass wasting in which rocks fall through the air.

rock-forming mineral   A mineral common in rocks, which is important in their identification and classification.

rock slide   A type of rapid mass movement in which rocks move downslope along a more or less planar surface.

rounding   The process by which the sharp corners and edges of sedimentary particles are abraded during transport.

runoff   The surface flow of streams and rivers.

R-wave   See *Rayleigh wave*.

S

salt crystal growth   A mechanical weathering process in which rocks are disaggregated by the growth of salt crystals in crevices and pores.

scientific method   A logical, orderly approach that involves gathering data, formulating and testing hypotheses, and proposing theories.

seafloor spreading   The theory that the seafloor moves away from spreading ridges and is eventually consumed at subduction zones.

seamount   A submarine volcanic mountain rising at least 1 km above the seafloor.

secondary wave See *S-wave*.

sediment Loose aggregate of solids derived from preexisting rocks, or solids precipitated from solution by inorganic chemical processes or extracted from solution by organisms.

sedimentary facies Any aspect of a sedimentary rock unit that makes it recognizably different from adjacent sedimentary rocks of the same, or approximately the same, age (e.g., a sandstone facies).

sedimentary rock Any rock composed of sediment (such as sandstone and limestone).

sedimentary structure Any structure in sedimentary rock formed at or shortly after the time of deposition (such as cross-bedding, mud cracks, and animal burrows).

seismic profiling A method in which strong waves generated at an energy source penetrate the layers beneath the seafloor. Some of the energy is reflected from the various layers to the surface, making it possible to determine the nature of the layers.

seismic risk map A map based on the distribution and intensity of past earthquakes; such maps indicate the potential severity of future earthquakes and are useful in planning.

seismic tomography A method of analyzing numerous seismic waves to develop a three-dimensional image of Earth's interior.

seismograph An instrument that detects, records, and measures the various waves produced by an earthquake.

seismology The study of earthquakes.

shear strength The resisting forces helping to maintain slope stability.

shear stress The result of forces acting parallel to one another but in opposite directions; results in deformation by displacement of adjacent layers along closely spaced planes.

sheet erosion Erosion that is more or less evenly distributed over the surface and removes thin layers of soil.

sheet joint A large fracture more or less parallel to a rock surface resulting from pressure released by expansion of the rock.

shield A vast area of exposed ancient rocks on a continent; the exposed part of a craton.

shield volcano A dome-shaped volcano with a low, rounded profile built up mostly of overlapping basalt lava flows.

shoreline The area between mean low tide and the highest level on land affected by storm waves.

silica A compound of silicon and oxygen atoms.

silica tetrahedron The basic building block of all silicate minerals, it consists of one silicon atom and four oxygen atoms.

silicate A mineral containing silica (such as quartz [SiO_2]).

sill A tabular or sheetlike concordant pluton.

sinkhole A depression in the ground that forms in karst regions by the solution of the underlying carbonate rocks or by the collapse of a cave roof.

slide A type of mass wasting involving movement of material along one or more surfaces of failure.

slow mass movement Mass movement that advances at an imperceptible rate and is usually only detectable by the effects of its movement.

slump A type of mass wasting that takes place along a curved surface of failure and results in the backward rotation of the slump mass.

soil Regolith consisting of weathered material, water, air, and humus that can support plants.

soil degradation Any process leading to a loss of soil productivity; may involve erosion, chemical pollution, or compaction.

soil horizon A distinct soil layer that differs from other soil layers in texture, structure, composition, and color.

solar nebula theory A theory for the evolution of the solar system from a rotating cloud of gas.

solifluction A type of mass wasting involving the slow downslope movement of water-saturated surface materials; especially the flow at high elevations or high latitudes where the flow is underlain by frozen soil.

solution A reaction in which the ions of a substance become dissociated in a liquid and the solid substance dissolves.

sorting A term referring to the degree to which all particles of sediment or sedimentary rock are about the same size.

specific gravity The ratio of a substance's weight, especially a mineral, to the weight of an equal volume of water; for example, the specific gravity of quartz is 2.65.

spheroidal weathering A type of chemical weathering in which corners and sharp edges of rocks weather more rapidly than flat surfaces, thus yielding spherical shapes.

spit A continuation of a beach forming a point of land that projects into a body of water, commonly a bay.

spring A place where groundwater flows or seeps out of the ground.

stock An irregularly shaped discordant pluton with a surface area less than 100 km^2.

stoping A process in which rising magma detaches and engulfs pieces of the surrounding country rock.

strain Deformation caused by stress. See also *elastic strain* and *plastic strain*.

strata (stratification) Strata (singular *stratum*) are the layers in sedimentary rocks, whereas stratification is the layered aspect of sedimentary rocks. See also *bed* (*bedding*).

stratified drift Glacial drift displaying both sorting and stratification.

stratovolcano See *composite volcano*.

stream terrace An erosional remnant of a floodplain that formed when a stream was flowing at a higher level.

stress The force per unit area applied to a material such as rock.

strike The direction of a line formed by the intersection of a horizontal plane with an inclined plane, such as a rock layer.

strike-slip fault A fault involving horizontal movement so that blocks on opposite sides of a fault plane slide sideways past one another. See *dip-slip fault*.

submarine canyon A steep-walled canyon best developed on the continental slope but some extend well up onto the continental shelves.

submarine fan A cone-shaped sedimentary deposit that accumulates on the continental slope and rise.

submarine hydrothermal vent A crack or fissure in the seafloor through which superheated water issues. See *black smoker*.

submergent coast A coast along which sea level rises with respect to the land or the land subsides.

superposed stream A stream that once flowed on a higher surface and eroded downward into resistant rocks while still maintaining its course.

surface wave Earthquake waves that travel along Earth's surface. Rayleigh (R-) and Love (L-) waves are both surface waves.

suspended load The smallest particles carried by a stream, such as silt and clay, which are kept suspended by fluid turbulence.

sustainable development The concept of satisfying basic human needs while safeguarding the environment to ensure continued economic development.

S-wave A shear wave that moves material perpendicular to the direction of travel, thereby producing shear stresses in the material it moves through; also known as a secondary wave. An S-wave travels only through solids.

S-wave shadow zone Those areas more than 103 degrees from an earthquake focus where no S-waves are recorded.

syncline A down-arched fold in which the youngest exposed rocks coincide with the fold axis and all strata dip toward the axis.

system A combination of related parts that interact in an organized fashion. Earth systems include the atmosphere, hydrosphere, biosphere, and solid Earth.

T

talus Weathered material that accumulates at the base of slopes.

tension A type of stress in which forces act in opposite directions but along the same line, thus tending to stretch an object.

terminal moraine A type of end moraine; the outermost moraine marking the greatest extent of a glacier. See also *end moraine* and *recessional moraine*.

terrane A block of rock with characteristics quite different from those of surrounding rocks. Terranes probably represent seamounts, oceanic rises, and other seafloor features that accreted to continents during orogenies.

terrestrial planet Any of the four innermost planets (Mercury, Venus, Earth, and Mars). They are all small and have high mean densities, indicating that they are composed of rock and metallic elements. See also *Jovian planet*.

theory An explanation for some natural phenomenon that has a large body of supporting evidence. To be considered scientific, a theory must be testable—for example, plate tectonic theory.

thermal convection cell A type of circulation of material in the asthenosphere during which hot material rises, moves laterally, cools and sinks, and is reheated and continues the cycle.

thermal expansion and contraction A type of mechanical weathering in which the volume of rock changes in response to heating and cooling.

thrust fault A type of reverse fault with a fault plane dipping less than 45 degrees.

tide The regular fluctuation in the sea's surface in response to the gravitational attraction of the Moon and Sun.

till All sediment deposited directly by glacial ice.

tombolo A type of spit that extends out into the sea or a lake and connects an island to the mainland.

transform fault A type of fault along which one type of motion is transformed into another; commonly displaces oceanic ridges, but movement on opposite sides of the fault between displaced ridge segments is the opposite of the apparent displacement; on land recognized as a strike-slip fault, such as the San Andreas fault.

transform plate boundary Plate boundary along which plates slide past one another and crust is neither produced nor destroyed; on land recognized as a strike-slip fault.

transport The mechanism by which weathered material is moved from one place to another, commonly by running water, wind, or glaciers.

transverse dune A long ridge of sand perpendicular to the prevailing wind direction.

tree-ring dating The process of determining the age of a tree or wood in structures by counting the number of annual growth rings.

trough The lowest point between wave crests.

tsunami   A destructive sea wave that is usually produced by an earthquake but can also be caused by submarine landslides or volcanic eruptions.

turbidity current   A sediment–water mixture, denser than normal seawater, that flows downslope to the deep seafloor.

U

unconformity   An erosion surface separating younger strata from older rocks. See also *angular unconformity, disconformity,* and *nonconformity.*

U-shaped glacial trough   A valley with steep or vertical walls and a broad rather flat floor; formed by the movement of a glacier through a stream valley.

V

valley   A linear depression bounded by higher areas such as ridges, hills, or mountains. Most are eroded by streams, although mass wasting is also important in their origin and evolution.

valley glacier   A glacier confined to a mountain valley or to an interconnected system of mountain valleys.

valley train   A long, narrow deposit of stratified drift confined within a glacial valley.

velocity   A measure of the downstream distance water travels per unit of time. Velocity varies considerably among streams and even within the same stream.

ventifact   A stone whose surface has been polished, pitted, grooved, or faceted by wind abrasion.

vesicle   A small hole or cavity formed by gas trapped in cooling lava.

viscosity   A fluid's resistance to flow.

volcanic explosivity index (VEI)   A semiquantitative scale for the size of a volcanic eruption based on evaluation of such criteria as volume of material explosively ejected and height of eruption cloud.

volcanic neck   An erosional remnant of the material that solidified in a volcanic pipe.

volcanic pipe   The conduit connecting the crater of a volcano with an underlying magma chamber.

volcanic (extrusive igneous) rock   An igneous rock formed when magma is extruded onto Earth's surface where it cools and crystallizes, or when pyroclastic materials become consolidated.

volcanic tremor   Ground motion lasting from minutes to hours resulting from magma moving below the surface, as opposed to the sudden jolts produced by most earthquakes.

volcanism   The process whereby magma and its associated gases rise through the crust and are extruded onto the surface or into the atmosphere.

volcano   A mountain formed around a vent as a result of the eruption of lava and pyroclastic materials.

W

water table   The surface separating the zone of aeration from the underlying zone of saturation.

water well   A well made by digging or drilling into the zone of saturation.

wave   An undulation on the surface of a body of water, resulting in the water surface rising and falling.

wave base   A depth of about one-half wavelength, where the diameter of the orbits of water in waves is essentially zero; the depth below which water is unaffected by surface waves.

wave-cut platform   A beveled surface that slopes gently seaward; formed by the retreat of a sea cliff.

wave height   The vertical distance from wave trough to wave crest.

wavelength   The distance between successive wave crests or troughs.

wave period   The time required for two successive wave crests (or troughs) to pass a given point.

wave refraction   The bending of waves so that they more nearly parallel the shoreline.

weathering   The physical breakdown and chemical alteration of rocks and minerals at or near Earth's surface.

Z

zone of accumulation   The part of a glacier where additions exceed losses and the glacier's surface is perennially covered with snow. Also refers to horizon B in soil where soluble materials leached from horizon A accumulate as irregular masses.

zone of aeration   The zone above the water table that contains both water and air within the pore spaces of the rock or soil.

zone of leaching   Another name for horizon A of a soil; the area in the upper part of a soil where soluble minerals are removed by downward moving water.

zone of saturation   The area below the zone of aeration in which all pore spaces are filled with groundwater.

zone of wastage   The part of a glacier where losses from melting, sublimation, and calving of icebergs exceed the rate of accumulation.

Credits

Chapter 1

CO1: NASA. **Figure 1.4:** Reed Wicander. **Figure 1.3:** Collection of the New York Public Library, Astor, Lenox and Tilden Foundations. **Figure 1.6:** Superstock. **Geo-Profile, 1.1, 1.2:** David Wunsch. **Figure 1.8:** Courtesy of Dana Berry. **Art Spread 1:** NASA. **Figure 1.15a-f:** Sue Monroe. **Geology in Unexpected Places, Figures 1, 2, 3:** Reed Wicander.

Chapter 2

CO2: Sue Monroe. **Figure 2.1a:** Los Angeles County Museum specimen, © Harold and Erica Van Pelt. **Figure 2.1b:** National Museum of Natural History Specimen #G7101. **Figure 2.1c:** Jerry Jacka Photography. **Figure 2.1d:** Jeffrey A. Scovil. **Figure 2.2:** Layne Kennedy/Corbis. **Figure 2.8a-b:** Sue Monroe. **Geology in Unexpected Places, Figure 1:** PhotoDisc Green/Getty Images. **Figure 2.13:** Sue Monroe. **Figure 2.14:** Sue Monroe. **Figure 2.15a-d:** Sue Monroe. **Art Spread 2:** National Museum of Natural History, Specimen #R12197, Photo by D. Penland, Smithsonian Institution; Bettmann/Corbis; J. C. H. Grabill/Corbis; Ken Lucas/Visuals Unlimited; James S. Monroe; Sue Monroe. **Figure 2.16a-c:** Sue Monroe. **Geo-Focus 2.1, Figure 1:** Richard D. Fisher. **Figure 2.19a-b:** Sue Monroe. **Figure 2.20:** Sue Monroe. **Geo-Profile, 2.1, 2.2:** Terry S. Mollo. **Figure 2.21:** Sue Monroe.

Chapter 3

CO3: Courtesy of Richard L. Chambers. **Figure 3.1a:** National Park Service. **Figure 3.1b:** Robb DeWall © Crazy Horse Memorial Foundation. **Figure 3.2:** P. Mouginis-Mark. **Figure 3.6b:** Sue Monroe. **Figure 3.8:** Sue Monroe. **Figure 3.10:** Sue Monroe. **Figure 3.11a-b:** Sue Monroe. **Figure 3.12a-b:** Sue Monroe. **Figure 3.13a-b:** Sue Monroe. **Figure 3.14a:** Courtesy of Steve Stahl. **Figure 3.14b:** Sue Monroe. **Figure 3.14c:** Wendell E. Wilson. **Figure 3.16a:** Courtesy of David J. Matty. **Figure 3.16b-d:** Sue Monroe. **Geology in Unexpected Places, Figure 1:** Marcus Kazmierczak, www.mkaz.com. **Art**

Spread 3: James S. Monroe, Sue Monroe, Martin G. Miller/Visuals Unlimited, http://formontana.net/cb.html. **Geo-Focus 3.1, Figure 1:** Richard List/Corbis. **Geo-Focus 3.1, Figure 2:** Courtesy of Frank Hanna. **Geo-Focus 3.1, Figure 3:** James S. Monroe. **Figure 3.17:** James S. Monroe.

Chapter 4

CO4: Stone/Getty Images. **Figure 4.1b-c:** Sue Monroe. **Figure 4.2:** JPL/NASA. **Figure 4.3a-c:** Sue Monroe. **Geo-Focus 4.1, Figure 1a:** Images of the World. **Geo-Focus 4.1, Figure 1b:** AFP/Corbis. **Figure 4.4a:** Hawaii Volcanoes National Park, USGS. **Figure 4.4b:** J. B. Judd/USGS. **Figure 4.4c:** Sue Monroe. **Figure 4.5a-b:** J. D. Griggs/USGS. **Figure 4.5c:** Robert Tilling/USGS. **Figure 4.5d:** James S. Monroe. **Figure 4.6a:** T. J. Takahashi/USGS. **Figure 4.6b:** J. D. Griggs/USGS. **Figure 4.7a-b:** James S. Monroe. **Figure 4.8c:** Sue Monroe. **Figure 4.9a:** Reproduced by permission of Marie Tharp, 1 Washington Ave., South Nyack, NY, 10960. **Figure 4.9a:** James S. Monroe. **Figure 4.10a:** Sue Monroe. **Figure 4.10b:** Reuters/Corbis. **Figure 4.11e:** James S. Monroe. **Figure 4.12b:** Robert Tilling/USGS. **Figure 4.12c:** James S. Monroe. **Geo-Focus 4.2, Figure 1:** Sue Monroe. **Figure 4.13a-c:** James S. Monroe. **Figure 4.13d:** Solarfilma/GeoScience Features. **Figure 4.14b:** R. Solkowski/Consulting Geologists, Vancouver, WA. **Figure 4.14c:** Sue Monroe. **Figure 4.14d:** Courtesy of Wayne E. Moore. **Geology in Unexpected Places, Figure 1:** Courtesy of Frederick A. Belton. **Figure 4.15a:** U.S. Department of the Interior/USGS/David Johnson, Cascades Volcano Observatory, Vancouver, WA. **Figure 4.15b:** Darrell G. Herd/USGS. **Figure 4.16a:** James S. Monroe. **Figure 4.16b:** T. P. Miller/USGS. **Figure 4.17a:** Reuters/Corbis. **Figure 4.17b:** neg. # 256108, E.O Hovey/Dept. of Library Services/American Museum of Natural History. **Art Spread 4:** D. R. Crandel/USGS, Courtesy of Keith Ronnholm, PhotoDisc/Getty Images, Lynn Topinka/USGS, Sue Monroe. **Figure 4.18a:** Frank Kujawa, University of Central Florida/GeoPhoto. **Figure 4.18c:** W. E. Scott/USGS.

Chapter 5

CO5: James S. Monroe. **Figure 5.1a-b:** James S. Monroe. **Figure 5.2a:** Courtesy of Frank Hanna. **Figure 5.2b:** Courtesy of Gary Rees. **Figure 5.2c:** James S. Monroe. **Figure 5.3b:** James S. Monroe. **Figure 5.4a-b:** Sue Monroe. **Figure 5.4c:** Mark Gibson/Visuals Unlimited. **Figure 5.5:** Courtesy of W. D. Lowry. **Figure 5.6a-b:** James S. Monroe. **Figure 5.8:** James S. Monroe. **Figure 5.9:** Bill Beatty/Visuals Unlimited.

Courtesy of Victor Royer, USGS. **Figure 12.16b:** Robert Caputo/Aurora. **Figure 12.17b:** Courtesy of John Faivre. **Figure 12.18b:** Japan Satellite/Getty Images. **Figure 12.19b:** Harvey Lloyd/Getty Images. **Figure 12.20b:** NASA. **Figure 12.23:** USGS. **Figure 12.28b:** Courtesy of R. V. Dietrich.

Chapter 13
CO13: Sue Monroe. **Figure 13.1a:** Courtesy of David J. Matty. **Figure 13.1b:** Reed Wicander. **Geology in Unexpected Places, Figure 1:** Jim Winkley/Ecoscene/ Corbis. **Figure 13.5a:** Sue Monroe. **Figure 13.5b:** James S. Monroe. **Figure 13.7b:** Courtesy of Kevin O'Brien. **Figure 13.6b:** Sue Monroe. **Figure 13.8:** Reed Wicander. **Figure 13.10b:** Martin F. Schmidt, Jr. **Figure 13.12d:** Sue Monroe. **Figure 13.13c:** John S. Shelton. **Figure 13.15a:** Galen Rowell/ Peter Arnold. **Figure 13.15b:** C. G. Tillmann. **Figure 13.16b:** Courtesy of David J. Matty. **Figure 13.16c:** James S. Monroe. **Art Spread 13:** James S. Monroe, Martin Miller/ Visuals Unlimited, John S. Shelton, Lindie R. Brewer/ USGS, David J. Matty. **Figure 13.18:** James S. Monroe. **Figure 13.19:** Sue Monroe. **Figure 13.21b:** Sue Monroe.

Chapter 14
CO14: AP/Wide World Photos. **Figure 14.3c:** Reed Wicander. **Figure 14.4:** James S. Monroe. **Figure 14.5d:** Courtesy of R. V. Dietrich. **Geology in Unexpected Places, Figure 1:** White Mountains Attractions Association. **Figure 14.6:** Boris Yaro/*Los Angeles Times*. **Figure 14.8:** Rod Rolle/Liaison Agency, Inc. **Figure 14.10:** Tonya Paul/*Oroville Mercury Register*. **Geo-Focus 14.1, Figure 1:** T. Spencer/ Colorific. **Figure 14.11a:** Sue Monroe. **Figure 14.11b:** James S. Monroe. **Geo-Profile, 14.1:** Sean Brown. **Figure 14.13:** John S. Shelton. **Art Spread 14:** Eleanora Robbins/USGS, Reed Wicander. **Figure 14.15:** Steven R. Lower, GeoPhoto Publishing Company. **Figure 14.16b:** B. Bradley and the University of Colorado's Geology Department, National Geophysical Data Center. **Figure 14.17:** James S. Monroe. **Figure 14.18:** B. Pipkin, University of Southern California. **Figure 14.19:** Reed Wicander. **Figure 14.20:** Courtesy of the Canadian Air Force. **Figure 14.21b:** Photo from "Alaska Earthquake Collection" no. 43ct/USGS. **Figure 14.22b:** B. Bradley and the University of Colorado's Geology Department. **Figure 14.23:** O. J. Ferrains, Jr./USGS. **Figure 14.24b-c:** B. Bradley and the University of Colorado's Geology Department. **Figure 14.24d:** Courtesy of David J. Matty. **Figure 14.26:** George Plafker/USGS. **Figure 14.28b:** Reed Wicander. **Figure 14.30b:** John D. Cunningham/Visuals Unlimited. **Figure 14.31b:** Dell R. Foutz/Visuals Unlimited. **Figure 14.32b:** Reed Wicander.

Chapter 15
CO15: Sue Monroe. **Figure 15.1a:** Schenectaday Museum; Hall of Electrical History/Corbis. **Figure 15.1b:** Bettmann/Corbis. **Figure 15.2a:** James S. Monroe. **Figure 15.2b:** Sue Monroe. **Figure 15.2c:** James S. Monroe. **Art Spread 15:** Landsat Imagery Courtesy of Earth Observation Satellite Co., Chris Stewart/Black Star, Michael Lawton, Sue Monroe. **Geo-Focus 15.1, Figure 1:** Sue Monroe. **Figure 15.9a:** Sue Monroe. **Figure 15.9b:** Courtesy of R. V. Dietrich. **Figure 15.10a-b:** James S. Monroe. **Figure 15.10c:** Sue Monroe. **Figure 15.12a-b:** James S. Monroe. **Figure 15.14b-c:** James S. Monroe. **Figure 15.15e:** James S. Monroe. **Figure 15.18b:** James S. Monroe. **Geology in Unexpected Places, Figure 1:** JPL/NASA. **Figure 15.20b:** John S. Shelton. **Figure 15.23a:** James S. Monroe. **Figure 15.23b:** Sue Monroe. **Figure 15.24a:** James S. Monroe. **Figure 15.26b:** Courtesy of the Kentucky Department of Parks. **Figure 15.26d:** James S. Monroe. **Figure 15.29a-b:** James S. Monroe. **Figure 15.31d:** J. R. Stacey/USGS. **Figure 15.32d:** James S. Monroe. **Figure 15.32e:** Sue Monroe.

Chapter 16
CO16: David Muench/Corbis. **Figure 16.4b:** Sue Monroe. **Figure 16.5a-b:** Sue Monroe. **Figure 16.9a:** Frank Kujawa/University of Central Florida, GeoPhoto Publishing Company. **Figure 16.9b:** James S. Monroe. **Figure 16.11a:** Reed Wicander. **Figure 16.11b:** John S. Shelton. **Art Spread 16:** Reed Wicander. **Figure 16.13:** Stone/Getty Images. **Figure 16.16:** USGS. **Figure 16.17:** Sarah Stone/Getty Images. **Figure 16.18:** Courtesy of R. V. Dietrich. **Figure 16.19:** City of Long Beach, Department of Oil Properties. **Geology in Unexpected Places, Figure 1:** Reed Wicander. **Figure 16.21a:** Reed Wicander. **Figure 16.21b-d:** James S. Monroe. **Figure 16.22:** British Tourist Authority. **Figure 16.23a-b:** James S. Monroe. **Figure 16.25a-b:** Reed Wicander. **Figure 16.26:** Julie Donnelly-Nolan/USGS.

Chapter 17
CO17: Sue Monroe. **Figure 17.1a:** Reprinted with permission from T. H. Van Andel, *New Views on an Old Planet*, p. 175 (table 11.1) © 1989 Cambridge University Press. **Figure 17.1b:** Sue Monroe. **Figure 17.2b:** Sue Monroe. **Figure 17.4a:** Engineering Mechanics, Virginia Polytechnic Institute and State University. **Figure 17.4b:** Sue Monroe. **Figure 17.5b-c:** James S. Monroe. **Figure 17.7:** James S. Monroe. **Figure 17.9a:** Courtesy of R. V. Dietrich. **Figure 17.9b:** Sue Monroe. **Figure 17.10:** AP/Wide World Photos. **Figure 17.11b:** James S. Monroe. **Figure 17.12a-c:** James S. Monroe. **Figure 17.13:** Sue Monroe. **Art Spread 17:** James S. Monroe, Swiss National Tourist Office. **Geo-Focus 17.1, Figure 1a:** James S. Monroe. **Geo-Focus 17.1, Figure 1b:** Ron Watts/Corbis. **Geo-Focus 17.1, Figure 2:** Carl H. Key, Northern Rocky Mountain Science Center, Glacier Field Station/USGS. **Figure 17.15:** Alan Kellehein/Mary Pat Ziter, JLM Visuals. **Figure 17.16a:** Courtesy of R. V. Dietrich. **Figure 17.16b:** James S. Monroe. **Figure 17.18a-b:** James S. Monroe. **Figure 17.19a:** Engineering Mechanics, Virginia Polytechnic Institute and State University. **Figure 17.19b:** Peter Kresan. **Figure 17.21:** Courtesy of Carl Guell Slide Collection, Department of Geography, University of Wisconsin, Oshkosh. **Figure 17.22a:** Courtesy of David J. Matty. **Figure 17.22b:** James S. Monroe. **Figure 17.23a:** James S. Monroe. **Figure 17.23b:** Courtesy of B. M. C. Pape. **Figure 17.23c:** Tom Bean/Corbis. **Geology in Unexpected Places, Figure 1:** D. D. Trent. **Figure 17.24a:** Canadian Geological Survey. **Figure 17.24b:** James S. Monroe. **Figure 17.28a-b:** P. Weiss/USGS.

Chapter 18

CO18: George Gerster/The National Audubon Society/Photo Researarchers, Inc. **Figure 18.2:** O. Alamany & W. Vicens/Corbis. **Geo-Focus 18.1, Figure 1:** United States Department of Energy. **Geology in Unexpected Places, Figure 1:** NASA. **Geology in Unexpected Places, Figure 2:** Reed Wicander. **Figure 18.3b:** Martin G. Miller/Visuals Unlimited. **Figure 18.4:** Courtesy of Marion A. Whitney. **Figure 18.5:** Martin G. Miller/Visuals Unlimited. **Figure 18.6c:** Courtesy of David J. Matty. **Figure 18.7:** PhotoDisc/Getty Images. **Figure 18.9:** Reed Wicander. **Figure 18.10b:** John S. Shelton. **Figure 18.11b:** © 1994 CNES. Provided by SPOT Image Corporation. **Figure 18.12b:** W. J. Weber/Visuals Unlimited. **Figure 18.13b:** Reed Wicander. **Figure 18.14b:** Nigel J. Dennis/Photo Researchers, Inc. **Figure 18.15:** Lowell Georgia/Corbis. **Figure 18.19:** Charlie Ott, The National Audubon Society Collection/Photo Researchers, Inc. **Geo-Profile, 18.1:** Nicolas Lancaster. **Art Spread 18:** Reed Wicander, Thomas Abercrombie/National Geographic/Getty Images, Tom Bean/Corbis. **Figure 18.20:** James S. Monroe. **Figure 18.21a-b:** John S. Shelton. **Figure 18.22:** Martin G. Miller/Visuals Unlimited. **Figure 18.23:** Alan L. and Linda D. Mayo, GeoPhoto Publishing Company. **Figure 18.24b:** Reed Wicander. **Figure 18.25a:** Royalty Free/Corbis. **Figure 18.25b:** Tom Bean/Corbis.

Chapter 19

CO19: Sue Monroe. **Figure 19.1a-b:** James S. Monroe. **Figure 19.2a-b:** James S. Monroe. **Figure 19.4b-c:** James S. Monroe. **Geo Focus 19.1, Figure 2:** Tom Till. **Figure 19.5a-b:** James S. Monroe. **Figure 19.6a-b:** Tom Servais/*Surfer Magazine*. **Figure 19.7:** Sue Monroe. **Figure 19.8b-c:** James S. Monroe. **Art Spread 19:** James S. Monroe, John S. Shelton, Sue Monroe. **Figure 19.9b-c:** Sue Monroe. **Figure 19.10b-c:** James S. Monroe. **Figure 19.11b:** James S. Monroe. **Figure 19.11c:** John S. Shelton. **Figure 19.12a:** NASA. **Figure 19.12b:** Valerie Bates. **Figure 19.14a:** James S. Monroe. **Figure 19.14b:** Sue Monroe. **Figure 19.15b:** John S. Shelton. **Figure 19.15c:** James S. Monroe. **Figure 19.16b:** Courtesy of Nick Harvey. **Figure 19.16c:** Sue Monroe. **Figure 19.20a-b:** U. S. Army Corps of Engineers. **Figure 19.21:** Rosenberg Library, Galveston, Texas. **Geology in Unexpected Places, Figure 1a:** NASA. **Geology in Unexpected Places, Figure 1b:** AP/Wide World Photo. **Figure 19.22:** Courtesy of Dr. Stanley R. Riggs, Department of Geology, Graham Bldg., East Carolina University, Greenville, NC, 27858, 252-328-4391. **Figure 19.23:** NASA. **Figure 19.24:** Sue Monroe.

Chapter 20

CO20: James S. Monroe. **Figure 20.1a-b:** James S. Monroe. **Figure 20.2a:** NASA. **Figure 20.2b:** Courtesy of David J. Matty. **Figure 20.2c:** Sue Monroe. **Figure 20.2d:** James S. Monroe.

Index